Microbial Products

Microbial Products: Applications and Translational Trends offers complete coverage of the production of microbial products, including biopolymers, biofuels, bioactive compounds, and their applications in fields such as bioremediation, agriculture, medicine, and other industrial settings. This book focuses on multiple processes including upstream procedures and downstream processing, and the tools required for their production. Lab-scale development processes may not be as efficient when aiming for large-scale industrial production, so it is necessary to utilize in-silico modeling tools such as graph theory and Petri net modeling for bioprocess design to ensure success at translational levels. Therefore, this book presents in-silico and mathematical simulations and approaches used for such applications. Further, it examines microbial products produced from bacteria, fungi, and algae. These major microbial categories have the capacity to produce various diverse secondary metabolites, bioactive compounds, enzymes, biopolymers, biofuels, probiotics, and more. The bioproducts examined in this book are of great social, medical, and agricultural benefit, and include examples of biodegradable polymers, biofuels, biofertilizers, and drug delivery agents.

- Presents approaches and tools that aid in the design of eco-friendly, efficient, and economic bioprocesses.
- Utilizes in-silico and mathematical simulations for optimal bioprocess design.
- Examines approaches to be used for bioproducts from the lab scale to widely applied microbial biotechnologies.
- Presents the latest trends and technologies in production approaches for microbial bioproduct manufacture and application.

This book is ideal for both researchers and academics, as it provides up-to-date knowledge of applied microbial biotechnology approaches for bioproducts.

Microbial Biotechnology for Food, Health, and the Environment Series
Series Editor: Ashok Kumar Nadda

Plant-Microbial Interactions and Smart Agricultural Biotechnology
Edited by Swati Tyagi, Robin Kumar, Baljeet Saharan, Ashok Kumar Nadda

Microbial Products
Applications and Translational Trends
Edited by Mamtesh Singh, Gajendra Pratap Singh, Shivani Tyagi

For more information about this series, please visit: www.routledge.com
https://www.routledge.com/Microbial-Biotechnology-for-Food-Health-and-the-Environment/book-series/CRCMBFHE

Microbial Products

Applications and Translational Trends

Edited by
Mamtesh Singh
Gajendra Pratap Singh
Shivani Tyagi

CRC Press is an imprint of the
Taylor & Francis Group, an **informa** business

Cover image: Dr. Gajendra Pratap Singh

First edition published 2023
by CRC Press
6000 Broken Sound Parkway NW, Suite 300, Boca Raton, FL 33487-2742

and by CRC Press
4 Park Square, Milton Park, Abingdon, Oxon, OX14 4RN

CRC Press is an imprint of Taylor & Francis Group, LLC

ISBN: 978-1-032-30820-3 (hbk)
ISBN: 978-1-032-30839-5 (pbk)
ISBN: 978-1-003-30693-1 (ebk)

DOI: 10.1201/9781003306931

Typeset in Times
by SPi Technologies India Pvt Ltd (Straive)

To my late loving father Mr Sher Singh,
to whom (along with my mother Mrs Jagroshni)
I owe everything I am today and will be.
M.S.

To my late grandparents and to my parents
(Mr Suraj Singh & Mrs Krishna Devi)
G.P.S.

Contents

Preface

Microbial products are one of the most rapidly expanding sectors to be explored today, and their capabilities are seeing them establish standards in every field. With advances in science and technology, an extensive range of techniques for exploitation of these substances has been emerging for the benefit of mankind. With the rise in environmental consciousness among young scientists, people are working hard to reduce the harmful effects of anthropogenic activities. Microorganisms are our great friends in this connection. Research in this field is opening up new avenues for the improvement of the environment. Development of new drugs and utilization of microbial products for such purposes in day-to-day life is important for sustainable living. Microorganisms provide a variety of substances for human use and help in the degradation of pollutants. Around 23,000 microbial products are utilized by humans. Secondary metabolites produced by microbes are "wonder molecules" for immunotherapy, vaccines and drugs. In the food industry microbial products have also had a significant impact, for example, microbial probiotics in the marine ecosystem have the ability to be harnessed as sources of microbial food products.

This book examines the ability of microorganisms to provide useful and eco-friendly natural products, discusses new findings, and provides insights into their applications and translational trends. This book is divided into four Parts: Environment, Agriculture, Medicine, and In Silico and Mathematical Tools. Microbial products have found their most successful significance in environmental biotechnology, such as in bioremediation, biodegradation, enzyme production industrial application, biopolymer production from biowaste, biosurfactants and many more. Many industries are based on microbial enzymes, including the food industry, waste management, leather industry, detergent industry, photographic industry, chemical industry, silk degumming, silver recovery, pharmaceuticals and the medical industry. Similar impacts of microbial products have been manifested in agriculture. Sustainable agriculture is the need of the hour, and microorganisms provide a sustainable approach. The use of microbes or their metabolites in agriculture improves nutrient uptake and reduces plant stress responses, thereby improving yield. Microbial products can also be used in biologically compatible formulations in agriculture, for example as biopesticides, biofertilizers, aquatic probiotics, and more. Interestingly the field of medicine is not an exception to the theme of this book. The use of products from genetically modified microorganisms, including hormones such as insulin, vaccines, and enzymes, has enhanced the lives and life expectancy of people with certain medical conditions. Similarly, fermented foods and beverages have played key roles in the maintenance of good health as they are good reservoirs of macro-nutrients and micro-nutrients, and have several other therapeutic properties. In view of increasing awareness of healthy food, probiotics have become a natural and promising immunity booster and may also improve gut microbiome disparities.

Nevertheless, having collected an enormous amount of information over several decades, our current era needs the enhanced exploration and exploitation of microbial products. The conventional lab approach, although it provides ultimate validation, cannot achieve what is required on its own. Silico and mathematical tools have therefore earned an important place in research and development in all fields. The integration of machine learning approaches can play a part in the display and optimization of bioprocesses. Integration of machine learning approaches speeds up the progress of bioprocesses. Computational machine learning techniques have high predictive accuracy and are suitable for various types of bioprocess modelling. Descriptions of studies and the future prospects of this field have also been incorporated for a holistic view.

Thus, this book provides a learning experience for those interested in working towards the application of microbial products. With contributions from scientists working in these diverse research fields, this book, covering the latest trends and applications, will nurture young enthusiastic minds to pursue research and development in the field of utilization of microbial products to the benefit of all of humanity. We would, therefore, like to thank all the contributors for investing their thoughts, ideas and time in our book. We will always be indebted to them for this great venture. In addition to this, we would like to express our gratitude to our parents and grandparents – late Mr Sher Singh (whom we lost before completion of this book), Mrs Jagroshni, Mr Suraj Singh, Mrs Krishna Devi, Mr Anil Kumar Tyagi, Mrs Prem Lata Tyagi, Mr Yashveer Singh Tyagi, and Mrs Sudha Tyagi – who have supported and inspired us throughout and made us the people we are today. Guidance and support from Dr V. C. Kalia (CSIR-IGIB), Dr Sangita Kansal (DTU), Dr Mukti Acharya (DTU), the late Dr B. D. Acharya and Dr D. K. Singh (DU) have been instrumental in shaping our careers and aptitudes. Hence, this book also reflects the expertise they have invested in us. We will always be gtrateful for the help and support we received from members of our immediate families: Anshuman Tyagi and our loving children Takshira, Suryanshi, Raghav and Dhruv. We must also acknowledge the support of our friends and colleagues: Dr Divya M. Gnaneswari, Dr Smriti Sharma, Dr Thoudam Regina Devi, Dr Rashmi Saini, Dr Chaitali Ghosh, Dr Neena Kumar and all those whom we are unable mention here, but whose support was invaluable to us. This acknowledgment would not be complete without thanking our student friends for their support: Rukhsar Afreen, Madhuri Jha, Sujit K. Singh, Riddhi Jangid, and Ramnayan Verma. Finally, we would like to express our deepest gratitude to Gargi College (University of Delhi); School of Computational and Integrative Sciences, Jawaharlal Nehru University; Science and Engineering Research Board (SERB) (project funds: ECR/2017/001130 & ECR/2017/003480/PMS), India for providing necessary funds and facilities for the completion of this book.

Mamtesh Singh
Gajendra Pratap Singh
Shivani Tyagi

Contributors

Preeti Agarwal
Department of Botany
Gargi College, University of Delhi
New Delhi, India

Archana Aggarwal
Department of Zoology
Maitreyi College, University of Delhi
New Delhi, India

Roman Kumar Aneshwari
Institute of Pharmacy
Pt. Ravishankar Shukla University
Raipur, India

Swati Bajaj
Department of Zoology
Gargi College, University of Delhi
Delhi, India

Rameshwari A. Banjara
Department of Chemistry
Rajeev Gandhi Government Post Graduate College
Ambikapur, India

Manju Bhaskar
Department of Zoology
D. B. S. College Govind Nagar
Kanpur, India

Chandrasekaran Binuramesh
PG and Research Department of Zoology
Thiagarajar College
Madurai, India

Nagendra Kumar Chandrawanshi
School of Studies in Biotechnology
Pt. Ravishankar Shukla University
Raipur, India

Archna Chaudhary
Department of Environmental Sciences, Faculty of Science
SGT University
Gurugram, India

Sonal Chaudhary
Amity Institute of Microbial Technology
Amity University
Noida, India

Divyanshi Chauhan
Department of Zoology
Gargi College, University of Delhi
New Delhi, India

Sakshi Dawer
Department of Botany
Gargi College, University of Delhi
New Delhi, India

Deepali
School of Studies in Biotechnology,
Pt. Ravishankar Shukla University,
Raipur, India

Rakhi Dhankhar
Department of Microbiology
Maharshi Dayanand University
Rohtak, India

Sonika Dhillon
Department of Microbiology
Maharshi Dayanand University
Rohtak, India

Neha Dhingra
Department of Zoology
University of Delhi
New Delhi, India

M. Divya Gnaneswari
Department of Zoology
Gargi College, University of Delhi
New Delhi, India

Neelam Gandhi
Department of Zoology
Hansraj College, University of Delhi
New Delhi, India

Paushali Ghosh
School of Biotechnology
Institute of Science, Banaras Hindu University
Varanasi, India

Pooja Gulati
Department of Microbiology
Maharshi Dayanand University
Rohtak, India

Rakhi Gupta
Department of Zoology
Maitreyi College, University of Delhi
New Delhi, India

Sadhna Gupta
Department of Zoology
Dayal Singh College, University of Delhi
New Delhi, India

Dharmesh Harwani
Department of Microbiology
Maharaja Ganga Singh University
(University of Bikaner)
Rajasthan, India

Anina James
Department of Zoology, Deen Dayal
Upadhyaya College, University of Delhi
Delhi, India

Riddhi Jangid
MathSciIntR Lab, School of Computational
and Integrative Sciences
Jawaharlal Nehru University
New Delhi, India

Leisan Judith
Department of Botany
Gargi College, University of Delhi
New Delhi, India

Seema Kalra
School of Sciences
Indira Gandhi National Open University
New Delhi, India

Vera Yurngamla Kapai
Department of Botany
Gargi College, University of Delhi
New Delhi, India

Rajeev Kumar Kapoor
Medical Microbiology and Bioprocess
Laboratory, Department of Microbiology
MD University
Rohtak, India

Twinkle Kathuria
Department of Zoology
University of Delhi
New Delhi, India

Prabhleen Kaur
Department of Zoology
University of Delhi
New Delhi, India

Anubhuti Kawatra
Department of Microbiology
Maharshi Dayanand University
Rohtak, India

Anjali Kosre
School of Studies in Biotechnology
Pt. Ravishankar Shukla University
Raipur, India

Ashish Kumar
Department of Biotechnology
Sant Gahira Guru Vishwavidyalaya
Sarguja Ambikapur, India

Jai Kumar
Department of Zoology, Gargi College
University of Delhi
New Delhi, India

Jay Kumar
Molecular and Structural Biology Division
CSIR-Central Drug Research Institute
Lucknow, India

Mohit Kumar
Department of Zoology
Hindu College, University of Delhi
New Delhi, India

Sanjay Kumar
Medical Microbiology and Bioprocess
Laboratory, Department of Microbiology
MD University
Rohtak, India

Sunil Kumar
Department of Biological Sciences
Sungkyunkwan University
Suwon, South Korea

Tarika Kumar
Department of Environmental Studies, Faculty of Science
The Maharaja Sayajirao University
Vadodara, India

Usha Kumari
Department of Zoology
Gargi College, University of Delhi
New Delhi, India

Kuntal
Department of Zoology
Gargi College, University of Delhi
New Delhi, India

Jyoti Lakhani
Department of Computer Science
Maharaja Ganga Singh University
Bikaner, India

Indra Mani
Department of Microbiology
Gargi College, University of Delhi
New Delhi, India

Reema Mishra
Department of Botany
Gargi College, University of Delhi
New Delhi, India

Aparajita Mohanty
Department of Botany
Gargi College, University of Delhi
New Delhi, India

Gladys Muivah
Department of Botany
Gargi College, University of Delhi
New Delhi, India

Shalini Porwal
Amity Institute of Microbial Technology
Amity University
Noida, India

Preeti
Division of Germplasm Evaluation
National Bureau of Plant Genetic Resources
New Delhi, India

Rita Rath
Department of Zoology, Dayal Singh College
University of Delhi
New Delhi, India

Vijay Rayasam
Department of Biotechnology
REVA University, Rukmini Knowledge Park
Bangalore, India

Smriti Sharma
Department of Zoology
Gargi College, University of Delhi
New Delhi, India

Sushma Sharma
University of Delhi
New Delhi, India

Pallee Shree
Department of Zoology
Bhaskaracharya College of Applied Sciences
University of Delhi
New Delhi, India

Aashita Singh
Department of Zoology
University of Delhi
New Delhi, India

Ayushi Singh
Amity Institute of Microbial Technology
Amity University
Noida, India

Dileep K. Singh
Department of Zoology, University of Delhi
Delhi, India

Divya Singh
School of Biotechnology
Institute of Science
Banaras Hindu University
Varanasi, India

Gajendra Pratap Singh
MathSciIntR Lab, School of Computational and Integrative Sciences
Jawaharlal Nehru University
New Delhi, India & Special Centre for Systems Medicine
Special Centre for Systems Medicine
Jawaharlal Nehru University
New Delhi, India

Garvita Singh
Department of Botany
Gargi College, University of Delhi
New Delhi, India

Jyotsna Singh
Department of Zoology
Daulat Ram College
University of Delhi
New Delhi, India

Mamtesh Singh
Department of Zoology
Gargi College, University of Delhi
New Delhi, India

Nirmal Singh
Department of Seed Science and Technology, College of Agriculture
CCS Haryana Agricultural University
Hisar, India

Sawraj Singh
Medical Microbiology and Bioprocess Laboratory, Department of Microbiology
MD University
Rohtak, India

Tanvi Singh
University of Delhi
New Delhi, India

Vijai Singh
Department of Biosciences, School of Science
Indrashil University
Raipur, India

Renu Soni
Department of Botany
Gargi College, University of Delhi
New Delhi, India

Neeraja Sood
Department of Zoology
Dayal Singh College, University of Delhi
New Delhi, India

Sourabh
ICAR - Central Arid Zone Research Institute
Jodhpur, Rajasthan, India

Nidhi Srivastava
Department of Zoology
School of Basic and Applied Sciences
Maharaja Agrasen University
Solan, India

Parasuraman Aiya Subramani
Centre for Fish Immunology
Vels Institute of Science, Technology, and Advanced Studies
Chennai, India

Yamini Tiwari
Vivekananda Global University
Jaipur, India

Shivani Tyagi
Department of Zoology
Gargi College, University of Delhi
New Delhi, India

Ajit Varma
Amity Institute of Microbial Technology
Amity University
Noida, India

Aarti Venkatesan
Department of Zoology
Gargi College, University of Delhi
New Delhi, India

Vinod Kumar Verma
Department of Life Science
C. S. J. M. University
Kanpur, India

Deepika Yadav
Department of Zoology
Shivaji College, University of Delhi
New Delhi, India

Madhu Yashpal
Department of Zoology
Gargi College, University of Delhi
New Delhi, India

About the Editors

Mamtesh Singh, PhD, is presently working as Assistant Professor, Gargi College (University of Delhi), Delhi, India. She holds a bachelor's degree in zoology from University of Delhi and PhD in Biotechnology from CSIR-Institute of Genomics and Integrative Biology (IGIB). She has over eight years' teaching experience in life sciences and zoology. Her research area of interest is microbial biotechnology for generation of bio-products utilizing biological wastes as cheap renewable resources, and she has been working on biopolymer and biofuel production since 2006. She has also worked on applications of in-silico tools and Petri nets for analysis and modelling in biotechnology. She received a Young Scientist Award from the National Environment Science Academy (NESA) in 2018 for her significant work in this field.

Lt. Gajendra Pratap Singh, PhD, is presently working as Assistant Professor, School of Computational and Integrative Sciences (SCIS), Jawaharlal Nehru University, Delhi, India and concurrent faculty at Special Centre for Systems Medicine, Jawaharlal Nehru University, Delhi, India. Here, he also serves as an Associate National Cadet Corps (NCC) Officer. He received his PhD in applied mathematics in the area of graph theory and Petri Nets from Delhi College of Engineering (now Delhi Technological University), Faculty of Technology, University of Delhiand has eight years of teaching and research experience. His teaching areas of interest are applied mathematics, biostatistics, discrete mathematics, graph theory, Petri Nets, network analytics, optimization, mathematical biology and their applications. He has been conducting research work in these prime areas and published around 40 papers in reputed journals. He has been awarded young scientist award in 2018 in the area of mathematics-Petri Nets theories. Dr Singh is an associate editor and member of the expert committee of many national/international journals of repute. He is also a life member of several mathematical societies.

Shivani Tyagi, PhD, is presently working as Assistant Professor, Gargi College (University of Delhi), Delhi, India. She has degrees including an M. Phil in botany, zoology, chemistry, immunology and research methodology from C.C.S University and a PhD in bacterial taxonomy and soil health from the Department of Zoology, University of Delhi. She was awarded a gold medal for gaining top marks in post-graduation. She has nearly ten years of teaching experience in life sciences and zoology, specifically teaching cell biology, ecology, environmental science, evolution, animal diversity and bioinformatics courses to undergraduate students. Her research areas of interest are bacterial taxonomy, biopolymers and bioplastics.

Part I

Environment

1 Microbial Products
Applications in the Field of Biotechnology and Bioremediation

Shalini Porwal, Sonal Chaudhary, Ayushi Singh and Ajit Varma
Amity Institute of Microbial Technology, Amity University, Noida, India

CONTENTS

1.1 INTRODUCTION

The advent of genetic engineering in the 1970s opened a new arena for transdisciplinary research in modern biotechnology, and today, with recent advances, it encompasses various disciplines and has wide fields of application. The term biotechnology was coined by the 19th-century Hungarian economist, Karoly Ereky, during the discovery of innovative techniques for obtaining more useful products by converting various raw materials. In 1988, the European Federation of Biotechnology defined biotechnology as an "integrated approach of biochemistry, microbiology, and engineering sciences to achieve the application of capabilities of microorganisms, cultured animal cells, or plant cells, or parts thereof, in industries, agriculture, health care, and environmental processes". The most prominent era for biotechnology as a key technology is the 21st century [1].

The global population explosion has caused high consumption demand which has had a negative impact on the feedback process for the purification and recycling of nutrients in the ecosystem. Environmental contamination and pollution, and climate change, is the major consequence of anthropogenic and industrial activity [2, 3], which are regarded as the major source of persistent toxic chemicals detected in our ecosystem. Because of the increased use of natural resources, particularly non-renewable ones, a rapid rate of production and extraction of natural resources may be required to fulfil the demands of the world's growing population. The trend and intensity of pollution in our living environment and ecosystem cannot be reversed sooner even with a huge increase in education levels and living standards. Pristine ecological niches are rapidly declining in number, and human activities are adding persistent and toxic chemicals, including polymeric materials or plastics, to our ecosystem [4]. Industrial processes designed to improve the quality

DOI: 10.1201/9781003306931-2

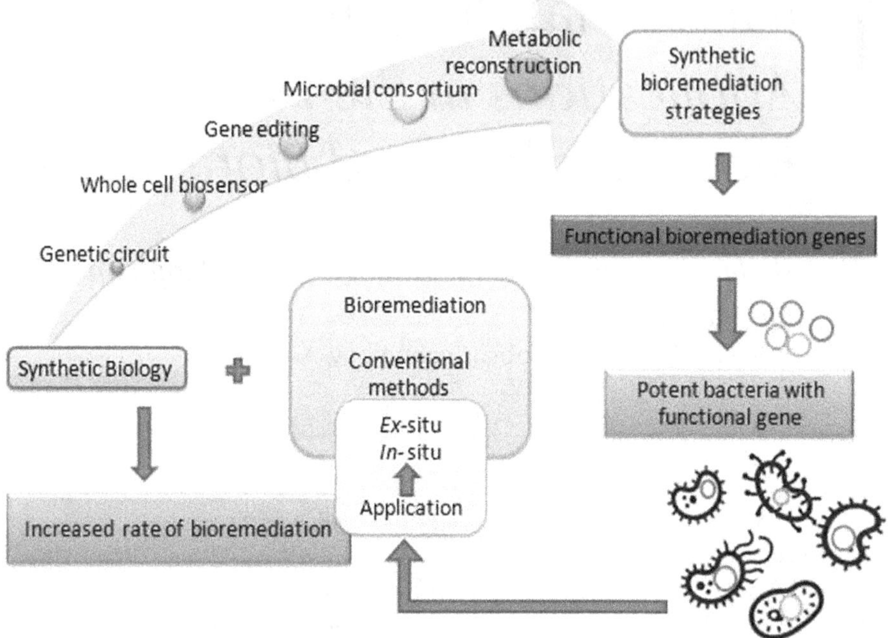

FIGURE 1.1 The strategies of synthetic biology applicable for bioremediation [13].

of human life have caused the natural ecosystem to become contaminated with a large quantity of various types of xenobiotics, such as anthropogenic halogenated hydrocarbon [5–7], which contains a wide range of biocidal chemicals in, for example, agriculture and household products, and invisible microplastics [4, 8].

The heterogeneity of the ecosystem, which consists of the lithosphere, hydrosphere, atmospheres, biospheres, and anthroposphere, does help provide different niches for sequestration and storage of toxic chemicals and materials [9]. Chemicals are extremely resistant to biodegradation, persisting for an extended period which results in accumulations of toxicity, ecologically and biologically [4, 8, 10–12]. While polymeric materials have remarkable application properties due to their chemical composition, they have also caused continuous xenobiotic and associated recalcitrant compound contamination of the environment.

The key to removing persistent contaminants from the environment is bioremediation. Traditional bioremediation processes have limitations, so new bioremediation technologies must be developed to achieve better results. Researchers are investigating several synthetic biological models of microbial bioremediation, including the conditions for constructing synthetic biological models of microbial bioremediation (Figure 1.1) [13]. With recent advances in analytical methods and developing understanding of the free and bioavailable fraction of total concentration, the actual concentration of the specific toxicant can be detected.

1.2 THE ROLE OF MICROBES IN BIODEGRADATION AND BIOREMEDIATION

During evolution organisms, whether small or large, have acquired the genetic traits and biochemical capabilities for dealing with unfavorable environmental conditions. Such capabilities in microbes including fungi (yeasts), bacteria, and viruses have laid the foundation for basic as well as translational research in environment biotechnology. Capabilities enabling them to detoxify and adapt via enzymes or biochemical reactions for extracting energy, and develop resistance to toxicity are

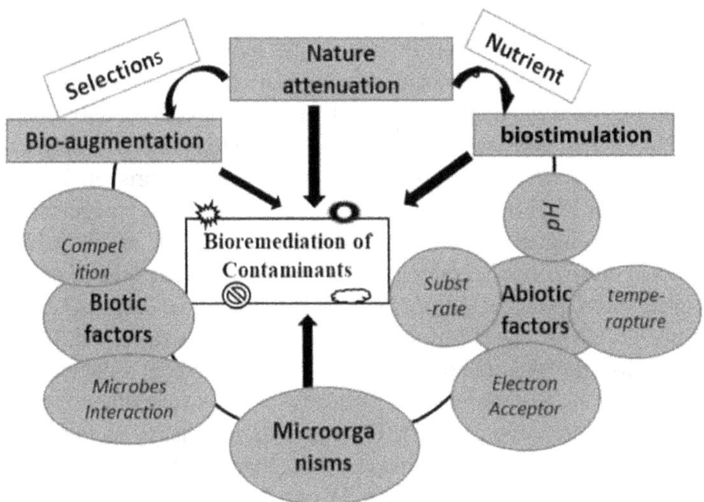

FIGURE 1.2 Bioremediation of contaminants and associated biotic and abiotic factors [14].

of prime interest in bioremediation and biodegradation which are accomplished with bioaugmentation and biostimulation (Figure 1.2). As we all know, highly toxic organic compounds such as fuels, PCBs, PAHs, pesticides, and dyes [9] have been synthesized and released into the environment for long-term (direct or indirect) applications. Other synthetic chemicals, such as radionuclides and metals, are much more resistant to biodegradation by native flora than naturally occurring organic compounds, which degrade quickly after being introduced into the environment, whereas there are some other toxic agrochemicals, such as atrazine, which metabolize through aerobic microorganisms. However, the deprivation of anaerobic conditions has not yet been proved [4, 5, 15, 16].

The biodegradation and bioremediation processes are not scientifically equivalent. Bioremediation uses living organisms, particularly microorganisms, to degrade pollutants and convert them into less toxic or nontoxic forms. Bacteria, fungi, and plants are examples of suitable organisms because they have the physiological abilities to degrade or detoxify the harmful elements. On the other hand, biodegradation is a natural process that recycles biologically essential elements in the Earth's biogeochemical cycles. It is mainly mediated by microbes and enzymes that are arranged in pathways to convert chemicals via a series of intermediates into end products which usually catalyze biodegradation processes [5, 15, 16]. These methods aim to use the incredible natural microbial catabolic diversity to degrade, transform, or accumulate a wide range of compounds, including hydrocarbons (e.g., oil), PCBs, PAHs, radionuclides, and metals [12]. Bioremediation of polluted places is an entirely separate problem, although there are many factors assisting in making a bridge between the two. The claim for on-site bioremediation based on laboratory demonstration of selective toxicant degradation highlights the huge gap between on-site bioremediation and biodegradation. Many of the successful bioremediation scenarios are in bioreactors or in field conditions where natural diminution is a major contributor. Nevertheless, these selective laboratory demonstrations have significant implications for on-site bioremediation and clean-up, but in conjunction with knowledge and data of the chemical, bioavailability, size characteristics, and bioactivities of potential degrading microorganisms, as well as changes in climate conditions.

The degradation of environmental pollutants by various microbes has been the subject of numerous studies. For instance, hydrocarbons, polychlorinated biphenyls (PCBs), polycyclic aromatic hydrocarbons (PAHs), pesticides, dyes, heavy metals, and many others are all degraded with the help of microorganisms. Bacteria that help in the degradation of hydrocarbons are known as hydrocarbon-degrading bacteria. The nitrate-reducing bacterial strains *Pseudomonas* sp., *Bacillus, Corynebacterium, Staphylococcus, Streptococcus, Shigella, Alcaligenes, Acinetobacter, Escherichia,*

Klebsiella, and *Enterobacter* isolated from petroleum-contaminated soil biodegrade hydrocarbons under aerobic and anaerobic conditions. PCBs can be biotransformed by anaerobic and aerobic bacteria alike. Anaerobic microorganisms dehalogenate higher chlorinated PCBs by reductive dehalogenation. Aerobic bacteria oxidize lower chlorinated biphenyls. Gram-negative strains of *Pseudomonas*, *Burkholderia, Ralstonia, Achromobacter, Sphingomonas*, and *Comamonas* have so far been the focus of research on aerobic bacteria. Several reports on PCB-degrading activity and characterization of PCB-degrading genes, on the other hand, suggested that some gram-positive strains could also degrade PCBs (*Rhodococcus, Janibacter, Bacillus, Paenibacillus*, and *Microbacterium*). The chlorpyrifos-degrading bacterium *Providencia stuartii* is isolated from agricultural soil [17] and isolates *Bacillus, Staphylococcus*, and *Stenotrophomonas* capable of degrading dichlorodiphenyltrichloroethane (DDT) from cultivated and uncultivated soil. Heavy metals cannot be destroyed biologically (no "degradation," or change in the element's nuclear structure), but can only be transformed from one oxidation state or organic complex to another to reduce their toxicity [14]. Furthermore, bacteria are effective in the bioremediation of heavy metals. Adsorption, uptake, methylation, oxidation, and reduction are some of the mechanisms that are used by microorganisms to protect themselves from heavy metal toxicity. Dissimilatory metal reduction is a technique for reducing metals: for example, under aerobic [18] or anaerobic conditions, the reduction of Cr(VI) to Cr(III), the reduction of Se(VI) to elemental Se, and the reduction of U(VI) to U(IV) [14], and the reduction of Hg(II) to Hg(IV).

Fungi metabolize dissolved organic matter and are the primary organisms responsible for carbon decomposition in the biosphere. Fungi have significant degradative capabilities, which have implications for the recycling of recalcitrant polymers (such as lignin and laccase) and the removal of hazardous wastes from the environment [19].

Yeasts are known for removing toxic heavy metals from the body. Biosorption of heavy metals by yeasts has been documented in numerous studies. Several studies have shown that yeasts can accumulate heavy metals like Cu(II), Ni(II), Co(II), Cd(II), and Mg (II) and are better metal accumulators than bacteria [20]. Researchers have discovered that *P. anomala* can remove hexavalent chromium Cr(VI), and have investigated Cr(VI) biosorption by live and dead cells of three yeast species: *Cyberlindnera fabianii, Wickerhamomyces anomalus*, and *C. tropicalis*. Several yeast strains have been shown to reduce Cr(VI) to Cr(III): *S. cerevisiae, P. guilliermondii, Rhodotorula pilimanae, Yarrowiali polytica*, and *Hansenula polymorpha* [21]. *P. guilliermondiis* tolerance to chromate was also found to be dependent on its ability to chelate Cr(VI) and Cr (III) outside the cell. The efficiency of immobilized yeast cells in metal removal has been reported in most studies; one example is S*chizosaccharomyces* pombe for copper removal.

1.3 ENVIRONMENTAL BIOTECHNOLOGY

Environmental biotechnology is based on fundamental science integrated with its application in solving environmental issues, e.g., in manufacturing, engineering, and the ecosystem. Two important pillars of fundamental sciences, biology and chemistry, provide support to environmental biotechnology with engineering and management thrown in for good measure, as shown in Figure 1.3. Similar to any biotechnology establishment, environmental biotechnology is first illustrated with organisms, biochemical processes, and specific chemical reactions in order to pave the way for the trial of prototypes to be tested using innovation theory [22]. Genuine operation and applications may also reach different scales, ranging from laboratory microtubes and various-sized controllable reactors to full-scale applications for pollutant elimination, [18] to purify wastewater, [20, 23] entire scales of polluted areas, or to restore and clean up a lake, rive, or wetland in a coastal environment. Thus, both fundamental biology and chemistry are required to lay solid foundations for the process involved. It includes the use of specific microorganisms (such as bacteria, archaea, insects, fungi, and plants), the selective enzyme and biochemical capability of the organism, the transformation of a biochemical product or intermediate for

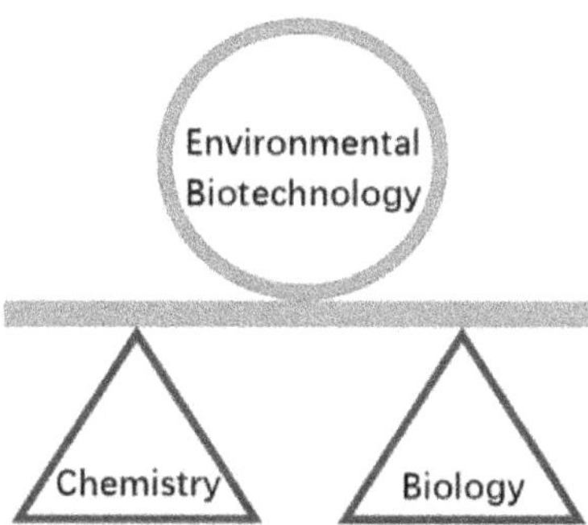

FIGURE 1.3 Chemistry and biology as the foundation of environmental biotechnology.

environmental pollution or toxics clean-up. Some of the successful implementations involve: i) activated sludge, which uses activated microorganisms to achieve removal and deprivation of organic and inorganic pollutant ranges or toxicants; ii) converting residual oil in reservoirs; iii) improving nutrition and food production; iv) protecting sustainable environments from eutrophication; v) extracting conventional energy from reservoirs; or vi) producing new energy supplies in a variable form [19, 21, 24].

1.4 BIOLOGICAL ACTIVITIES OF NATURAL PRODUCTS AND BIOLOGICS

Some natural products have biological properties that are also relevant for the health of humans, including antifungal, antibiotic, anticancer, anti-inflammatory, and biofilm inhibitory activities. These products can be grouped into different categories.

1.4.1 Antibiotics

Natural products are rich resources for the development of antibiotic drugs, but the maximum usage is made of those that can also be classified as non-ribosomal peptides, polyketides, and aminoglycosides [24]. When combined with polyketide synthases, polyketides produce the most diverse class of chemically diverse natural compounds, which are also among the most significant secondary metabolites for use in medicine, industry, and agriculture. Puromicin, which is highly potent against multi-drug-resistant respiratory pathogens, was the first polyketide antibiotic, produced from *S. venezuelae* in 1950. Erythromycin, discovered in 1952, is another polyketide antibiotic with significant clinical application, from *S. erythraea*. It is used to treat bacterial infections, such as gastrointestinal and respiratory, for acne, and in patients allergic to penicillin.

Vancomycin is a glycopeptide antibiotic that prevents the formation of cell walls. The hydrophilic part of vancomycin can bind to the alanine residues of NAM/NAG-peptides, preventing the cell wall linking enzyme from joining. Vancomycin was discovered in 1953 by Edmund Kornfeld from a soil sample of *Amycolatopsis orientalis*, and it was first marketed in 1954. It treated gram-positive bacterial infections that were resistant to other antibiotics. Vancomycin is inefficient against gram-negative bacteria due to its mechanism [25].

1.4.2 Antifungal Agents

In the year 1950, the primary active polyene antifungal reagent obtained was Nystatin from *Streptomyces noursei*, which was highly effective against *Aspergillus* species. It is used in the treatment of oral and genital candidosis, and gastrointestinal infections [26, 27]. Amphotericin B is a traditional antifungal polyene product that helps in the treatment of life-threatening fungal infections caused by *Aspergillus* species and is especially beneficial to patients who have had organ transplants, are undergoing rigorous chemotherapy, or have acquired immunodeficiency syndrome.

The use of *B. licheniformis* to protect ornamental plants against fungal diseases has been authorized. It produces an antibiotic that kills fungi and may also make an antifungal enzyme. Many species of fungus, particularly those that cause leafspot and blight diseases, are resistant to *B. licheniformis*. A flavus CM5 growth was reduced by 88% in in-vivo experiments on maize ears, with total prevention of fungal sporulation and aflatoxin build-up. 3-methyl-1-butanol was the most abundant component in the GC–MS-based volatile profile. The findings imply that *B. licheniformis* BL350-2 is an effective biocontrol agent for mycotoxigenic fungus, at least during cereal grain storage [17].

1.4.3 Anticancer Agents

Many microbe-derived anticancer agents are evaluated through clinical trials. The polyketide actinomycin was discovered from *Streptomyces parvulus* in 1940, and it was also the first antibiotic to be shown to have anticancer action. FDA-approved Actinomycin D, which is also known as dactinomycin, is also used widely in clinical practice as the anticancer drug for the treatment of tumor-like childhood rhabdomyosarcoma, Wilms Tumor, metastatic, non-seminomatous testicular cancer.

Antibiotics generated from the microorganism *Streptomyces peucetius* are known as anthracyclines. Doxorubicin, a hydroxylated daunorubicin derivative, is used to treat lymphoma, sarcomas, and carcinomas in people and animals. It has no cell cycle specificity and produces cytotoxicity by a number of methods, including free radical production, DNA intercalation, and protein synthesis suppression. It also inhibits topoisomerase, resulting in the formation of cleavable complexes, DNA damage, and cell death [28]

1.4.4 Immunosuppressive Agent

FK506 (Tacrolimus) and Rapamycin, also called Sirolimus, are microbial natural products with immunosuppressive properties. Rapamycin inhibits cell proliferation in response to stimulation by IL-2, IL-3, platelet-derived growth factor, insulin, and epidermal growth factor [29]. Rapamycin works in conjunction with other immunosuppressants, such as cyclosporin, to reduce kidney damage and acute renal allograft rejection. The chemical is being developed for the purpose of coating coronary stents and inhibiting organ transplant rejection and lymphangioleiomyomatosis. It received FDA approval in 1999. Rapamycin also possesses various other biological attributes, including antitumor, lifespan extension activity, antineoplastic, etc.

1.4.5 Anti-Inflammatory Agents

Many natural products are also concerned with anti-inflammatory activities. FK506 showed efficacy for treating refractory rheumatoid arthritis, a chronic inflammatory disease [29]. By reducing the activation and multiplication of inflammatory cells and cytokines, Rapamycin also inhibits the inflammatory response after spinal cord injury, and thus there is a reduction in injuries of the spinal cord, further providing a neuroprotective effect. Salinamides A and B from *Streptomyces sp.* CNB-091 have strong anti-inflammatory effects in a phorbol ester-induced mouse-ear edoema experiment. The peptides can inhibit the NF-KB pathways in vitro and have anti-inflammatory properties in vivo in the colitis model produced by dinitrobenzene sulfate.

1.5 CONCLUSION AND FUTURE PROSPECTS

Environmental biotechnology redefines solutions for numerous environmental hazards in cases where solutions in the form of biodegradation and bioremediation are identified. This chapter has underlined the problems observed in identifying and designing solutions for environmentally friendly approaches, where the limitations rest in the viability of metabolically dynamic behavior and the ability of the microorganism to degrade the pollutants under consideration. All types of life can be expected to have applications in environmental biotechnology. The classification of hazardous pollutants and their relationship with microorganisms producing natural bioremediate plays a vital role where antifungal, antibiotic, anticancer, anti-inflammatory, and immunosuppressive agents are employed as solutions. With the assistance of genetic engineering and recombinant DNA technologies, researchers are able to manipulate at the genetic levels of living organisms as well. As environmental biotechnology is the translational outcome of fundamental biological and chemical knowledge, equal priority should be given to both from the beginning. Moreover, no single method can be considered the optimal solution for environmental problems. A collective approach that includes the application of biotechnology will be more prominent and effective, and it is in this area that ongoing research is anticipated.

REFERENCES

1. Alpat, S.K., Alpat, Ş., Kutlu, B., Özbayrak, Ö., Büyükışık, H.B. 2007. Development of a biosorption-based algal biosensor for Cu (II) using Tetraselmis chuii. *Sensors and Actuators B: Chemical*. Dec 12;128(1):273–278.
2. Ackerman, D. 2014. *The Human Age: The World Shaped by Us*. W.W. Norton, New York, p. 352
3. Carson, R. 1962. *Silent Spring*. Penguin Books, Ltd, London, UK,. p. 336
4. Gu, J.-D. 2020. Anthroposphere, a new physical dimension of the ecosystems. *Applied Environmental Biotechnology*, 5(1): 1–3. http://doi.org/10.26789/AEB.2020.01.001
5. Schwarzenbach, R.P., Escher, B.I., Fenner, K., Hofstetter, T.B., Jonson, C.A., von Gunten, U., Wehrli, B. 2006. The challenge of micropollutants in aquatic systems. *Science*, 313: 1072–1077.
6. Cheung, K.H., Gu, J.-D. 2007. Mechanisms of hexavalent chromium detoxification by bacteria and bioremediation applications. *International Biodeterioration and Biodegradation*, 59: 8–15. http://doi.org/10.1016/j.ibiod.2006.05.002
7. Han, X., Gu, J.-D. 2010. Sorption and transformation of toxic metals by microorganisms. In: R. Mitchell and J.-D. Gu (eds) *Environmental Microbiology* (2nd ed.), 153–176, John Wiley, New York.
8. Gu, J.-D. 2021. Biodegradability of plastics: the issues, recent advances and future perspectives. *Environmental Science and Pollution Research*, 28(2): 1278–1282. http://doi.org/10.1007/s11356-020-11501-9
9. Gu, J.-D. 2018a. The endocrine-disrupting plasticizers will stay with us for a long time. *Applied Environmental Biotechnology*, 3(1): 61–64. http://doi.org/10.26789/AEB.2018.01.008
10. Gu, J.-D. 2003. Microbiological deterioration and degradation of synthetic polymeric materials: recent research advances. *International Biodeterioration and Biodegradation*, 52: 69–91.
11. Gu, J.-D. 2019a. Microbial ecotoxicology as an emerging research subject. *Applied Environmental Biotechnology*, 4(1): 1–4. http://doi.org/10.26789/AEB.2019.01.001
12. Gu, J.-D. 2019b. On applied toxicology. *Applied Environmental Biotechnology*, 4(2): 1–4. http://doi.org/10.26789/AEB.2019.02.001
13. Jaiswal, S., Shukla, P. 2020. Alternative strategies for microbial remediation of pollutants via synthetic biology. *Frontiers in Microbiology*, 11: 808.
14. Aragaw, T.A. 2020. Functions of various bacteria for specific pollutants degradation and their application in wastewater treatment: A review. *International Journal of Environmental Science and Technology*, 18(7): 2063–2076.
15. Atlas, R.M. 1995. *Bioremediation. Chemical and Engineering News*, American Chemical Society, Washington DC, 32–42.
16. Alexander, M. 1999. *Biodegradation and Bioremediation* (2nd ed.). Academic Press, San Diego, California.
17. Ul Hassan, Z., Al Thani, R., Alnaimi, H., Migheli, Q., Jaoua, S. 2019. Investigation and application of bacillus licheniformis volatile compounds for the biological control of *Toxigenic Aspergillus* and *Penicillium spp. ACS Omega*, 4(17): 17186–17193. http://doi.org/10.1021/acsomega.9b01638
18. Meng, L., Li, W., Zhang, S., Wu, C., Jiang, W., Sha, C. 2016. Effect of different extra carbon sources on nitrogen loss control and the change of bacterial populations in sewage sludge composting. *Ecological Engineering*, Sep 1; 94: 238–243.

19. Irfan, M., Zhou, L., Ji, J.-H., Chen, J., Yuan, S., Liang, T.-T., Liu, J.-F., Yang, S.-Z., Gu, J.-D., Mu, B.-Z., 2020a. Enhanced energy generation and altered biochemical pathways in an enrichment microbial consortium amended with natural iron minerals. *Renewable Energy*, 159: 585–594. http://doi.org/10.1016/j.renene.2020.05.036
20. Irfan, M., Zhou, L., Ji, J.H., Yuan, S., Liu, J.F., Yang, S.Z., Gu, J.D., Mu, B.Z. 2020. Energy recovery from the carbon dioxide for green and sustainable environment using iron minerals as electron donor. *Journal of Cleaner Production*, Dec 20; 277: 124134.
21. Irfan, M., Bai, Y., Zhou, L., Yuan, S., Mbadinga, S.M., Yang, S.-Z., Yang, J.-F., Sand, W., Gu, J.-D., Mu, B.-Z., 2019. Direct microbial transformation of carbon dioxide to value-added chemicals: A comprehensive analysis and application potential. *Bioresource Technology*, 288: 121401. http://doi.org/10.1016/j.biotech.2019.121401
22. Liu, L., Corma, A. 2018. Metal catalysts for heterogeneous catalysis: from single atoms to nanoclusters and nanoparticles. *Chemical Reviews*, Apr 16; 118(10): 4981–5079.
23. Bai, Y., Zhou, L., Irfan, M., Liang, T.-T., Cheng, L., Liu, Y.-F., Liu, J.-F., Yang, S.-Z., Sand, W., Gu, J.-D., Mu, B.-Z. 2020. Bioelectrochemical methane production from CO_2 by Methanosarcina barkeri via direct and H_2-mediated indirect electron transfer. *Energy*, 210: 118445. http://doi.org/10.1016/j.energy.2020.118445
24. Irfan, M., Zhou, L., Ji, J.-H., Yuan, S., Liu, J.-F., Yang, S.-Z., Gu, J.-D., Mu, B.-Z., 2020b. Energy recovery from CO_2 for green and sustainable environment using iron minerals as electron donor. *Journal of Cleaner Production*, 277: 124–134. http://doi.org/10.1016/j.jclepro.2020.124134
25. Cui, Q., Bian, R., Xu, F., Li, Q., Wang, W., Bian, Q. 2021. New molecular entities and structure–activity relationships of drugs designed by the natural product derivatization method from 2010 to 2018. In: Atta-ur-Rahman (ed), *Studies in Natural Products Chemistry*, (Vol. 69). 371–415, Elsevier, Netherlands.
26. Wright, G. 2014. Synthetic biology revives antibiotics: Re-engineering natural products provides a new route to drug discovery. *Nature*, May 1; 509(7498 SI): S13.
27. Vezina, C., Kudelski, A., Sehgal, S.N. 1975. Rapamycin (AY-22, 989), a new antifungal antibiotic I. Taxonomy of the producing streptomycete and isolation of the active principle. *The Journal of Antibiotics*, 28(10): 721–726.
28. Borgatti, A. 2013. Chemotherapy. In: Washabau, R.J., Day, M.J. (eds), *Canine and Feline Gastroenterology-e-Book*. 494–499, Elsevier Saunders, St. Louis, Missouri
29. Migita, K., Eguchi, K. 2003. FK506: Anti-inflammatory properties. *Current Medicinal Chemistry-Anti-Inflammatory and Anti-Allergy Agents*, Sep 1; 2(3): 260–264.

2 Microbial Proteases

A Significant Tool for Industrial Applications

Tanvi Singh and Sushma Sharma
University of Delhi, New Delhi, India

Yamini Tiwari
Vivekananda Global University, Jaipur, India

CONTENTS

2.1 INTRODUCTION

Enzymes act as the root system for catalyzing the metabolic activities of living organisms. They are essential for survival, and their malfunctioning can lead to various ailments. Enzymes are classified into six groups: hydrolases, oxidoreductases, lyases, transferases, ligases and isomerases.

Proteases belong to the hydrolases group of enzymes (International Union of Biochemistry and Molecular Biology (IUBMB) nomenclature, group 3, subgroup 4). They are degrading enzymes

DOI: 10.1201/9781003306931-3

which act as catalysts to cleave peptide bonds in proteins, thereby hydrolyzing them. They show selectivity and specificity in protein modification. Proteases perform various roles, from cellular to organ and organism level, producing a cascade of systems vital for life. According to the International Protease Network, each human has approximately 500 proteases that account for 2% of their entire genome (Gurumallesh et al., 2019).

Proteases are involved in synthesis regulation and activation as well as the turnover of proteins to control biological processes. They are also responsible for processes associated with the normal non-pathological as well as pathological conditions of cell physiology. Physiological processes involved in birth, development, aging, and death are controlled by these enzymes. Proteases play a significant part in the life-cycle of infectious diseases and have therefore been of vital importance in discovery of drugs (Poza et al., 2007).

Proteases are one of the three key enzyme groups used in industries, carbohydrases and lipases being the other two. There has been a dramatic increase in the global market for proteases, which account for 60% of the total enzyme market (Razzaq et al., 2019).

2.2 CATEGORIES OF PROTEASES

Based on the type of reaction, proteases are categorized into exopeptidases and endopeptidase (Rao et al., 1998). Based on their isolation source, proteases can be grouped into animal proteases, plant proteases and microbial proteases (Figure 2.1) (Singh et al., 2016).

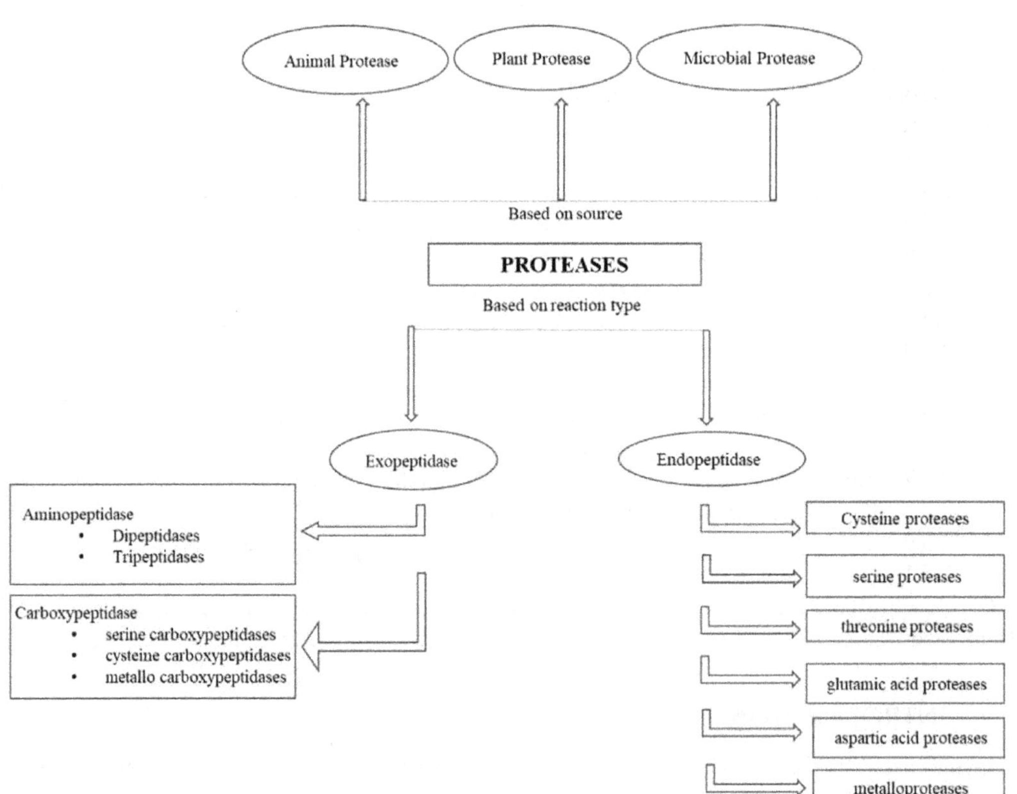

FIGURE 2.1 Types of proteases.

2.2.1 Reaction Type

Exopeptidases (E.C. No 3.4.11-3.4.19) catalyze the hydrolysis only on the carbon or nitrogen terminal of the polypeptide chain. They are called aminopeptidase and carboxypeptidase, depending on their activity. Aminopeptidases act on the free amino-terminal (N-terminal) of the polypeptide chain and release either one amino acid residue or a dipeptide or a tripeptide. Carboxypeptidases act on the carboxy terminal (C-terminal) of the protein. According to the nature of the amino acid residue at the active site they are further classified into three categories:

- Serine carboxypeptidases, which use serine residues at their active site.
- Cysteine carboxypeptidases, which use cysteine residues at their active site.
- Metallo carboxypeptidases, which use metal at their active site.

Endopeptidases (E.C. No 3.4.21-3.4.24, 3.4.99) catalyze the hydrolysis on non-terminal regions of the polypeptide chain and are categorized into six classes, based on the presence of the chemical group responsible for the catalysis.

- Cysteine proteases possess cysteine at their active site.
- Serine proteases contain residual serine at the active site.
- Glutamic acid proteases possess glutamic acid residue at their active site.
- Threonine proteases possess threonine residue at their active site.
- Aspartic acid proteases are acidic proteases which utilize activated molecules of water bound to aspartic acid residues at the site of catalysis.
- Metalloproteases use a divalent metal ion for proteolysis.

2.2.2 Source

2.2.2.1 Animal Proteases

Livestock is the primary source of the production of animal proteases. The most familiar animal proteases include chymotrypsin, pancreatic trypsin, rennin and pepsin. Trypsin acts as the chief enzyme of the intestine and is responsible for protein hydrolysis of food. It is extracted from bovine pancreas and pancreatic juices. Trypsin has been used in food industries, for medical applications, and for preparing bacterial culture media. Chymotrypsin is obtained from the pancreatic gland of animals. It is very expensive in its pure state and therefore is only utilized for analytical or diagnostic applications. It has been used extensively for reducing the antigenicity of hydrolysates, making them non-allergenic. Pepsin, being an acidic protease, is present within the gastric juices of stomach lumen of nearly all vertebrates. Pepsin is used in food industry, leather industry and in biomedical research. Rennin, also known as chymosin, is produced in the stomach of ruminant animals and is also extensively utilized in the food industry to form curds and in cheese making (Poza et al., 2007).

2.2.2.2 Plant Proteases

Bromelain, papain, and ficin are a few of the best-known plant proteases (Silva-López and Gonçalves, 2019). Papain is obtained from the dried latex of the papaya fruit (*Carica papaya*). It is commonly used in the food industry as a meat tenderizer, in cell culturing to dissociate cells, as stabilizing treatment in brewing, in therapeutic areas for cleaning infected tissues or cells, and for treating tooth decay. Bromelain is extracted from the stems of pineapples and pineapple juice. It has long been used for medicinal purposes to reduce swelling and inflammation, to remove dead skin from burns, to reduce soreness and muscle ache, and in the food industry as a meat tenderizer. Ficin, obtained from the latex of the fig tree (*Ficus carica*), is used in the food industry as a meat tenderizer, collagen hydrolyzer for obtaining gelatin, and for clotting of milk. In the biotechnology industry, it has been used as a biocatalyst for dipeptide synthesis and to provide immunoreactivity (Tantamacharik et al., 2018).

2.2.2.3 Microbial Proteases

Microorganisms form an admirable source of proteases due to their rapid growth and limited space requirement for their culture. Microorganisms are also easier to manipulate genetically and offer a limitless source of novel enzymes with diverse biochemical profiles, resulting in entirely new proteases or modifications to existing ones to meet new requirements from continuously growing industries. The diversity of microorganisms makes it possible to find a protease beneficial for all types of biotechnological processes (Beg and Gupta, 2003). Microbial proteases fulfill a variety of physiological, regulatory, and biochemical roles.

Using animals and plants as a source of proteases depends on numerous factors, such as space for raising the desired plants and animals, and climatic conditions. Protease extraction from plants and animals is a long and laborious process. The amount of enzyme extracted depends on the stock and on the approaches used for extraction and purification (Poza et al., 2007). Ethical issues associated with slaughtering animals to extract proteases can also delay the process. Microorganisms have therefore attracted research attention as a source of proteases on a large scale and at high rates. Of all marketable proteases produced worldwide, nearly two-thirds are microbial proteases (Gurumallesh et al., 2019).

2.3 CLASSIFICATION OF MICROBIAL PROTEASES

Based on the active site present on long chain polypeptides, microbial proteases are categorized into two types:

1. *Exopeptidases.* These enzymes exhibit their action at the end terminus of the polypeptide chains. Depending on their action site, they are sub-divided into two categories: aminopeptidases and carboxypeptidases.
2. *Endopeptidases.* These enzymes cleave peptide bonds by acting on the inner sides of the polypeptide chains and at a distance from both ends of the chain (Rao et al., 1998).

Depending on their optimum pH reactions, microbial proteases are graded as alkaline, neutral or acidic (Gupta et al., 2002).

1. *Alkaline proteases* are those which work at alkaline pH 9–11. They are essential enzymes, so they are ubiquitous in nature.
2. *Acidic proteases* are more stable and active at pH ranging from 3.8 to 5.6.
3. *Neutral proteases* are enzymes that work at neutral, weak acidic or weak alkaline pH. They have low thermotolerance and medium reaction rates (Razzaq et al., 2019). They are identified and characterized based on their elevated affinity against hydrophobic amino acids. Thermolysin is a known neutral protease, having molecular weight of 34 kDa. It possesses a single peptide and has no disulfide bridges. In between the two folds of protein lobes, a vital Zn atom and four Ca atoms are present. It is very active at 80°C and has half-life of 1 hour (Dawson and Kent, 2000).

Based on the presence of catalytic residue in the active site, microbial proteases are further divided into six categories: aspartate, serine, cysteine, glutamic acid, threonine and metallo proteases (Rao et al., 1998).

1. *Aspartate proteases* are endopeptidases and belong to acid proteases. For catalytic activity two aspartic residues are present. They exhibit optimal activity at acidic pH (pH 3 to 4) and retain isoelectric points from pH 3 to 4.5. Aspartate proteases preferentially disrupt peptide

bonds linking residues of non-polar amino acids. Pepstatin and diazoketone are inhibitors of aspartate proteases.

2. *Serine proteases* are most widely present in microbes and have reactive serine residue at the active site. Optimum activity of serine proteases is at a broad range of pH 7 to 11 and has wide substrate specificity with esterolytic and amidase activity. Serine alkaline proteases are used on a commercial scale as they are highly active and stable at critical conditions (Singh et al., 2016).
3. *Cysteine proteases* constitute a catalytic dyad having cysteine and histidine at their active site. They exhibit activity at extensive temperature and pH ranges, and are mainly used in food and pharmaceuticals. Of the various cysteine proteases in use, papain is most extensively used in the food industry. Cysteine proteases are mostly involved in degradation of proteins and scrapie protein metabolically (usually degrading neuronal and dendritic cells).
4. *Glutamic acid proteases*. Glutamic acid residue is present at their active acid.
5. *Threonine proteases* have threonine residue at catalytic site like acyltransferases and proteasome.
6. *Metallo proteases* need divalent metal ions like zinc, manganese or cobalt to exhibit their catalytic activity. They are sensitive to chelating agents such as EDTA owing to their sequestering effect on the metal ions. Metallo proteases possess a wide range of substrate specificities.

2.4 SOURCE OF MICROBIAL PROTEASES

2.4.1 Bacterial Proteases

The main bacterial genera that are the source of proteases include *Alcaligenes, Aeromonas, Arthrobacter, Halomonas, Bacillus, Serratia* and *Pseudomonas* (Table 2.1).

Neutral bacterial proteases show activity at near neutral pH values between 5 and 8. These are sparingly thermotolerant and this property gives them an advantage for generation of partial protein hydrolysates. These proteases generate protein hydrolysates which taste less bitter than animal proteases. A few of them fall into the metalloprotease group, and need divalent metal ions for activation, whereas neutral serine-type proteases are not affected by chelators.

Alkaline bacterial proteases are most active at alkaline pH and temperature of 60°C.

Most of the commercialized proteases, primarily alkaline or neutral, are obtained from bacteria belonging to the genus *Bacillus* and make up 35% of total microbial proteases (Jayakumar et al., 2012). Proteases derived from *Bacillus* spp. are favorable to the industrial process conditions. They exhibit the property of poly-extremotolerance, i.e., the potential to function in unfavorable process settings such as the presence of solvents, extreme temperature, extreme pH, presence of soaps, presence of other enzyme inhibitors. They are thus widely utilized in leather, textiles, food, waste water treatment and organic synthesis industries. For heterologous expression of genes, *Bacillus* spp. can also act as a model system (Sadeghi et al., 2009).

B. subtilis, B. licheniformis, and *B. amyloliquefaciens* are the most accepted species for bioprocessing related to protein synthesis, owing to their exceptional fermentation properties, high yield, and absence of toxic by-products.

2.4.2 Fungal Proteases

A wide variety of proteases are produced by fungi. These proteases exhibit activity at a broad pH range of 4–11 and show broad substrate specificity, but they exhibit a slow rate of reaction and less thermotolerance than bacterial proteases. Fungal proteases are of interest to researchers due to their broad substrate specificity, high diversity, stability in extreme environments, mycelial separation by simple filtration method, and convenient production by solid-state fermentation. The fungus *Aspergillus oryzae* has been reported to be capable of producing acid, neutral, and alkaline proteases (Poza et al., 2007)

TABLE 2.1
Protease-Producing Fungi, Bacteria and Viruses

Fungi	Bacteria		Virus
Aspergillus sp.	***Alteromonas* sp.**	***Pseudomonas sp.***	**Alphaviruses**
A. candidus	*Arthrobacter protophormiae*	*P. aeruginosa*	Aleutian disease virus (ADV)
A. clavatus	*Bacillus* sp.	*P. fluorescens*	Flaviviruses
A. fumigatus	*B. alcalophilus* subsp. *halodurans*KP1239, *B. alcalophilus* ATCC 21522	*P. maltophilia*	Human immunodeficiency virus (HIV)
A. flavus	*B. amyloliquefaciens, B. amyloliquefaciens S94*	*Pseudomonas sp. SJ320*	Herpes viruses
A. melleus	*B. cereus* strain CA15	*Staphylothermus marinus*	Hepatitis C virus (HCV)
A. oryzae	*B. intermedius*	*Salinivibrio* sp. Strain AF-2004	Picornaviruses
A. niger	*B. firmus*	*Streptomyces* sp.	
A. sydowi	*B. coagulans*PB-77	*Streptomyces sp.* EGS-5, *Streptomyces* sp. YSA-130	
A. sojae	*B. megaterium*	*S. microflavus*	
Beauveria felina	*B. licheniformis, B. licheniformis UV-9 Mutant*	*S. moderatus*	
Cephalosporium sp. KSM 388	*B. lentus*	*S. nogalator AC 80*	
Chrysosporium keratinophilum	*B. sphaericus*	*S. rectus, S. rectus var. proteolythicus*	
Conidiobolus coronatus	*B. pumilus, B. pumilus CBS*	*S. rimosus*	
Engyodontium album	*B. proteolyticus*	*Thermo bacteroides proteolyticus*	
Entomophthora coronata	*B. stearothermophilus*	*Thermoactinomyces* sp.	
Fusarium eumartii	*B. subtilis, B. subtilis DKMNR, B. subtilis var. amylosacchariticus, B. subtilis RTSBA*	*Thermococcus* sp.	
Lentinus citrinus	*B. thermoruber*BT2T	*T. stetteri*	
Mucor sp.	*B. circulans*	*T. celer*	
Paecilomyces lilacinus	*B. laterosporus*	*T. litoralis*	
Penicillium sp.	*Bacillus* sp. B001, *Bacillus* sp. B21-2, *Bacillus* sp. CW-1121, *Bacillus* sp. KSM-K16, *Bacillus* sp. *NKS-21, Bacillus* sp. *Y, Bacillus* sp. *Ya-B, Bacillus* sp. JB-99	*Thermomonospora fusca*	
P. chrysogenum	*Brevibacterium linens*	*Thermus sp.*	
P. godlewskii SBSS 25	*Geobacillus caldoproteolyticus*	*Thermus* sp. strain Rt41A	
Rhizomucor nainitalensis	*Hyphomonas jannaschiana* VP 3	*T. aquaticus* YT-1	
Rhizopus oligosporus	*Lactobacillus helveticus*	*Torula thermophila*	
Scedosporium apiospermum	Listeria monocytogenes	*Vibrio sp.*	
Tritirachium album Limber	*Malbranchea pulchella* var. *sulfurea*	*V. alginolyticus*	
	Microbacterium sp.	*V. metschnikovii* RH 530	
	Nocardiopsis dassonvillei	*Xanthomonas maltophila*	
	*Oerskovia xanthine olytica*TK-1	*Xenorhabdus nematophila*	
	Pimelobacter sp. 2483		

Source: Poza et al., 2007; Veloorvalappil et al., 2013; Sharma and Gupta, 2017; dos Santos Aguilar & Sato, 2018; Gurumallesh et al., 2019.

Fungal acid proteases are stable at pH range 2.5–6 and optimal at pH 4–4.5.

Fungal neutral proteases show activity at pH 7. These are usually metalloproteases and show inhibition through chelating agents. They are specifically functional towards hydrophobic amino acid bonds which permits the generation of protein hydrolysates having a slight bitter taste.

Fungal alkaline proteases are hardly used in food protein modification. Yeasts like *Candida caseinolytica* and *Yarrowia lipolytica* produce very active alkaline proteases in large quantities, which can be used in detergents as well as food industries (Gurumallesh et al., 2019). Table 2.1 shows a list of fungi commonly employed in production of proteases.

2.4.3 Viral Proteases

All viral proteases are endopeptidases. Serine, cysteine, and aspartic proteases have been found in viruses, while metalloproteases have not yet been found. Viral endopeptidases possess convincing features including small size, less sequence similarity (limited to active sites), adaptability to various roles, and specificity in cutting requirements. This specificity makes them potential tools for biotechnology and antiviral therapies (Shamsi et al., 2016; Tran et al., 2017). Retroviral aspartic proteases are involved in biological roles associated with assembly of viral particles and replication. Most importantly, viral proteases have achieved significant status because of their role in processing the viral proteins involved in development of a fatal diseases like AIDS or cancer. Table 2.1 shows important viruses that are the source of proteases exploited in drug design.

2.5 INDUSTRIAL APPLICATIONS OF MICROBIAL PROTEASES

Microbial proteases form the most imperative class of hydrolytic enzymes and alkaline microbial proteases are the highest rated in the enzymes market (Sharma et al., 2017). Microorganisms engendering a huge quantity of extracellular proteases are considered as of utmost importance. There are numerous industrial applications of microbial proteases (Figure 2.2).

2.5.1 Food Industry

In the food industry, microbial proteases are widely employed. Proteases are capable of fortifying food proteins to develop enhanced dietary value, digestibility, solubility, taste, and piquancy, and to curtail allergens (Deniz, 2019; Ismail et al., 2019). Apart from these basic functions, proteases are

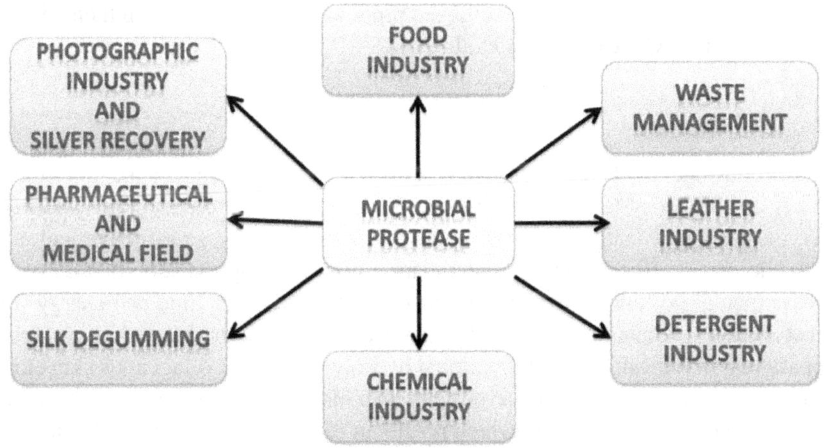

FIGURE 2.2 Industrial applications of microbial proteases.

also employed to alter certain functional properties like gel strength, protein coagulation, foaming, emulsification, binding of fat to proteins, etc. in food (Tavano et al., 2018).

Proteases are one of the vital enzymes used in the dairy industry for cheese making. Specific peptide bonds are hydrolyzed to yield kappa casein and macro-peptides to avert coagulation by alleviating micelle formation (Moschopoulou, 2018). In baking, proteases augment uniformity, consistency, and quicker preparation of dough, and maintain the strength of gluten during bread preparation by partially hydrolyzing it using a heat-labile protease (Mamo and Assefa, 2018).

Protein hydrolysate, having high nutritional value, is prepared by employing the catalytic function of microbial alkaline proteases for use in medicinal dietary products, baby food, fortified drinks and juices, to reduce the turbidity complex arising due to protein in alcohol-based liquors and fruit juices, and for production of soy products including soy sauce (Gao et al., 2019; Golkar et al., 2019; Ashaolu, 2020; Samaei et al., 2020). Proteases are capable of easily hydrolyzing connective tissue and muscle fiber proteins that can be employed to make meat tenderer and to recover meat proteins during gelatin hydrolysis (Philipps-Wiemann, 2018).

The bioactive peptide formed during hydrolysis of a variety of food proteins act as an antioxidant to restrain auto-oxidation of linoleic acid and the scavenging effects of free radicals in the cell. Exceptional solubility of protein hydrolysates is one of the reasons why antioxidant activities of protein hydrosylates were found to be improved (Horax et al., 2017). These bioactive peptides are further used in the formation of a range of pharmaceutical drugs (Lorenzo et al., 2018). Mackerel hydrolysate interferes during hydrolysis of proteins, subsequently forming free amino acids like anserine, carnosine, and some petite peptides by making use of proteases (Wang et al., 2017).

2.5.2 Waste Management

Today, the use of chemicals is not only increasing process input costs but is also threatening the ecosystem, leading to an urgent need for an alternative eco-friendly way to treat waste. Microbial proteases are one such alternative, and they play an important part managing domestic, medical and manufacturing waste. Proteinaceous waste can easily be solubilized by proteases, reducing the biological oxygen demand (BOD) of marine systems. Alkaline proteases effectively degrade household and food-processing industry waste (Naveed et al., 2021). Poultry feathers contain very stiff keratin, a plentiful source of natural waste that can be converted into useful biomass with the help of alkaline protease enzymes procured from microorganisms. Using the keratinolytic process, large amounts of poultry waste can be disintegrated into highly digestible feedstuffs and food. Alkaline protease extracted from *Bacillus subtilis* has been found effective in waste processing of feathers obtained from poultry slaughterhouses (Emon et al., 2020). *Bacillus amyloliquefacines, B. subtitlis*, and *Streptomyces* sp. are brought together with thioglycolate as a formulation and employed as a commercial surfactant to clean drains and clogged pipes and remove hair from them. This combination has been patented as Genex (Thakur et al., 2018).

2.5.3 Leather Industry

Toxic and hazardous chemicals used for conventional leather processing cause pollution and a detrimental effect on the ecosystem due to effluent disposal. Owing to elastolytic and keratinolytic activity, enzyme-mediated leather processing has been found to be an effective and eco-friendly alternative to chemical methods. Enzyme application not only improves the quality of leather but reduces environmental pollution, helps to minimize waste, and saves time and energy (Fang et al., 2017; Zhou et al., 2018; Khambhaty, 2020). Proteases are utilized in bating, dehairing, soaking junctures of preparation of skin and pelts, in fractional hydrolysis of non-collagenous skin elements, and also are reported to eradicate non-fibrillar proteins like albumins and globulins. Microbial alkaline proteases check rapid water absorption which results in reduced soaking time (Akhter et al.,2020). Alkaline proteases, in addition to hydrated lime and sodium chloride or sodium sulfide, reduce waste during dehairing and dewooling processes (Vitolo, 2019). Microbial alkaline protease accelerates dehairing as alkaline conditions cause hair roots to swell, making it easy for protease enzymes to attack hair follicle proteins. . Bating is the process of removing redundant inter-fibrillary proteins to produce soft and silky leather with increased elasticity (Sasia et al., 2019). Alkaline proteases extracted from certain microorganisms have exhibited effectual bating. Bating substantiates internal separation of collagens by hydrolyzing keratin, revealing maximum skin area for tanning.

2.5.4 Detergent Industry

Crude pancreatic extract along with sodium carbonate was the first enzyme to be used with a detergent named Brunus (Banerjee and Ray, 2017). Later, under the label BIO-40, microbial enzymes were employed commercially in enzyme-detergent preparation (Nasre Taheri et al., 2019). Protease extracted from *B. cereus* (BM1) was found to be a good cleaner, exhibiting robust activity in 10% (w/v) solution of commercial detergent (Illanes, 2008). Detergents account for up to 20–30% of the overall enzyme trade, which is expected to expand rapidly, with a compound annual growth rate of around 11.5% between 2015 and 2020 (Al-Ghanayem and Joseph 2020). Protease enzymes are helpful for digesting food and blood stains (Zhang et al., 2019). They are added as the crucial element in the formulation of domestic washing detergents, and for cleaning and storage solutions for contact lenses and dentures (Gürkök, 2019; Sangeetha and Arulpandi, 2019). The use of enzymes in detergent formulations reduces soaking and washing time, resulting in more eco-friendly laundering. Serine proteases are more often added in detergent formulations as the microbial protease additives that are extracted from *Bacillus* strains (Rekik et al., 2019). Alkaline proteases from fungal sources are also employed in detergent formulations due to easy downstream processing (Savitha et al., 2011). A mixture of various enzymes consisting of protease, lipase, amylase and cellulase are added to most domestic detergent formulations (Nathan and Rani, 2020).

Recombinant DNA technology is now being incorporated in the manufacture of bioengineered detergent proteases with oxidizing agents producing consistent high-temperature performance superior to that of existing commercial proteases (Al-Ghanayem et al., 2020). Stable proteases have been synthesized by replacing a few specific amino acid residues via protein engineering (Liu et al., 2019). The use of proteases is not limited to washing detergents; they are also used in dishwashing and for refining detergents in various other industries (Sharma ct al., 2019).

2.5.5 Photographic Industry

Proteases obtained from *Bacillus subtilis*, *B. lehensis*, *Purpureocillium lilacinum*, *Aspergillus versicolor*, *Conidiobolus coronatus* and *Streptomyces avermectnus* have been found effective for retrieval of silver from X-ray films, as they possess gelatinolytic activity which is much eco-friendlier than chemical processes (Rakotonirainy et al., 2016; Singh and Bajaj, 2017; Varia et al., 2019). Recovery of silver by making use of thermostable mutant alkaline proteases obtained from *Bacillus* spp. is reported to be effective as the elevated temperature and somewhat basic conditions favor easy removal of the gelatin coat. Proteolytic enzymes produced through nutrition-rich media were also competent to hydrolyze X-ray-bound gelatin efficiently (Kumaran et al., 2013).

2.5.6 Chemical Industry

In the chemical industry, highly stable enzymes in organic solvents are required for peptide synthesis in non-aqueous medium during biocatalysis. *Pseudomonas aeruginosa*, *Bacillus pseudofirus* and *Aspergillus flavus* producing alkaline proteases exhibited significant outcomes in synthesis of peptides, as these are much more stable in organic solvents than other enzymes (Sundus et al., 2016). For peptide and organic synthesis, some of the *Bacillus* and *Streptomyces* species that produce alkaline proteases are preferred to others (Yadav et al., 2015).

2.5.7 Silk Degumming

Silk is a multifaceted substance with two fibroin filaments cemented by coating of sericin (More et al., 2013). Processing of silk involves several steps including degumming. In this process sericin is removed using any soap or alkali to obtain silk with required characteristics. Using chemicals not only damages the quality of the silk, but also causes environmental pollution (Da Silva et al., 2017). An alternative is enzyme-based silk degumming involving use of proteolytic enzymes that are able to break the peptide bonds present in sericin exceptionally well without damaging the fibroin filaments (Radha et al., 2012). The bacterial enzyme alkalase advertised by Novo has been reported to hydrolyze sericin well. The combination of protease and lipase has been found to enhance wettability of silk fibers and is effective for degumming and dewaxing. According to Joshi and Satyanarayana (2013), a recombinant alkaline serine protease obtained from *B. lehensis* enhances the softness and luster of silk fibers. Moreover, the nitrogen-rich discharges obtained from the enzyme-based silk-degumming process are used as potential nutrient substrates for production of more protease from *B. licheniformis* and *Aspergillus flavus*. The liquor left as waste after degumming, loaded with plenty of sericin, is used as the raw material for producing sericin powder. Sericin powder is used in the cosmetics industry in moisturizers, for hair care products and also as a dietary additive in the food industry (Chongjun et al., 2016; Ghosh et al., 2017).

2.5.8 Pharmaceuticals and the Medical Field

A broad-spectrum use of microbial proteases is now to be seen for diagnostic and therapeutic purposes in the pharmaceutical and medical fields. Microbial proteases (specifically bacterial and fungal) are contributing to produce effectual therapies for clot disintegration, and as anticancer, antimicrobial and anti-inflammatory agents (Fernández de Ullivarri et al., 2020). Treatment of acute and chronic inflammation via protease has been proved to be cost-effective and to have no ill-effects. A protease enzyme produced by the *Serratia* species, serratiopeptidase, is reported to reduce pain. Serratiopeptidase is also reported to be the most efficient protease for treatment of inflammation (Tiwari, 2017). Protease obtained from *Aspergillus oryzae*is is used to treat lytic enzyme deficiency during digestion (He et al., 2019). Asparaginase extracted from *Escherichia coli is* used for the treatment of lymphocytic leukemia. Likewise, clostridial collagenase produced by *Clostridium histolyticum* is used for treating wounds and burns. Nattokinase obtained from *Bacillus subtilis* is used in the same way as a nutraceutical for cardiovascular disease, dissolving thrombus and preventing coagulation of blood (Weng et al., 2017). Various bacterial and fungal proteases have also been recognized for their antimicrobial properties (Yang et al., 2017). Degradation of keratin on skin, and preparation of vaccines for dermatophytosis therapy are also performed using proteases. Hydrolytic enzymes are used for regeneration of epithelia, removal of scars, and in wound healing.

Various formulations such as ointments, non-woven tissues, patches and gauze contain alkaline proteases obtained from *Bacillus subtilis*, and provide competent healing properties. Some lytic enzyme deficiency syndromes are treated by oral administration of alkaline proteases (Joshi and Satyanarayana, 2013). Alkaline fibrinolytic proteases have been reported to degrade fibrin, a potent molecule used in manufacturing of anticancer drugs and in thrombolytic therapy.

2.5.9 Silver Recovery

Silver has various industrial applications in photographic films, X-ray films, jewelry, silverware and electronics. A large amount of silver is used in the photographic industry. During processing of photographs, silver is not eliminated but can be recovered and recycled. X-ray and photographic films contain about 1.5–2% (w/w) of silver in their gelatin layers. Recovery of silver via traditional processes involving burning of X-ray films causes environmental pollution. Alkaline proteases are involved in the bioprocessing and recovery of silver from X-ray and photographic films. The enzyme removes the gelatin layer entangled with silver in X-ray films in a slow, economic and pollution-free way. Alkaline proteases obtained by *Conidiobolus coronatus, Bacillus subtilis* and *Streptomyces avermectinus* have been reported to be effective in silver recovery (Varia et al., 2019).

2.5.10 Other Uses

Besides acting as important industrial tools, proteases have other significant uses. They are used in basic research to cleave peptide bonds selectively, which explains the relationship between structure and function in peptide synthesis. Proteases can also be used in place of other chemicals involved in physiological and biochemical processes. For example, in DNA isolation, in place of proteinase K, alkaline proteases obtained by *Vibrio metschnikovii* can be utilized (Vijayaraghavan and Vincent, 2015; Razzaq et al., 2019). These have also been used in extraction and purification of non-protein products from plants or animals, such as mucopolysaccharides and carbohydrate gums. Research continues to prepare protease formulations applicable in various fields, such as skin care formulations involved in desquamation, as skin cleansers, human digestion-improving formulations, toothpastes, contact lens cleaning solutions and wellness medication (Gurumallesh et al., 2019). Biodegradable sucrose polyester is synthesized by *Bacillus* alkaline protease, commercially known as Proleather (Poza et al., 2007)

2.6 CONCLUSION

Proteases are a strongly growing sector establishing standards in each and every field of life.

With advances in science and technology, a wide range of techniques for exploitation of protease for the benefit of mankind have been emerging. Most enzyme-driven industrial methods are suitable substitutes for expensive, tedious and polluting conventional processes. Generating protease can produce genes to support product attribute improvement and increased production. Microbial proteases have been widely utilized and established as the backbone for various industries. Enzyme application in the detergent industry ensures an eco-friendly approach, reducing the use of chemicals in surfactants, soaps, bleach, chelating and oxidizing agents, as well as enhancing washing performance particularly for dirt or dust having biological origin. It also contributes to reducing washing temperatures, saving energy and protecting the environment. A wide range of ailments are extremely successfully treated using protease enzymes

It is possible that applications of microbial proteases may be extended to include precise curatives for prion-related diseases in animals and humans. In view of the adverse side-effects and high cost of existing chemotherapeutics, microbial proteases are being assessed for generation of promising thrombolytic agents. Protease enzymes are extremely widely used in the food industry, and precautions are being taken to ensure food quality and safety. In the textile industry, protease application improves product quality as well as making processes more efficient and eco-friendly.

The use of protease in wool and silk manufacturing units guarantees production of fabric of increasing quality with a harmless and eco benevolent approach. The application of proteases helps advance the green extraction of silver from X-ray films, reducing the massive ecological contamination that occurs in traditional silver extraction. Pollution from chemicals found in pesticides may be mitigated by application of proteases as potential biocontrol agents.

Successful and effective commercialization of proteases is required to pave the way for various cleaner, greener and more sustainable processes. The latest advances in the field of biotechnology must be used to produce novel/anthropogenic enzymes with improved efficacy within existing industrial process microenvironments.

REFERENCES

Akhter, M., Wal Marzan, L., Akter, Y., & Shimizu, K. (2020). Microbial bioremediation of feather waste for keratinase production: An outstanding solution for leather dehairing in tanneries. *Microbiology Insights*, *13*, 1178636120913280. http://doi.org/10.1177%2F1178636120913280

Al-Ghanayem, A. A., & Joseph, B. (2020). Current prospective in using cold-active enzymes as eco-friendly detergent additive. *Applied Microbiology and Biotechnology*, *104*(7), 2871–2882. http://doi.org/10.1007/s00253-020-10429-x

Ashaolu, T. J. (2020). Applications of soy protein hydrolysates in the emerging functional foods: A review. *International Journal of Food Science & Technology*, *55*(2), 421–428. http://doi.org/10.1111/ijfs.14380

Banerjee, G., & Ray, A. K. (2017). Impact of microbial proteases on biotechnological industries. *Biotechnology and Genetic Engineering Reviews*, *33*(2), 119–143. http://doi.org/10.1080/02648725.2017.1408256

Beg, Q. K., & Gupta, R. (2003). Purification and characterization of an oxidation-stable, thiol-dependent serine alkaline protease from Bacillus mojavensis. *Enzyme and Microbial Technology*, *32*(2), 294–304. http://doi.org/10.1016/S0141-0229(02)00293-4

Chongjun, Y., Bing, L., & Fusheng, C. (2016). Extraction of sericin and its application in cosmetics. *Animal Husbandry and Feed Science*, *8*(4), 223.

Da Silva, O. S., Gomes, M. H. G., de Oliveira, R. L., Porto, A. L. F., Converti, A., & Porto, T. S. (2017). Partitioning and extraction protease from Aspergillus tamarii URM4634 using PEG-citrate aqueous two-phase systems. *Biocatalysis and Agricultural Biotechnology*, *9*, 168–173. http://doi.org/10.1016/j.bcab.2016.12.012

Dawson, P. E., & Kent, S. B. (2000). Synthesis of native proteins by chemical ligation. *Annual Review of Biochemistry*, *69*(1), 923–960. http://doi.org/10.1146/annurev.biochem.69.1.923

Deniz, I. (2019). Production of microbial proteases for food industry. In *Green Bio-Processes* (pp. 9–14). Springer, Singapore.

dos Santos Aguilar, J. G., & Sato, H. H. (2018). Microbial proteases: Production and application in obtaining protein hydrolysates. *Food Research International*, *103*, 253–262. http://doi.org/10.1016/j.foodres.2017.10.044

Emon, T. H., Hakim, A., Chakraborthy, D., Bhuyan, F. R., Iqbal, A., Hasan, M., Aunkor T. H., & Azad, A. K. (2020). Kinetics, detergent compatibility and feather-degrading capability of alkaline protease from Bacillus subtilis AKAL7 and Exiguobacterium indicum AKAL11 produced with fermentation of organic municipal solid wastes. *Journal of Environmental Science and Health, Part A*, *55*(11), 1339–1348. http://doi.org/10.1080/10934529.2020.1794207

Fang, Z., Yong, Y. C., Zhang, J., Du, G., & Chen, J. (2017). Keratinolytic protease: A green biocatalyst for leather industry. *Applied Microbiology and Biotechnology*, *101*(21), 7771–7779.

Fernández de Ullivarri, M., Arbulu, S., Garcia-Gutierrez, E., & Cotter, P. D. (2020). Antifungal peptides as therapeutic agents. *Frontiers in Cellular and Infection Microbiology*, *10*, 105.

Gao, X., Yin, Y., Yan, J., Zhang, J., Ma, H., & Zhou, C. (2019). Separation, biochemical characterization and salt-tolerant mechanisms of alkaline protease from Aspergillus oryzae. *Journal of the Science of Food and Agriculture*, *99*(7), 3359–3366.

Ghosh, S., Rao, R. S., Nambiar, K. S., Haragannavar, V. C., Augustine, D., & Sowmya, S. V. (2017). Sericin, a dietary additive: Mini review. *Journal of Medicine, Radiology, Pathology and Surgery*, *4*(2), 13–17.

Golkar, A., Milani, J. M., & Vasiljevic, T. (2019). Altering allergenicity of cow's milk by food processing for applications in infant formula. *Critical Reviews in Food Science and Nutrition*, *59*(1), 159–172.

Gupta, R., Beg, Q., Khan, S., & Chauhan, B. (2002). An overview on fermentation, downstream processing and properties of microbial alkaline proteases. *Applied Microbiology and Biotechnology*, *60*(4), 381–395.

Gürkök, S. (2019). Microbial enzymes in detergents: a review. *International Journal of Scientific and Engineering Research*, *10*, 75–81.

Gurumallesh, P., Alagu, K., Ramakrishnan, B., & Muthusamy, S. (2019). A systematic reconsideration on proteases. *International Journal of Biological Macromolecules*, *128*, 254–267.

He, B., Tu, Y., Jiang, C., Zhang, Z., Li, Y., & Zeng, B. (2019). Functional genomics of Aspergillus oryzae: Strategies and progress. *Microorganisms*, *7*(4), 103.

Horax, R., Vallecios, M. S., Hettiarachchy, N., Osorio, L. F., & Chen, P. (2017). Solubility, functional properties, ACE-I inhibitory and DPPH scavenging activities of Alcalase hydrolysed soy protein hydrolysates. *International Journal of Food Science & Technology*, *52*(1), 196–204.

Illanes, A. (2008). *Enzyme Biocatalysis. Principles and Applications*. New York, NY: Springer-Verlag New York Inc http://doi.org/10.1007/978-1-4020-8361-7

Ismail, B., Mohammed, H., & Nair, A. J. (2019). Influence of proteases on functional properties of food. In *Green Bio-processes* (pp. 31–53). Springer, Singapore.

Jayakumar, R., Jayashree, S., Annapurna, B., & Seshadri, S. (2012). Characterization of thermostable serine alkaline protease from an alkaliphilic strain Bacillus pumilus MCAS8 and its applications. *Applied Biochemistry and Biotechnology*, *168*(7), 1849–1866.

Joshi, S., & Satyanarayana, T. (2013). Characteristics and applications of a recombinant alkaline serine protease from a novel bacterium Bacillus lehensis. *Bioresource Technology*, *131*, 76–85.

Khambhaty, Y. (2020). Applications of enzymes in leather processing. *Environmental Chemistry Letters*, *18*(3), 747–769.

Kumaran, E., Mahalakshmipriya, A., & Sentila, R. (2013). Effect of fish waste based Bacillus protease in silver recovery from waste X-ray films. *International Journal of Current Microbiology and Applied Sciences*, *2*(3), 49–56.

Liu, Q., Xun, G., & Feng, Y. (2019). The state-of-the-art strategies of protein engineering for enzyme stabilization. *Biotechnology Advances*, *37*(4), 530–537.

Lorenzo, J. M., Munekata, P. E., Gómez, B., Barba, F. J., Mora, L., Pérez-Santaescolástica, C., & Toldrá, F. (2018). Bioactive peptides as natural antioxidants in food products–A review. *Trends in Food Science & Technology*, *79*, 136–147.

Mamo, J., & Assefa, F. (2018). The role of microbial aspartic protease enzyme in food and beverage industries. *Journal of Food Quality*, *2018*.

More, S. V., Khandelwal, H. B., Joseph, M. A., & Laxman, R. S. (2013). Enzymatic degumming of silk with microbial proteases. *Journal of Natural Fibers*, *10*(2), 98–111.

Moschopoulou, E. (2018). Microbial non-coagulant enzymes used in cheese making. In *Microbial Cultures and Enzymes in Dairy Technology* (pp. 204–221). IGI Global, Hershey, PA.

Nasre Taheri, M., Ebrahimipour, G., & Sadeghi, H. (2019). Isolation and identification of alkaline protease-producing bacterium in the presence of washing-powder from Geinarje hot spring, Ardabil, Iran. *Nova Biologica Reperta*, *6*(3), 292–301.

Nathan, V. K., & Rani, M. E. (2020). A cleaner process of deinking waste paper pulp using Pseudomonas mendocina ED9 lipase supplemented enzyme cocktail. *Environmental Science and Pollution Research*, *27*(29), 36498–36509.

Naveed, M., Nadeem, F., Mehmood, T., Bilal, M., Anwar, Z., & Amjad, F. (2021). Protease—a versatile and ecofriendly biocatalyst with multi-industrial applications: An updated review. *Catalysis Letters*, *151*(2), 307–323.

Philipps-Wiemann, P. (2018). Proteases – Animal feed. In Nunes, C. S., Kuamr, V. (eds), *Enzymes in Human and Animal Nutrition* (pp. 279–297). Academic Press, London, UK

Poza, M., Ageitos, J. M., Vallejo, J. A., De Miguel, T., Veiga-Crespo, P., & Villa, T. G. (2007). Industrial applications of microbial proteases and genetic engineering. *Hot Spots in Applied Microbiology*, 661, 91–125

Radha, S., Sridevi, A., HimakiranBabu, R., Nithya, V. J., Prasad, N. B. L., & Narasimha, G. (2012). Medium optimization for acid protease production from Aspergillus sp. under solid state fermentation and mathematical modelling of protease activity. *Journal of Microbiology and Biotechnology Research*, *2*(1), 6–16.

Rakotonirainy, M. S., Vilmont, L. B., & Lavédrine, B. (2016). A methodology for detecting the level of fungal contamination in the French Film Archives vaults. *Journal of Cultural Heritage*, *19*, 454–462.

Rao, M. B., Tanksale, A. M., Ghatge, M. S., & Deshpande, V. V. (1998). Molecular and biotechnological aspects of microbial proteases. *Microbiology and Molecular Biology Reviews*, *62*(3), 597–635.

Razzaq, A., Shamsi, S., Ali, A., Ali, Q., Sajjad, M., Malik, A., & Ashraf, M. (2019). Microbial proteases applications. *Frontiers in Bioengineering and Biotechnology*, *7*, 110.

Rekik, H., Jaouadi, N. Z., Gargouri, F., Bejar, W., Frikha, F., Jmal, N., Bejar, S., & Jaouadi, B. (2019). Production, purification and biochemical characterization of a novel detergent-stable serine alkaline protease from Bacillus safensis strain RH12. *International Journal of Biological Macromolecules*, *121*, 1227–1239.

Sadeghi, H. M. M., Rabbani, M., & Naghitorabi, M. (2009). Cloning of alkaline protease gene from Bacillus subtilis 168. *Research in Pharmaceutical Sciences*, *4*(1), 43–46.

Samaei S. P., Ghorbani M., Tagliazucchi D., Martini S., Gotti R., Themelis T., Tesini F., Gianotti A., Toschi T. G., & Babini, E. (2020). Functional, nutritional, antioxidant, sensory properties and comparative peptidomic profile of faba bean (Vicia faba, L.) seed protein hydrolysates and fortified apple juice. *Food Chemistry*, *330*, 127120.

Sangeetha, R., & Arulpandi, I. (2019). Anti-inflammatory activity of a serine protease produced from Bacillus pumilus SG2. *Biocatalysis and Agricultural Biotechnology*, *19*, 101162.

Sasia, A. A., Sang, P., & Onyuka, A. (2019). Recovery of collagen hydrolysate from chrome leather shaving tannery waste through two-step hydrolysis using magnesium oxide and bating enzyme. *Journal of Bacteriology & Mycology: Open Access*, *3*(1), 191–194.

Savitha, S., Sadhasivam, S., Swaminathan, K., & Lin, F. H. (2011). Fungal protease: Production, purification and compatibility with laundry detergents and their wash performance. *Journal of the Taiwan Institute of Chemical Engineers*, *42*(2), 298–304.

Shamsi, T. N., Parveen, R., & Fatima, S. (2016). Characterization, biomedical and agricultural applications of protease inhibitors: A review. *International Journal of Biological Macromolecules*, *91*, 1120–1133.

Sharma, K. M., Kumar, R., Panwar, S., & Kumar, A. (2017). Microbial alkaline proteases: Optimization of production parameters and their properties. *Journal of Genetic Engineering and Biotechnology*, *15*(1), 115–126.

Sharma, A., & Gupta, S. P. (2017). Fundamentals of viruses and their proteases. In Gupta, S.P. (ed), *Viral Proteases and Their Inhibitors* (pp. 1–24). Academic Press, London, UK.

Sharma, M., Gat, Y., Arya, S., Kumar, V., Panghal, A., & Kumar, A. (2019). A review on microbial alkaline protease: An essential tool for various industrial approaches. *Industrial Biotechnology*, *15*(2), 69–78.

Silva-López, R. E., & Gonçalves, R. N. (2019). Therapeutic proteases from plants: biopharmaceuticals with multiple applications. *Journal of Applied Biotechnology and Bioengineering*, *6*, 101–109.

Singh, R., Kumar, M., Mittal, A., & Mehta, P. K. (2016). Microbial enzymes: Industrial progress in 21st century. *3 Biotech*, *6*(2), 1–15.

Singh, S., & Bajaj, B. K. (2017). Agroindustrial/forestry residues as substrates for production of thermoactive alkaline protease from Bacillus licheniformis K-3 having multifaceted hydrolytic potential. *Waste and Biomass Valorization*, *8*(2), 453–462.

Sundus, H., Mukhtar, H., & Nawaz, A. (2016). Industrial applications and production sources of serine alkaline proteases: A review. *Journal of Bacteriology & Mycology: Open Access*, *3*, 191–194.

Tantamacharik, T., Carne, A., Agyei, D., Birch, J., & Bekhit, A. E. D. A. (2018). Use of plant proteolytic enzymes for meat processing. In *Biotechnological Applications of Plant Proteolytic Enzymes* (pp. 43–67). Springer, Cham.

Tavano, O. L., Berenguer-Murcia, A., Secundo, F., & Fernandez-Lafuente, R. (2018). Biotechnological applications of proteases in food technology. *Comprehensive Reviews in Food Science and Food Safety*, *17*(2), 412–436.

Thakur, N., Goyal, M., Sharma, S., & Kumar, D. (2018). Proteases: Industrial applications and approaches used in strain improvement. *Biological Forum–An International Journal*, *10*(1), 158–167).

Tiwari, M. (2017). The role of serratiopeptidase in the resolution of inflammation. *Asian Journal of Pharmaceutical Sciences*, *12*(3), 209–215.

Tran, P. T., Widyasari, K., Park, J. Y., & Kim, K. H. (2017). Engineering an auto-activated R protein that is in vivo activated by a viral protease. *Virology*, *510*, 242–247.

Varia, A. D., Shukla, V. Y., & Tipre, D. R. (2019). Alkaline protease-a versatile enzyme. *International Journal of Research and Analytical Reviews*, *6*(2), 208–217.

Veloorvalappil, N. J., Robinson, B. S., Selvanesan, P., Sasidharan, S., Kizhakkepawothail, N. U., Sreedharan, S., Prakasan, P., Moolakkariyil, S. J., & Sailas, B. (2013). Versatility of microbial proteases. *Advances in Enzyme Research*, *1*, 39–51

Vijayaraghavan, P., & Vincent, S. P. (2015). A low cost fermentation medium for potential fibrinolytic enzyme production by a newly isolated marine bacterium, *Shewanella sp.* IND20. *Biotechnology Reports*, *7*, 135–142.

Vitolo, M. (2019). Miscellaneous use of enzymes. *World Journal of Pharmaceutical Research*, *9*(2), 199–224.

Wang, X., Yu, H., Xing, R., Chen, X., Liu, S., & Li, P. (2017). Optimization of the extraction and stability of antioxidative peptides from mackerel (Pneumatophorus japonicus) protein. *BioMed Research International*, *2017*, 1–14

Weng, Y., Yao, J., Sparks, S., & Wang, K. Y. (2017). Nattokinase: An oral antithrombotic agent for the prevention of cardiovascular disease. *International Journal of Molecular Sciences*, *18*(3), 523.

Yadav, S. K., Bisht, D., Tiwari, S., & Darmwal, N. S. (2015). Purification, biochemical characterization and performance evaluation of an alkaline serine protease from Aspergillus flavus MTCC 9952 mutant. *Biocatalysis and Agricultural Biotechnology*, *4*(4), 667–677.

Yang, J., Lee, K. S., Kim, B. Y., Choi, Y. S., Yoon, H. J., Jia, J., & Jin, B. R. (2017). Anti-fibrinolytic and antimicrobial activities of a serine protease inhibitor from honeybee (Apis cerana) venom. *Comparative Biochemistry and Physiology Part C: Toxicology & Pharmacology*, *201*, 11–18.

Zhang, H., Li, H., Liu, H., Lang, D. A., Xu, H., & Zhu, H. (2019). The application of a halotolerant metalloprotease from marine bacterium Vibrio sp. LA-05 in liquid detergent formulations. *International Biodeterioration & Biodegradation*, *142*, 18–25.

Zhou, C., Qin, H., Chen, X., Zhang, Y., Xue, Y., & Ma, Y. (2018). A novel alkaline protease from alkaliphilic Idiomarina sp. C9-1 with potential application for eco-friendly enzymatic dehairing in the leather industry. *Scientific Reports*, *8*(1), 1–18.

3 Microbial Melanin

Role, Biosynthesis, and Applications

Sanjay Kumar, Sawraj Singh and Rajeev Kumar Kapoor
MD University, Rohtak, India

CONTENTS

3.1 INTRODUCTION

Melanin is a biocompatible and eco-friendly natural pigment produced in almost all forms of life, including plants, animals, and microorganisms (Gosset, 2017). It is a negatively charged high molecular weight compound that exhibits multiple activities. Owing to its ubiquitous nature, melanin plays an important role as a life-supporting agent in various organisms. In humans and animals, melanin synthesis takes place in unique cells known as melanocytes. Melanin contributes to physical attributes such as eye, hair, and skin color, and is of significance as it protects DNA from harmful UV radiation, preventing undesirable mutations. Melanin granules are stored in specialized cell organelles known as melanosomes.

Microbes such as bacteria and fungi also synthesize melanin which serves diverse purposes in cells. The well-established role of melanin in bacteria as a virulence factor explains how various pathogenic bacteria such as *Burkholderia cepacia, Proteus mirablis, Klebsiella pneumonia, E. Coli, Mycobacterium leprae*, and *Vibrio cholera* defend themselves against host defense mechanisms and cause severe infections (Plonka & Grabacka, 2006; Singh et al., 2018; Pavan et al., 2020). In fungi, melanin contributes to virulence in melanotic parasitics by neutralizing reactive oxygen species

DOI: 10.1201/9781003306931-4

and providing protection against the lytic action of phagocytes. In leguminous plants, phenolic compound wastes are converted into life-supporting melanin pigments via tyrosinase activity of microbes residing in root nodules, which enhance atmospheric nitrogen fixation thereby promoting better growth of the plant. In general, microorganisms synthesize melanin by oxidation of phenolic or indolic compounds such as L-tyrosine, using them as substrates (Solano, 2014). The major pathways through which fungi and bacteria synthesize melanin are DHN (1,8-dihydroxynaphthalene) and DOPA (3,4-dihydroxyphenylalanine) pathways (Pal et al., 2014). Different enzymes such as polyketide synthase, tyrosinase, and laccase play a role in melanin synthesis in microbes. Microbial production of melanin is eco-friendly and can be more economical than chemical synthesis. This chapter focuses on the role of melanin in various microorganisms, its biosynthesis and applications in diverse fields.

3.2 TYPES OF MELANIN

The five types of melanin vary in color and structure. Neuromelanin is found in nerve cells located in the substantia nigra pars compacta (SNpc) and locus coeruleus in the brain. Chemically, neuromelanin contains both indole and benzothiazine units. Pyomelanin is a dark-colored pigment formed from the oxidation of homogentisic acid in microorganisms (Pralea et al., 2019). Pheomelanins are yellow, red, or brown sulfur-containing melanin pigments synthesized by oxidative polymerization of cysteine conjugate of l-3,4-dihydroxyphenylalanine (L-DOPA) (Ito et al., 2019). Eumelanin is a black or brown-colored melanin that is synthesized by the action of tyrosinase on tyrosine to produce L-DOPA and subsequently convert it to DOPA quinone, which later undergoes polymerization to yield melanin (Eisenman & Casadevall, 2012). Allomelanins are nitrogen-free heterogeneous groups produced by catechol precursors (Tran-Ly et al., 2020).

3.3 ROLE OF MELANIN IN MICROBES

Melanin plays an important role in pathogenesis in various microorganisms and helps them survive against host defense mechanisms (Lee et al., 2019). It can alter several cytokine responses of cells to reduce cell phagocytosis. It also changes responses to antifungals, removes and neutralizes the accumulated reactive oxygen species (ROS) and other free radicals in the cell. Melanin production in free-living microorganisms gives protection from extreme weather conditions.

3.3.1 Role of Melanin in Bacteria as a Virulence Factor

Melanin synthesis in pathogenic bacteria enhances their virulence against host defense mechanisms in two ways: by weakening the host immune response and by minimizing the susceptibility of the disease-causing microorganisms to host defense system (Pavan et al., 2020). *Burkholderia cepacia*, a pathogenic bacterium that causes severe pulmonary infection by developing sepsis in patients with cystic fibrosis, gives the best explanation of the role of melanin as a virulence factor (Zughaier et al., 1999). The lipopolysaccharide (LPS) synthesized by the bacteria increases the production and release of pro-inflammatory cytokines in the host. LPS stimulates the immune system and also accelerates oxidative response to different stimuli causing oxygen burst. The oxygen burst is an important mechanism for evading pathogens in blood cells. The harmful effect of reactive oxygen produced in leucocytes during oxygen burst can be neutralized by the melanin present in *B. cepacia*. Melanin acts as an antioxidant; however, it does not affect the kinetics of ROS production and release of O_2. Thus, *B. cepacia* can survive phagocytosis because they can neutralize the action of host phagocytes by removing superoxide anion with the help of melanin. Due to melanin, it can survive in alveolar macrophages, easily proliferate inside the host cells, infect secondary macrophages, and cause host cell damage (Plonka & Grabacka, 2006).

In *Proteus mirabilis*, the level of ROS is decreased by melanin, which makes the bacteria more resistant to oxygen bursts triggered by the host's immunity (Agodi et al., 1996). In *Klebsiella pneumonia*, melanin is synthesized by an unusual melanogenesis pathway from 4-hydroxyphenyl acetic acid, imparting virulence (Singhet al., 2018).

E.coli and *Proteus mirabilis* responsible for urinary tract infection in humans also utilize melanin by tyrosinase activity (Agodi et al., 1996). *Mycobacterium leprae* has the unique property of oxidizing diphenols to o-quinones, so in such bacteria, oxidation of L-DOPA is a convenient diagnostic feature (Prabhakaran & Harris 1985). Melanin is a polymer of phenolic or indolic compounds that can accept or donate electrons to act as a shuttle or final acceptor in the electron exchange with insoluble iron compounds. In the last step of the respiratory chain, assimilating bacteria use a variety of electron acceptors replacing O_2 in anaerobic respiration. In bacteria, melanin acts as a reducing agent that converts insoluble Fe(III) to Fe(II), boosting energy by electron transfer (Turick et al., 2002). In *Vibrio cholera*, melanin stimulates pilus expression, increases their colonization in the host cells, and enhances cholera toxin production (Pavan et al., 2020)

3.3.2 Atmospheric Nitrogen Fixation

Tyrosinase plays a vital role in leguminous plants by removing phenolic compounds by oxidation because the concentration of phenolic compounds increases in root nodules over time. *Azotobacter chroococcum* is known for its potential to fix nitrogen (Shivprasad & Page, 1989). It contains polyphenol oxidase (tyrosinase), which catalyzes 1,2-dihydroxybenzene (catechol) to produce melanin. The demand for oxygen increases due to melanogenesis, which helps remove phenolic compounds necessary for atmospheric N_2 binding (Hynes et al., 1988). Polyphenol oxidase catalase produces melanin as well as protecting from hydroxyl radicals. This is necessary for providing catechol and binding metal ions (Cu and Fe ions) in the structure of melanin.

3.3.3 Virulence Mechanism in Melanotic Parasitic Fungi

Melanin renders virulence to melanotic parasitic fungi by neutralizing the oxidants. It neutralizes ROS and protects fungal cells from the lytic action of phagocytes. Melanin can act as both an electron acceptor and electron donor due to its electrochemical properties. In *Colletotrichum gloeosporioides*, a fungus that contains the laccase gene (*LAC1*) plays a crucial role in its development and virulence on various fruit plants. This gene is involved in virulence and participates in conidiation, enzyme activity, nutrition, and melanin biosynthesis.

Activated host macrophages produce oxygen bursts to release reactive nitrogen species (RNS) and ROS against parasitic fungi. *LAC1* stimulates tyrosinase activity and triggers melanin synthesis, neutralizing RNS and ROS (NO, H_2O_2 and UV) to protect them against host defense mechanisms. Melanotic parasitic fungi can defend themselves against the acidic solution of $NaNO_2$ due to the presence of melanin (Plonka & Grabacka, 2006).

3.3.4 Role of Melanin in Nematophagous Activity in Fungi

Nematophagous fungi can kill nematodes by trapping and infection through adhesion or without adhesion. These fungi are the natural predators of nature controlling the vast diversity of nematodes (Freitas et al., 2021). About 700 species of such fungi are distributed among the phylum Ascomycota, Basidiomycota, Calcarisporiellomycota, Entomophthoromycota, and Zoopagomycota.

The habitat of these nematophagous fungi is soil-containing organic matter, and the fungus helps in the decomposition of agricultural and animal waste. These fungi infect nematodes and kill their eggs by capturing them with zoospores, conidia formation, adhesive nodules, non-differentiated adhesive hyphae, forming constructive and non-constructive rings, and forming two- and three-dimensional networks. Melanin plays a significant role in providing all requirements for fighting against host defense mechanisms. In melanosomes, melanogenesis takes place, and it stores melanin for further use. In their structure and functions, fungal melanosomes show analogy with mammalian melanosomes. Further analysis and more advanced study in this field can contribute to more effective development of "melanin as a biocontrol agent" (Freitas et al., 2021).

3.4 BIOSYNTHESIS PATHWAYS OF MELANIN IN MICROBES

Fungi and bacteria can synthesize melanin through two pathways: the DHN (1,8-dihydroxynaphthalene) pathway and DOPA (3,4-dihydroxyphenylalanine) pathway (Pal et al., 2014). Under extreme and adverse conditions, these organisms produce melanin that helps them survive. The melanin synthesis pathways in these microorganisms start from two substrates, malonyl-CoA and L-tyrosine, and use various enzymes such as polyketide synthase, laccase, or tyrosinase.

3.4.1 DHN Pathways

Generally, fungi use this pathway for the production of melanin by using malonyl-CoA or acetyl-CoA as the monomeric units. Polyketide synthase catalyzes to produce intermediate molecules, i.e., 1,3,6,8-tetrahydroxynaphthalene (THN). In the production of THN, sequential decarboxylative condensation of 5- molecules of malonyl-CoA occurs. THN is converted into 1, 8-dihydroxy naphthalene (DHN) by serial dehydration and reduction reactions. Further polymerization of DHN gives allomelanin (Eisenman & Casadevall, 2012). Polyketide synthase is involved in the polymerization of DHN to produce DHN melanin. Some ascomycetes fungi such as *Neurospora crassa, Magnaporthe grisea* use the DHN pathway for melanin formation (Ao*etal.*, 2019). Allomelanin lacks nitrogen which makes it different from other melanin. This melanin has different polymer groups—catechols, dihydrofolate and homogentisic acid—which make its heterogeneous structure (Plonka & Grabacka, 2006).

3.4.2 DOPA-Pathway

This pathway uses either of the two types of enzymes, laccase or tyrosinase. For the production of DOPAmelanin, various parent molecules such as L-tyrosine, L-DOPA, catecholamines, and laccase and tyrosinase-produced derivatives are used. First, monophenolase activity of tyrosinase transforms L-tyrosine into L-DOPA by hydroxylation. In the next step L-DOPA quinones are synthesized by further oxidation of L-DOPA by diphenolase activity of tyrosinase (Claus & Decker, 2006). DOPAquinones further autopolymerize to synthesize melanin by spontaneous oxidation (See Figure 3.1).

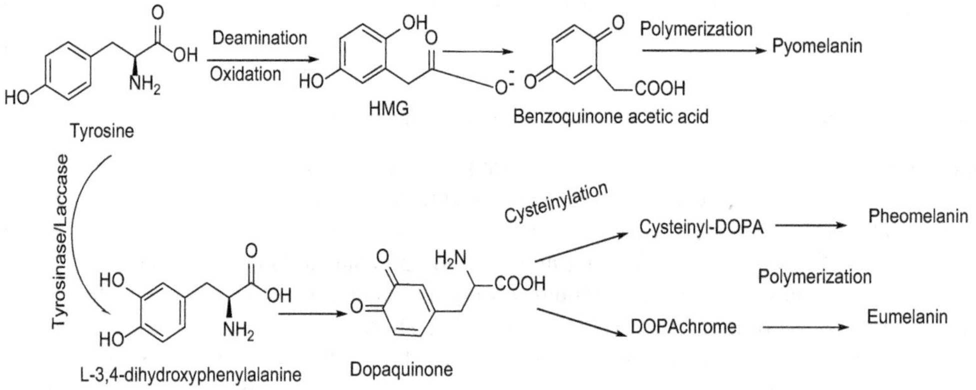

FIGURE 3.1 Synthesis of pyomelanin, pheomelanin, and eumelanin in bacteria.

DOPA quinone also produces DOPAchrome and by polymerization reaction forms eumelanin (Plonka & Grabacka, 2006). In this pathway, it is the enzyme tyrosinase that catalyzes the reaction. In a laccase catalyzed pathway DOPA quinone undergoes cysteinylation to give Cysteinyl-DOPA, which undergo polymerization to synthesize pheomelanin (Eisenman and Casadevall, 2012). (See Figure 3.1). To synthesize pyomelanin, tyrosine undergoes deamination reaction to give 4-HPP/4-HPA (Hydroxyphenylpyruvate/Hydroxyphenylacetate), which is converted into 3-hydroxy-3-methyl-gluterate (HMG), and its subsequent enzymatic oxidation produces benzoquinone. This benzoquinone undergoes polymerization to produce pyomelanin (Ahmad et al., 2017). In the catabolic process, tyrosine breakdown takes place with the help of 4-hydroxyphenylpyruvic acid dioxygenase (4-HPPD) and homogentisic acid oxidase (HGA-oxidase) to produce fumarate and acetoacetate.

3.4.3 Enzymes in Melanin Synthesis Pathways

3.4.3.1 Polyketide Synthase (PKS)

This enzyme catalyzes in the DHN pathway to the synthesis of allomelanin. Various microorganisms, fungi, bacteria, and plants synthesize this enzyme. Various biocompounds such as allomelanin, toxins, polyketide, and antibiotics are produced by this enzyme (Plonka & Grabacka, 2006). Many PKS genes are present in the genome of a pathogenic fungus *Penicillium marneffei*. Gene *pks1* is involved in the synthesis of Type I PKS for melanin biosynthesis in *Pestalotiopsis microspora*. There are three types of PKS, depending on structure and functions. Modular PKS/Type I PKS is a multifunctional protein complex organized into different modules, each capable of carrying out a specific function. Acyl carrier protein (ACP) domains present in them activate acyl-CoA substrate. Malonyl-CoA, methylmalonyl-CoA, or ethylmalonyl-CoA are the common elongation groups. Type I PKS helps in the assembling of secondary metabolites by C-C condensation. Type II/Discrete PKS is a multienzyme complex with different catalytic proteins with defined functions. They help in the biosynthesis of aromatic molecules by the Clasien condensation reaction. Type III PKS/Ketosynthase PKS/is a homodimeric enzyme with each monomer capable of catalyzing priming, elongation, and cyclization reactions involved in the biosynthesis of polyketides. Despite their structural simplicity, they play a role in synthesizing a variety of natural products (Gokulan et al., 2014). These PKSs are involved in the aromatic polyketide biosynthetic pathway used by various fungi to synthesize fungal melanin (Yu et al., 2015).

3.4.3.2 Tyrosinase (EC 1.14.18.1, monophenol, o-diphenol: oxygen oxidoreductase)

This enzyme can be produced by bacteria, fungi, insects, animals, and plants. It is a copper (Cu) containing multifunctional enzymes (Claus & Decker, 2006). Tyrosinase has both monophenolase and diphenolase activity and is involved in DOPA-pathway to synthesize DOPA melanin (eumelanin) (Zaidi et al., 2014). In monophenolase activity it causes the hydroxylation of monophenols (tyrosine) to o-diphenol (L-DOPA), whereas in diphenolase activity, it catalyzes the oxidation of diphenols into o-quinones (Fairhead & Thöny-Meyer, 2012).

3.4.3.3 Laccase (EC 1.10.3.2)

In bacteria, *S. meliloti, and Bacillus subtilis* laccase are involved in melanin biosynthesis. It is a Cu containing glycoprotein that can catalyze a large range of substrates such as methoxy-substituted phenols, diamines, 4-hydroxy-3,5-dimethoxybenzaldehyde azine and polyphenols (Upadhyay et al., 2016).

It plays a significant role in the synthesis of melanin via the DHNand DOPApathways (Eisenman et al., 2007). Tyrosinase has monophenolase and diphenolase activity, so it catalyzes the two-step conversion from L-tyrosine to L-dopaquinone. In contrast, laccase catalyzes in single-step oxidation from L-DOPA to L-DOPAquinone, indicating a difference of substrate specificities between tyrosinase and laccase (Langfelder et al., 2003).

3.5 VARIOUS STUDIES ON MELANIN PRODUCTION FROM MICROORGANISMS

Microbial production of melanin is superior to its chemical synthesis and extraction from other sources because of its high yield and eco-friendly nature. Microbial production can be cost-effective, and the yields are free from seasonal variations. There are various influencing factors in microbial melanin production. Temperature, pH, aeration, irradiation, stress, flask volume, inoculums age, and media composition need to be optimized depending upon the nature and type of the organisms. It is mandatory to select proper productive culture strains and favorable culture conditions for product formation, which helps in microbial fermentation. In one study, *Streptomyces kathirae* SC-1 was tested for melanin production and yield was maximized upto 13.7g/L by optimizing media and culture conditions (Guo et al., 2014). The addition of Cu and yeast extract with tyrosine enhanced melanin production. In another study melC operon and its promoter, P_{skmel} were PCR amplified from *S. kathirae* and melanin level was enhanced by plasmid-based expression of melanin genes (Guo et al., 2015). *Vibrio alginolyticus MMRF534 and Vibrio harveyi MMRF 535* obtained from sponges could yield 50 g/L and 40 g/L melanin in marine broth (Vijayan et al., 2017). *Burkholderia cenocepacia* is a pathogenic bacterium that produces unique ochre-colored melanin that provides a defense against the host immune system, enabling itto survive in the phagocytic cell. Other bacteria like *Pseudomonas* sp., *E. coli, Streptomyces glaucescens, and Rhizobium etl* have also been explored for melanin production (see Table 3.1). Several fungi, including *Armillaria cepistipes, Gliocephalotrichum simplex*, and *Auricularia auricular* have been studied for the production of melanin (see Table 3.2).

TABLE 3.1
Melanin Production in Bacteria Using Different Substrates

Source	Substrates	Incubation	Yield (g/L)	References
Bacillus safensis	Fruits waste extract	24 h	6.9	Tarangini & Mishra (2014)
Nocardiopsis alba MSA10	Tyrosine, sucrose	7 days	3.4	Kiran et al. (2014)
Pseudomonas sp. WH001 55	Tyrosine, starch & yeast extract	6 days	7.6	Kiran et al. (2017)
Streptomyces glaucescens NEAE-H	Tyrosine, protease peptone	6 days	0.4	El-Naggar & El-Ewasy (2017)
Streptomyces sp. ZL-24	Tyrosine, soy peptone	5 days	4.2	Wang et al. (2019)
Streptomyces kathirae SC-1	Tyrosine, amylodextrin& yeast	5 days	13.7	Guo et al. (2014)
Streptomyces lusitanus DMZ-3	Tyrosine, beef extract	6 days	5.3	Madhusudhan et al. (2014)

TABLE 3.2
Melanin Production in Fungi Using Different Substrates

Source	Substrates	Incubation Time	Yield (g/L)	References
Armillaria borealis	Tyrosine, glucose, yeast extract	97 h	11.58	Ribera et al. (2019)
Armillaria ostyae	Tyrosine, glucose, yeast extract	153 h	24.8	Ribera et al. (2019)
Armillaria cepistipes	Tyrosine, glucose, yeast extract	161 h	27.98	Ribera et al. (2019)
Aspergillus fumigates	Dextrose, peptone	10 days	0.01	Raman et al. (2015)
Auricularia auricular	Tyrosine, lactose, yeast extract	8 days	2.97	Sun et al. (2016)
Daldinia concentric	Tyrosine, glucose, yeast extract	73 h	1.78	Ribera et al. (2019)

3.6 APPLICATIONS OF MELANIN

1. Melanin can be used in energy production from dye-sensitized solar cells as melanin is a photosensitizing pigment that can be an alternative to platinum complexes or organic dyes that pollute the environment (Silva et al., 2019).
2. Microbial melanin can be used for water purification by making composite material with polyurethane and polycaprolactone polymer. Heavy metals such as mercury (Hg) and lead (Pb) can be removed from polluted water using melanin-based hybrid materials (Tran-Ly et al., 2020).
3. Melanin acts as a sunscreen agent to protect skin from harmful UV radiations (Brenner & Hearing, 2008).
4. In nanotechnology, microbial melanin can be used to synthesize metal nanoparticles that have diverse applications in cosmetics, food processing and packaging, paint additives, and antimicrobial agents (Kiran et al., 2014).
5. Melanin can be used in implantable devices with minimum side-effects in the biomedical field because of its biodegradability, bioavailability, and biocompatibility (Vahidzadeh et al., 2018).
6. Melanin acts as an antioxidant and radical scavenger and prevents biomolecular damage by scavenging harmful free radicals produced through various metabolic pathways. Due to various functional groups present in melanin, it can neutralize ROS (Ju et al., 2011; Le Na et al., 2019; Oh et al., 2020).
7. It can be used as a coloring agent in textile industries that are more eco-friendly compared to synthetic dyes (Singh et al., 2021).
8. Melanin can be used in the treatment of cancer. Its anticancer activity *in vitro* has been tested on cancerous cell lines (El-Naggar & El-Ewasy, 2017).
9. It has been explored as a potential antiviral agent against SARS-CoV-2 (Vijayababu & Kurian, 2021; Paria et al., 2020).
10. Its hydrogel can be used in skin and wound healing agents and in bone and cartilage tissue engineering (Cavallini et al., 2020).

3.7 FUTURE PROSPECTS

Melanins, the unique natural dark-colored pigments conventionally known for their role in skin pigmentation and related disorders in humans, have recently emerged as biomolecules with many potential industrial applications with research potential in areas such as biomedicine, biotechnology, bioelectronics, and bioinspired nanotechnology. In microbes, melanin plays a crucial role in ensuring their survival in harsh conditions. Recent understanding of the role of melanin in pathogenesis in fungi can impact future research designing therapies against infectious diseases. Their unique potential to exhibit multiple properties can be applied only when their structure, structure activity relationship and mechanisms involved in their biosynthesis are deciphered. The structure of melanin is difficult to predict at present due to the limitations of analytical techniques. The use of next-generation analytical techniques for developing an understanding about related structure and activity can pave the way for advanced melanin-based technologies.

The diverse functions and biocompatibility of melanin have motivated researchers to design melanin-inspired nanoparticles which have potential applications in drug delivery and bioimaging. Future nanotechnology research to develop melanin-like nanoparticles, explore their applications and improve safety and effectiveness is anticipated. The molecular mechanisms of melanogenesis in myxomycetes in response to light exposure in plasmodium are close to melanogenesis in primordial melanogenic Eukaryota. Studies of melanogenesis in microbes such as protozoa can be helpful in understanding the future evolution of this phenomenon in mammals.

In view of the significance of its diverse applications, future research should also focus on designing a cost-effective bioprocess for efficient large-scale production. The microbial production of melanin can be more economical than chemical synthesis. The search for novel native cells and engineering cells for optimum production of soluble melanin, coupled with the development of ideal production techniques and downstream processes, may make melanin-based technologies a reality of the future.

REFERENCES

Agodi, A., Stefani, S., Corsaro, C., Campanile, F., Gribaldo, S., & Sichel, G. (1996). Study of a melanic pigment of *Proteus Mirabilis*. *Research in Microbiology*, *147*(3), 167–174.

Ahmad, S., Lee, S. Y., Khan, R., Kong, H. G., Son, G. J., Roy, N., ... & Lee, S. W. (2017). Identification of a gene involved in the negative regulation of pyomelanin production in Ralstonia solanacearum. *Journal of Microbiology and Biotechnology*, *27*(9), 1692–1700.

Ao, J., Bandyopadhyay, S., & Free, S. J. (2019). Characterization of the Neurospora crassa DHN melanin biosynthetic pathway in developing ascospores and peridium cells. *Fungal Biology*, *123*(1), 1–9.

Brenner, M., & Hearing, V. J. (2008). The protective role of melanin against UV damage in human skin. *Photochemistry and Photobiology*, *84*(3), 539–549. http://doi.org/10.1111/j.1751-1097.2007.00226.x

Cavallini, C., Vitiello, G., Adinolfi, B., Silvestri, B., Armanetti, P., Manini, P., Pezzella, A., D'ischia, M., Luciani, G., & Menichetti, L. (2020). Melanin and melanin-like hybrid materials in regenerative medicine. *Nanomaterials*, *10*(8), 1–32. http://doi.org/10.3390/nano10081518

Claus, H., & Decker, H. (2006). Bacterial tyrosinases. *Systematic and Applied Microbiology*, *29*(1), 3–14.

Eisenman, H. C., & Casadevall, A. (2012). Synthesis and assembly of fungal melanin. *Applied Microbiology and Biotechnology*, *93*(3), 931–940.

Eisenman, H. C., Mues, M., Weber, S. E., Frases, S., Chaskes, S., Gerfen, G., & Casadevall, A. (2007). Cryptococcus neoformans laccase catalyses melanin synthesis from both D-and L-DOPA. *Microbiology*, *153*(12), 3954–3962.

El-Naggar, N. E. A., & El-Ewasy, S. M. (2017). Bioproduction, characterization, anticancer and antioxidant activities of extracellular melanin pigment produced by newly isolated microbial cell factories *Streptomyces glaucescens* NEAE-H. *Scientific Reports*, *7*(January), 1–19. http://doi.org/10.1038/srep42129

Fairhead, M., & Thöny-Meyer, L. (2012). Bacterial tyrosinases: Old enzymes with new relevance to biotechnology. *New Biotechnology*, *29*(2), 183–191. http://doi.org/10.1016/j.nbt.2011.05.007

Freitas, D. F., da Rocha, I. M., Vieira-da-Motta, O., & de Paula Santos, C. (2021). The role of melanin in the biology and ecology of nematophagous fungi. *Journal of Chemical Ecology*, *47*(7), 597–613. http://doi.org/10.1007/s10886-021-01282-x

Gokulan, K., Khare, S., & Cerniglia, C. (2014). Metabolic pathways. Production of Secondary Metabolites of Bacteria. In Batt, C. A. (Ed.) *Encyclopedia of Food Microbiology*, 2, 561–569. Elsevier Ltd, Academic Press.

Gosset, G. (2017). Biotechnological production of melanins with microorganisms. In Singh O. V. (Ed.) *Biopigmentation and Biotechnological Implementations*, 161–171. John Wiley & Sons, Inc.

Guo, J., Rao, Z., Yang, T., Man, Z., Xu, M., & Zhang, X. (2014). High-level production of melanin by a novel isolate of *Streptomyces kathirae*. *FEMS Microbiology Letters*, *357*(1), 85–91. http://doi.org/10.1111/1574-6968.12497

Guo, J., Rao, Z., Yang, T., Man, Z., Xu, M., Zhang, X., & Yang, S. T. (2015). Cloning and identification of a novel tyrosinase and its overexpression in Streptomyces kathirae SC-1 for enhancing melanin production. *FEMS Microbiology Letters*, *362*(8), 1–7. http://doi.org/10.1093/femsle/fnv041

Hynes, M. F., Brucksch, K., & Priefer, U. (1988). Melanin production encoded by a cryptic plasmid in a Rhizobium leguminosarum strain. *Archives of Microbiology*, *150*(4), 326–332.

Ito, S., Kolbe, L., Weets, G., & Wakamatsu, K. (2019). Visible light accelerates the ultraviolet A-induced degradation of eumelanin and pheomelanin. *Pigment Cell & Melanoma Research*, *32*(3), 441–447.

Ju, K. Y., Lee, Y., Lee, S., Park, S. B., & Lee, J. K. (2011). Bioinspired polymerization of dopamine to generate melanin-like nanoparticles having an excellent free-radical-scavenging property. *Biomacromolecules*, *12*(3), 625–632.

Kiran, G. S., Dhasayan, A., Lipton, A. N., Selvin, J., Arasu, M. V., & Al-Dhabi, N. A. (2014). Melanin-templated rapid synthesis of silver nanostructures. *Journal of Nanobiotechnology*, *12*(1), 1–13.

Kiran, G. S., Jackson, S. A., Priyadharsini, S., Dobson, A. D. W., & Selvin, J. (2017). Synthesis of Nm-PHB (nanomelanin-polyhydroxy butyrate) nanocomposite film and its protective effect against biofilm-forming multi drug resistant Staphylococcus aureus. *Scientific Reports*, *7*(1), 1–13. http://doi.org/10.1038/s41598-017-08816-y

Langfelder, K., Streibel, M., Jahn, B., Haase, G., & Brakhage, A. A. (2003). Biosynthesis of fungal melanins and their importance for human pathogenic fungi. *Fungal Genetics and Biology*, *38*(2), 143–158.

Lee, D., Jang, E. H., Lee, M., Kim, S. W., Lee, Y., Lee, K. T., & Bahn, Y. S. (2019). Unraveling melanin biosynthesis and signaling networks in Cryptococcus neoformans. *MBio*, *10*(5), e02267–e02319.

Le Na, N. T., Loc, S. D., Tri, N. L. M., Loan, N. T. B., Son, H. A., Toan, N. L., Thu, H. P., Nhung, H. T. M., Thanh, N. L., Van Anh, N. T., & Thang, N. D. (2019). Nanomelanin potentially protects the spleen from radiotherapy-associated damage and enhances immunoactivity in tumor-bearing mice. *Materials*, *12*(10). http://doi.org/10.3390/MA12101725

Madhusudhan, D. N., Mazhari, B. B. Z., Dastager, S. G., & Agsar, D. (2014). Production and cytotoxicity of extracellular insoluble and droplets of soluble melanin by *Streptomyces lusitanus* DMZ-3. *BioMed Research International*, *2014*. http://doi.org/10.1155/2014/306895

Oh, J. J., Kim, J. Y., Kwon, S. L., Hwang, D. H., Choi, Y. E., & Kim, G. H. (2020). Production and characterization of melanin pigments derived from Amorphothecaresinae. *Journal of Microbiology*, *58*(8), 648–656. http://doi.org/10.1007/s12275-020-0054-z

Pal, A. K., Gajjar, D. U., & Vasavada, A. R. (2014). DOPA and DHN pathway orchestrate melanin synthesis in Aspergillus species. *Medical Mycology*, *52*(1), 10–18.

Paria, K., Paul, D., Chowdhury, T., Pyne, S., Chakraborty, R., & Mandal, S. M. (2020). Synergy of melanin and vitamin-D may play a fundamental role in preventing SARS-CoV-2 infections and halt COVID-19 by inactivating furin protease. *Translational Medicine Communications*, 5(1), 1–14.

Pavan, M. E., López, N. I., & Pettinari, M. J. (2020). Melanin biosynthesis in bacteria, regulation and production perspectives. *Applied Microbiology and Biotechnology*, *104*(4), 1357–1370.

Plonka, P. M., & Grabacka, M. (2006). Melanin synthesis in microorganisms - Biotechnological and medical aspects. *Acta Biochimica Polonica*, *53*(3), 429–443. https://doi.org/10.18388/abp.2006_3314

Prabhakaran, K., & Harris, E. B. (1985). A possible metabolic role for o-diphenoloxidase in Mycobacterium leprae. *Experientia*, *41*(12), 1571–1572.

Pralea, I. E., Moldovan, R. C., Petrache, A. M., Ilieş, M., Hegheş, S. C., Ielciu, I., … & Iuga, C. A. (2019). From extraction to advanced analytical methods: The challenges of melanin analysis. *International Journal of Molecular Sciences*, *20*(16), 3943.

Raman, N. M., Shah, P. H., Mohan, M., & Ramasamy, S. (2015). Improved production of melanin from *Aspergillus fumigatus* AFGRD105 by optimization of media factors. *AMB Express*, *5*(1), 1–9. http://doi.org/10.1186/s13568-015-0161-0

Ribera, J., Panzarasa, G., Stobbe, A., Osypova, A., Rupper, P., Klose, D., & Schwarze, F. W. M. R. (2019). Scalable biosynthesis of melanin by the basidiomycete *Armillaria cepistipes*. *Journal of Agricultural and Food Chemistry*, *67*(1), 132–139. http://doi.org/10.1021/acs.jafc.8b05071

Shivprasad, S., & Page, W. J. (1989). Catechol formation and melanization by Na+-dependent Azotobacter chroococcum: A protective mechanism for aeroadaptation?*Applied and Environmental Microbiology*, *55*(7), 1811–1817.

Silva, C., Santos, A., Salazar, R., Lamilla, C., Pavez, B., Meza, P., & Barrientos, L. (2019). Evaluation of dye sensitized solar cells based on a pigment obtained from Antarctic Streptomyces fildesensis. *Solar Energy*, *181*, 379–385.

Singh, D., Kumar, J., & Kumar, A. (2018). Isolation of pyomelanin from bacteria and evidences showing its synthesis by 4-hydroxyphenylpyruvate dioxygenase enzyme encoded by hppD gene. *International Journal of Biological Macromolecules*, 119, 864–887.

Singh, S., Nimse, S. B., Mathew, D. E., Dhimmar, A., Sahastrabudhe, H., Gajjar, A., Ghadge, V. A., Kumar, P., & Shinde, P. B. (2021). Microbial melanin: Recent advances in biosynthesis, extraction, characterization, and applications. *Biotechnology Advances*, February, 107773. http://doi.org/10.1016/j.biotechadv.2021.107773

Solano, F. (2014). Melanins: Skin pigments and much more—types, structural models, biological functions, and formation routes. *New Journal of Science*, *2014*.

Sun, S., Zhang, X., Chen, W., Zhang, L., & Zhu, H. (2016). Production of natural edible melanin by Auricularia auricula and its physicochemical properties. *Food Chemistry*, *196*, 486–492. http://doi.org/10.1016/j.foodchem.2015.09.069

Tarangini, K., & Mishra, S. (2014). Production of melanin by soil microbial isolate on fruit waste extract: Two step optimization of key parameters. *Biotechnology Reports*, *4*(1), 139–146. http://doi.org/10.1016/j.btre.2014.10.001

Tran-Ly, A. N., Reyes, C., Schwarze, F. W. M. R., & Ribera, J. (2020). Microbial production of melanin and its various applications. *World Journal of Microbiology and Biotechnology*, *36*(11), 1–9. http://doi.org/10.1007/s11274-020-02941-z

Turick, C. E., Tisa, L. S., & Caccavo Jr, F. (2002). Melanin production and use as a soluble electron shuttle for Fe (III) oxide reduction and as a terminal electron acceptor by Shewanella algae BrY. *Applied and Environmental Microbiology*, *68*(5), 2436–2444.

Upadhyay, P., Shrivastava, R., & Agrawal, P. K. (2016). Bioprospecting and biotechnological applications of fungal laccase. *3 Biotech*, *6*(1), 15.

Vahidzadeh, E., Kalra, A. P., & Shankar, K. (2018). Melanin-based electronics: From proton conductors to photovoltaics and beyond. *Biosensors and Bioelectronics*, *122*, 127–139.

Vijayababu, P., & Kurian, N. K. (2021). Melanin and its precursors as effective antiviral compounds: With a special focus on SARS CoV2. *Molecular Biology*, 10(5), 290.

Vijayan, V., Jasmin, C., Anas, A., ParakkaparambilKuttan, S., Vinothkumar, S., PerunninakulathSubrayan, P., & Nair, S. (2017). Sponge-associated bacteria produce non-cytotoxic melanin which protects animal cells from photo-toxicity. *Applied Biochemistry and Biotechnology*, *183*(1), 396–411. http://doi.org/10.1007/s12010-017-2453-0

Wang, X., Sheng, J., & Yang, M. (2019). Melanin-based nanoparticles in biomedical applications: From molecular imaging to treatment of diseases. *Chinese Chemical Letters*, *30*(3), 533–540. http://doi.org/10.1016/j.cclet.2018.10.010

Yu, X., Huo, L., Liu, H., Chen, L., Wang, Y., & Zhu, X. (2015). Melanin is required for the formation of the multi-cellular conidia in the endophytic fungus Pestalotiopsismicrospora. *Microbiological Research*, *179*, 1–11.

Zaidi, K. U., Ali, A. S., Ali, S. A., & Naaz, I. (2014). Microbial tyrosinases: Promising enzymes for pharmaceutical, food bioprocessing, and environmental industry. *Biochemistry Research International*, *2014*.

Zughaier, S. M., Ryley, H. C., & Jackson, S. K. (1999). A melanin pigment purified from an epidemic strain of Burkholderiacepacia attenuates monocyte respiratory burst activity by scavenging superoxide anion. *Infection and Immunity*, *67*(2), 908–913.

4 Cyanobacteria as Natural Biofactories

Garvita Singh
Gargi College, New Delhi, India

Paushali Ghosh and Divya Singh
Banaras Hindu University, Varanasi, India

Jay Kumar
CSIR-Central Drug Research Institute, Lucknow, India

CONTENTS

4.1 INTRODUCTION

Cyanobacteria are among the morphologically diverse and ubiquitous group of photosynthetic prokaryotes that have modified Earth's atmosphere suitable for aerobic metabolism. They are the most primitive living organisms and have played an important role in the transformation and evolution of complex life on Earth (Chittora et al., 2020). Fossil records of cyanobacteria (Schirrmeister et al., 2016) enabling molecular clock studies provide insights into their evolutionary history. Initially, researchers found that at the organismal level, oxygenic photosynthesis emerged from ancestral anoxygenic phototrophs, 'protocyanobacteria', over 2.5 billion years ago (Hohmann-Marriott and

DOI: 10.1201/9781003306931-5

Blankenship, 2011). From the outset, the two photosystems (PSI and PSII) have evolved the ability to carry out oxidation of water to a remarkable degree, and have specialized further, from gene duplication events, leading to oxygenic photosynthesis (Sánchez-Baracaldo and Cardona, 2019). According to the 2019 forecast from the United Nation's Population Division, global population was estimated around 7 billion in 2011 and is expected to increase to 9.7 billion in 2050. With the ever-expanding population, our demand for resources is increasing, coupled with the increase in reliance on agricultural crops (Guihéneuf et al., 2016). The increased demand has led to over-exploitation of agricultural crops for food, chemicals, and energy resources (Kumar et al., 2019a). The challenges arising from improper management of food and energy resources calls our continuing existence into question (Patel et al., 2021a; Patel et al., 2021b). Cyanobacteria represent the most attractive contenders to meet future demand due to two inherent properties: their photosynthetic capability and potential for genetic engineering (Ducat et al., 2011). Cyanobacteria have been broadly used as model organisms for investigating photosynthetic mechanisms, nitrogen fixation, and production of primary and secondary metabolites, and in pharmaceutical applications. They can colonize in almost all kinds of aquatic and terrestrial ecosystems, and adapt to extreme environmental conditions (Kini et al., 2020). Cyanobacteria owe their adaptability and versatility to complex morphologies enabling them to survive in extreme habitats (Kumar et al., 2019b). Another feature which makes them promising candidates is their capacity to grow on non-arable lands. Their growth on residuary nutrients with soaring productivity has been established, along with enrichment in carbohydrates, lipids and proteins,which could alleviate the soaring prices of various products acquired from agriculture-based industries (Rittmann, 2008; Guihéneuf et al., 2016) and can be used as a source of several other metabolites and products beneficial to life.

The cyanobacteria metabolic pathway could prove promising for the production of value-added compounds that has boosted the biotechnology industry in the past few decades (Noreña-Caro and Benton, 2018). Researchers have recently identified the immense potential of cyanobacteria as natural biofactories for the production of pharmaceuticals and chemicals of interest (Figure 4.1). Among the valuable pharmacologically vital compounds are products that have antibacterial (Volk and Furkert, 2006), anticancerous (Srinivasan and Sivasubramanian, 2014), antifungal (Rath and Priyadarshani, 2013), antiviral and immunosuppressive (Vijayakumar and Menakha, 2015) properties, leading to new prospects in the field of medicine. The use of the inexpensive raw materials needed to fuel the cyanobacterial metabolism – sunlight, CO_2, and water – has converted them into powerful

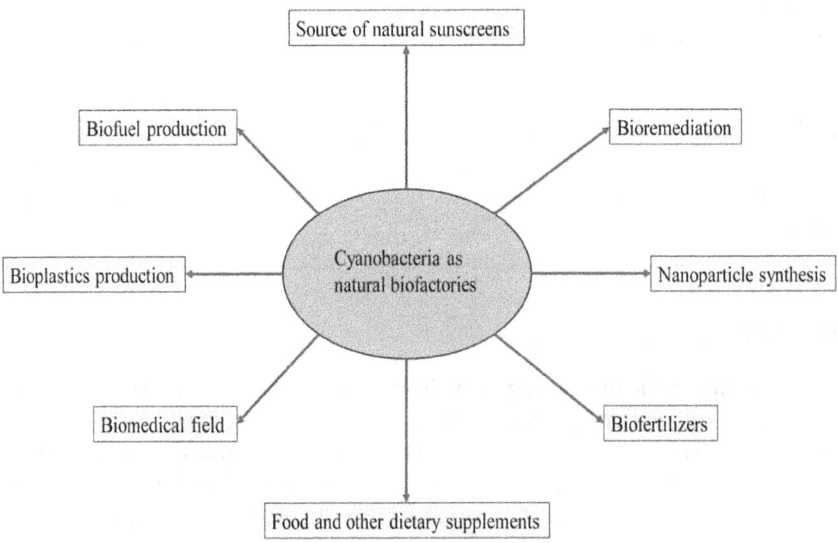

FIGURE 4.1 Diagrammatic representation depicting cyanobacteria as natural biofactories.

photoautotrophic cell biofactories (Ducat et al., 2011). To make the leap from laboratory work to industrial and commercial production of valuable chemicals, certain essential factors should be present. These include strain improvement to achieve large-scale productivity to increase the rate of cell survival under adverse conditions, increased efficiency in the harvesting and isolation of products from cyanobacteria, and gaining new insights into the cyanobacterial metabolism (Zhou et al., 2016).

This chapter focuses on general applications of cyanobacteria as natural cell biofactories, highlighting their role in the synthesis of diverse bioactive metabolites like photoprotective compounds (MAAs, scytonemin and carotenoids), biofuels, bio-polymers, high-value chemicals, and pharmaceuticals, and their roles in the agriculture and food sectors as depicted in Table 4.1.

TABLE 4.1
Important Applications of Cyanobacteria

Representative Cyanobacteria	Applications	References
	Biofuels	
Synechococcus elongates PCC 7942	Isobutanol	Atsumi et al. (2009)
Spirulina	Biogas	Mussgnug et al. (2010)
Anabena	Biogas	Mussgnug et al. (2010)
	Food Supplements	
Nostoc	Nutritional supplements	Nicoletti (2016)
Anabena	Animal food additives	Nicoletti (2016)
Oscillatoria	Nutraceuticals	Nicoletti (2016)
Phormidium	Nutraceuticals	Nicoletti (2016)
Spirulina platensis	Dietary supplements	Belay (2002)
Spirulina maxima	Dietary supplements	Belay (2002)
Aphanotheca sacrum	Edible product	Fujishiro et al. (2004)
	Bioplastics	
Chlorogloea fritschii	Polyhydroxybutyrate	Capon et al. (1983)
Spirulina	Polyhydroxybutyrate	Miyake et al. (1996)
Oscillatoria	Polyhydroxybutyrate	Fernandez-Nava (2008)
Gleocapsa	Polyhydroxybutyrate	Fernandez-Nava (2008)
Nostoc	Polyhydroxybutyrate	Miyake et al. (1996)
Synechococcus sp. PCC6803	Polyhydroxybutyrate	Osanai et al. (2013)
	Bioremediation	
Spirulina indica, Spirulina platensis and *Spirulina maxima*	Biosorption of zinc and nickel	Balaji et al. (2014)
Synechocystis	Biodegradation of chlorinated hydrocarbons	Abed et al. (2009)
Oscillatoria	Biodegradation of chlorinated hydrocarbons	Abed et al. (2009)
Pleurocapsa	Biodegradation of chlorinated hydrocarbons	Abed et al. (2009)
Chroococcus sp. HH-11	Biosorption of chromium	Anjana et al. (2007)
*Nostoccalciola*HH-12	Biosorption of chromium	Anjana et al. (2007)
*Synechocystis*sp. PUPCCC 64	Biodegradation of organochlorine or organophosphorous pesticides	Singh et al. (2011a)
Nostoc hatei	Biodegradation of organochlorine or organophosphorous pesticides	Jha and Mishra (2005)
Anabena sphaerica	Biodegradation of organochlorine or organophosphorous pesticides	Singh et al. (2011b)
Oscillatoria sp.	Oxidation of naphthalene to 1-napthol	Al-Hasan et al. (1998)
Phormidium corium	Biodegradation of n-alkanes	Al-Hasan et al. (1998)

(Continued)

TABLE 4.1 *(Continued)*

Microcoleuschthonoplastes	Biodegradation of n-alkanes	Al-Hasan et al. (1998)
Oscillatoria salina	Biodegradation of aliphatics	Raghukumar et al. (2001)
	Biomedicals	
Lyngby lagerheimeii	Anti-HIV compounds	Gustafson et al. (1989); Raghukumar et al. (2001)
Phormidium tenue	Anti-HIV compounds	Gustafson et al. (1989)
Spirulina	Immunomodulatory compounds	Ratha et al. (2021)
Nostoc ellipsosporum	Virucidal compound	Matei et al. (2010)
Microcystis aeruginosa	Anti-HIV compounds	Chaves et al. (2018)
Scytonemavarium strain HG-24-1	Virucidal compound	Bokesch et al. (2003)
Oscillatoria agardhii	Anti-HIV activity	Férir et al. (2014)
Oscillatoria acuminate MHM-632MK014210	Anticancer and antiviral activity	Saad et al. (2020)
Cyanothece sp. PCC7424	Anti-HIV activity	Matei et al. (2016)
Spirulina sp.	Antiviral drugs	Heo et al. (2017); Anekthanakul et al. (2019)
Lyngbya majuscule	Anticancer metabolites	Gutiérrez et al. (2008)
*Schizothrix*sp.	Anticancer metabolites	Gutiérrez et al. (2008)
*Symploca*sp.	Antiproliferative agents	Kurisawa et al. (2020)
Nostoc sp. GSV 224, *Nostoc linckia* and *Nostoc spongiaeforme*	Anticancerous drug	Hemscheidt et al. (1994), Torres et al. (2014)
Anabaena doliolum	Anti-inflammatory and anticancer properties	Rajeshkumar et al. (2014)
Nostoc commune	Antibacterial activity	Blom et al. (2006)
Microcystis aeruginosa PCC 7806	Antileishmanial and antimalarial compounds	Portmann et al. (2008
Blennothrixcantharidosmum	Antimalarial activity	Clark et al. (2008)
Lyngbya majuscule	Antileishmanial and antimalarial activity	Shah et al. (2017), Morone et al. (2020)
	Nanobiotechnology	
Oscillatoria limnetica	Synthesis of silver nanoparticles	Hamouda et al. (2019)
Cylindrospermumstagnale	Synthesis of silver nanoparticles	Sudha et al. (2013)

4.2 THE ROLE OF CYANOBACTERIA IN BIOFUEL PRODUCTION

With the two crucial problems to be addressed by the human race, global warming and the energy crisis, there is an urgent need for identification of alternative, economical, and environmentally-friendly renewable energy sources. Biofuelis widely regarded as a low-cost and environmentally friendly substitute for petroleum and other fossil fuels. Biofuels like biohydrogen, biodiesel, bioethanol and biogas are regarded as clean and green fuelswith greater energy conversion efficiency than petroleum (Chen et al., 2014). Of many possible solutions, the production through biological routes of ethanol and hydrogen as future fuels is an eco-friendly process(Patel et al., 2012). Cyanobacterial species have been recognized as promising alternative energy sources replacing fossil-based fuels for economical, sustainable energy production.

Currently, the fermentation of agricultural crops and residues are the primary sources for bioethanol production. However, limitations on agricultural land/grain use for fuel, along with the high energy input linked to traditional sugar/starch and lignocellulosic biomass-based fermentation systems, are regarded as considerable challenges for ethanol production (Silva and Bertucco, 2016). Bioindustrial approaches depend on photosynthetic microorganisms due to their capacity to grow in diverse locations, their eco-friendly and renewable nature, low transportation costs, ease of cultivation and capacity for genetic engineering (Shimizu, 2003). The inexpensive substrates required to fuel the metabolism of cyanobacteria vis-à-vis sunlight, CO_2, water and other minerals are convenient in contrast to carbohydrate feedstocks. The cyanobacterial biomass provides feedstock for the production of biofuels like bioethanol, biodiesel, biohydrogen, biomethane, etc. (Ducat et al., 2011).

4.2.1 Alcohols: Ethanol and Isobutanol

The light-converting reactions carried out by Photosystems I and II are the engines driving the manufacture of all organic matter in cyanobacteria. In the majority of cyanobacterial production pathways, the metabolites that are an essential component of biofuels,generated directly downstream of the Calvin cycle, are redirected towards the desired endproduct (Brenner et al., 2006). Dexter et al. (2015) reported that cyanobacteria synthesize ethanol directly in a photosynthetic process referred to as the "photofermentative" route. Some cyanobacterial strains like *Synechocystis* sp. have been genetically adapted to perform the photofermentative route of bioethanol synthesis (Dexter and Fu, 2009; Gaoet al., 2012). Deng and Coleman (1999) have introduced alcohol dehydrogenase II (ADHII) and pyruvate decarboxylase (PDC) genes from *Zymomonas mobilis* into the cyanobacterium *Synechococcus* sp. PCC 7942. Ingram et al. (1987) previously demonstrated the heterologous expression of these genes during the establishment of a similar PDC/ADHII cassette in *E. coli* to that generated in this model system. In the current metabolic pathway, utilizing the reducing power of photosynthesis, carbon dioxide is fixed for the synthesis of pyruvate which forms acetaldehyde and ethanol via the inserted ADHII and PDC. To reduce the cost of production and escalate productivity, scientists have now focused on enhancement of cyanobacterial strains. Namakoshi et al. (2016) have reported studies that include genetic modifications along with the combinatorial excisions of some parts of the glycogen synthesis pathway and poly (3-hydroxybutyrate) (PHB) synthesis pathway to enhance the ethanol production rate in *Synechocystis* sp. PCC 6803.

Isobutanol, a favoredalternative to gasoline, is extensively manufactured by synthetic approaches due to its high-octane value, compatibility and greater energy density (Atsumi et al., 2008; Machado and Atsumi, 2012). Atsumi et al. (2008) have developed a 2-ketoacid-based pathway to test the capacity of cyanobacteria for increased production of isobutanol. Atsumi et al. (2009) investigated the production of isobutaraldehyde, a precursor of isobutanol, in the mutant strain of *Synechococcus elongatus* PCC7942 via heterologous co-expression of valine biosynthesis enzymes and ketoacid decarboxylase gene (kivD). Alcohol dehydrogenase gene (*yqh*D) from *E.coli* showed efficient conversion from isobutaraldehyde to isobutanol yielding 3.12 mg L^{-1} h^{-1} (Atsumi et al., 2009).

4.2.2 Biodiesel

Biodiesel production has been a designated energy alternative for conventional-based diesel fuel. Production of biodiesel is reliant on plant-derived oil feedstock for transesterification of lipids to yield fatty acid alkyl esters (Machado and Atsumi, 2012). However, a more environmentally friendly and economical downstream process for recovery of lipids from cyanobacteria has been developed. High-yield oil-producing cyanobacterial strains have been recently demonstrated by genetic engineering (Schirmer et al., 2010; Liu et al., 2011a). The algal crude lipids comprise free fatty acids (FFA) and triacylglycerols which can be transformed into fatty acid alkyl esters. Generally, the process involved in the recovery of algal lipids includes degradation of cell membrane lipids into FFA and then cell lysis leading to extraction and separation of the desired product. The cyanobacteria have the ability to sequester and fix CO_2, hence FFA produced is converted into biodiesel by esterification. This procedure accounts for nearly 50% of the total production cost of biofuel synthesis. The engineered cyanobacterium was capable of yielding 36.1×10^{-12} mg/cell of fatty acid under CO_2 fixation (Liu et al., 2011b).

4.2.3 Biomethane

Being ubiquitous, cyanobacteria have been found to produce methane (CH_4) biogas at significant rates under both oxic and anoxic, and light and dark conditions. Bižić et al. (2020) investigated the conversion of fixed inorganic carbon into CH_4 by cyanobacteria under both light and dark conditions. Their potential synthesis of CH_4 via distinct pathways depending on the surroundings makes them key players in the global CH_4 cycle (Visser et al., 2016). Additionally, the biochemical composition of cyanobacterial biomass, the absence of hard polysaccharide-based cell walls and high C/N ratio makes them suitable for higher digestibility (Bohutskyi and Bouwer, 2013). The two most studied genera,*Spirulina* and *Anabena*, have been explored for biogas production, where methane fraction accounted for nearly 60–75% in the biogas (Mussgnug et al., 2010).

4.2.4 Biohydrogen

Among the different methods being explored, biological hydrogen (H_2) production has gained remarkable importance due to its production under ambient physiological conditions and the discovery of many H_2 producers (Patel et al., 2014). Cyanobacteria can also produce molecular hydrogen (H_2) which could be an ideal alternative to fossil fuels. Solar energy can be converted into hydrogen, i.e., chemical energy, by using photosynthetic bacteria that are either oxygenic (O_2-producing)-deriving electrons from water or anoxygenic-deriving electrons from substrate other than water. Hydrogen production by microbes can either be generated by light (e.g., photosynthetic bacteria and cyanobacteria) or synthesized in the dark from the fermentation of organic sources (e.g., members of Enterobacteriaceae). Due to partial decomposition in dark fermentation, the yield is only 10–20% of the hydrogen stored in the substrate although the reaction rate is high (Dutta et al., 2005; Sharma and Stal, 2014). There are diverse intrinsic (genetic components and cellular proteins in cyanobacteria) and extrinsic factors (light, temperature, salinity, micronutrient levels, oxygen, carbon and nitrogen sources) that affect cyanobacterial hydrogen production. H_2 production in cyanobacteria occurs either during photosynthesis by the activity of hydrogenase enzymes or as a byproduct of nitrogen fixation (Pinzon-Gamez et al., 2005). Photosynthetic H_2 production can be effectively scaled up by metabolically engineered cyanobacterial strains with low pigment production, resulting in a 1.4-fold increase in H_2 conversion efficiency and innovation of bioreactor designs with increased surface area/volume ratio which can further be used as a source of clean energy.

4.3 CYANOBACTERIA AS A SOURCE OF NATURAL SUNSCREENS AND ITS APPLICATIONS IN COSMETICS

A significant range of evolutionary adaptations, from biochemical to physiological and behavioral, have been extensively studied in cyanobacteria. Among these, the production of putative secondary metabolites for their value as sunscreens and antioxidants has high potential for commercial application (Soule and Garcia-Pichel, 2014). Cyanobacteria possess a wide range of defense mechanisms to protect/minimize the photodamage caused to the metabolic and physiological machinery (Castenholz and Garcia-Pichel, 2012). The major cellular processes impaired in cyanobacteria include photosynthesis, nitrogen fixation, and cell differentiation(Blakefield and Harris, 1994; Castenholz and Garcia-Pichel, 2012). Cellular mechanisms to repair damaged molecules or other processes are exhausting and mainly not preventive. Preventive strategies require the capability to produce and gather photoprotective compounds that can act as either UV-absorbing sunscreens or as antioxidant scavengers of reactive oxygen radicals induced under UV radiation (Garcia-Pichel and Castenholz, 1991, 1993).

Since cyanobacteria possess various protection and repair strategies, they can overcome the damaging effects of UV radiation. To survive under both high UV radiation and severe dessication, large numbers of cyanobacteria live in adverse habitats and synthesize metabolites. (Fleming and

Castenholz, 2007). One of the interesting applications of cyanobacteria is their ability to produce photoprotective compounds such as MAAs (Mycosporine-like amino acids) and scytonemin (Richa et al., 2011), which are very stable and can counteract UV radiation without forming reactive oxygen species (ROS) inside cells. These attributes make cyanobacteria an attractive biological source of cosmetic ingredients. MAAs are low molecular weight (< 400 Da) colorless molecules with absorption maxima in the range of 310–362 nm. MAAs are photo-stable and heat-stable, with the ability to protect membrane lipids and reduce sunburn (Soule and Garcia-Pichel, 2014). Misonou et al. (2003) investigated the mitigation strategies of MAAs against DNA damage and discovered their ability to efficiently obstruct thymine dimer formation induced under UV radiation. Oyamada et al. (2008) also studied the defense mechanisms of three different kinds of MAAs (shinorine, porphyra-334 and mycosporine-glycine) and discovered their protective role against UV-induced cell death on human fibroblast cells. MAAs also act as multifunctional secondary metabolites with various cellular functions, for example as antioxidant molecules to quench oxidative radicals and lower cellular oxidative stress (Suh et al., 2003). These natural metabolites are potential substitutes for commercially available sunscreens (White et al., 2011). The use of MAAs in the treatment of photo-aging of skin induced by reactive free radicals upon UV radiation exposure has been explored. Daniel et al. (2004) reported that the application of a cream containing 0.005% Porphyra-334 encapsulated in liposomes potentially improved skin texture after 4 weeks compared to other creams containing synthetic UV filters.

Another crucial sunscreen pigment, scytonemin, has been isolated from cyanobacterial species growing under intense solar radiation with an absorption maximum of 380 nm. Scytonemin usually present in the extracellular sheath of cyanobacterial colonies provides protection against short-wavelength UV radiation. Besides showing great potential in photoprotection in medical and industrial fields, these photoprotective compounds also exhibit other applications. Initial studies determined the application of scytonemin in neovascularization, which is common in inflammatory disorders, indicating its activity as a powerful anti-inflammatory agent. Its implementation in the treatment of human umbilical vein endothelial cell (HUVEC) proliferation was analyzed. Stevenson et al. (2002) discovered a gradual decrease in edema in the mouse ear through topical application of scytonemin. Similarly, scytonemin was found to be antiproliferative in nature as it turned out to be a significant inhibitor of polo-kinase 1 (PLK1), a serine/threonine kinase found to be a potential regulator of the G2/M cell cycle transition that is dominant in many tumor cells (Stevenson et al., 2002; McInnes et al., 2005). PLK1 has been confirmed as critical for t melanoma cell survival, though its functional role in melanoma is unknown (Schmit et al., 2009) Moreover, Duan et al. (2010) have reported a greater sensitivity of human osteosarcoma cells to scytonemin than normal osteoblast cells. Carotenoids are potent natural antioxidants produced from plants and photosynthetic microorganisms like cyanobacteria (Singh and Kumar, 2018). The important carotenoids, such as ß-carotene, lutein, and lycopene, are natural lipophilic pigments which are distributed through all of nature and have various biological functions and health benefits including properties such as antioxidants, hormonal precursors, dyes and several components essential for photosynthesis. These compounds are most effective against ROS and prevent singlet oxygen radicals from attacking different molecules including those of the skin tissue (Godwin et al., 2006).

4.4 FOOD AND OTHER DIETARY SUPPLEMENTS

With the rising burden of population explosion, limitations on production of traditional crops, and dwindling energy resources, alternative sources of functional foods (proteins, lipids, dietary fibers, etc.) have become a priority (Vanthoor-Koopmans et al., 2014). Over the last decade, cyanobacteria have garnered huge interest as a sustainable source of food supplements as they are rich in carbohydrates, lipids, proteins, antioxidants, and vitamins. Certain attributes, such as their worldwide distribution, ease of digestibility,cultivation conditions and product stability make them suitable sources of functional foods (Panjiar et al., 2017). Many species of *Spirulina, Nostoc, Anabaena* and *Oscillatoria* are grown on an enormous scale as nutritional supplements for humans, as animal food additives, and as neutraceuticals (Nicoletti, 2016).

The commercial use of *Spirulina platensis* and *S. maxima* began in the 1970s due to their high nutritional content. *Spirulina* belongs to the list of substances under the category GRAS (Generally Recognized as Safe) by the US Food and Drug Administration (Belay, 2002). In comparison to regular protein foods, crude protein content is 50–60% higher in *Spirulina* (Ali and Saleh, 2012). Lipid content of the filamentous cyanobacterial species has been found to contain large amounts of polyunsaturated fatty acids (25–60%) and essential fatty acids,thus having medical implications (Moreira et al., 2013). Lipids obtained from *Spirulina* are free from cholesterol which is metabolically supportive for people with ailments like diabetes and obesity. However, *Scytonema bohneri* was found to have a higher carbohydrate content (28.4%) compared to 15–20% for other *Spirulina* species (Rajeshwari and Rajashekhar, 2011).

Cyanobacteria have the potential to provide a richer source of vitamin B12 and provitamin A than other dietary sources (Cohen, 1986; Shetty et al., 2006). Substantial amounts of minerals like iron, magnesium, calcium, and potassium having strong antioxidant properties are also very high in *Spirulina* (Shetty et al., 2006). Gamma linolenic acid (GLA), abundantly found in *Spirulina*, regulates lipid metabolism in humans after conversion into prostaglandin E2. C-phycocyanin (C-PC), one of the major biliproteins extensively found in *Spirulina*, has both antioxidant and anti-inflammatory characteristics (Romay et al., 2003). Pulz and Gross (2004) have reported that the biomass of cyanobacteria and microalgae can also be used as additives in beverages and bakery products. *Spirulina is* designated a "superfood" by NASA and has been successfully used as a dietary supplement on space missions (Karkos et al., 2008). *Nostoc*, widely used as a food supplement, can be consumed raw as dried powder, stir-fried, and as a thickener for other foods (Facciola, 1998). The ball-shaped colonies of *N. punctiforme*, referred to as "lakeplum" by local people in South America, have been traditionally used in human diets (Trainor, 1978). Besides the filamentous variety *Aphanotheca sacrum*, a unicellular form is another consumable microbe in Japan enjoyed as a special delicacy called *suizenji-nori* (Fujishiro et al., 2004).

4.5 THE ROLE OF CYANOBACTERIA IN BIOPLASTICS PRODUCTION

The use of plastics has become universally important in computer hardware, medical equipment, home appliances, packaging materials, and automobiles (Kumar et al., 2019a). However, the constant use of non-degradable plastics manufactured from 80% of non-renewable chemicals produced by the petrochemical industry has focused attention on biodegradable plastics (bioplastics). Kalia et al. (2021) presented innovative strategies to increase the commercial viability and sustainability of bioplastics. Plastic pollution has generated particles with a diameter of 1.01–4.75 mm, classed as microplastics (i.e., diameter less than 5 mm) (Eriksen et al., 2014). Due to their biomagnifications in the food chain, health-related concerns are increasing in humans. Exposure to microplastics in beverages, salt, sugar, honey, and drinking water has caused an increase in toxic effects and complications in various diseases (Wright and Kelly, 2017).

Bioplastics can be classified as: bio-based and biodegradable (polyhydroxyalkanoates-PHAs), bio-based (starch/lignocellulose-based plastics), and fossil-based biodegradable plastics (e.g., bionylons, bio-polyurethanes and polylactic acid-PLA) (Karan et al., 2019). Currently, difficulties in production, processibility of polymers and economic output are experienced in most bioplastics produced from carbohydrate-rich plant-based feedstock such as corn or sugarcane. Next-generation cyanobacterial-based bioplastics have the potential to address many of these issues. Cyanobacteria-based systems have the benefit of biochemical conversion of greenhouse gas like carbon dioxide to eco-friendly plastics using solar energy. Cyanobacterial species accumulate storage materials like polysaccharides, cyanophycin, polyphosphates and poly-β-hydroxybutyrate (PHB) which supply carbon and energy reserves (Koutinas et al., 2007).

PHB, a homopolymer identified among the 150 varied types of polyhydroxy-alkanoids, is an extensive storage reserve in different taxonomic groups of prokaryotes like cyanobacteria. Cyanobacteria are reported to be the most efficient microbes in the synthesis of a huge family of polymers (PHAs) under photoautotrophic conditions, having minimal nutrient requirements (Singh and Mallick, 2017). The carbon sources assimilated photoautotrophically are converted to hydroxy-alkanoate monomer units, polymerized, and further stored as water-insoluble granules in the cell cytoplasm. *Chlorogloea fritschii* was the first cyanobacterial species in which the presence of PHB was reported. To date, the presence of PHB has been shown in diverse cyanobacteria such as *Spirulina, Oscillatoria, Gleocapsa, Nostoc* and *Synechococcus* (Capon et al., 1983; Miyake et al., 1996; Fernandez-Nava et al., 2008). PHB serves as an excellent feedstock for plastic production and can efficiently replace conventional plastics to reduce carbon dioxide emissions and manage fossil resources (Dias et al., 2006). PHB homopolymer is a highly crystalline, sturdy but brittle substance and behaves as a hard elastic material when spun into fibers (Sangkharak and Prasertsan, 2007). Hai et al. (2001) reported the hydrophobicity, biodegradability, and biocompatibility properties of PHB as comparable to those of polypropylene. It has gained momentum for its application in the biomedical and biopharmaceutical fields (Sudesh et al., 2000). For the past two decades, PHAs have been widely regarded as a potential alternative to non-biodegradable petrochemical-based plastics. They are actively biodegraded by the action of naturally occurring microorganisms into monomeric units to be utilized as a carbon source for growth.

Cyanobacteria that do not accumulate PHB have been engineered by transformation with genes encoding PHB synthesis, leading to the accumulation of the biopolymer (Sangkharak and Prasertsan, 2007; Balaji et al., 2013). Biotechnological methods like fermentation and genetic engineering have been extensively explored in pursuit of new metabolic pathways to enhance PHB synthesis in cyanobacteria. Osanai et al. (2013)reported the overexpression of sigma factor SigE in *Synechococcus* sp. PCC6803 during nitrogen starvation, leading to an increase in the synthesis of PHB. Algal-based bioplastics require future research aiming at increasing and optimizing the feasibility of bioplastic production rather than studying aspects of internal storage reserves or optimizing nutrient removal.

4.6 CYANOBACTERIA-MEDIATED BIOREMEDIATION

A well-balanced and natural ecosystem in terms of atmospheric gases, water vapor, nutrient cycle, and natural resources for the sustenance of diverse flora and fauna has been badly disrupted by human activities (Acho-Chi, 1998). There has been a significant increase in contaminated sites globally; therefore, maintenance of a productive and fertile ecosystem is the urgent priority of current environmental biotechnology research. Since a wide variety of physicochemical approaches have been both expensive and incompatible with supporting the potential demand for the remediation of polluted sites, bioremediation using natural and economic sources of biomass have gained momentum. Ruffing (2011) efficiently tested the potential of genetically engineered or native microbes as potential candidates for reducing the burden of pollutants from various sites

Cyanobacteria are gaining considerable importance due to their natural ability to degrade and detoxify heavy metals, radionuclides, antibiotics, pesticides, organic wastes and recalcitrant compounds from contaminated sites (Mohamed, 2001). The presence of different functional and other charged groups, namely hydroxyl, phosphate, and sulfate, on the biomass of *Spirulina* are important for biosorption of toxic heavy metals. Recently, the immense potential of several strains of *Spirulina* in the biosorption of zinc and nickel metals was discovered (Balaji et al., 2014). It is interesting to note that cyanobacteria potentially form symbiotic associations with other microorganisms in the environment. Cyanobacterial blooms including *Synechocystis, Oscillatoria*, and *Pleurocapsa* facilitate the degradation process of chlorinated hydrocarbons in oil by providing nutrients to the related oil-degrading bacteria (Abed et al., 2009). Anjana et al. (2007) reported the biosorption of heavy metal Cr (VI) using *Chroococcus* sp. HH-11 and *Nostoc calciola* HH-12 immobilized in Ca-alginate. The presence of metallothioneins and intracellular polyphosphates in cyanobacteria has been reported to be involved in the sequestration of toxic metal ions (Gardea-Torredey et al., 1998). *Anabena sphaerica, Synechocystis* sp. PUPCCC 64 and *Nostochatei* efficiently degrade organochlorine or organophosphorous pesticides in aquatic regions (Jha and Mishra, 2005; Singh et al. 2011a). Degradation of hazardous organic waste, for example, oxidation of polycyclic aromatic hydrocarbons like benzo(a)pyrene, anthracene, perylene, pyrene, and fluorine by cyanobacteria, have also been reported. Raghukumar et al. (2001) reported the degradation of crude oil-containing aromatics (anthracene and perylene), waxes and aliphatics by cyanobacterial consortium viz., *Oscillatoria salina, Aphanocapsa* and *Plectonema* sp. Similarly, Al-Hasan et al. (1998) have reported that *Phormidium corium* and *Microcoleus chthonoplastes* degrade n-alkanes while *Oscillatoria* sp. can effectively oxidize naphthalene to 1-napthol.

Other potential applications of cyanobacteria in the field of environmental detoxification include: maintaining the physicochemical properties of the soil by mobilization of nutrients like phosphorous and nitrogen in nutrient-deprived soils and soil-aggregation properties in saline-sodic soils, thereby improving air and water movement in the soil (Kaushik, 1983); phycoremediation of textile wastewaters (Pathak et al., 2014); biodegradation of polymers like polyethylene by the production of oxidative and lignolytic enzymes (Gupta et al., 2017); and reducing the emission of greenhouse gases via photosynthetic machinery (Kaplan et al., 1994).

4.7 CYANOBACTERIA IN THE BIOMEDICAL FIELD

As phototrophic organisms, cyanobacteria have prokaryotic cell organization, perform photosynthesis and respiration similar to eukaryotes, and have considerable potential as innovative sources of effective unexplored bioactive compounds. The study of those pharmaceutically active compounds could fill the information gap and lead to the eradication and control of several diseases caused by viruses, bacteria, fungi, and other microbes as potential antiviral, antibacterial, antifungal and other multitarget compounds.

4.7.1 Antivirals

Various lethal diseases caused by such viruses as HIV, AIDS, dengue, avian influenza (H5N1 virus), SARS, and COVID-19 have emerged as epidemics and pandemic with serious consequences for human beings. To tackle these diseases several therapies and drugs have been tested,such as the anti-HIV antiretroviral therapy (HAAR) which was found to be effective in controlling HIV, although a number of studies suggested it to be toxic under several conditions and treatment lines(Luescher-Mattli, 2003). Many antiviral compounds isolated and purified from cyanobacterial species are reported by several researchers as showing an antiviral response by blocking the viral penetration and limiting replication of viruses. Gustafson et al. (1989)reported the effective role of *Phormidium tenue* extract against the cytopathic effect of HIV. Another study suggested a role of cyanobacterial extract as inhibitor with sulfonic acid-containing glycolipid against HIV virus. The elaborative usage of *Spirulina* extracts in various health and dietary products was recognized by the Food and Drug Administration (FDA) and a similar study showed very promising results when a concoctions made from *Spirulina* sp. used in treatment lines of HIV patients showed exceptionally low viral load along with a boost in the activity of macrophages and all other immunomodulatory responses (Hirahashi et al., 2002; Ngo-Matip et al., 2015; Ratha et al. 2021).One such virucidal compound, cyanovirin-N (CV-N) made of 11-kDa protein, was isolated from extracts of the cyanobacterium *Nostoc ellipsosporum* (Matei et al. 2010). Studies have shown this virucidal extract to be to be effective and promising when applied on different strains of HIV.

The role of cyanovirin-N as a potent microbicide in the field of HIV transmission is being developed by Cellegy Pharmaceuticals, Inc.. Similarly, another virucidal compound isolated from *Microcystis aeruginosa* (Chaves et al., 2018), microvirin (MVN) inhibits the infection of many HIV-1 strains. A4-fold stronger anti-HIV activity was reported from cyanobacterium isolated from rice fields, *Cyanothece* sp. PCC7424 (Matei et al., 2016). Similarly, active metabolites from *Oscillatoria* sp. were found to be effective in carcinogenic and viral treatment lines (Saad et al., 2020).

4.7.1.1 Role of Cyanobacteria in Treating SARS and COVID-19

The study of cyanobacterial metabolites as a remedy for SARS-CoV2 infections, while still in its infancy, shows many promising responses. A continuous effort is required to identify their pharmaceutical role in the fight against COVID-19 and its after-effects. With the continuously mutating virus and new strains now emerging, there is an immediate need to develop broad-spectrum antivirals without developing resistance. A number of cyanobacterial members have been studied for their virucidal properties, among them *Spirulina* sp., which alongside other benefits is a rich source of antiviral compounds. Phycobilin-derived peptides from *Spirulina* sp. as enzymatic hydrolysate were found to be effective in suppressing the angiotensin-converting enzyme (ACE) obtained from phycobilin-derived peptides (Heo et al., 2017; Anekthanakul et al., 2019).

These phycobilin peptides show remarkable control of cardiovascular abnormalities, including blood pressure issues, by resolving blood vessel constriction with the aid of ACE, reducing the formation of angiotensin II when studied in several patients. Another class of cyanobacterial metabolites includes sulfoglycolipids, which were among the first antiviral compounds from cyanobacteria to be discovered (Gustafson et al., 1989). They are present in the cyanobacterial thylakoid membrane and heterocystcell wall. The role of sulfoglycolipids mainly concerned regulating the function of DNA polymerase in HIV-1,and their role in ribonuclease H was not widely reported (Gustafson et al., 1989; Loya et al., 1998).

4.7.2 Anticancer

In order to develop drugs against lethal cancers, various research centers employed cyanobacterial extracts, first reported by Moore and Gerwick in the 1990s (Moore et al., 1988; Gerwick et al., 1994; Moore et al., 1996; Gerwick et al., 2008). Studies have shown the active role of these cyanobacterial extracts in targeting tubulin or actin filaments in eukaryotic cells, which has made researchers interested in isolating and characterizing compounds from cyanobacteria for their potential role in controlling carcinogenic cells (Jordan and Wilson, 1998). Several anticancer metabolites have been reported in the cyanobacterium *Leptolyngbya*. Studies conducted by pharmaceutical companies seeking to develop a better drug have resulted in the identification of a group of secondary metabolites known as lipopeptides showing strong results against tumor cells and found to be effective in apoptosis. The lipopeptide Somocystinamide A (Van Damme et al., 1998), isolated from *Lyngbya majuscule*, is a conjugate of marine cyanobacteria with exceptional biological properties. Another anticancer compound, Apratoxin A (Gutiérrez et al., 2008), isolated from *Lyngbya majuscule*, showed an impressive cytotoxic effect on cancer cell line*in vitro*. Another group of peptides known as dolastatins which have an antiproliferative effect was obtained from marine cyanobacterium *Symploca* sp. and was found to be effective in controlling cancer cell growth by apoptosis. Similarly, another class of anticancer compounds called cryptophycins, isolated and purified from *Nostoc* sp., were effective cytotoxic agents, and have been explored for their application as anticancerous drug. Approximately 30 anticarcinogenic cryptophycins obtained from *Nostoc* sp. were effective cytotoxic agents studied by the Moore group. Among the different types studied to date, the compound cryptophycin 52 showed promising results in clinical trials in the treatment of ovarian and lung cancer (Edelman et al., 2003).

4.7.3 Antibacterial and Antifungal Activity

With recent advances in microbial studies, a number of strains of cyanobacteria have been studied and reported for their potential antibiotic activity against many bacterial and fungal growths. In somecases, although the exact nature of the extract is not known, researchers are pursuing the characterization of cyanobacterial compounds from their extracts as antibacterial agents (Skulberg, 2000; Biondi et al., 2008). With the increased application of antibiotics, several multidrug-resistant bacteria have emerged, for example, methicillin-resistant *Staphylococcus aureus* and vancomycin-resistant *Enterococci*. This has led to therapeutic challenges that require immediate attention (Reinert et al., 2007). Noscomin obtained from *Nostoc commune* and nostocarboline from *Nostoc* showed antibacterial activity against several bacteria and the unwanted growth and development of certain types of toxic algae belonging to different classes (Jaki et al., 2000; Blom et al., 2006), respectively. Nostocine A purified from *Nostoc spongiaeforme* are effective in controlling growth of green algae (Hirata et al., 2003; Asthana et al., 2009). Another compound, ahapalindole (alkaloids) obtained from Nostoc CCC537 and *Fischerella* sp., were reported to be effective antimicrobial agents against various drug-resistant bacterial species such as *Mycobacterium tuberculosis* H37Rv, *Staphylococcus aureus* ATCC25923, *Salmonella typhi* MTCC3216, *E. coli* ATCC25992, and *Enterobacter aerogenes* MTCC2822.

The antifungal activity of these cyanobacterial cell extracts was studied on *Curvularia lunata*, *Fusarium sporotrichoids*, *Macrophomina phaseolina*, *Rhizoctoni solani*, *Sclerotium rolfsii* and *Trichoderma harzianum* (Ghazala et al., 2004). Various cyanobacterial compounds including polysulfides (cyclic), amino acids, and halogenated metabolites are well known for their antibiotic activity (Wratten and Faulkner, 1976; Singh et al., 2011b).

4.7.4 Antiparasitic Agents

In order to develop effective treatments against pathogenic parasitic agents such as *Plasmodium*, *Leishmania*, and*Schistosoma* that cause various diseases of concern like malaria and leishmaniasis, a thorough study of natural and effective compounds that can act against resistant protozoans is needed (Lanzer and Rohrbach, 2007; Simmons et al., 2008). To promote effective and economical treatment of disease caused by parasitic agents, cyanobacteria like *Nostoc* sp. were reported to be active with the aid of protease inhibitor nostocarboline, an alkaloid against *Trypanosoma brucei*, *Leishmania donovani*, and *Plasmodium falciparum* (Barbaraus et al., 2008). *Microcystis aeruginosa* PCC 7806 has been reported to be an effective source of aerucyclamide which was effective in treatment against *T. brucei*, followed by aerucyclamide B, which was studied for its active role in targeting *P. Falciparum* (Portmann et al., 2008). The marine cyanobacterium *Lyngbya majuscule* has long been studied for its role as an antiparasitic agent. Some antimalarial activities have been isolated from the cyanobacterium with such bioactive compounds as alkynoiclipopeptides, carmabin A (Uzair et al., 2012), dragomabin (Agrawal, 2016), dragonamide A (Shah et al., 2017), and dragonamide B (Hossain et al., 2016). In 2010, several researchers characterized many cytotoxic almiramids A-C, lipopeptides obtained from *L. majuscle*(Ananyev et al., 2012; Bajpai et al., 2018; Zahra et al., 2020), which showed antileishmanial activity, principally against *L. donovani*, *L. infantum*, and *L. chagasi*, which were of much concern.

4.7.5 Protease Inhibition and Immunomodulatory Activity of Cyanobacteria

A new class of protease inhibitors, aeruginosins, microginins, and cyanopeptolins studied from cyanobacteria, has been discovered for the treatment of various disorders. Microginins play an important role in controlling high blood pressure. Cyanopeptolin, as a serine protease inhibitor, is applied in the treatment of asthma and viral infections. Cyanobacteria have been well reported for their role as immunomodulant when various products synthesized from *Spirulina* affect immune response through the rising phagocytic activity by macrophages, increased cytokine synthesis activating T and B cells, and substantial elevation of natural killer cells (NK) (Khan et al., 2005).

4.8 CYANOBACTERIA AS BIOFACTORIES FOR NANOPARTICLE SYNTHESIS

Advances in cyanobacterial studies have paved the way for understanding the importance of these microorganisms in nanoparticle synthesis, utilizing either intracellular or extracellular methods. Cyanobacteria could be a preferred microbe as they require simple growth conditions along with a rapid growth rate, supplemented with bioactive compounds in their various pharmaceutical products (Sharma et al., 2016).Studies have demonstrated the role of various species of cyanobacteria in the synthesis of different types of nanoparticles such as silver nanoparticles (Ag-NPs), along with various other nanoparticles such as zinc oxide nanoparticles (ZnONPs), titanium dioxide (TiO2), copper oxide nanoparticles (CuONPs), etc. (Mubarak Ali et al., 2013; Mohamed et al., 2017; Ebadi et al., 2019). Several characterization techniques have demonstrated the role of the bioreduction pathway of silver nitrate and the metabolic active compounds of microorganisms, including nitrogen fixation pathway and enzymes. In order to synthesize Ag-NPs, biomass of *Oscillitoria limnetica* was utilized as aqueous extract at room temperature for approximately 18 hrs (Hamouda et al., 2019). Other stress factors including time, pH, and concentration of metabolically active cellular compounds responsible for silver nanoparticle synthesis have also been studied. Husain et al. (2015) used more than 30 cyanobacteria in order to form Ag-NPs extracellularly in varied size distribution ranges from various precursors. Another study included screening of Ag-NP fabrication potential among several cyanobacteria species tested and studied from mangroves (Muthupet) (Sudha et al., 2013). Of the various forms studied, only a few were capable of reduction and only *Microcoleus* sp. was capable of reduction and formed spherical Ag-NPs (size 55 nm).The results obtained showed that all species of cyanobacteria reduced silver nitrate to Ag-NPs under direct light. The ability of cyanobacteria to formnanoparticles covers a range of nanoparticles including silver, gold, titanium dioxide, and zinc oxide. Several researchers have screened and found the potential of *P. boryanum* UTEX 485 in forming many types of nanoparticles such as platinum and palladium with varying temperature ranges (Lengke et al., 2006; Lengke et al., 2007).

Nanotechnology has now crossed the barriers of basic science and is widely accepted in the medical and pharmaceutical fields as well as for sustainable agriculture focusing on more varied and diversified applications (Dejous et al., 2017). Several reports have suggested the application of cyanobacterial nanoparticles for treatment in various therapies including treatment against cancer, microbes of drug-resistant bacteria, and viruses. Various studies in the nanotechnology field have been reported to effectively control tumor growth and metastasis. Another study published promising results in which Ag-NPs synthesized from cyanobacteria were found to be effective against Dalton's Lymphoma and colo205 cancer cells (Singh et al., 2014). The role of Ag-NPs even at low concentration synthesized silver nanoparticles was reported to be able to control proliferation of both cell types with ROS formation leading to DNA damage and ultimately leading to cell death. Another study projected the effectiveness of silver nanoparticles obtained from *Nostoc* sp. HKAR-2 against tumor cell lines MCF-7 (Sonker et al., 2017). The nanoparticles synthesized killed tumor cells and were reported for their potential role in treating cancer and tumor diseases. The role of cyanobacteria in nanobiotechnology in the form of different metals and alloys or by bioconjugates with bioactive metabolites is being tested by various pharmaceutical companies in, for example, medicine, fertilizers, sensors, and cosmetics, nanoformulated as antioxidative creams with strong anti-inflammatory properties utilizing cyanobacterial metabolites (Amendola and Meneghetti, 2009; Rajeshkumar et al., 2014; Bin-Meferij and Hamida, 2019).

4.9 CYANOBACTERIA AS BIOFERTILIZERS

Biofertilizers are a strong alternative to minimize the use of synthetic fertilizers that not only adversely affect soil health but are also very costly. Biofertilizers are living microorganisms that utilize natural processes like nitrogen fixation, phosphate solubilization, and plant growth hormone production to provide nutrients and promote plant growth (Rai, 2006; Kumar et al., 2017).

Cyanobacteria fix atmospheric dinitrogen (N_2), enriching soil fertility which results in increased rice yields. Apart from fixing N_2 and nutrient cycling, several vitamins, growth hormones, organic acids, andbioactive compounds are also produced by cyanobacteria (Singh et al., 2016). By increasing water retention capacity and soil pore size, they increase soil fertility and can tolerate pesticides above suggested levels in the field (Cohen, 2006; Kaushik, 2013; Singh et al., 2016).

Several heterocystous as well as non-heterocystous species aid nitrogen fixation. Various unicellular species, mainly *Aphanothece, Gloeocapsa* and several filamentous non-heterocystous species such as *Oscillatoria* and many others are the central flora of rice fields that fix N_2 (Bergman et al., 1997; Whitton and Potts, 2000). A symbiotic relationship is formed between cyanobacteria and several eukaryotic hosts including plants. The hosts thereafter utilize the N_2 fixed by the cyanobiont (Adams, 2000). Heterocysts contain many biochemical and physiological attributes, which helps in the reduction of N_2 to NH_3 via nitrogenase enzyme; heterocysts are hence referred to as the biological factory of N_2 fixation (Wolk, 1996). Certain marine species like *Anabaena* and symbiotic cyanobacteria, mainly *Nostoc cycadaes* and *Anabaena azollae*, show increased heterocyst frequency which correlates directly with nitrogenase activity, indicating a crucial role for heterocysts in fixing N_2 (Kumar et al., 1982; Kumar and Kumar, 1988; Rai, 1990; Adams, 2000). Many researchers have reported increased formation and frequency of heterocysts either by adding some non-toxic inhibitors or by raising mutant strains with greater heterocyst frequency (Kumar and Kumar, 1980; Meeks, 1998). Fixed N_2 is excreted as amino acids, ammonia, several hormones, vitamins, and other bioactive compounds. Many cyanobacteria sps. acts as biofertilizers in different crops (Karthikeyan et al., 2007; Singh et al., 2016).N_2-based studies have shown that cyanobacteria can fix up to 20–40 kg N ha^{-1} per crop (Ladha and Reddy, 2000).

Apart from fixing N_2, cyanobacteria promote plant growth by the secretion of certain plant growth-promoting hormones [gibberellins, auxins, indole-3-acetic acid (IAA) iron chelators and vitamins (folic acid, Vitamin B12, pantothenic acid, nicotinic acid)]. In fact, the presence of other plant growth promoters (PGRs) like jasmonic acid, ethylene, and abscisic acid have been found in cyanobacteria (Ordog and Pulz, 1996) by researchers. A more detailed and critical study is required using other advanced analytical techniques for validating the presence of PGRs in other species of cyanobacteria.

4.10 CONCLUSION

Cyanobacteria comprise a varied group of prokaryotes that leverage solar energy and assimilate CO_2 to produce multiple biological value-added products. Thus, they have the potential to be industrialized as solar-cell factories because of their photoautotrophic nature. Evolutionary processes have led to the development of standard mechanisms in cyanobacteria that enhance carbon availability to drive sustainable growth and production of vital metabolic products. These native metabolic processes in cyanobacterial organisms result in the synthesis of various key compounds that can act as precursor metabolites to industrial organic products. In this respect, metabolic engineering in cyanobacteria has been instrumental in accelerating the production of organic compounds like alcohols (mono-, di- and polyhydric alcohols), acids (organic acids and carboxylic acid derivatives) and bioplastics. Peptide-like secondary metabolites is another class of compounds generated by cyanobacteria that can potentially be used in various pharmaceutical formulations. Although bioengineering of cyanobacteria can lead to biotechnologically beneficial systems, the yield of these organisms under photoautotrophic growth is nascent. The combined application of systems biology, metabolic engineering, bioinformatics, and genetic engineering is recommended to remediate these inefficiencies. The natural carbon sink pathway facilitated by cyanobacteria that leads to the production of peptides and carbohydrates can be evolved into productive and high-quality industrial biofactories. The move to such a biotechnologically advanced system will require scientific intervention in the form of novel metabolic pathway designs that catalyze utilization of amino acids and carbohydrates.

ACKNOWLEDGEMENT

PG is grateful to the University Grants Commission (UGC), New Delhi for the award of a Senior Research Fellowship ((F.16-6 (DEC. 2016)/2017(NET). DS is the recipient of a DST-INSPIRE Fellowship (DST/INSPIRE Fellowship/2014/296, IF140707). The authors are grateful to the coordinators at the School of Biotechnology, and the Head of the Department of Botany, Gargi College, for providing facilities as and when required.

REFERENCES

Abed, R.M.M., Dobretsov, S., Sudesh, K. (2009). Applications of cyanobacteria in biotechnology. *J. Appl. Microbiol.*, 106 (1), 1–12.

Acho-Chi (1998). Human interference and environmental instability: addressing the environmental consequences of rapid urban growth in Bamenda, Cameroon. *Environ Urban*, 10, 161–174.

Adams, D.G. (2000). Symbiotic interactions. In B.A. Whitton, M. Potts (Eds). *The Ecology of Cyanobacteria* (1st ed., pp. 523–561). Boston: Kluwer Academic Publishers. doi:10.1007/0-306-46855-7

Agrawal, M.K. (2016). Antimicrobial activity of *Nostoccalcicola* (Cyanobacteria) isolated from central India against human pathogens. *Asian J. Pharm.*, 10 (4).

Al-Hasan, R., Al-Bader, D., Sorkhoh, N., et al., (1998). Evidence for *n*-alkane consumption and oxidation by filamentous cyanobacteria from oil-contaminated coasts of the Arabian Gulf. *Mar. Biol.*, 130, 521–527.

Ali, S.K., Saleh, A.M. (2012). *Spirulina* – an overview. *Int. J. Pharm. Pharmaceut. Sci.*, 4(3), 9–15.

Amendola, V., Meneghetti, M. (2009). Laser ablation synthesis in solution and size manipulation of noble metal nanoparticles. *Phys. Chem. Chem. Phys.*, 11(20), 3805–3821. doi:10.1039/b900654k

Ananyev, G.M., Skizim, N.J., Dismukes, G.C. (2012). Enhancing biological hydrogen production from cyanobacteria by removal of excreted products. *J. Biotechnol.*, 162, 97–104.

Anekthanakul, K., Senachak, J., Hongsthong, A., Charoonratana, T., Ruengjitchatchawalya, M. (2019).Natural ACE inhibitory peptides discovery from *Spirulina (Arthrospira platensis)* strain C1. *Peptides*, 118, 170107.

Anjana, K., Kaushik, A., Kiran, B., Nisha, R. (2007). Biosorption of Cr (VI) by immobilized biomass of two indigenous strains of cyanobacteria isolated from metal contaminated soil. *J. Hazard. Mater.*, 148, 383–386.

Asthana, R.K., Tripathi, M.K., Deepali, A. et al., (2009). Isolation and identification of a new antibacterial entity from the Antarctic cyanobacterium *Nostoc* CCC 537. *J. Appl. Phycol.*, 21, 81–88.

Atsumi, S., Hanai, T., Liao, J.C., (2008). Non-fermentative pathways for synthesis of branched-chain higher alcohols as biofuels. *Nature*, 451, 86–89.

Atsumi, S., Higashide, W., Liao, J. (2009). Direct photosynthetic recycling of carbon dioxide to isobutyraldehyde. *Nat. Biotechnol.*, 27, 1177–1180. doi:10.1038/nbt.1586

Bajpai, V., Shukla, S., Kang, S.M., Hwang, S., Song, X., Huh, Y., Han, Y.K.J.M.D. (2018). Developments of cyanobacteria for nano-marine drugs: relevance of nanoformulations in cancer therapies. *Mar. Drugs*, 16, 179.

Balaji, S., Gopi, K., Muthuvelan, B. (2013). A review on production of poly β hydroxybutyrates from cyanobacteria for the production of bio plastics. *Algal Research*, 2, 278–285.

Balaji, S., Kalaivani, T., Rajasekaran, C. (2014). Biosorption of zinc and nickel and its effect on growth of different *Spirulina* strains. *CLEAN—Soil, Air, Water*, 42 (4), 507–512.

Barbaraus, D., Kaiser, M., Brun, R., Gademann, K. (2008). Potent and selective antiplasmodial activity of the cyanobacterial alkaloid nostocarboline and its dimers. *Bioorg. Med. Chem. Lett.*, 18, 4413–4415.

Belay, A. (2002) The potential application of *Spirulina* (*Arthrospira*) as a nutritional health and therapeutic supplement in health management. *J. Am. Nutraceut. Ass.*, 5, 27–48.

Bergman, B., Gallon, J.R., Rai, A.N., Stal, L.J., (1997). N_2 fixation by non-heterocystous cyanobacteria. *FEMS Microbiol. Rev.*, 19, 139–185.

Bin-Meferij, M.M., Hamida, R.S. (2019). Biofabrication and antitumor activity of silver nanoparticles utilizing novel *Nostoc* sp. Bahar M. *Int. J. Nanomedicine*, 14, 9019–9029. doi:10.2147/IJN.S230457

Biondi, N., Tredici, M.R., Taton, A. et al. (2008). Cyanobacteria from benthic mats of Antarctic lakes as a new source of bioactivities. *J. Appl. Microbiol.*, 105, 105–115.

Bižić, M., Klintzsch, T., Ionescu, D., et al., (2020). Aquatic and terrestrial cyanobacteria produce methane. *Sci. Adv.*, 6, eaax5343.

Blakefield, M.K., Harris, D.O. (1994). Delay of cell differentiation in *Anabaena aequalis*caused by UV-B radiation and the role of photoprotection and excision repair. *Photochem. Photobiol.*, 59, 204–208.
Blom, J.F., Brutsch, T., Barbaras, D.et al. (2006). Potent algicides based on the cyanobacterial alkaloid nostocarboline. *Org. Lett.*, 8, 737–740.
Bohutskyi, P., Bouwer, E. (2013). Biogas production from algae and cyanobacteria through anaerobic digestion: a review, analysis, and research needs. In J.W. Lee (Ed.). *Advanced Biofuels and Bioproducts* (pp. 873–975). New York, London: Springer.
Bokesch, H.R., O'Keefe, B.R., McKee, T.C., Pannell, L.K., Patterson, G.M., Gardella, R.S., et al., (2003). A potent novel anti-HIV protein from the cultured cyanobacterium *Scytonema varium. Biochemistry*, 42, 2578–2584.
Brenner, M.P., Bildsten, L., Dyson, F., et al. (2006) Engineering microorganisms for energy production. Report JSR-05-300, U. S. Department of Energy, (Washington, DC).
Capon, R.J., Dunlop, R.W., Ghisalberti, E.L., Jefferies, P.R. (1983). Poly-3-hydroxyalkanoates from marine and freshwater cyanobacteria. *Phytochemistry*, 22, 1181–1184.
Castenholz, R.W. and Garcia-Pichel, F. (2012). Cyanobacterial responses to UV radiation. In B.A. Whitton (Ed.). *Ecology of CYANOBACTERIAII* (pp. 481–502). Dordrecht: Springer.
Chaves, R.P., da Silva, J.P.F.A., Carneiro, R.F. et al. (2018). *Meristiella echinocarpa* lectin (MEL): a new member of the OAAH-lectin family. *Appl. Phycol.*, 30, 2629–2638.
Chen, M., Li, J., Zhang, L., Chang, S., Liu, C., Wang, J., Li, S. (2014). Auto-flotation of heterocyst enables the efficient production of renewable energy in cyanobacteria. *Sci. Rep.*, 4, 3998.
Chittora, D., Meena, M., Barupal, T., Swapnil, P., Sharma, K. (2020). Cyanobacteria as a source of biofertilizers for sustainable agriculture. *Biochem. Biophys. Rep.*, 22, 100737. doi:10.1016/j.bbrep.2020.100737
Clark, I.A., Alleva, L.M., Budd, A.C., Cowden, W.B. (2008). Understanding the role of inflammatory cytokines in malaria and related diseases. *Travel Med. Infect. Dis.*, 6, 67–81.
Cohen, R.R.H. (2006). Use of microbes for cost reduction of metal removal from metals and mining industry waste streams. *J. Clean. Prod.*, 14, 1146–1157. doi:10.1016/j.jclepro.2004.10.009
Cohen, Z. (1986). Products from microalgae. In A. Richmond(Ed.). *Handbook of Microalgal Mass Culture* (pp. 421–454). Boca Raton, Florida: CRC Press.
Daniel, S., Cornelia, S., Fred, Z. (2004). UV-A sunscreen from red algae for protection against premature skin aging. *Cosmet. Toiletries Manuf. Worldwide*, 204, 139–143.
Dejous, C., Hallil, H., Raimbault, V., Rukkumani, R., Yakhmi, J.V. (2017). Using microsensors to promote the development of innovative therapeutic nanostructures. In. D. Ficai and A. M. Grumezescu (Eds.). *Nanostructures for Novel Therapy* (pp. 539–566). Elsevier.
Deng, M., Coleman, J.R. (1999). Ethanol synthesis by genetic engineering in cyanobacteria. *Appl. Environ. Microbiol.*, 65(2), 523–528.
Dexter, J., Armshaw, P., Sheahan, C., Pembroke, J.T. (2015). The state of autotrophic ethanol production in Cyanobacteria. *J. Appl. Microbiol.* doi:10.1111/jam.12821
Dexter, J., Fu, P. (2009). Metabolic engineering of cyanobacteria for ethanol production. *Energ. Environ. Sci.*, 2, 857–864.
Dias, J.M., Lemos, P.C., Serafim, L.S., Oliveira, C., Eiroa, M., Albuquerque, M.G. (2006). Recent advances in polyhydroxyalkanoate production by mixed aerobic cultures: from the substrate to the final product. *Macromol. Biosci.*, 6, 885–906.
Duan, Z.F., Ji, D.N., Weinstein, E.J., et al. (2010). Lentiviral shRNA screen of human kinases identifies PLK1 as a potential therapeutic target for osteosarcoma. *Cancer Lett.*, 293, 220–229.
Ducat, D.C., Way, J.C., Silver, P.A. (2011). Engineering cyanobacteria to generate high-value products. *Trends Biotechnol.*, 29, 95–103.
Dutta, D., De, D., Chaudhuri, S. (2005) Hydrogen production by cyanobacteria. *Microb. Cell Fact.*, 4, 1–11.
Ebadi, M., Zolfaghari, M.R., Aghaei, S.S., et al. (2019). A bio-inspired strategy for the synthesis of zinc oxide nanoparticles (ZnO NPs) using the cell extract of cyanobacterium *Nostoc* sp. EA03: from biological function to toxicity evaluation. *RSC Adv.*, 9(41), 23508–23525. doi:10.1039/C9RA03962G
Edelman, M.J., Gandara, D.R., Hausner, P., et al. (2003). Phase 2 study of cryptophycin 52 (LY355703) in patients previously treated with platinum based chemotherapy for advanced non-small cell lung cancer. *Lung Cancer*, 39, 197–199. doi:10.1016/S0169-5002(02)00511-1
Eriksen, M, Lebreton, L.C.M., Carson, H.S., et al. (2014). Plastic pollution in the world's oceans: more than 5 trillion plastic pieces weighing over 250,000 tons afloat at sea. *PLoS One*, 9(12), e111913.
Facciola, S. (1998). *Cornucopia II: a Sourcebook of Edible Plants* (2nd ed.). Vista: Kampong Publications.
Férir, G., Huskens, D., Noppen, S., Koharudin, L.M., Gronenborn, A.M., Schols, D. (2014). Broad anti-HIV activity of the *Oscillatoria agardhii* agglutinin homologue lectin family. *J. Antimicrob. Chemother.*, 69(10), 2746–2758.

Fernandez-Nava, Y., Maranon, E., Soons, J., Castrillon, L. (2008). Denitrification of wastewater containing high nitrate and calcium concentration. *Bioresour. Technol.* 99(17), 7976–7981.

Fleming, E.D., Castenholz, R.W. (2007). Effects of periodic desiccation on the synthesis of the UV-screening compound, scytonemin, in cyanobacteria. *Environ. Microbiol.*, 9, 1448–1455.

Fujishiro, T., Ogawa, T., Matsuoka, M., Nagahama, K., Takeshima, Y., Hagiwara, H. (2004). Establishment of a pure culture of the hitherto unicellular cyanobacterium *Aphanothecasacrum*, and phylogenetic position of the organism. *Appl. Environ. Microbiol.*, 70, 3338–3345.

Gao, Z., Zhao, H., Li, Z., Tan, X., Lu, X. (2012). Photosynthetic production of ethanol from carbon dioxide in genetically engineered cyanobacteria. *Energ. Environ. Sci.*, 5, 9857–9865.

Garcia-Pichel, F., Castenholz, R.W. (1991). Characterizationand biological implications of scytonemin, a cyanobacterial sheath pigment. *J. Phycol.*, 27, 395–409.

Garcia-Pichel, F., Castenholz, R.W. (1993). Occurrenceof UV-absorbing, mycosporine-like compounds amongcyanobacterial isolates and an estimate of their screening capacity. *Appl. Environ. Microbiol.*, 59, 163–169.

Gardea-Torredey, J.L., Arenas, J.L., Francisco, N.M.C., Tiemann, K.J., Webb, R. (1998). Ability of immobilized cyanobacteria to remove metal ions from solution and demonstration of the presence of metallothionein genes in various strains. *J Hazard Subst Res.*, 1, 1–18.

Gerwick, W.H., Coates, R.C., Engene, N. et al. (2008). Giant marine cyanobacteria produce exciting potential pharmaceuticals. *Microbe*, 3, 277–284.

Gerwick, W.H., Roberts, M.A., Proteau, P.J., Chen, J.L. (1994). Screening cultured marine microalgae for anticancer-type activity. *J. Appl. Phycol.*, 6, 143–149. doi:10.1007/BF02186068

Ghazala, B., Shameel, M., Choudhary, M.I., Shahzad, S., Leghari, S.M. (2004). Phycochemistry and bioactivity of *Tetraspora* (volvocophyta) from Sindh. *Pak. J. Bot.*, 36, 531–548.

Godwin, D.A., Wiley, C.J., Felton, L.A. (2006).Using cyclodextrin complexation to enhance secondary photoprotection of topically applied ibuprofen. *Eur. J. Pharm. Biopharm.*, 62(1), 85–93.

Guihéneuf, F., KhanA., TranL.S.P. (2016). Genetic engineering: a promising tool to engender physiological, biochemical, and molecular stress resilience in green microalgae. *Front. Plant Sci.*, 7(400), 1–8. doi:10.3389/fpls.2016.00400

Gupta, V.K., Zeilinger, S., Filho, E.X.F., Duran-Dominguez-de-Bazua, M.C., Purchase, D., (2017). *Microbial applications - recent advancements and future developments*. Retrieved from https://www.worldcat.org/title/microbial-applications-recent-advancements-and-future-developments/oclc/931648826

Gustafson, K.R., Cardellina, J.H., Fuller, R.W., Weislow, O.S., Kiser, R.F., Snader, K.M., Patterson, G.M., Boyd, M.R. (1989).AIDS-antiviral sulfolipids from cyanobacteria (blue-green algae). *J. Natl. Cancer Inst.*, 81, 1254–1258.

Gutiérrez, M., Suyama, T.L., Engene, N., Wingerd, J.S., Matainaho, T., Gerwick, W.H. (2008). Apratoxin D, a potent cytotoxic cyclodepsi peptide from Papua New Guinea collections of the marine cyanobacteria *Lyngbyamajuscula* and *Lyngbyasordida*. *J. Nat. Prod.*, 71, 1099–1103.

Hai, T., Hein, S., Steinbuchel, A. (2001). Multiple evidence for widespread. and general occurrence of type-III PHA synthases in cyanobacteria and molecular characterization of the PHA synthases from two thermophilic cyanobacteria: *Chlorogloeopsisfritschii* PCC6912 and *Synechococcus* sp. strain MA19. *Microbiology*, 147, 3047–3060.

Hamouda, R.A., Hussein, M.H., Abo-Elmagd, R.A., Bawazir, S.S. (2019). Synthesis and biological characterization of silver nanoparticles derived from the cyanobacterium *Oscillatoria limnetica*. *Sci. Rep.*, 9(1), 13071. doi:10.1038/s41598-019-49444-y

Hemscheidt, T., Puglisi, M.P., Larsen, L.K. et al., (1994). Structure and biosynthesis of borophycin, a new boeseken complex of boreic acid from a marine strain of the blue-green alga *Nostoc linckia*. *J. Org. Chem.*, 59, 3467–3471.

Heo, S.Y., Ko, S.C., Kim, C.S., et al. (2017).A heptameric peptide purified from *Spirulina* sp. gastrointestinal hydrolysate inhibits angiotensin I-converting enzyme- and angiotensin II-induced vascular dysfunction in human endothelial cells. *Int. J. Mol. Med.*, 39, 1072–1082.

Hirahashi, T., Matsumoto, M., Hazeki, K., Saeki, Y., Ui, M., Seya, T. (2002).Activation of the human innate immune system by Spirulina: augmentation of interferon production and NK cytotoxicity by oral administration of hot water extract of *Spirulina platensis*. *Int. Immunopharmacol.*, 2, 423–434.

Hirata, K., Yoshitomi, S., Dwi, S. et al. (2003). Bioactivities of nostocine A produced by a freshwater cyanobacterium *Nostocspongiaeforme* TISTR8169. *J. Biosci. Bioeng.*, 95, 512–517.

Hohmann-Marriott, M.F., Blankenship, R.E. (2011). Evolution of photosynthesis. *Annu. Rev. Plant Biol.*, 62, 515–548.

Hossain, M.F., Ratnayake, R.R., Meerajini, K., Wasantha Kumara, K.L. (2016). Antioxidant properties in some selected cyanobacteria isolated from fresh water bodies of Sri Lanka. *Food Sci. Nutr.*, 4, 753–758.

Husain, S., Sardar, M., Fatma, T. (2015). Screening of cyanobacterial extracts for synthesis of silver nanoparticles. *World J. Microbiol. Biotechnol.*, 31(8), 1279–1283. doi:10.1007/s11274-015-1869-3

Ingram, L.O., Conway, T., Clark, D.P., Sewell, G.W., Preston, J.F. (1987). Genetic engineering of ethanol production in Escherichia coli. *Appl. Environ. Microbiol.*, 53, 2420–2425.

Jaki, B., Orjala, J., Heilmann, J., Linden, A., Vogler, B. and Sticher, O. (2000). Novel extracellular diterpenoids with biological activity from the cyanobacterium *Nostoc commune*. *J. Nat. Prod.*, 63, 339–343.

Jha, M.N., Mishra, S.K. (2005). Biological responses of cyanobacteria to insecticides and their insecticide degrading potential. *Bull. Environ. Contam. Toxicol.*, 75(2), 374–381.

Jordan, M.A., Wilson L. (1998). Microtubules and actin filaments: dynamic targets for cancer chemotherapy. *Curr. Opin. Cell Biol.*, 10, 123–130.

Kalia, V.C., Patel, S.K.S., Shanmugam, R., Lee, J.K. (2021). Polyhydroxyalkanoates: trends and advances toward biotechnological applications. *Bioresour. Technol.*, 326, 124737.

Kaplan, A., Schwarz, R., Lieman-Herwitz, J., Reinhold, L. (1994). Physiological and molecular studies on the response of cyanobacteria to changes in the ambient inorganic carbon concentration. In D.A. Brynat (Ed.). *The Molecular Biology of Cyanobacteria. Advances in Photosynthesis* (1st ed., pp. 469–485). Dordrecht: Springer.

Karan, H., Funk, C., Grabert, M., Oey, M., Hankamer, B. (2019). Green bioplastics as part of a circular bioeconomy. *Trends Plant Sci.*, 24(3), 237–249. doi:10.1016/j.tplants.2018.11.010

Karkos, P.D., Leong, S.C., Karkos, C.D., Sivaji, N., Assimakopoulos, D.A. (2008). Spirulina in clinical practice: evidence-based human applications. *Hindawi Publ. Corporation*, 531053, 1–4. doi:10.1093/ecam/nen058

Karthikeyan, N., Prasanna, R., Nain, L., Kaushik, B.D. (2007). Evaluating the potential of plant growth promoting cyanobacteria as inoculants for wheat. *Eur. J. Soil Boil.*, 43, 23–30.

Kaushik, B.D. (1983). Effect of native algal flora on nutritional and physico-chemical properties of sodic soils. *J Biol Res.*, 3, 99–103.

Kaushik, B.D. (2013). Developments in cyanobacterial biofertilizer. *Proc. Indian Natn. Sci. Acad.*, 80, 379–388. doi:10.16943/ptinsa/2014/v80i2/55115

Khan, Z., Bhadouria, P., Bisen, P.S. (2005). Nutritional and therapeutic potential of *Spirulina*. *Curr. Pharm. Biotechnol.*, 6, 373–379.

Kini, S., Divyashree, M., Mani, M.K., Mamatha, B.S. (2020). Algae and cyanobacteria as a source of novel bioactive compounds for biomedical applications. In P.K. Singh, A. Kumar, V.K. Singh, A.K. Shrivistava (Eds). *Advances in Cyanobacterial Biology* (1st ed., pp 173–193). India: Academic Press. doi:10.1016/B978-0-12-819311-2.00012-7

Koutinas, A.A., Xu, Y., Wang, R.H., Webb, C. (2007). Polyhydroxybutyrate production from a novel feedstock derived from a wheat-based biorefinery. *Enzyme Microb. Technol.*, 40, 1035–1044.

Kumar, A., Kumar, H.D. (1980). Differential effects of amino acid analogs on growth and heterocyst differentiation in two nitrogen-fixing blue-green algae. *Curr. Microbiol.*, 3, 213–218.

Kumar, A., Kumar, H.D. (1988). Nitrogen-fixation by blue-green algae. In S.P. Sen (Ed.). *Proceedings of the Plant Physiology Research, Socieryfor Plant Physiology and Biochemistry* (pp. 85–103). 1st International Congress of Plant Physiology, New Delhi, India.

Kumar, A., Tabita, F.R., van Baalen, C. (1982). Isolation and characterization of heterocysts from *Anabena* strain CA. *Arch. Microbiol.*, 133, 103–109.

Kumar, J., Singh, D., Ghosh, P., Kumar, A., (2017). Endophytic and epiphytic modes of microbial interactions and benefits. In D. Singh, H. Singh, R. Prabha (Eds). *Plant-microbe Interactions in Agro-Ecological Perspectives* (pp. 227–253). Singapore: Springer Nature.

Kumar, J., Singh, D., Tyagi, M.B., Kumar, A. (2019a). Cyanobacteria: applications in biotechnology. In A.K. Mishra, D.N. Tiwari, A.N. Rai (Eds). *Cyanobacteria From Basic Sci. to Appl* (pp. 327–346). Academic Press. doi:10.1016/B978-0-12-814667-5.00016-7

Kumar, V., Patel, S.K.S., Gupta, R.K., Otari, S.V., Gao, H., Lee, J.K., Zhang, L. (2019b). Enhanced saccharification and fermentation of rice straw by reducing the concentration of phenolic compounds using an immobilized enzyme cocktail. *Biotechnol. J.*, 14, 1800468.

Kurisawa, N., Iwasaki, A., Jeelani, G., Nozaki, T., Suenaga, K. (2020). Iheyamides A–C, antitrypanosomal linear peptides isolated from a Marine *Dapis* sp. Cyanobacterium. *J. Nat. Prod.*, 83 (5), 1684–1690.

Ladha, J.K., Reddy, P.N. (Eds). (2000). *The Quest for Nitrogen Fixation in Rice*. Retrieved from https://books.irri.org/9712201120_content.pdf

Lanzer, M.R., Rohrbach, P. (2007). Subcellular pH and Ca^{2+} in *Plasmodium falciparum*: implications for understanding drug resistance mechanisms. *Curr. Sci.*, 92, 1561–1570.

Lengke, M.F., Fleet, M.E., Southam, G. (2006). Synthesis of platinum nanoparticles by reaction of filamentous cyanobacteria with platinum (IV)-chloride complex. *Langmuir*, 22(17), 7318–7323. doi:10.1021/la060873s

Lengke, M.F., Fleet, M.E., Southam, G. (2007). Synthesis of palladium nanoparticles by reaction of filamentous cyanobacterial biomass with a palladium (II) chloride complex. *Langmuir*, 23(17), 8982–8987. doi:10.1021/la7012446

Liu, X., Fallon, S., Sheng, J., Curtiss, R. (2011a). CO_2-limitation-inducible Green Recovery of fatty acids from cyanobacterial biomass. *Proc. Natl. Acad. Sci. U. S. A.*, 108(17), 6905–6908.

Liu, X., Sheng, J., Curtiss, R. (2011b). Fatty acid production in genetically modified cyanobacteria. *Proceedings of the National Academy of Sciences*, 108(17), 6899–6904.

Loya, S., Reshef, V., Mizrachi, E., Silberstein, C., Rachamim, Y., Carmeli, S., Hizi, A. (1998). The inhibition of the reverse transcriptase of HIV-1 by the natural sulfoglycolipids from cyanobacteria: contribution of different moieties to their high potency. *J. Nat. Prod.*, 61, 891–895.

Luescher-Mattli, M. (2003). Algae as a possible source of new antiviral agents. *Curr. Med. Chem. Anti-Infect. Agents*, 2, 219–225.

Machado, I.M.P., Atsumi, S. (2012). Cyanobacterial biofuel production. *J. Biotechnol.*, 162, 50–56. doi:10.1016/j.jbiotec.2012.03.005

Matei, E., Basu, R., Furey, E., Shi, J., Calnan, C., Aiken, C., Gronenborn, A.M. (2016). Structure and glycan binding of a new cyanovirin- N homolog. *J. Biol. Chem.*, 291, 18967–18976.

Matei, E., Zheng, A., Furey, W., Rose, J., Aiken, C., Gronenborn A.M. (2010). Anti-HIV activity of defective cyanovirin-N mutants is restored by dimerization. *J. Biol. Chem.*, 285, 13057–13065.

McInnes, C., Mezna, M., Fischer, P.M. (2005). Progress in the discovery of polo-like kinase inhibitors. *Curr. Top. Med. Chem.*, 5, 181–197.

Meeks, J.C. (1998). Symbiosis between nitrogen-fixing cyanobacteria and plants. *Bioscience*, 48, 266–276.

Misonou, T., Saitoh, J., Oshiba, S., et al. (2003).UV-absorbing substance in the red alga *Porphyrayezoensis* (Bangiales, Rhodophyta) block thymine photodimer production. *Marine Biotechnol.*, 5, 194–200.

Miyake, M., Erata, M., Asada, Y. (1996). A thermophilic cyanobacterium, *Synechococcus*sp. MA19, capable of accumulating poly-b-hydroxybutyrate. *J. Ferment. Bioeng.*, 82, 512–514.

Mohamed, A., Fouda, A., Elgamal, M., et al. (2017). Enhancing of cotton fabric antibacterial properties by silver nanoparticles synthesized by new Egyptian strain *Fusarium keratoplasticum* A1-3. *Egypt J Chem.*, 60, 63–71. doi:10.21608/ejchem.2017.1626.113

Mohamed, Z.A. (2001). Removal of cadmium and manganese by a non-toxic strain of the freshwater cyanobacterium *Gloeothece magna*. *Water Res.*, 35, 4405–4409.

Moore, R.E., Corbett, T.H., Patterson, G.M.L., Valeriote, F.A. (1996). The search for new antitumor drugs from blue green algae. *Curr. Pharm. Design*, 2, 317–330.

Moore, R.E., Patterson, G.M.L., Carmichael, W.W. (1988). New pharmaceuticals from cultured blue green algae. In D.G. Fautin (Ed.). *Biomedical Importance of Marine Organisms*. San Francisco, CA: California Academy of Sciences.

Moreira, L.M., Ribeiro, A.C., Duarte, F.A., Morais, M.G., Soares, L.A.S. (2013). *Spirulina platensis* biomass cultivated in Southern Brazil as a source of essential minerals and other nutrients. *Afr. J. Food Sci.*, 7(12), 451–455.

Morone, J., Alfeus, A., Vasconcelos, V., Martins, R. (2020). Revealing the potential of cyanobacteria in cosmetics and cosmeceuticals: a new bioactive approach. *Algal Res.*, 41, 101–141.

Mubarak Ali, D., Arunkumar, J., Nag, K.H., et al. (2013). Gold nanoparticles from Pro and eukaryotic photosynthetic microorganisms- Comparative studies on synthesis and its application on biolabelling. *Colloids Surf. B Biointerfaces*, 103, 166–173. doi:10.1016/j.colsurfb.2012.10.014

Mussgnug, J.H., Klassen, V., Schlüter, A., Kruse, O. (2010). Microalgae as substrates for fermentative biogas production in a combined biorefinery concept. *J. Biotechnol.*, 150 (1), 51–64.

Namakoshi, K., Nakajima, T., Yoshikawa, K., Toya, Y., Shimizu, H. (2016). Combinatorial deletions of glgC and phaCE enhance ethanol production in *Synechocystis* sp. PCC 6803. *J. Biotechnol.* 239(3). doi:10.1016/j.jbiotec.2016.09.016

Ngo-Matip, M.E., Pieme, C.A., Azabji-Kenfack, M., Moukette, B.M., Korosky, E., Stefanini, P., Ngogang, J.Y., Mbofung, C.M. (2015).Impact of daily supplementation of *Spirulina platensis* on the immune system of naïve HIV-1 patients in Cameroon: a 12-months single blind, randomized, multicenter trial. *Nutr. J.*, 14, 70.

Nicoletti, M. (2016). Microalgae nutraceuticals. *Foods.*, 5(54). doi:10.3390/foods5030054

Noreña-Caro, D., Benton, M.G. (2018). Cyanobacteria as photoautotrophic biofactories of high-value chemicals. *J. CO_2 Util.*, 28, 335–366.

Ordog, V., Pulz, O. (1996). Diurnal changes of cytokine- like activity in a strain of *Arthronemaafricanum* (cyanobacteria), determined by bioassays. *Algol. Stud.*, 82, 57–67.

Osanai, T., Numata, K., Oikawa, A., Kuwahara, A., Iijima, H., Doi, Y., Tanaka, K., Saito, K., Hirai, M.Y. (2013). Increased bioplastic production with an RNA polymerase sigma factor SigE during nitrogen starvation in Synechocystis sp. PCC 6803. *DNA Res.*, 20(6), 525–535. doi:10.1093/dnares/dst028

Oyamada, C., Kaneniwa, M., Ebitani, K. et al. (2008). Mycosporine-like amino acids extracted from scallop (*Patinopectenyessoensis*) ovaries: UV protection and growth stimulation activities on human cells. *Marine Biotechnol.*, 10, 141–150.

Panjiar, N., Mishra, S., YadavA.N., Verma, P. (2017). Functional foods from cyanobacteria: an emerging source for functional food products of pharmaceutical importance. In V. K. Gupta, H. Treichel, V. Shapaval, L.A. deOliveira, M.G. Tuohy(Eds). *Microbial Functional Foods and Nutraceuticals*(1st ed., pp. 21–37). John Wiley & Sons Ltd doi:10.1002/9781119048961.ch2

Patel, S.K.S., Kumar, P., Kalia, V.C. (2012). Enhancing biological hydrogen production through complementary microbial metabolisms. *Int. J. Hydrogen Energy*, 37(14), 10590–10603.

Patel, S.K.S., Kumar, P., Mehariya, S., Purohit, H.J., Lee, J.K., Kalia, V.C. (2014). Enhancement in hydrogen production by co-cultures of Bacillus and Enterobacter. *Int. J. Hydrogen Energy*, 39(27), 14663–14668.

Patel, S.K.S., Gupta, R.K., Das, D., Lee, J.K., Kalia, V.C. (2021a) Continuous biohydrogen production from poplar biomass hydrolysate by a defined bacterial mixture immobilized on lignocellulosic materials under non-sterile conditions. *J. Clean. Prod.*, 287, 125037.

Patel, S.K.S., Das, D., Kim, S.C., Cho, B.K., Kalia, V.C., Lee, J.K. (2021b). Integrating strategies for sustainable conversion of waste biomass into dark-fermentative hydrogen and value-added products. *Renew. Sustain. Energy Rev.*, 150, 111491.

Pathak, V.V., Singh, D.P., Kothari, R., Chopra, A.K. (2014). Phytoremediation of textile wastewater by unicellular microalga *Chlorella pyrenoidosa*. *Cell. Mol. Biol.* 60, 35–40.

Pinzon-Gamez, N.M., Sundaram, S., Ju, L.K. (2005). Heterocyst differentiation and H_2 production in N_2-fixing cyanobacteria. In *AIChE Annual Meeting, Conference Proceedings*, Cincinnati, OH, pp. 8949–8951.

Portmann, C., Blom, J.F., Kaiser, M., Brun, R., Juttner, F., Gademann, K. (2008). Isolation of aerucyclamides C and D and structure revision of microcyclamide 7806A, heterocyclic ribosomal peptides from *Microcystis aeruginosa* PCC 7806 and their antiparasite evaluation. *J. Nat. Prod.*, 71, 1891–1896.

Pulz, O., Gross, W. (2004). Valuable products from biotechnology of microalgae. *Appl. Microbiol. Biotechnol.* 65, 635–648.

Raghukumar, C., Vipparty, V., David, J.J., Chandramohan, D. (2001). Degradation of crude oil by marine cyanobacteria. *Appl. Microbiol. Biotechnol.*, 57, 433–436.

Rai, A.N. (Ed.) (1990). *CRC Handbook of Symbiotic Cynanobacteria.* Retrieved from https://www.routledge.com/CRC-Handbook-of-SymbioticCyanobacteria/Rai/p/book/97813

Rai, M.K. (2006). *Handbook of Microbial Biofertilizers.* New York: Haworth Press.

Rajeshkumar, S., Malarkodi, C., Paulkumar, K., Vanaja, M., Gnanajobitha, G., Annadurai, G. (2014). Algae mediated green fabrication of silver nanoparticles and examination of its antifungal activity against clinical pathogens. *Int J Metals.*, 2014.

Rajeshwari, K.R. and Rajashekhar, M. (2011). Biochemical composition of seven species of cyanobacteria isolated from different aquatic habitats of Western Ghats, southern India. *Brazil. Arch. Biol. Technol.*, 54(5), 849–857.

Rath, B., Priyadarshani, I. (2013). Antibacterial and Antifungal activity of marine cyanobacteria from Odisha Coast. *Int. J. Curr. Trends Res*, 2, 248–251.

Ratha, S.K., Renuka, N., Rawat, I., Tech, M., Bux, F. (2021). Prospective options of algae-derived nutraceuticals as supplements to combat COVID-19 and human coronavirus diseases. *Nutrition*, 83, 111089.

Reinert, R., Donald, E.L., Rosi, F.X., Watal, C., Dowzicky, M. (2007). Antimicrobial susceptibility among organisms from the Asia/Pacific Rim, Europe and Latin and North America collected as part of TEST and thein vitro activity of tigecycline. *J. Antimicrob. Chemother.*, 60, 1018–1029.

Richa, R.P., Rastogi, S., Kumari, K.L., Singh, V.K., Kannaujiya, G., Singh, M., Kesheri, R.P. (2011). Biotechnological potential of mycosporine-like amino acids and phycobiliproteins of cyanobacterial origin. *Biotechnol. Bioeng.*, 1, 159–171.

Rittmann, B.E. (2008). Opportunities for renewable bioenergy using microorganisms. *Biotechnol. Bioeng.*, 100(2), 203–212.

Romay, C., Gonzales, R., Ledon, N., Remirez, D., Rimbau, V. (2003). C-phycocyanin: a biliprotein with antioxidant, anti-inflammatory and neuroprotective effects. *Curr. Protein Pept. Sci.*, 4, 207–216.

Ruffing, A.M. (2011). Engineered cyanobacteria: teaching an old bug new tricks. *Bioeng. Bugs*, 2, 136–149.

Saad, M.H., El-Fakharany, E.M., Salem, M.S., Sidkey, N.M. (2020). In vitro assessment of dual (antiviral and antitumor) activity of a novel lectin produced by the newly cyanobacterium isolate, *Oscillatoria* acuminate MHM-632 MK0142101. *J. Biomol. Struct. Dyn.*, 17, 1–21.

Sánchez-Baracaldo, P., Cardona, T. (2019). On the origin of oxygenic photosynthesis and Cyanobacteria. *New Phytol.*, 225(4), 1440–1446.

Sangkharak, K., Prasertsan, P. (2007). Optimization of polyhydroxybutyrate production from a wild type and two mutant strains of *Rhodobactersphaeroides* using statistical method, *J. Biotechnol.*, 132, 331–340.
Schirmer, A., Rude, M.A., Li, X., Popova, E., del Cardayre, S.B. (2010). Microbial biosynthesis of alkanes. *Science*, 329, 559–562.
Schirrmeister, B.E., Sánchez-Baracaldo, P., Wacey, D. (2016). Cyanobacterial evolution during the Precambrian. *Int. J. Astrobiol.* 15, 187–204.
Schmit, T.L., Zhong, W., Setaluri, V. (2009). Targeteddepletion of polo-like kinase (Plk) 1 through lentiviral shRNA or a small-molecule inhibitor causes mitotic catastrophe and induction of apoptosis in human melanomacells. *J. Invest. Dermatol.*, 129, 2843–2853.
Shah, S., Akhter, N., Auckloo, B., Khan, I., Lu, Y., Wang, K., Wu, B., Guo, Y.W.J.M.D. (2017). Structural diversity, biological properties and applications of natural products from cyanobacteria. *Mar. Drugs*, 15, 354.
Sharma, A., Sharma, S., Sharma, K., et al. (2016). Algae as crucial organisms in advancing nanotechnology: a systematic review. *J. Appl Phycol.*, 28(3), 1759–1774. doi:10.1007/s10811-015-0715-1
Sharma, N.K., Stal, L.J. (2014). The economics of cyanobacteria-based biofuel production: challenges and opportunities. In N.K. Sharma, L.J. Stal, A.K. Rai (Eds). *Cyanobacteria: an Economic Perspective* (pp 167–180). UK: John Wiley & Sons.
Shetty, K., Paliyath, G., Pometto, A., Robert, E.L. (2006). *Food Biotechnology* (2nd ed.). Boca Raton: CRC Press.
Shimizu, Y. (2003). Microalgal metabolites. *Curr. Opin. Microbiol.*, 6, 236–243.
Silva, C.E.F., Bertucco, A. (2016). Bioethanol from microalgae and cyanobacteria. *Process Biochem.*, 51, 1833–1842.
Simmons, L.T., Engene, N., Urena, L.D. et al. (2008). Viridamides A and B, lipodepsipeptides with antiprotozoal activity from marine cyanobacterium *Oscillatorianigroviridis*. *J. Nat. Prod.*, 71, 1544–1550.
Singh, A.K., Mallick, N. (2017). Advances in cyanobacterial polyhydroxyalkanoates production. *FEMS Microbiol. Lett.*, 364, (20), fnx189.
Singh, G., Babele, P.K., Shahi, S.K., Sinha, R.P., Tyagi, M.B., Kumar, A. (2014). Green synthesis of silver nanoparticles using cell extracts of *Anabaena doliolum* and screening of its antibacterial and antitumor activity. *J. Microbiol. Biotechnol.*, 24(10), 1354–1367. doi:10.4014/jmb.1405.05003
Singh, G., Kumar, J. (2018). Artificial and natural photoprotective compounds. In R.P. Rastogi (Ed.). *Sunscreens*(1st ed., pp. 153–199). India: Nova Science Publishers.
Singh, J.S., Kumar, A., Rai, A.N., Singh, D.P. (2016). Cyanobacteria: a precious bio-resource in agriculture, ecosystem, and environmental sustainability. *Front. Microbiol.*, 7, 1–19. doi:10.3389/fmicb.2016.00529
Skulberg, O.M. (2000). Microalgae as a source of bioactive molecules: experience from cyanophyte research. *J. Appl. Phycol.*, 12, 341–348.
Sonker, A.S., Pathak, J., Kannaujiya, V.K., Sinha, R.P. (2017). Characterization and in vitro antitumor, antibacterial and antifungal activities of green synthesized silver nanoparticles using cell extract of *Nostoc* sp. strain HKAR-2. *Can. J. Biotechnol.*, 1(1), 26. doi:10.24870/cjb.2017-00010398
Soule, T., Garcia-Pichel, F. (2014). Ultraviolet photoprotective compounds from cyanobacteria in biomedical applications. In N.K. Sharma, A.K. Rai, L.J. Stal (Eds). *Cyanobacteria: An Economic Perspective*, 119–143. New York: John Wiley & Sons, Ltd.
Srinivasan, M., Sivasubramanian, V. (2014). Anticancer activity of *Oscillatoria terebriformis* cyanobacteria in human lung cancer cell line a549. *Int J Appl Biol Pharm.*, 5(2), 2–34.
Stevenson, C.S., Capper, E.A., Roshak, A.K. (2002). Scytonemin, a marine natural product inhibitor of kinases key in hyperproliferative inflammatory diseases. *Inflamm. Res.*, 51, 112–114.
Sudesh, K., Abe, H., Doi, Y. (2000). Synthesis, structure and properties of polyhydroxyalkanoates: biological polyesters. *Prog. Polym. Sci.*, 25, 1503–1555.
Sudha, S.S., Rajamanickam, K., Rengaramanujam, J. (2013). Microalgae mediated synthesis of silver nanoparticles and their antibacterial activity against pathogenic bacteria. *Indian J. Exp. Biol.*, 51(5), 393–399.
Suh, H.J., Lee, H.W., Jung, J. (2003). Mycosporine glycine protects biological systems against photodynamic damage by quenching singlet oxygen with a high efficiency. *Photochem. Photobiol.*, 78, 109–113.
Torres, F.A.E., Passalacqua, T.G., Velásque, A.M.A., de Souza, R.A., Colepicolo, P., Graminha, M.A.S. (2014). New drugs with antiprotozoal activity from marine algae: a review. *Rev. Bras. Farmacogn.*, 24, 265–276.
Trainor, F.R. (1978). *Introductory Phycology*. New York, NY: John Wiley & Sons.
Uzair, B., Tabassum, S., Rasheed, M., Rehman, S.F. (2012). Exploring marine cyanobacteria for lead compounds of pharmaceutical importance. *Sci.World J.*, 2012, 10.
Van Damme, E.J.M., Peumans, W.J., Barre, A., Rougé, P. (1998). Plant lectins: a composite of several distinct families of structurally and evolutionary related proteins with diverse biological roles. *Crit. Rev. Plant Sci.*, 17, 575–692.

Vanthoor-Koopmans, M., Cordoba-Matson, M.V., Arredondo-Vega, B.O., Lozano-Ramírez, C., Garcia-Trejo, J.F., Rodriguez-Palacio, M.C. (2014). Microalgae and cyanobacteria production for feed and food supplements. In R. Guevara-Gonzalez, I. Torres-Pacheco (Eds.). *Biosystems Engineering: Biofactories for Food Production in the Century XXI* (pp. 253–275). Switzerland: Springer.

Vijayakumar, S., Menakha, M. (2015). Pharmaceutical applications of cyanobacteria-A review. *J. Acute Med.*, 5(1), 1–9. doi:10.1016/j.jacme.2015.02.004

Visser, P.M., Verspagen, J.M.H., Sandrini, G., Stal, L.J., Matthijs, H.C.P., Davis, T.W., Paerl, H.W., Huisman, J. (2016). How rising CO_2 and global warming may stimulate harmful cyanobacterial blooms. *Harmful Algae*, 54, 145–159.

Volk, R.B., Furkert, F.H. (2006). Antialgal, antibacterial and antifungal activity of two metabolites produced and excreted by cyanobacteria during growth. *Microbiol. Res.*, 161(2), 180–186. doi:10.1016/j.micres.2005.08.005

White, D.A., Polimene, L., Llewellyn, C.A. (2011). Effects of ultraviolet-A radiation and nutrient availability on the cellular composition of photoprotective compounds in *Glenodiniumfoliaceum* (Dinophyceae). *J. Phycol.*, 47, 1078–1088.

Whitton, B.A., Potts, M. (Eds). (2000). *The Ecology of Cyanobacteria; Their Diversity in Time and Space.* Retrieved from https://www.springer.com/gp/book/9780792347354

Wolk, C.P. (1996). Heterocyst formation. *Annu. Rev. Genet.*, 30, 59–78.

Wratten, S.J., Faulkner, D.J. (1976). Cyclic polysulfides from the red alga *Chondriacalifornica. J. Org. Chem.*, 41, 2465–2467.

Wright, S.L., Kelly, F.J. (2017). Plastic and human health: a micro issue? *Environ. Sci. Technol.*51, 6634–6647.

Zahra, Z., Choo, D.H., Lee, H., Parveen, A. (2020). Cyanobacteria: review of current potentials and applications. *Environments*, 7, 13.

Zhou, J., Zhu, T., Cai, Z., Li, Y. (2016). From cyanochemicals to cyanofactories: a review and perspective. *Microb. Cell Fact.*, 15(1), 1–9. doi:10.1186/s12934-015-0405-3

5 Microbial Production of Polyhydroxyalkanoate from Biological Waste

Tarika Kumar
The Maharaja Sayajirao University, Vadodara, India

Vijay Rayasam
REVA University, Bangalore, India

CONTENTS

5.1 INTRODUCTION

The advantages offered by polymeric materials like plastics have led to continuous demand since they first came onto the market. Their inertness to physical, chemical and biological attacks, ease of handling, low cost, and ability to mold in different shapes have kept them in demand for the past five to six decades (Siracusa, 2019). The other side of the coin brings us face to face with the reality of the pollution caused by excessive use of these petrochemical-based plastics. These are among the most difficult polymers to manage when it comes to waste management, because they are non-biodegradable and in widespread use in packaging films, containers/bottles, household products, etc. (Preethi et al., 2020).

In 1926, Maurice Lemoigne, a French scientist, observed inclusion bodies inside the bacterium *Bacillus megaterium* (Lemoigne, 1926), which were later named polyhydroxybutyrates (PHB) and are now the most widely studied type of biopolymers (Keshavarz and Roy, 2010). Not all the biopolymers known are biodegradable, as in addition to naturally occurring polymers, the term also covers those which are polymerized into high molecules by chemical/biological methods (Sudesh and Iwata, 2008). On the other hand, polymers that are biodegradable are produced by aerobic fermentation, composting using renewable raw materials or by waste management, or by anaerobic fermentation (Zhong et al., 2009).

This has motivated researchers across the globe to produce a polymer that has similar potential to petrochemical-derived plastics, but is biodegradable (Gabor (Naiaretti) and Tita, 2012). Currently, the polyhydroxyalkanoates (PHAs) are the most widely researched and studied family of polymers and are among the most suitable commodities to substitute for synthetic plastics in the market. These biopolymers are produced by microorganisms (usually under unfavorable conditions), in a medium that has surplus carbon and limited/complete absence of other important macronutrients such as nitrogen, phosphorous, and sulfur (Aslan et al., 2016). The production and commercialization of

DOI: 10.1201/9781003306931-6

PHAs has been difficult due to the complexity of the process, which is influenced by the composition and biodegradability of the biopolymer which is also used to classify these biopolymers. The biosynthesis of PHA activates an energy storage mechanism in bacteria that plays a fundamental role in promoting long-term survival of both sporulating/non-sporulating bacteria in stressful conditions (Kadouri et al., 2005). These are synthesized intracellularly and stored as energy storage compounds in Gram-positive and Gram-negative bacteria under severe nutrient-deficient conditions such as limited nitrogen, oxygen, phosphorous or pH shifts (Keshavarz and Roy, 2010; Philip and Keshavarz, 2007; Shah et al., 2008). PHA-producing bacteria have generally been observed to be more tolerant of environmental stress such as ultra-violet (UV) radiation, osmotic stress, or heat than other bacterial cultures (Kadouri et al., 2005).

Polylactic acid (PLA) has been studied extensively due to its physiochemical properties (Chanprateep, 2010; Sudesh and Iwata, 2008). These cytoplasmic granules, typically 0.2–0.5 m in diameter, can be visualized either by simple staining methods using dyes like the Sudan Black B or Nile Red, or by using phase contrast microscopy, as the granules are highly refractive in nature. Once the microbes are exposed to nutrient limiting conditions, the energy storage compounds are degraded to derive energy as a substitute for the carbon source (Khanna and Srivastava, 2005).

A complete lack of control over the polymer content makes it very unsuitable with respect to the thermal and mechanical properties of the resultant polymer (Chen et al., 2020). Finished products made of PHA are often associated with high costs due to the steps involved in sterilization, bioprocessing, the carbon source as raw materials, poor growth of the microorganisms, and downstream processing (Chen et al., 2020). The high manufacturing cost of these biopolymers is a reflection of the pure raw materials used. Generally, pure carbohydrates such as glucose, glycerol, sucrose, starch, maltose, fatty acids, and methanol, etc. are used as raw materials for microbial growth contributing to PHA production which accounts for 50% of the cost of the entire procedure. The price of the finished products ultimately shoots up, making it unaffordable for consumers (Bugnicourt et al., 2014; Shen et al., 2009). Moreover, the Food and Agriculture Organization (FAO) of the United Nations (UN) reports that of the total food produced for human consumption globally, around one-third of the food (1.3 billion tons) is wasted or lost every year. The food comes from industrial and household waste, agricultural residues and commercial waste (Xue et al., 2017). To overcome these problems, researchers are now targeting microorganisms that are resistant to any amount of contamination or which can resist unfavorable environments better than other bacteria. This would help obtain better microbial growth and consequently lead to improved PHA production. This method is termed "next-generation industrial biotechnology" (NGIB) (Chen et al., 2020). This chapter discusses the research challenges of achieving high PHA production, ways in which this is currently being addressed, and suggestions for possible ways in which it can be addressed in future.

5.2 METABOLISM OF POLYHYDROXYALKANOATES

PHAs generally comprises hydroxy acid (HA) monomers in a linear chain linked by an ester bond, which acts as a link between a carboxyl group and a hydroxyl group of two adjacent monomers (Philip and Keshavarz, 2007). These are classified as short chain length (scl) comprising 3–5 carbon atoms, mostly biosynthesized by *Cupriavidus necator;* and medium chain length (mcl) comprising 6–14 carbon atoms, mostly biosynthesized by *Pseudomonas species*. Poly(3-hydroxybutyrate) [P(3HB)] is the most common type of scl-PHA, comprising HB monomers (Możejko-Ciesielska and Kiewisz, 2016).

Of all the biosynthetic pathways that are currently known, acetyl-CoA is an important precursor to the synthesis of scl/mcl PHAs (Schubert et al., 1991), involving three steps catalyzed by acetyl-CoA acetyltransferase (β-ketothiolase; PhaA), acetoacetyl-CoA reductase (PhaB), and PHA

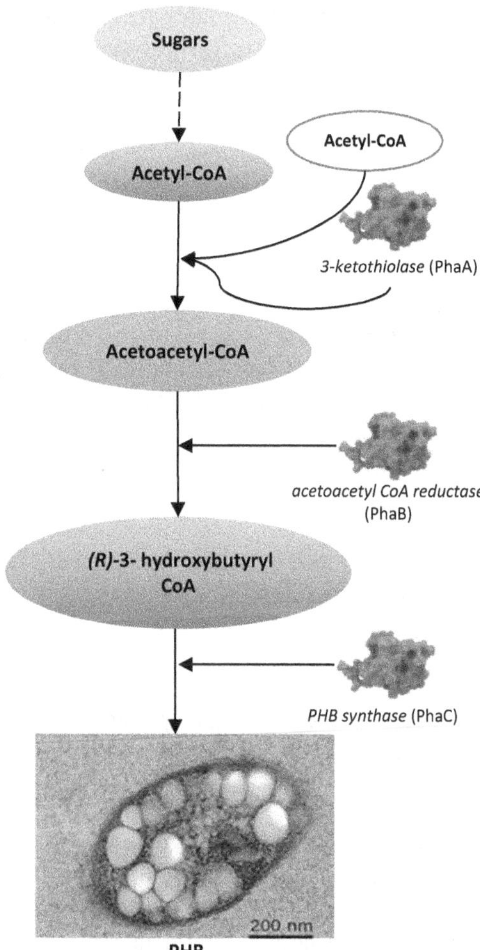

FIGURE 5.1 PHB biosynthetic pathway. The images shown in the figure are only for representational purpose.

CoA, coenzyme A; PHB, polyhydroxybutyrate

synthase (PhaC) (Slater et al., 1988). Two acetyl-CoA molecules are condensed into acetoacetyl-CoA, which is further transformed into *(R)*-3-hydroxybutyryl-CoA (*[R]*-3-HB-CoA) by PhaB, which gets polymerized into the growing chain of PHB by PhaC (Figure 5.1) (Sagong et al., 2018). These three genes, PhaA, PhaB, and PhaC, constitute a phaCAB operon (Mozes-Koch et al., 2017). The crucial enzyme for the use of *(R)*-3-hydroxyacyl-CoA as a substrate is PHA synthase (PhaC), which again is classified into four classes (class I to IV) depending on the specificity of the substrate, amino acid sequence and the subunits. Class I, III and IV PHA synthases polymerizes scl monomers while class II polymerizes mcl monomers. PHA synthases found in *Cupriavidus necator* qualifies under class I and that found in *Pseudomonas putida* qualifies under class II, both of which consist of a single subunit (PhaC; molecular weight: 61–73 kDa) (Qi and Rehm, 2001). On the other hand, two types of subunits are required for the PHA synthases belonging to class III (PhaC and PhaE; molecular weights: 40.3 and 20–40 kDa, respectively) and IV (PhaC and PhaR; molecular weights: 41.5 and 22 kDa, respectively). Class III PHA synthases are found in *Allochromatium vinosum* and class IV PHA synthases in *Bacillus megaterium* (Chek et al., 2017; Kim et al., 2017a; Kim et al., 2017b; Wittenborn et al., 2016).

In addition to acetyl-CoA, many other intermediates are also shared among the PHA biosynthesis pathways and central metabolic pathways (glycolysis, β-oxidation, Krebs cycle, amino acid catabolism, *de novo* fatty acids synthesis, serine pathway and Calvin cycle) (Madison and Huisman, 1999; Shimizu et al., 2013). The PHA metabolism depends very much on the production of acetyl-CoA from the tricarboxylic acid (TCA) cycle. When nutrients are in short supply, PHAs are depolymerized into monomers which ultimately produce acetyl-CoA, which is further metabolized into carbon and energy sources in the TCA cycle. Acetic acid, the main source of acetyl-CoA, is converted into pyruvate and later into lipids and serine which are utilized in the synthesis of cell membranes and also promotes synthesis of PHAs via TCA cycle. The phosphoenolpyruvate (PEP) pathway produces the precursor to PHA which is catalyzed by the enzyme pyruvate dehydrogenase complex. Another pathway leading to the production of acetyl-CoA includes threonine synthesis/degradation, PEP carboxylation, and formation of serine from glycine/serine deamination (Wang et al., 2019). Polymers that contain more than one substrate are known as copolymers. Bacteria convert the carbon sources into scl copolymers like poly (3-hydroxybutyrate-co-3-hydroxyvalerate) [P(3HB-co-3HV)] or poly (3-hydroxybutyrate-co-4-hydroxybutyrate) [P(3HB-co-4HB)] and mcl copolymers like poly (3-hydroxyhexanoate-co-3-hydroxyoctanoate) [P(3HHx-co-3HO)] (Możejko-Ciesielska and Kiewisz, 2016) or a combination of both scl and mcl polymers like poly (3-hydroxybutyrateco-3-hydroxyhexanoate) (P[3HB-co-3HHx]) (Surendran et al., 2020).

5.3 FERMENTATION

The process of fermentation breaks the complex compounds into simple compounds/metabolites. This is usually done either by bacteria, fungi, algae or actinomycetes (Amanullah et al., 2010; Chakravarty et al., 2017). Global PHA production is currently continuously increasing, and currently represents about one percent of the >368 million tonnes of plastic produced annually. Data from European Bioplastics in collaboration with the nova-Institute states that global bioplastics production will rise from 2.11 million tonnes in 2020 to ~2.87 million tonnes in 2025 (Bioplastics Market Data, 2020). Bacteria produce PHA either by using cultures or by bioprocess strategies such as batch, fed-batch or continuous fermentation techniques (Coats et al., 2016; Rodriguez-Perez et al., 2018), which increases the biomass and PHA yield and reduces overall process cost (Gómez-Cardozo et al., 2020). Depending on the culture conditions, PHA-producing bacteria can be classified into two groups: one that require limitation of essential nutrients such as nitrogen and oxygen, and the presence of excess carbon sources for the efficient synthesis of PHA (for e.g., *Cupriavidus necator*, *Protomonas extorquens* and *Protomonas oleovorans*); and one that does not require nutrient limitation for synthesis and can accumulate PHA while growing exponentially (for e.g., *Alcaligenes latus*, mutant *Azotobacter vinelandii* and a recombinant *Escherichia coli with* PHA biosynthetic operon of *Cupriavidus necator*). The culture conditions are very important criteria to be considered during the PHA biosynthesis for the development of better cultivation techniques to be used in large-scale production of PHA. In the industrial fermentation processes, fed-batch cultivation is more efficient than batch cultivation in terms of achieving high product and cell concentration because the composition of the medium can be controlled by substrate inhibition. This helps avoid the initial high concentration of substrates to be fed. The main limitation of the fed-batch is its high operation cost which is due to the long down-time between two batches, which makes it suitable for bacteria belonging to the first group. Imperial Chemical Industries (ICI) initially adopted the two-stage cultivation method for the production of P(3HB-*co*-3HV), with a minimal change in the technology since then. During the first stage, the bacterial cells are allowed to grow until they reach a pre-determined cell mass concentration in a medium without any nutrient limitation. During the second stage, these cells are then transferred to a medium with limited nutrients in which cells are unable to multiply and remain almost constant. The cells then utilize the carbon substrate which is fed in order to produce and accumulate PHA as intracellular storage material, due to which the cells increase in its size and weight (Chee et al., 2010).

The three main types of fermentation process for the production of PHA are:

1. *Batch process:* This is a simple operating system in which all the media requirements for production of the biomass and its metabolites during either the primary or secondary phase of ultivation are added in a closed vessel at the beginning of the process (Ohno et al., 1995; Ruiz-Sanchez et al., 2010; Robinson et al., 2001; Singh et al., 2010a).
2. *Fed-batch process:* This phase usually starts when in batch fermentation, one or more substrate(s)/nutrient(s) is consumed and a fresh medium is introduced into the system (Amanullah et al., 2010; Krishna and Nokes, 2001; Rossi et al., 2009; Tang et al., 2011). This medium is introduced either in a fixed or a variable volume (Hewitt and Nienow, 2007; Hölker et al., 2004). The volume of the fresh medium to be added is controlled using a feedback mechanism which monitors the process and prevents high substrate concentration in the medium and catabolism repression (Singh et al., 2010b).
3. *Continuous process:* In this industrial-scale process (Aguilar et al., 2001), there is a continuous removal from the bioreactor of microbial cells, which are replaced by fresh nutrients and new microbial cells that are already in the exponential phase of their life cycle (Sen and Roychoudhury, 2013; Vogel et al., 2012). The pH and dissolved oxygen are controlled by the auxostat, the chemostat monitors the availability of the limiting substrate, and the turbidostat monitors the turbdity of the medium (Aguilar et al., 2001).

5.4 POLYHYDROXYALKANOATE PRODUCTION USING BIOWASTES

The oil crisis of the early 1970s drove the search for a new substitute to replace petroleum-based plastics. It was not till 50 years after its discovery that PHAs were produced commercially (Philip and Keshavarz, 2007). 3HB as a monomer of PHA was reported in activated sludge (Wallen et al., 1974) and the presence of 3HV and 3HHx as the major and minor constituents were also reported (Anderson and Dawes, 1990). As bacteria possess the ability to break down complex organic wastes into essential intermediates which participate in biopolymer production, biowastes naturally become the first choice for feed (Patel et al., 2012; Singh et al., 2013). The sustainability and safe disposal of organic wastes have sparked interest in research on reducing our dependence on fossil fuels by converting biowaste into bio-energy and bio-materials. Different microorganisms that can produce biopolymers are used in converting wastes from agriculture, municipal sources, and food processing units. *Alcaligenes latus*, *Alcaligenes eutrophus*, *Azotobacter vinelandii*, *Azotobacter chroococcum*, *Methylotrophs*, *Azotobacter beijerincki*, *Nocardia* spp., *Bacillus* spp., *Pseudomonas* spp., *Rhizobium* spp., and recombinant *Escherichia coli* are examples of microorganisms that have been successfully tested for their ability to produce PHAs on an industrial scale by utilizing organic wastes (Pagliano et al., 2017). One such study reports the use of biological hydrogen (H_2) as a potential substitute for petrochemical fuels. Properties such as its high conversion efficiency and high specific content make it a stronger candidate. The purple non-sulfur bacteria (*Rhodopseudomonas sp.*) were investigated for the production of PHBs with a simultaneous photo-evolution of H_2. Three different media combinations with different sources of carbon (succinate, acetate, and malate) were prepared that contained a nitrogen source (glutamate) under the batch cultivation system and continuous irradiation. The resultant production of the biomass, PHB, and H_2 indicated that the type of carbon source in the medium is a factor. Under nutrient sulfur-deficient conditions, PHB (mg/L [%]); H2 (mL/L) produced were: acetate (68.99 [18.28]; 2286); succinate ([12.50]; nil); malate (47.52 [17.9]; nil). Of the three carbon sources, acetate proved to be the best feed material for PHB and H_2 production (Touloupakis et al., 2021).

Researchers have also reported specific substrates like organic acids, valeric acid, and propionic acid required for synthesis of PHA/polyhydroxyvalerate (PHV) monomers (Rocha et al., 2008; Steinbüchel and Lütke-Eversloh, 2003) which can sometimes be as high as 80% accumulation of the biomass. This is because, as per the life cycle assessment, the energy required for PHB production is generally lower than that required for production of conventional high-density plastic polymers (Harding et al., 2007).

Bacterial isolates were also isolated from contaminated sources, such as molasses from the sugar industry and oil spills from the oil industry and were screened for their ability to produce PHA. The isolates were grown on nitrogen limitation media with 2% glucose and PHA accumulation was detected using the Nile blue staining method. Of the 120 positive isolates, the amount of PHA produced was quantified using a UV spectrophotometer at 235nm. Post quantification, 15 high PHA-producing isolates were selected and the accumulation was re-quantified using gas chromatography (GC). The PHA accumulation was observed in: *Psuedomonas* species (27.2%–66.5%); *Acinetobacter* species (36.8%) and *Caulobacter* species (25.3%–25.9%) (Reddy and Thirumala, 2012).

Inexpensive substrates such as cane molasses were used to grow *Bacillus subtilis* and *Escherichia coli* isolated from contaminated industrial soil samples, and the amount of PHA production was studied. The amount of PHA accumulation by *Bacillus subtilis* was 54.1% (6% molasses) and by *Escherichia coli* 47.16% (8% molasses). An addition of 1% ethanol to the molasses medium increased biomass growth and PHA accumulation. Further addition of nitrogen sources (1g/L of ammonium sulfate and ammonium nitrate) increased production up to 62.21% and 58.7% respectively. Functional groups of the accumulated PHA granules were identified using Fourier transform infrared (FTIR) spectroscopy analysis (Gomaa, 2014).

Previous studies have also reported the use of biowastes such as pea-shell slurry in the production of H_2 and PHB, using defined mixed microbial culture (MMC4: *Enterobacter*, *Proteus*, *Bacillus* spp.). Different ratios (1:1, 3:7, 7:3, 9:1) of the pea-shell slurry were prepared with the GM-2/M-9 medium combined with glucose (0.5% w/v) supplementation, the effect of which on H_2 and PHB production was investigated. The resulting supernatant, after centrifuging effluent at 6000 rpm, 4°C for 20 min, from the batch and continuous fermentation process was inoculated with the PHB-producing strain by addition of nutrients to GM-2 media in continuous culture stage with/without glucose (Table 5.1) (Patel et al., 2015). In addition to all the substrates discussed, production of PHAs using the organic fraction from municipal waste has also been reported. Organic acid-rich substrate was obtained from municipal waste using a pilot-scale anaerobic percolation biocell reactor. From the 151 g/kg organic acid produced from the municipal waste, the total PHA produced was 223±28 g/kg. Thus, the study reported that optimizing acid fermentation to obtain organic acids from municipal waste and converting it into PHA could very well act as a starting point towards obtaining PHAs that are economically feasible (Colombo et al., 2017).

5.5 ACKNOWLEDGMENT

We are grateful to Prof. K.D. Vachhrajani, Head of the Department of Environmental Studies and Prof. H.R. Kataria, Dean of the Faculty of Science, The Maharaja Sayajirao University of Baroda, for their encouragement.

TABLE 5.1
Studies Reporting the Conversion of Biowastes Into PHA

Bacteria	Polymer	Carbon/Nitrogen Sources	HOI	Cell Concentration (g/L)	Polymer Concentration (g/L)	Polymer Content (%)	Productivity (g/L)	References
Alcaligenes eutrophus	PHB	Glucose	150	164	121	76	2.42	Kaewkannetra, (2012)
Alcaligenes eutrophus	PHB	Carbon dioxide	40	91.3	61.9	67.8	1.55	
Alcaligenes eutrophus	PHB	Tapioca hydrolysate	59	106	61.9	57.5	1.03	
Alcaligenes eutrophus	P(3HB-co-3HV)	Glucose + propionic acid	46	158	117	74	2.55	
Alcaligenes latus	PHB	Sucrose	18	143	71.4	50	3.97	
Azotobacter vinelandii	PHB	Glucose	47	40.1	32	79.8	0.68	
Chromobacterium violaceum	PHV	Valeric acid	-	39.5	24.5	62	-	
Methylobacterium organophilium	PHV	Methanol	70	250	130	52	1.68	
Protomonas extorquens	PHB	Methanol	170	233	149	64	0.88	
Pseudomonas oleovorans	P(3HBx-co-3HO)	n-octane	38	37.1	12.1	33	0.32	
Recombinant *Escherichia coli*	PHB	Glucose	39	101.4	81.2	80.1	2.08	
Recombinant *Klebsiella aerogenes*	PHB	Molasses	32	37	24	65	0.75	
C.necator H16	PHA	WCO/Urea	72	6.1	N/A	64	N/A	Kamilah et al. (2013)
C.necator PHB-4	PHA	WCO/Urea	72	5.3	N/A	61	N/A	
C.necator H16	PHA	WCO/Ammonium Chloride	72	4.1	N/A	83	3.4	
C.necator PHB-4	PHA	WCO/Ammonium Chloride	72	4	N/A	79	3.2	
Bacillus thuringiensis	PHA	Glucose	N/A	N/A	N/A	58.5	3.3	Gowda & Shivakumar (2014)
Bacillus thuringiensis	PHA	Starch	N/A	N/A	N/A	41.5	2.8	
Bacillus thuringiensis	PHA	Starch (10g/L)	N/A	N/A	N/A	72.5	2.6	
Bacillus thuringiensis	PHA	Straw hydrolysate	N/A	N/A	N/A	-	-	
Bacillus thuringiensis	PHA	Jack fruit powder	N/A	N/A	N/A	51.7	8.03	
Bacillus thuringiensis	PHA	Bagasse	N/A	N/A	N/A	39.6	4.2	
Bacillus thuringiensis	PHA	Mango peel	N/A	N/A	N/A	45.6	3.3	
Bacillus thuringiensis	PHA	Ragi husk	N/A	N/A	N/A	29.6	1.6	
Bacillus thuringiensis	PHA	Ragi husk	N/A	N/A	N/A	29.4	1.46	
Bacillus thuringiensis	PHA	Waste glycerol	N/A	N/A	N/A	45.7	4.23	

(Continued)

TABLE 5.1 *(Continued)*

Bacteria	Polymer	Carbon/Nitrogen Sources	HOI	Cell Concentration (g/L)	Polymer Concentration (g/L)	Polymer Content (%)	Productivity (g/L)	References
Pseudomonas corrugata	PHA	Soy molasses	N/A	N/A	N/A	5.0–17.0	N/A	Solaiman et al. (2006)
Cupriavidus necator NCIMB 11599	PHA	Oil palm frond juice	N/A	N/A	N/A	75	N/A	Zahari et al. (2014)
Burkholderia sacchari DSM 17165	PHA	Wheat straw hydrolysate	N/A	N/A	N/A	72	N/A	Cesário et al. (2014)
Cupriavidus necator DSM 545	PHA	Olive mill wastewater	N/A	N/A	N/A	55.00 (11% 3HV co-polymer)	N/A	Agustín Martinez et al. (2015)
Azohydromonas lata	PHA	Sugarbeet juice	N/A	N/A	N/A	65.6	N/A	Wang et al. (2013)
Azohydromonas lata	PHA	Maple sap	N/A	N/A	N/A	77.6	N/A	Yezza et al. (2007)
Bacillus cereus SPV	PHA	Molasses (sugarcane)	N/A	N/A	N/A	61.07	N/A	Akaraonye et al. (2010)
Cupriavidus necator H16	PHA	Waste rapeseed oil	N/A	N/A	N/A	76.00 (8% 3-HV co-polymer)	N/A	Obruca et al. (2010)
Haloferax mediterranei	PHA	Extruded rice bran and wheat bran	N/A	N/A	N/A	56	N/A	Huang et al. (2006)
Activated sludge (MMC)	PHA	Rice grain distillery wastewater with nutrient added	N/A	N/A	N/A	67	N/A	Khardenavis et al. (2007)
Cupriavidus necator DSM545	PHA	Crude glycerol	N/A	N/A	N/A	62.7	N/A	Mozumder et al. (2014)
Bacillus sp.	PHA	Pineapple juice	N/A	N/A	N/A	48.89	N/A	Suwannasing et al. (2015)
6 strains bacterial cultures1	PHB/PHV	Pea-shell slurry	48	4.85	0.85/0.02	18.1	N/A	Kumar et al. (2014)
5 strains bacterial cultures 1	PHB/PHV	Pea-shell slurry	48	4.51	1.62/0.03	36.2	N/A	
5 strains bacterial cultures 2	PHB/PHV	Pea-shell slurry	48	4.64	1.60/0.02	34.7	N/A	
4 strains bacterial cultures 1	PHB/PHV	Pea-shell slurry	48	3.33	0.43/0.03	13.8	N/A	
4 strains bacterial cultures 2	PHB/PHV	Pea-shell slurry	48	4.67	0.96/0.02	21	N/A	
4 strains bacterial cultures 3	PHB/PHV	Pea-shell slurry	48	4.59	0.57/0.03	13	N/A	
3 strains bacterial cultures 1	PHB/PHV	Pea-shell slurry	48	4.39	0.87/0.03	21	N/A	

Bacteria	Polymer	Carbon/Nitrogen Sources	HOI	Cell Concentration (g/L)	Polymer Concentration (g/L)	Polymer Content (%)	Productivity (g/L)	References
2 strains bacterial cultures 1	PHB/PHV	Pea-shell slurry	48	4.54	0.78/0.04	18.1	N/A	
2 strains bacterial cultures 2	PHB/PHV	Pea-shell slurry	48	4.2	0.84/0.04	21	N/A	
Bacillus thuringiensis EGU45	PHA-co-	Apple pomace	48	0.72	0.12	17	N/A	Ray et al. (2018)
Mixed culture	PHV		48	0.6	0.12	18	N/A	
Bacillus thuringiensis EGU45	PHA-co-	Apple pomace + 1% crude	48	0.82	0.13	16	N/A	
Mixed culture	PHV	glycerol	48	1.85	0.2	10	N/A	
Bacillus thuringiensis EGU45	PHA-co-	Potato peels	48	1.04	0.04	2	N/A	
Mixed culture	PHV		48	1.5	0.12	11	N/A	
Bacillus thuringiensis EGU45	PHA-co-	Potato peels + 1% crude	48	2.34	0.15	6	N/A	
Mixed culture	PHV	glycerol	48	1.72	0.2	5	N/A	
Bacillus thuringiensis EGU45	PHA-co-	Pea-shells	48	1.2	0.07	6	N/A	
Mixed culture	PHV		48	0.85	0.07	8	N/A	
Bacillus thuringiensis EGU45	PHA-co-	Pea-shells + 1% crude	48	1.23	0.09	6	N/A	
Mixed culture	PHV	glycerol (v/v)	48	1.7	0.45	26	N/A	
Bacillus thuringiensis EGU45	PHA-co-	Onion peels	48	3.59	0.35	8	N/A	
Mixed culture	PHV		48	1.9	0.45	23	N/A	
Bacillus thuringiensis EGU45	PHA-co-	Onion peels + crude	48	5.4	0.63	11	N/A	
Mixed culture	PHV	glycerol	48	3.4	2.04	60	N/A	
Ralstonia eutropha H16	PHB	Glucose, fructose, aceticacid, valeric acid	24	N/A	N/A	70–80	N/A	Du et al. (2001), Poirier et al. (1995)
	PHV		24	N/A	N/A	90	N/A	
Recombinant *Ralstonia eutropha* H16	PHB	Sucrose orgluconate	24	N/A	N/A	80	N/A	de Andrade Rodrigues et al. (2000)
Recombinant *Echerichia coli*	P(3HPE)		24	N/A	N/A	4–60	N/A	
Burkholderia cepacia ATCC 17759	P(3HB-co-3HV)	Xylose: levulinicacid	74	N/A	2.4	45–49	N/A	Keenan et al. (2004), Verlinden et al. (2007)

(Continued)

TABLE 5.1 *(Continued)*

Bacteria	Polymer	Carbon/Nitrogen Sources	HOI	Cell Concentration (g/L)	Polymer Concentration (g/L)	Polymer Content (%)	Productivity (g/L)	References
Pseudomonas sp.	PHA	Corn oil	24	12.53	N/A	35.63	N/A	Chaudhry et al. (2010)
			48	10.68	N/A	29.45	N/A	
			72	9.81	N/A	26.47	N/A	
	PHA	Spent mash	24	8.56	N/A	25.46	N/A	
			48	7.98	N/A	16.45	N/A	
			72	6.83	N/A	14.36	N/A	
	PHA	Fermented mash	24	7.02	N/A	23.56	N/A	
			48	6.52	N/A	15.69	N/A	
			72	6.23	N/A	13.98	N/A	
	PHA	Molasses	24	10.54	N/A	20.63	N/A	
			48	9.62	N/A	18.79	N/A	
			72	9.05	N/A	16.45	N/A	
Bacillus cereus EGU43	PHA	Pea-shell slurry in M-9 medium with 1% TS	48	1.02	N/A	3.4	0.03	Patel et al. (2015)
	PHA	Pea-shell slurry in M-9 medium with 2% TS	48	1.21	N/A	5.8	0.07	
	PHA	Pea-shell slurry in M-9 medium with 3% TS	48	1.35	N/A	6.3	0.08	
	PHA	Pea-shell slurry in M-9 medium with 5% TS	48	1.52	N/A	6.6	0.1	
	PHA	Pea-shell slurry in M-9 medium with 7% TS	48	1.7	N/A	5.3	0.09	
	PHA	Pea-shell slurry in GM-2 medium with 1% TS	48	0.1	N/A	4.4	0.04	
	PHA	Pea-shell slurry in GM-2 medium with 2% TS	48	1.24	N/A	12.4	0.15	
	PHA	Pea-shell slurry in GM-2 medium with 3% TS	48	1.28	N/A	12.1	0.15	

Bacteria	Polymer	Carbon/Nitrogen Sources	HOI	Cell Concentration (g/L)	Polymer Concentration (g/L)	Polymer Content (%)	Productivity (g/L)	References
	PHA	Pea-shell slurry in GM-2 medium with 5% TS	48	1.61	N/A	10.2	0.16	
	PHA	Pea-shell slurry in GM-2 medium with 7% TS	48	1.86	N/A	8.1	0.15	
	PHA	Pea-shell slurry with M-9 media in 9:1 ratio	48	2.45	N/A	8.1	0.2	
	PHA	Pea-shell slurry with M-9 media in 7:3 ratio	48	2.12	N/A	12	0.25	
	PHA	Pea-shell slurry with M-9 media in 1:1 ratio	48	1.91	N/A	13.1	0.25	
	PHA	Pea-shell slurry with M-9 media in 3:7 ratio	48	1.72	N/A	12.2	0.21	
	PHA	Pea-shell slurry with GM-2 media in 9:1 ratio	48	2.84	N/A	15	0.42	
	PHA	Pea-shell slurry with GM-2 media in 7:3 ratio	48	2.18	N/A	17.4	0.38	
	PHA	Pea-shell slurry with GM-2 media in 1:1 ratio	48	1.95	N/A	18.4	0.36	
	PHA	Pea-shell slurry with GM-2 media in 3:7 ratio	48	1.81	N/A	19.1	0.34	

Co, copolymer; N/A, not available; PHA, polyhydroxyalkanoate; PHB, polyhydroxybutyrate; 3HO, 3-hydroxyoctanoate; P(3-HB-co-3HV), copolymer of poly (3-hydroxybutyrate and 3-hydroxyvalerate); P(3HPE), poly (3-hydroxy-4-pentenoic acid); TS, total solids; WCO, waste cooking oil

REFERENCES

Aguilar, C. N., Augur, C., Favela-Torres, E., & Viniegra-González, G. (2001). Production of tannase by *Aspergillus niger* Aa-20 in submerged and solid-state fermentation: Influence of glucose and tannic acid. *Journal of Industrial Microbiology and Biotechnology*, *26*(5), 296–302. https://doi.org/10.1038/sj.jim.7000132

Agustín Martinez, G., Bertin, L., Scoma, A., Rebecchi, S., Braunegg, G., & Fava, F. (2015). Production of polyhydroxyalkanoates from dephenolised and fermented olive mill wastewaters by employing a pure culture of *Cupriavidus necator*. *Biochemical Engineering Journal*, *97*, 92–100. https://doi.org/10.1016/j.bej.2015.02.015

Akaraonye, E., Keshavarz, T., & Roy, I. (2010). Production of polyhydroxyalkanoates: The future green materials of choice. *Journal of Chemical Technology and Biotechnology*, *85*(6), 732–743. https://doi.org/10.1002/jctb.2392

Amanullah, A., Otero, J. M., Mikola, M., Hsu, A., Zhang, J., Aunins, J., Schreyer, H. B., Hope, J. A., & Russo, A. P. (2010). Novel micro-bioreactor high throughput technology for cell culture process development: Reproducibility and scalability assessment of fed-batch CHO cultures. *Biotechnology and Bioengineering*, *106*(1), 57–67. https://doi.org/10.1002/bit.22664

Anderson, A.J. & Dawes, E.A. (1990). Occurrence, metabolism, metabolic role, and industrial uses of bacterial polyhydroxyalkanoates. *Microbiological Reviews*, *54*(4), 450–472.

Aslan, A. K. H. N., Ali, M. D. M., Morad, N. A., & Tamunaidu, P. (2016). Polyhydroxyalkanoates production from waste biomass. *IOP Conference Series: Earth and Environmental Science*, *36*(1). https://doi.org/10.1088/1755-1315/36/1/012040

Bioplastics Market Data. (2020). European Bioplastics. Retrieved from https://www.european-bioplastics.org/market/

Bugnicourt, E., Cinelli, P., Lazzeri, A., & Alvarez, V. (2014). Polyhydroxyalkanoate (PHA): Review of synthesis, characteristics, processing and potential applications in packaging. *Express Polymer Letters*, *8*(11), 791–808. https://doi.org/10.3144/expresspolymlett.2014.82

Cesário, M. T., Raposo, R. S., de Almeida, M. C. M. D., van Keulen, F., Ferreira, B. S., & da Fonseca, M. M. R. (2014). Enhanced bioproduction of poly-3-hydroxybutyrate from wheat straw lignocellulosic hydrolysates. *New Biotechnology*, *31*(1), 104–113. https://doi.org/10.1016/j.nbt.2013.10.004

Chakravarty, I., Kundu, K., Ojha, S., & Kundu, T. (2017). Development of various processing strategies for new generation antibiotics using different modes of bioreactors. *JSM Biotechnol Bioeng*, *4*(1), 1073. https://pdfs.semanticscholar.org/4cf8/91c8477af9ff455f73647917c4b8bff15f9f.pdf

Chanprateep, S. (2010). Current trends in biodegradable polyhydroxyalkanoates. *Journal of Bioscience and Bioengineering*, *110*(6), 621–632. https://doi.org/10.1016/j.jbiosc.2010.07.014

Chaudhry, W. N., Jamil, N., Ali, I., Ayaz, M. H., & Hasnain, S. (2010). Screening for polyhydroxyalkanoate (PHA)-producing bacterial strains and comparison of PHA production from various inexpensive carbon sources. *Annals of Microbiology*, *61*(3), 623–629. https://doi.org/10.1007/s13213-010-0181-6

Chee, J., Yoga, S., Lau, N., Ling, S., & Abed, R. M. M. (2010). Bacterially Produced Polyhydroxyalkanoate (PHA): Converting renewable resources into bioplastics. current research, technology and education topics in applied microbiology and microbial biotechnology. A. Mendez-Vilas, 1395–1404.

Chek, M. F., Kim, S. Y., Mori, T., Arsad, H., Samian, M. R., Sudesh, K., & Hakoshima, T. (2017). Structure of polyhydroxyalkanoate (PHA) synthase PhaC from *Chromobacterium sp*. USM2, producing biodegradable plastics. *Scientific Reports*, *7*(1), 1–15. https://doi.org/10.1038/s41598-017-05509-4

Chen, G. Q., Chen, X. Y., Wu, F. Q., & Chen, J. C. (2020). Polyhydroxyalkanoates (PHA) toward cost competitiveness and functionality. *Advanced Industrial and Engineering Polymer Research*, *3*(1), 1–7. https://doi.org/10.1016/j.aiepr.2019.11.001

Coats, E. R., Watson, B. S., & Brinkman, C. K. (2016). Polyhydroxyalkanoate synthesis by mixed microbial consortia cultured on fermented dairy manure: Effect of aeration on process rates/yields and the associated microbial ecology. *Water Research*, *106*, 26–40. https://doi.org/10.1016/j.watres.2016.09.039

Colombo, B., Favini, F., Scaglia, B., Sciarria, T. P., D'Imporzano, G., Pognani, M., Alekseeva, A., Eisele, G., Cosentino, C., & Adani, F. (2017). Enhanced polyhydroxyalkanoate (PHA) production from the organic fraction of municipal solid waste by using mixed microbial culture. *Biotechnology for Biofuels*, *10*(1), 1–15. https://doi.org/10.1186/s13068-017-0888-8

de Andrade Rodrigues, M. F., Valentin, H. E., Berger, P. A., Tran, M., Asrar, J., Gruys, K. J., & Steinbüchel, A. (2000). Polyhydroxyalkanoate accumulation in *Burkholderia sp*.: A molecular approach to elucidate the genes involved in the formation of two homopolymers consisting of short-chain-length 3-hydroxyalkanoic acids. *Applied Microbiology and Biotechnology*, *53*(4), 453–460. https://doi.org/10.1007/s002530051641

Du, G., Chen, J., Yu, J., & Lun, S. (2001). Continuous production of poly-3-hydroxybutyrate by *Ralstonia eutropha* in a two-stage culture system. *Journal of Biotechnology*, *88*(1), 59–65. https://doi.org/10.1016/S0168-1656(01)00266-8

Gabor (Naiaretti), D., & Tita, O. (2012). Biopolymers Used in Food Packaging: A Review. *Acta Universitatis Cibiniensis Series E: Food Technology*, *16*(2), 3–19.

Gomaa, E. Z. (2014). Production of polyhydroxyalkanoates (PHAs) by *Bacillus subtilis* and *Escherichia coli* grown on cane molasses fortified with ethanol. *Brazilian Archives of Biology and Technology*, *57*(1), 145–154. https://doi.org/10.1590/S1516-89132014000100020

Gómez-Cardozo, J. R., Velasco-Bucheli, R., Marín-Pareja, N., Ruíz-Villadiego, O. S., Correa-Londoño, G. A., & Mora-Martínez, A. L. (2020). Fed-batch production and characterization of polyhydroxybutyrate by *Bacillus megaterium* LVN01 from residual glycerol. *DYNA (Colombia)*, *87*(214), 111–120. https://doi.org/10.15446/DYNA.V87N214.83523

Gowda, V., & Shivakumar, S. (2014). Agrowaste-based Polyhydroxyalkanoate (PHA) production using hydrolytic potential of *Bacillus thuringiensis* IAM 12077. *Brazilian Archives of Biology and Technology*, *57*(1), 55–61. https://doi.org/10.1590/S1516-89132014000100009

Harding, K. G., Dennis, J. S., von Blottnitz, H., & Harrison, S. T. L. (2007). Environmental analysis of plastic production processes: Comparing petroleum-based polypropylene and polyethylene with biologically-based poly-β-hydroxybutyric acid using life cycle analysis. *Journal of Biotechnology*, *130*(1), 57–66. https://doi.org/10.1016/j.jbiotec.2007.02.012

Hewitt, C. J., & Nienow, A. W. (2007). the scale-up of microbial batch and fed-batch fermentation processes. *Advances in Applied Microbiology*, *62*(07), 105–135. https://doi.org/10.1016/S0065-2164(07)62005-X

Hölker, U., Höfer, M., & Lenz, J. (2004). Biotechnological advantages of laboratory-scale solid-state fermentation with fungi. *Applied Microbiology and Biotechnology*, *64*(2), 175–186. https://doi.org/10.1007/s00253-003-1504-3

Huang, T. Y., Duan, K. J., Huang, S. Y., & Chen, C. W. (2006). Production of polyhydroxyalkanoates from inexpensive extruded rice bran and starch by *Haloferax mediterranei*. *Journal of Industrial Microbiology and Biotechnology*, *33*(8), 701–706. https://doi.org/10.1007/s10295-006-0098-z

Kadouri, D., Jurkevitch, E., Okon, Y., & Castro-Sowinski, S. (2005). Ecological and agricultural significance of bacterial polyhydroxyalkanoates. *Critical Reviews in Microbiology*, *31*(2), 55–67. https://doi.org/10.1080/10408410590899228

Kaewkannetra, P. (2012). Fermentation of Sweet sorghum into added value biopolymer of Polyhydroxyalkanaotes (PHAs). In: Casparus, Johannes, Reinhard, Verbeek Eds. *Products and Applications of Biopolymers*, 41–60. https://doi.org/10.5772/32985

Kamilah, H., Tsuge, T., Yang, T. A., & Sudesh, K. (2013). Waste cooking oil as substrate for biosynthesis of poly(3-hydroxybutyrate) and poly(3-hydroxybutyrate-co-3-hydroxyhexanoate): Turning waste into a value-added product. *Malaysian Journal of Microbiology*, *9*(1), 51–59. https://doi.org/10.21161/mjm.45012

Keenan, T. M., Tanenbaum, S. W., Stipanovic, A. J., & Nakas, J. P. (2004). Production and characterization of poly-β-hydroxyalkanoate copolymers from *Burkholderia cepacia* utilizing xylose and levulinic acid. *Biotechnology Progress*, *20*(6), 1697–1704. https://doi.org/10.1021/bp049873d

Keshavarz, T., & Roy, I. (2010). Polyhydroxyalkanoates: bioplastics with a green agenda. *Current Opinion in Microbiology*, *13*(3), 321–326. https://doi.org/10.1016/j.mib.2010.02.006

Khanna, S., & Srivastava, A. K. (2005). Recent advances in microbial polyhydroxyalkanoates. *Process Biochemistry*, *40*(2), 607–619. https://doi.org/10.1016/j.procbio.2004.01.053

Khardenavis, A. A., Suresh Kumar, M., Mudliar, S. N., & Chakrabarti, T. (2007). Biotechnological conversion of agro-industrial wastewaters into biodegradable plastic, poly β-hydroxybutyrate. *Bioresource Technology*, *98*(18), 3579–3584. https://doi.org/10.1016/j.biortech.2006.11.024

Kim, J., Kim, Y. J., Choi, S. Y., Lee, S. Y., & Kim, K. J. (2017a). Crystal structure of *Ralstonia eutropha* polyhydroxyalkanoate synthase C-terminal domain and reaction mechanisms. *Biotechnology Journal*, *12*(1), 1–29. https://doi.org/10.1002/biot.201600648

Kim, Y. J., Choi, S. Y., Kim, J., Jin, K. S., Lee, S. Y., & Kim, K. J. (2017b). Structure and function of the N-terminal domain of *Ralstonia eutropha* polyhydroxyalkanoate synthase, and the proposed structure and mechanisms of the whole enzyme. *Biotechnology Journal*, *12*(1), 1–25. https://doi.org/10.1002/biot.201600649

Krishna, C., & Nokes, S. E. (2001). Predicting vegetative inoculum performance to maximize phytase production in solid-state fermentation using response surface methodology. *Journal of Industrial Microbiology and Biotechnology*, *26*(3), 161–170. https://doi.org/10.1038/sj.jim.7000103

Kumar, P., Singh, M., Mehariya, S., Patel, S. K. S., Lee, J. K., & Kalia, V. C. (2014). Ecobiotechnological approach for exploiting the abilities of bacillus to produce co-polymer of Polyhydroxyalkanoate. *Indian Journal of Microbiology*, *54*(2), 151–157. https://doi.org/10.1007/s12088-014-0457-9

Lemoigne, M. (1926). Produits de dehydration et de polymerisation de l'acide ß-oxobutyrique. *Bulletin de La Société de Chimie Biologique*, *8*, 770–782.

Madison, L. L., & Huisman, G. W. (1999). Metabolic engineering of poly(3-hydroxyalkanoates): From DNA to plastic. *Microbiology and Molecular Biology Reviews*, *63*(1), 21–53. https://doi.org/10.1128/mmbr.63.1.21-53.1999

Możejko-Ciesielska, J., & Kiewisz, R. (2016). Bacterial polyhydroxyalkanoates: Still fabulous? *Microbiological Research*, *192*(2016), 271–282. https://doi.org/10.1016/j.micres.2016.07.010

Mozes-Koch, R., Tanne, E., Brodezki, A., Yehuda, R., Gover, O., Rabinowitch, H. D., & Sela, I. (2017). Expression of the entire polyhydroxybutyrate operon of *Ralstonia eutropha* in plants. *Journal of Biological Engineering*, *11*(1), 1–9. https://doi.org/10.1186/s13036-017-0062-7

Mozumder, M. S. I., de Wever, H., Volcke, E. I. P., & Garcia-Gonzalez, L. (2014). A robust fed-batch feeding strategy independent of the carbon source for optimal polyhydroxybutyrate production. *Process Biochemistry*, *49*(3), 365–373. https://doi.org/10.1016/j.procbio.2013.12.004

Obruca, S., Marova, I., Snajdar, O., Mravcova, L., & Svoboda, Z. (2010). Production of poly(3-hydroxybutyrate-co-3-hydroxyvalerate) by *Cupriavidus necator* from waste rapeseed oil using propanol as a precursor of 3-hydroxyvalerate. *Biotechnology Letters*, *32*(12), 1925–1932. https://doi.org/10.1007/s10529-010-0376-8

Ohno, A., Ano, T., & Shoda, M. (1995). Production of a lipopeptide antibiotic, surfactin, by recombinant *Bacillus subtilis* in solid state fermentation. *Biotechnology and Bioengineering*, *47*(2), 209–214. https://doi.org/10.1002/bit.260470212

Pagliano, G., Ventorino, V., Panico, A., & Pepe, O. (2017). Integrated systems for biopolymers and bioenergy production from organic waste and by-products: A review of microbial processes. *Biotechnology for Biofuels*, *10*(1), 1–24. https://doi.org/10.1186/s13068-017-0802-4

Patel, S. K. S., Kumar, P., Singh, M., Lee, J. K., & Kalia, V. C. (2015). Integrative approach to produce hydrogen and polyhydroxybutyrate from biowaste using defined bacterial cultures. *Bioresource Technology*, *176*, 136–141. https://doi.org/10.1016/j.biortech.2014.11.029

Patel, S. K. S., Singh, M., Kumar, P., Purohit, H. J., & Kalia, V. C. (2012). Exploitation of defined bacterial cultures for production of hydrogen and polyhydroxybutyrate from pea-shells. *Biomass and Bioenergy*, *36*, 218–225. https://doi.org/10.1016/j.biombioe.2011.10.027

Philip, S., Keshavarz, T. (2007). Polyhydroxyalkanoates: biodegradable polymers with a range of applications. *Journal of Chemical Technology and Biotechnology*, *82*, 233–247.

Poirier, Y., Nawrath, C., & Somerville, C. (1995). Production of polyhydroxyalkanoates, a family of biodegradable plastics and elastomers, in bacteria and plants. *Nature Biotechnology*, *13*(2), 142–150. https://doi.org/10.1038/nbt0295-142

Preethi, R., Maria Leena, M., Moses, J. A., & Anandharamakrishnan, C. (2020). Biopolymer nanocomposites and its application in food processing. In *Green Nanomaterials* (pp. 283–317). https://doi.org/10.1007/978-981-15-3560-4_12

Qi, Q., & Rehm, B. H. A. (2001). Polyhydroxybutyrate biosynthesis in *Caulobacter crescentus*: Molecular characterization of the polyhydroxybutyrate synthase. *Microbiology*, *147*(12), 3353–3358. https://doi.org/10.1099/00221287-147-12-3353

Ray, S., Sharma, R., & Kalia, V. C. (2018). Co-utilization of crude glycerol and biowastes for producing polyhydroxyalkanoates. *Indian Journal of Microbiology*, *58*(1), 33–38. https://doi.org/10.1007/s12088-017-0702-0

Reddy, S. V., & Thirumala, M. (2012). Isolation of polyhydroxyalkanoates (PHA) producing bacteria from contaminated soils. *International Journal of Environmental Studies*, *2*(3), 104–107. http://urpjournals.com/tocjnls/13_12v2i3_1.pdf

Robinson, T., Singh, D., & Nigam, P. (2001). Solid-state fermentation: A promising microbial technology for secondary metabolite production. *Applied Microbiology and Biotechnology*, *55*(3), 284–289. https://doi.org/10.1007/s002530000565

Rocha, R. C. S., da Silva, L. F., Taciro, M. K., & Pradella, J. G. C. (2008). Production of poly(3-hydroxybutyrate-co-3-hydroxyvalerate) P(3HB-co-3HV) with a broad range of 3HV content at high yields by *Burkholderia sacchari* IPT 189. *World Journal of Microbiology and Biotechnology*, *24*(3), 427–431. https://doi.org/10.1007/s11274-007-9480-x

Rodriguez-Perez, S., Serrano, A., Pantión, A. A., & Alonso-Fariñas, B. (2018). Challenges of scaling-up PHA production from waste streams. A review. *Journal of Environmental Management*, *205*(January), 215–230. https://doi.org/10.1016/j.jenvman.2017.09.083

Rossi, S. C., Vandenberghe, L. P. S., Pereira, B. M. P., Gago, F. D., Rizzolo, J. A., Pandey, A., Soccol, C. R., & Medeiros, A. B. P. (2009). Improving fruity aroma production by fungi in SSF using citric pulp. *Food Research International*, *42*(4), 484–486. https://doi.org/10.1016/j.foodres.2009.01.016

Ruiz-Sanchez, J., Flores-Bustamante, Z. R., Dendooven, L., Favela-Torres, E., Soca-Chafre, G., Galindez-Mayer, J., & Flores-Cotera, L. B. (2010). A comparative study of Taxol production in liquid and solid-state fermentation with *Nigrospora sp.* a fungus isolated from *Taxus globosa*. *Journal of Applied Microbiology*, *109*(6), 2144–2150. https://doi.org/10.1111/j.1365-2672.2010.04846.x

Sagong, H. Y., Son, H. F., Choi, S. Y., Lee, S. Y., & Kim, K. J. (2018). Structural insights into polyhydroxyalkanoates biosynthesis. *Trends in Biochemical Sciences*, *43*(10), 790–805. https://doi.org/10.1016/j.tibs.2018.08.005

Schubert, P., Kruger, N., & Steinbuchel, A. (1991). Molecular analysis of the *Alcaligenes eutrophus* poly(3-hydroxybutyrate) biosynthetic operon: Identification of the N-terminus of poly(3-hydroxybutyrate) synthase and identification of the promoter. *Journal of Bacteriology*, *173*(1), 168–175. https://doi.org/10.1128/jb.173.1.168-175.1991

Sen, S., & Roychoudhury, P. K. (2013). Development of optimal medium for production of commercially important monoclonal antibody 520C9 by hybridoma cell. *Cytotechnology*, *65*(2), 233–252. https://doi.org/10.1007/s10616-012-9480-z

Shah, A. A., Hasan, F., Hameed, A., & Ahmed, S. (2008). Biological degradation of plastics: A comprehensive review. *Biotechnology Advances*, *26*(3), 246–265. https://doi.org/10.1016/j.biotechadv.2007.12.005

Shen, L., Haufe, J., & Patel, M. K. (2009). *Product overview and market projection of emerging bio-based plastics. PRO-BIP; Final Report, Report No: NWS-E-2009-32*. 243. https://www.uu.nl/sites/default/files/copernicus_probip2009_final_june_2009_revised_in_november_09.pdf

Shimizu, R., Chou, K., Orita, I., Suzuki, Y., Nakamura, S., & Fukui, T. (2013). Detection of phase-dependent transcriptomic changes and Rubisco-mediated CO_2 fixation into poly (3-hydroxybutyrate) under heterotrophic condition in *Ralstonia eutropha* H16 based on RNA-seq and gene deletion analyses. *BMC Microbiology*, *13*(1), 1. https://doi.org/10.1186/1471-2180-13-169

Singh, H. B., Singh, B. N., Singh, S. P., & Nautiyal, C. S. (2010a). Solid-state cultivation of *Trichoderma harzianum* NBRI-1055 for modulating natural antioxidants in soybean seed matrix. *Bioresource Technology*, *101*(16), 6444–6453. https://doi.org/10.1016/j.biortech.2010.03.057

Singh, M., Kumar, P., Patel, S. K. S., & Kalia, V. C. (2013). Production of Polyhydroxyalkanoate Co-polymer by *Bacillus thuringiensis*. *Indian Journal of Microbiology*, *53*(1), 77–83. https://doi.org/10.1007/s12088-012-0294-7

Singh, R. K., Mishra, S. K., & Kumar, N. (2010b). Optimization of α-amylase production on agriculture byproduct by *Bacillus cereus* MTCC 1305 using solid state fermentation. *Research Journal of Pharmaceutical, Biological and Chemical Sciences*, *1*(4), 867–876.

Siracusa, V. (2019). Microbial degradation of synthetic biopolymers waste. *Polymers*, *11*(6). https://doi.org/10.3390/polym11061066

Slater, S. C., Voige, W. H., & Dennis, D. E. (1988). Cloning and expression in *Escherichia coli* of the *Alcaligenes eutrophus* H16 poly-β-hydroxybutyrate biosynthetic pathway. *Journal of Bacteriology*, *170*(10), 4431–4436. https://doi.org/10.1128/jb.170.10.4431-4436.1988

Solaiman, D. K. Y., Ashby, R. D., Hotchkiss, A. T., & Foglia, T. A. (2006). Biosynthesis of medium-chain-length Poly(hydroxyalkanoates) from soy molasses. *Biotechnology Letters*, *28*(3), 157–162. https://doi.org/10.1007/s10529-005-5329-2

Steinbüchel, A., & Lütke-Eversloh, T. (2003). Metabolic engineering and pathway construction for biotechnological production of relevant polyhydroxyalkanoates in microorganisms. *Biochemical Engineering Journal*, *16*(2), 81–96. https://doi.org/10.1016/S1369-703X(03)00036-6

Sudesh, K., & Iwata, T. (2008). Sustainability of biobased and biodegradable plastics. *Clean - Soil, Air, Water*, *36*(5–6), 433–442. https://doi.org/10.1002/clen.200700183

Surendran, A., Lakshmanan, M., Chee, J. Y., Sulaiman, A. M., Thuoc, D. van, & Sudesh, K. (2020). Can polyhydroxyalkanoates be produced efficiently from waste plant and animal oils? *Frontiers in Bioengineering and Biotechnology*, *8*(March), 1–15. https://doi.org/10.3389/fbioe.2020.00169

Suwannasing, W., Imai, T., & Kaewkannetra, P. (2015). Cost-effective defined medium for the production of polyhydroxyalkanoates using agricultural raw materials. In *Bioresource Technology* (Vol. 194). Elsevier Ltd. https://doi.org/10.1016/j.biortech.2015.06.087

Tang, Y. J., Zhang, W., Liu, R. S., Zhu, L. W., & Zhong, J. J. (2011). Scale-up study on the fed-batch fermentation of *Ganoderma lucidum* for the hyperproduction of ganoderic acid and Ganoderma polysaccharides. *Process Biochemistry*, *46*(1), 404–408. https://doi.org/10.1016/j.procbio.2010.08.013

Touloupakis, E., Poloniataki, E. G., Ghanotakis, D. F., & Carlozzi, P. (2021). Production of biohydrogen and/or poly-β-hydroxybutyrate by *Rhodopseudomonas sp*. Using various carbon sources as substrate. *Applied Biochemistry and Biotechnology*, *193*(1), 307–318. https://doi.org/10.1007/s12010-020-03428-1

Verlinden, R. A. J., Hill, D. J., Kenward, M. A., Williams, C. D., & Radecka, I. (2007). Bacterial synthesis of biodegradable polyhydroxyalkanoates. *Journal of Applied Microbiology*, *102*(6), 1437–1449. https://doi.org/10.1111/j.1365-2672.2007.03335.x

Vogel, J. H., Nguyen, H., Giovannini, R., Ignowski, J., Garger, S., Salgotra, A., & Tom, J. (2012). A new large-scale manufacturing platform for complex biopharmaceuticals. *Biotechnology and Bioengineering*, *109*(12), 3049–3058. https://doi.org/10.1002/bit.24578

Wallen, L. L., Rohwedder, W. K., Regional, N., & District, P. S. (1974). Poly-β-hydroxyalkanoate from activated.pdf. *Environmental Science & Technology*, *8*(6), 576–579.

Wang, B., Sharma-Shivappa, R. R., Olson, J. W., & Khan, S. A. (2013). Production of polyhydroxybutyrate (PHB) by *Alcaligenes latus* using sugarbeet juice. *Industrial Crops and Products*, *43*(1), 802–811. https://doi.org/10.1016/j.indcrop.2012.08.011

Wang, P., Qiu, Y. Q., Chen, X. T., Liang, X. F., & Ren, L. H. (2019). Metabolomic insights into polyhydroxyalkanoates production by halophilic bacteria with acetic acid as carbon source. *Bioscience, Biotechnology and Biochemistry*, *83*(10), 1955–1963. https://doi.org/10.1080/09168451.2019.1630252

Wittenborn, E. C., Jost, M., Wei, Y., Stubbe, J. A., & Drennan, C. L. (2016). Structure of the catalytic domain of the class I polyhydroxybutyrate synthase from *Cupriavidus necator*. *Journal of Biological Chemistry*, *291*(48), 25264–25277. https://doi.org/10.1074/jbc.M116.756833

Xue, L., Liu, G., Parfitt, J., Liu, X., van Herpen, E., Stenmarck, Å., O'Connor, C., Östergren, K., & Cheng, S. (2017). Missing food, missing data? A critical review of global food losses and food waste data. *Environmental Science and Technology*, *51*(12), 6618–6633. https://doi.org/10.1021/acs.est.7b00401

Yezza, A., Halasz, A., Levadoux, W., & Hawari, J. (2007). Production of poly-β-hydroxybutyrate (PHB) by *Alcaligenes latus* from maple sap. *Applied Microbiology and Biotechnology*, *77*(2), 269–274. https://doi.org/10.1007/s00253-007-1158-7

Zahari, M. A. K. M., Abdullah, S. S. S., Roslan, A. M., Ariffin, H., Shirai, Y., & Hassan, M. A. (2014). Efficient utilization of oil palm frond for bio-based products and biorefinery. *Journal of Cleaner Production*, *65*, 252–260. https://doi.org/10.1016/j.jclepro.2013.10.007

Zhong, Z. W., Song, B., & Huang, C. X. (2009). Environmental impacts of three polyhydroxyalkanoate (PHA) manufacturing processes. *Materials and Manufacturing Processes*, *24*(5), 519–523. https://doi.org/10.1080/10426910902740120

6 Exopolysaccharides for Heavy Metal Remediation

A Review of Current Trends and Future Prospects

Anina James
Deen Dayal Upadhyaya College, University of Delhi, Delhi, India

Deepika Yadav
Shivaji College, University of Delhi, New Delhi, India

Mohit Kumar
Hindu College, University of Delhi, New Delhi, India

CONTENTS

6.1 INTRODUCTION

Heavy metals are natural elements, some of which (including Cu^{2+}, Mg^{2+}, Zn^{2+}, $Fe^{2+/3+}$) are essential for optimal functioning of living organisms. However, they can be toxic even at slightly higher concentrations. The hazardous impact of heavy metals on the environment has been apparent in the last few decades as a result of their widespread application in of the pharmaceutical, tanneries, organic chemicals, pesticide formulations, wood processing and rubber industries (Lakherwal, 2014). Free forms of heavy metals have contaminated every system of the biosphere, both terrestrial and aquatic. Their highly persistent nature enhances the severity of their harmful impact. Heavy metals cannot be biodegraded, and undergo bioaccumulation and biomagnification. Unfortunately, some heavy metals like mercury can transform to more toxic states under certain environmental conditions (Wang and Chen, 2006). On the other hand, heavy metals as a source of raw materials for industry are growing scarcer day by day. Heavy metals vary in their toxic manifestation in living organisms. It is now well established that lead causes oxidative stress due to free radical imbalance; mercury, chromium and arsenic result in formation of harmful thiol and methyl derivatives; cadmium and aluminum replace cofactors and metal ions; chromium (IV) causes ion channel disruption, cell membrane leak and DNA and protein damage; cadmium binds to cellular proteins; iron has corrosive effects and causes lipid peroxidation (Jaishankar et al., 2014). Lead is harmful to various systems, from blood

DOI: 10.1201/9781003306931-7

cell generation to liver and kidneys, and neuronal networks. Its chronic toxicity at 400–600 μg L^{-1} causes encephalopathy, convulsions, delirium and even coma if not treated (Flora et al., 2012). Chronic toxicity of cadmium via air leads to proteinurea and emphysema; acute toxicity in higher doses causes diarrhea, headache and nausea, and osteomalacia. Arsenic accumulation also causes cancer, dermatitis, cirrhosis of the liver, ulcer, bronchitis and neural insufficiencies at concentrations beyond the World Health Organization (WHO) recommendation, that is, 10 μg L^{-1} in drinking water (Alluri et al., 2007; Rajendran et al., 2003; Wang and Zhao, 2009).

Since the recognition of the hazardous nature of heavy metals in the environment, scientists have devised many methods to alleviate their concentration at the source of contamination. Some of the commonly used techniques are chemical precipitation, ultrafiltration, coagulation, flocculation, electrodialysis, ion exchange, evaporative recovery, reverse osmosis, and nanofiltration (Volesky, 1987; Lakherwal, 2014). These methods are effective, but expensive and energy-intensive; they also produce toxic sludge and harmful by-products which contaminate the environment. Sometimes, these processes may lead to incomplete metal ion removal. Thus, it is pertinent to develop effective, environment-safe and economical methods to remediate heavy metals from the environment.

Sorption is a phenomenon that involves the association of one material on to another on the basis of chemical and/or physical interactions between them. The interactions could be due to multiple factors such as covalent bonding, and electrostatic interactions between the functional groups of the sorbent and sorbate. Sorption involving a biological sorbent is called biosorption. Living or dead prokaryotic and eukaryotic microorganisms such as bacteria and fungi have been the front-running candidates for biosorption of heavy metals (Alluri et al., 2007; Rani et al., 2010). Heavy metals in particulate or soluble form can accumulate in intact bacterial cells (living or dead) or products derived from them.

Biopolymers are high molecular weight molecules made by enzymes in living organisms using building blocks like sugars, amino acids and fatty acids. Several classes of biopolymers including polysaccharides (sugars or sugar acids linked by glycosidic bonds), polyamides (amino acids linked by peptide bonds), polyesters (hydroxy fatty acids bonded by ester linkages), and polyphosphates (inorganic phosphates bonded via anhydride linkages) can be synthesized by bacteria. These biopolymers act as a protective coating on bacterial cells forming biofilms. Biofilms facilitate bacteria to persist and grow even under unfavorable conditions. The functions are modulated by environmental stimuli and include adhesion, nutrient storage, and protection (Rehm, 2010). They are produced primarily for self-defense, protecting cells against desiccation and toxins, and they may also provide a source of carbon (Sheng et al., 2010). Biofilms may also dictate bacterial behaviors like fixation onto surfaces, movement and invasion. The biosynthetic polymers can be bound as capsular polysaccharide (CPS) or as slime on cell surfaces (Whitfield, 1988). Biofilms are extracellular polymeric substances or exopolysaccharides (EPS) that are essentially structured microbial communities (Flemming and Wuertz, 2019). Researchers have been interested in the structure, function and synthesis of these water-soluble biopolymers for their potential role in environmental, clinical and industrial applications due to their physiochemical and rheological properties (Comte et al., 2008). In nature, EPS is secreted to coat the cell surface to prevent infiltration of toxins such as metals ions. EPS have metal ion sequestration potential, due to which they are of interest for metal bioremediation. The structure and composition of EPS promotes the sequestration of metal ions onto its coating, preventing metal ions' entry into cells. EPS comprises high molecular weight polysaccharide (hexose sugar moieties existing as homo or heteropolysachhrides) and small proportions of protein and uronic acid (Comte et al., 2008; Czaczyk and Myszka, 2007; Lau et al., 2005).

This review attempts to briefly discuss the synthesis of EPS by several bacteria, and further, details the structural and chemical characteristics that form the basis of potential applications of EPS. Lastly, the implementation of EPS from different bacteria inhabiting varied environments for heavy metal remediation is explained. The review aims to provide a holistic view of the use of EPS for decontamination of toxic heavy metals from the environment.

6.2 BIOSYNTHESIS OF BACTERIAL EPS

Many bacterial species have been reported so far that have the potential for adsorption via secretion of EPS (Figure 6.1). Besides the polysaccharide backbone, EPS have diverse structural characteristics due to different side chains, various bondings and linkages, functional groups and non-carbohydrate substituents (Whitfield, 1988). Theis structural and chemical diversity depends on the type and amount of the available nutrient sources, environmental stressors like temperature and pH, and the growth phase of the microorganism (Czaczyk and Myszka 2007; Sheng et al., 2010). Heavy metal remediation is primarily based on the utilization of anionic charges on EPS. Examples of bacterial EPS commercially available with net negative charge are gellan (*Sphingomona spaucimobilis*), galactopol (*Pseudomonas oleovorans*), alginate (*Pseudomonas aeruginosa, Azotobacter vinelandii*), hyaluronan (*Pasteurella multocida, Streptococco* attenuated strains, *Pseudomonas aeruginosa*), fucopol (*Enterobacter* A47), and xanthan (*Xanthomonas campestris*) (Czaczyk and Myszka 2007; Freitas et al., 2009; Freitas et al., 2011a; Freitas et al., 2011b). Biosysnthesis of EPS can be categorized according to the procedure and site of synthesis, and whether intracellular or extracellular (Rehm, 2009).

6.2.1 EXTRACELLULAR SYNTHESIS OF EP

Polysaccharides that are synthesized extracellularly are mostly homopolymers. The enzymes involved transfer precursor monopolymers to/from polymers from appropriate substrates. The

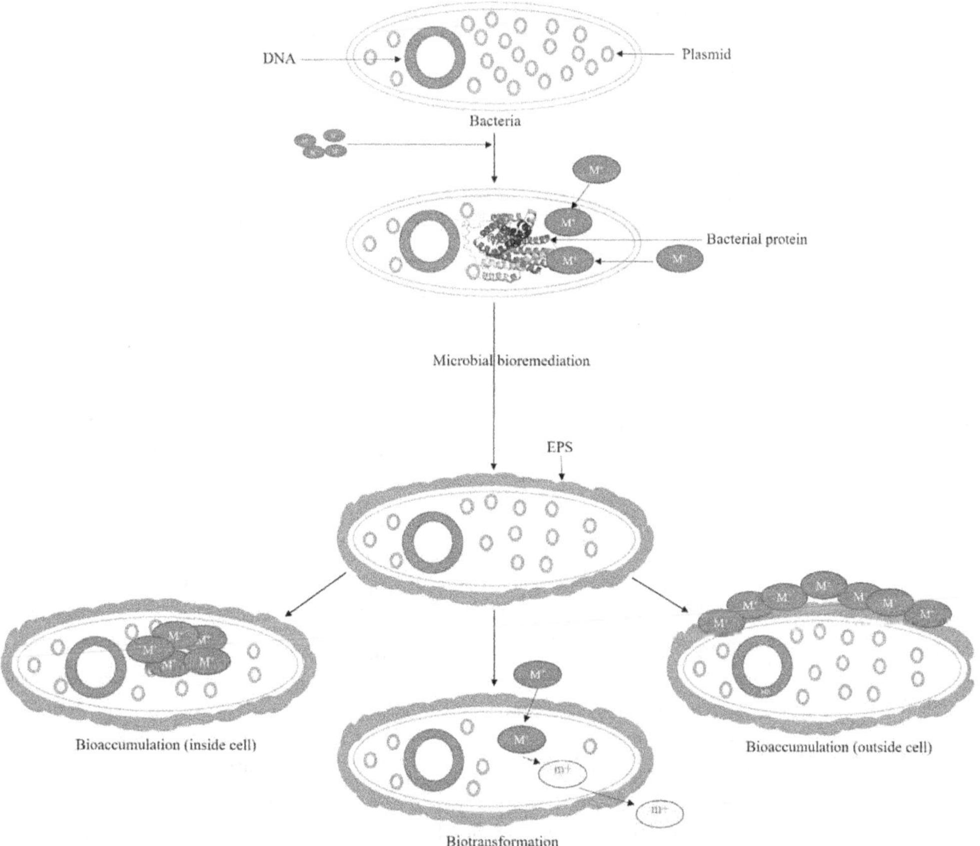

FIGURE 6.1 Illustration of heavy metal interaction with bacteria, EPS production and fate of metal ions.

polymers assume the particular structure with a specific linkage pattern and branching. Enzymes dextran sucrase and levan sucrase act on the sucrose (substrate) to form dextran, levan and mutan (Boels et al., 2001; Whitfield, 1988).

6.2.2 Intracellular Synthesis of EPS

Biopolymer synthesis inside the cell involves linking to sugar nucleotide precursors, which is then exported outside the cell. Several enzymes and other cellular components participate in multiple metabolic pathways to secrete EPS. A substrate (mostly a hexose sugar) enters the microorganism actively or passively and is catabolized via intracellular phosphorylation or periplasmic oxidation (Freitas et al., 2011a). Several molecules taken up by the microorganism may not participate in the central metabolic pathway but may take part in biosysnthesis of EPS. The metabolic pathway for EPS synthesis requires charged, energy-dense monosaccharides and nucleotide diphosphate/monophosphate (NDP/NMP) as precursor molecules. However, before this step, phosphorylated sugar (sugar-1P/-2P/-6P) is formed via an independent pathway, to provide an activated primary residue (Madhuri and Prabhakar, 2014; Whitfield, 1988). EPS synthesis uses central metabolic pathway participants such as fructose-6-phosphate and glucose-1-phosphate for making precursor molecules uridine diphosphate N acetyl glucasamine (UDP-GlcNAc), uridine diphosphate N acetyl galactosamine (UDP-GalNAc) and dTDP-rhamnose (Boels et al., 2001). Sugar-6-phosphate is converted to sugar-1-phosphate, the primary glycosyl donor for nucleotide diphosphate/monophosphate sugar synthesis, by the catalysis of phosphoglucomutase and further, sugar-1-phosphate is converted to UDP-glucose and dTDP-glucose via the catalysis of UDP-glucosepyrophosphorylase and dTDPglucosepyrophosphorylase (Madhuri and Prabhakar, 2014). In *Lactobacillus rhamnosus* UDP-GlcNAc is converted to UDP-GalNAc by catalytic activity of UDP-N-acetylglucosamine 4-epimerase (Boels et al., 2001). In this bacterium, there are four genes that synthesize dTDP-L-rhamnose NDP sugar, after which the sugar from activated NDP/NMP sugar moieties gets transferred to undecaprenyl phosphate by the catalysis of enzyme glycosyltransferase (Sutherland, 2001); undecaprenyl phosphate (C55-P) is a cytoplasmic membrane isoprenoid lipid carrier. Boels et al. (2001) reported that in *Lactobacillus rhamnosus* six glycosyltransferase genes were responsible for formation of heptasaccharide units comprising rhamnose moieties. Whitfield (1988) surmised that the lipid carrier molecule established the formation of monomeric molecules and modulation of extracellular transport of elongating chain polymerized monomers. Lipid metabolic pathways are utilized by gram-negative bacteria for EPS assembly and transport and rarely, gram-positive bacteria too may utilize these pathways for the same purpose. In gram-positive bacteria the assembled monomers on the lipid carrier are translocated to the cell surface, whereas in gram-negative bacteria, a Wzx-Wzy or ABC transporter-dependent pathway is followed (Freitas et al., 2011a). However, *Pseudomonas aeruginosa*, synthase-dependent pathway occurs. In Wzx-Wzy pathway, the monomers assemble in cytoplasmic space following which assembly and elongation occurs within the periplasm; the completed polymer is then translocated to the surface of the bacterium. This is the mechanism followed by the gram-negative bacterium *E. coli* (Whitfield, 1988). Several enzymes take part in the activities beginning with monomer assembly and elongation length as well as translocation, export and detachment from carrier; transportation of lipid-linked monomers is catalyzed by flippase, and precursor coupling is catalyzed by polymerase (Madhuri and Prabhakar, 2014). Central metabolic pathway enzymes like phosphoenol pyruvate and acetyl CoA are donors for additional non-carbohydrate substitutes such as pyruvate, acetate, succinate, sulfate and hydroxybutyrate. These substitutes are added onto the elongating chain of monomers at the lipid carriers' position (Coplin and Cook, 1990). Some reports indicate that biosynthesis of EPS proceeds only via NDP-sugar monomers where no lipid carriers are involved in the periplasmic region. Aloni et al. (1983) reported the production of cellulose by *Acetobacter xylinum* by this method. In some bacteria, modifications occur after monomer assembly as well, such that homopolysaccharides are modified after being translocated to the cell surface. For example, in *Azotobacter vinelandii* the enzyme mannuronan-C5 epimerase converts mannuronic acid residue to glucuronic acid at cell surface from the polymannuronic acid molecule (Whitfield, 1988). Through one of the mechanisms explained above, the EPS polymerization and assembly is complete and is attached to the surface of cells as CPS or extracellularly secreted as slime.

6.3 METAL BINDING TO EPS

Although there are several studies that report the use of bacteria for metal decontamination and recovery (Ayangbenro and Babalola, 2017; Choudhary et al., 2017; Kratochvil and Volesky, 1998), not much work has been done to understand the mechanisms involved (Volesky and Holan, 1995). Similarly, much remains to be discovered about the role of metal interaction with substances secreted extracellularly (Volesky, 1987). Essentially, the binding of metals occurs through the secretion of several metabolites such as polysaccharides outside the cell surface when there has been uptake or accumulation of ions (Beveridge, 1989; Brierley, Goyak and Brierley, 1986; Kelly and Pekar, 1979Wong and Kwok, 1992). The excellent adsorption potential of EPS has been reported by many scientists as the primary basis of metal ion removal (Loaëc, Olier and Guezennec, 1997; Ozdemir et al., 2003). It is now accepted that exopolymers are mainly composed of EPS which in turn are repeating sugar monomers. The adsorption of metal by EPS is essentially a physiochemical process that involves the interaction of metal ions with the negatively charged functional groups; the processes may further be classified as ion exchange, precipitation, complexation, andphysical adsorption. EPS is predominantly anionic owing to the negatively charged ionizable carboxylate, phosphate, acetate and infrequently, sulfate groups that interact with the metal cations (Liu and Fang, 2002; Ozdemir et al., 2003). Every microorganism may have a unique EPS composition based on the types of functional groups present; studies elucidating the structural and chemical composition of the EPS are important for understanding the mechanism of metal bioadsortion (Sharma et al., 2008). Studies indicate that among functional groups, the amino group in polysaccharides are responsible for chelation of heavy metal (Loaëc, Olier and Guezennec, 1997; Mittelman and Geesey, 1985). On the other hand, highly stable coordination complexes are formed by carboxyl and hydroxyl groups and heavy metals (Cozzi, Desideri and Lepri, 1969). Negatively charged functional groups such as those in uronic acid and phosphoryl groups in membrane and carboxylic groups of amino acids have electrostatic interactions with metal cations. There could also be cationic bond formation by polymers that are positively charged or coordination with hydroxyl groups (Gutnick and Bach, 2000). Researchers surmise that the reason microbes are resistant to heavy metals is their ability to prevent the intracellular entry of heavy metals, and this may be achieved by bioadsorption of the toxic heavy metals onto the cell membrane and the associated cell wall materials (Das et al., 2009; De Philippis, Colica and Micheletti, 2011; Naja and Volesky, 2011; Xue, Stumm and Sigg, 1988). Another mechanism involves the secretion of organic compounds such as EPS into its surroundings (Berg et al., 1979). Some microbes may be called metal tolerant, in that they can withstand large amounts of heavy metals inside the cells via uptake and accumulation. Hence, there is a distinction between heavy metal resistance and heavy metal tolerance; while one is a surface phenomenon, the other is intracellular. In *Lysinibacillus* sp. strain HG17, the removal of 100 μM $HgCl_2$ employed two mechanisms; it secreted EPS to bind the heavy metal and its cell wall also showed adsorption (François et al., 2012). Various heavy metals have different potentials with respect to affinity and specificity for binding to the EPS, for example, zinc and copper may have a lower affinity to EPS than cadmium (Gadd, 1992). The structure of EPS can be modified to bioadsorb and accommodate different types of heavy metals. Studies show that when a concoction of several metal ions is adsorbed on EPS it may hinder adsorption of other metal ions at an adjacent site (Parker et al., 2000; Sağ, Kaya and Kutsal, 2000; Tien, 2002). There may also be a synergistic interaction between metal adsorption at adjacent sites, which may be due to the alteration of binding sites due to interaction of the metals at the first binding site (Chen and Yang, 2005). Studies indicate that the degree of acetylation affects the affinity of EPS for binding to metal (Sutherland, 1983). Micheletti et al. (2008) reported that when several metal ions were involved, in cyanobacteria, the EPS–metal interaction could be synergistic, non-interactive or competitive, depending on the chemical nature and structure of EPS and the type of metal ions. Bioadsorption of metals onto the EPS-coated cell surface may have an effect on the metal homeostatsis of the microbe. Such an effect is best elucidated by Pearson's hard-soft acid base theory, according to which the EPS would act as a Lewis base and the metal ions would act as Lewis acid. The affinity of the metal ions to the EPS (ligand) can be used to classify them as class A (hard acids), class B (soft acids) and borderline (Pearson, 1963, 1968). The affinity of metal ions to EPS is influenced by the radius of the metal ions and their electronegativity (Can and Jianlong, 2007; Neiboer and Rishardson, 1980).

Pal and Paul (2008) discussed diverse types of EPS from microorganisms with various structural and chemical variations; they also compiled the types of biopolymers obtained from sundry biofilms, activated sludge, planktons and other biomass for metal remediation via adsorption and transformation.

Enzymes in the EPS play a significant role in decontamination of heavy metal via transformation and precipitation. In aquatic microbes, the EPS chelates the metal ions to itself such that both remain attached to the cell surface (Santamaría et al., 2003). The EPS, particularly those containing uronic acid, act as polyanions and form a salt bridge with negatively charged carboxyl groups. They can also form electrostatic interactions with neutral carbohydrates having hydroxyl groups. EPS can use any type of macromolecule, nucleic acid, uronic acid, protein, amino acid, for electrostatic interactions with no bearing on the selectivity of EPS towards metals (Geesey and Jang, 1989). The anionic properties of EPS can also be due to proteins rich in aspartic and glutamic amino acids that are acidic in nature (Mejáre and Bülow, 2001). On the other hand, nucleic acids are polyanionic due to the presence of phosphates imparting a negative charge to the EPS. The electrostatic interactions of EPS with multivalent cations may be due to the presence of acidic amino acids, uronic acids and nucleotides (Beech and Sunner, 2004). The bacteria may also have the potential to transform a toxic heavy metal into a less toxic or non-toxic form, as well as the EPS-mediated chelation providing added detoxification.

6.4 APPLICATION OF EPS FOR HEAVY METAL REMEDIATION

The application of EPS for metal remediation has been reported over the last four decades (Kurita et al., 1979; Norberg and Enfors, 1982). Kim et al. (1996) reported the removal of copper and lead ions by the EPS produced by *Methylobacterium organophilum*, a methylotrophic bacterium, within 30 minutes of reaction incubation at pH 7. *Herminiimonas arsenicoxydans*, a gram-negative bacterium, was reported to produce biofilm in response to arsenic in its surrounding environment and this bacterium has the potential to bind arsenic at 5mM concentration of the heavy metal (Marchal et al., 2010). Similarly, *Thiomonas* sp CB2, a betaproteobacteria, was also reported to secrete biofilm in response to the heavy metal, trapping it within its EPS (Marchal et al., 2011). Studies such as these suggest the potential of bacteria not only to produce biofilm due to heavy metal stress but also to bind the heavy metal to their EPS, removing them from their immediate environment.

Several studies report the presence of bacteria resistant to heavy metals in activated sludges (Lau et al., 2005; Ozdemir et al., 2005). Lau et al. (2005) studied the metal ion uptake potential of *Pseudomonas* sps. They reported that whole bacterial cells showed more adsorption of copper ions than isolated pure EPS. There are several other sources that harbor EPS-producing bacteria. The marine environment is one such source. Bhaskar and Bhosle (2006) reported that EPS from *Marinobacter* adsorbed copper and lead ions under neutral pH, with more propensity of adsorption for copper ions. Marine bacteria have more uronic acid in their EPS, leading to enhanced anionicity. Iyer et al. (2005) studied the potential of the marine bacterium *Enterobacter cloaceae* to chelate cadmium and copper and to some extent cobalt ions. In their study, it was observed that *E. cloaceae* EPS could adsorb 2 to 16 mg cadmium and copper, reducing the 100 mg L^{-1} load of the heavy metals by up to 65%. The bacterial cells together with the EPS contributed to 75% removal of 100 mg L^{-1} of the heavy metal (Iyer et al., 2004).

Many cyanobacteria also produce EPS that can chelate metal ions. *Anabaena spiroides* showed significant affinity towards Mn^{2+}. The Langmuir and Freundlich models for electron paramagnetic resonance (EPR) studies estimated the potential maximum Mn^{2+} complexation at 8.92 mg g^{-1} (Freire-Nordi et al., 2005). Raungsomboon et al. (2006) studied the bacterium *Gloeocapsa gelatinosa* for lead sequestration. They reported that the bacterial EPS showed higher potential for lead sequestration (82 to 86 mg) than whole cells due to the presence of high acidic sugar in their CPSs. Structure and composition of the EPS varied based on the bacterial growth phase; consequently, maximum lead decontamination was observed at the stationary phase when the net acidic sugar incorporation was at its highest. In another study, Raungsomboon et al., (2007) reported the potential of *Calothrix marchica*, a cyanobacteria, for lead removal of up to 65 mg per gram of the CPS secreted. De Philippis et al. (2007) compared the potential of CPSs of cyanobacteria *Cyanospira capsulate* and *Nostoc* sps for copper, nickel and zinc ion removal. When bacterial cells were deployed with their CPS intact, maximum adsorption was recorded for copper ions. In another study, *Gloeocapsa calcarea* and *Nostoc punctiforme* EPS showed chromium decontamination; although *Gloeocapsa calcarea* showed higher production of EPS, *Nostoc punctiforme* showed higher sequestration of chromium metal ions (Sharma et al., 2008). Kiran and Kaushik (2008) reported another cyanobacterium, *Lyngbya putealis*, whose exopolysacharides showed excellent chromium ion uptake of up to 157 mg g^{-1} of EPS at 40 mg L^{-1} of the metal ion concentration. Studies using response surface methodology (RSM) predicted optimum temperature and pH of 45°C and 2 respectively. Brunaur Emmett Teller (BET) and Langmuir models suggest the adsorption to be physically a multilayer phenomenon with coefficient of determination of 0.9925. In an interesting study, Mota et al. (2016) compared the potential of isolated EPS, live cell-bound EPS and dead cell-bound EPS of *Cyanothece* sp CCY0110 for removal of copper, cadmium and lead ions. Of the 10 mg L^{-1} initial concentration of heavy metal, isolated EPS showed the highest metal decontamination efficiency. Similarly, in *Ensifer meliloti*, the metal ion uptake was higher by the polysaccharide compared with dried cells. At an initial concentration of 50 mg L^{-1}, Langmuir isotherm estimated a maximum adsorption of 89, 85 and 66% of Pb^{2+}, Ni^{2+} and Zn^{2+} respectively. EPS at 0.02 g showed adsorption equilibrium within 20 minutes of the initiation of the reaction (Lakzian, 2008).

Ha et al. (2010) studied a comparison between the EPS production of wild and mutant *Shewenella oneidensis* using potentiometric titration. They observed that the wild type was more adept at EPS production, hence had greater potential for adsorption of zinc and lead. In the mutants, the metal ions showed more affinity towards the phosphoryl groups, potentially suggesting that a reduction in EPS may lead to metal diffusion to cell walls that are rich in phosphoryl groups. Wang et al. (2013) elucidated the bioadsorption potential of EPS secreted by *Rhizobium radiobacter* for lead and zinc ions using kinetic studies and adsorption isotherms. The bioadsorption fitted the Langmuir model and was observed to be pseudo-first order; the optimum pH for the process was estimated to be 5 and 6 for lead and zinc respectively, and additionally, the system could undergo five cycles of adsorption-desorption efficiently. EPS from another *Rhizobium* sps, *Sinorhizobium meliloti*, displayed arsenic and mercury resistance. The bacterium showed an interesting phenomenon called rescuing in which the EPS-producing, hence resistant species, showed a protective nature towards non-EPS-producing strains in a co-culture experiment (Nocelli et al., 2016). Through this study it may be deduced that even a single EPS secreting strain can confer its defensive properties on other strains, potentially creating a consortium of metal-resistant bacteria useful for application for bioremediation. Kiliç et al. (2015) isolated from a local stream *Stenotrophomonas maltophilia* EPS that displayed 40 to 70% chromium and copper ions decontamination on a starting concentration of 25 mg L^{-1}. They reported 0.74 g L^{-1} and 1.05 g L^{-1} of EPS production on account of copper and chromium exposure. Yang et al. (2015) studied a binary system using *Klebsiella* sp J1 EPS for removal of copper and zinc ions and found that copper ions were decontaminated more than zinc by the bacterium. They

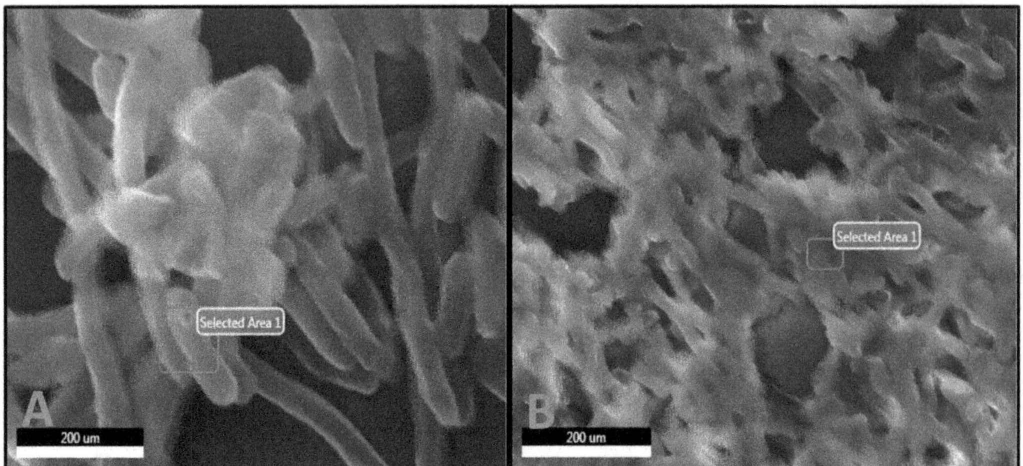

FIGURE 6.2 The SEM images of bacteria (A) under normal conditiond, and (B) under the stress of 100 ppm cadmium. The shape of bacteria, under stress of cadmium, is disturbed and bacteria form EPS to reduce/prevent the contact with metal.

found the reason behind this competitive adsorption was the creation of complex by the metal ion with some protein or similar substances potentially useful for the bacterium.

Bacillus firmus, soil bacterium, isolated by Salehizadeh and Shojaosadati (2003), showed the highest uptake of lead among copper and zinc under the influence of an acidic pH. Langmuir and Freundlich models estimated 1103, 860 and 722 mg of lead, copper and zinc adsorption per gram of EPS corresponding to 98.3%, 74.9% and 61.8% decontamination of an initial concentration of 1000 mg L^{-1} of metal ions, reaching an equilibrium within ten minutes of initiation of the reaction. Another study reported the significance of pH in EPS-mediated metal remediation; Rasulov et al. (2013) isolated *Azotobacter chroococum* from soil, and the EPS showed chelation with lead and mercury ions in a pH-dependent manner. Maximum adsorption of nearly 50% was observed at a low pH of 4 to 5. In a study conducted in our lab, scanning electron microscopy (SEM) images indicated cadmium-induced stress in bacterial structure, resulting in generation of EPS (Figure 6.2).

6.5 CONCLUSION

The potential of several bacteria to secrete EPS has created much interest in the scientific community due to its physiological significance and prospective applications. Currently bacterial EPS is being commercially used in food, cosmetics, pharmaceuticals and textiles, among other fields. The structure of EPS and their associated physical and chemical properties have proved them to be good candidates for heavy metal and organic compounds bioremediation for water treatment. Furthermore, EPS is biodegradable, biocompatible and eco-friendly. The valency of the cation and charge density governs the interaction with negatively charged EPS. In both *ex-situ* and *in-situ* studies, application of single metal or multi-metal solutions regulate the efficacy of heavy metal removal by the EPS. Furthermore, parameters such as adsorbent dosage of EPS and time duration of interaction between EPS and the heavy metal/s also play significant roles in remediation. Varying the amount of EPS secretion may have a bearing on the amount of metal ions that can be bound to it. Affinity and specificity of EPS can be modified to alter the type of metal that binds to it. High metal concentration, extreme temperature and pH may prove to be limiting factors since they can hamper the growth and sustenance of most bacteria. To overcome this challenge in such abiotic conditions, use of extracted EPS or dead cells may be effective. Additionally, bacteria that naturally inhabit extreme environments may be used. Hence, application of EPS polymers as bioadsorbents for heavy metal remediation requires stringent management of EPS secretion via bacterial multiplication. In view of the discussion in this review, it may be concluded that EPS has a promising future sustainable remediation of heavy metals from the environment.

REFERENCES

Alluri, H. K., Ronda, S. R., Settalluri, V. S., Bondili, J. S., Suryanarayana, V., & Venkateshwar, P. (2007). Biosorption: An eco-friendly alternative for heavy metal removal. *African Journal of Biotechnology 6*, 2924–2931.

Aloni, Y., Cohen, R., Benziman, M., & Delmer, D. (1983). Solubilization of the UDP-glucose: 1, 4-beta-Dglucan 4-beta-D-glucosyltransferase (cellulose synthase) from Acetobacter xylinum. A comparison of regulatory properties with those of the membrane-bound form of the enzyme. *Journal of Biological Chemistry 258*, 4419–4423.

Ayangbenro, A. S., & Babalola, O. O. (2017). A new strategy for heavy metal polluted environments: A review of microbial biosorbents. *International Journal of Environmental Research and Public Health 14*(1), 94.

Beech, I. B., & Sunner, J. (2004). Biocorrosion: towards understanding interactions between biofilms and metals. *Current Opinion in Biotechnology 15*(3), 181–186.

Berg, C. V. D., Wong, P. T. S., & Chau, Y. K. (1979). Measurement of complexing materials excreted from algae and their ability to ameliorate copper toxicity. *Journal of the Fisheries Board of Canada 36*(8), 901–905.

Beveridge, T. J. (1989). Role of cellular design in bacterial metal accumulation and mineralization. *Annual Reviews in Microbiology 43*(1), 147–171.

Bhaskar, P., & Bhosle, N. B. (2006). Bacterial extracellular polymeric substance (EPS): A carrier of heavy metals in the marine food-chain. *Environment International 32*, 191–198.

Boels, I. C., van Kranenburg, R., Hugenholtz, J., Kleerebezem, M., & de Vos, W. M. (2001) Sugar catabolism and its impact on the biosynthesis and engineering of exopolysaccharide production in lactic acid bacteria. *International Dairy Journal 11*, 723–732.

Brierley, J. A., Goyak, G. M., & Brierley, C. L. (1986) Considerations for commercial use of natural products for metal recovery. In: Eccles H., Hunt S. (Eds) *Immobilization of Ions by Biosorption*, pp. 105–117, Ellis Horwood Chichester, New York

Can, C., & Jianlong, W. (2007). Correlating metal ionic characteristics with biosorption capacity using QSAR model. *Chemosphere 69*(10), 1610–1616.

Chen, J. P., & Yang, L. (2005). Chemical modification of *Sargassum* sp. for prevention of organic leaching and enhancement of uptake during metal biosorption. *Industrial & Engineering Chemistry Research 44*(26), 9931–9942.

Choudhary, M., Kumar, R., Datta, A., Nehra, V., & Garg, N. (2017). Bioremediation of heavy metals by microbes. In: Arora, S., Singh, A., Singh, Y. (Eds) *Bioremediation of Salt Affected Soils: An Indian Perspective*, pp. 233–255 Springer International Publishing, New York.

Comte, S., Guibaud, G., & Baudu, M. (2008). Biosorption properties of extracellular polymeric substances (EPS) towards Cd, Cu and Pb for different pH values. *Journal of Hazardous Materials 151*, 185–193.

Coplin, D., & Cook, D. (1990). Molecular genetics of extracellular polysaccharide biosynthesis in vascular phytopathogenic bacteria. *Molecular Plant-Microbe Interactions 3*, 271–279.

Cozzi, D., Desideri, P. G., & Lepri, L. (1969). The mechanism of ion exchange with algenic acid. *Journal of Chromatography A 40*, 130–137.

Czaczyk, K., & Myszka, K. (2007). Biosynthesis of extracellular polymeric substances (EPS) and its role in microbial biofilm formation. *Polish Journal of Environmental Studies 16*, 799–806.

Das, B. K., Roy, A., Koschorreck, M., Mandal, S. M., Wendt-Potthoff, K., & Bhattacharya, J. (2009). Occurrence and role of algae and fungi in acid mine drainage environment with special reference to metals and sulfate immobilization. *Water Research 43*(4), 883–894.

De Philippis, R., Colica, G., & Micheletti, E. (2011). Exopolysaccharide-producing cyanobacteria in heavy metal removal from water: Molecular basis and practical applicability of the biosorption process. *Applied Microbiology and Biotechnology 92*(4), 697.

De Philippis, R., Paperi, R., & Sili, C. (2007). Heavy metal sorption by released polysaccharides and whole cultures of two exopolysaccharide-producing cyanobacteria. *Biodegradation 18*, 181–187.

Flemming, H. C., & Wuertz, S. (2019). Bacteria and archaea on earth and their abundance in biofilms. *Nature Reviews. Microbiology 17*, 247–260.

Flora, G., Gupta, D., & Tiwari, A. (2012). Toxicity of lead: A review with recent updates. *Interdisciplinary Toxicology 5*, 47–58.

François, F., Lombard, C., Guigner, J. M., Soreau, P., Brian-Jaisson, F., Martino, G., Vandervennet, M., Garcia, D., Molinier, A. L., Pignol, D., Peduzzi, J., & Peduzzi, J. (2012). Isolation and characterization of environmental bacteria capable of extracellular biosorption of mercury. *Applied and Environmental Microbiology 78*(4), 1097–1106.

Freire-Nordi, C. S., Vieira, A. A. H., & Nascimento, O. R. (2005). The metal binding capacity of *Anabaena spiroides* extracellular polysaccharide: an EPR study. *Process Biochemistry 40*, 2215–2224.

Freitas, F., Alves, V. D., & Reis, M. A. (2011a). Advances in bacterial exopolysaccharides: from production to biotechnological applications. *Trends in Biotechnology 29*, 388–398.

Freitas, F., Alves, V. D., Pais, J., Costa, N., Oliveira, C., Mafra, L., Hilliou, L., Oliveira, R., & Reis, M. A. (2009). Characterization of an extracellular polysaccharide produced by a *Pseudomonas* strain grown on glycerol. *Bioresource Technology 100*, 859–865.

Freitas, F., Alves, V. D., Torres, C. A., Cruz, M., Sousa, I., Melo, M. J., Ramos, A. M., & Reis, M. A. (2011b). Fucose-containing exopolysaccharide produced by the newly isolated *Enterobacter strain* A47 DSM 23139. *Carbohydrate Polymers 83*, 159–165.

Gadd, G. M. (1992). In: Herbert, R.A., Sharp, R.J. (Eds.), *Molecular Biology and Biotechnology of Extremophiles*. Blackie and Son Ltd, Glasgow (United Kingdom). ISBN 0-216-93153-3. 225–257.

Geesey, G. G., & Jang, L. (1989). Interactions between metal ions and capsular polymers. In: Terrance J. Beveridge, R. J. Doyle (Eds) *Metal ions and Bacteria*, pp. 325–357 Wiley, New York.

Gutnick, D. L., & Bach, H. (2000). Engineering bacterial biopolymers for the biosorption of heavy metals; new products and novel formulations. *Applied Microbiology and Biotechnology 54*(4), 451–460.

Ha, J., Gélabert, A., Spormann, A. M., & Brown, G. E. (2010). Role of extracellular polymeric substances in metal ion complexation on *Shewanella oneidensis*: batch uptake, thermodynamic modeling, ATR-FTIR, and EXAFS study. *Geochimica et Cosmochimica Acta 74*, 1–15.

Iyer, A., Mody, K., & Jha, B. (2004). Accumulation of hexavalent chromium by an exopolysaccharide producing marine *Enterobacter Cloaceae*. *Marine Pollution Bulletin 49*, 974–977.

Iyer, A., Mody, K., & Jha, B. (2005). Biosorption of heavy metals by a marine bacterium. *Marine Pollution Bulletin 50*, 340–343.

Jaishankar, M., Tseten, T., Anbalagan, N., Mathew, B. B., & Beeregowda, K. N. (2014). Toxicity, mechanism and health effects of some heavy metals. *Interdisciplinary Toxicology 7*, 60–72.

Kelly, R. A., & Pekar, H. G. (1979). U.S. Patent No. 4,165,242. Washington, DC: U.S. Patent and Trademark Office.

Kiliç, N. K., Kürkçü, G., Kumruoğlu, D., & Dönmez, G. (2015). EPS production and bioremoval of heavy metals by mixed and pure bacterial cultures isolated from Ankara Stream. *Water Science and Technology 72*, 1488–1494.

Kim, S.-Y., Kim, J.-H., Kim, C.-J., & Oh, D.-K. (1996). Metal adsorption of the polysaccharide produced from *Methylobacterium organophilum*. *Biotechnology Letters 18*, 1161–1164.

Kiran, B., & Kaushik, A. (2008). Chromium binding capacity of *Lyngbya putealis* exopolysaccharides. *Biochemical Engineering Journal 38*, 47–54.

Kratochvil, D., & Volesky, B. (1998). Advances in the biosorption of heavy metals. *Trends in Biotechnology 16*(7), 291–300.

Kurita, K., Sannan, T., & Iwakura, Y. (1979). Studies on chitin. VI. binding of metal cations. *Journal of Applied Polymer Science 23*, 511–515.

Lakherwal, D. (2014). Adsorption of heavy metals: A review. *International Journal of Environmental Research and Development 4*, 41–48.

Lakzian, A. (2008). Adsorption capability of lead, nickel and zinc by exopolysaccharide and dried cell of *Ensifer meliloti*. *Asian Journal of Chemistry 20*, 6075–6080.

Lau, T., Wu, X., Chua, H., Qian, P., & Wong, P. (2005). Effect of exopolysaccharides on the adsorption of metal ions by *Pseudomonas* sp. CU-1. *Water Science and Technology 52*, 63–68.

Liu, H., & Fang, H. H. (2002). Characterization of electrostatic binding sites of extracellular polymers by linear programming analysis of titration data. *Biotechnology and Bioengineering 80*(7), 806–811.

Loaëc, M., Olier, R., & Guezennec, J. (1997). Uptake of lead, cadmium and zinc by a novel bacterial exopolysaccharide. *Water Research 31*(5), 1171–1179.

Madhuri, K., & Prabhakar, K. V. (2014). Microbial Exopolysaccharides: Biosynthesis and Potential Applications. *Oriental Journal of Chemistry 30*, 1401–1410.

Marchal, M., Briandet, R., Koechler, S., Kammerer, B., & Bertin, P. N. (2010). Effect of arsenite on swimming motility delays surface colonization in *Herminiimonas arsenicoxydans*. *Microbiology 156*, 2336–2342.

Marchal, M., Briandet, R., Halter, D., Koechler, S., DuBow, M. S., Lett, M.-C., & Bertin, P. N. (2011). Subinhibitory arsenite concentrations lead to population dispersal in *Thiomonas sp*. *PloS One 6*, e23181.

Mejáre, M., & Bülow, L. (2001). Metal-binding proteins and peptides in bioremediation and phytoremediation of heavy metals. *Trends in Biotechnology 19*(2), 67–73.

Micheletti, E., Colica, G., Viti, C., Tamagnini, P., & De Philippis, R. (2008). Selectivity in the heavy metal removal by exopolysaccharide-producing Cyanobacteria. *Journal of Applied Microbiology 105*(1), 88–94.

Mittelman, M. W., & Geesey, G. G. (1985). Copper-binding characteristics of exopolymers from a freshwater-sediment bacterium. *Applied and Environmental Microbiology 49*(4), 846–851.

Mota, R., Rossi, F., Andrenelli, L., Pereira, S. B., De Philippis, R., & Tamagnini, P. (2016). Released polysaccharides (RPS) from *Cyanothece* sp. CCY 0110 as biosorbent for heavy metals bioremediation: interactions between metals and RPS binding sites. *Applied Microbiology and Biotechnology 100*(17), 7765–7775.

Naja, G., & Volesky, B. (2011). The mechanism of metal cation and anion biosorption. In: Kotrba, P., Mackova, M., Macek, T. (Eds) *Microbial Biosorption of Metals*, pp. 19–58, Springer, Dordrecht.

Neiboer, E., & Rishardson, D. H. (1980). The replacement of the nondescript term 'heavy metals' by a biologically and chemically significant classification of metal ions. *Environmental Pollution. Series B: Chemical and Physical 1*(1), 3–26.

Nocelli, N., Bogino, P. C., Banchio, E. & Giordano, W. (2016). Roles of extracellular polysaccharides and biofilm formation in heavy metal resistance of *Rhizobia*. *Materials 9*, 418.

Norberg, A. B., & Enfors, S.-O. (1982). Production of extracellular polysaccharide by *Zoogloea ramigera*. *Applied and Environmental Microbiology 44*, 1231–1237.

Ozdemir, G., Ceyhan, N., & Manav, E. (2005). Utilization in alginate beads for Cu (II) and Ni (II) adsorption of an exopolysaccharide produced by *Chryseomonas luteola* TEM05. *World Journal of Microbiology and Biotechnology 21*, 163–167.

Ozdemir, G., Ozturk, T., Ceyhan, N., Isler, R., & Cosar, T. (2003). Heavy metal biosorption by biomass of *Ochrobactrum anthropi* producing exopolysaccharide in activated sludge. *Bioresource Technology 90*(1), 71–74.

Pal, A., & Paul, A. K. (2008). Microbial extracellular polymeric substances: central elements in heavy metal bioremediation. *Indian Journal of Microbiology 48*(1), 49.

Parker, D. L., Mihalick, J. E., Plude, J. L., Plude, M. J., Clark, T. P., Egan, L., et al. (2000). Sorption of metals by extracellular polymers from the cyanobacterium *Microcystis aeruginosa* fo. flos-aquae strain C3-40. *Journal of Applied Phycology 12*(3–5), 219–224.

Pearson, R. G. (1963). Of the American Chemical Society. *Journal of the American Chemical Society 85*(22), 3533–3539.

Pearson, R. G. (1968). Hard and soft acids and bases. HSAB, part II: Underlying theories. *Journal of Chemical Education 45*(10), 643.

Rajendran, P., Muthukrishnan, J., & Gunasekaran, P. (2003). Microbes in heavy metal remediation. *Indian Journal of Experimental Biology 41*, 935–944.

Rani, M. J., Hemambika, B., Hemapriya, J., & Kannan, V. R. (2010). Comparative assessment of heavy metal removal by immobilized and dead bacterial cells: A biosorption approach. *African Journal of Environmental Science and Technology 4*, 077–083.

Raungsomboon, S., Chidthaisong, A., Bunnag, B., Inthorn, D., & Harvey, N. W. (2006). Production, composition and Pb2+ adsorption characteristics of capsular polysaccharides extracted from a cyanobacterium *Gloeocapsa gelatinosa*. *Water Research 40*, 3759–3766.

Rasulov, B., Yili A., & Aisa H. (2013) Biosorption of Metal Ions by Exopolysaccharide Produced by *Azotobacter chroococcum* XU1. *Journal of Environmental Protection* 4 (9), 989–993.

Rehm, B. (2009). *Microbial Production of Biopolymers and Polymer Precursors: Applications and Perspectives.* Caister Academic, Norwich.

Rehm, B. H. A. (2010). Bacterial polymers: Biosynthesis, modifications and applications. *Nature Reviews. Microbiology 8*, 578–592.

Ruangsomboon, S., Chidthaisong, A., Bunnag, B., Inthorn, D., & Harvey, N. W. (2007). Lead (Pb^{2+}) adsorption characteristics and sugar composition of capsular polysaccharides of cyanobacterium *Calothrix marchica*. Songklanakarin. *Journal of Science and Technology 29*, 529–541.

Sağ, Y., Kaya, A., & Kutsal, T. (2000). Lead, copper and zinc biosorption from bicomponent systems modelled by empirical Freundlich isotherm. *Applied Microbiology and Biotechnology 53*(3), 338–341.

Salehizadeh, H., & Shojaosadati, S. (2003). Removal of metal ions from aqueous solution by polysaccharide produced from *Bacillus firmus*. *Water Research 37*, 4231–4235.

Santamaría, M., Díaz-Marrero, A. R., Hernández, J., Gutiérrez-Navarro, A. M., & Corzo, J. (2003). Effect of thorium on the growth and capsule morphology of *Bradyrhizobium*. *Environmental Microbiology 5*(10), 916–924.

Sharma, M., Kaushik, A., Bala, K., & Kamra, A. (2008). Sequestration of chromium by exopolysaccharides of *Nostoc* and *Gloeocapsa* from dilute aqueous solutions. *Journal of Hazardous Materials 157*, 315–318.

Sheng, G.-P., Yu, H.-Q., & Li, X.-Y. (2010). Extracellular polymeric substances (EPS) of microbial aggregates in biological wastewater treatment systems: a review. *Biotechnology Advances 28*, 882–894.

Sutherland, I.W. (1983). Extracellular polysaccharides. In: Rehm, H.J., Reed, G. (Eds), *Biotechnology: Biomass, Microorganisms for Special Applications, Microbial Products I, Energy from Renewable Resources*, pp. 531–574, John Wiley & Sons, New York.

Sutherland, I. W. (2001) Microbial polysaccharides from Gram-negative bacteria. *International Dairy Journal 11*, 663–674.

Tien, C. J. (2002). Biosorption of metal ions by freshwater algae with different surface characteristics. *Process Biochemistry 38*(4), 605–613.

Volesky, B. (1987). Biosorbents for metal recovery. *Trends in Biotechnology 5*(4), 96–101.

Volesky, B., & Holan, Z. R. (1995). Biosorption of heavy metals. *Biotechnology Progress 11*(3), 235–250.

Wang, J., & Chen, C. (2006). Biosorption of heavy metals by *Saccharomyces cerevisiae*: A review. *Biotechnology Advances 24*, 427–451.

Wang, L., Yang, J., Chen, Z., Liu, X., & Ma, F. (2013). Biosorption of Pb (II) and Zn (II) by extracellular polymeric substance (Eps) of *Rhizobium radiobacter*: Equilibrium, kinetics and reuse studies. *Archives of Environmental Protection 39*, 129–140.

Wang, S., & Zhao, X. (2009). On the potential of biological treatment for arsenic contaminated soils and groundwater. *Journal of Environmental Management 90*, 2367–2376.

Whitfield, C. (1988). Bacterial extracellular polysaccharides. *Canadian Journal of Microbiology 34*, 415–420.

Wong, P. K., & Kwok, S. C. (1992). Accumulation of nickel ion by immobilized cells of *Enterobacter* species. *Biotechnology Letters 14*(7), 629–634.

Xue, H. B., Stumm, W., & Sigg, L. (1988). The binding of heavy metals to algal surfaces. *Water Research 22*(7), 917–926.

Yang, J., Wei, W., Pi, S., Ma, F., Li, A., Wu, D., & Xing, J. (2015). Competitive adsorption of heavy metals by extracellular polymeric substances extracted from *Klebsiella* sp. J1. *Bioresource Technology 196*, 533–539.

7 Biosurfactants
A Greener Alternative for a Sustainable Future

Neha Dhingra
University of Delhi, New Delhi, India

CONTENTS

DOI: 10.1201/9781003306931-8

7.1 INTRODUCTION

Biosurfactants are amphiphilic compounds produced on the surfaces of microbial cells that contain distinct polar and non-polar moieties which facilitate formation of micelles that accumulate at the interface between fluids of different polarities, such as water and oil, and then have the ability to reduce surface pressure (Kumar and Das, 2018). Surface-active chemicals produced naturally have gained in popularity over the years, with the number of publications devoted to the extraction, characterization, and refinement of biosurfactant producers steadily growing (due to their potential applications in a variety of disciplines (Açıkel, 2011 Franzetti et al., 2014b; Akbari et al., 2018; Fenibo et al., 2019). Because of their hydrophobic and hydrophilic characteristics, these natural biomolecules have excellent emulsion-generating capabilities (Shah et al., 2016).

Surfactants are widely used in farming, beauty products, pharmaceuticals, nutrition and engineering; due to their refractory and defined character, the majority of the mixtures are synthetically concocted which may lead to environmental and toxicological issues (Vijayakumar and Saravanan, 2015). The microbial cell surface, polysaccharide-lipid composite, mycolic corrosive, phospholipid, lipoprotein/lipopeptide, or glycolipids can be accompanying components (Shekhar et al., 2015). Tiny organisms such as microbes and yeast are capable of providing biosurfactants.

Biosurfactants were first recognized as hydrocarbon disintegration agents in the 1960s, and in the last five decades their uses have expanded dramatically as a superior alternative to compound surfactants, notably in the food, pharmaceutical, and oil industries. Every year, it is estimated that well over 10 million tonnes of complex surfactants and microbial biosurfactants are supplied. Biosurfactants have several advantages over their chemical counterparts: their composition is simpler than that of synthetic surfactants; synthetic surfactants cannot withstand high salt, but biosurfactants may withstand up to 10% salinity (Shekhar et al., 2015).

As a result, they are used in a number of commercial items, including medicines, cosmetics, cleaning agents, and the food sector. Due to their chemical origin, artificial surfactants are not recommended for use in food and cosmetics (Edwards et al., 2003; Santos et al., 2016). Regardless of the kind and quantity of microbial surfactants used, production is mostly determined by the living being that creates them; other variables such as nitrogen and carbon, temperature, air circulation, and follow-up constituents also impact the life form's production (Kumar and Das, 2018). All commercial surfactants are made from finite oil-based resources, and because of their volatile nature, these are much more costly as well as raising environmental concerns. Microbial surfactants represent the best solutions for surfactant creation. Biosurfactants have unique characteristics that allow them to be substituted for or added to manufactured surfactants. Biosurfactants have a greater surface action, lower toxicity, and better ecological compatibility than synthetic surfactants, as well as having compound diversity and being more biodegradable. Biosurfactants are commonly used in environmental applications, as anticancer agents, as antibacterial medication in the cosmetics sector, and as adhesives to tiny organisms and yeasts for therapeutic purposes, due to their high (Santos et al., 2016) surface movement and natural resemblance. Biosurfactant-producing microorganisms are typically found in oil-depleted soil. Hydrocarbons with complex compound structures, such as aliphatic and sweet-smelling hydrocarbons, are present in significant amounts in oil-polluted conditions. Microorganisms utilize biosurfactants for their emulsifying action and utilize hydrocarbons like a substrate to render them harmless.

Several individuals have attempted to produce biosurfactants using unusual and agro-based raw materials, but this technique has yet to be commercialized. To save costs, these lipopeptides may be made from low-cost raw materials such as sugarcane molasses (Hippolyte et al., 2018), maize steep liquor (López-Prieto et al., 2019), agro-industrial wastes (Kumar and Das, 2018), and some that are readily accessible.

Worldwide demand for biosurfactants is expected to exceed US$2.4 billion by 2025, according to research published by World Market Insights, Inc. in 2019. The biosurfactants market is likely to develop as the cosmetics industry expands and people become more health-conscious and concerned about the environment. Consumers prefer organic or bio-based skincare over synthetic skincare, therefore there will be plenty of potential market (Rivera et al., 2019).

Extremophile biosurfactants recently attracted widespread attention due to their multiple persistent characteristics (Mehetre et al., 2019), which have been found to be preserved even under harsh environmental circumstances. They can withstand extreme temperatures (30–100°C), a pH range of 2–12, and salt levels of up to 10% (Cameotra and Singh, 2004; Santos et al., 2016).

Due to their advantageous characteristics, biosurfactants have been found to be great candidates for application in the food sector (Sanchita and Pritisnigdha, 2019), agriculture (Chen et al., 2017; Singh et al., 2019; Naughton et al., 2019), soil (Lal et al., 2018; Karlapudi et al., 2018) and water remediation (Karlapudi et al., 2018; Olasanmi and Thring, 2018; Ravindran et al., 2020), enhancement of oil recovery (Putra and Hakiki, 2019), the biomedical field (Naughton et al., 2019), nanoscience (Pasternak et al., 2020) and various sectors involving applications in detergent and cleaning solutions (Fei et al., 2019).

Following recent advances in biotechnology, alternative environmentally friendly methods for the generation of several types of biosurfactants from microbes have been studied (Drakontis and Amin, 2020). The aim of this study is to provide recent data on types and uses of biosurfactants in a variety of sectors, as well as a leap towards renewable technology.

7.2 PROPERTIES OF BIOSURFACTANTS

Biosurfactants have been found to offer better properties than their chemically generated counterparts, as well as a wide spectrum of substrate availability, making them appropriate for commercial usage. Microbial surfactants have properties such as pH, low emulsifying and demulsifying capacity, surface movement, temperature, biodegradability, hazard resistance, ionic quality resilience, and antimicrobial activity. (Singh et al., 2019). The most important aspects of each biosurfactant characteristic are discussed below.

7.2.1 Surface and Interface Activity

Surfactant contributes to the reduction of surface tension and interfacial pressure. Surfactin generated by *B. subtilis* may reduce water surface tension to 25 mN m^{-1} and water/hexadecane interfacial strain to less than 1 mN m^{-1} (Roy, 2018). Biosurfactants are much more powerful and efficient than chemical surfactants, and their critical micelle concentration is many times lower, meaning that less surfactant is needed for maximal surface strain reduction (Sanchita and Pritisnigdha, 2019).

7.2.2 Temperature and pH Tolerance

Much attention has been paid to extremophile biosurfactant manufacturing in the last ten years because of its economic potential. Many biosurfactants can be used at high temperatures and pH values ranging from 2 to 12. Biosurfactants can withstand a salt content of up to 10%, but manufactured surfactants are inactivated by just 2% NaCl. Biosurfactants are resilient across a wide range of temperatures and pH levels. Lichenysin from *B. licheniformis* JF-2 is stable at 50°C, pH 4.5–9.0, and 50 g/L and 25 g/L NaCl and Ca concentrations, respectively. Natural factors such as temperature and pH have little effect on biosurfactants and their surface activity. *Arthrobacter protophormiae* biosurfactant has been shown to be thermostable (30–100°C) and pH stable (2 to 12). Because industrial operations involve temperature, pH, and weight extremes, it is critical to isolate innovative microbial things that are ready to perform in these situations (Roy, 2018).

7.2.3 Biodegradability

Microbially generated molecules are quickly degraded, especially in comparison to chemical surfactants, and suitable bioremediation and biosorption are examples of natural uses. Growing environmental concerns are forcing us to seek alternatives, such as biosurfactants (Sanchita and Pritisnigdha, 2019). Degradable biosurfactants produced from microbial sources have already been used, while synthetic chemical surfactants have been linked to environmental difficulties in the biosorption of poorly solvent polycyclic sweet-smelling hydrocarbon phenanthrene contaminated in aqueous surfaces (Olasanmi and Thring, 2018).

7.2.4 Low Toxicity

There is little documentation on the toxic nature of these biosurfactants, which are believed to be from low to mostly non-harmful substances that can be suitable for medicinal, remedial, and nutritional purposes. Biosurfactant sophorolipids from *Candida bombicola* have a reduced toxicity profile, making them useful in nutrition (Roy, 2018). They have no detrimental effects on the heart, lung, kidney and liver. They pose no threat to blood coagulation. Studies have found the toxicity of chemically produced surfactant with LC 50 against *Photobacterium phosphoreum* is ten times lower than that of rhamnolipids.

7.2.5 Emulsion Framing and Emulsion Breaking

Biosurfactants include emulsifiers and de-emulsifiers. A heterogeneous environment made up of one immiscible fluid scattered as beads in another, having a diameter of at least 0.1 mm, appears to be an emulsion. Emulsions are divided into two types: oil-in-water and water-in-oil. They have a low degree of stability which can be accounted for with other chemicals like biosurfactants, and they can be kept stable for months or even years as emulsions (Hippolyte et al., 2018).

7.2.6 Anti-adhesive Agents

A biofilm is a bacterial or other organic material coating that has formed on a surface (Roy, 2018). Bacterial adhesion to the exterior is the first process in biofilm development, and this is affected by a multitude of variables including microorganism characteristics. Their capacity to supply extracellular polymers that help cells grapple to substrates, external hydrophobicity and electrical charges, environmental conditions, and ability to deliver extracellular polymers that help cells grapple to substrates are all factors to consider. Biosurfactants might be utilized to alter the environment, or the hydrophobicity of surface, which affects how microorganisms engage with it. (Kubicki et al., 2019).

7.3 TYPES OF BIOSURFACTANTS

7.3.1 Glycolipids

Glycolipids are microbiological surface-active molecules made up of a carbohydrate moiety coupled to fatty acids and generated by a variety of bacteria. They can minimize surface and interfacial tension at the surface and interface and have wide structural variation. They can also be utilized as antibacterial, antifungal, and hemolytic agents because of their ability to generate holes and damage cellular membranes. They can be employed as therapeutic agents in pharmaceuticals because of their antiviral and anticancer characteristics. Glycolipids can also inhibit the bioadhesion of harmful microorganisms, making them useful as anti-adhesive agents and also for disrupting biofilm development in the cosmetics sector. (Roy, 2018).

7.3.2 Rhamnolipids

Rhamnolipids appear to be glycolipids that contain between one and two rhamnose molecules and one or two hydroxydecanoic acid molecules. The main glycolipids generated by *P. aeruginosa* are biosurfactants, which have been extensively investigated (Roy, 2018).

7.3.3 Sophorolipids

Yeasts produce these glycolipids, which are made up of a dimeric carbohydrate called sophorose that is glycosidically linked to a long-chain hydroxyl fatty acid. Sophorolipids, which are generally a mixture of at least six to nine different hydrophobic sophorolipids and the lactone form of the sophorolipid, are better suited to a variety of applications (Roy, 2018).

7.3.4 Trehalolipids

When liquid hydrocarbons are present inside the growing media, species of the *Rhodococcus* genus generate biosurfactants. Cell-bound glycolipids with trehalose as the carbohydrate are by far the most common biosurfactants. Certain glycolipids have a variety of physiological functions, including absorption of water-insoluble substrates, cell attachment to hydrophobic surfaces, and enhanced *Rhodococcal* resilience to physicochemical effects. *Rhodococcus* biosurfactants interact favorably with the other microbial and synthetic surfactants in respect of surfactant properties (e.g., surface and interfacial tension, emulsifying activity, critical micelle concentration). Biological activity of *rhodococci* trehalolipids was also found to comprise immunomodulating, anticancer, and anti-adhesive capabilities (Roy, 2018).

7.3.5 Surfactin

Bacillus subtilis incorporates this as a standout amongst the most promising biosurfactants. Surfactin is a lipopeptide antibiotic comprised of seven amino acids and one-hydroxy fatty acid with a lengthy fatty acid component. D-configurations are seen in three of those. It's a biosurfactant that induces erythrocytes and microorganisms to lyse. It also acts as a coagulation blocker. Surfactin's physicochemical interactions with the membrane surface are considered to destabilize or disassemble membranes. (Roy, 2018).

7.3.6 Lichenysin

Bacillus licheniformis incorporates many biosurfacants that work together to provide exceptional salt, temperature, and pH security. They are also comparable to surfactin in terms of auxiliary along with their physio-synthetic characteristics (Roy, 2018).

7.3.7 Fatty Acids, Phospholipids and Neutral Lipids

During the development of n alkanes, some bacteria and yeast produce large amounts of unsaturated fats as well as phospholipid surfactants. Phosphatidyl ethanolamine-rich vesicles are supplied by *Acinetobacter spp.* 1-N, and in water, they produce optically clear smaller-sized alkane emulsions. For therapeutic uses, these biosurfactants are required. The lack of phospholipid protein complex, is the fundamental cause of breathing difficulties in premature babies. They also suggested that the traits that determine the synthesis of these surfactants be separated and cloned and employed in the fermentation process.

7.3.8 Bio-emulsifiers

Bacteria, yeast, and fungus generate bio-emulsifiers, which are high-molecular-weight molecules made up of a complicated combination of lipopolysaccharides, heteropolysaccharides, proteins and lipoproteins, that can stick to the cell membrane or be discharged. Different microorganisms that generate bio-emulsifiers provide a wide range of physicochemical characteristics. Emulsan is a D-galactose-amine-containing lipoheteropolysaccharide polymer formed even during the stationary phase. The highest concentration was reached when a culture medium containing 12 carbon-based fatty acids was used as the carbon source. Emulsan may be made using a variety of fermentation processes, including batch, chemo-stat, immobilized cell system, and self-cycling fermentation (Saimmai et al., 2019).

7.3.9 Particulate Biosurfactants

Extracellular film vesicles are vital in microbial alkane absorption because they partition hydrocarbons from a microemulsion. Protein, phospholipids, and lipopolysaccharide are found in *Acinetobacter sp.* vesicles with a diameter of 20–50 nm and a light thickness of 1.158 cg/cm3 (Roy, 2018).

7.4 APPLICATIONS OF BIOSURFACTANTS

7.4.1 Pharmaceuticals and Therapeutics

Biosurfactants can be used in a wide range of medicinal applications. Biosurfactants appear as antibacterial, antifungal, and antiviral operators, as well as safe modulator particles, antibodies, and quality therapy. Several biosurfactants have been shown to have antibacterial activity against a variety of microscopic organisms, as well as green growth, parasites, and diseases (Kubicki et al., 2019).

7.4.2 Antimicrobial Action

Biosurfactants have a harmful impact on the ozone layer penetrability due to their distinct structures. Some have powerful antifungal, antibacterial, and antiviral effects; these surfactants act as anti-cement experts to microorganisms, making them beneficial for treating a wide range of diseases as well as medicinal and probiotic applications. The marine *B. circulans* biosurfactant had a good antibacterial impact against Gram-positive and Gram-negative pathogens, as well as semi-pathogenic microorganisms including MDR strains (Nikolova and Gutierrez¸2019).

7.4.3 Anticancer Property

Certain microorganisms with extracellular glycolipids cause cell separation instead of cell growth in the human promyelocytic leukemia cultured cells. Likewise, MEL enhanced acetylcholine esterase movement and infiltrated into the cell cycle at the G1 stage, culminating in an excess of neurites as well as inadequate cell separation in PC 12 cells (Ohadi et al., 2020).

7.4.4 Antiviral Activity

The basic analogs of *C. bombicola* sophorolipid spermicidal, anti-HIV, and cytotoxic activities were all investigated. Diacetate ethyl ester derivative has been found to be by far the most effective spermicidal and virucidal speciality (Nikolova and Gutierrez, 2019).

7.4.5 Cosmetics Industry

Sophorolipids are frequently employed in beauty products. They contain anti-oxidant qualities, stimulate dermal fibroblast metabolism, and have hygroscopic properties that help maintain good skin physiology. Various cosmetics, facial creams, beauty washes, and hair products are among the long-term prospects for sophorolipid-based products (Roy, 2018).

7.4.6 Oil Industry

Biosurfactants and bio-emulsifiers are unique atom gatherings that are among the most effective and adaptable outcomes that sophisticated microbial technology can provide in fields such as bio-consumption and hydrocarbon biofouling inside oil storage tanks, catalysts and biocatalysts for oil. Biosurfactants also play an important role in oil extraction, transportation, overhauling and refining, and in the production of petrochemicals (Nikolova and Gutierrtz, 2020).

7.4.7 Agriculture

Biosurfactants are also utilized in agriculture and are important in microbiological biocontrol components including parasitism, competition, and antibiosis-induced basic protection. Surfactants can be used as preparation specialists to aid in the solubilization of bio-hazardous chemical mixtures such as PAH, enhancing solvency of hydrophobic organic contaminants (HOCs). Surfactants are also thought to make it easier for organisms to adsorb to surroundings affected by toxins, reducing the gap between source of administration and the location of microbe biouptake. Surfactants are, however, utilized in horticulture to promote high flowability and even manure absorption by hydrophilizing contaminated soils. They also inhibit solidifying of specific compost during capacity and hasten the diffusion and entrance of herbicide toxic elements (Nikolova and Gutierrtz, 2020). The antibacterial effects of the rhamnolipid biosurfactant, which is primarily generated by *Pseudomonas* species, are well established. Moreover, rhamnolipid biosurfactants are not thought to have any detrimental impact on humans and the ecosystem as a whole. Antifungal activities were also discovered in fengycins, indicating that they might be used in the biocontrol of phytopathogens.

7.4.8 Commercial Laundry Detergents

Artificial surfactants are an important constituent in today's commercial detergents. However, they are damaging to young aquatic life. The search for environmentally safe, characteristic alternatives to complex surfactants in detergents has been re-energized by growing public awareness of the environmental problems and risks associated with them (Nikolova and Gutierrtz, 2020). Biosurfactants like cyclic lipopeptide (CLP) are robust over a wide pH range (7.0–12.0) and retain their surface-dynamic characteristics when subjected to heat. They showed superb emulsion production capacity with vegetable oils, as well as exceptional strength and similarity to professional laundry detergents.

7.4.9 Phytoremediation

Heavy metals are the most problematic of the inorganic contaminants, owing to the excessive contamination. When consumed in excessive amounts, they release free radicals and induce oxidative stress. Researchers substituted several fundamental proteins and colors, which affected their usual function. In any event, utilizing both biosurfactant-producing and tiny metal-safe organisms can boost the plant's phytoremediation capability. The biosurfactant-producing *Bacillus sp.* J119 strain, for example, can boost the productivity of assault, sundagrass, tomato, and maize plants as well as cadmium uptake. It is obvious from this analysis that the species selected for this purpose has a root colonization effect. As a result, an organism that aids the phytoremediation process is established for cleaning up toxic metals (Roy, 2018).

7.4.10 Soil Washing

This is an ex-situ remedial procedure that involves washing polluted soil with a solution that is typically laced with chemicals to remove hazardous substances from (excavated) soil. Soil washing is used to remove pollutants from fine-grained soils such as clay, sand, silt and gravels (Abdel-Moghny et al., 2012). The wastewater may subsequently be cleaned and discarded, with the washed soil used as backfill at the excavated location. Fuels, semi-volatile organic chemicals, pesticides, and metals may all be removed by this method. Water, water/surfactants, bases or acids, water/chelating agents, or organic solvents can all be used as washing fluids (Hazrina et al., 2018), depending on the pollutant in question. Soil washing is the most effective way to eliminate organic compound waste and chlorinated hydrocarbon pollutants. Surfactants reduce the surface tension between the contaminant and the soil particles, improving the solubility of non-aqueous-phase liquids (NAPLs). Tween 80 (a non-anionic model), alkylbenzyldimethylammonium chloride and sodium dodecyl sulfate (an anionic model) are all common synthetic surfactants used for dirt cleaning (a cationic model). Biosurfactants, on the other hand, have recently been employed in soil-cleaning technology due to their environmental benefits (Mao et al., 2015).

Rhamnolipids have indeed been proven to be effective soil-washing reagents for removing hydrocarbons and metals. Mobilization and solubilization (Zhong et al., 2016) lead in rhamnolipid-enhanced soil washing aimed at hydrocarbons, allowing for greater contaminant partitioning in the aqueous phase and simpler separation of contaminants from solid particles (Lai et al., 2009).

Concentration affects monorhamnolipid sorption on soil matrix components, according to Ochoa-Loza et al. (2007), and the monorhamnolipid produced sorbs more effectively alone than mixed rhamnolipids. In a comparison study, Lai et al. (2009) found that rhamnolipid eliminated total petroleum hydrocarbons from severely contaminated soil up to 63 percent more effectively than surfactin (62 percent), Triton-100 (40 percent), and Tween 80 (35 percent). Although significant recognition has been reported with biosurfactant administration in soil cleaning, soil sorption remains a significant barrier to biosurfactant application.

An anionic biosurfactant will work much better as just a cleaning agent than just a cationic or nonionic surfactant, and anionic effluent would be easier to destabilize via the charge neutralization procedure. A desorbing medium other than a biosurfactant (with a comparable structure and composition) could be used in combination with the biosurfactant to promote hydrocarbon desorption from polluted soils (Yu et al., 2014). In the case of a loss of concentration, a calculated homogeneous biosurfactant can be introduced to the washing system on a regular basis to fulfill the remediation goal (Sandrin and Maier, 2002). Rhamnolipids are one of the biosurfactants that have met the bulk of these requirements for soil cleansing. These conclusions are supported by a significant body of literature on the use of rhamnolipid for dirt cleansing.

7.4.11 Metal Bioremediation

Metals are long-lasting soil pollutants that pose a variety of health risks to animals and people. Metal toxicity has been related to physical and mental retardation, birth abnormalities, cancer, kidney and liver damage, learning difficulties, and other health problems (Wuana and Okieimen, 2011). Landfilling has been used to remediate soil polluted with hazardous metals such as cadmium, lead, chromium and zinc (Santos et al., 2016). Because conventional remediation is becoming more expensive, microorganisms are being used to effect in-situ remediation of metal-contaminated surface and subsurface soils. Surfactant use for both organics and metals has the same goal: to make the pollutant more soluble so that it may be removed more easily by decomposition or flushing. However, there are significant distinctions between metal-contaminated and organic-contaminated soils. Heavy metals, with the exception of organic pollutants, are not degradable and are commonly found as cationic species (Ochoa-Loza et al., 2001). Metal contaminants can be eliminated or immobilized in two ways, altering their mobility and toxicity potency as a result of being changed from one chemical state to another, either through a redox process or alkylation (Mulligan et al., 1999; Zaghloul and Saber, 2019).

Similar to other pollutant remediation, eco-friendly techniques in metal remediation using renewable resources such as plants, microbes, and biosurfactants, is presently being studied.

Metal remediation caused by biosurfactants uses a variety of processes, comprising sorption, desorption, and complexation (Franzetti et al., 2014a) To aid the dispersion and solubilization of metals along with desorption in contaminated groundwater, microbial surfactants are being used in pump-and-treat procedures and soil washing. Precipitation–dissolution, electrostatic contact, counter-ion association, and ion exchange are all processes that drive biosurfactant–metal binding (Bhaskar and Bhosle, 2006). Because pollutant sorption is dependent both on soil and on the metal's chemical composition, the surfactant employed for contaminant complexation is critical (Nikolova and Gutierrez, 2019). The inclusion of a biosurfactant might help heavy metals desorb from their solid phases. In theory, an anionic biosurfactant forms electrovalent bonds with metals, resulting in stronger nonionic complexes than those generated between soil and metal. The biosurfactant-formed complexes desorb from the soil matrix and move to the soil solution, where they are incorporated into micelles (Burakov et al., 2017). Surfactins and rhamnolipids have already been identified as attractive biosurfactants for metal remediation. The inclusion of appropriate microbes in the metal remediation process would help improve effectiveness of the procedure. Microorganisms can indirectly affect metal mobility by changing pH or stimulating chemicals, both of which can alter metal movement. A number of species, including yeast, filamentous fungus, and bacteria, have been found to decrease nickel toxicity by raising pH (Rufino et al., 2012). The high pH circumstances and the ability of microbes to acquire or adsorb a high amount of metal ions in a metabolism-dependent manner are two explanations for this detoxification process (Ohadi et al., 2020).

7.5 FUTURE ASPECTS

The expense and large-scale manufacturing of biosurfactants are the primary bottlenecks. Whey, a waste product from the agricultural, culinary, and dairy industries, might be used as a substrate. Invert sugar, vitamins, sucrose, amino acids, and minerals are all abundant in whey. Potato plant effluent, banana peel, bagasse, coconut shells, straw, and other food wastes are examples of agricultural waste. The fermentation process and material recycling are the main operational expenses in their manufacture. Production under nonsterile conditions, if possible, would greatly lower production costs. Furthermore, because the downstream process and purification phase are expensive, development in this sector is critical for lowering manufacturing costs. Technical advances in bioreactor design and process management are needed to enable full-scale manufacturing. Examining the longevity and assessing the life cycle of biosurfactants is another topic that has been highlighted for further investigation. In this case, the research effort must focus on all phases, from generation and consumption to utilization. Research also should take into account significant aspects including toxicity, carbon footprint and resource depletion. To boost output, the raw material can be low cost with readily accessible stimulators or commercial components, and can be used to boost microbiological growth and yield. It's also possible to employ mutant strains with documented genomes and metabolic data (Singh et al., 2019). Furthermore, using biosurfactant and nanoparticle mixtures for ecological cleaning (Amani, 2017) and medication delivery has achieved impressive outcomes; hence, nanoparticles and biosurfactants mixtures can also be used in various domains (Zgorova and Ananikov, 2017). The creation of nano-scale biosurfactants to improve their activity could be another amazing future project (Rawat and Kumar, 2020). These are versatile biomolecules with the potential to enhance environmental sustainability.

7.6 CONCLUSION

Biosurfactants have a number of characteristics that might be useful in a variety of industries. Their anti-adhesive characteristics make them a viable option for preventing and disrupting biofilm formation on contact surfaces. They are indicated as multifunctional fixes or additional compounds because of their unique properties, such as anti-adhesive, emulsifying and antibacterial activity. Insufficient knowledge of the hazards and high production costs are reportedly the main reasons for the limited use of biosurfactants in the basic subsistence zone.

However, the utilization of agro-industrial waste can lower biosurfactant production prices and also waste processing expenditure, giving food- and nutrition-related businesses an alternative way of monetizing their losses along with becoming recognized microbiological surfactant producers. Biosurfactants produced from microorganisms that are generally viewed as safe (GRAS), such as *lactobacilli* and yeasts, provide excellent protection for food and solution purposes, but much more study is needed in this area. While the potential for new types of surface-dynamic mixes derived from microorganisms can aid in the identification of various atoms in terms of characteristics and composition, toxicological aspects of new and existing biosurfactants must be emphasized in order to confirm the safety of such mixes for human consumption. Despite multiple lab-based successes in the production of biosurfactants and their widespread commercial implementations, the production of biosurfactants on a large scale remains a challenge, as the configuration of the final product is affected by supplements and micronutrients, along with natural components. The use of biosurfactants in diverse regions should be governed by specific rules and guidelines.

REFERENCES

Abdel-Moghny, T., Mohamed, R. S., El-Sayed, E., Aly, S. M., & Snousy, M. G. (2012). Effect of soil texture on remediation of hydrocarbons-contaminated soil at El-Minia district, upper Egypt. *ISRN Chemical Engineering*, 2012, 1–13. doi:http://dx.doi.org/10.5402/2012/406598 [CrossRef]

Açıkel, Y. S. (2011). Use of biosurfactants in the removal of heavy metal ions from soils. In: Khan, M.S., Zaidi, A., Goel, R., Musarrat, J. (eds) *Bio-management of Metal Contaminated Soils*. Springer, Dordrecht, pp 183–223.

Akbari, S., Abdurahman, N.H., Yunus, M.R., Fayaz, F., & Alara, O.R. (2018). Biosurfactants-a new frontier for social and environmental safety: a mini review. *Biotechnology Res Innovations*, 2, 81–90. doi:http://dx.doi.org/10.1016/j.biori.2018.09.001

Amani, H. (2017). Synergistic effect of biosurfactant and nanoparticle mixture on microbial enhanced oil recovery. *Journal of Surfactants and Detergents*, 20, 589–597.

Bhaskar, P., & Bhosle, N. B. (2006). Bacterial extracellular polymeric substance (EPS): A carrier of heavy metals in the marine food-chain. *Environment International*, 32(2), 191–198. doi:http://dx.doi.org/10.1016/j.envint.2005.08.010

Burakov, A., Galunin, E., Burakova, I., Memetova, A., Agarwal, S., Tkachev, A., & Gupta, V. (2017). Adsorption of heavy metals on conventional and nanostructured materials for wastewater treatment purposes: A review. *Ecotoxicology and Environmental Safety*, 148, 702–712. doi:http://dx.doi.org/10.1016/j.ecoenv.2017.11.034

Cameotra, S. S., & Singh, P. (2004). Potential applications of microbial surfactants in biomedical sciences. *Journal of Trends in Biotechnology*, 22, 142–146.

Chen, J., Wu, Q., Hua, Y., Chen, J., Zhang, H., & Wang, H. (2017). Potential applications of biosurfactant rhamnolipids in agriculture and biomedicine. *Applied Microbiology and Biotechnology*, 101, 8309–8319. doi:http://dx.doi.org/10.1007/s00253-017-8554-4

Drakontis, C. E., & Amin, S. (2020). Biosurfactants: Formulations, properties, and applications. *Current Opinion in Colloid & Interface Science*. doi:http://dx.doi.org/10.1016/j.cocis.2020.03.013

Edwards, K. R., Lepo, J. E., & Lewis, M. A. (2003). Toxicity comparison of bio-surfactants and synthetic surfactants used in oil spill remediation to two estuarine species. *Marine Pollution Bulletin*, 46, 1309–1316.

Fei, D., Zhou, G. W., Yu, Z. Q., Gang, H. Z., Liu, J. F., Yang, S. Z., Ye, R. Q., & Mu, B. Z. (2019). Low-toxic and nonirritant biosurfactant surfactin and its performances in detergent formulations. *Journal of Surfactants and Detergents*, 23(1), 109–118.

Fenibo, E. O., Ijoma, G. N., Selvarajan, R., & Chikere, C. B. (2019). Microbial surfactants: the next generation multifunctional biomolecules for applications in the petroleum industry and its associated environmental remediation. *Microorganisms*, 7(11), 581. doi:http://dx.doi.org/10.3390/microorganisms7110581

Franzetti, A., Gandolfi, I., Fracchia, L., Hamme, J. V., Gkorezis, P., Marchant, R., & Banat, I. M. (2014a). Biosurfactant use in heavy metal removal from industrial effluents and contaminated sites. In: Kosaric, N., Sukan, F.V. (eds) *Biosurfactants: Production and Utilization-Processes, Technologies and Economics*, Taylor & Francis Group, 361–369.

Franzetti, A., Gandolfi, I., Fracchia, L., van Hamme, J., Gkorezis, P., Marchant, R., Banat, I. M. (2014b). Biosurfactant use in heavy metal removal from industrial Effluents and contaminated sites. *Biosurfactants*, 372–381. doi:http://dx.doi.org/10.1201/b17599-19

Hazrina, H. Z., Noorashikin, M. S., Beh, S. Y., Loh, S. H., & Zain, N. N. (2018). Formulation of chelating agent with surfactant in cloud point extraction of methylphenol in water. *Royal Society Open Science*, 5(7), 180070. doi:http://dx.doi.org/10.1098/rsos.180070 [CrossRef]

Hippolyte, M. T., Augustin, M., Hervé, T. M., Robert, N., & Devappa, S. (2018). Application of response surface methodology to improve the production of antimicrobial biosurfactants by *Lactobacillus paracasei* subsp. tolerans N2 using sugar cane molasses as substrate. *Bioresources and Bioprocessing*, 5, 48. doi:http://dx.doi.org/10.1186/s40643-018-0234-4

Karlapudi, A. P., Venkataswaralu, T. C., Tammineedi, J., Kanumuri, L., Ravuru, B. K., Dirisala, V. R., & Kodali, V. P. (2018). Role of biosurfactants in bioremediation of oil pollution-a review. *Petroleum*, 4(3), 241–249.

Kubicki, S., Bollinger, A., Katzke, N., Jaeger, K. E., Loeschcke, A., & Thies, S. (2019). Marine Biosurfactants: Biosynthesis, structural diversity and biotechnological applications. *Journal of Marine Drugs*, 17, 408. doi:http://dx.doi.org/10.3390/md17070408

Kumar, R., & Das, A. J. (2018). Utilization of agro-industrial waste for bio-surfactant production under submerged fermentation and its application in oil recovery from sand matrix. *Bioresource Technology*, 260, 233–240.

Lai, C., Huang, Y. C., Wei, Y., & Chang, J. (2009). Biosurfactant-enhanced removal of total petroleum hydrocarbons from contaminated soil. *Journal of Hazardous Materials*, 167(1–3), 609–614. doi:http://dx.doi.org/10.1016/j.jhazmat.2009.01.017 [CrossRef] [PubMed]

Lal, S., Ratna, S., Said, O. B., & Kumar, R. (2018). Biosurfactant and exopoly-saccharide-assisted rhizobacterial technique for the remediation of heavy metal contaminated soil: An advancement in metal phytoremediation technology. *Environmental Technology and Innovation*, 10, 243–263.

López-Prieto, A., Martínez-Padrón, H., Rodríguez-López, L., Moldes, A. B., & Cruz, J. M. (2019). Isolation and characterization of a microorganism that produces biosurfactants in corn steep water. CyTA. *Journal of Food*, 17(1), 509–516. doi:http://dx.doi.org/10.1080/19476337.2019.1607909

Mao, X., Jiang, R., Xiao, W., & Yu, J. (2015). Use of surfactants for the remediation of contaminated soils: A review. *Journal of Hazardous Materials*, 285, 419–435. doi:http://dx.doi.org/10.1016/j.jhazmat.2014.12.009 [CrossRef]

Mehetre, G. T., Dastager, S. G., & Dharne, M. S. (2019). Biodegradation of mixed polycyclic aromatic hydrocarbons by pure and mixed cultures of biosurfactant producing thermophilic and thermo-tolerant bacteria. *Science of the Total Environment*, 679, 52–60. doi:http://dx.doi.org/10.1016/j.scitotenv.2019.04.376

Mulligan, C. N., Yong, R. N., Gibbs, B. F., James, S., & Bennett, H. P. (1999). Metal removal from contaminated soil and sediments by the Biosurfactant Surfactin. *Environmental Science & Technology*, 33(21), 3812–3820. doi:http://dx.doi.org/10.1021/es9813055

Naughton, P. J., Marchant, R., Naughton, V., & Banat, I. M. (2019). Microbial biosurfactants: Current trends and applications in agricultural and biomedical industries. *Journal of Applied Microbiology*, 127, 12–28.

Nikolova, C., & Gutierrez, T. (2019). Use of microorganisms in the recovery of oil from recalcitrant oil reservoirs: Current state of knowledge, technological advances and future perspectives. *Frontiers in Microbiology*, 10, 2996. doi:http://dx.doi.org/10.3389/fmicb.2019.02996

Nikolova, C., & Gutierrtz, T. (2020). Use of microorganisms in the recovery of oil from recalcitrant oil reservoirs. Current state of knowledge, technological advances and future perspectives. *Frontiers in Microbiology*, 10, 2996. doi:http://dx.doi.org/10.3389/fmicb.2019.02996

Ochoa-Loza, F. J., Artiola, J. F., & Maier, R. M. (2001). Stability constants for the Complexation of various metals with a Rhamnolipid Biosurfactant. *Journal of Environmental Quality*, 30(2), 479–485. doi:http://dx.doi.org/10.2134/jeq2001.302479x

Ochoa-Loza, F. J., Noordman, W. H., Jannsen, D. B., Brusseau, M. L., & Maier, R. M. (2007). Effect of clays, metal oxides, and organic matter on rhamnolipid biosurfactant sorption by soil. *Chemosphere*, 66(9), 1634–1642. doi:http://dx.doi.org/10.1016/j.chemosphere.2006.07.068 [CrossRef] [PubMed]

Ohadi, M., Shahravan, A., Dehghannoudeh, N., Eslaminejad, T., Banat, I. M., & Dehghannoudeh, G. (2020). Potential use of microbial surfactant in microemulsion drug delivery system: A systematic review. *Drug, Design, Development and Therapy*, 14, 541–550.

Olasanmi, I. O., & Thring, R. W. (2018). The role of biosurfactants in the continued drive for environmental sustainability. *Sustainability*, 10, 4817. doi:http://dx.doi.org/10.3390/su10124817

Pasternak, G., Askitosari, T. D., & Rosenbaum, M. A. (2020). Biosurfactants and synthetic surfactants in bioelectrochemical systems: A mini-review. *Frontiers in Microbiology*, 11, 358. doi:http://dx.doi.org/10.3389/fmicb.2020.00358

Putra, W., & Hakiki, F. (2019). Microbial enhanced oil recovery: Interfacial tension and biosurfactant-bacteria growth. *Journal of Petroleum Exploration and Production Technology*, 9, 2353–2374. doi:http://dx.doi.org/10.1007/s13202-019-0635-8

Ravindran, A., Sajayan, A., Priyadarshani, G. B., Selvin, J., & Kiran, G. S. (2020). Revealing the efficacy of thermo-stable biosurfactant in heavy metal bioremediation and surface treatment in vegetables. *Frontiers in Microbiology*. doi:http://dx.doi.org/10.3389/fmicb.2020.00222

Rawat, G., & Kumar, V. (2020). Contributions of biosurfactants in environment: A green and clean approach. In: Gallegos, A.C.F., Jasso, R.M.R., Aguilar, C.N. (eds.) *Bioprocessing of Agri-food Residues for Production of Bioproducts*. Apple Academic Press, Palm Bay.

Rivera, D. Á., Urbina, M., López, M. A., & López, V. E. (2019). Advances on research in the use of agro-industrial waste in biosurfactant production. *World Journal of Microbiology Biotechnology*, 35, 155. doi:http://dx.doi.org/10.1007/s11274-019-2729-3

Roy, A. (2018). A review on the biosurfactants: properties, types and its applications. *Journal of Fundamentals of Renewable Energy and Applications*, 08(01). doi:http://dx.doi.org/10.4172/2090-4541.1000248

Rufino, R. D., Luna, J. M., Campos-Takaki, G. M., Ferreira, S., & Sarubbo, L. A. (2012). Application of the biosurfactant produced by Candida lipolytica in the remediation of heavy metals. *Chemical Engineering Transactions*, 27, 61–66.

Saimmai, A., Riansa-Ngawong, W., Maneerat, S., & Dikit, P. (2019). Application of biosurfactants in the medical field. Walailak. *Journal of Science and Technology*, 17(2), 154–166.

Sanchita, K., & Pritisnigdha, P. (2019). Production and functional characterization of food compatible biosurfactants. *Applied Food Science Journal*, 3(1), 1–4.

Sandrin, T. R., & Maier, R. M. (2002). Effect of pH on cadmium toxicity, speciation, and accumulation during naphthalene biodegradation. *Environmental Toxicology and Chemistry*, 21(10), 2075–2079. doi:http://dx.doi.org/10.1002/etc.5620211010

Santos, D. K. F., Rufino, R. D., Luna, J. M., Santos, V. A., & Sarubbo, L. A. (2016). Biosurfactants: Multifunctional biomolecules of the 21st century. *International Journal of Molecular Sciences*, 17(3), 401. doi:http://dx.doi.org/10.3390/ijms17030401

Shah, N., Nikam, R., Gaikwad, S., Sapre, V., & Kaur, J. (2016). Biosurfactant: Types, detection methods, importance and applications. *Indian Journal of Microbiol Research*, 3(1), 5–10.

Shekhar, S., Sundaramanickam, A., & Balasubramanian, T. (2015). Biosurfactant producing microbes and their potential applications: a review. *Journal of Environmental Science and Technology*, 45(14), 1522–1554. doi:http://dx.doi.org/10.1080/10643389.2014.955631

Singh, P., Patil, Y., & Rale, V. (2019). Biosurfactant production: Emerging trends and promising strategies. *Journal of Applied Microbiology*, 126(1), 2–13.

Vijayakumar, S., & Saravanan, V. (2015). Biosurfactants-types sources and applications. *Research Journal of Microbiology*, 10(5), 181–192. doi:http://dx.doi.org/10.3923/jm.2015.181.192

Wuana, R. A., & Okieimen, F. E. (2011). Heavy metals in contaminated soils: A review of sources, chemistry, risks and best available strategies for remediation. *ISRN Ecology*, 2011, 1–20. doi:http://dx.doi.org/10.5402/2011/402647 [CrossRef]

Yu, H., Xiao, H., & Wang, D. (2014). Effects of soil properties and biosurfactant on the behavior of PAHs in soil-water systems. *Environmental Systems Research*, 3(1), 6. doi:http://dx.doi.org/10.1186/2193-2697-3-6 [CrossRef]

Zaghloul, A., & Saber, M. (2019). Modern technologies in remediation of heavy metals in soils. *International Journal of Environment and Pollution*, Model.2, 10–19.

Zgorova, K., & Ananikov, V. (2017). Toxicity of metal compounds: Knowledge and myths. *Organometallics*, 36(21), 4071–4090.

Zhong, H., Yang, X., Tan, F., Brusseau, M. L., Yang, L., Liu, Z., Zeng, G., & Yuan, X. (2016). Aggregate-based sub-CMC solubilization of N-alkaNes by monorhamnolipid biosurfactant. *New Journal of Chemistry*, 40(3), 2028–2035. doi:http://dx.doi.org/10.1039/c5nj02108a [CrossRef] [PubMed]

8 Biosurfactants
Versatile Molecules with Potential Applications

Manju Bhaskar
D. B. S. College Govind Nagar, Kanpur, India

Vinod Kumar Verma
C. S. J. M. University, Kanpur, India

CONTENTS

8.1 INTRODUCTION

Studies of biosurfactants date back to the 1960s and their use has increased in recent decades (Roy, 2017). Biosurfactants have been found to be effective and efficient in various industries such as cosmetics, medicine, food, petroleum, agriculture, textiles, and wastewater treatment (Akbari et al., 2018). Carbon and energy are the sources of the growth of microorganisms which facilitate the intracellular diffusion and production of different substances (Vijayakumar and Saravanan, 2015; Vandana and Singh, 2018). Great biosurfactant producers include *Pseudomonas* and *Bacillus* (Olasanmi and Thring, 2018). Some bacterial biosurfactants may be pathogenic in nature so cannot be used in the food industry. *Candida bombicola* and *C. lipopolytica* are the most common biosurfactants produced by yeast (Rufino et al., 2007; Ocampo, 2016). *Yarrowia lipolytica*, *Saccharomyces cerevisiae*, and *Kluyveromyces lactis* are in the "generally regarded as safe" (GRAS) category which does not have any toxicity or pathogenicity (Santos et al., 2016). Sen et al. (2017) reported good

DOI: 10.1201/9781003306931-9

emulsification activity of sophorolipids produced by *Rhodotorula babjevae*, isolated from Assam, India.

There is plenty of evidence that biosurfactants exhibit significant activities, such as detergency, emulsification, and foaming, that are pertinent to human and animal health (Fu et al., 2008; Shao et al., 2012; Fracchia et al., 2015). Commercial biosurfactants are manufactured from microorganisms, animal fat, plants and petrochemicals. Shortage of potential feedstock and raw materials increase the cost of biosurfactants (Chaprao et al., 2015; Rufino et al., 2014). Environmental protection and climate change are serious issues for society; eco-friendly biosurfactants are generated from various natural and sustainable sources (Marchant and Banat, 2012a). Various plants and some marine organisms manufacture natural surfactants (Kubicki et al., 2019).

This chapter focuses on potential applications of biosurfactants that substitute for and complement synthetic molecules.

8.2 COMPOSITION AND STRUCTURE OF BIOSURFACTANTS

Biosurfactants are classified according to their microbial origin or chemical structure. They have different components including glycolipids, phospholipids, lipopeptides, fatty acids, peptides and antibiotics (Usman et al., 2016). Biosurfactants are generated by a large range of bacteria, fungi and yeast. Most biosurfactants are considered to be secondary metabolites. The amphipathic nature of biosurfactants explains their broad use in environmental applications. Rahman and Gakpe (2008) explain that all surfactants have two ends – hydrophobic and hydrophilic. The hydrophobic region comprises a long chain of fatty acids. The attachment to and removal of microorganisms from the surface is expanded by the surface area and bioavailability of hydrophobic organic substrates (Vijayakumar and Saravanan, 2015).

According to Shu et al. (2021), the most common biosurfactants are glycolipids which can also be categorized as high molecular weight polymers or bio-emulsions, and low molecular weight molecules called biosurfactants. Biosurfactants with low molecular weight and low surface and interfacial tension are considered to be more effective as emulsion-stabilizing agents. Low molecular weight biosurfactants include surfactin and rhamnolipids (Mulligan and Gibbs, 1990; Whang et al., 2008; Souza et al., 2014). Rahman and Gakpe (2008) highlight that extracellular lipopolysaccharide biosurfactants that are high molecular weight emulsifiers are polyanionic heteropolysaccharides. They also emphasized that the hydrophilic region is soluble in water and consists of carbohydrate, amino acid, cyclic protein-peptide, carboxylic acid, phosphate, and alcohol. Karanth et al. (1999) described the amphiphilic moiety of biosurfactants as able to reduce further friction and surface interactions between isolated molecules in space and visual cues respectively.

8.3 TYPES OF BIOSURFACTANTS

Biosurfactants are classified as lipopeptides, glycolipids, polysaccharides, peptides, phospholipids, and fatty acids (Table 8.1).

8.3.1 Polypeptides

A polypeptides chain attached to lipid makes lipopeptide (Rosenberg and Ron, 1999). These molecules have high surface activity and antibiotic potential. All strains of *Bacillus subtilis* manufacture the lipopeptide iturin. Surfactin is more efficient than iturin. It contains a fatty acid chain alongside a lactonic linkage integrated with several amino acids (Arima et al., 1968) which can inactivate herpes and retroviruses. Arthrofactin produced by the *Arthrobacter* species was identified and characterized by Morikawa et al. (1993). Lichenysins produced by *Bacillus licheniformis* are more stable under extreme temperature, pH, and salt conditions (Vijayakumar and Saravanan, 2015).

TABLE 8.1
Template of Biosurfactants

Biosurfactants		Microorganisms	References
Type	Subtype		
Glycolipids	Rhamnolipids	*Pseudomonas aeruginosa*	Jadhav et al. (2011)
		Candida bombicola	Casas et al. (1997)
	Trehalolipids	*Nocardia erythropolis*	Krishnaswamy et al. (2008)
	Sophorolipids	*Candida batistae*	Konishi et al. (2008)
Lipopeptides	Surfactin	*Bacillus subtilis*	Kim et al. (2010)
	Iturin		
	Fengycin		
	Lichenysin	*Bacillus licheniformis*	Jenny et al. (1991)
	Mannosylerythritol	*Candida antartica*	Kitamoto et al. (1993)
	Viscosin	*Pseudomonas fluorescens*	Abouseoud et al. (2008)
	Polymyxin	*Bacillus polymyxia*	Vandana & Singh (2018)
	Gramicidin	*Bacillus brevis*	
Polymeric surfactants	Emulsan	*Acinetobacter calcoaceticus*	
	Liposan	*Candida lipolytica*	
	Mannan	*Candida tropicalis*	
Fatty acids	Fatty acid	*Corynebacterium lepus*	
	Neutral Lipids	*Nocardia erythropolis*	
	Phospholipids	*Thiobacillus thioxidans*	

8.3.2 Fatty Acids, Phospholipids, and Neutral Lipids

Several yeast and bacteria such as Acinetobacter sp. produce phosphatidyl ethanolamine-rich vesicles which form optically clear micro-emulsions of alkanes in water (Vijayakumar and Saravanan, 2015). Fatty acids are produced by *Corynebacterium lepus*. Phospholipids are produced by *Thiobacillus thiooxidans*. Neutral lipids are produced by *Nocardia erythropolis* (Vandana and Singh 2018).

8.3.3 Polymeric Biosurfactants

Alasan, liposan, emulsan, lipomannana, and other polysaccharides are the best known polymeric biosurfactants. Liposan contain 83% carbohydrate and 17% protein and exhibits extracellular water-soluble emulsifier properties (Cooper and Paddock, 1984). Emulsan is a successful emulsifying agent for hydrocarbon in water (Hatha et al., 2007).

8.3.4 Particulate Biosurfactants

Micro-emulsions take part in alkane uptake by microbial cells which are structured by extracellular membrane vesicles. Vesicles of *Acinetobactor* spp. with a diameter of 20–50 nm and buoyant density of 1.158 cubic gcm are composed of protein, phospholipids, and lipopolysaccharide (Kaeppeli and Finnerty, 1979; Chakrabarti, 2012).

8.3.5 Glycolipids

The most common type of biosurfactants include sophorolipids, trehalolipids, and rhamnolipids. They are composed of sugars with long-chain aliphatic acids or hydroxy aliphatic acids (Drakontis and Amin, 2020). *Pseudomonas* and *Burkholderia* species produce rhamnolipids (Abdel-Mawgoud et al., 2010). There are many reports suggesting that petroleum hydrocarbons are efficient in biodegradation (Chrzanowski et al., 2011 & 2012; Szulc et al., 2014). *C. bombicola* and *C. apicola* produce sophorolipids consisting of a glycosidic bond making a connection between dimeric carbohydrate sophorose and long-chain hydroxyl fatty acids (Cortes-Sanchez et al., 2013). *Mycobacterium*, *Nocardia*, and *Corynebacterium* are producers of trehalolipids which can lower surface and interfacial tensions (Vijayakumar & Saravanan, 2015).

8.4 APPLICATIONS OF BIOSURFACTANTS

8.4.1 Antimicrobial Activity

Many biosurfactants are important antimicrobial agents that have come to be regarded as safe and effective therapeutic agents and acceptable alternatives to synthetic medicines. *Pseudomonas* spp. produces rhamnolipids (Guerra-Santos et al., 1986) and *Torulopsis* spp. manufactures sophorolipids (Cooper and Paddock 1984). Both these biosurfactants show antimicrobial activities (Elshikh et al., 2017). *R. erythropolis* and various *Mycobacterium* spp. can change cell wall complexity (Ristau and Wanger, 1983). Gram-negative pathogens, gram-positive pathogens, and semi-pathogenic microbial strains are effectively countered by biosurfactants produced by *B. circulans* (Kugler et al., 2015). Antimicrobial activity is also shown by *P. fluorescens* and *Pseudomonas aeruginosa* against some of the bacteria causing disease in humans (Vandana and Peter, 2014). *Bacillus* is known for producing lipopeptide biosurfactants. The first biosurfactant produced by *B. subtilis* is surfactin (Gudina et al., 2015). Jenny et al. (1991) determined the structure and activities of biosurfactants produced by *Bacillus licheniformes*. Grangemard et al. (2001) described the chelating property of lichenysin which causes lipopeptide membrane disruption. Surfactin persuades antibiotic and hemolytic action of lipopeptides due to formation of pores (Carrillo et al., 2003). Other lipopeptides such as fengycin, iturin, bacillomycin, and mycosubtilin are combined (Das et al., 2008).

In India, Thenmozhi and Krishnan (2011) reported antifungal activity of biosurfactants from Streptomyces against *Aspergillus fumigatus*. Surfactants have been found to be effective against human immuno-deficiency virus (HIV 1) (Itokawa et al., 1994). *C. lipolytica* (Sarubbo et al., 2007), *C. ishiwadae* (Thamnomsub et al., 2004), *C. batistae* (Konishi et al., 2008) *Aspergillus ustus* (Alejandro et al., 2011), and *Trichosporon ashii* (Chandran and Das, 2010) are some of the fungi known to produce biosurfactants. Thimon et al. (1995) reported that iturin produced by *B. subtilis* had antifungal properties affecting the morphology and membrane structure of yeast cells. Adu et al. (2020) claimed skin irritation and allergic reactions were due to synthetic personal care products. Biosurfactants are composed of lipid and proteins, and are reconcilable with skin-cell membranes (Vecino et al., 2015). *Lactobacillus pentosus* produces biosurfactants that manifest antimicrobial and anti-adhesive properties (Vecino et al., 2018). Gudina et al. (2015) did not find any antimicrobial activity with *L. agilis* against *Entamoeba coli* or *C. albicans*. Singh et al. (2014) reported that biosurfactants produced from different carbon sources and fermenting conditions reveal different antimicrobial properties.

8.4.2 Pharmaceuticals/Cosmetics

Biosurfactants are widely used as detergents and as wetting, emulsifying, foaming and solubilizing agents in medicine and the cosmetics industry (Marchant and Banat, 2012b). Biosurfactants are found in shampoo, hair conditioner, soap, shower gel, toothpaste, cream, moisturizer, cleansers, and many other skincare and health care products (Chakraborty et al., 2015; Satpute et al., 2010). The use of synthetic surfactants in cosmetics can cause major skin allergy and irritation problems (Sil et al., 2017, Vecino et al., 2017). These surfactants are also toxic to freshwater living organisms due to their chemical synthetization (Vandana and Singh, 2018). Ferreira et al. (2017) reported that these biosurfactants greatly improve these products. Biosurfactants are very useful for skin moisturizing such as ceramides (Kitgawa et al., 2011). It is known that ceramides are effective in treatment of damaged skin, preventing roughness and dryness. Microorganism-based surfactants have a good emulsifying capacity and can therefore can be used as ceramides to improve rough skin and eliminate ceramides deficiency in the skin (Kitgawa et al., 2011). Vecino et al. (2017) and Kitgawa et al. (2011) have reported that a combination of biosurfactants helps reduce skin roughness. Biosurfactants are reported to have good results as emulsifiers due to their low toxicity and higher biodegradability in different industries such as cosmetics and pharmaceuticals (Akbari et al., 2018).

8.4.3 Food Industry

In the food industries, biosurfactants are used to lower surface and interfacial tensions as well as in formalin stabilization. According to Campos et al. (2014), emulsifiers contribute to disposal of immiscible substances, solubilization of aromas and maintaining consistency and texture formulation in food. Emulsions contain one immiscible liquid which is dispersed in another in the form of droplets. Emulsifiers can also be used to improve the texture and shelf life of starch products, maintain agglomeration of fat globules, and stabilize aerated systems (Krishnaswamy et al., 2008). Emulsifier enhancement in the stability of dough, and in the volume, texture, and conservation of bakery products is done with the assistance of emulsifiers (Van Haesendonck and Vanzeveren, 2004). *C. utilis* and *S. cerevisiae* are used for stabilization of emulsion in food products and salad dressings (Torabizadeh et al., 1996; Campos et al., 2015). The use of rhamnolipids was insisted by Vijayakumar and Saravanan (2015) to improve the properties of butter, cream, and frozen confectionery products.

8.4.4 Textiles

One of the major contributors to our heavily polluted environment is industrial effluents, including those from textile manufacturing (Toprak and Anis, 2017). Biosurfactants are natural compounds that can balance these environmental challenges. Surfactants are used as antistatic untangling and softening agents in various textile manufacturing processes such as scouring, lubrication, dyeing, and finishing (Kesting et al., 2007). Textile surfactants contain both hydrophilic (water-loving) and hydrophobic (water-hating) regions. Due to their easy availability and cost-effectiveness, synthetic dispersants have been exclusively used in various textile processes (Fracchia et al., 2012). However, biosurfactants are replacing synthetic surfactants due to environmental concerns. Biosurfactants do not damage the surface and color characteristics of fabric. Biosurfactants can remove oil stains (Lima et al., 2017). Due to their low toxicity, tolerance of extreme temperature and pH, high biodegradability, and the fact they are produced from renewable sources, biosurfactants are recommended over chemical surfactants (Ricon-Fontan et al., 2017). Biosurfactants of *C. echinulate* can eliminate 86% of burned engine oil from cotton fabric – without having any toxic effects – as compared to synthetic detergents which removed 92% (Andrade et al., 2018).

8.4.5 Agriculture

Biosurfactants offer great advantages in agriculture (Karanth et al., 1999). Agricultural activities and crop production require adequate soil quality (Sarubbo et al., 2015a). The presence of inorganic and organic pollutants negatively affect agricultural soil (Sachdev and Cameotra, 2013). Plants and microbes, especially bacteria, need appropriate plant–microbe interaction (Hong et al., 2019). Ma et al. (2016) show how low metal phytotoxicity, induction of defense mechanisms against pathogens, and metal bioavailability of soil are achieved by plant growth-promoting microorganisms (PGPMS). According to Fenibo et al., 2019 (as reported by Neethu et al., 2019; Kaczorek et al.,

2018 and Demir and Koleli, 2018), biosurfactants participate in bioremediation of hydrocarbons, metal detoxification and/or removal, and soil washing technology. Sachdev and Cameotra (2013) describe how biosurfactants help in root cell differentiation which contributes to nutrient uptake. They also explain antimicrobial activity against plant pathogens. The zoospores of *Phytophthora capsica* cause damping-off in cucumber, lysed by biosurfactants produced by *Pseudomonas putida* (Krujit et al., 2009). *P. fluorescens* produces biosurfactants that are effective against plant pathogens, namely *Pythium ultimum*, *Fusarium oxysporum*, and *Phytophthora crypto* gear. Lipopeptide from bacteria is used as a biopesticide against the fruit fly *Drosophila melanogaster* (Mulligen, 2005).

8.4.6 Bioremediation

Bioremediation is an important biological element in the transformation or mineralization of organic contaminants by living organisms, leading to less harmful substances (Cameotra and Bollag, 2003; Sobrinho et al., 2013). According to Olasanmi and Thring (2018), a major source of environmental pollution is drilling waste and drilling cuttings and hydrocarbon in the oil and gas exploration industry. The discharge of oil waste into soils and water bodies causes massive environmental pollution (Urum et al., 2006), which is being felt at global level. Contamination of the environment remains an immediate concern (Luna et al., 2015).

According to Batista et al. (2006), biosurfactants are suitable for bioremediation of pollutants. Sen (2008) described the influential role of biosurfactants in microbial enhanced oil recovery. The recovery of oil from oily sludge before disposal reduces its environmental impact (Olasanmi and Thring, 2018). *P. aeruginosa* F-2, a rhamnolipid-producing strain, is used for the recovery of refinery oil sludge (Yan et al., 2012). Chirwa et al. (2013) revealed that recovered oil also serves as an energy source, which has led to renewed and intensified efforts to treat oily sludge. Silva et al. (2014) reported that oil spill accidents cause contamination of ocean and shoreline environments. The use of chemical surfactants in oil reservoirs causes pollution of water and soil, which poses serious health risks to humans, plants, and animal species (Jha et al., 2016). The use of biosurfactants facilitates the mobilization of oil and further oil recovery (Hosseininoosheri et al., 2016). Biosurfactants can be generated for both ex-situ and naturally positioned applications (Geetha et al., 2018) through aerobic fermentation of microbes. In-situ applications enhance oil recovery due to introduction of bacteria and their nutrients (Geetha et al., 2018).

Physical, chemical, or biological methods can reform contaminated soil(Rufino et al., 2013). Destruction and immobilization of the contaminants occurs through bioremediation (Sarubbo et al., 2015b). Biosurfactants show higher potency and advantages with excellent surface activity and lower surface tension (Sobrinho et al., 2013). Profitable remediation of soil has been carried out by species of *Pseudomonas* (Silva et al., 2013), *Candida* (Rufino et al., 2011; Rufino et al., 2013, 2014) and *Bacillus* (Vijayakumar and Saravanan, 2015; Chaprao et al., 2015).

8.5 FACTORS AFFECTING THE PRODUCTION OF BIOSURFACTANTS

Extracellularly, part of the cell membrane in yeast, bacteria, or filament gives rise to biosurfactants (Karanth et al., 1999). The concentration of different elements like nitrogen, iron, and phosphorus ions, and also the carbon source and different environmental temperature, pH, and agitation all modulate biosurfactant production (Vijayakumar & Saravanan, 2015). Rahman and Gakpe (2008) reported that pH, calcium, and magnesium have no effect on the production of biosurfactants by *Pseudomonas* strains. Duvnjak et al. (1982) describe how biosurfactant production from *Arthrobacter paraffines* ATCC 19558 is increased after utilization from ammonium as inorganic nitrogen. The pH and temperature also influence biosurfactant production, stimulating production from *Pseudomonas* sp. DSM 2874 and *A. paraffineus* ATCC 19558 (Rahman and Gakpe, 2008; Muller et al., 2011). Production of biosurfactants by *Ustilago maydis*, *Pseudomonas* sp., and *T. bombicola* is positively affected by pH (Hewald et al., 2005; Cooper & Paddock 1984).

8.6 CONCLUSION

This chapter provides information about various types of biosurfactant from different microorganisms. Bacteria, fungi and yeast manufacture biosurfactants, which are deemed more advantageous for restoring the environment, synthetic surfactants being more hazardous to nature. The use of synthetic surfactants is associated with massive environmental input and uninvited interference with the environment. Biosurfactants are thought to be the best alternative as their effect on the environment is less toxic. Potential applications of biosurfactants are valuable to human health, and they have a wide range of uses in different industries including medicine, pharmaceuticals, textiles, food processing, agriculture, paper, and cosmetics, etc. However, their use is found to be limited because of their high production costs (Vandana and Singh, 2018). Currently, very few industries are involved in producing microbial surfactants. Large industries should be required to incorporate microbial biosurfactants in commercial products to enhance their position in the global market (Fenibo et al., 2019). Further research should be focused on the production of cost-effective biosurfactant yields. There is a need to explore biosurfactant production processes so that they can benefit humankind.

REFERENCES

Abdel-Mawgoud, A. M., Lepine, F. & Deziel, E. (2010). Rhamnolipids: Diversity of structures, microbial origins, and roles. *Applied Microbiology and Biotechnology*, 86(5), 1323–1336. doi:10.1007/s00253-010-2498-2

Abouseoud, M., Yataghene, A. Amrane, A. & Maachi, R. (2008). Biosurfactant production by free and alginate entrapped cells of *Pseudomonas fluorescens*. *Journal of Industrial Microbiology*, 35, 1303–1308.

Adu, S., Naughton, P., Marchant, R. & Banat, I. (2020). Microbial biosurfactants in cosmetic and personal skincare pharmaceutical formulations. *Pharmaceutics*, 12(11), 1099.

Akbari, S., Abdurhman, N. H., Yunus, R. M., Fayaz, F. & Alara O. R. (2018). Biosurfactants- a new frontier for social and environmental safety: A mini-review. *Biotecnology Research & Innovation*, 2, 81–90. http://www.journals.elsevier.com/biotechnology-research-and-innovation/.

Alejandro, C. S., Humberto, H. S. & Maria, J. F. (2011). Production of glycolipids with antimicrobial activity by *Ustilago maydis* FBD12 submerged culture. *African Journal of Microbiology Research*, 5, 2512–2523. doi: 10.5897AJMR10.814

Andrade, R. F. S., Silva, T. A. L., Ribeaux, D. R., Rodriguez, D. M., Souza, A. F., Lima, M. A. B., Lima, R. A., da Silva, C. A. A. & Campos-Takaki, G. M. (2018). Promising biosurfactant produced by *Cunninghamella echinulata* UCP 1299 using renewable resources and its application in the cotton fabric cleaning process. *Advances in Materials Science and Engineering*, 2018. doi:10.1155/2018/1624573

Arima, K., Kakinuma, A. & Tamura, G. (1968). Surfactin, a crystalline peptide lipid surfactant produced by *Bacillus subtilis*: Isolation, characterization and its inhibition of fibrin clot formation. *Biochemical and Biophysical Research Communications*, 31(3), 488–494.

Batista, S. B., Mounteer, A. H., Amorim, F. R. & Totola, M. R. (2006). Isolation and characterization of biosurfactants/emulsifier-producing bacteria from petroleum contaminated sites. *Bioresource Technology*, 97(6), 868–875. doi:10.1016/j.biortech.2005.04.020

Cameotra, S. S. & Bollag, J. (2003). Biosurfactant enhanced bioremediation of polycyclic aromatic hydrocarbons. *Chemosphere*, 144, 635–644.

Campos, J. M., Stamford, T. L. & Sarubbo, L. A. (2014). Production of a bioemulsifier with potential applications in the food industry. *Applied Biochemistry and Biotechnology*, 172(6), 3234–3252.

Campos, J. M., Stamford, T. L. M., Rufino, R. D., Luna, J. M., Stamford, T. C. M. & Sarubbo, L. A. (2015). Formulation of mayonnaise with the addition of a bioemulsifier isolated from *Candida utilis*. *Toxicology Reports*, 2, 1164–1170.

Carrillo, C., Teruel, J. A., Aranda, F. J. & Ortiz A. (2003). Molecular mechanism of membrane permeabilization by the peptide antibiotic Surfactin. *Biochimica et Biophysica Acta*, 1611(1–2), 91–97. doi:10.1016/S0005-2736(03)00029-4

Casas, J. A., De Lara, S. G. & Garcia-Ochoa, F. (1997). Optimization of a synthetic medium for *Candida bombicola* growth using factorial design of experiments. *Enzyme and Microbial Technology*, 21, 221–229. doi:10.1016/S0141-0229(97)00038-0

Chakrabarti, S. (2012). *Bacterial biosurfactants: Characterization, antimicrobial, and metal remediation properties*. Ph.D. Thesis. National Institute of Technology.

Chakraborty, S., Ghosh, M., Chakraborti, S., Jana, S., Sen, K. K., Kokare, C. & Zhang, L. (2015). Biosurfactants produced from *Actinomycetes nocardiosis* A17: Characterization and its biological evaluation. *International Journal of Biological Macromolecules*, 79, 405–412. doi:10.1016/j.ijbiomac.2015.04.068

Chandran, P. & Das, N. (2010). Biosurfactants production and diesel oil degradation by yeast species *Trichosporon ashii* isolated from petroleum hydrocarbon contaminated soil. *International Journal of Engineering Science and Technology*, 2(12), 6942–6953.

Chaprao, M. J., Ferreira, I. N. S., Correa, P. F., Rufino, R. D., Luna, J. M., Silva, E. J. & Sarubbo, L. A. (2015). Application of bacterial and yeast biosurfactants for enhanced removal and biodegradation or motor oil from contaminated sand. *Electronic Journal of Biotechnology*, 18(6), 471–479. doi:10.1016/j.ejbt.2015.09.005

Chirwa, E. M. N., Mampholo, T. & Fayemiwo, O. (2013). Biosurfactants as demulsifying agents for oil recovery from oily sludge-performance evaluation. *Water Science and Technology*, 67, 2875–2881. doi:10.2166/wst.2013.207

Chrzanowski, L., Lawniczak, L. & Czaczyk, K. (2012). Why do microorganisms produce rhamnolipids? *World Journal of Microbiology and Biotechnology*, 28(2), 401–419. doi:10.1007/s11274-011-0854-8

Chrzanowski, L., Owsianiak, M., Szulc, C., Marecik, R., Piotrowska-Cyplik, A., Olejnik Schmidt, A. K., Staniewski, J., Ciesielczyk, F. & Jesionowski, T. (2011). Interaction between rhamnolipid biosurfactants and toxic chlorinated phenols enhance biodegradation of a model hydrocarbon rich effluent. *International Journal of Biodeterioration and Biodegradation*, 65, 605–611. doi:10.1016/j.ibiod.2010.10.015

Cooper, D. G. & Paddock, D. A. (1984). Production of biosurfactants from *Torulopsis bombicola*. *Applied and Environmental Microbiology*, 47, 173–176.

Cortes-Sanchez, A. J., Sanchez, H. H. & Jaramillo-Flores, M. E. (2013). Biological activity of glycolipids produced by microorganisms: New trends and possible therapeutic alternatives. *Microbiological Research*, 168, 22–32 doi:10.1016/j.micres.2012.07.002

Das, P., Mukherjee, S. & Sen, R. (2008). Genetic regulations of Biosynthesis of microbial surfactants: An overview. *Biotechnology and Genetic Engineering News*, 25, 165–186. doi:10.5661/bger-25-165

Demir, A. & Koleli, N. (2018). The removal of Pb and Cd from heavily contaminated soil in Kayseri, Turkey by a combined process of soil washing and electrodeposition. *Soil and Sediment Contamination (formerly Journal of Soil Contamination)*, 27, 469–484. https://www.researchgate.net/publication/326051606.

Drakontis, C. E. & Amin, S. (2020). Biosurfactants: Formulations, properties, and applications. *Current Opinion in Colloid & Interface Science*, 48, 77–90. doi:10.1016/j.cocis.2020.03.013

Duvnjak, Z., Cooper, D. G. & Kosaric, N. (1982). Production of surfactants by *Arthrobacter paraffineus* ATCC 19558. *Biotechnology and Bioengineering*, 24, 165–175.

Elshikh, M., Moya-Ramrez, I., Moens, H., Roelants, S., Soetaert, W., Marchant, R. & Banat, I. M. (2017). Rhamnolipids and lactonic sophorolipids: Natural antimicrobial surfactants for oral hygiene. *Journal of Applied Microbiology*, 123, 1111–1123.

Fenibo, E. O., Grace, N. I., Ramganesh, S. & Chioma, B. C. (2019). Microbial surfactants: The next generation of multifunctional biomolecules for applications in the petroleum industry and its associated environmental remediation. *Microorganisms*, 7, 581. doi:10.3390/microorganisms7110581

Ferreira, A., Vecino, X., Ferreira, D., Cruz, J. M., Moldes, A. B. & Rodrigues, L. R. (2017). Novel cosmetic formulations containing a biosurfactants from *Lactobacillus paracasei*. *Colloids and Surfaces B: Biointerfaces*, 155(4), 522–529. doi:10.1016/j.colsurfb.2017.04.026

Fracchia, L., Banat, J. J., Cavallo, M., Ceresa, C. & Banat, I. M. (2015). Potential therapeutic applications of microbial surface-active compounds. *AIIMS Bioengineering*, 2, 144–162. doi:10.3934/bioeng.2015.3144

Fracchia, L., Cavallo, M. & Giovanna, M. M. I. (2012). Biosurfactants and bioemulsifiers biomedical and related applications-Present status and future potential. *Biomedical Science Engineering and Technology*, 325–370. doi:10.5722/23821

Fu, S. L., Wallner, S. R., Bowne, W. B., Hagler, M. D., Zenilman, M. D., Gross, R. & Bluth, M. H. (2008). Sophorolipids and their derivatives are lethal against human pancreatic cancer cells. *The Journal of Surgical Research*, 148, 77–82. doi:10.1016/j.jss.2008.03.005

Geetha, S. J., Banat, I. M. & Joshi S. J. (2018). Biosurfactants: Production and potential applications in microbial enhanced oil recovery (MEOR). *Biocatalysis and Agricultural Biotechnology*, 14, 23–32. doi:10.1016/j.bcab.2018.01.010

Grangemard, I., Wallach, J., Maget-Dana, R. & Peypoux, F. (2001). Lichenysin: a more efficient cation chelator than surfactin. *Applied Biochemistry and Biotechnology*, 90(3), 199–210. doi:10.1385/abab:90:3:199

Gudina, E. J., Fernandes, E. C., Rodrigues, A. I., Teixeira, X. A. & Rodrigues, L. R. (2015). Biosurfactants production by *Bacillus subtilis* using corn steep liquor as culture medium. *Frontiers in Microbiology*, 6, 59. doi:10.3389/fmicb.2015.00059

Guerra-Santos, L. H., Kappeli, O. & Fiechter, A. (1986). Dependence of *Pseudomonas aeruginosa* continuous culture biosurfactants production on nutritional and environmental factors. *Applied Microbiology and Biotechnology*, 24, 443–448.

Hatha, A. A. M., Edward, G. & Rahman, K. S. M. P. (2007). Microbial biosurfactants-review. *Journal of Marine and Atmospheric Research*, 3, 1–17.

Hewald, S., Josephs, K. & Bolkar, M. (2005). Genetic analysis of biosurfactant production in *Ustilago maydis*. *Applied and Environmental Microbiology*, 71, 3033–3040.

Hong, E., Jeong, M. S., Kim, T. H., Lee, J. H. Cho, J. H. & Lee, K. S. (2019). Development of coupled biokinetic and thermal model to optimize cold water Microbial Enhanced Oil Recovery (MEOR) in homogenous reservoir. *Sustainability*, 11, 1652.

Hosseininoosheri, P., Lashgari, H. R. & Sepehrnoori, K. (2016). A novel method to model and characterize in-situ biosurfactant production in microbial enhanced oil recovery. *Fuel*, 183, 501–511. doi:10.1016/j.fuel.2016.06.035

Itokawa, H., Miyashita, T., Morita, H. (1994). Structural and confirmational studies of the [Ile7] and [Leu7] surfactins from *Bacillus subtilis*. *Chemical and Pharmaceutical Bulletin*, 42(3), 604–607. doi:10.1248/cpb.42.604

Jadhav, M., Kalme, S. Tamboli, D. & Govindwar, S. (2011). Rhamnolipid from *Pseudomonas desmolyticum* NCIM-2112 and its role in the degradation of brown 3REL. *Journal of Basic Microbiology*, 51, 385–396. doi:10.1002/jobm.201000364

Jenny, K., Kappeli, O. & Fietcher, A. (1991). Biosurfactants from *Bacillus licheniformis*: Structural analysis and Characterization. *Applied Microbiology and Biotechnology*, 36,5–13. doi:10/1007/BF00164690

Jha, S. S., Joshi, S. J. & Geetha, S. J. (2016). Lipopeptide production by *Bacillus subtilis* R1 and its possible applications. *Brazilian Journal of Microbiology*, 47(4), 955–964. doi:10.1016/j.bjm.2016.07.056

Kaczorek, E., Pachholak, A., Zdatra, A. & Smulek, W. (2018). The impact of biosurfactants on microbial properties leading to hydrocarbon bioavailability. *Colloids Interfaces*, 2, 35.

Kaeppeli, O. & Finnerty, W. R. (1979). Partition of alkane by an extracellular vesicle derived from hexadecane-grown *Acinetobacter*. *Journal of Bacteriology*, 140, 707–712.

Karanth, N., Deo, P. & Veenanadig, N. (1999). Microbial production of biosurfactants and their importance. *Current Science*, 77(1), 116–126.

Kesting, W., Tummuscheit, M., Schacht, H. & Schollmeyer, E. (2007). Ecological washing of textiles with microbial surfactants. *Progress in Colloid and Polymer Science*, 101, 125–130. doi:10.1007/BFb0114456

Kim, P. I., Ryu, J., Kim, Y. H. & Chi Y. T. (2010). Production of biosurfactant lipopeptide Iturin A, Fengycin, and Surfactin A from *Bacillus subtilis* CMB32 for control of *Colletotrichum gloeosporioides*. *Journal of Microbial Technology*, 20(1), 138–145.

Kitamoto, D., Yanagishita, H., & Shinbo, T. (1993). Surface active properties and antimicrobial activities of mannosylerythritol lipids as biosurfactants produced by *Candida antarctica*. *Journal of Biotechnology*. 29, 91–96.

Kitgawa, M., Suzuki, M., Yamamoto, S., Sogabe, A. Kitamoto, D., Imura, T. & Morita, T. (2011). *U. S.Ppatent Application No. 13/170,432*.

Konishi, M., Fukuoka, T., Morita, T., Imura, T. & Kitamoto, D. (2008). Production of new types of sophorolipids by *Candida batistae*. *Journal of Oleo Science*, 57, 359–369. doi:10.5650/JOS.57.359

Krishnaswamy, M., Subbuchettiar, G., Ravi, T. K. & Panchaksharam, S. (2008). Biosurfactants properties, commercial production, and application. *Current Science*, 94, 736–747.

Krujit, M., Tran, H. & Raaijmakers, J. M. (2009). Functional genetic and chemical characterization of biosurfactants produced by plant growth-promoting *Pseudomonas putida*. *Journal of Applied Microbiology*, 107(2), 546–556. doi:10.1111/j.1365-2672.2009.04244.x

Kubicki, S., Bollinger, A., Katzke, N., Jaeger, K. E., Loeschcke, A. & Thies, S. (2019). Marine biosurfactants: Biosynthesis, structural diversity and biotechnological applications. *Marine Drugs*, 17(7), 408. doi:10.3390/md17070408

Kugler, J. H., Le Rose-Hill, M., Syldatk, C. & Hausmann, R. (2015). Surfactants tailored by the class Actinobacteria. *Frontiers of Microbiology*, 6, 212.

Lima, R. A., Andrade, R. F. S., Rodriguez, D. M., Aroujo, H. W. C., Santos, V. P. & Campos-Takaki, G. M. (2017). Production and characterization of a biosurfactant isolated from *Candida glabrata* using renewable substrates. *African Journal of Microbiology Research*, 11(6), 237–244.

Luna, J. M., Rufino, R. D., Jara, A. M. A. T., Braeileiro, P. P. F. & Sarubbo, L. A. (2015). Environmental applications of the biosurfactant produced by *Candida sphaerica* cultivated in low-cost substrates. *Colloid and Surfaces A Physicochemical and Engineering Aspects*, 480, 413–418. doi:10.1016/j.colsurfa.2014.12.014

Ma, Y., Oliveira, R. S., Freitas, H. & Zhang, C. (2016). Biochemical and molecular mechanisms of plant-microbe-metal interactions: Relevance for phytoremediation. *Frontiers in Plant Science*, 7, 918. doi:10.3389/fpls.2016.00918

Marchant, R. & Banat, I. M. (2012a). Microbial biosurfactants: Challenges and opportunities for future exploitation. *Trends in Biotechnology*, 30, 558–565.

Marchant, R. & Banat, I. M. (2012b). Biosurfactants: A sustainable replacement chemical surfactants? *Biotechnology Letters*, 34, 1597–1605.

Morikawa, M., Daido, H., Takao, H. Murata, S., Shimonishi, Y. & Imanaka, T. (1993). A new lipopeptide biosurfactant produced by *Arthrobacter* sp. Strain MIS38. *Journal of Bacteriology*, 175(20), 6459–6466. doi:10.1128/jb.175.20.6459-6466.1993

Muller, M. M., Hormann, B., Kugel, M., Syldatk, C. & Hausmann, R. (2011). Evaluation of rhamnolipid production capacity of *Pseudomonas aeruginosa* PAO1 in comparison to the rhamnolipid over producer strains DSM 7108 and DSM 2874. *Applied Microbiology and Biotechnology*, 89(3), 585–592.

Mulligan, C. N. & Gibbs, B. F. (1990). Recovery of biosurfactants by ultrafiltration. *Journal of Chemical Technology and Biotechnology*, 47, 23–29.

Mulligen, C. N. (2005). Environmental applications for biosurfactants. *Environmental Pollution*, 133, 183–198.

Neethu, C. S., Sarvanakumar, C., Purvaja, R., Robin, R. S. & Ramesh, R. (2019). Oil spill triggered shift in indigenous microbial structure and functional dynamics in different marine environmental matrices. *Scientific Reports*, 9, 1354. doi:10.1038/s41598-018-37903-x

Ocampo, G. Y. (2016). Role of biosurfactants in nature and biotechnological applications. *Journal of Bacteriology and Mycology*, 2(4), 95–96. doi:10.15406/jbmoa.2016.02.00031

Olasanmi, I. O. & Thring, R. W. (2018). The role of biosurfactants in the continued drive for environmental sustainability. *Sustainability*, 10, 4817. doi:10.3390/su10124817

Rahman, P. K. & Gakpe, E. (2008). Production, characterization and applications of biosurfactants. *Review Biotechnology*, 7, 360–370. doi:10.3923/biotech.2008.360.370

Ricon-Fontan, M., Rodriguez-Lopez, L., Vecino, X., Cruz, J. M. & Moldes, A. B. (2017). Influence of micelle formation on the adsorption capacity of a biosurfactant extracted from corn on dyed hair. *RSC Advances*, 7(27), 16444–16452. doi:10.1039/C7RA01351E

Ristau, E. & Wanger, F. (1983). Formation of novel trehalose lipids from *Rhodococcus erythropolis* under-growth RR limiting conditions. *Biotechnology Letters*, 5, 95–100.

Rosenberg, E. and Ron, E. Z. (1999). High and low molecular mass microbial surfactants. *Applied Microbiology and Biotechnology*, 52(2), 154–162.

Roy, A. (2017). Review on the Biosurfactants: Properties, types and its applications. *Journal of Fundamentals of Renewable Energy and Applications*, 8, 248. doi:10.4172/20904541.1000248

Rufino, R. D., de Luna, J. M., de Campos Takaki, G. M. & Sarubbo, L. A. (2014). Characterization and properties of the biosurfactant produced by *Candida lipolytioca* UCP 0988. *Electronic Journal of Biotechnology*, 17(1), 34–38. doi:10.3390/ijms17030401

Rufino, R. D., Luna, J. M., Marinho, P. H. C., Farias, C. B. B., Ferreira, S. R. M. & Sarubbo, L. A. (2013). Removal of petroleum derivative adsorbed to soil by biosurfactant Rufisan produced by *Candida lypolytica*. *Journal of Petroleum Science and Engineering*, 109, 117–122. doi:10.1016/j.petrol.2013.08.014

Rufino, R. D., Luna, J. M., Rodrigues, G. I. B., Campos-Takaki, G. M., Sarubbo, L. A., & Ferreira, S. R. M. (2011). Application of yeast biosurfactant in the removal of heavy metals and hydrophobic contaminant in a soil used as slurry barrier. *Applied and Environmental Soil Science*, 2011. doi:10.1155/2011/939648

Rufino, R. D., Sarubbo, L. A. & Campos-Takaki, G. M. (2007). Enhancement of stability of biosurfactant produced by *Candida lipolytica* using industrial residue as substrate. *World Journal of Microbiology and Biotechnology*, 23, 729–734. doi:10.1007/s11274-006-9278-2

Sachdev, D. P. & Cameotra, S. S. (2013). Biosurfactants in agriculture. *Applied Microbiology and Biotechnology*, 97(3), 1005–1016. doi:10.1007/s00253-012-4641-8

Santos, D. K. F., Rufino, R. D., Luna, J. M., Santos, V. A. & Sarubbo, S. A. (2016). Biosurfactants: Multifunctional biomolecules of 21st Century. *International Journal of Molecular Sciences*, 17(3), 401. doi:10.3390/ijms17030401

Sarubbo, L., Luna, J. & Rufino, R. (2015a). Application of biosurfactant produced in a low-cost substrate in the removal of hydrophobic contaminants. *Chemical Engineering Transactions*, 43, 295–300. doi:10.3303/CET1543050

Sarubbo, L. A., Farias, C. B. B. & Campos-Takaki, G. M. (2007). Co-utilization of canola oil and glucose on the production of a surfactant by *Candida lipolytica*. *Current Microbiology*. 54(1), 68–73.

Sarubbo, L. A., Rocha, R. B., Jr. Luna, J. M., Rufino, R. D., Santos, V. A. & Banat. M. (2015b). Some aspects of heavy metals contamination remediation and role of biosurfactants. *Chemistry and Ecology*, 31(8), 1–17. doi:10.1080/02757540.2015.1095293

Satpute, S. K., Bhuyan, S. S., Pardesi, K. R., Majumdar, S. S., Dhakephalkar, P. K., Shete, A. M. & Chopade, B. A. (2010). Molecular genetics of biosurfactant synthesis in microorganisms. *Advances in Experimental Medicine and Biology*, 672, 14–41. doi:10.1007/978-1-4419-5979-2

Sen, R. (2008). Biotechnology in petroleum recovery: The Microbial EOR. *Progress in Energy and Combustion Science*, 34(6), 714–724. doi:10.1016/j.pecs.2008.05.001

Sen, S., Borah, S. N., Bora, A. & Deka, S. (2017). Production, characterization and antifungal activity of a biosurfactant produced by *Rhodotorula babjevae* YS3. *Microbial Cell Factories*, 16(95). doi:10.1186/s12934-017-0711-z

Shao, L., Song, X., Ma, X., Li, H. & Qu, Y. (2012). Bioactivities of sophorolipids with different structures against human oesophageal cancer cells. *The Journal of Surgical Research*, 173, 286–291. doi:10.1016/j.jss.2010.09.013

Shu, Q., Lou, H., Wei, T., Liu, X. & Chen, Q. (2021). Contributions of glycolipid biosurfactants and glycolipid modified materials to antimicrobial strategy: A review. *Pharmaceuticals*, 13(2), 227. doi:10.3390/pharmaceutics13020227

Sil, J., Dandapat, P. & Das S. (2017). Health care applications of different biosurfactants: Review. *International Journal of Science and Research*, 6, 41–50.

Silva, R. C. F. S., Almeida, D. G., Rufino, R. D., Luna, J. M., Santos, V. A. & Sarubbo, L. A. (2014). Applications of biosurfactants in the petroleum industry and remediation of oil spills. *International Journal of Molecular Sciences*, 15(7), 12523–12542. doi:10.3390/ijms150712523

Silva, R. C. F. S., Rufino, R. D., Luna, J. M., Farias, C. B. B., Filho, H. J. B., Santos, V. A. & Sarubbo, L. A. (2013). Enhancement of biosurfactant production from *Pseudomonas cepacian* CCT6659 through optimisation of nutritional parameters using Response Surface Methodology. *Tenside Surfactants Detergents*, 50(2), 137–142. doi:10.3139/113.110241

Singh, A. K., Rautela, R. & Cameotra, S. S. (2014). Substrate-dependent in vitro antifungal activity of Bacillus sp strain AR2. *Microbial Cell Factories*, 13, 67. doi:10.1186/1475-2859-13-67

Sobrinho, H. B. S., Luna, J. M., Rufino, R. D., Porto, A. L. & Sarubbo, A. L. F. (2013). Biosurfactants: Classification, properties and environmental applications. In *Recent Developments in Biotechnology*. 1st ed., Govil, J. N., Ed., Studium Press LLC: Houston, TX, 11, 11–29.

Souza, E. C., Vessoni-Penna, T. C. & De Souza Oliveira, R. P. (2014). Biosurfactants enhanced hydrocarbon bioremediation: An overview. *International Biodeterioration & Biodegradation*. 89, 88–94.

Szulc, A., Ambrozewicz, D., Sydow, M., Lawnickzak, L., Piotrowska-Cyplik, A., Marecik, R. & Chrzanowski, L. (2014). The influence of bioaugmentation and biosurfactant addition on bioremediation efficiency of diesel-oil contaminated soil: Feasibility during field studies. *Journal of Environmental Management*, 132, 121–128. doi:10.1016/j.jenvman.2013.11.006

Thamnomsub, B., Watcharachaipong, T., Chotelersak, K., Arunrattiyakorn, P. Nitoda, T. & Kanzaki, H. (2004). Monoacylglycerol: Glycolipid biosurfactants produced by a thermotolerant yeast *Candida ishiwadae*. *Journal of Applied Microbiology*, 96(3), 588–592. doi:10.1111/j.1365-2672.2004.02202.x

Thenmozhi, M. & Krishnan, K. (2011). Anti-Aspergillus activity of *Streptomyces* sp. VITSTK7 isolated from the Bay of Bengal coast of Puducherry, India. *Journal of Medical Mycology*, 23(2). doi:10.1016/j.mycmed.2013.04.005

Thimon, L., Peypoux, F. & Wallach, J. (1995). Effect of lipopeptide antibiotic, iturin A, on morphology and membrane ultrastructure of yeast cells. *FEMS Microbiology Letters*, 128(2), 101–106.

Toprak, T. & Anis. P. (2017). Textile industry's environmental effects and approaching cleaner production and sustainability, an overview. *Journal of Textile Engineering & Fashion Technology*, 2(4), 429–442. doi:10.15406/jteft.2017.02.00066

Torabizadeh, H., Shojaosadati, S. & Tehrani, H. (1996). Preparation and characterization of bioemulsifier from *Saccharomyces cerevisiae* and its application in food products. *LWT-Food Science Technology*, 29(8), 734–737.

Urum, K., Grigson, S., Pekdemir, T. & McMenamy, S. (2006). A comparison of the efficiency of different surfactants for removal of crude oil from contaminated soils. *Chemosphere*, 62, 1403–1410.

Usman, M. M., Dadrasnia, A., Lim, K. T., Mahmud, A. F. & Ismail, S. (2016). Application of biosurfactants in environmental biotechnology; remediation of oil and heavy metal. http://www.aimspress.com/journal/Bioengineering. 3(3), 289–304.

Van Haesendonck, I. P. H. & Vanzeveren, E. C. A. (2004). *Rhamnolipids in bakery lipids*. W. O. 2004/040984 International Application Patent (PCT). Washington, DC., USA.

Vandana, P. & Peter, J. K. (2014). Determination of antimicrobial activity and production of biosurfactant by *Pseudomonas aeruginosa. International Journal of Engineering Sciences and Research Technology*, 3(10), 254–260.

Vandana, P. & Singh, D. (2018). Review on Biosurfactant production and its application. *International Journal of Current Microbiology and Applied Science*, 7(8), 4228–4241. doi:10.20546/ijcmas.2018.708.443

Vecino, X., Bustos, G., Devesa Rey, R., Jose Manual, C. F. & Ana, M. (2015). Salt free aqueous extraction of cell bound biosurfactants: A kinetic study. *Journal of Surfactants and Detergents*, 18, 267–274.

Vecino, X., Cruz, J. M., Moldes, A. B. & Rodrigues, L. R. (2017). Biosurfactants in cosmetic formulations: Trends and challenges. *Critical Reviews in Biotechnology*, 37, 911–923. doi:10.1080/07388551.2016.1269053

Vecino, X., Rodrigues-Lopez, L., Ferreira, D., Cruz, J. M., Moldes, A. B. & Rodrigues, L. R. (2018). Bioactivity of glycol-lipopeptide cell bound biosurfactants against skin pathogens. *International Journal of Biological Macromolecule*, 109, 971–979. doi:10.1016/j.ijbiomac.2017.11.088

Vijayakumar, S. & Saravanan, V. (2015). Biosurfactants-Types, Source and Applications. *Research Journal of Microbiology*, 10(5), 181–192.

Whang, L. M., Liu, P. W. G., Ma, C. C., & Cheng S. S. (2008). Application of biosurfactants, rhamnolipid and surfactin for enhanced biodegradation of diesel-contaminated water and soil. *Journal of Hazardous Materials*, 151, 155–163.

Yan, P., Lu, M., Yang, Q. (2012). Oil recovery from refinery oily sludge using a rhamnolipid biosurfactants-producing *Pseudomonas. Bioresource Technology*, 116, 24–28.

9 Production of Microbial Enzymes Using Spent Mushroom Compost (SMC) and Its Application

Anjali Kosre and Deepali
Pt. Ravishankar Shukla University, Raipur, India

Ashish Kumar
Sant Gahira Guru Vishwavidyalaya, Sarguja Ambikapur, India

Nagendra Kumar Chandrawanshi
Pt. Ravishankar Shukla University, Raipur, India

CONTENTS

DOI: 10.1201/9781003306931-10

9.1 INTRODUCTION

Mushrooms have been consumed as sources of protein and energy since ancient times. They are popularly known as a functional food as they are low in calories, and high in proteins, minerals, and dietary fiber (Fazenda et al., 2008). Global production and consumption of mushrooms has been rising rapidly in recent decades, due to their nutritional value and flavor, and they are widely appreciated for their beneficial effect on human health (Nakajima et al., 2018). A huge quantity of waste product is generated during mushroom cultivation in the form of spent mushroom compost (SMC). Disposal of the SMC generated after the final harvest is a major problem for mushroom producers. SMC is presently disposed of by burning, composting with manure or land-fill, which is not eco-friendly (Phan and Sabaratnam, 2012). Researchers have been exploring and developing novel applications for SMC including bioremediation, energy feedstock (biogas, bioethanol), animal feed and crop production (Jordan and Mullen, 2007). Another efficient use of SMC, which has to date been little explored by scientists or the industry, is recovery of enzymes (Schimpf and Schulz, 2016) (Finney et al., 2009). This has the advantage of converting a cheap and easily available waste raw material into high value products of industrial importance (Zhu et al., 2018)

9.1.1 Spent Mushroom Compost (SMC)

Mushrooms belong to the basidiomycetes, which have the potential to generate a broad range of enzymes that degrade complex substances into simpler one. The enzymes secreted from these mushrooms are non-specific and they degrade a variety of structurally different compounds. Spent mushroom compost (SMC) is the agro-residues biologically degraded by fungal mycelium that are generated during mushroom cultivation and remain after the harvest of mushroom fruiting bodies (Phan and Sabaratnam, 2012). It is generally obtained from renewable agro-waste deposits such as sawdust, paddy straw, wheat straw, sugarcane bagasse, poultry manure, rice husk, groundnut shell, hay, corncobs, cocoa shells, gypsum, cottonseed meal, etc. (Jordan et al., 2008). This agro-waste biomass consists of cellulose, hemicellulose and lignin. After degradation of lignocellulosic substrates, it comprises mycelia biomass, enzymes, and bioactive compounds like protein and phenol (Lim et al., 2013).

9.1.2 Composition of Spent Mushroom Compost

A large amount of SMC is produced commercially worldwide. This compost is rich in a degraded form of naturally present biopolymers such as chitin, lignin, chitosan, cellulose and hemicelluloses (Alhujaily et al., 2018) and mycelium with a high amount of carbohydrate and crude protein etc. (Lin et al., 2016). Its organic content of 22–40% provides enriched quality of nutrition for plant growth, chiefly because of its high cation exchange capacity, the porosity of substrate as manure, and slow mineralization, though the characteristics of SMC may differ depending on its primary constituents as well as on the type of agro-waste. The mineral content is up to 78%. It is a good source of macro and micro nutrients such as nitrogen (N), potassium (K), magnesium (Mg), phosphorus (P), calcium (Ca), sodium (Na), and other trace elements such as copper, iron, manganese, zinc, etc. in the dry matter. It contains 45% water, even though it looks quite a complex structure compared to manure, due to its low bulk density. One cubic meter of spent mushroom compost is equivalent to several tonnes of manure in nutrient content (Dann, 1996). It is high in lignocellulosic substances. The fungal population in the SMC devours this lignocellulosic content and degrades it into a simpler form (Fen et al., 2014). The other microbial population present in it degrades this content and utilizes it for their growth by the secretion of lignocellulolytic enzymes (Singh et al., 2003). For that reason, SMC can be an excellent resource for the extraction of commercially essential enzymes for industrial applications.

9.2 ENZYMES PRESENT IN SPENT MUSHROOM COMPOST

The breakdown of cellulose and hemicellulose is a critical process that necessitates assorted enzymatic pathways and produces certain by-products, some of which are also present as insoluble crystalline fibers. SMC is rich in defiant lignocellulose, which is brittle in nature and can hinder the formation of compost (Hu et al., 2019). Several approaches contribute to the process of compost formation. Microbial inoculation is a technique that is extensively applied for improvement of lignocellulose degradation in the process of agro-waste compost formation. The microbial metabolism of compost using the cellulosic content of SMC can also have potential nutrients for enzyme production (Sun et al., 2021).

9.2.1 Ligninolytic Enzymes

The efficient breakdown of lignocellulosic waste in simple form is due to such microbial populations as bacteria and fungi, as previously reported by several researchers (Janusz et al., 2017). The lignin-modifying enzymes are collectively called ligninolytic enzymes. They are secreted extracellularly as mycelia consortium which remain in the waste mushroom compost after harvest and have the potential to produce groups of enzymes called lignocellulytic enzymes that degrade the agro-waste (Sanchez, 2009). The lignin-degrading enzymes produced by these microorganisms are categorized as heme-containing peroxidase, i.e., lignin peroxidase (LiP), manganese peroxidase (MnP) and laccases, etc. The breakdown macromolecules present in agro-waste, e.g., lignin, generate free radicals to break the bonds present in the macromolecule (Ball and Jackson, 1995). Subsequently, the free radical formed permits the secreted enzymes to degrade the complex substrates, providing non-specificity and good oxidizing power to the enzyme.

9.2.1.1 Laccase

Laccase is an extracellular copper containing blue oxidoreductase enzymes excreted by white-rot fungi. It belongs to an extensive category of enzymes called polyphenol oxidases containing copper as the catalytic center, also known as multicopper oxidases. It can be present in monomeric, dimeric or tetrameric form containing 60–80 kDa of molecular weight, along with 15–20% carbohydrate content. Laccase isoenzymes have diverse catalytic features and stability. They oxidize and degrade many environment pollutants such as xenobiotics, including phenols, dyes and chlorophenols, etc. (Tanyolac & Aktas, 2003). Laccase enzymes may potentially take action on lignocellulosic compounds of agro-waste substances, including phenolic and non-phenolic compounds (Mayolo-Deloisa et al., 2009), as well as on other compounds, for its detoxification and degradation (Kumar and Chandra, 2020). It has protective properties such as humification, cell wall protection, cellular oxidation, and fruit browning, and is used in several industrial applications like wine processing, dyeing, and pulp delignification, as well as in the treatment of polluted soil and water (Schlosser and Hofer, 1999).

9.2.1.2 Lignin Peroxidase (LiP)

Lignin peroxidase belongs to the oxidoreductase family, which in the presence of H_2O_2 is applicable in the degradation of lignin and its derivatives (Edwards, 1993). These are heme-containing enzymes primarily secreted by higher fungi including some bacteria that are helpful in degradation of the polymer present in lignolytic substrate using an oxidative pathway (Pothiraj et al., 2006). This enzyme has been employed for the oxidative breakdown of an assortment of organic compounds with the help of H_2O_2 as a mediator. It also plays a significant role in the degradation of β-O-4-linked lignin containing compounds and other non-phenolic derivatives which are homologous to ketones or aldehydes. Lignin peroxidases with ferric protoporphyrin ring as prosthetic group are heme proteins, often attached with Ca, and act as catalysts for reactions in the presence of hydrogen peroxide (H_2O_2) (Duran and Esposito, 2000). Its molecular weight is approx. 38–43 kDa. It can perform many functions such as aromatic ring cleavage, cleavage of alkyl-phenyl, hydroxylation demethylation, demethoxylation and polymerization, along with redox potentials (Field et al., 1996).

9.2.1.3 Manganese Peroxidase (MnP)

Manganese peroxidase (MnP) is an enzyme that contains heme. It also belongs to the oxidoreductase family. It is secreted by ligninolytic microorganisms in solid and liquid condition into their environments (Hatakka et al., 2003). The secretion of MnP isozymes with approximately 40–50kDa molecular mass has been reported in a number of bacteria, algae and basidiomycetes that utilize various genes after the process of coding and regulating for lignin degradation (Wong, 2009). Its molecular structure contains two Ca2þ ions and five disulfide-bridging elements, which accounts for maintenance of the active site of the enzyme (Kumar and Chandra, 2020). Ligninolytic microorganisms oxidize Mn2þ to Mn3þ in a combined process involving several steps. The Mn2þ triggers the functions and degradation of the substrate by MnP enzymes. Consequently, the Mn3þ is catalyzed through MnP enzymes which play a role as an intermediate in the oxidation process for numerous phenolic and non-phenolic compounds (Ten Have and Teunissen, 2001). The Mn3þ, by the process of chelation, diffuses oxalate into the catalytic site of the enzyme. Manganese peroxidase enzymes hold specific groups that perform like both oxidase and peroxidase (Singh et al., 2011).

9.2.2 Hydrolytic Enzymes

9.2.2.1 Cellulases

Cellulase enzymes help in the composting process by hydrolyzing polymeric chains of cellulose into smaller units, allowing effortless consumption of substrate by microorganisms. It can improve the metabolism of microorganisms, change the microbial population and help in the regulation of pH and permeability of the compost formed. This process will influence biological and physiochemical reactions such as metabolism of nitrogen and humus synthesis during the formation of compost (Ma et al., 2019). With the development of the enzyme industry and the application of molecular biotechnology, commercial enzyme preparation, including its use in composting, has become economical (Sun et al., 2021).

9.2.2.2 Xylanase

These enzymes are normally present in marine algae, protozoans, snails, insects, seeds, crustaceans, and several microorganisms and plants. But commercially, filamentous fungi are the main source of these enzymes (Polizeli et al., 2005). Xylan, also called wood gum, is a polysaccharide (da Silva et al., 2012) present in cell walls of plants. Structurally, it is a linear set of residues of 1, 4- linked D-xylopyranose with a 5-carbon reducing sugar producing the enzyme xylanase (Lim et al., 2012). The most important purpose of this enzyme is to break down hemicellulose into smaller units leading to the formation of a simple sugar called xylose. Recently, xylanase and its derivatives, hydrolytic enzymes, have become of huge industrial interest for its application as animal feed, in bakery, beverages, textiles, bleaching, ethanol and xylitol production (Lim et al., 2013).

9.2.2.3 β-glucosidases

These are a diverse assembly of phylogenetically arranged hydrolytic enzymes extensively found in the living world which play a fundamental role in some important biological processes in cellulolytic microbes. For instance they are involved in cellulase induction (as a result of their transglycosylation activities) and hydrolysis of cellulose. They play an efficient role in complete hydrolysis of lignocellulosic-rich biomass. Direct application of fungal-based fermentation broths can be a cost-effective method using commercial enzymes for manufacturing of biofuels and bioproducts (Sorensen et al., 2011).

9.3 MICROBIAL PRODUCTION OF ENZYMES THROUGH SPENT MUSHROOM COMPOST

The substrates on which mushrooms grow are generally agro-wastes with lignocellulosic content. First the mushroom in the form of mycelia secretes lignolytic enzymes that allow breakdown of compounds such as lignin and cellulose abundantly found in the agro-wastes (Jasinska, 2018). The enzymes involved depend upon the type of agro-biomass and its composition, which in turn depends on the geographic location where it is harvested (Beyer, 1996). The microorganisms secrete enzymes by solid or submerged fermentation. Microorganisms such as fungi and bacteria have been used for the industrial production of cellulase using lignocellulosic raw materials as substrates. Companies including Genencor and Novozymes produce commercial enzymes using fungal species such as *Trichoderma* (cellulase enzymes) (Paul et al., 2021). In general, the enzyme is produced as by-product through the microorganisms which are inoculated in the medium by the consumption of carbon sources (such as glucose) in the submerged fermentation medium. In the enzymatic saccharification process of lignocellulosic biomass, enzymes (commercial type) are applied to collect glucose as a final end product (Abdeshahian et al., 2011). The hydrolysate thus obtained from enzymatic hydrolysis is composed of glucose which is supplied as a carbon source (substrate) to microorganisms (bacteria and fungi) inoculated in the culture medium. Consequently, enzymes produced in the broth can be extracted as crude enzymes at the end. However, in case of solid-state fermentation (SSF), cheaply available substrates may be used, such as spent mushroom composts (Sukharnikov et al., 2012).

9.3.1 Extraction and Recovery

The extraction and recovery of lignocellulose-degrading enzymes from SMC requires different physicochemical treatments. The maximum recovery of active enzymes is in the extracts collected from spent compost which had been blended physically. From these blended compost certain enzymes are extracted, such as peroxidases, xylan-debranching enzymes such as arabinofuranosidase and acetyl

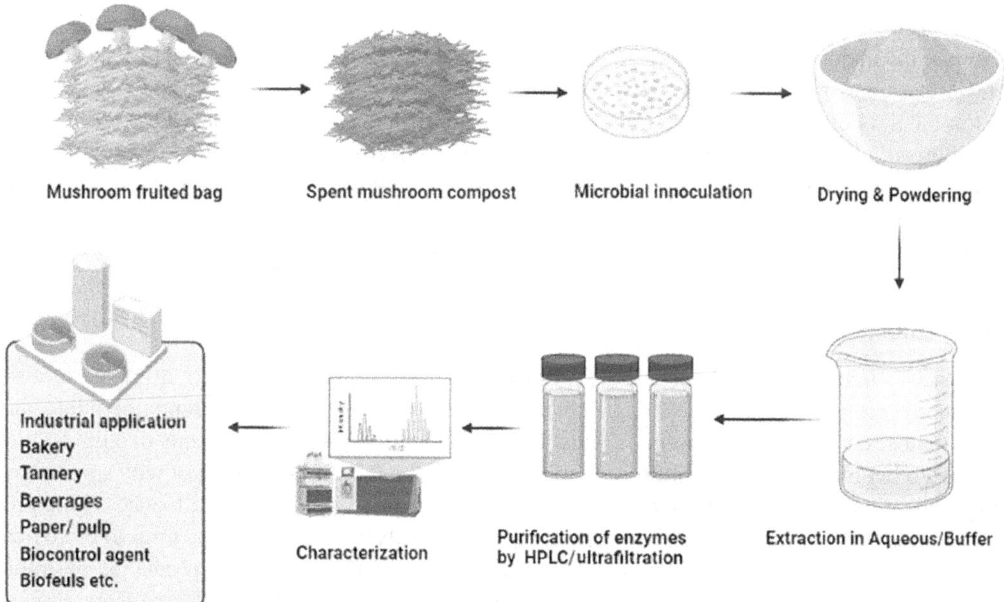

FIGURE 9.1 Extraction and purification of enzymes using spent mushroom composts (SMC).

esterase, and cellulose-degrading enzymes such as cellobiohydrase, endoglucanase and ﬂ-glucosidae. The compost extract also releases several reducing sugars (Ball and Jackson, 1995). The enzymatic activity and stability suggest a potential role of enzyme mixture extracted from compost for industrial application. Several physicochemical pretreatments, including acid/alkali pretreatment, steam explosion, organic solvents, ionic liquids, ionizing radiation, etc., have been employed to assist the enzymatic hydrolysis of lignocelluloses (Figure 9.1). These methods require high energy, have high production costs, form undesirable end products and generate inhibitors affecting the fermentation medium and enzymatic hydrolysis (Mosier et al., 2005, Himmel et al., 2007).

9.4 FACTORS AFFECTING ENZYMES IN SPENT MUSHROOM COMPOST

Several factors–temperature, pH, carbon and nitrogen ratios of substrates, incubation time, the moisture content of substrates – affect production of enzymes during mushroom cultivation using agro-industrial waste.

9.4.1 pH

pH plays a significant role in fungal cultivation but managing it in solid substrate culture is not easy. The production of lignolytic enzymes is influenced by the initial pH of the fermentation medium. Most fungi grow well at their favorable pH range 4.0–5.0 and reduce the acidity of the substrate (Agosin and Odier, 1985). Laccase enzyme production is drastically affected by pH of the medium. The optimum pH range for *P. ostreatus* species of mushroom is 5.0. A change in pH affects the three-dimensional structure of laccase which in turn decreases laccase activity (Patel et al., 2009). Specific enzymes have a specific pH on which they are active, so the pH of the substrate significantly affects enzyme productivity. For instance, the most favorable pH for production of xylanase enzymes during oyster mushroom (*Pleurotus* sp.) cultivation is 4.0 (Surendran et al., 2021).

9.4.2 Temperature

It is necessary to maintain a suitable incubation temperature at the time of biological pretreatment of the substrate. Optimum temperatures may vary with the type of microorganism inoculated. For example, the majority of the ascomycetes fungi grow optimally around 39°C, whereas basidiomycetes prefer temperatures between 25 and 30°C. The metabolic mechanism of these fungi produces heat and builds up temperature gradients in solid-state media. The accumulated heat formed may inhibit growth and metabolism of the fungus itself. The main challenges in scaling up production of solid-state cultivation are designing and developing an efficient bioreactor with low heat generation. Different temperature ranges for pretreatment of biomass are required because of the physiology and strain of the fungus, and the type of substrate (Millati et al., 2011).

9.4.3 Biomass Type

Agro-waste biomass rich in lignocellulosic content is a bioresource abundantly available in agricultural crops and forest residues. Lignocelluloses are composed mainly of lignin, cellulose and hemicelluloses, with a small amount of other organic and non-organic components like lipids, proteins and other extractives. The biological constituents of the feedstocks may vary with species and geographical location of the agro-wastes harvested, which will have a major influence on the pretreatment process to be followed. Hence a compositional analysis should be done prior to biological pretreatment. Further treatment can be carried out using microbial inoculums producing desirable enzymes, resulting in the hydrolysis of the lignocellulosic waste. Biomass collected from agricultural land should be harvested at a suitable stage of maturity so that it can produce different reducing sugars with biological pretreatment as the end product (Sindhu et al., 2016).

9.4.4 Type of Microorganism

Biological pretreatment (fungal) is one of the efficient methods for enzymatic saccharification of agro-wastes using wood rot fungus. *Gloeophyllumtrabeum* is a brown rot fungus that secretes enzymes which can breaks the polymeric chain of cellulose and hemicelluloses of wood containing modified lignin as brown residue (Gao et al., 2012). SMC is biologically pretreated with fungi by lignin degradation to enhance the enzymatic hydrolysis of feedstock. The pretreatment causes separation effect of substrate as well as partial removal of other products such as xylans. The altered structure of lignin resulting from disruption of the structure of the agro-waste cell wall improves accessibility of cellulase enzymes. Several studies have found that the application of fungal population achieves faster degradability of biomass (Sindhu et al., 2016).

9.5 INDUSTRIAL APPLICATIONS AND FUTURE PROSPECTS

Ligninolytic enzymes secreted by microbes have enormous potential to be applied for industrial purposes in context as diverse as cosmetics, food processing, degradation and detoxification of paper, textiles, distillery effluent, production of fine chemicals, and certain biofuels (Table 9.1, Malherbe and Cloete, 2002). For example, laccase from SMC has been investigated for breakdown of lignin polymers, and has significant applications in the pulp and paper industry (Widsten and Kandelbauer, 2008). Furthermore, lignin peroxidase (LiP) and manganese peroxidase (MnP) are used in deprivation of paper mill effluents. In the food and beverages industry, laccase is used as a coloring and flavoring agentand for the removal of phenolic compounds. In the textile industry, laccase is used for biodegradation of dye and bleaching of industrial waste (Gomes et al., 2009). Laccase, lignin peroxidases, and manganese peroxidases are used in biodegradation of a range of polycyclic aromatic hydrocarbons (PAHs) and xenobiotics compounds (Wen et al., 2009). They are also used in the manufacture of pharmaceutical products, polymer production, coupling of phenols and steroids, synthesis of complex natural products, personal hygienic products and biosensors (Barbosa et al., 2008). The most valuable application of these enzymes reported is the transformation of lignocellulose agro-waste substances into value-added products such as composite, pulp and paper, animal feeds, biofuels, and industrially important enzymes (Malherbe and Cloete, 2002). However, the potential of SMC as good-quality raw material for industrial purposes still requires further assessment.

TABLE 9.1
Spent Mushroom Substrate-Based Enzymes and Their Applications

Enzymes	Spent Mushroom Compost (SMC)	Species of Mushroom	Application	Reference
Amylase Cellulase β - Glucosidase	Paddy straw	*Pleurotusostreatus,* *Lentinula edodes,* *Flammulinavelutipes* *Hericiumerinaceum*	Biodegradation of dye	Ko et al. (2005) Gomes et al. (2009)
Laccase	Sugarcane bagasese	*Flammulinavelutipes* *Agaricusbisporus* *C. comatus*		
Xylanase Lignin peroxidase	Sawdust (Metroxylonsagu) Rice straw	*Hericiumerinaceum* *Pleurotussajor-caju*	Degradation of polycyclic aromatic hydrocarbons (PAHs)	Kumaran et al. (1997), Li et al. (2010) Singh et al. (2003)
Manganese peroxidase		*Pleurotusostreatus*	Pulp and paper, biofuels,	Wen et al. (2009)

9.6 CONCLUSION

Spent mushroom compost generated from mushroom cultivation is no longer treated as a waste product but is appreciated as a renewable resource with industrial applications. It can now be utilized with several new technologies to produce clean and green energy. The enzymes recovered from SMC are potentially useful for the bioremediation of pollutants and other industrial biotechnology purposes. The extraction of industrially important enzymes is affected by several factors, which may operate individually or in combination. Parameters such as temperature, pH, incubation period, type of biomass, carbon and nitrogen content of different substrates, microorganisms and pretreatment method provide a synergistic effect for optimization of enzyme production. This chapter has therefore suggested ideas for production of industrially important enzymes from mushroom cultivation, along with novel technology using agro-industrial waste residues.

REFERENCES

Abdeshahian, P., Samat, N., Hamid, A.A. & Yusoff, W.M.W. (2011). Solid substrate fermentation for cellulase production using palm kernel cake as a renewable lignocellulosic source in packed-bed bioreactor. *Biotechnology and Bioprocess Engineering.*, 16(2), 238–244.

Agosin, E. & Odier, E. (1985). Solid state fermentation, lignin degradation and resulting digestibility of wheat straw fermented by selected white rot fungi. *Applied Microbiology and Biotechnology*, 21, 393–403.

Alhujaily, A., Yu, H., Zhang, X. & Ma, F. (2018). Highly efficient and sustainable spent mushroom waste adsorbent based on surfactant modification for the removal of toxic dyes. *International Journal of Environmental Research* and *Public Health*, 15, 1421.

Ball, A. & Jackson, A. (1995). The recovery of lignocellulose-degrading enzymes from spent mushroom compost. *Bioresource Technology*, 54, 311–314.

Barbosa, E.S., Perrone, D., Vendramini, A.L.A. & Leite, S.G.F. (2008). Vanillin production by *Phanerochaetechrysosporium* grown on green coconut agro-industrial husk in solid state fermentation. *Bio Resources*, 3, 1042–1050.

Beyer, M. (1996). The impact of the mushroom industry on the environment. *Mushroom News*, 44(11), 6–13.

da Silva, A.E., Marcelino, H.R., Gomes, M.C., Oliveira, E.E., Nagashima, T.J. & Egito, E.S. (2012). Xylan, a promising hemicellulose for pharmaceutical use. In: Verbeek J, (ed.). *Products and Applications of Biopolymers*. Shanghai: In Tech China, pp. 61–84.

Dann, M.S. (1996). The many uses of spent mushroom substrate. *Mushroom News*, 44(8), 24–27.

Duran, N. & Esposito, E. (2000). Potential applications of oxidative enzymes and phenoloxidase-like compounds in wastewater and soil treatment: A review. *Applied Catalysis B: Environmental*, 28, 83–99.

Edwards, C. (1993). Isolation properties and potential applications of thermophilic actinomycetes. *Applied Biochemistry and Biotechnology*, *42*(2), 161–179.

Fazenda, M.L., Seviour, R., McNeil, B. & Harvey, L.M. (2008). Submerged culture fermentation of "Higher Fungi": The Macrofungi. *Advances in Applied Microbiology*, 63, 33–103.

Fen, L., Xuwei, Z., Nanyi, L., Puyu, Z., Shuang, Z. & Xue, Z. (2014). Screening of lignocellulose-degrading superior mushroom strains and determination of their CMCase and laccase activity. *The Scientific World Journal*, 6, 222–228. 763108.

Field, J.A., Vledder, R.H., Van Zelst, J.G. & Rulkens, W.H. (1996). The tolerance of lignin peroxidase and manganese-dependent peroxidase to miscible solvents and the in vitro oxidation of anthracene in solvent: Water mixtures. *Enzyme and Microbial Technology*, 18, 300–308.

Finney, K.N., Ryu, C., Sharifi, V.N. & Swithenbank, J. (2009). The reuse of spent mushroom compost and coal tailings for energy recovery: Comparison of thermal treatment technologies. *Bioresource Technology*, 100, 310–315.

Surendran, G., Karunakaran, S., Prabhu, N., Karthika, S.D. & Gajendran, T. (2021). Sequestration of Potential Enzymes from Mushrooms compost waste. *IOP Conference Series: Materials Science and Engineering*, 1145, (1), 012116. IOP Publishing.

Gao, Z., Mori, T. & Kondo, R., 2012. The pretreatment of corn stover with Gloeophyllumtrabeum KU-41 for enzymatic hydrolysis. *Biotechnology Biofuels*, 5, 1–11.

Gomes, E., Aguiar, A.P., Carvalho, C.C., Bonfa, M.R.B., Da Silva, R. & Boscolo, M. (2009). Ligninases production by Basidiomycetes strains on lignocellulosic agricultural residues and their application in the decolorization of synthetic dyes. *Brazilian Journal of Microbiology*, 40, 31–39.

Hatakka, A., Lundell, T., Hofrichter, M. & Maijala, P. (2003). Manganese peroxidase and its role in the degradation of wood lignin. In: Mansfield, S.D., Saddler, J.N. (Eds), *Applications of Enzymes to Lignocellulosic*, 855, 230–243. ACS Symposium Series. ACS Publications.

Himmel, M.E., Ding, S.Y., Johnson, D.K., Adney, W.S., Nimlos, M.R., Brady, J.W. & Foust, T.D. (2007). Biomass recalcitrance: Engineering plants and enzymes for biofuels production. *Science*, 315, 804–807.

Hu, T., Wang, X.J., Zhen, L.S., Gu, J., Zhang, K.Y., Wang, Q.Z., Ma, J.Y., Peng, H.L., Lei, L.S. & Zhao, W.Y., 2019. Effects of inoculating with lignocellulose-degrading consortium on cellulose-degrading genes and fungal community during cocomposting of spent mushroom substrate with swine manure. *Bioresource Technology*, 291, 121876.

Janusz, G., Pawlik, A., Sulej, J., Burek, U.S., Wilkołazka, A.J. & Paszczynski, A. (2017). Lignin degradation: Microorganisms, enzymes involved, genomes analysis and evolution. *FEMS Microbiology Reviews*, 41, 941–962.

Jasinska, A. (2018). Spent mushroom compost (SMC) – retrieved added value product closing loop in agricultural production. TY - JOUR *Acta Agraria Debreceniensis*. doi:10.34101/actaagrar/150/1715

Jordan, S.N. & Mullen, G.J. (2007). Enzymatic hydrolysis of organic waste materials in a solid–liquid system. *Waste Management*, 27(12), 1820–1828.

Jordan, S.N., Mullen, G.J. & Murphy, M.C. (2008). Composition variability of spent mushroom compost in Ireland. *Bioresource Technology*, 99, 411–418.

Ko, H.G., Park, S.H., Kim, S.H., Park, H.G. & Park, W.M. (2005). Detection and recovery of hydrolytic enzymes from spent compost of four mushroom species. *Folia Microbiologica*, 50, 103–106.

Kumar, A. & Chandra, R. (2020). Ligninolytic enzymes and its mechanisms for degradation of lignocellulosic-waste in environment. *Heliyon*, 6(2), e03170. doi:10.1016/j.heliyon.2020.e03170

Kumaran, S., Sastry, C.A. & Vikineswary, S. (1997). Laccase, cellulase and xylanase activities during growth of *Pleurotussajor-caju* on sago hampas. *World Journal of Microbiol Biotechnol*ogy, 13, 43–49.

Li, X., Lin, X., Zhang, J., Wu, Y., Yin, R., Feng, Y. & Wang, Y. (2010). Degradation of polycyclic aromatic hydrocarbons by crude extracts from spent mushroom substrate and its possible mechanisms. *Current Microbiology*, 60, 336–342.

Lim, S.H., Kim, J.K., Lee, Y.H. & Kang, H.W. (2012). Production of lignocellulytic enzymes from spent mushroom compost of *Pleurotuseryngii*. *Korean Journal of Mycology*, 40, 152–158.

Lim, S.H., Lee, Y.H. & Kang, H.W. (2013). Efficient recovery of lignocellulolytic enzymes of spent mushroom compost from oyster mushrooms, *Pleurotus*spp., and potential use in dye decolorization. *Mycobiology*, 41, 214–220.

Lin, L., Cui, F., Zhang, J., Gao, X., Zhou, M., Xu, N., Zhao, H., Liu, M., Zhang, C. & Jia, L. (2016). Antioxidative and renoprotective effects of residue polysaccharides from *Flammulinavelutipes*. *Carbohydrate Polymers*, 146, 388–395.

Ma, C., Hu, B., Wei, M.B., Zhao, J.H. & Zhang, H.Z. (2019). Influence of matured compost inoculation on sewage sludge composting: Enzyme activity, bacterial and fungal community succession. *Bioresource Technology*, 294, 122165.

Malherbe, A. & Cloete, T.E. (2002). Lignocellulose biodegradation: fundamentals and applications. *Review Environment Science Biotechnology*, 1, 105–114.

Mayolo-Deloisa, K., Trejo-Hernandez, M.D.R. & Rito-Palomares, M. (2009). Recovery of laccase from the residual compost of *Agaricusbisporus* in aqueous two-phase systems. *Process Biochemistry*, 44, 435–439.

Millati, R., Trihandayani, E., Cahyanto, M., Taherzadeh, M. J. & Niklasson, C. (2011). Ethanol from oil palm empty fruit bunch via dilute-acid hydrolysis and fermentation by *Mycur Indicus* and *Saccharomyces cerevisiae*. *Agricultural Journal*, *6*(2), 54–59.

Mosier, N., Wyman, C., Dale, B., Elander, R., Lee, Y.Y., Holtzapple, M. & Ladisch, M. (2005). Features of promising technologies for pretreatment of lignocellulosic biomass. *Bioresource Technology*, 96, 673–686.

Nakajima, V.M., de Freitas Soares, F.E. & de Queiroz, J.H. (2018). Screening and decolorizing potential of enzymes from spent mushroom composts of six different mushrooms. *Biocatalysis and Agricultural Biotechnology*, 13, 58–61.

Patel, H., Gupte, A. & Gupte, S. (2009). Effect of different culture conditions and inducers on production of laccase by basidiomycetes fungal isolate *Pleurotusostreatus* HP-1 under solid state fermentation. *Bioresources*, 4, 268–284.

Paul, M., Mohapatra, S., Mohapatra, P.K.D. & Thatoi, H. (2021). Microbial cellulases: An update towards its surface chemistry, genetic engineering and recovery for its biotechnological potential. *Bioresource Technology*, 340, 125710.

Phan, C.W. & Sabaratnam, V. (2012). Potential uses of spent mushroom substrate and its associated lignocellulosic enzymes. *Applied Microbiology and Biotechnology*, 96, 863–873. doi:10.1007/s00253-012-4446-9

Polizeli, M.L., Rizzatti, A.C., Monti, R., Terenzi, H.F., Jorge, J.A. & Amorim, D.S. (2005). Xylanases from fungi: Properties and industrial applications. *Applied Microbiology and Biotechnology*, 67, 577–591.

Pothiraj, C., Kanmani, P. & Balaji, P. (2006). Bioconversion of lignocellulose materials. *Mycobiology*, 34(4), 159–165.

Sindhu, R., Parameswaran, B. & Pandey, A. (2016). Biological pretreatment of lignocellulosic biomass:An overview. *Bioresource Technology*, 199, 76–82.

Sanchez, C. (2009). Lignocellulosic residues: Biodegradation and bioconversion by fungi. *Biotechnology Advances*, 27, 185–194.

Schimpf, U. & Schulz, R. (2016). Industrial by-products from white-rot fungi production. Part I: Generation of enzyme preparations and chemical, protein biochemical and molecular biological characterization. *Process Biochemistry*, 51, 2034–2046.

Schlosser, D. & Hofer, C. (1999). Novel enzymatic oxidation of Mn2+ to Mn catalyzed by a fungal laccase. *FEBS Letters*, 451, 186–190.

Singh, A.D., Abdullah, N. & Vikineswary, S. (2003). Optimization of extraction of bulk enzymes from spent mushroom compost. *Journal of Chemical Technology & Biotechnology*, 78, 743–752.

Singh, D., Zeng, J. & Chen, S., (2011). Increasing manganese peroxidase productivity of Phanerochaetechrysosporium by optimizing carbon sources and supplementing small molecules. *Letters in Applied Microbiology*, 53, 120–123.

Sorensen, A.L., Stephensen, P., Mette, L., Johan, T.P. & Ahring, B.K. (2011). β-Glucosidases from a new Aspergillus species can substitute commercial β-glucosidases for saccharification of lignocellulosic biomass. *Canadian Journal of Microbiology*, 57(8), 638–650. doi:10.1139/w11-05

Sukharnikov, L.O., Alahuhta, M., Brunecky, R., Upadhyay, A., Himmel, M.E., Lunin, V.V. & Zhulin, I.B. (2012). Sequence, structure, and evolution of cellulases in glycoside hydrolase family. *Journal of Biological Chemistry*, 287(49), 41068–41077.

Sun, C., Wei, Y., Kou, J., Han, Z., Shi, Q., Liu, L. & Sun, Z. (2021). Improve spent mushroom substrate decomposition, bacterial community and mature compost quality by adding cellulaseduring composting. *Journal of Cleaner Production*, 299, 126928. doi:10.1016/j.jclepro.2021.1269

Tanyolac, A. & Aktas, N. (2003). Reaction conditions for laccase catalyzed polymerization of catechol. *Bioresource Technology*, 87, 209–214.

Ten Have, R. & Teunissen, P.J.M. (2001). Oxidative mechanisms involved in lignin degradation by white-rot fungi. *Chemical Review*, 101(11), 3397–3413.

Wen, X., Jia, Y. & Li, J. (2009). Degradation of tetracycline and oxytetracycline by crude lignin peroxidase prepared from *Phanerochaetechrysosporium*-a white rot fungus. *Chemosphere*, 75, 1003–1007.

Widsten, P. & Kandelbauer, A. (2008). Laccase applications in the forest products industry: A review. *Enzyme and Microbial Technology*, 42, 293–307.

Wong, W.S.D. (2009). Structure and action mechanism of ligninolytic enzymes. *Applied Biochemistry and Biotechnology*, 157, 174–209.

Zhu, H., Tian, L., Zhang, L., Bi, J., Song, Q., Yang, H. & Qiao, J. (2018). Preparation, characterization and antioxidant activity of polysaccharide from spent *Lentinus edodes* substrate. *Internatioanl Journal of Biology and Macromolecules*, 112, 976–984.

10 Indigenous Fermented Food and Beverages of Manipur

Vera Yurngamla Kapai
University of Delhi, New Delhi, India

Leisan Judith
University of Delhi, New Delhi, India

Gladys Muivah
University of Delhi, New Delhi, India

CONTENTS

DOI: 10.1201/9781003306931-11

10.1 INTRODUCTION

Of the eight states of Northeast India, Manipur is one state with rich ethno-cultural diversity. It is home to a diverse population of approximately 30 ethnic groups, both tribal and non-tribal. The majority non-tribal Meitei community lives in the valley. The Naga and Kuki-Chin-Mizo tribes live in the hilly region (Wahengbam et al., 2020). Food plays a very significant role in distinguishing one ethnic group from another (Keishing and Banu, 2013), and this is reflected in their food habits. Depending on the availability of diverse local food sources such as bamboo, soybean, fish, grains, meat, and vegetables, foods are fermented (Das and Deka, 2012; Tamang et al., 2012) using various traditional techniques handed down through generations of practice. The fermentation processes carried out serve to preserve and enhance the taste, aroma, flavor, and texture of the food that are the result of biological enrichment and modification of food products by different microbial populations (Sekar and Mariappan, 2007). All ethnic communities' traditional diet systems and cultures include a variety of indigenous fermented foods. In Manipur the most commonly used ethnic plant-based fermented foods are fermented bamboo shoots (*soidon, soibum*), fermented soybean (*hawaijar/theishui*), fermented mustard leaves (*Inziangsang/ziang-dui/ziang-sang*) and animal-based fermented fish (*ngari, hentak*). Fermented alcoholic rice beverages use ethnic starters—*hamei, chamri, khai*—to obtain non-distilled beverages (*yu angouba, khor*) and distilled liquors (*atingpa, yu, acham*).

This chapter focuses on the indigenous fermented food and beverages consumed by the people of Manipur and prepared using age-old traditional processes, the associated microbes, and the nutritional benefits derived from them.

10.2 FERMENTED BAMBOO SHOOT

Bamboo shoot is one of the most widely consumed and marketed nutritious vegetables in Northeast India. Protein, carbohydrates, amino acids, minerals, fat, sugar, fiber, and inorganic salts are the nutrients present in bamboo shoots. Fresh shoots are a good source of thiamine, niacin, vitamin A, vitamin B6, and vitamin E (Visuphaka, 1985; Xia, 1989; Shi and Yang, 1992). They contain 17 amino acids, eight of which are essential for human health (Qiu, 1992). Bamboo shoots are consumed in fresh, canned, dried and fermented form (Nirmala et al., 2008).

Fermented bamboo shoot is extensively used in the states of Northeast India. It adds flavor and is an important ingredient in Manipuri cuisine. Using their traditional techniques, several ethnic communities produce a variety of fermented bamboo shoots, including the following.

10.2.1 *SOIBUM*

Soibum is an indigenous fermented food prepared from thin slices of the inner part of young bamboo shoot (Table 10.1). *Dendrocalamus hamiltonii, D. sikkimensis, D. giganteus, Melocana bambusoide, Bambusa tulda*, and *B. balcooa* are among the succulent bamboo shoot species used to prepare *soibum*. The fermentation process occurs during the months of June to September, when young bamboo shoots begin to grow. *Soibum* production takes place mostly in Manipur's hilly regions, although it is dependent on availability and quantity of raw material (Bhatt et al., 2003; Jeyaram et al., 2009, Figure 10.1).

Lactobacillus brevis, L. plantarum, L. coryniformis, L. delbrueckii, L. lactis, Leuconostoc falllax, L. mesenteroides, Enterococcus durans, Streptococcus lactis, Bacillus subtilis, B. licheniformis, B. coagulans, and the yeasts *Candida sp., Saccharomyces sp., Torulopsis sp.* have been identified as organisms involved in the fermentation of *soibum* (Tamang et al., 2008; Tamang and Tamang, 2009).

For *soibum* preparation, *Noney/Kwatha* and *Andro* are the two types of fermentation techniques used. The outside hard inedible sheaths of succulent bamboo sprouts are peeled off in both procedures, leaving just the inner soft succulent portions to be chopped up. These are pressed tightly into wooden or earthenware pots to ferment for 6–12 months.

TABLE 10.1

Types of Fermented Food Products, Nutritional Composition Associated Microbes and Their Uses

Local Name of Fermented Products	Substrate	Nutritional Composition		Microorganisms Associated with the Fermentation	Uses	Community	References
		Organic Nutrients	Inorganic Nutrients				
Soibum	Tender bamboo shoot *Dendrocalamus hamiltonii, D. sikkimensis, D. giganteus, Melocana bambusoide, Bambusa tulda* and *B. balcooa*	Fat: 3.2%, Protein: 36.3%, Carbohydrate: 47.2%, Food value: 362.8 kcal/100 gm,	Ca: 16.0 mg/100 gm, Na: 2.9 mg/100 gm, and K: 212.1 mg/100 gm	*Lactobacilus brevis, L. coryniformis, L. delbrueckii, L. lactis, L. plantarum Leuconostoc falllax, L. mesentroides, Enterococcus durans, Streptococcus lactis, Bacillus subtilis, B. licheniformis, B. coagulans, Candida sp., Saccharomyces sp., Torulopsis sp.*	Cooked in curries; chutney	Meitei	Jeyaram et al. (2010), Sarangthem and Singh (2003), Tamang et al. (2008), Tamang and Tamang (2009)
Soidon	Matured bamboo shoot *Teinostachyum wightii, Bambusa tulda, Dendrocalamus giganteus, Melocana bambusoide*	Fat: 3.1%, Protein: 37.2%, Carbohydrate: 46.6%, Food value: 363.1 kcal/100 gm,	Ca: 18.5 mg/100 gm, Na: 3.7 mg/100 gm, and K: 245.5 mg/100 gm	*Lactobacillus brevis, L. plantarum, L. acetotolera, L. fallax, L. citreum, Lactis subsp. cremoris, Weissella cibaria, W. ghanensis*	Eaten as curry and pickle;	Meitei	Tamang et al. (2008), Romi et al. (2015)
Hawaijar/ Theishui	Soybean (*Glycine max*)	Protein: 43.9%, Fat: 27.9%, Carbohydrate: 23.4%, Food value: 521.2 kcal/100 gm,	Ca: 357.8 mg/100 gm, Na: 88.7 mg/100 gm,Fe: 92.3 mg/100 gm, K: 835.1 mg/100 gm and Zn: 63.0 mg/100 gm	*Bacillus subtilis, B. licheniformis, B. amyloliquefaciens, B.cereus, Staph. aureus, Staph. sciuri, Alkaligenes sp., Providencia rettgers, Proteus mirabilis*	consumed as '*ametpa*' - a paste with chili and salt; also added in vegetable or meat stew.	Meitei, Tangkhul	Jeyaram et al. (2008a, 2008b), Singh et al. (2014), Wahengbam et al. (2020)
Ngari	Fish (*Puntius sophore, Tenualosa ilisha*)	Protein: 34.1%, Fat: 13.2%, Carbohydrate: 31.6%, Food value: 381.6 kcal/100 gm,	Ca: 41.7 mg/100 gm, Fe: 0.9 mg/100 gm, Mg: 0.8 mg/100 gm, Mn: 0.6 mg/100 gm, and Zn: 1.7 mg/100 gm	*Bacillus subtilis, B. licheniformis, B. amyloliquefaciens, B.cereus, Staph. aureus, Staph. sciuri, Alkaligenes sp., Providencia rettgers, Proteus mirabilis*	used in *eromba, ametpa, kangsoi, singju*, either after frying or steaming; also consumed with rice as a side dish	Meitei	Thapa et al. (2004), Thapa et al. (2007), Devi et al. (2015)
Hentak	Finger-sized fish *Esomus danricus* and petioles of *Alocasia macrorrhiza*	Protein: 32.7%, Fat: 13.6%, Carbohydrate: 38.7%, Food value: 408.0 kcal/100 gm,	Ca: 38.2 mg/100 gm, Fe: 1.0 mg/100 gm,Mg: 1.1 mg/100 gm, Mn: 1.4 mg/100 gm, and Zn: 3.1 mg/100 gm	*Lactobacillus fructosus, L. amylophilus, Enterococcus faecium, Bacillus cereus, B. subtilis, Staphylococcus aureus, Candida sp*	Used as the main ingredient or in curry. also consumed directly by roasting and steaming; and in chili-chutney	Meitei	Thapa et al. (2004), Thapa et al. (2007)
Ziang-sang/ Ziang-dui	Mustard leaves (*Brassica* sp.)	Protein: 38.7%, Fat: 3.2%, Carbohydrate: 41.2%, Food value: 348.4 kcal/100 gm	Ca: 240.4 mg/100 gm, Na: 133.7 mg/100 gm, and K: 658.4 mg/100 gm	*L. brevis, L. plantarum Pediococcus acidilactici*	Consumed as a condiment for flavoring local dishes and soups	Zeliangrong	Tamang et al. (2005)

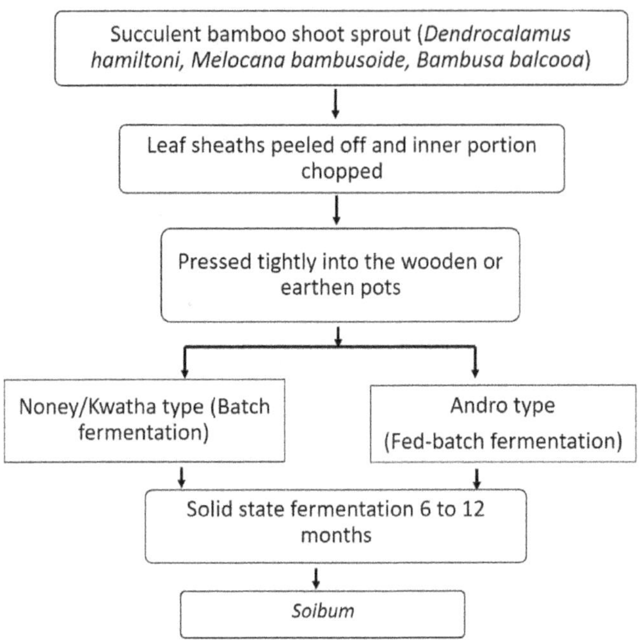

FIGURE 10.1 Flowchart for steps involved in preparation of *soibum*.

(Adapted from Jeyaram et al. 2009)

In the *Noney/Kwatha* type, batch-type fermentation is carried out in traditionally designed bamboo chambers perforated at the base to allow seepage of fluid. Bamboo chambers are packed tightly with slices of juicy and delicate bamboo shoots filled to capacity, sealed with a polythene sheet, then weights are placed on top of it to keep the seal intact. The bottom of this basket is perforated so that the acidic juices produced can flow through during fermentation. For solid-stage fermentation, the set-up is left for 6–12 months. The fermented product can be stored for a long time.

The *Andro* type of fermentation is conducted exclusively in Andro village. Fed-batch fermentation takes place in a large earthenware vessel. Part of the pot is first filled with bamboo shoot slices, which are allowed to ferment. More bamboo shoot slices are added once the volume has been reduced by fermentation. To condense the bulk, pressure is applied from above. This technique is repeated until the pot is completely filled, after which it is kept for a period of 6–12 months. In the *Andro* type of fermentation, the fluid is not allowed to drain (Jeyaram et al., 2008a). The product has a whitish color, a slight aroma, and a sour flavor. As a side dish, it is usually served with steamed rice. *Soibum* is also used to make Manipuri meals using corms of *Colocasia* sp., green peas, pumpkins, and potatoes. Some people also cook it with fish (Jeyaram et al., 2009; Tamang and Tamang, 2009.

10.2.1.1 Nutritional Value

The nutritional profile of *soibum* includes carbohydrate, dietary fiber, protein, fat, ash, and minerals (Ca, Na, K, Mg). In comparison to freshly picked young shoots, *soibum* has lower levels of all nutritional components except dietary fiber (Nirmala et al., 2008; Table 10.1). Fiber-rich diets lower body fat, serum and hepatic lipids, particularly cholesterol, while also promoting peristaltic activity in the intestines, which aids digestion and prevents constipation. Water-soluble vitamins (ascorbic acid, riboflavin, folic acid, cyanocobalamin, etc.) are abundant in *soibum* (Singh et al., 2011; Sonar et al., 2015). In edible bamboo shoots, fermentation reduces poisonous and anti-nutritive components such as cyanogenic glycoside, phytate, and saponin. As a result, their safety, nutritional quality, and health advantages have improved (Sarangthem and Singh 2013). Lactic acid bacteria, particularly *Lactobacillus* spp., give *soibum* its functional value. It has probiotic capabilities, high riboflavin production, hydrolytic enzyme production, anti-nutritive chemical degradation properties, protease activities, and broad-spectrum antibacterial activity against common food-borne pathogens (Sarangthem and Singh 2013; Sonar and Halami 2014; Romi 2015; Sonar et al. 2015; Thakur and Tomar 2015, 2016; Tamang, 2020).

10.2.2 *Soidon*

Soidon is a product made by fermenting only the apical portion of succulent bamboo shoots. It is made from bamboo species such as *Teinostachyum wightii, Bambusa tulda Roxb., Dendrocalamus giganteus Munro, Melocana bambusoide Trin.* (Jeyaram et al., 2009; Tamang and Tamang, 2009. *Lactobacillus brevis, Leuconostoc fallax*, and *Lactobacillus lactis* are microorganisms associated with the production of *soidon* (Tamang et al., 2008, 2016).

The tips of succulent bamboo shoots are cut transversely into small pieces or sometimes vertically into 2–4 pieces and submerged in water in an earthenware pot (Figure 10.2). As a starter culture, a 1:1 dilution of milky fermented sour liquid (*soijim*) from the previous batch is added to this mixture. As an acidifier and flavor enhancer, the leaves of the locally available heibung plant (*Garcinia pedunculata*) can be added to the fermenting mixture. To improve the color, rice washing water called *chenghi* is occasionally added in a 1:10 dilution. Fermentation is allowed to take place for 3–7 days once the pot is covered, after which the *soidon* is ready to be consumed (Figure 10.9B). When stored in plastic containers at room temperature, it can be kept for up to a year (Jeyaram et al., 2009; Tamang and Tamang, 2009). *Soidon* is consumed both as a curry and as a pickle. The liquid part (*soijim*) can be used as a condiment to give curry an acidic flavor (Tamang and Tamang, 2009).

10.2.2.1 Nutritional Values

Soidon contains about 55 volatile organic chemicals (VOCs), that are responsible for its odor, flavor, and texture, including alkanes, alkenes, esters, aldehydes, ketones, alcohols, and phenolic aromatic compounds (Sonar et al., 2015). P-cresol, 2-methyl naphthalene, 2-heptanol, acetic acid, linalool, and phenyl acetaldehyde are some of the most common VOCs. Among the fermented bamboo shoots of Manipur, *soidon* is considered to have the least glucosidase inhibitory action. As a result, it reduces glucose absorption from the gut into the bloodstream, lowering blood sugar levels and

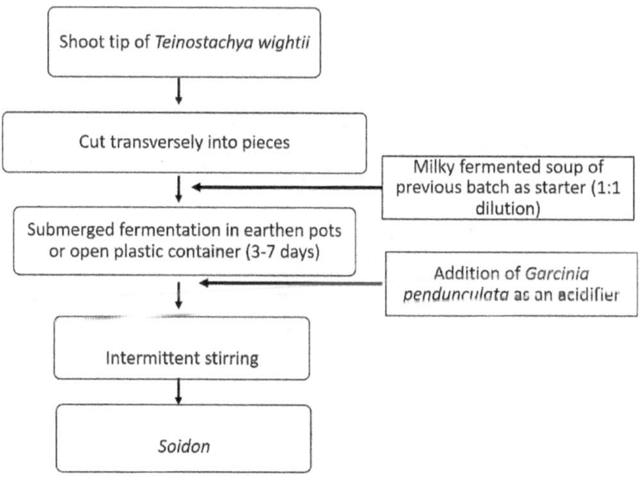

FIGURE 10.2 Flowchart for steps involved in preparation of *soidon*.

(Adapted from Jeyaram et al. 2009)

aiding diabetes control. Ascorbic acid, folic acid, and cyanocobalamin are the most prevalent vitamins in *soidon* (Sonar et al., 2015). *Soidon* has the least amount of cyanogenic glycosides, making it the safest fermented bamboo shoot to eat (Sarangthem and Singh 2013). *Lactobacillus* spp responsible for the functional value of *soidon* have been linked to probiotic qualities, hydrolytic enzyme synthesis, anti-nutritive chemical degradation, and protease activity.

10.3 FERMENTED FISH

Many tribes in Northeast India prepare fermented fishes, which are usually made from locally available small species of freshwater fish. Extreme perishability and small size are important factors limiting the use of freshwater fish. In this region, traditional fish processing methods including salting, drying, and smoking are routinely used to preserve fish. In Manipur, the fish that are traditionally fermented and preserved are *ngari* and *hentak* (Figures 10.3 and 10.4). These products have a

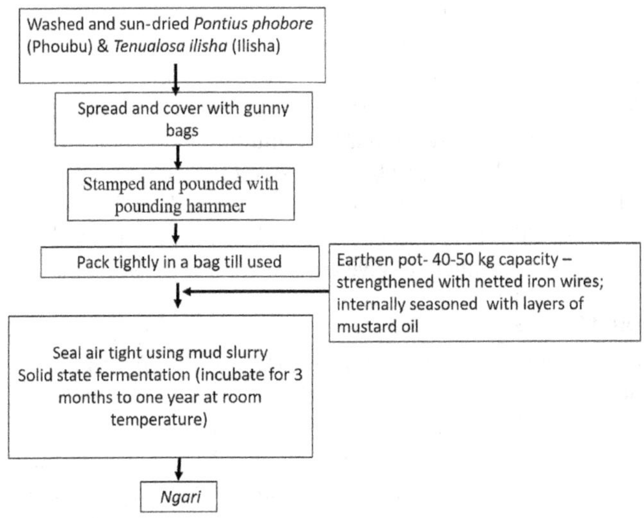

FIGURE 10.3 Flowchart for steps involved in preparation of *ngari*.

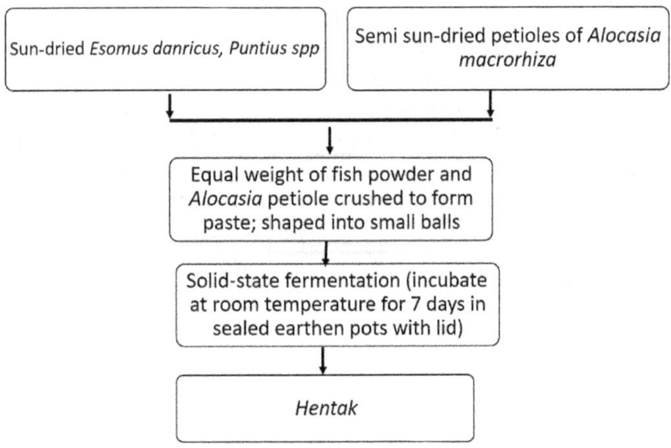

FIGURE 10.4 Flowchart for steps involved in preparation of *hentak*.

(Adapted from Jeyaram et al. 2009)

unique taste and smell and are sought after for making various culinary preparations. They are also a source of protein.

10.3.1 *Ngari*

Ngari is an intrinsic component mainly in the diet of the Meitei communities. The fish species mainly used for its preparation are *Puntius sophore* and *Tenualosa ilisha*, locally called *Phabou nga* and *Ilisha* (Jeyaram et al., 2009; Soibam and Ayam, 2018). Lactic acid bacterium (LAB) is the most common fermenting organism found in *ngari* samples. *Lactococcus plantarum* and *Lactobacillus plantarum* have been identified. *Ngari* has also yielded *Bacillus subtilis*, *Bacillus pumilus*, and *Miocrococcus* sp. (Tamang et al., 2016). *Candida* sp. has been identified as the fungal isolate (Thapa et al., 2004; Tamang et al., 2016).

Ngari preparation is mostly confined to small village households. The fish are either sourced directly from local fishponds or imported from the Brahmaputra valley and Bangladesh. Processing time is during the fishing season, from October to January, when the fish are readily accessible.

Whole fishes are sorted, washed with water and drained followed by drying in the sun for 3–4 days (Figure 10.3). They are then spread on gunny bags, covered with another layer of gunny bags and stamped and pounded with a traditional long-handled pounding hammer locally known as a *droomboos* (tamper) to crush or soften the head and bones for faster and better fermentation. After that, the dried fish are either utilized immediately for fermentation or stored in gunny bags until needed.

The fish are put in earthenware pots called *ngari chaphu* which have been seasoned with multiple layers of mustard oil. The oil coatings help to provide anaerobic conditions inside the pot, which is key to the process of fermentation. The outer wall of the *ngari chaphu* has netting with thick, strong iron wires to strengthen the pots. After the fish are packed tightly (airtight), the pots are sealed with polythene sheet, fish scales, oil slurry, and finally smeared with mud/sand and cow dung slurry. The pots are then kept in the dark at room temperature, where fermentation takes 3 to 6 months to complete and maturation takes 12 months (Figure 10.9D). It has a characterestic odor and a shelf life of 12 to 18 months (Jeyaram et al., 2009; Singh et al., 2010). The majority of LAB bacteria recovered from *ngari* had a high degree of hydrophobicity, demonstrating its probiotic properties (Thapa et al., 2004).

It is used for preparing different delicacies and signature dishes of the Meitei people such as *eromba, ametpa, kangsoi, singju*, where it is added either after frying or steaming (Soibam and Ayam, 2018). It is also consumed with rice as a side dish (Singh et al., 2010).

10.3.1.1 Nutritional Value

The traditional meal has significant nutritional value due to the high protein content of *ngari*. It contains a high amount of essential and non-essential amino acids, lipids (Omega-3 and Omega-6 fatty acids), and minerals (calcium, salt, potassium, and magnesium) and has a food value of 381.6 kcal/100 g. (Devi et al. 2015; Majumdar et al. 2015; Thapa, 2016; Table 10.1). *Ngari* contains 39.6% of the total amino acid content in the form of essential amino acids. Compared to other similar fish products, *ngari* has a higher concentration of volatile nitrogenous compounds, which causes its organoleptic characteristics to develop during fermentation (Tamang, 2010). *Ngari* has been shown to have antioxidant, fibrinolytic, and probiotic activities in several investigations. The antioxidant properties of *ngari* were discovered to be dependent on the protein concentration and fermentation period, as indicated by its DPPH (2, 2-diphenyl-1-picrylhydrazyl) scavenging activity. (Phadke et al., 2014). The angiotensin converting enzyme (ACE) inhibitory action increases as the fermentation period and protein concentration increase. *Ngari* are of non-microbial endogenous origin and have stronger fibrinolytic activity than most fermented soybean products, including *hawaijar*. Regular consumers are thought to be protected from hypertension and cardiovascular disease due to the ACE inhibitory action of *ngari*, in combination with its endogenous fibrinolytic activity (Phadke et al., 2014; Singh et al., 2014; Keishing and Banu, 2015). *Ngari* also has probiotic qualities, as *Enterococcus faecium* BDU7 isolated from it shows bile acid tolerance, auto-aggregation, and hydrophobicity, among other characteristics (Abdhul et al., 2014).

10.3.2 *Hentak/Khaiti*

Another fermented fish product in the form of paste, used in Manipur is *hentak* also known as *khaiti* by the Tangkhul Naga ethnic group (Jeyaram et al., 2009). *Lactobacillus fructosus, L. amylophilus, Enterococcus faecium, Bacillus cereus, B. subtilis, Staphylococcus aureus, Enterococcus faecium*and *Candida* sp have been identified from *hentak* (Thapa et al., 2004).

For preparing *hentak* (Figure 10.4), fishes of the species *Puntius sophore, Esomus danricus, Amblypharyngodon mola, Puntius manipurensis* and *Trichogaster fasciata* are obtained from local sources (Soibam and Ayam, 2018). They are sorted, washed and dried in the sun and pulverized with a mortar and pestle. Meanwhile petioles of the aroid plant (*Alocasia macrorhiza*) are chopped into small pieces and semi-sundried. The fish powder and the chopped plant material are blended and crushed in a 1:1 ratio and made into a paste (Figure 10.9G). These pastes are shaped into small balls, placed in earthenware jars, sealed, and allowed to ferment at room temperature. The entire procedure takes around 2 weeks, following which the product is ready for consumption.

These balls become hardened during storage and can then be pounded again to form paste by adding water and stored for future use (Jeyaram et al., 2009). *Hentak* is consumed directly either by roasting, steaming or as a supplement with boiled rice. It is sometimes given to women in the final stages of pregnancy or patients who are recovering from illness or injury (Sarojnalini and Singh, 1988).

10.3.2.1 Nutritional Value

Hentak has a food value of 408 kcal/100 g. It is a good source of proteins, lipids, and minerals (Majumdar et al., 2015; Thapa, 2016). The use of sun-dried fish as a raw material explains the low moisture content of *hentak* (35%). Glycine (5.72% dry weight), alanine (4.09%), proline (4.45%), aspartic acid (3.84%), glutamic acid (3.35%), and essential amino acids (44% of total amino acids) are all abundant in *hentak* (Table 10.1). *Hentak* possesses antioxidant, fibrinolytic, and probiotic effects, similar to *ngari*. The antioxidant property of *hentak* was determined to be around 36% based on its DPPH scavenging activity (Singh et al., 2018a). *Debaryomyces fabryi*, a yeast found in *hentak*, shows significant fibrinolytic activity. As a result, *hentak* has higher fibrinolytic activity than *ngari*, which could be micro- and endogenous (Singh et al., 2014). Probiotic bacteria isolated from *hentak, Levilactobacillus brevis, Levilactobacillus pentosus*, and *Bacillus subtilis* have high antioxidant and antibacterial properties (Aarti et al., 2016, 2017; Singh et al., 2018a).

10.4 FERMENTED BEANS

In many developing countries, including India, fermented foods made from legumes constitute an important part of the human diet (Sandhu and Soni, 1989). In almost all the states of Northeast India, fermented soybean products are widely used. It is similar to the Chinese *tou-shi, hamanatto, chiang-you, shi-tche, chiang* and *tofu*, Indonesian *tempeh kedele, kecap* and *taoco*, and Japanese *shoyu* and *miso*.

Soyabean seeds contain flavonoids, terpenoids and other natural antioxidants like carotene, ascorbic acid, and tocipherol. Some traditionally fermented bean products have shown an increase in crude protein content and a decrease in fat content (Dajanta et al., 2011, Appaiah et al., 2011). Also, the total soluble sugar of fermented seeds is seen to decrease (Premarani and Chhetry, 2011). The sticky fermented soyabean product used in Manipur is known as *hawaijar* or *theishui* by Tangkhul Naga people.

10.4.1 *Hawaijar/Theishui*

Hawaijar is derived from *hawai* meaning pulses and *jar* which is a shortened form of *achar*, meaning pickle (Jeyaram et al., 2009; Premarani and Chhetry, 2010).

Traditionally, in the preparation of *hawaijar*, small and medium-sized soyabean (*Glycine max* L.) seeds are used which are cleaned and sorted (Devi and Kumar, 2012). The product is brown in color and has a white slimy substance covering it (Jeyaram et al., 2009; Figure 10.9C).

The species *Bacillus subtilis, B. licheniformis, B. cereus, Staphylococcus aureus, S. sciuri, Alkaligenes sp.* and *Providencia rettgeri* have been found to be predominantly present in fermented *hawaijar* (Jeyaram et al., 2008a: Table 10.1).

Soybean seeds are prepared by soaking for a few hours before being washed thoroughly with water and cooked until the seeds soften (Figure 10.5) They are then rinsed in hot water and packed tightly in a small bamboo basket (*lubak*) base-lined with fig (*Ficus hispida*), banana (*Musa* sp.) or bamboo leaves at the base. The basket is covered with a jute cloth and left out in the sun, kept near the fireplace or buried in paddy. This helps maintain the optimal temperature (> 40°C) for fermentation. After 3–5 days, the fermented product is ready for consumption. The final product is brown in color, sticky in texture, and emits an ammonia-like odor. For storage, the fermented product is wrapped in banana leaves (Jeyaram et al., 2009; Premarani and Chhetry, 2010; Tamang et al., 2012).

Without refrigeration, *hawaijar* has a very short shelf life of roughly 7 days (Tamang, 2015). Hence, the product is sometimes sun-dried to enable long-term storage (Premarani and Chhetry, 2010). It is known for its unique organoleptic properties. *Chagempomba*, which is a special delicacy of the Meitei people, is prepared using *hawaijar*, rice and other vegetables. It is consumed as a paste with chili and salt known as *ametpa*. It is also added in vegetable stew (Premarani and Chhetry, 2010) or cooked with pork.

10.4.1.1 Nutritional Value

Hawaijar is consumed commonly in the local diet as a low-cost source of high protein food (Devi and Kumar, 2012). Various health benefits have been reported in many studies, such as anticancer, anti-osteoporosis and hypocholesterolemic effects (Keishing and Banu, 2013). Protein hydrolysis is the most significant biochemical change that occurs during *hawaijar* fermentation (Appaiah et al., 2011). Proteolysis raises the protein content of *hawaijar* during fermentation; protease enzymes

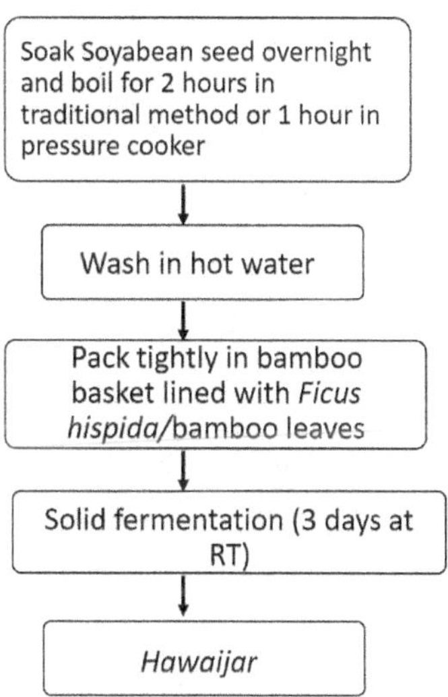

FIGURE 10.5 Flowchart for steps involved in preparation of *hawaijar*.

(Adapted from Jeyaram et al. 2009)

produced by microorganisms results in hydrolysis of protein and releases free amino acids (Dajanta et al., 2011). Benefits of fermented soybean include fibrinolytic and thrombolytic activity, antiviral activity, antioxidant activity, antidiabetic activity, ACE inhibitory activity, anti-glucosidase activity, and acetyl-choline esterase (AChE) inhibitory activity. As a result, it is a healthy diet for people with heart disease, high blood pressure, and diabetes (Singh et al., 2014; Rai and Jeyaram, 2015).

10.5 FERMENTED ALCOHOLIC BEVERAGES

Since time immemorial, the tribal people of Northeast India have been known to consume fermented alcoholic beverages of different forms. These products are similar to *shaosingiju* and *laochao* in China, *sake* in Japan, *chongju* and *takju* in Korea, *brem bali, tapuy* and *tape-ketan* in Indonesia, *khaomak* in Thailand and *tapai pulul* in Malaysia. The commonly consumed alcoholic beverage in Manipur is prepared from rice. Any kind of rice can be used for the preparation of alcoholic beverages but glutinous sticky rice is preferred by some. Three different kinds of alcoholic beverages are discussed below.

10.5.1 Yu angouba/khor

For the preparation of *yu angouba* the rice is soaked in water for around 2–3 hrs along with some germinated paddy (Figure 10.6). For 1 kg rice around 100g germinated paddy is used. After water has been drained, the soaked rice is pounded into powder with the help of a mortar. In another container, water is boiled and added to the crushed powdered rice, followed by continuous stirring and cooling. The container is covered with a muslin cloth and kept aside for 2–3 days without any disturbance. During these few days, due to the setting in of fermentation and microbial activity, a typical flavor and odor develop. This indicates that *yu angouba*, also known as *khor*, is now ready for consumption. *Yu angouba* can be stored for about a week, but not for a long time.

10.5.2 Atingba

An alcoholic beverage called *atingba* is prepared from glutinous rice. The starter culture used for this preparation is called *hamei/khai*. The secret recipes for its preparation are passed down through

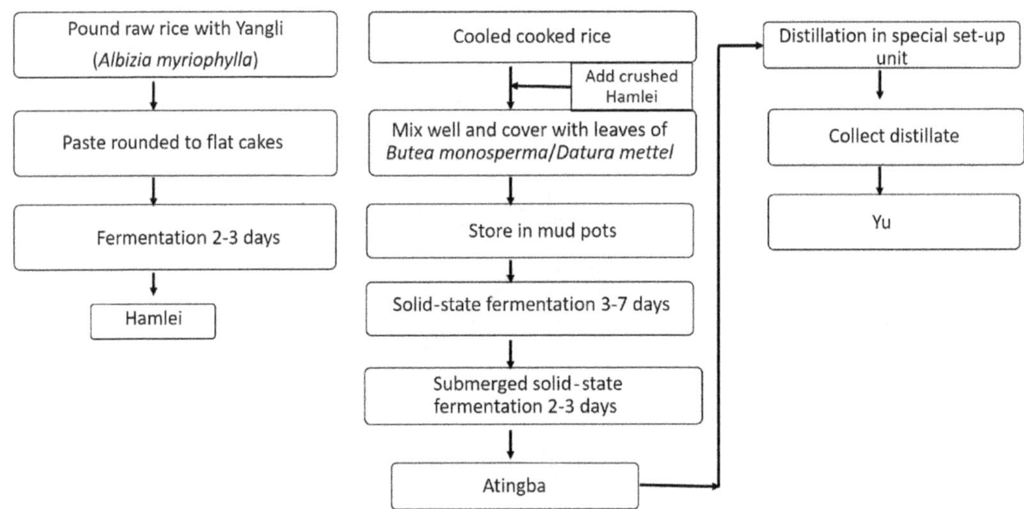

FIGURE 10.6 Flowchart for steps involved in preparation of *yu angouba*.

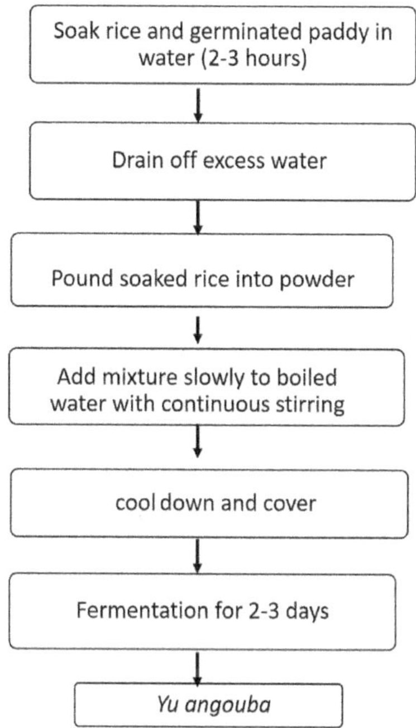

FIGURE 10.7 Flowchart for steps involved in preparation of *yu*.

the generations (Jeyaram et al., 2008b). To prepare *hamei* (Figure 10.7), raw rice is crushed with bark powder from the yangli plant (*Albizia myriophylla*) (0.25 kg per kg of rice) along with water to form a dough-like mass. The powdered *hamei* from the previous batch is added to this freshly prepared dough and mixed thoroughly. Thereafter it is molded into flat cakes with a diameter of 2–7 cm and a thickness of 0.6–1.5 cm. For 2 to 3 days at room temperature, they are spread over rice husks in the floor or bamboo baskets without overlapping. The cakes swell after fermentation, producing an alcoholic flavor and a yellowish hue. They can be dried and kept for up to a year (Tamang et al., 2007; Jeyaram et al., 2008b; Jeyaram et al., 2009). The LAB isolated from samples of *hamei* have been identified as *Lactobacillus plantarum and Pediococcus pentosaceus* (Tamang and Nikkuni, 1998). *Saccharomyces cerevisiae, Pichia anomala, P. guilliermondi, P. fabianii, Candida tropicalis, C. parapsilosis, C. montana, Torulaspora delbrueckii and Trichosporon sp*. are the fungal agents associated with its fermentation (Jeyaram et al., 2008b; Sha et al., 2018, 2019).

Glutinous rice is cooked, cooled, and then mixed with crushed *hamei* to make *atingba* (5 cakes for 10 kg). The mixture is then fermented in solid state for 3–4 days in the summer and 6–7 days in the winter in mud pots covered with *Butea monosperma, Alocasia macrorhyza*, or *Datura mettel* leaves. After mixing with water at a 1:1 ratio, the fermentation is submerged in earthenware pots for 2–3 days. *Atingba* is the beverage obtained following filtration of the fermented product (Jeyaram et al., 2009; Tamang, 2020).

10.5.3 Yu

Yu or *acham* is prepared from *atingba* through the process of distillation. In this preparation, *atingba* is poured into an aluminum pot which is cooked over a low flame. An aluminum funnel is placed above this pot and from this, a pipe is connected to another pot to collect the distillate which is known as *yu*. The pot is covered tightly with an aluminum plate. On top of this, another aluminum pot is placed containing cold water. Cow dung paste is used to seal all the connecting points. Continuous distillation occurs until all the alcohol present in the content is collected. The first collected one is the strongest and is known as *machin*. The remains after the extraction of *yu* are used as pig feed. Meiteis use *machin* for quick healing of vaginal stitches following childbirth.

10.5.4 Nutritional Benefits

The principal ingredient, *Albizia myriophylla* bark (yangli) in the starter culture *hamei* or *khai* is high in polyphenols, soluble sugars, terpenoids, flavonoids, and anthraquinone. It has antioxidant and antibacterial properties, and inhibits lipid peroxidation.

Atingba brewed from Manipur's glutinous black rice has an alcohol content of 5.71 to 7.75%. There is also lactic acid and anthocyanin, but no biogenic amines

(Mangang et al., 2017a, 2017b). The consumption of *atingba* is supposed to aid in the urination-based elimination of kidney stones. *Yu angouba* or *khor* improves skin tone and overall wellbeing. *Machin* is used for medicinal purposes and contains a high percentage of alcohol. It is reported that massaging the lower surface of the foot has a soothing effect, eases pain from arthritis and muscular tightness, improves wound stitch healing, and so on. Women with menstrual abnormalities, infertility, obesity, appetite loss, and postpartum health issues can benefit from herbal therapies made by traditional healers. (Singh and Singh, 2006; Soibam and Ayam, 2018; Singh et al., 2018b).

10.6 OTHER FERMENTED PRODUCTS

Still lesser-known fermented products used in Manipur are fermented perishable vegetables, meat etc. They are processed and fermented to increase shelf life for storage and also to develop flavor when used as additives in the preparation of various dishes (Tamang et al., 2012). Notable among these are:

10.6.1 *Inziangsang/Ziang-dui/Ziang-sang*

During the winter season, fresh leaves of *hangam* (*Brassica* sp.) are allowed to wither for 2 to 3 days, and crushed using traditional wooden mortar and pestle and soaked in warm water (Figure 10.8).

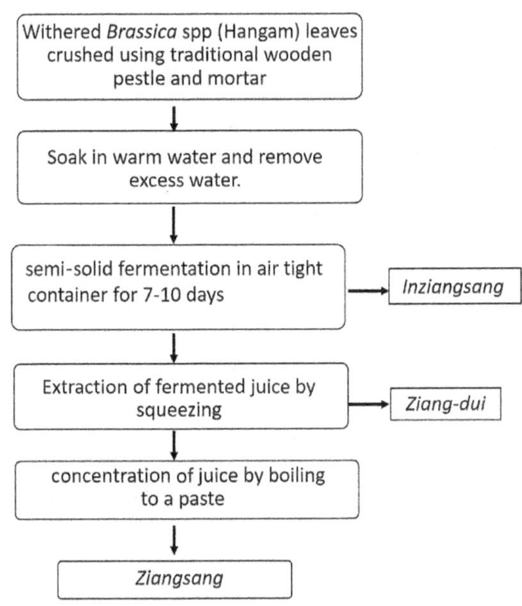

FIGURE 10.8 Flowchart for steps involved in preparation of *ziang-sang*.

FIGURE 10.9 Some fermented foods of Manipur: (A) *soibum*, (B) *soidon*, (C) *hawaijar*, (D) *ngari*, (E) *aphurthao*, (F) *sāyung* (G) *hentak* (H) *ziang-sang*.

The excess water is removed by squeezing, placed in an airtight container and left to ferment in semi-solid state for 7–10 days at room temperature (20–30°C). After fermentation, the juice and paste are separated by squeezing by hand. The paste is sun-dried for 4–5 days to produce *ziang-dui*, a product that may be preserved for up to a year. The liquid obtained is referred to as *inziangsang*. Boiling concentrates the liquid component to form *ziang-sang* (Figure 10.9H), which is preserved in traditional bamboo vessels with a one-year shelf life. It is predominantly prepared by Zeliangrong Naga ethnic people.

Lactobacillus brevis, Lactobacillus plantarum, Pediococcus pentosaceus, Pediococcus acidilactici, and *Leuconostoc fallax* are the most common LAB species found in fermented *ziang-dui/ziang-sang* (Tamang et al., 2005).

10.6.1.1 Nutritional Value

The protein, carbohydrate, minerals and moisture composition of *ziang-dui/ziang-sang* contribute to its nutritional value (Tamang et al., 2012). The probiotic qualities (hydrophobicity, adhesion to intestinal cells) of the lactic acid bacteria present in *ziang-dui/ziang-sang* contribute to its functional value by degrading anti-nutritive compounds and oligosaccharides. It also makes bacteriocin, which is used to fight food-borne infections (Tamang et al., 2009).

10.6.2 *Khaireoshui*

A fermented food product made from crab (*Scylla sp.*), is called *khaireoshui* and used by Tangkhul Naga peoples. It is similar to the *japangangnagtsu'* prepared by the Ao Nagas in Nagaland. The black species of crabs with hard shells are preferably used as they develop pleasant aroma and taste. Crabs are first washed thoroughly, and shredded into pieces leaving the hard appendages and entrails. They are then ground with *hanshi* (*Perilla frutescens*) and wrapped in banana leaves, *Macaranga indica*, or *Phrynium pubinerve* leaf or kept in a pot and fermented for a week over the fire in a traditional Naga hearth to yield a powerful aromatic product ready to be used in cooking or preparation of other delicacies.

10.6.3 *Sāyung*

The Tangkhul Nagas produces *sāyung*, a type of fermented pork, in late December, usually after the Christmas festival. During the festive season, extra pork is processed into *sāyung*, which is fermented by seasoning raw pork with or without salt and spices, followed by hanging it on wooden beams of the roof for 3–4 months or more. The fermentation process imparts a distinctive taste and also changes its texture and color (Figure 10.9F). *Sāyung* can be kept for a long time and consumed after several months. The people believe that the longer it is kept, the more therapeutic it becomes. *Sāyung* is given to pregnant women and breastfeeding mothers since it is traditionally believed that it provides them with increased energy and nutrients (Wahengbam et al., 2020).

10.6.4 *Sathu*

Sathu is the fermented pork fat often prepared and used by the Kuki, Hmar, Paite, and Vaiphei ethnic groups. Comparable products used by Tangkhul Nagas and Rongmei Nagas are known as *aphurthao* and *gwag ruum*, respectively. *Sathu* is also similar to *sa'um* in Mizoram. It has a distinct odor and is greasy and somewhat acidic (pH 6.3–6.5). It is added to impart a distinct flavor to many traditional signature dishes. *Sathu* is prepared by washing and cutting pork fat into small pieces, then half-boiling in water for 15 to 20 minutes. The boiled pork fats are then fermented for 3–5 days in an airtight container. It is possible to keep the fermented products for up to a year (Figure 10.9E). These can be eaten as a pickle or served as a side dish along with other dishes. (Chakrabarty et al., 2009). These communities also prepare and consume *bongthu*, which are similar products made from cow or buffalo fat.

Literature on the microbiology and nutritional composition of some of these fermented products are not available and hence can be areas for further investigations.

10.7 CONCLUSION

Fermented foods and beverages are prepared in a variety of ways that are often passed down through the generations as traditional knowledge. These foods have played key roles in maintaining the overall health of the people who consume them as they are found to be very good reservoirs of macro-nutrients and micronutrients and have several other therapeutic properties. Therefore, their consumption across regions and cultures should be encouraged by creating more awareness about their high nutritional values. The first step would be to increase their appeal to non-traditional consumers, which can be challenging, as to relish these foods one first needs to acquire a taste for them. In addition, the threats posed by socio-economic and cultural shifts leading to many traditional foods being overshadowed by commercial foods need to be addressed. Promotion of these indigenous products should be taken up actively by the government, as besides the numerous nutritional and health benefits, they also play an important role in sustaining the economic livelihood of the local people and empowering the women, who are the prime movers in sustaining and promoting traditional food products. An effort should be made to streamline production methods which should be more systematic and sustainable while not compromising on the standard of the products.

REFERENCES

Aarti, C., Khusro, A., Arasu, M.V., Agastian, P. & Al-Dhabi, N.A. (2016). Biological potency and characterization of antibacterial substances produced by *Lactobacillus pentosus* isolated from *Hentak*, a fermented fish product of North-East India. *Springerplus 5*, 1743. doi:https://doi.org/10.1186/s40064-016-3452-2.

Aarti, C., Khusro, A., Varghese, R., Arasu, M.V., Agastian, P., Al-Dhabi, N.A., Ilavenil, S. & Choi, K.C. (2017) *In vitro* studies on probiotic and antioxidant properties of *Lactobacillus brevis* strain LAP2 isolated from *Hentak*, a fermented fish product of North-East India. *LWT Food Sci Technol., 86*, 438–446.

Abdhul, K., Ganesh, M., Shanmughapriya, S., Kanagavel, M., Anbarasu, K. & Natarajaseenivasan, K. (2014). Antioxidant activity of exopolysaccharide from probiotic strain *Enterococcus faecium* (BDU7) from *Ngari. Int. J. Biol. Macromol., 70*, 450–454.

Appaiah K.A.A., General, T. & Koijam, K. (2011). Process Improvement as Influenced by Inoculums and Product Preservation in the Production of Hawaijar. A Traditional Fermented Soybean. *African J. Food Sci., 5* (2), 63–68.

Bhatt, B.P., Singha, L.B., Singh, K. & Sachan, M.S. (2003). Some commercially available edible bamboo species of North-East India: Production, indigenous uses, cost benefit and management strategies. *J. Am. Bamboo Soc., 17*(1), 4–20.

Chakrabarty, J., Sharma, G.D. & Tamang, J.P. (2009). Substrate utilisation in traditional fermentation technology practiced by tribes of North Cachar Hills District of Assam. *Assam University J. Sci. Technol.: Biol. Sci., 4* (I), 66–72.

Dajanta, K., Apichartrangkoon, A., Chukeatirote, E. & Frazier, R.A. 2011. Free-amino acid profiles of thau nao, a Thai fermented Soybean. *Food Chem., 125*, 342–347.

Das, A.J. & Deka, C. (2012). Fermented foods and beverages of Northeast India. *Int. Food Res. J., 19*, 377–392.

Devi, K.R., Deka, M. & Jeyaram, K. (2015). Bacterial dynamics during yearlong spontaneous fermentation for production of ngari, a dry fermented fish product of Northeast India. *Int. J. Food Microbiol., 199*, 62–71. doi:https://doi.org/10.1016/j.ijfoodmicro.2015.01.004

Devi, P. & Kumar, S.P. (2012). Traditional ethnic and fermented foods of different tribes of Manipur. *Indian J. Trad. Knowledge, 11*(1), 70–77.

Jeyaram, K., Romi, W., Singh, T.A., Devi, A.R. & Devi, S.S. (2010). Bacterial species associated with traditional starter cultures used for fermented bamboo shoot production in Manipur state of India. *Int. J. Food Microbiol., 143*, 1–8.

Jeyaram, K., Singh, T.A., Romi, W., Devi, R.A., Singh, W.M., Dayanidhi, H. Singh, N.R. & Tamang, J.P. (2009). Traditional fermented foods of Manipur. *Indian J. Trad. Knowledge, 8* (1), 115–121.

Jeyaram, K., Singh, W.M., Capece, A. & Romano, P. (2008b). Molecular identification of yeast species associated with 'Hamei': A traditional starter used for rice wine production in Manipur, India. *Int. J. Food Microbiol., 124*, 115–125. doi:https://doi.org/10.1016/j.ijfoodmicro.2008.02.029

Jeyaram, K., Singh, W.M., Premarani, T., Ranjita Devi, A., Chanu, K.S., Talukdar, N.C. & Singh, M.R. (2008a). Molecular identification of dominant microflora associated with 'Hawaijar': A traditional fermented soybean (*Glycine max* (L.)) food of Manipur, India. *Int J. Food Microbiol., 122*, 259–268.

Keishing, S. & Banu, A.T. (2013). Hawaijar—a fermented soya of Manipur, India: Review. *IOSR J. Environ Sci. Toxicol Food Technol., 4*(2), 29–33.

Keishing, S. & Banu, A.T. (2015). Fermented fish (*ngari*) of Manipur–preparation technique and its potential as a functional food ingredient. *Elixir Food Sci., 85*, 34502–34507.

Majumdar, R.K., Bejjanki, S.K., Roy, D., Shitole, S., Saha, A. & Narayan, B. (2015). Biochemical and microbial characterization of *Ngari* and *Hentaak*-traditional fermented fish products of India. *J. Food Sci. Technol., 52*, 8284–8291.

Mangang, K.C.S., Das, A.J. & Deka, S.C. (2017a), Comparative shelf-life study of two different rice beers prepared using wild-type and established microbial starters. *J. Inst. Brew., 123*, 579–586.

Mangang, K.C.S., Das, A.J. & Deka, S.C. (2017b). Shelf-life improvement of rice beer by incorporation of *Albizia myriophylla* extracts. *J. Food Process Preserv., 41*, 1–13.

Nirmala, C., Sharma, M. L. & David, E. (2008). A comparative study of nutrient component of freshly harvested, fermented and canned bamboo shoots of *Dendrocalamus giganteus* Munro. *J. Am. Bamboo Soc., 21* (1), 33–39.

Phadke, G., Elavarasan, K. & Shamasundar, B. (2014). Angiotensin-I converting enzyme (ACE) inhibitory activity and antioxidant activity of fermented fish product *ngari* as influenced by fermentation period. *Int. J. Pharm. Bio. Sci. 5*, 134–142.

Premarani, T. & Chhetry, G.K.N. (2010). Evaluation of traditional fermentation technology for the preparation of Hawaijar in Manipur. *Assam University J. Sci. Technol.: Biol. Environ. Sci. 6*(I), 82–88.

Premarani, T. & Chhetry, G.K.N. (2011). Nutritional analysis of fermented soyabean (Hawaijar). *Assam University Journal of Science & Technology*, 7 (1), 96–100.

Qiu, F.G. (1992). The recent development of bamboo foods. *Proceedings of the International Symposium on Industrial Use of Bamboo. International Timber Organization and Chinese Academy of Forestry*, Beijing, China: Bamboo and its Use. pp 333–337.

Rai, A.K. & Jeyaram, K. (2015). Health benefits of functional proteins in fermented foods. In Tamang, J.P. (Ed.) *Health Benefits of Fermented Foods and Beverages*. Boca Raton: CRC Press, Taylor & Francis Group, pp 455–474. ISBN: 9781466588097

Romi, W. (2015). Metagenomic and culture-dependent analyses of microbial communities associated with natural fermentation of different indigenous bamboo shoot products in Northeast India. PhD thesis, Department of Biotechnology, Gauhati University, Guwahati, Assam, India, p. 347. http://hdl.handle.net/10603/98978

Romi, W., Ahmed, G. & Jeyaram, K. (2015). Three-phase succession of autochthonous lactic acid bacteria to reach a stable ecosystem within 7 days of natural bamboo shoot fermentation as revealed by different molecular approaches. *Mol. Ecol., 13*, 3372–3389. doi:https://doi.org/10.1111/mec.13237

Sandhu, D.K. & Soni, S.K. (1989). Microflora associated with Indian Punjabi warri fermentation. *J. Food Sci. Technol., 26*, 21–25.

Sarangthem, K. & Singh, T.N. (2003). Microbial bioconversion of metabolites from fermented succulent bamboo shoots during fermentation. *Curr. Sci., 84*, 1544–1547.

Sarangthem, K. & Singh, T.N. (2013). Fermentation decreases the antinutritional content in bamboo shoots. *Int. J. Curr. Microbiol. App. Sci., 2*, 361–369.

Sarojnalini, C. & Singh, W.V. (1988). Composition and digestibility of fermented fish foods of Manipur. *J. Food Sci. Technol., 25*, 349–351.

Sekar, S. & Mariappan, S. (2007). Usage of traditional fermented products by Indian rural folks and IPR. *Indian J. Trad. Knowledge, 6* (1), 111–120.

Sha, S.P., Suryavanshi M.V. & Tamang, J. P. (2019). Mycobiome diversity in traditionally prepared starters for alcoholic beverages in India by high-throughput sequencing method. *Front. Microbiol., 10*, Article 348 doi:https://doi.org/10.3389/fmicb.2019.00348.

Sha, S.P., Suryavanshi, M.S., Jani, K., Sharma, A., Shouche, Y. & Tamang, J.P. (2018). Diversity of yeasts and molds by culture-dependent and culture-independent methods for mycobiome surveillance of traditionally prepared dried starters for the production of Indian alcoholic beverages. *Front. Microbiol., 9*, 2237. https://doi.org/10.3389/fmicb.2018.02237

Shi, Q.T. & Yang, K.S. (1992). Study on relationship between nutrients in bamboo shoots and human health. *Proceedings of the International Symposium on Industrial Use of Bamboo. International Tropical Timber Organization and Chinese Academy*, Beijing, China: Bamboo and its Use, pp. 338–346.

Singh, H.D., Singh, A.S., Singh, N.R. & Singh, N.R. (2011). Biochemical composition of *soibum*—a fermented bamboo shoot and its dynamics during fermentation in real time model. *Int. Conf. Food Eng. Biotechnol.*, May 7–9, *Int. Proc. Chem. Biol. Environ. Eng. 9*, 198–202.

Singh, P.K. & Singh, K.I. (2006). Traditional alcoholic beverage, *Yu* of Meitei communities of Manipur. *Ind. J. Tradit. Knowl. 5*, 184–190.

Singh, S.K., Singh, C.A., Singh, Y.J. & Das, P. (2010, January 2). *Ngari: An indigenous fermented fish product from Manipur, Part 2*. Retrieved from: https://e-pao.net.

Singh, S.S., Mandal, S.E., Lalnunmawii, E. & Kumar, N.S. (2018a). Antimicrobial, antioxidant and probiotics characterization of dominant bacterial isolates from traditional fermented fish of Manipur, North-East India. *J. Food Sci. Technol. 55*, 1870–1879.

Singh, T. A., Devi, K. R., Ahmed, G. & Jeyaram, K. (2014). Microbial and endogenous origin of fibrinolytic activity in traditional fermented foods of Northeast India. *Food Res. Int., 55*, 356–362. doi:https://doi.org/10.1016/j.foodres.2013.11.028

Singh, T.A., Sarangi, P.K. & Singh, N.J. (2018b). Traditional process foods of the ethnic tribes of western hills of Manipur, India, *Int. J. Curr. Microbiol. App. Sci. 7*, 1100–1110.

Soibam, H. & Ayam, V.S. (2018). The traditional fermented foods of meiteis of Manipur, India: A case study. *J. Pharmacognosy Phytochem., 7*(4), 535–539.

Sonar, N.R. & Halami, P.M. (2014). Phenotypic identification and technological attributes of native lactic acid bacteria present in fermented bamboo shoot products from North-East India. *J. Food Sci. Technol. 51*, 4143–4148.

Sonar, N.R., Vijayendra, S.V.N., Prakash, M., Saikia, M., Tamang, J.P. & Halami, P.M. (2015). Nutritional and functional profile of traditional fermented bamboo shoot based products from Arunachal Pradesh and Manipur states of India. *Int Food Res J*, *22*, 788–797.

Tamang, J. P. (2020). Ethnic fermented foods and beverages of India: Science history and culture. https://doi.org/10.1007/978-981-15-1486-9.

Tamang, J.P. & Nikkuni, S. (1998). Effect of temperatures during pure culture fermentation of kinema. *World J. Microbiol. Biotechnol.*, *14*, 847–850.

Tamang, B. & Tamang, J.P. (2009). Traditional knowledge of biopreservation of perishable vegetables and bamboo shoots in Northeast India as food resources. *Indian J. Trad. Knowledge*, *8*(1), 89–95

Tamang, B., Tamang, J.P., Schillinger, U., Franz, C.M.A.P., Gores, M. & Holzapfel W.H. (2008). Phenotypic and genotypic identification of lactic acid bacteria isolated from ethnic fermented bamboo tender shoots of North East India. *Int. J. Food Microbiol.*, *121*, 35–40.

Tamang, J.P., Dewan, S., Tamang, B., Rai, A., Schillinger, U. & Holzapfel, W.H. (2007). Lactic acid bacteria in Hamei and Marcha of North East India. *Indian Journal of Microbiology*, 47, 119–125.

Tamang, J.P., Tamang, B., Schillinger, U., Guigas, C. & Holzapfel, W.H. (2009). Functional properties of lactic acid bacteria isolated from ethnic fermented vegetables of the Himalayas. *International Journal of Food Microbiology*, 135, 28–33

Tamang, J. P. (2015). Naturally fermented ethnic soybean foods of India. *J. Ethnic Foods*, *2*, 8–17

Tamang, J. P., Watanabe, K. & Holzapfel, W. H. (2016). Review: Diversity of microorganisms in global fermented foods and beverages. *Front. Microbiol.*, *7*, 377. doi:https://doi.org/10.3389/fmicb.2016.00377

Tamang, J.P. (2010). *Himalayan Fermented Foods: Microbiology, Nutrition And Ethnic Values*. CRC Press, Taylor & Francis Group, New York, p. 295. ISBN: 978-1-4200-9324-7

Tamang, J.P., Tamang, B., Schillinger, U., Franz, C.M., Gores, M. & Holzapfel, W.H. (2005). Identification of predominant lactic acid bacteria isolated from traditionally fermented vegetable products of the Eastern Himalayas. *Int. J. Food Microbiol.*, *105*(3), 347–356.

Tamang, J.P., Tamang, N., Thapa, S., Dewan, S., Tamang, B., Yonzan, H., Rai, A.P., Chettri, R., Chakrabarty, J. & Kharel, N. (2012). Microorganism and nutritive value of ethnic fermented foods and alcoholic beverages of North-East India. *Indian J. Trad. Know.*, *11*, 7–25.

Thakur, K. & Tomar, S.K. (2015). Exploring indigenous *Lactobacillus* species from diverse niches for riboflavin production. *J. Young Pharm.*, *7*, 122–127.

Thakur, K. & Tomar, S.K. (2016). *In vitro* study of riboflavin producing lactobacilli as potential probiotic. *LWT Food Sci. Technol. 68*, 570–578.

Thapa, N. (2016). Ethnic fermented and preserved fish products of India and Nepal. *J Ethnic Foods*, 3, 69–77.

Thapa, N., Pal, J. & Tamang, J.P. (2004). Microbial diversity in ngari, hentak and tungtap, fermented fish products of North East India. *World J. Microbiol. Biotechnol.*, *20*, 599–607.

Thapa, N., Pal, J. & Tamang, J.P. (2007). Microbiological profile of dried fish products of Assam. *Indian J. Fisheries*, *54*, 121–125.

Visuphaka, K. (1985). The role of bamboo as a potential food source in Thailand. Proceedings of the International Bamboo Workshop; Oct 6–14; Hangzhou, China: Recent Research on Bamboos. pp. 301–303.

Wahengbam, R., Thangjam, A.S., Keisam, S., Asem, I.D., Ningthoujam, D.S. & Jeyaram, K. (2020). Ethnic fermented foods and alcoholic beverages of Manipur. In J. Tamang (Ed.). *Ethnic Fermented Foods and Beverages of India: Science History and Culture* (1st ed., pp. 349–419). Singapore: Springer. https://doi.org/10.1007/978-981-15-1486-9_14

Xia, N.H. (1989). Analysis of nutritive constituents of bamboo shoots in Guangdong. *Acta Botanica Austro Sinica*, *4*, 199–206.

11 Microorganisms in Cosmetology

Rita Rath, Neeraja Sood and Sadhna Gupta
University of Delhi, New Delhi, India

CONTENTS

11.1 INTRODUCTION

Cosmetics or toiletries are preparations which are used on skin, eyes, mouth, hair or nails to augment and enhance appearance or odor or provide protection from dryness, UV rays, etc. (de Groot & White, 1995). These may include cleansing and moisturizing creams, soaps, shampoos, toothpastes, lipsticks, eyeshadows, nail varnish, hair coloring and styling agents, fragrances and UV light protectants. The cosmetics industry is an important and fast-growing industry worldwide. Income from the global cosmetics market is likely to increase to US$429.8 billion by 2022, and predicted to reach an annual growth rate of about 4.3% for the period 2016–2022 (Research and Markets, 2016). In the past few decades, changing consumer choices favoring cosmetics of natural origin over chemical-based synthetically produced ones has been apparent (El-Enshashy et al., 2016).The presence of

DOI: 10.1201/9781003306931-12

low levels of phathalates in cosmetics, toiletries, perfumes, etc. and botulinum toxins, mainly in anti-aging creams, are claimed to be safe by the cosmetics industry but controversy prevails as environmentalists hold that they are still at chronic levels of exposure. There have been reports of reproductive and developmental problems due to exposure to these cosmetic toxins (Barrett, 2005). Many of these cosmetics-based chemicals are linked to birth defects in males, decreased sperm counts, changes in the consequences of pregnancy in animals, but as yet there is no definite evidence in the case of humans. It has been suggested by Bouslimani et al. (2019) that the effects of cosmetic products may last for several weeks but may produce individualized responses such as alterations in pheromone and steroid levels. A large number of potentially harmful chemicals are found in many skincare products that may cause allergic reactions, skin rashes, irritation, dermatitis, etc. Hence consumers are trying to avoid any potentially harmful ingredients that may be applied to skin. In India, we have traditionally using many herbal-based natural ayurvedic products for cosmetic purposes. These natural products were obtained from plants or animal extracts. Natural cosmetics are considered to be safer, biocompatible and less toxic (Brunt & Burgess, 2018). For a long time, natural cosmetics using plant extracts were in use as herbal cosmetics. However, lately the role and potential of microorganisms such as algae, bacteria and fungi as a source of active ingredients for the cosmetics industry is being realized.

In the Asia-Pacific region, India is an emerging market for cosmetics. The Indian cosmetics market was valued at US$11.16 billion in 2017 and is likely to grow at a compound annual growth rate (CAGR) of 5.19% during 2017–2030 (Goldstein Market Intelligence Report, 2017). Domestic demand in India has grown by 60% since 2010, making India's one of the fastest-growing cosmetics industries in the world. This is possibly due to increased awareness of body aesthetics and increased disposable income among the Indian population. A preference for herbal, ayurvedic and biological-origin products due to the known side-effects of cosmetics with chemical formulations has given a further boost and impetus to the use of naturally originating products in the cosmetics industry.

Some of India's top cosmetics-producing companies that are using ingredients of biological origin are Himalaya, Biotique, Dabur, Lotus, Patanjali, Kama ayurveda, and Shahnaz Husain. In the last few years, many international brands have entered the Indian market, such as Coach, LVMH, Puig, Shiseido, the Estée Lauder Co., Ralph Lauren, L'Oréal, Coty, Revlon, Avon, Hermes, Oriflame, Proctor and Gamble, and Unilever (Golden Market Intelligence Report, 2017). The cosmetics products made from biological ingredients by these companies have found widespread acceptance among our urban, rural and mature clients because of our traditional knowledge and experience with herbal and ayurvedic ingredients (Patwardhan et al., 2017).

Due to growth in the consumption of cosmetic products, use of harmless and efficient biologically safe ingredients has become a necessity. In this respect, the need for the use of microorganisms as producers of novel compounds has become very important and a significant potential alternative which has to be explored and exploited.

11.2 MICROORGANISMS AS POTENTIAL PRODUCERS FOR THE COSMETICS INDUSTRY

The role of microorganisms as the cheapest, most easily renewable and novel source for biomolecules production has only recently gained significant attention (Balkrishna et al., 2018). Microorganisms are omnipresent and exist in several natural habitats such as soils, oceans, glaciers, ponds and other extreme environments (Oren, 2002). Out of the diverse microorganisms occurring naturally, many cannot be cultured artificially. Hence, so far, only a few microorganisms have been successfully utilized commercially in the cosmetics industry.

Microorganisms are good and efficient candidates for large-scale production of metabolites as these can be produced economically (Waites et al., 2001). They are also capable of surviving and adapting to different conditions and producing compounds which are unique. Microorganisms are known to produce fatty acids, peptides, enzymes, lipopolysaccharides, vitamins and various pigments which have wide applications in cosmetology (Brunt & Burgess, 2018). Therefore, this underutilized microbial biodiversity should be explored as potential producers for cosmetic applications. Over the last decade, attempts have been made to extract compounds and study their effectiveness for cosmetics and personal care products (Andersen & Williams, 2000).

11.3 BACTERIA AS POTENTIAL PRODUCERS FOR THE COSMETICS INDUSTRY

Bacteria are single-celled prokaryotes found in soil, air and water. Some bacteria can survive in extreme temperatures, pH and salinity. Bacteria and their metabolites have been exploited in almost every field. Now, bacteria are also considered a potential sustainable source of cosmetic products that can replace existing chemical products. Bacteria produce a variety of low molecular weight compounds, including organic acids, alcohols, proteins and polysaccharides, biosurfactants, etc. which can be used in cosmetic formulations. Such products are used as binding, thickening, coloring, and stabilizing agents.

11.3.1 Polysaccharides from Bacteria

Many bacterial species produce polysaccharides which are water-soluble polymers used as binders, coagulants, emulsifiers, film-forming and gelling agents, lubricants, stabilizers and thickening agents (Freitas et al., 2015). Bacterial cellulose is a biodegradable, biocompatible and non-toxic biopolymer (Balkrishna et al., 2018). It is used in face masks, in skin treatments like skin tighteners, and in wound care including burns, as it provides a moist environment and good conditions for skin regeneration (Bianchet et al., 2020). Numerous cosmetic formulations use petroleum derivatives that are difficult to degrade and are harmful to health (Sharma et al., 2019). These also result in accumulation of microplastics, mainly in oceans, which adversely affects marine biodiversity (Pacheco et al., 2018).

Oligosaccharides of bacterial origin have contributed diversely in cosmetic formulations. Cyclodextrin are produced by enzymatic transformations using the different bacterial strains of *Bacillus subtilis*, *Microbacterium terrae* and *Brevibacillus* sp. These are used in perfumes and room freshener gels to reduce the volatility of esters, and as aromatic agents in soaps, talcum powders, diapers, menstrual discs, etc. (Rajput et al., 2016).

Exopolysaccharides or extra-polymeric substances (EPS), secreted by *Streptococcus mutans*, which are biocompatible and less toxic, are exploited in the manufacturing of cosmetics. Some hydrophilic EPS, like dextran, have the ability to retain water and are used as skin-smoothing and brightening agents, and wrinkle-reducing and skin-tightening products. Usually, alginates are obtained from seaweed, however *Pseudomonas aeruginosa* and *Azobacter vinelandii* also produce copious amounts of alginates as EPS, which are used in gelling agents as thickeners (Gupta et al., 2019). Xanthun gum produced by *Xanthomonas* sp. is widely used as a viscosity controller (Freitas et al., 2015). Xanthan also has water-retention properties and hence is used in moisturizers and various skin formulations (Gupta et al., 2019).

11.3.2 Bacterial Metabolites as Biosurfactants

Biosurfactants are amphiphilic compounds. They possess both hydrophobic and hydrophilic groups, which decrease surface tension. They have low toxicity, high biodegradability and are environment-friendly. Biosurfactants are good emulsifiers, solubilizers and wetting agents. These properties make them a preferred choice in cosmetics formulations. Most of the biosurfactants are glycolipids, lipopeptides, phospholipids and fatty acids (Mukherjee et al., 2006). Sophorolipids, rhamnolipids, mannosyloerythritol (MEL) and lipolipids are commonly used biosurfactants. Surfactins produced by *Bacillus* species, like *B. subtilis*, *B. licheniforme,s* etc., are used as anti-wrinkle and cleansing agents (Farias et al., 2021). Due to their low CMC (critical micelles concentration) and good foaming features, surfactants are used as oil formulations in dermatology and cosmetology. Rhamnolipids are produced by *Pseudomonas aeruginosa* and are used in anti-wrinkle formulations. They also possess antimicrobial properties and are used in deodorants, nail care products and toothpastes (Varvaresou & Lakovou, 2015).

11.3.3 Role of Bacteria-Derived Hyaluronic Acid in Cosmetic Formulations

Bacterial strains of *Streptococci*, particularly S. *equii*, are a potential source of hyaluronic acid (HA) (Bettenhausen, 2021). HA is a glycosaminoglycan devoid of sulfate group and is a biodegradable polymer which has huge potential and numerous applications in the cosmetics industry. HA of different molecular weights exhibits different properties and is utilized in making different formulations like dermal fillers to uplift sagging skin, moisturizing creams and gels. HA of approximately 300 kDa molecular weight forms a protective layer on the skin without being absorbed. It forms a barrier and retains the moisture in the skin. HA of low molecular weight penetrates the skin, holds water in deeper tissues and reduces wrinkles (Essendoubi et al., 2015). Brands like Johnson and Johnson and Neutrogena use HA of low molecular weights, from 50–1700 kDa, in their moisturizers. In cosmetic formulations, micelles formed by combining HA with linoleic acid are used in skin whiteners (Bettenhausen, 2021). HA along with stem cell-derived exosomes are used for healing of wound (Vasvani et al., 2019). Dissolvable HA microneedles combined with bacterial nanocellulose confer hydrating and regenerative properties and are used to treat skin patches (Fonseca et al., 2021; Saha & Rai, 2021).

11.3.4 Enzymes and Proteins from Bacteria

Since ancient times, the use of proteins in the form of curd, milk, egg whites, etc., for cosmetic purposes has been a part of every civilization. Soluble proteins are appropriate for cosmetic formulations but insoluble proteins are also used in facial masks. Insoluble collagen is used in making facial masks to provide hydration, shine and smoothness to the skin. High molecular weight proteins and hydrolysates provide a smoothing and softening effect for the skin by making a colloidal film on its surface (Secchi, 2008).

Superoxide dismutase (SOD) and peroxidase work in synergy as exfoliators (Gupta et al., 2019). These enzymes and lactate dehydrogenase can also effectively remove free radicals, thereby protecting the skin from harmful UV radiation. Earlier, *Marinomonas* sp., *Sulpholobus acidocladarius* and a few other extremophiles were used to produce SOD and peroxidases.

The peptide bonds of keratin, elastin and collagen of the skin are hydrolyzed by proteases and thus help in the treatment of several skin disorders like xerosis, psoriasis, ichthyoses, etc. (Gupta et al., 2019). Proteases such as keratinases are used in the treatment of scars, stretch marks and for regeneration of epithelial cells. Keratin hydrolysates are used in tropical ointment creams for heels and elbows. *Bacillus licheniformes* is used for commercial production of keratinase. Other potential bacterial sources of keratinase are *Thermoanaerobacter, Thermococcus, Microbacterium* and *Xanthomonas*. Along with enzymes, several *Bacillus* sp. are used to produce peptides, which are used in gels, emulsions and powders (Gupta et al., 2013; Bindal & Gupta, 2017).

Table 11.1 lists some important cosmetic ingredients obtained from bacteria.

11.4 ALGAE AS POTENTIAL PRODUCERS FOR THE COSMETICS INDUSTRY

Algae are found in aquatic environments and wetlands. They are either unicellular or multicellular eukaryotic oxygenic organisms which can photosynthesize sugar/starch and other biomass using chlorophyll, sunlight, carbon dioxide and water (Croce & Amerongen, 2014). They can withstand severe conditions of temperature, salinity, UV rays, osmotic pressure, pH and anaerobiosis by producing moieties like folic acid, vitamin B_{12}, E, zeaxanthin and lutein (de Morais et al., 2015). These metabolites are known to have antibiotic and antimicrobial properties.

TABLE 11.1
Cosmetic Ingredients from Bacteria and Their Cosmetic Applications

Molecule Class	Cosmetic Ingredient	Cosmetic Application	Bacteria
Polysaccharides	Dextran Alginate Xanthun gum Cellulose	Gelling agent, UV protectant, emulsifier, film-forming antioxidant, skin hydration, brightening agent, smoothing agent, moisture retainer	*Xanthomonas* sp., *Cornybacterium autotrophicum*, *Streptococcus mutans*
Oligosaccharides	Cyclodextrine	Reduce volatility of esters in perfumes and room fresheners. Aromatic agents in soaps, talcum etc.	*Microbacterium terrae*, *Bacillus subtilis*, *Brevibacillus* sp.
Biosurfactants	Glycolipids Lypoproteins Phosphor lipids like sophorolipids, Mannosyloerythritol (MEL)	Emulsifiers Solubilizers and wetting agents Anti-wrinkle and cleansing agents	*Pseudomonas aeruginosa*, *Bacillus licheniformes*, *Acinetobacter calcoaceticus*,
Enzymes	Superoxide dismutases Peroxidases Glutathione Peroxidases	Exfoliators, remove free radicals, hair removal, anti-aging agents	*Xanthomonas* sp., *Bacillus licheniformes*, *Thermobacter* sp., *Sulphobolus acidocladarius*. *Clostridium histolyticum*
Pigments	Astaxanthin	Antioxidant and improve skin texture	*Agrobacterium aurantiacum*

Source: Balakrishna et al. (2018); Gupta et al. (2019); Farias et al. (2021); Saha & Rai, (2021).

The use of photosynthetic algae in the cosmetics industry has shown increasing and very promising trends. According to several reports, algal products have been found to be an efficient alternate and substitute to synthetic compounds for the cosmetics industry (Corinaldesi et al., 2017; Joshi et al., 2018; Future Market Insight, 2021).

The algae belong to two major groups: micro-algae and macro-algae.

Micro-algae (blue green algae or cyanobacteria). These are prokaryotes, microscopic and unicellular phytoplankton which grow heterotrophically. They are present in freshwater or marine ecosystems either singly, in chains or in groups. They have the ability to photosynthesize similar to terrestrial plants utilizing CO_2 in the presence of sunlight (Wolkers et al., 1989). These are a rich source of phosphorus, Ca, Fe, folic acid, biotin, beta-carotene, pantothenic acid, vitamins A, B, C, E, B_{12}, etc. (Fabregas & Herrero, 1990).

Macro-algae (seaweeds). These are marine, benthic, eukaryotic, macroscopic, multicellular algae. These are important in the cosmetics industry because of their ability to produce rich bioactive molecules such as phlorotannins, sulfated polysaccharides and tyrosine inhibitors (Thomas & Kim, 2013). The macro-algae of commercial importance belong to three classes: Rhodophyceae (red algae), Phaeophyceae (brown algae), and Chlorophyceae (green algae). They are also a rich source of health-promoting molecules like dietary fiber, vitamins A, B, C, E, omega fatty acids and essential amino acids (Rajapakse & Kim, 2011; Kim, 2014) which are important cosmetic ingredients (Kim et al., 2008; Xing et al., 2013).). The metabolites derived from many of the macro-algae (Blunt et al., 2011) have anti-inflammatory (Wijesakara et al., 2012, Dore et al., 2013), antioxidant

TABLE 11.2
Main Cosmetic Ingredients from Algae and Their Application in Cosmetic Products

Class	Pigment Present	Cosmetic Ingredient	Cosmetic Applications
Rhodophyceae	Phycoerythrin	Porphyra 334	Sunscreen lotion
(Red algae)		Shinorin	Sunscreen lotion
		Asthaxanthin	Anti-aging, lipsticks
		Carrageenan	Thickening and moisturizing agent in skin creams
Chlorophyceae	Chlorophyll a	Lutein	Sunscreen lotions
(Green algae)	Chlorophyll b	Neoxanthin	Anti-aging creams
	Beta-carotene	Beta-carotene	UV protectants
		Violaxanthin	
		Linoleic acid	
		Palmitic acid	
Phaeophyceae	Fucoxanthin	Potassium alginate	Skin care serum, hair gel
(Brown algae)	Chlorophyll c	Phlorotannin	Anti-aging cream
	Cantaxanthin	Laminarin	Skin protectant
		Fucoxanthin	De-pigmenting
		Fucoidan	Sunscreen
		Phloroglucinol	UV protectant, skin whitening
		Mistric acid	Antioxidant
		Oleic acid	
Cyanophyceae	Phycocyanin	Phycocyanobilin	Anti-wrinkle Collagen synthesis
(Blue green algae/Cyanobacteria)		Phycoerythrobilin	Anti-inflammatory,
		Gamma-linoleic acid	antioxidant

Source: Gupta et al. (2019).

(Wijesakara et al., 2012), anti-allergic (Vo et al., 2012), anti-tanning, UV protecting and moisturizing (Wang et al., 2015) properties in addition to their neuroprotective role, hyaluronidase inhibition and bone-related diseases in pharmaceutical industry as well as in cosmetics (Thomas & Kim, 2013). Several studies indicate the significance of biological activities of marine algae and their role in promoting skin health and beauty products (Kim et al., 2007; Shibata et al., 2008; Ariede et al., 2017; Malakar & Mohanty, 2021).

Macro-algae are currently being commercially grown in lagoons or huge open ponds in Asia and Europe to provide food and additives for the cosmetics, pharmaceutical and chemical industries (Vigani et al., 2015; Arauj et al., 2021).[1]

Some of the important ingredients produced by algae that have applications in the cosmetics industry are shown in Table 11.2.

11.4.1 Anti-Skin Aging Role of Algae

Aging of skin results in the appearance of lines, ridges and creases, loss of elasticity and discoloration (Gancevicience et al., 2012). The maintenance of youthful, vibrant and flexible skin is due the presence of collagen and elastin fiber (Peytavi et al., 2016). Insufficient hydration of skin, exposure to heavy metals and nutritional deficiency expedite the aging and wrinkling process. Several scientific studies have shown the efficacy of many algal products containing vitamin E and beta-carotene on rejuvenating and delaying skin aging or protecting from skin cancers (Schagen et al., 2012; Keen & Hassan, 2016). Among well-studied bioactive ingredients from algae are phlorotannin from

Eisenia bicyclisis and brown algae *Ecklonia* (Shibata et al., 2008), and beta-carotene from green and red algae (Schagen et al., 2012). Other algal species used in developing anti-aging products are *Turbinaria ornate, Gracilaria, Padina* and *Hydroclathrus* (Kelman et al., 2012). Many amino acids, such as Mycosporine-like amino acids (MAAs) extracted from the algae *Porphyra umbilicatis*, have been shown to prevent premature aging (Daniel et al., 2004). Extracts from green micro-algae *Chlorella* help restore the firmness of skin by inhibiting the enzymes collagenase and elastase which degrade skin (Wang et al., 2015). It is commercially available under the brand name Dermochlorella. Other products, such as Astaxanthin from *Haematococcus pluviali*; hexadecatetraenoic acid from diatom, *Stauroneis amphyioxys*; and hexadecapentaenoic acid from algae *Anadyomene stellata* are also known to have anti-aging properties (Tominaga et al., 2012).

11.4.2 Skin Whitening and Depigmenting Role of Algae

Melanin pigment is a complex polymer which imparts color to the skin, protects and shields it from damage by absorbing the UV radiation in the atmosphere (Thomas & Kim, 2013). Exposure of skin to the sunlight for long increases melanocytes production resulting in skin tanning due to hyperpigmentation. With aging, the melanocyte distribution becomes irregular and dark spots may appear. Inhibition of the tyrosinase enzyme responsible for melanin pigment formation can reduce this hyperpigmentation. Phlorotannin and 7-pholoroeckol present in the brown algae *Ecklonia cava* are known to inhibit tyrosinase and hence have potential as skin whitening agents (Yoon et al., 2009). Fucoxanthin from *Laminaria japonica, Macrocystis* and *Alaria chorda* is also known to inhibit melanogenesis (Shimoda et al., 2010). Asthaxanthin extracted from *Haematococcus. pluvialis*; zeaxanthin from micro-algae, *Nannochloropsis ovulate*; and fucodan from *Undaria pinnatifida* and *Fucus vesiculosus* have also been found to play a significant role in skin lightening and protection (Shen et al., 2011; Fitton et al., 2015; Kim, 2016).

11.4.3 Role of Algae As Antioxidant

An antioxidant inhibits the damage caused by oxidation by reacting with reactive oxygen species (ROS), by transferring electrons to an oxidizing agent or by preventing the ability to produce free radicals. In the dermal cells, the aging process is hastened due to lipid peroxidation caused by OH^- and H_2O_2 radicals (Kim, 2016). The well-known antioxidant compounds abundantly present in the red algae, such as carotenoids, phycocyanins, xanthophylls and allophycocyanins (Ariede et al., 2017), and the bioactive lipids with high omega-3 fatty acids levels extracted from micro-algae (Conde et al., 2021) can be exploited for preventing the scavenging effects of these radicals. Commercial products containing these active ingredients are already available in the market, such as the pure clay red algae mask by L'Oréal Paris. MAAs such as polythine, scytonenin, etc. extracted from cyanobacteria also help in protecting skin. The algae *Spirulina maxima* and *Chlorella vulgaris* are known to be rich sources of vitamin A and C which are natural antioxidants helping hair growth, skin toning and cleansing (Mourelle et al., 2017).

11.4.4 Role of Algae in Hair Care

The protein sericin, an important ingredient in hair care and skin treatment, is obtained from the silkworm, *Bombyx mori*. However, some algal species such as *Arthrospire platensis* and *Chlorella vulgaris* are also good source of such proteins which should be considered (Bari et al., 2017). They are known to produce 7-phloroeckol which stimulates hair growth. Another compound, omega-3 fatty acid, which is present in Argan oil, prevents drying of hair, hair loss, brittleness, dandruff and itchy scalp. Eicosapentaenoic acid (EPA) and docosahexaenoic acid (DHA) extracted from micro-algae provide nourishment to the hair and are important ingredients in hair oils, hair serum and other hair products.

11.4.5 Role of Algae As Moisturizing Agents

A moisturizer is used for making the epidermis of the skin softer and supple which, in turn, prevents drying, bruising and wrinkling of skin and also acne and eczema. Acids, such as hyaluronic acids, and polysaccharides, such as agar, carrageenan, fucoidans, and alginates obtained from algae can regulate water distribution in skin (Wang et al., 2013). These compounds isolated from algae are non-hazardous, cheap and available in substantial quantities, and therefore have enormous potential as ingredients in cosmetic products.

11.4.6 Role of Algae As Thickening Agent

Thickening agents such as polyethylene glycol and vegetable gum are used in products and formulations containing high water content to maintain their consistency. However, agar obtained from red algal species, *Gracillaria* and *Gellidium*; and cartagena from *Chondrus* can also be a good alternative (Dita et al., 2020).

11.4.7 Algae As Photo Protector/UV Protector

A range of UV protection compounds are produced by algae such as MAAs, flavonoids, phycobiliproteins, carotenoids, scytonemins (Rastogi, 2014; Bedoux et al., 2014). These substances are capable of absorbing UV light in the range of 300–365 mm without generating free radicals. MAAs act as a protective sunscreen preventing cell dedication and flavonoids and help in dermatitis treatments.

Table 11.3 lists some of the algal species (micro and macro) from which active biomolecules are extracted through different processes and incorporated as active ingredients in many cosmetic products (Xing et al., 2013).

TABLE 11.3
Some Important Algal Species As Active Ingredient Producers for the Cosmetics Industry

Algal Species	Active Biomolecules	Function
Fucus vesiculosus	Laminaran, fucoidan, alginate	Antioxidant
Gracilaria chilensis	Polyphenols	Antioxidant
Laminaria digilata	Phlorotannins	Anti-melanogenesis, anti-aging, antioxidant
Undaria pinnatifida	Wakamine	Whitening/lightening agent
Ulva lactuca	Polyphenols	Macromolecular antioxidants
Ecklonia stolonifera	Oxylipins, Phlorotannins	Matrix metalloproteinase inhibition activity
Enteromorpha sp.	Polyphenols	Macromolecular antioxidants
Porphyra umbilicalis	Mycosporine-like amino acids	Photo-protection
Turbinaria conoides	Laminaran, fucoidan, alginate	Antioxidant
Isochrysis	Fucoxanthin, cantaxanthin, mistiric acid, oleic acid	Antioxidant, sunscreen, soothing cream
Alaria esculenta	Lipophilic extract	Progerin reduction in aged fibroblast
Dunaliella salina	Beta-carotene	Photo-protection against UV, antioxidant
Spirulina platensis	Crude extract for skin creams	Wound healing of keratinocyte cells
Haematococcus pluviallis	Astaxanthin	Antioxidant and scavenger of free radicals
Scytonema sp	Scytonemin	UV-A sunscreen
Sargassum macrocarpum	Sargafuran	Anti-acne activity
Muriellopsis sp.	Lutein	Ant-oxidant
Chlorella species	Sporopollenin	Anti-wrinkle potential

Source: Gupta et al. (2019); Joshi et al. (2018).

TABLE 11.4
Important Companies Producing/Extracting Algal Biomolecules for Use in the Cosmetics Industry

Company	Algae	Product Name	Application
L'Oréal Paris	Red algae	Pure face mask	Mixed with clay, it exfoliates, refines and cleanses using algal properties.
Nykaa	Green algae & *Spirulina*	Iraya algae body serum	For softening of skin by moisturizing
Algatech	Green algae		Astaxanthin from *Haematococcus* used in cosmetics and food industry.
La Pairie	Snow algae, *Chlamydomonas nivalis*	Cellular Swiss ice crystal dry oil	Mixed with sea almond oil and made into anti-aging cream.
Dove	Red algae	Dove regenerative repair shampoo	Nourishment and damage prevention of hair thereby restoring hair strength
Aubrey Organics	Blue green algae	Blue green algae hair rescue conditioning mask	Algal protein helps in hair strengthening preventing breakage and split ends
Algenist	Green algae, *Dunaliella salina, Haematococcus pluviallis*	Reveal color correcting eye serum brightener	Used as a concealer that covers dark circles and reduces uneven skin around the eyes.
Jenelt	Red algae, *Porphyra umbicalis*	Ultra UV defenses brightening cream with SPF 30	Used as sunscreen with antioxidants and anti-aging.
Osea	Red algae, *Chondrus crispus*	Osea eyes and lips	Hydrates the skin of lips and around eyes.

Source: Joshi et al. (2018).

Recently many companies have been exploiting the potential of algae as a source for production of important biomolecules (Martin et al., 2014). Table 11.4 lists some companies which are commercially producing active ingredients from algae to be used in cosmetic products.

According to Future Market Insights (FMI) 2021, the demand for algae in the personal care and cosmetics sector will be US$50.59 million in 2021, growing by CAGR 4.2% for 2021–2031. The market grew by CAGR 2.6% between 2016 and 2020. Various patents have been granted to skin care manufacturing companies for the use of algae.

11.5 FUNGI AS POTENTIAL PRODUCERS FOR THE COSMETICS INDUSTRY

Fungi are the group of eukaryotes that include molds, yeast and mushrooms (Moore, 1980). They are heterotrophs, cannot photosynthesize and are present all over the world in a wide range of habitats (Hawksworth and Lücking, 2016). They have been a part of human food (mushrooms), leavening agents (bread making) and for the fermentation of several food products such as soy sauce, beer and wine. Mushrooms were among the first fungi to be used in medicinal products (Wasser, 2002). For thousands of years, mushrooms were known for their beneficial effects on human health and were thus used as an important nutrient in the diet, so much so that their consumption was restricted to royal families (Wasser, 2002). In the past few decades increasing customer awareness of the use of chemicals in cosmetics and their harmful effect has led to examination of the use of fungi and its products in the cosmetics industry (Wu et al., 2016).

Some fungal products derived from wild and edible mushrooms have been found to have beneficial applications for both cosmetic and therapeutic effects (cosmeceuticals), and are reported to have nutricosmetic benefits too as these products can be ingested orally (Chang & Buswell, 1996). Since such products do not require any legal provision to prove their claims, many have become popular in the cosmetics industry (Wu et al., 2016; El-Enshansy et al., 2016).

Compounds or metabolites derived from mushrooms and other fungi, and their roles in the cosmetics industry, are discussed further in the following sections.

11.5.1 Kojic Acid from Fungi

Kojic acid (KA) is a chelating agent and fungal by-product produced by *Aspergillus oryzae* (called *koji* in Japanese) while fermenting rice for the purpose of making Japanese rice wine (Yabuta, 1924). It can undergo many reactions vis-à-vis oxidation reduction, alkylation, acylation, substitution, etc. because of the presence of a polyfunctional heterocyclic ring. Moreover, the free hydroxyl group of it can also react with various metal ions such as sodium, zinc, calcium, cadmium (Brtko et al., 2004). KA is aqueous and soluble in some organic solvents like ethanol, ethyl acetate, chloroform and acetone, but insoluble in benzene. It has many uses, such as a mild inhibitor in the production of pigment in plant and animal tissue, as a food additive and as preservatives in cosmetics (Beelik, 1956; Lim, 2009; Wakisaka, 1998). Kojic acid is a very important and popular skin lightening agent as it inhibits the tyrosinase activity by binding to copper and thus inhibits the formation of melanin (Chang, 2005). It is often used in combination with glycolic acid to increase the effect of skin lightening. Earlier hydroquinone was used in the cosmetics industry for the same purpose, but because of its side-effects, it has been completely replaced by KA (Garcia & Fulton, 1996; Lim, 2009; Mohammad et al., 2010). It also acts as an antioxidant since it can counteract free radicals in air and hence may prevent oxidative skin damage. KA also shows antibacterial activity and its role in curing blemishes caused by bacterial infection has been recommended by dermatologists (Arbab & Eltahir, 2010).

11.5.2 Lactic Acid from Fungi

Lactic acid is a valuable product obtained during fermentation from many fungal species of *Rhizopus* genus, which is more economical than production by bacteria due to the low substrate cost. It is often used as an exfoliating agent on skin for lightening purposes. It also slows down the eruption of acne (Hyde & Bahkali, 2010).

11.5.3 Ceramides from Fungi

Ceramide is a complex lipid made up of sphingosine and fatty acid. Many fungal species such as *Candida albicans, Agaricus bisporus* and *Armillaria tabescens* are sources of glycosyl ceramides which is an important ingredient in topical skin medications to treat eczema or dermatitis (Prasad, 1996; Gao et al., 2004; Murakami et al., 2015).

11.5.4 Terpenoids from Fungi

Terpenoids or isoprenoids are organic compounds made up of a large number of terpenes units. These are isolated from *Ganoderma* species and are used in sun lotions to protect the skin from UV radiation. They also show immunomodulating and anti-infective activities (El-Enshahy et al., 2016).

11.5.5 L-Ergothioneine from Fungi

L-Ergothioneine is an amino acid that occurs naturally in some species of *Portabella* and *Criminis*. Its role as an antioxidant in anti-aging creams and lotions has been widely reported. Some derivatives, such as trehalose and gallic acid from *Aspergillus niger* and *Lentinula edodes* also show very high antioxidant activity and are very important ingredients in skin moisturizing cream (Bazela et al., 2014).

11.5.6 Polysaccharides from Fungi

Polysaccharides derived from mushrooms are both homo and heteroglycans and are able to form peptidoglycan or protein polysaccharide complexes with various other proteins. Many polysaccharides have been isolated from a variety of mushrooms and have been shown to have beneficial effects in many cosmetic products (El-Enshasy et al., 2016).

Pullulan is an aqueous, non-ionic, non-toxic white polymer of maltotriose units produced by the the yeast *Aureobasidium pullulance*. It has a molecular weight ranging from 100 to 150 kDa. It is widely used as an important ingredient in cosmetics as it has no immunogenic, mutagenic or carcinogenic effects (El-Enshasy et al., 2016). It is used for tissue regeneration, grafts and healing of wounds, and for target drug and gene delivery (Prajapati et al., 2013).

Chitin glucan is another polysaccharide present in the cell wall of fungi. It is a copolymer made up of covalently bound chitin and branched glucose. It is used for thickening purposes in different formulations to provide stability to the products and act as preservative. Its use in anti-aging products and skin care cream has been appreciated (Gautier et al., 2008; Aranaz et al., 2018).

Another polysaccharide, beta-glucan polymer secreted by *Grifola frondosa* and known as dancing mushroom in Japanese, has been in use for centuries in food in Japan and China. It is known to enhance the process of synthesis of collagen and provide protection from light, hence its used in the cosmetics industries. Kim et al. (2007) have shown the effect of beta-glucan in reducing matrix metalloproteinase activity in the skin after photo-exposure from sun, making it a very valuable product for skin care (Lee et al., 2003; Bae et al., 2005).

Another polymer, beta-13-D glucan, derived from *Pleurotus*, has been shown to have anticancer properties and its role in skin whitening cosmetics has long been known (Wu et al., 2016).

Polysaccharides from *Tremella* mushroom have shown an inhibitory effect on melanogenesis, lightning skin blemishes. It also has a moisturizing effect and thus is a potential ingredient in moisturizing lotions (Gupta et al., 2019).

Some edible mushrooms such as *Agaricus bisporus* and *Lentinus edodes* contain selenium, which is an essential element needed in traces for different enzymes in mammals and can strengthen teeth, hair and nails. It is widely used in shampoos (Ogra et al., 2004).

Terpenes and terpenoids have been extracted from some species of *Ganoderma* and show anti-infective and immuno-modulating effects. Their potential to be used in cosmetics is being explored (El-Enshasy et al., 2016).

Table 11.5 presents a list of some important chemicals produced by mushrooms. Many cosmetic products using species of mushrooms or their extracts as active ingredients are now available commercially (Table 11.6).

11.6 CONCLUSION

Growing awareness of the hazardous impact of synthetic compounds in the cosmetics industry has led to increased demand for biological and eco-friendly products. Therefore, the cosmetics industry is constantly working towards increasing their use of botanicals and microorganisms as natural sources for their products. Microorganisms represent a good alternative, due to their immense diversity, adaptability to different environments and low production costs. Since cosmetic products are applied directly on the skin and penetrate into deeper layers, it is imperative that all precautions, care and efforts should be directed towards replacing chemical-based products available in the market with biologically derived compounds from microorganisms.

ACKNOWLEDGMENTS

The authors gratefully acknowledge the support and motivation provided by Dyal Singh College, University of Delhi during the preparation of this manuscript.

TABLE 11.5
Mushroom Species Involved in Production of Important Chemicals and Their Applications

Mushroom Species	Biomolecule	Application
Agaricus bisporus, Lyophyllum shimeiji, Pleurotus ostreatus, Termitomyces eurhizus, Volvareilla volvacea	Carbohydrates	Antimicrobial
Agaricus bisporus, Lyophyllum shimeiji, Pleurotus ostreatus, Termitomyces eurhizus, Volvareilla volvacea	Alkaloids	Antimicrobial, Anti-inflammatory, Antioxidant
Agaricus bisporus, Coprinus comatus, Lentinus edodes, Pleurotus ostreatus, Sparassis crispa, Volvareilla volvacea	Protein and amino acids	Antimicrobial, Anti-inflammatory
Agaricus subrufescens, Marasmius oreades, Panellus serotinus, Pleurotus eryngii, Stropharia rugosoannulata	Steroids	Anti-inflammatory
Lactarius deliciosus, Lentinus edodes, Macrolepiota mastoidea, Russula griseocarnosa	Flavonoids	Antimicrobial, Anti-inflammatory, Antioxidant
Flammulina velutipes, Grifola frondosa, Hypsizigus mamoreus, Lentinus edodes, Pholiota nameko	Glycosides	Anti-inflammatory, Antioxidant
Agaricus bisporus, Lentinus edodes, Phellinus linteus, Pleurotus ostreatus, Sparassis crispa, Tricholoma equestre	Phenols and Polyphenols	Antimicrobial, Anti-inflammatory, Antioxidant
Agaricus bisporus, Lentinus edodes, Lentinus sajor-caju, Volvareilla volvacea	Tannins	Antimicrobial, Antioxidant
Agaricus bisporus, Ganoderma lucidum, Pleurotus ostreatus, Termitomyces albuminosus, Wolfiporia cocos	Saponins	Anticancer, Antioxidant
Ganoderma colossum, Lepista nuda, Naematoloma sublateritium, Panellus serotinus, Scleroderma citrinum, Tricholoma matsutake	Triterpenoids	Antibacterial, Anti-inflammatory

Source: Wu et al. (2016).

TABLE 11.6
Some Important Commercially Available Cosmetic Products Obtained from Mushroom Species

Mushroom Species	Function	Product Name
Ganoderma lucidum	Anti-aging of skin	Menard Embellir Refresh Massage, France
Ganoderma lucidum	Boosts collagen formation, improves elasticity and provides hydration	Kat Burki Form Control Marine Collagen Gel, UK
Ganoderma lucidum	Provides hair with sun protection and prevents fading of color	Tela Beauty Organics Encore Styling Cream, UK
Ganoderma lucidum	Anti-aging	Yves Saint-Laurent Temps Majeur Elixir de Nuit, France
Ganoderma lucidum and *Pleurotus ostreatus*	Skin tightening and vitalization	Hankook Sansim Firming Cream (Tan Ryuk SANG), Korea
Lentinula edodes	Lift away dirt, oil and make-up and fight signs of aging	Aveeno Positively Ageless Daily Exfoliating Cleanser, US
Agaricus subrufescens (also known as *A. brasiliensis*)	Renew and revitalize skin	Vitamega Facial Moisturizing Mask, Brazil
Cordyceps sinensis	Moisturizer and suppress melanin production	Kose Sekkisei Cream, Japan
Inonotus obliquus	Anti-inflammatory to help soothe irritated skin	Root Science RS Reborn Organic Face Mask, U.S.
Tremella	Skin improvement around eyes	Surkran Grape Seed Lift Eye Mask, U.S.

Mushroom Species	Function	Product Name
Schizophyllum commune	Anti-aging and lifting	Alqvimia Eternal Youth Cream Facial Máxima Regeneración, Spain
Tremella fuciformis	Moisturizer which nourishes, revitalizes and hydrates skin	La Prairie Advanced Marine Biology Night Solution, Switzerland
Ganoderma lucidum and *Lentinula edodes*	Antioxidants and vitamin D	La Bella Figura Gentle Enzyme Cleanser, Italy

Source: Wu et al. (2016).

NOTE

1. Training manual source: http://www.fao.org/3/AB730E/AB730E03.htm

REFERENCES

Andersen, D. E., & William, R. J. (2000). Pharmaceuticals from the sea. In: R. E. M. Nester & R. E. Harrison, (Eds) *Chemistry in the Marine Environment*, (pp 55–79). The Royal Society of Chemistry, Cambridge. https://doi.org/10.1039/9781847550453-00055

Aranaz, I., Acosta, N., Civera, C., Elora, B., Mingo, J., Castro, C., Handia, M. deL, & Caballero, A.G. (2018). Cosmetics and cosmeceutical applications of chitin, chitosan and their derivatives. *Polymer*, 10(2) https://doi.org/10.3390/polym10020213

Arauj, O. R., Lopez, J., Calderon, F.V., Azevedo, I. C., Bruhn, A., Fluch, S., Trasenda, M.G., Ghaderiardakani, F., Ilmjarv, T., Laurans, M., Mac Managail, M., Manginin, S., Peteiro, C., Rebours, C., Steffanson, T., & Ullmann, J. (2021). Current status of the algal production industry in Europe: An Emerging Sector of the Blue Bioeconomy. *Frontiers Marine Science*, 7, 1–24. https://doi.org/10.3389/fmars.2020.626389

Arbab, A. H., & Eltahir, M.M. (2010). Review on skin whitening agents. *Khartoum Pharmacy Journal*, 13 (1), 5–10. https://www.researchgate.net/publication/249650287_Review_on_Skin_Whitening_Agents

Ariede, M. B., Candido, T. M., Jacome, A. L. M., Velasco, M.V. R., de Carvello, J. C. M., & Baby A.R. (2017). Cosmetic attributes of algae: A review. *Algal Research*, 25, 483. https://doi.org/10.1016/j.alg

Bae, J. T., Sim, D. H., Lee, B. C., Pyo, H. B., Choe, T. B., & Yun, J. W. (2005). Production of exopolysaccharide from mycelial culture of *Grifola frondosa* and its inhibitory effect on matrix metalloproteinase -1 expression in ultraviolet A irradiated human dermal fibroblasts. *FTMS Microbiology Letters*, 251, 347–354. https://doi.org/10.1016/j.femsle.2005.08.021

Balkrishna, A., Agarwal, V., Kumar, G., & Gupta, A.K. (2018). Applications of bacterial polysaccharides with special reference to the cosmetic industry. In: J. Singh, D. Sharma, G. Kumar, N. R. Sharma (Eds). *Microbial Bioprospecting for Sustainable Development*, (pp, 189–202). Springer, Singapore. https://doi.org/10.1007/978-981-13-0053-0

Bari, E., Arciola, C.R., Vigani, B., Crivelli, B., Moro, P., Marubini, G., Sorrenti, M., Catenacci, L., Bruni, G., Chlapanidas, T., Lucarelli, E., Perteghella, S., & Torre, M. L. (2017). In vitro effectiveness microspheres based on silk sericin and *Chlorella vulgaris* or *Arthrospira platensis* for wound healing applications. *Materials* (Basel), 10(9), 983. https://doi.org/10.3390/ma10090983

Barrett, J. R. (2005) Chemical exposures: The ugly side of beauty products. *Environmental Health Perspective*, 113(1). https://doi.org/10.1289/ehp.113-a24

Bazela, K., Solygya-Zurek, A., Debowska, K., Rogiewicz, E., & Bartnik, I. E. (2014). 1 Ergothioneine porotects skin cells against UV induced damage – A preliminary study. *Cosmetics*, 1, 51–57 https://doi.org/10.3390/cosmetics1010051

Bedoux, G., Hardouin, K., Bulot, A. S., & Bourgougnon, N. (2014). Bioactive compounds from seaweeds. In: N. Bourgougnon (Ed.), *Advances Botanical Research*, (pp. 345–378). Academic Press, Elsevier, Netherlands. https://doi.org/10.1016/b978-0-12-408062-1.00012-3

Beelik, A. (1956). Kojic acid. *Advances in Carbohydrate Chemistry and Biochemistry*, 11, 145–183 https://pubmed.ncbi.nlm.nih.gov/13469630/

Bettenhausen, C. (2021). Hyaluronic acid is just getting started. *Chemical and Engineering News*, 99 (16). https://doi.org/10.47287/cen-09916-cover

Bianchet, R. T., Cubas, A. L.V., Machado, M. M., & Moecke, E. H. S. (2020). Applicability of bacterial cellulose in cosmetics-bibliometric review. *Biotechnology Reports*, 27 https://doi.org/10.1016/j.btre.2020.e00502

Bindal, S., & Gupta, R. (2017) Hyperproduction of gamma-glutamyl transpeptidase from *Bacillus licheniformes* ER 15 in the presence of high salt concentration. *Preparative Biochemistry and Biotechnology*, 47(2) https://doi.org/10.1080/10826068.2016.1188314

Blunt, J. W., Copp, B. R., Munro, M. H. G., Northiote, P. T., & Prinsep, M. R. (2011). Marine natural products. *Natural Products Reports*, 28, 196–268. https://doi.org/10.1039/c2np20112g

Bouslimani, A., da Silva, R., Kosciolek, T., Janssen, S., Callewaert, C., Amir, A., Dorrestein, K., Melnik, A., Zaramela, L.S., Kim, J.N., Humphrey, G., Schartz, T., Sanders, K., Brennan, C., Luzzatto-Knaan, T., Ackermann, G., McDonald, D., Zenger, K., Knight, R., & Dorrestein, P. C. (2019) The impact of skin care products on skin chemistry and microbiome dynamics. *BMC Biology*, 17, 47 https://doi.org/10.1186/s12915-019-0660-6

Brtko, J., Rondahl, L., Fickova, M., Hudecova, D., Eybl, V., & Uher, M. (2004). Kojic acid and its derivatives; history and present state of art. *Central European Journal of Public Health*, 12, 16–18 https://pubmed.ncbi.nlm.nih.gov/15141965/

Brunt, E. G., & Burgess, J. W. (2018). The promise of marine molecules as cosmetic active ingredients. *International Journal of Cosmetic Sci*ence, 40 (1). https://doi.org/10.1111/ics.12435

Chang, S. T. (2005). An update review on tyrosinase inhibitors. *International Journal of Molecular Science*, 10, 2400–2475 10.3390/ijms10062440

Chang, S. T., & Buswell, J.A. (1996) *World Journal of Microbiology and Biotechnology*, 12, 473–476 https://doi.org/10.1007/BF00419460

Conde, T. A., Neves, B. F., Couto, D., Melo, T., Neves, B., Costa, M., Silva, J., Domingues, P., & Domingues, M.R. (2021). Microalgae as sustainable bio-factories of healthy lipids: Evaluating fatty acid content and anti-oxidant activity. *Marine Drugs*, 19(7), 357. https://doi.org/10.3390/md19070357

Corinaldesi, C., Barone, G., Marcellini, F., Dell'Anno, A., Danovaro, R. (2017). Marine microbial-derived molecules and their potential use in cosmeceutical and cosmetic products. *Marine Drugs*, 15, (4), 118. https://doi.org/10.3390/md15040118

Croce, R., & Amerongen, H.V. (2014). Natural strategies for photosynthetic light harvesting. *Nature Chemical Biology*, 10, 492–501. https://doi.org/10.1038/nchembio.1555

Daniel, S., Cornelia, S., Fred, Z., & Mibelte, A.G. (2004). Biochemistry, UV-A sunscreen from red algae for protection against premature skin ageing. *Cosmetics and Toiletries Manufacture Worldwide*, 139–143. https://www.mib.bio.com

de Groot, A. C., & White, I. R. (1995). Cosmetics and skin care products. In: R. J. G. Rycroft; T. Menni, & P. J. Frosch (Eds) *Textbook of contact Dermatitis*, (pp. pp. 461–476). Springer, Berlin, Heidelberg. https://doi.org/10.1007/978-3-662-03104-9_23

de Morais, M. G., de Silva Vaz, B., de Morais, E. G., & Viera Costa, J.A. (2015). Biologically active metabolites synthesized by microalgae. *Biomedical Research International*, 2015, 15. https://doi.org/10.1155/2015/835761

Dita, L., & Sudarno Triastuti, J. (2020). Utilization of agar *Gracilaria* sp. As a natural thickener on liquid bath soap formulation. *IOP Conference Series Earth and Environmental Scienc*e, 441. https://doi.org/10.1088/1755-1315/441/1/012021

Dore, C. M. P., Alves, M. G. D. C. F., Will, L. S. E., Costa, T. G., Sabry, D. A., Rego, L. A. R. S., Accardo, C. M., Rocha, O. H. A., Filgueira, L. G. A., & Leite, E. L. (2013). A sulfates polyssacchride, fucans, isolated from brown algae, *Sargassum vulgare* with anticoagulant, antithrombotic, antioxidant and anti-inflammatory effects. *Carbohydrate Polymers*, 91(1), 467–475. https://doi.org/10.1016/j.carbpol.2012.07.075

El-Enshasy, H. A., Hamid, M. A., Malek, R. A., Elmarzugi, N., & Sarmidi, M.R. (2016). Microbial metabolites in the cosmetic industry. In: V.K. Gupta, G.D. Sharma, M.G. Tuohy & R. Gaur (Eds) *The Handbook of Microbial Bioresources*, (pp. pp. 388–405). CABI, Oxfordshire, United Kingdom. https://doi.org/10.1079/9781780645216.0388

Essendoubi, M., Gobinet, C., Reynaud, R., Angiboust, J.F., Manfait, M., & Piot, O. (2015). Human skin penetration of Hyaluronic acid of different molecular weights as probed by Raman Spectroscopy. *Skin Research Technology*, 22(1), 55–62. https://doi.org/10.11/srt.12228

Fabregas, J., & Herrero, C. (1990.) Vitamin content of four marine microalgae. Potential use as source of vitamins in nutrition. *Journal of Industrial Microbiology*, 5, 259–264. https://doi.org/10.1007/bf01569683

Farias, C. B. B., Almeida, F. C. G., Silva, I. A., Souza, T. C., Meira, H. M., de Cassia, R., da Silva, F. S., Luna, J. M., Santos, V. A., Converti, A., Banat, I. M., & Sarubbo, L. A. (2021). Production of green surfactants: Market prospects. *Electronic Journal of Biotechnol*ogy, 51 https://doi.org/10.1016/j.ejbt.2021.02.002

Fitton, J. H., Dell'Acqua, G., Gardiner, V. A., Karpiniec, S. S., Stringer, D. N., & Davis, E. (2015). Topical benefits of two fucoidan-rich extracts from marine macroalgae. *Cosmetics*, 2(2), 66–81. https://doi.org/10.3390/cosmetics3010012

Fonseca, D. F. S., Vilela, C., Pinto, R. J. B., & Bastos, V., (2021). Bacterial nanocellulose-Hyaluronic acid microneedle patches for skin applications: *In vitro* and *in vivo* evaluations. *Material Science and Engineering*, 118, 111350. https://doi.org/10.1016/j.msec.2020.111350

Freitas, F., Alves, V. D., & Reis, M. A. M. (2015). Bacterial polysaccharides: Production and applications in cosmetic industry. In: K. Ramawat & J. M. Merillon (Eds) *Polysaccharides*, (pp. 2017–2043). Springer. https://doi.org/10.1007/978-3-319-16298-0_63

Future Marketing Insights (FMI) (2021) futuremarketinginsights.com: Demand for microalgae in the Personal care and Cosmetic Sector, Global sales Analysis and Opportunities 2031 by source-Marine Water and Fresh Water for 2021–2031. https://www.futuremarketinsights.com/reports/microalgae-personal-care-and-cosmetics-sector

Ganceviciene, P., Liakou, A. I., Theodoridis, A., Makrantonaki, E, & Zouboulis, C.C. (2012). Skin anti-ageing strategies-*Dermato – Endocrinology*, 4, 308–319. https://doi.org/10.4161/derm.22804

Gao, J. M., Zhang, A. L., Chen, H., & Liu, J.K. (2004) Molecular species of ceramides from ascomycete, *Tuber indicum. Chemistry and Physics of Lipids*, 131(2), 205. 10.1016/j.chemphyslip.2004.05.004

Garcia, A., & Fulton, J. E. (1996).The combination of glycolic acid and hydroquinone or kojic acid for the treatment of melasma and related conditions. *Dermatology Surgery*, 22, 443–447. https://doi.org/10.1111/j.1524-4725.1996.tb00345.x

Gautier, S., Xhauflaire-Uhoda, E., Gonry, P., & Pirrard, G. E. (2008). Chitin-glucan, a natural cell scaffold for skin moisturization and rejuvenation. *International Journal of Cosmetic Science*, 30, 459–469. https://doi.org/10.1111/j.1468-2494.2008.00470.x

Goldstein Market Intelligence 2017. *Global Cosmetics Industry Analysis Report 2017-2030*. https://www.pr-inside.com/global-cosmetics-industry-analysis-report-2017-2030-by-goldstein-market-intelligence-r4808286.htm

Gupta, P. L., Rajput, M., Oza, T., Trivedi, U., & Sanghvi, G. (2019). Eminence of microbial products in cosmetic industry. *Natural Products and Bioprospecting*, 9, 267–278. https://link.springer.com/article/10.1007/s13659-019-0215-0

Gupta, R., Sharma, R., & Beg, Q. (2013). Revisiting microbial keratinases: Next generation proteases for sustainable biotechnology, *Critical Review of Biotechnology*, 33 (2), 216–228. https://doi.org/10.3109/07388551.2012.685051

Hawksworth, D. L., & Lücking, R. (2016). Fungal diversity revisited: 2.2 to 3.8 million species. *Microbiology Spectrum*, 5(4). 10.1128/microbiolspec.FUNK-0052-2016

Hyde, K. D., Bahkali, A. H., & Moslem, M. A. (2010). Fungi-an unusual source for cosmetics. *Fungal Diversity*, 43, 1–29. https://doi.org/10.1007/s13225-010-0043-3

Joshi, S., Kumari, R., & Vivek, N. U. (2018) Application of algae in cosmetics: An overview. *International Journal of Innovative Research in Science, Engineering and Technology*, 7(2), 1269–1278. https://doi.org/10.15680/IJIRSET.2018.0702038

Keen, M. A., & Hassan, I. (2016). Vitamin E in dermatology. *Indian Dermatology Online Journal*, 7, 311–315. https://doi.org/10.4103/2229-5178.185494

Kelman, D., Posner, E. K., McDermid, K. J., Tabendera, N. K., & Wright, A.D. (2012). Antioxidant activity of Hawaiian marine algae. *Marine Drugs*, 10, 403–416. https://doi.org/10.3390/md10020403

Kim, S. K. (2014). Marine cosmeceutical. *Journal of Cosmetic Dermatology*, 13 (1), 56–67. https://doi.org/10.1111/jocd.12057

Kim, S. K. (2016). *Marine Cosmeceuticals. Trends and Prospects*, CRC Press. Boca Raton. https://doi.org/10.1201/b10120

Kim, S. K., Ravichandran, Y. D., Khan, S. B., & Kim, Y. T. (2008). Prospective of the cosmeccutical derived from marine organisms. *Biotechnology and Bioprocessing Engineering*, 13, 511–523. https://doi.org/10.1007/s12257-008-0113-5

Kim, S. W., Hwang, H. J., Lee, B. C., & Yun, J. W. (2007) Submerged production and characterization of *Grifola frondosa* polysaccharides- a new application to cosmeceuticals. *Food Technology and Biotechnology*, 45 (3), 295–305. https://hrcak.srce.hr/24177

Lee, B. C., Bar, J. T., Pyo, H. B., Choe, T. B., Kim, S. W., Hwang, H. J., & Yun, J. W. (2003). Biological activities of the polysaccharides produced from submerged culture of the edible basidiomycetes *Grifola frondosa. Enzyme and Microbial Technology*, 32, 574–581. https://doi.org/10.1016/S0141-0229(03)00026-7

Lim, J. T. (2009). Treatment of melasma using kojic acid in gel containing hydroquinone and glycolic acid. *Dermatologic Surgery*, 25, 282–284. 10.1046/j.1524-4725.1999.08236.x

Malakar, B., & Mohanty, K. (2021). The budding potential of algae in cosmetics. In: S. K. Mandota, A.K. Upadhyay, A. S. Ahluwalia (Eds) *Algae; Multifarious Applications for Sustainable World*, (pp. 181–199). Springer. https://doi.org/10.1007/978-981-15-7518-1_8

Martin, A., Viera, H., Gaspar, H., & Santos, S. (2014). Marketed marine natural products in the pharmaceutical and cosmeceutical industries: Tips for success. *Marine Drugs*, 12(2). https://doi.org/10.3390/md12021066

Mohamad, R., Mohamed, M. S., Suhaili, N., & Mohamad, S., & Arif, A. B. (2010). Kojic acid: Applications and development of fermentation process for production. *Biotechnology and Molecular Biology Reviews*, 5(2), 24–37. https://doi.org/10.5897/BMBR2010.0004

Moore, R.T. (1980). Taxonomic proposals for classification of marine yeasts and other yeast like fungi including the smuts. *Botanica Marina*, 23, 361–373. https://www.mycoguide.com/guide/fungi

Mourelle, M. L., Gomez, C. P., & Legido, J. L. (2017) The potential use of marine microalgae and cyanobacteria in cosmetics and thalassotherapy. *Cosmetics*, 4 (4), 46. https://doi.org/10.3390/cosmetics4040046

Mukherjee, S., Das, P., & Sen, R. (2006). Towards commercial production of microbial surfactants. *Trends in Biotechnology*, 24, 509–515. https://doi.org/10.1016/j.tibtech.2006.09.005

Murakami, S., Shimamoto, T., Nagano, H., Tsuruno, M., Okuhara, H., Hatanka, H., Tojo, H., Kodama, Y., & Funato, K. (2015). Producing human ceramides- NS by metabolic engineering using yeast *Saccharomyces cerevisiae. Science Reports*, 5, 16319. 10.1038/srep16319

Ogra, Y., Ishiwaka, K., & Suzuki, K. T. (2004). Speciation of selenium in selenium enriched shiitake mushroom *Lentulus edodes. Analytical and Bioanalytical Chemistry*, 379, 861–866. https://doi.org/10.1007/s00216-004-2670-6

Oren, A. (2002). Diversity of halophylic microorganisms: Environments, phylogeny, physiology and applications. *Journal of Industrial Microbiology and Biotechnology*, 28(1), 56–63. https://doi.org/10.1038/sj/jim/7000176

Pacheco, G., de Mello C. V., Chiari-Andreo, B. G., Issac, V. L. B., Ribeiro, S. J. L., Pecoraro, E., & Trovatti, E. (2018). Bacterial cellulose skin masks-Properties and sensory tests. *Journal of Cosmetic Dermatology*, 17(5), 840–847. https://doi.org/10.1111/jocd.12441

Patwardhan, K., Pathak, J., & Acharya, R. (2017). Ayurvedic formulations: A roadmap to address the safety concerns. *Journal of Ayurveda Integrated Medicine*, 8(4), 279–282. https://doi.org/10.1016/j.jaim.2017.08.010

Peytavi, U. B., Kottner, J., Sterry, W., Hodin, M.W., Griffiths, T.W., Watson, R. E., Hay, R. J., & Griffiths, C.E. (2016). Age-associated skin conditions and diseases: Current perspectives and future options. *The Gerantologist*, 56, 242–248. https://doi.org/10.1093/geront/gnw003

Prajapati, V. D., Jani, G. K., & Khanda, S. M. (2013). Pullulan: An exopolysaccharide and its various applications. *Carbohydrate Polymers*, 95, 540–549. https://doi.org/10.1016/j.carbpol.2013.02.082

Prasad, R., & Ghannoum, M. A. (1996). *Lipids of Pathogenic Fungi*, CRC Press https://www.routledge.com/Revival-Lipids-of-Pathogenic-Fungi-1996/Prasad-Ghannoum/p/book/9781138560581

Rajapakse, N., & Kim, S. K. (2011). Nutritional and digestive health benefits of seaweed. In S.K. Kim (Ed.) *Advances in Food and Nutritional Research*, 64, (pp. 17–28). Academic Express, San Diego, CA USA.

Rajput, K. N., Patel, K. C., & Trivedi, U. B. (2016). Beta-cyclodextrin production by cyclodextrin glucanotransferase from an alkaliphile *Microbacterium terrae* KNR 9 using different starch substrates. *Biotechnology Research International*, 2016. https://doi.org/10.1155/2016/2034359

Rastogi, R.P., & Incharoensakdi, A. (2014). Characterization of UV-screening compounds, cyclosporine-like amino acids and scytonemin in the cyanobacterium *Lyngbya* sp. CU2555 *FEMS Microbiology Ecology*, 87 (1), 244–256. https://doi.org/10.1111/1574-6941.12220

Research and Markets (2016). World Cosmetic Market-Opportunities and Forecasts, 2014–2022. https://www.researchandmarkets.com/reportd/3275915/world-cosmetics-mart-opportunities-and. Accessed September, 2021.

Saha, I., & Rai, V.K. (2021). Hyaluronic acid based microneedle array: Recent applications in drug delivery and cosmetology. *Carbohyrate Polymers*, 267, 118168. https://doi.org/10.1016/j.carbpol.2021.118168

Schagen, S. K., Zampeli, V. A., Makrantonaki, E., & Zouboulis, C.C. (2012) Discovering the link between nutrition and skin ageing. *Dermato Endocrinology*, 4, 298–307. https://www.ncbi.nlm.nih.gov/pmc/articles/PMC3583891/

Secchi, G. (2008). Role of proteins in cosmetics. *Clinics in Dermatology*, 26(4), 321–325. https://doi.org/10.1016/j.clindermatol.2008.04.004

Sharma, A., Thakur, M., Bhattacharya, M., & Mandal, T. (2019). Commercial application of cellulose nanocomposites: A review. *Biotechnology Reports*, 21, e00316. https://doi.org/10.1016/j.btre.2019.e00316

Shen, C.T., Chen, P.Y., Wu, J. J. Lee, T.M., Hsu, S. E., Chang, M. J., Young, C. C., & Sheih, C. J. (2011). Purification of algal anti-tyrosinase zeaxanthin from *Nannochlorpsis ovulate* using supercritical anti solvent precipitation. *Journal of Supercritical Fluids*, 55, 955–962. https://doi.org/10.1016/j.supflu.2010.10.003

Shibata, T., Ishimaru, K., Kawaguchi, S., Yoshikawa, H., & Hama, Y. (2008). Antioxidant activities of phlorotannins isolated from Japanese Laminariacea. *Journal of Applied Phycology*, 20, 705. https://doi.org/10.1007/s10811-007-9254-8

Shimoda, H., Tanaka, J., Shan, S. J., & Maoka, T. (2010). Anti-pigmenting activity of fucoxanthin and its influence on skin mRNA expression of melanogenic molecules. *Journal of Pharmacy and Pharmacology*, 62, 1137–1145. https://doi.org/10.1111/j.2042-7158.2010.01139.x

Thomas, N.V., & Kim, S.K. (2013). Beneficial effects of marine algal compounds in cosmeceuticals. *Marine Drugs*, 11, 146–164. https://doi.org/10.3390/md11010146

Tominaga, K., Hongo, N., Karato, M., & Yamashita, E. (2012). Cosmetic benefits of asthaxanthin on human subjects. *Acta Biochimica Polonica*, 59(1), 43. https://doi.org/10.18388/ABP.2012_2168

Varvaresou, A., & Lakovou, K. (2015). Biosurfactants in cosmetics and biopharmaceuticals. *Letters in Applied Microbiology*, 61(3), 214–223. https://doi.org/10.1111/lam.12440

Vasvani, S., Kulkarni, P., & Rawtani, D. (2019). Hyaluronic acid: A review on its biology, aspects of drug delivery, route of administrations and a special emphasis on its approved marketed products and recent clinical studies. *International Journal of Biological Molecules*, 151, 1012–1029. https://doi.org/10.1016/j.ijbiomac.2019.11.066

Vigani, M., Parisi, C., & Rodriguez-Cerezo, E. (2015). Food and feed products from microalgae: Markets opportunities and challenges for the EU. *Trend in Food Science and Technology*, 42 (1), 81–92. https://doi.org/10.1016/j.tifs.2014.12.004

Vo, T. S., Ngo, D. H., & Kim, S. K. (2012). Marine algae as a potential pharmaceutical source for anti-allergic therapeutics. *Process Biochemistry*, 47, 386–394. https://doi.org/10.1016/j.procbio.2011.12.014

Waites, M. J., Morgan, N. L., Rockey, J. S., & Higton, G. (2001). Industrial microbiology: An introduction. *The Quarterly Review of Biology*, 78 (1) https://doi.org/10.1086/377850

Wakisaka, Y., Segawa, T., Imamura, K., Sakiyama, T., & Nakanishi, K. (1998). Development of a cylindrical apparatus for membrane-surface liquid culture and production of kojic acid using *Aspergillus oryzae* and NRRL 484, *Journal of Fermentation and Bioengineering*, 85, 488–494. https://doi.org/10.1016/S0922-338X(98)80067-6

Wang, H. M. D., Chen, C. C., Huynh, P., & Chang, J. S. (2015). Exploring the potential of using algae in cosmetics. *Bioresources Technology*, 184, 355–362. https://doi.org/10.1016/j.biortech.2014.12.001

Wang, J., Jin, W., Hou, Y., Niu, X., Zhang, H, & Zhang, Q. (2013). Chemical composition and moisture absorption/retention ability of polysaccharides extracted from 5 algae. *International Journal of Biological Macromolecules*, 57, 26–29. https://doi.org/10.1016/j.ijbiomac.2013.03.001

Wasser, S.P. (2002) Medicinal mushroom as a source of anti-tumour and immunomodulating polysaccharides. *Applied Microbiology and Biotechnology*, 60 (3), 258–357. https://doi.org/10.1007/s00253-002-1076-7

Wijesakara, I., Senevirathne, M., Li, Y.Y., & Kim, S. K. (2012). Functional ingredients from marine algae as antioxidants in the food industry. In: S.K. Kim (Ed.) *Handbook of Marine Macro Algae: Biotechnology and Applied Phycology*, (pp 398–402). John Wiley & Sons Ltd, UK. https://doi.org/10.1002/9781119977087.ch23

Workers, H., Barbose, M. J., Kleinegris, D. M. M., Bosma, R., Wijffels, R. H., & Harman, P. (1989). *Microalgae the Green Gold of the Future? Large Scale Sustainable Cultivation of Microalgae for the Production of Bulk Commodities*. Wageningen, Wageningen UR, Netherlands

Wu, Y.W., Choi, M. H., Li, J., Yang, H., & Hyun, J. S. (2016). Mushroom cosmetics: The present and future. *Cosmetics*, 3 (3), 22. https://www.researchgate.net/publication/305078746_Mushroom_Cosmetics_The_Present_and_Future

Xing, Z. Q., Wang, J. F., Hao, Y.Y., & Wang, Y. (2013). Recent advances in the discovery and development of marine microbial natural products. *Marine Drugs*, 11(3), 700–717. https://doi.org/10.3390/md11030700

Yabuta, T. (1924). The constitution of kojic acid, a gamma-pyrone derivative formed by *Aspergillus oryzae* from carbohydrates. *Journal of Chemical Society*, 125, 575–587. https://doi.org/10.1039/CT9242500575

Yoon, N.Y., Eom, T. K., Kim, M. M., & Kim, S. K. (2009). Inhibitory effects of phlorotannins isolated from *Ecklonia cava* on mushroom tyrosinase activity and melanin formation in mouse B 16F10 melanoma cells. *Journal of Agriultural and Food Chemistry*, 57(10), 4124–4129. https://doi.org/10.1021/jf900006f

Part II

Agriculture

12 Microbes and Their Products in Sustainable Agriculture

Smriti Sharma
Gargi College, University of Delhi, New Delhi, India

Jyotsna Singh
Daulat Ram College, University of Delhi, New Delhi, India

CONTENTS

12.1 INTRODUCTION

The Brundtland Commission's report defined sustainable development as "meeting the needs of the present without compromising the ability of future generations to meet their needs" (Michelle, 2016). In the past, measures to increase agricultural yields were introduced without evaluating their long-term impact on humans, non-target organisms and the environment. We have all witnessed the adverse impact of single and disproportionate use of synthetic pesticides. The situation demands alternative and sustainable ways to improve our agricultural yield to meet the requirements of our ever-increasing population.

Microbes have always been an integral part of the agricultural ecosystem and are known to inhabit almost all parts of the plant. A plethora of evidence exists that supports the beneficial role of microbes in agriculture. These beneficial microbes, referred to as plant growth-promoting microbes (PGPMs) promote plant growth mainly by facilitating nutrient uptake or by regulating plant hormone levels (Glick, 2012). Many of these microorganisms are also great decomposers which contribute to soil fertility and possible remediation of contaminated soils. The PGPMs are also known

DOI: 10.1201/9781003306931-14

to help plants to recover from stressful environmental conditions (Stamenković et al., 2018). The microbes may indirectly promote plant growth by acting as biocontrol agents protecting plants from pests, parasites or pathogens. Many bacterial, fungal, algal, actinomycetes and yeast groups are potentially effective in agriculture.

This chapter provide a comprehensive account of various beneficial roles of microbes and their products in agriculture and their successful implementation in sustainable agriculture practices.

12.2 MICROBES PROMOTING NUTRIENT MINERALIZATION AND AVAILABILITY

Many agricultural soils lack an adequate amount of one or more nutrients required by plants, including nitrogen (N), phosphorus (P), potassium (K) and iron (Fe), which are essential for their optimal growth and development. This lack of nutrients is met by adding chemical fertilizers to soil. However, in most instances it leads to over-reliance and over-use of chemical fertilizers, inflicting long-term environmental damage by depleting soil fertility, ultimately leading to a reduction in crop yield and eutrophication of water bodies. Microbes present in soil have been providing N, P and many other nutrients naturally to plants. Ways of enhancing and exploiting these biological processes to fulfill plants' nutrient requirements will go a long way towards reducing reliance on chemical fertilizers.

Bacteria are the most abundant microbes present in the rhizosphere, a thin film of soil surrounding the roots of plants. These types of bacteria are designated as plant growth-promoting rhizobacteria (PGPR) or simply plant growth-promoting bacteria (PGPB). PGPR/PGPB promote growth of most plants by employing different mechanisms, for example biological N fixation (BNF), P and K solubilization, phytohormone production, siderophore production etc. (Glick, 2012). Some of the successful PGPR includes *Azotobacter, Azospirillum, Bacillus, Burkholderia, Pseudomonas, Rhizobium, Bradyrhizobium, Mesorhizobium* etc. (Wani and Gopalakrishnan, 2019). Apart from PGPR, other microorganisms found in rhizospheric soil and promoting plant growth include actinomycetes, fungi, and algae (Prasad et al., 2017).

Microbial inoculants or biofertilizers, consisting of artificially multiplied cultures of beneficial soil microorganisms, are gaining prominence as an environment-friendly and cost-effective alternatives to chemical fertilizers. They are also non-toxic and easy to apply (Nosheen et al., 2021). Some of the most commonly used biofertilizers worldwide to replenish soil nutrients include N-fixing bacteria (*Rhizobium*), N-fixing cyanobacteria (*Anabaena*), P-solubilizing bacteria (*Pseudomonas* sp.), and arbuscular mycorrhizal fungi (AM fungi).

12.2.1 Nitrogen-Fixing Biofertilizers

Significant amounts of nitrogen are fixed as ammonia, nitrites, and nitrates by the microorganisms present in soil which plants can readily use for their growth and development (Table 12.1). This process of converting atmospheric dinitrogen (N_2) into a plant-accessible form by the microorganisms is known as biological nitrogen fixation (BNF). These nitrogen-fixing microbes are either present in symbiotic association with plants, e.g., *Rhizobium*, associated with leguminous plants, or may be free living (non-symbiotic) which includes *Azospirillum* sp., *Azotobacter* spp., *Acetobacter diazotrophicus*, *Clostridium* sp., *Herbaspirillum* sp., *Bacillus* sp., *Azoarcus* sp. (Gupta, 2012).

The use of BNF in agriculture offers a renewable source of nitrogen, reducing reliance on chemical fertilizers (Peoples et al. 1995). The most widely used microbial inoculants for the purpose of nitrogen fixation are rhizobia which currently constitute around 79% of worldwide demand (Owen et al., 2015). These symbiotic bacterial species invade roots of leguminous crops and form nodules where they fix atmospheric nitrogen into ammonia through the enzyme nitrogenase (Murray, 2011), which is used as a source of nitrogen by plants. The plant in turn provides organic acids which are used by the bacteria to produce energy. Several rhizobia, such as *Rhizobium, Bradyrhizobium*, and

TABLE 12.1
Microbial Inoculants/Biofertilizers Reported to Promote Plant Growth by Facilitating Nutrient Uptake

Microorganisms	Host Plant	Reference
I) **NITROGEN-FIXING BIOFERTILIZERS**		
A. ***Symbiotic nitrogen-fixing biofertilizers***		
Rhizobium isolates	Common bean	Argaw (2016)
Bradyrhizobia	Soybean	Sanginga et al. (2000)
Rhizobium leguminosarum bv. phaseoli LCS0306	Common bean	Pastor-Bueis et al. (2019)
Rhizobium strains	Common bean	Safikhani et al. (2013)
Native + exotic rhizobia	Kabuu bean	Menge et al. (2018)
Rhizobium strains, IRC 1045 and IRC 1050	*Leucaena leucocephala*	Sanginga et al. (1994), Ojo & Fagade (2002)
Azolla-Anabaena azollae	Rice	Kannaiyan (1993) Choudhury and Kennedy (2004)
B. **Non-symbiotic free-living *nitrogen-fixing biofertilizers***		
Gluconacetobacter diazotrophicus and *Herbaspirillum* sp.	Sugarcane	Muthukumarasamy et al. (2006)
G. diazotrophicus and *Burkholderia vietnamiensis*	Sugarcane plants	Govindarajan et al. (2006)
Burkholderia spp.	Rice seedlings	Baldani et al. (2000).
Azospirillum spp.; *Azotobacter* sp.	Rice, Maize	Choudhury and Kennedy (2004), Biari et al. (2008)
Nostoc spp. strain 2S9B	Wheat plant	Gantar and Elhai, (1999)
Nostoc 2S6B, 2S9B or *A nabaena* C 5	Wheat	Obreht et al. (1993)
Bacillus thuringiensis and *Achromobacter spanius*	Sugarcane	Santos and Rigobelo, (2021)
Arbuscular Mycorrhiza Fungi (AMF)	Berseem clover seeds	Saia et al. (2014)
II) **PHOSPHATE-SOLUBILIZING BIOFERTILIZERS**		
A. ***Phosphate-Solubilizing Bacteria (PSB)***		
Azotobacter chroococcum	Wheat Strawberries	Kumar et al. (2001) Negi et al. (2021)
Pseudomonas species	Wheat Walnut Strawberries	Elhaissoufi et al. (2020); Afzal et al. (2005) Yu et al. (2011) Negi et al. (2021)
Pseudomonas Sp. DN 13–01	Maize	Benbrik et al. (2020)
Pseudomonas fluorescens	Peanut	Dey et al. (2004)
Pseudomonas striata	Potato & Soybean	Munda et al. (2018)
Bacillus species	Wheat	Afzal et al. (2005)
Bacillus cereus	Walnut	Yu et al. (2011)
B. megaterium	Tomato	El-Yazeid and Abou-Aly, (2011)
Bacillus pimilus X22, *Bacillus cereus* 263AG5	Maize, *Zea mays*	Benbrik et al. (2020)
Bacillus subtilis-320	Maize, *Zea mays*	Lobo et al. (2019)
Streptomyces griseus	Wheat	Hamdali et al. (2008)
Paenibacillus polymyxa	Tomato	El-Yazeid and Abou-Aly, (2011)
Azotobacter chroococcum	Strawberries	Negi et al. (2021)
Enterobacter 15S	Cucumber & Maize, *Zea mays*	Alzate Zuluaga et al. (2021)

(Continued)

TABLE 12.1 *(Continued)*

Microorganisms	Host Plant	Reference
Rhizobium meliloti.	Cotton	Egamberdiyeva et al. (2004)
Mesorhizobium mediterraneum (strain PECA21)	Chickpea and barley plants	Peix et al. (2001)
Rhizobium leguminosarum bv. Trifolii	*Phaseolus vulgaris*	Abril et al. (2007)
Sphingobacterium suaedae T47	Maize, *Zea mays*	Benbrik et al. (2020)
Burkholderia anthina	Mung bean (*Vigna radiata* L.)	Walpola and Yoon (2013)
Pantoea agglomerans	Mung bean (*Vigna radiata* L.)	Walpola and Yoon (2013)
Actinomycetes,	Wheat	Hamdali et al. (2008)
Cyanobacteria	Rice	Choudhury and Kennedy (2004)
B. ***Phosphate-Solubilizing Fungi (PSF)***		
Aspergillus awamori	Tomato	Khan and Khan (2002)
Arbuscular mycorrhizae (*Glomus fasciculatum*);	Potato & Soybean	Munda et al. (2018)
Chaetomium globosum	Wheat, Pearl millet	Tarafdar & Gharu (2006)
Penicillium radicum	Walnut	Whitelaw et al. (1997)
Penicillium digitatum	Tomato	Khan and Khan (2002)
III) POTASSIUM SOLUBILIZING BIOFERTILIZERS		
Bacillus edaphicus	Cotton & rape	Sheng (2005)
Bacillus mucilaginosus,	Maize & wheat; Sudan grass (*Sorghum vulgare*)	Singh et al. (2010); Basak and Biswas (2009)
Bacillus subtilis ANctcri3	Elephant foot yam	Anjanadevi et al. (2016)
Bacillus megaterium ANctcri7	Elephant foot yam	Anjanadevi et al. (2016)
Rhizobium spp.	Maize & wheat	Singh et al. (2010)
Azotobacter chroococcum	Maize & wheat	Singh et al. (2010)
Arthrobacter sp.	Ryegrass	Xiao et al. (2017)
Paenibacillus sp.	Ryegrass	Xiao et al. (2017)
Mesorhizobium sp.	Ryegrass	Xiao et al. (2017)
Pantoea ananatis (KM977993)	Rice	Bakhshandeh et al. (2017)
Rahnella aquatilis (KM977991)	Rice	Bakhshandeh et al. (2017)
Enterobacter cloacae	Tobacco	Zhang and Kong (2014)
Enterobacter sp. (KM977992),	Rice	Bakhshandeh et al. (2017)
Klebsiella variicola	Tobacco	Zhang and Kong (2014)
Pisolithus sp	*Eucalyptus dunnii*	Luciano et al. (2010)
IV) SULFUR-OXIDIXING BIOFERTILIZERS		
Thiobacillus sp.	Maize	Pourbabaee et al. (2020)
	Groundnut and canola	Anandham et al. (2007)
Thiobacillus thioxidans	Mustard	Abhijit et al. (2014)
Pandoraeasputorum ATSB28	Groundnut and canola	Anandham et al. (2008)

Sinorhizobium (Ensifer) are used as biofertilizers in agriculture and have been reported to significantly increase nodulation as well as nodule dry weight (Argaw, 2016; Babalola and Glick, 2012; Carareto Alves et al., 2014). Inoculants containing single or diverse strains of rhizobia have been developed and have shown great potential to improve crop growth and yield (Koskey et al., 2021, Pastor-Bueis et al., 2019; Safikhani et al., 2013). Kawaka et al. (2014) showed superior performance of native rhizobia strains compared to the commercial ones in common beans. Menge et al. (2018), however, reported a significant increase in nitrogen content with an inoculation mix of native consortium and commercial rhizobia in common bean. Nitrogen built up in soil from legume sources persists for longer and is less susceptible to loss compared to chemical N fertilizer. Strains of rhizobium such as IRC 1045 and IRC 1050 have been reported to persist and effectively nodulate their host plant even 10 years after their introduction (Sanginga et al., 1994; Ojo & Fagade, 2002). Rhizobium species can fix nitrogen in non-leguminous crops as well, such as *Parasponia*, oilseed rape and maize (Saikia and Jain, 2007).

Soybean (*Glycine max* (L.) Merr.) is one of the most rhizobia inoculant-consuming crops the world over, especially in Brazil, mostly comprising bacterial species belonging to the genus *Bradyrhizobium*. Some commercial inoculants containing *Bradyrhizobium* spp. include *Bradyrhizobium japonicum* SEMIA 5079, *B. diazoefficiens* SEMIA 5080 (Santos et al., 2021) and *B. elkanii* SEMIA 5019 (Table 12.5).

Another symbiotic microorganism, *Anabaena azollae* (a blue green algae), is used as a potent biofertilizer for its contribution of nitrogen to rice, especially wetland rice. In South-East Asia, rice is grown with *Azolla*, an aquatic fern which in association with *Anabaena azollae* is able to fix atmospheric nitrogen for the crop (Kannaiyan, 1993). This arrangement has been recorded to increase rice yield by 10–25% over corresponding *Azolla*-free rice fields (Kour et al., 2020).

Other diazotrophic microorganisms, such as *Azospirillum* sp., *Azotobacter* sp., *Herbaspirillum* sp., *Burkholderia sp*. and cyanobacteria can also meet plant nitrogen requirements by fixing atmospheric nitrogen (Choudhury and Kennedy, 2004). The amounts of nitrogen fixed by these free-living diazotrophs is significantly less (10–160 kg N per ha) compared to rhizobia (13–360 kg N ha_1) species, yet they are of special significance for non-leguminous plants (Atieno et al., 2020; Wani and Gopalakrishnan, 2019). *Azotobacteria* inoculation in crops like rice, wheat, sorghum, maize, pearl, millet, cotton, sesame and vegetables has effectively resulted in increased yield (Mazid & Khan, 2014). In addition to its BNF activity, bacteria of the genus *Azotobacter* also contribute to hydrolysis of organic and inorganic phosphate compounds and produce IAA which is also responsible for its plant growth-promoting attributes (Farajzadeh et al., 2012). Use of *Gluconacetobacter diazotrophicus*, *Herbaspirillum* sp. and *Burkholderia vietnamiensis* in sugarcane fields mitigated N-fertilizer application and also considerably improved its yield (Govindarajan et al., 2006; Muthukumarasamy et al., 2006). Recently, Santos and Rigobelo (2021) reported the potential of *Bacillus thuringiensis* and *Achromobacter spanius* for use as future inoculants for sugarcane plants for their ability to fix nitrogen and show high levels of cellulolytic activity and potassium solubilization respectively.

More than 113 plant species have been reported to be benefited by the ability of *Azospirillum* species to enhance nitrogen availability and acquisition (Zeffa et al., 2019). *Azospirillum* species in combination with rhizobial strains improve growth and productivity of legume crops as well. Early nodulation and increase in yield have been reported in soybean when *Azospirillum brasilense* was used in combination with *Bradyrhizobium* spp. (Chibeba et al., 2015; Hungria et al., 2013) (Table 12.4). *Azospirillum*-based N-fixing bioinoculants are used worldwide specifically for the maize plant. Mexico (2002) was one of the first countries to commercialize *Azospirillum*-based inoculant for maize crop followed by Argentina, which uses the commercial strain *A. brasilense* Az39 in maize and wheat (Santos et al., 2019). *Azospirillum*-based inoculants such as Azo-N R and Azo-N Plus R for cultivation of grain and cover crop legumes are popularly used in South Africa (Raimi et al., 2017) (Table 12.5).

Nitrogen-fixing ability of free-living photosynthetic cyanobacteria, viz. *Nostoc*, *Anabaena* spp., has been widely reported in the wheat plant (Gantar and Elhai, 1999; Obreht et al., 1993). Other crops also reported to benefit from cyanobacteria inoculation include maize, barley, chili, tomato, lettuce, oats, radish and cotton (Thajuddin and Subramanian, 2005).

Arbuscular mycorrhizal fungi (AMF or AM fungi) belonging to the phylum Glomeromycota have established a symbiotic relationship with almost 90% of known agricultural plants, particularly cereals, vegetables and horticultural plants (Diagne et al., 2020). AMF forms an extensive hyphal network which can extend beyond the root surface and thus are able to explore a large volume of soil and acquire immobile nutrients resulting in improvement in plant growth (Rouphael et al. 2015).

AMF inoculation is especially beneficial to plants under stress condition such as drought, salinity, etc. Inoculation of berseem clover seeds with arbuscular mycorrhizal spores has been reported to result in increases in N content under water stress conditions (Saia et al., 2014). Inoculation with AMF, *Diversispora versiformis*, enhanced growth and root nitrogen uptake of *Chrysanthemum morifolium* (Hangbaiju) plants under moderate soil saline levels (Wang et al., 2018). Co-inoculation of AMF and *Bradyrhizobium* enhances nitrogen fixation and growth in green grams (Musyoka et al., 2020) (Table 12.4).

12.2.2 Phosphate-Solubilizing Biofertilizers

Phosphorus (P) is the second most essential nutrient required for plant's growth and development. A large quantity of phosphorus is present in soil in both organic and inorganic forms (somewhere between 400–1, 200 mg kg^{-1} of soil), but most of it is in insoluble form and thus cannot be taken up by the plant's root (Glick, 2012; Sharma et al., 2013). To meet this limitation, P fertilizers are added to the agricultural soil often disproportionately causing more damage than repair. Application of microbial inoculants/biofertilizers with P-solubilizing activities to agricultural soils or crops is reported to reduce usage of phosphate fertilizers by half without compromising crop yield (Jilani et al., 2007; Mahanta et al., 2014; Yazdani et al., 2009). The demand for phosphate-solubilizing biofertilizers has increased the global market share of these fertilizers to approximately 15% (Owen et al., 2015).

Many species of bacteria, fungi actinomycetes and some algae naturally found in soil and rhizosphere are capable of solubilizing and mineralizing insoluble soil phosphate for the growth of plants (Table 12.1). These microorganisms are referred to as phosphate-solubilizing microorganisms (PSMs) (Alori et al., 2017). PSMs employ various mechanisms to make phosphorus accessible for absorption by plants. These include lowering of soil pH, due to the production of organic acids and release of protons by microorganisms (Kalayu, 2019). Phosphate may further be converted into plant-usable form by the action of hydroxyl and carboxyl groups of acids (Satyaprakash et al., 2017). Mineralization of soil organic phosphate is also one of the important mechanisms by which production of phosphatases like phytase by microorganisms brings about hydrolysis of organic phosphate compounds releasing inorganic phosphorus which is taken up by the plants (Kalayu, 2019).

Many plant growth-promoting rhizobacteria (PGPR) having P-solubilizing abilities are specifically referred to as phosphate-solubilizing bacteria (PSB). PSB represent 1 to 50% of the soil's micro biota and are more effective in promoting plant growth and crop yield than phosphate-solubilizing fungi (PSF) (Alam et al., 2002), which constitutes only 0.1–0.5% of the total respective population (Kalayu, 2019; Sharma et al., 2013; Walpola & Yoon, 2012). Bacterial species belonging to generas such as *Pseudomonas*, *Bacillus*, *Agrobacterium*, *Paenibacillus* and *Burkholderia* are widely used as inoculants to increase P availability in agricultural soil (Atieno et al., 2020; Glick, 2012). Lobo et al. (2019) reported increase in maize yield and phosphorus concentration under field conditions on inoculation with *Bacillus subtilis*-320 isolates, strongly suggesting its use as a biological inoculant for maize crops. Alzate Zuluaga et al. (2021) reports two different mechanisms employed by *Enterobacter* 15S to improve plant growth and P nutrition in maize and cucumber plants. Under P deficiency, the bacterium alone was effective in improving the root system of maize, while in cucumber the bacterium when used with $Ca_3(PO_4)_2$ produced a better effect. Increased growth due to enhanced P uptake in wheat plant is reported with *Azotobacter chroococcum* (Kumar et al., 2001) and *Pseudomonas* species (Elhaissoufi et al., 2020).

A consortium comprising two or more bacterial species is reported to be better at phosphate solubilization than single bacterial isolate. Efficacy of co-inoculation of two PSB, viz. *Pantoea agglomerans* and *Burkholderia anthina*, in enhancing P uptake and growth of mung bean (*Vigna radiata* L.) has been shown (Walpola and Yoon, 2013) (Table 12.4). Benbrik et al. (2020) reported growth enhancement of *Zea mays* when inoculated with bacterial consortium containing *Pseudomonas Sp.* DN 13–01, *Sphingobacterium suaedae* T47, *Bacillus pimilus* X22 and *Bacillus cereus* 263AG5. Similar potential to increase crop yield of biofertilizers (*Azotobacter chroococcum* and *Pseudomonas fluorescens*) when used in combination with manures has been reported for strawberry crops (Negi et al., 2021) (Table 12.4).

Rhizobium that fixes atmospheric nitrogen for the host plants also show P-solubilizer activity. Peix et al. (2001) pointed out that selection of rhizobia for soil inoculation should not be based entirely on their nitrogen-fixing ability, since these microorganisms can contribute to plant growth using other mechanisms as well, for example, phosphate solubilization. They further showed the efficacy of *Mesorhizobium mediterraneum* (strain PECA21) to increase phosphorous content and enhance growth of chickpea and barley plants. Similar increase in phosphorus content on inoculation with *Rhizobium meliloti* was reported in cotton plants (Egamberdieva et al., 2004). Dual inoculation of *Rhizobium leguminismarum* Thal-8/SK8 and PSB, *Pseudomonas* sp. strain 54RB resulted in yield increase of wheat (*Triticum aestivum*) by 29% and 25% with and without fertilizer respectively (Afzal and Bano, 2008).

The P-solubilizing ability of actinomycetes is of special interest as these organisms can survive extreme environmental conditions such as drought and fire. In addition to this they also produce antibiotics and phytohormones that could additionally benefit plant growth. The P-solubilizing potential of actinomycete, *Streptomyces griseus*, has been reported (Hamdali et al., 2008).

AMF are also very effective in aiding plants in nutrient uptake (P in particular) from soils deficient in nutrients (Kayama and Yamanaka, 2014). Fungi are able to traverse long distances in soil (Kucey, 1983), and produce more acids compared to the more prevalent bacterial species, consequently exhibiting greater P-solubilizing activity (Venkateswarlu et al. 1984). AMF is an obligate symbiont which is difficult to cultivate in pure cultures away from its host plants. Thus, the soil from the root zone of a plant hosting AMF is mostly used as inoculum. AMF spores and hyphae are also sometimes used as inoculum (Berruti et al., 2016). Widada et al. (2007) demonstrated the potential of co-inoculation of AMFs *Glomus manihotis* with rhizobacteria such as *Entrophospora colombiana* and PSB, *Pseudomonas* sp., in increasing yield in sorghum plant under acid and low phosphate level soil (Table 12.4). Application of biofertilizers, [PSB (*Pseudomonas striata*) and arbuscular mycorrhizae (*Glomus fasciculatum*)] has been reported to increase P availability to both soybean and potato crops (Munda et al., 2018).

Other than AMF, fungal species of the genera *Aspergillus, Emericella, Gliocladium, Penicillium,* and *Trichoderma* are known to release huge amount of phytase or phosphatase enzymes, efficiently mobilizing unavailable phosphate for plants. Members of the *Aspergillus* sp. are the most efficient in producing phosphatases. Tarafdar and Gharu (2006) reported enhancement in the production of wheat and pearl millet crops due to phosphate mobilization by *Chaetomium globosum*.

Some organic acids produced by fungi such as acetate, lactate, oxalate, tartarate, succinate, citrate, gluconate, ketogluconate, glycolate etc. are also reported to be very effective in P solubilization (Tarafdar, 2019). Fungal species reported to solubilize phosphate include *Aspergillus, Penicillium, Trichoderma, Rhizoctonia solani* (Jacobs et al., 2002) etc. Root-dip treatment with the phosphate-solubilizing microorganisms *Aspergillus awamori* and *Penicillium digitatum* resulted in significant increase in the yield of tomato (Khan and Khan 2002). A successful demonstration of plant growth promotion by *Penicillium radicum* inoculation in both glasshouse and field trials resulted in the development of a commercial *P. radicum* inoculant known as Pr70RELEASE (BioCare Pty Ltd, Somersby, Australia) in Australia (Wakelin and Ryder, 2004). Another *Penicillium* sp. reported to increase P uptake and grain yields under P-limiting conditions is *P. bilaiae* which has been used successfully as biofertilizer under the name JumpStart® for many years in North America (PhilomBios Ltd., Saskatoon, Canada) (Leggett et al., 2007) (Table 12.5).

12.2.3 Potassium-Solubilizing Biofertilizers

Potassium (K) is an essential macronutrient that plays an important role in various plant metabolic processes e.g., protein synthesis, photosynthesis and enzyme activation (inadequate supply of potassium results in poor development of roots, slow growth, production of small seeds and overall lower yields).

Although most soils are rich in K minerals, yet a very small fraction of it (0.1–0.2%) is available in forms that can be used by the plants. Leaching and run-off from the upper layers of soil further contribute to K deficiencies increasing the demand for K fertilizer worldwide. Potassium-solubilizing microorganisms, popularly known as KSMs, are fungal and bacterial species promoting K-solubilization from silicate present in soil, enhancing the fertility status of soils (Table 12.1).

Various bacterial species, e.g., *Pseudomonas* sp., *Burkholderia* sp., *Acidothiobacillus ferrooxidans, Bacillus mucilaginosus, B. edaphicus, B. circulans, B. megaterium* and *Paenibacillus* sp., have been reported to release K from K-bearing minerals in soil (Parmar and Sindhu, 2013). India and China extensively use potassium-solubilizing bacteria (KSB), as biofertilizer due to the deficient levels of available K in soil. A major mechanism employed by KSB is the production of organic and inorganic acids and protons which convert insoluble K from mica, muscovite, illite (Sheng and He, 2006) and feldspar (Badr et al., 2006) to soluble forms of K. Inoculation with KSB *Bacillus edaphicus* increased root and shoot growth in cotton and rape (Sheng, 2005). In addition to K, increase in N and P content was also seen on bacterial inoculation. Ability to mobilize potassium from waste mica has been reported for three PGPR species, namely *Azotobacter chroococcum, Bacillus mucilaginosus* and *Rhizobium* in maize and wheat plant (Singh et al., 2010) and *Bacillus mucilaginosus* in Sudan grass (Basak and Biswas, 2009). Four KSB strains belonging to *Klebsiella* spp. and *Enterobacter* spp. were found to significantly increase plant dry weight as well as nitrogen and potassium uptake by tobacco plant (Zhang and Kong, 2014).

There are a number of reports of phosphate-solubilizing bacteria (PSB) possessing potassium solubilization potential as well and vice versa. Bakhshandeh et al. (2017) reported the potential of three bacterial strains, viz. *Pantoea ananatis* (KM977993), *Rahnella aquatilis* (KM977991) and *Enterobacter* sp. (KM977992), to be used as both PSB and KSB for enhancing rice growth. In a field experiment, Anjanadevi et al. (2016) found that approximately one-third of K fertilizer could be replaced by use of potent KSB such as *Bacillus subtilis* ANctcri3 and *Bacillus megaterium* ANctcri7 in the cultivation of elephant foot yam. Both the bacterial species were also capable of P solubilization and production of a phytohormone, indole acetic acid (IAA). Co-inoculation of a PSB, *Bacillus megaterium* var. *phosphaticum* and a KSB, *Bacillus mucilaginosus*, increased P and K availability in cucumber and pepper plants (Han et al., 2006) (Table 12.4). Increase in plant biomass in Sudan grass was recorded when K-solubilizing bacteria *Bacillus mucilaginosus* and N-fixing bacteria *Azotobacter chroococcum* A-41 were co-inoculated with waste mica (Basak and Biswas, 2010) (Table 12.4).

K mineral-solubilizing ability of a rhizobial strain, *Mesorhizobium* sp., *Paenibacillus* sp. and *Arthrobacter* sp., has been reported by Xiao et al. (2017). In a pot experiment, inoculation of these three strains into K-deficient soil increased ryegrass growth, yield and K uptake.

Luciano et al. (2010) showed the efficacy of ectomycorrhizal fungi (*Pisolithus* sp.) in promoting growth of *Eucalyptus dunnii* when inoculated in combination with alkaline breccia, a source of phosphorus and potassium for the plants.

12.2.4 Sulfur-Oxidixing Biofertilizers

Sulfur (S) is an essential macro-element necessary for plant growth. It is one of the major elements for the growth of oilseed crops, some important vegetables like, onion, oat, cauliflower, etc. and in some spices, e.g., ginger, garlic, etc. The deficiency of S in agricultural soils has increased in the past few decades due to the reduction in atmospheric levels of S and crop varieties that remove S from soil very rapidly. Approximately 95% of S present in soil is bound organically either as sulfate-esters or sulphonates. These organic forms of sulfur are not directly available to plants and thus they depend on microbes (mainly bacterial and fungal species) present in rhizospheric soil for organo-S mobilization (Gahan and Schmalenberger, 2014) (Table 12.1).

Bacterial species belonging to the genus *Thiobacillus, Thiomicrospira, Thiosphaera, Xanthobacter, Pseudomonas, Paracoccus* and *Alcaligens* are reported to mobilize sulfate-esters

(Kuenen and Beudeker, 1982). Pourbabaee et al. (2020) observed increase oxidation of elemental sulfur when maize plants were inoculated with *Thiobacillus*. Likewise, increase in the yield of mustard was reported on inoculation with *Thiobacillus thioxidans* under field conditions (Abhijit et al., 2014). Growth of two more oil seed plants viz. groundnut and canola, was also reported to be positively affected on inoculation with *Thiobacillus* sp. (Anandham et al., 2007) and *Pandoraeasputorum* ATSB28 (Anandham et al., 2008).

12.3 MICROORGANISMS SYNTHESIZING PLANT HORMONES AND INDUCING STRESS TOLERANCE

Plant hormones are chemical messengers that may stimulate or inhibit plant growth. Microorganisms can synthesize phytohormones that can alter the rhizosphere chemistry affecting physiological responses in plants, such as growth or fruit ripening (Table 12.2). Five major groups of plant hormones have been identified, viz, auxins, gibberellins, ethylene, cytokinins, and abscisic acid (Saharan and Nehra, 2011). The role of rhizobium-derived phytohormone, auxin, in legume root nodule initiation and morphogenesis was identified as far back as 1936 (Thimann, 1936). Later its role in production of other plant growth-promoting phytohormones, like indole-3-acetic acids (IAA), cytokinins, gibberellins, etc. were also identified. Etesami et al. (2009) suggested the use of rhizobia as a biofertilizer in wheat crop, as it facilitate nutrient uptake by producing IAA and enhancing the plant root system. Other PGPR reported to produce phytohormones include *Bacillus, Pseudomonas, Azotobacter, Azospirillum* (Bottini et al., 1989; Joseph et al., 2007; Khakipour et al., 2008; Malhotra and Srivastava, 2008). The ability of bacterial species *Enterobacter lignolyticus* strain TG1, *Burkholderia* sp. strain TT6, *Bacillus pseudomycoides* strain SN29 and *Pseudomonas aeruginosa* strain KH45 (Dutta et al., 2015) to solubilize phosphate, produce IAA, siderophore, ACC deaminase, protease, cellulose and antifungal metabolite was observed. Inoculation with desiccation-tolerant bacteria *Micrococcus luteus*-chp37 resulted in growth promotion in *Zea mays* (Raza and Faisal, 2013). The growth promotion was attributed to the production of plant growth-promoting substances, cytokinin and hydrogen cyanide by the bacteria.

Another phytohormone namely ethylene regulates growth, senescence and stress in plant at low concentrations but at higher concentration adversely affects plant root growth (Souza et al., 2015). Ethylene in the plant is produced from ACC, i.e., 1-aminocyclopropane-1-decarboxylate, which in turn is broken down by ACC deaminase enzyme, thereby regulating ethylene production (Naik et al., 2019). Plant growth-promoting bacteria (Glick, 2014) and actinomycetes like *Microbispora* sp. and *Streptomyces* sp. (Kruasuwan and Thamchaipenet, 2016) produce ACC deaminase alleviating any adverse effects of ethylene. A possible synergistic interaction between ACC deaminase, plant and bacterial auxin and indole-3-acetic acid (IAA) is required for the optimal functioning of these plant growth-promoting bacteria (Glick, 2014) and actinomycetes (Kruasuwan and Thamchaipenet, 2016). A significant reduction in stress ethylene production in rice seedlings treated with *Microbacterium* sp. AR-ACC2, *Paenibacillus* sp. ANR-ACC3 and *Methylophaga* sp. AR-ACC3 has been observed (Bal and Adhya, 2021). Decrease in the production of ethylene brought about by the ACC deaminase-producing microbes has been found to enhance the formation of the nodule as well. Co-inoculation of mung bean (*Vigna radiate* L.) with *Bradyrhizobium* and *Pseudomonas* spp. with ACC deaminase activity remarkably enhanced nodule formation in comparison to inoculation with *Bradyrhizobium* alone (Shaharoona et al., 2006) (Table 12.4).

Plants are often subjected to one or many abiotic stresses, such as water unavailability, drought, alkalinity, acidity, salinity, presence of heavy metals in soils, etc. that inhibit plant growth, reducing overall crop yield. Phytohormones produced by root-associated microbes help plants to tolerate different stress conditions (Egamberdieva et al., 2017). Improved tolerance to salt and drought stress in cucumber plant due to gibberellins (GAs) and indoleacetic acid (IAA)-producing endophytic fungi *Phoma glomerata* and *Penicillium* sp. has been reported (Waqas et al., 2012). Another gibberellin-producing endophytic fungus, *Aspergillus fumigatus*, associated with soybean roots was

TABLE 12.2
Microbial Inoculants/Biofertilizers Reported to Enhance Plant Growth by Synthesizing Plant Hormones and Alleviating Abiotic Stress

A) Microorganisms Synthesizing Plant Hormones

Microorganisms	Host Plant	Plant Hormones Secreted	Reported Effect	Reference
Rhizobia	Wheat	IAA	Uptake more nutrients by increasing plant root system	Etesami et al. (2009)
Bacillus megaterium	Tea plants	IAA	Solubilize phosphate, produce siderophore, ACC deaminase, protease, cellulose and antifungal metabolite	Chakraborty et al. (2006)
Enterobacter lignolyticus strain TG1, *Burkholderia* sp. strain TT6, *Bacillus pseudomycoides* strain SN29 and *Pseudomonas aeruginosa* strain KH45	Tea plants	IAA	Solubilize phosphate, produce siderophore, ACC deaminase, protease, cellulose and antifungal metabolite	Dutta et al. (2015)
Aspergillus fumigatus	Soybean plants	Gibberellins	Increase of photosynthetic pigments & shoot biomass under salt stress	Khan et al. (2011)
Micrococcus luteus-chp37.	*Zea mays*	Cytokinin	Production of growth-promoting substances and hydrogen cyanide.	Raza and Faisal, (2013)
Actinomycetes like *Microbispora* sp. and *Streptomyces* sp.		Ethylene regulators (ACC deaminase) + IAA	Lowers plant ethylene levels thus alleviating adverse effects of ethylene.	Kruasuwan and Thamchaipenet (2016)
Plant growth-promoting bacteria		Ethylene regulators (ACC deaminase) + IAA	Lowers plant ethylene levels thus alleviating adverse effects of ethylene	Glick (2014)
Microbacterium sp. AR-ACC2, *Paenibacillus* sp. ANR-ACC3 and *Methylophaga*	Rice	Ethylene regulators (ACC deaminase)	Reduction in stress due to ethylene production and enhanced formation of the nodule	Bal and Adhya, (2021)

B) **Microorganisms inducing stress tolerance**

Microorganisms	Host plant	Abiotic stress	Possible mechanism	Reference
Bacillus sp. NBRI YN4.4	Maize (*Zea mays* L.)	Alkaline stress	Increased photosynthetic pigments & soluble sugar, decreased proline level	Dixit et al. (2020)
Pseudomonas aeruginosa GGRJ21	Mung bean	Water stress	Fast accumulation of antioxidant enzymes, cell osmolytes & up-regulating the stress responsive genes	Sarma and Saikia (2014)
PGPR strains (*Pantoea agglomerans* and *Bacillus* sp.) alone or in combination with AMF (*Rhizophagus fasciculatus*)	*Casuarina obesa*	Saline stress	Increase in chlorophyll content & accumulation of plant proline.	Diagne et al. (2020)
Micrococcus luteus-chp37	*Zea mays*	Water stress	Production of plant growth-promoting substances, cytokinin and hydrogen cyanide.	Raza and Faisal (2013)

C) **Phytohormones from microorganisms inducing stress tolerance in plants**

Microorganisms	Host plant	Plant Hormones secreted	Abiotic stress	Reference
(Ni)-tolerant *Bacillus subtilis* strain SJ-101	Mustard, *Brassica juncea* (L.).	IAA	Ni phytotoxicity	Zaidi et al. (2006)
B. licheniformis HSW-16	Wheat	IAA	Salt stress	Singh and Jha (2016)
Pseudomonas spp	Maize	IAA	Salt stress	Mishra et al. (2017)
Serratia plymuthica RR-2-5-10, *Stenotrophomonas rhizophila* e-p10, *P. fluorescens* SPB2145, *P. extremorientalis* TSAU20, and *P. fluorescens* PCL1751	Cucumber	IAA	Salt stress	Egamberdieva, (2011)
Bacillus licheniformis Rt4M10 and *Pseudomonas fluorescens* Rt6M10	*Vitis vinifera*	ABA	Water stress	Salomon et al. (2014)

shown to increase photosynthetic pigments and shoot biomass under conditions of salt stress (Khan et al., 2011). Significant increase in the growth of *Brassica juncea* (L.), the Indian mustard plant, inoculated with nickel (Ni)-tolerant *Bacillus subtilis* strain SJ-101was attributed to the production of indole acetic acid (IAA) and solubilization of inorganic phosphate by the bacteria (Zaidi et al., 2006). Similar growth enhancement by the production of IAA in saline soil conditions has been reported in wheat by *B. licheniformis* HSW-16 (Singh and Jha, 2016), in maize by *Pseudomonas* sp. (Mishra et al., 2017) and in cucumber by *Serratia plymuthica* RR-2-5-10, *Stenotrophomonas rhizophila* e-p10, *P. fluorescens* SPB2145, *P. extremorientalis* TSAU20, and *P. fluorescens* PCL1751 (Egamberdieva et al., 2011). ABA synthesis by *Bacillus licheniformis* Rt4M10 and *Pseudomonas fluorescens* Rt6M10 in *Vitis vinifera* was found to alleviate stress induced by water losses (Salomon et al., 2014).

Microorganisms may also alleviate plant stress by improving soil health (Dixit et al., 2020). Co-inoculation of plant growth-promoting *Rhizobium* and *Pseudomonas* species acted synergistically resulting in positive adaptive responses in maize plants under saline conditions (Bano and Fatima, 2009) (Table 12.4). Arbuscular mycorrhizal fungi (AMF) are reported to improve tolerance in crop plants like maize, rice, potato (Chifetete and Dames, 2020) and tomato (Balliu et al., 2015) to drought and salinity stress. PGPR strains (*Pantoea agglomerans* and *Bacillus* sp.) when used alone or in combination with AMF (*Rhizophagus fasciculatus*) improved *Casuarina obesa* performance under saline stress (Diagne et al., 2020).

12.4 MICROORGANISMS USED FOR PLANT PEST CONTROL

Microbial pesticide is among the three major classes of biopesticides being employed for the control of pests. It exclusively consists of microscopic living organisms such as bacteria, fungi, viruses, protozoa or nematodes or even their toxins and metabolites as its active ingredient. These are readily available as fungicides, herbicides, insecticides, dust, sprays or granules and even growth regulators. Microbial pesticides can be used as an effective controlling agent against several pests, although each of these products are highly specific for their target pests. This acute target specificity makes them an effective option to be included in integrated pest management strategies as they are not harmful to the non-target organisms such as natural enemies of insect pests, beneficial insects and humans. Moreover, being biodegradable, they pose no threat to the environment.

The most widely used biopesticides are biofungicides (consisting of *Trichoderma*), bioherbicides (like *Phytopthora*) and bioinsecticides (like *Bacillus thuringiensis, B. sphaericus*). Their use may potentially benefit both agriculture and public health programs (Gupta and Dikshit, 2010). However, the use of these pesticides in agriculture for pest management is not a novel approach and dates back as far as the 1930s, when spores from *Bacillus popilliae*, a naturally occurring ring bacterium, were first isolated from an infected larva of a Japanese beetle, *Popillia japonica*, and were found to be pathogenic to scarab larvae. Later in 1948, *Bacillus siae*, now known as *Paenibacillus popilliae*, was registered as the first microbial pesticide. Since then, it has been commercialized and used for the control of these beetles in the United States (Koppenhöfer & Wu, 2017).

The mode of action of most of the microbial entomopathogens is invasion through the insect integument or gut, followed by multiplication of the pathogens which ultimately results in the death of the host insect (Usta, 2013).

12.4.1 Types of Microbial Biopesticides Used Worldwide

12.4.1.1 Entomopathogenic Fungi-Based Microbial Pesticide

The approach of replacing conventional insecticide completely with fungal entomopathogens or using their combination can prove beneficial for managing insecticide-induced resistance in pests (Hajeck and Leger, 1994) (Table 12.3). Such biopesticides with fungi as their active ingredient are

TABLE 12.3
Microbial Insecticides and Their Uses

Microbial Type	Commercial Preparation Name	Types of Pests Controlled	Uses & Applications
		BACTERIA	
Bacillus thuringiensis var. *kurstaki* (Bt)	Caterpillar Killer®, Dipel®, Futura®, Javelin®, SOK-Bt®, Thuricide®, Bactur®, Bactospeine®, Topside®, Tribactur®, Worthy Attack®, Bioworm®.	Foliage feeding caterpillars	• Effective in controlling Indian meal moths in stored grains also • Must be applied in the evening or on cloudy days as it gets rapidly deactivated in sunlight • Should also be sprayed on the lower surfaces of leaves • Effective only when ingested • Available either as wettable powders or liquid concentrates or ready to use dusts and granules
Bacillus thuringiensis var. *israelensis* (Bt)	LarvX®, Mosquito Attack®, Skeetal®, Teknar®, Vectobac®, Aquabee®, Bactimos®, Gnatrol®,	Larvae of mosquitoes like *Aedes* and *Psorophora*, black flies and fungus gnats	• Must be ingested • Exhibits reduced activity in turbid or polluted water • Generally applied over wider areas for mosquito and blackfly control
Bacillus thuringiensis var. *tenebrinos*	M-Track®, Foil® M-One®, Trident®, Novardo®	Grubs of beetles like Colorado potato beetle and adults of elm leaf beetle	• Should be ingested for efficacy • Breaks down in UV light • Not very persistent in the environment
Bacillus thuringiensis var. *aizawai*	Certan®	Caterpillars of wax moth	• Effective against wax moths infesting honeybee hives
Bacillus popilliae and *Bacillus lentimorbus*	Milky Spore Disease®, Grub Attack®, Doom®, Japidemic®	Grubs of Japanese beetles	• Persistent in soil for years
Bacillus sphaericus	Vectolex CG®, Vectolex WDG®	Larvae of mosquitoes like *Culex*, *Psorophora*, *Culiseta* some *Aedes* species.	• Must be ingested • Ineffective in turbid or stagnant water
		FUNGI	
Beauveria bassiana	Naturalis®, Botanigard®, Mycotrol®	Most hemipteran pests like mites, thrips, whiteflies, aphids, fungus, mealy bugs and gnats	• Requires high moisture • Has limitations like lack of proper storage techniques • Longevity is reduced • Has to compete with other microbes present in soil.
Lagenidium giganteum	Laginex®	Larvae of most mosquito species	• Stays effective even during dry periods in field • Unable to survive high temperatures
		PROTOZOA	
Nosema locustae	Grasshopper Attack®, NOLO Bait®	Grasshoppers and mormon crickets, caterpillars of European cornborer	• Must be ingested • Not very effective to use in backyard gardens as it is slow acting while grasshoppers are highly mobile

(Continued)

TABLE 12.3 *(Continued)*

Microbial Type	Commercial Preparation Name	Types of Pests Controlled	Uses & Applications
		VIRUSES	
Gypsy moth nuclear polyhedrosis (NPV)	Gypchek® virus	Gypsy moth caterpillars	Not much information available regarding use
Tussock moth NPV	TM Biocontrol-1®	Tussock moth caterpillars	Not much information available regarding use
Pine sawfly NPV	Neochek-S®	Pine sawfly larvae	Not much information available regarding use
Codling moth granulosis virus (GV)		Codling moth caterpillars	• Unavailable currently • Must be ingested • Breaks down rapidly in UV rays.
	ENTOMOGENOUS NEMATODES		
Steinernema feltiae (=*Neoaplectana carpocapsae*), *S. riobravis*, *S. carpocapsae* and other *Steinernema* species	Ecomask®, Scanmask®, Biosafe®, also sold generically (wholesale and retail), Vector®	Larvae of most soil-dwelling and boring insects	• Requires moisture • Can't be stored for long period • Extreme low or high temperatures also debilitate the nematodes • Must be stored and used in cool temperatures
Heterorhabditis heliothidis	Available in bulk or only for research or large-scale commercial uses	Larval instars of many soil-dwelling and boring insects	• Used mostly in European countries
Steinernema scapterisci	Nematac®S	Late nymphal instars and adults of mole crickets	• Good irrigation required both at the time of application and after application

Source: Usta (2013).

called mycoinsecticides. These provide effective control against diverse agricultural insect pests. They enter their host insects via the cuticle into the hemolymph, produce toxins and grow by using up the nutrients present in haemocoel in order to avoid the immune responses of the host (Meadows, 1993). They can be applied in the form of mycelium or conidia which sporulates thereafter. Some examples of commercially produced mycoinsecticide include Boverin, based on *Beauveria bassiana*, and lower doses of trichlorophon used to suppress the second-generation outbreaks of *Cydia pomonella* (Usta, 2013) and *Metarhizium anisopliae* used to control *Aedes aegypti* and *Aedes albopictus* (Shia and Feng, 2004).

12.4.1.2 Virus-Based Microbial Pesticides

Around 1600 different viruses have been reported to infect about 1100 insects and mites. Out of these around 100 insect species are alone susceptible to a specialized virus category, the baculovirus (Usta, 2013). These rod-shaped particles form a virus inclusion body bearing DNA within a protein coat. Upon entering the host insect's midgut, the inclusion body releases the viral particles due to the dissolution of the protein cover in the alkaline environment of the gut. These particles then multiply rapidly by fusing with the epithelial cells of midgut and ultimately killing the host. They are quite expensive due to their highly specific host action. Another limitation is that their action is too slow to satisfy farmers. Moreover, they are unstable when exposed to ultraviolet rays. In order to resolve this, research is being carried out to enclose baculoviruses within UV protectants, ensuring their longer viability in field (Usta, 2013).

Successful examples of introducing baculovirus (NPV) into the environment date back to the pre-World War II period, when it was released along a parasitoid imported from Canada, which led to effective suppression of spruce sawfly, *Diprion hercyniae*. Since then no other control measures have been required against this hymenopteran species. In 1975, another viral insecticide called Elcar™ was introduced by Sandoz Inc., containing *Heliothis zea* NPV. This broad-range baculovirus provided control against pests belonging to *Helicoverpa* and *Heliothis* genera which attack maize, beans, soybean, sorghum and tomato (Usta, 2013). Another successful example is the use of *Anticarsia gemmatalis nucleopolyhedrovirus* (AgMNPV) against the velvet bean caterpillar infesting soybean since the early 1980s in Brazil. These success stories of baculovirus pesticides prove the numerous advantages of microbial over synthetic pesticides (Mettenmeyer, 2002).

12.4.1.3 Protozoa-Based Microbial Pesticides

Protozoan pathogens have a natural potency to infect a variety of insect hosts. An added advantage of using them to control pest populations is that they also reduce the number of offspring produced, besides killing their insect hosts. But despite playing a significant role in lowering the insect populations, only a few of these preparations have proved suitable for use as insecticides. One such example is the Microsporidia, which includes species that inflict common infections in insects, eventually leading to low to moderate insect mortality. Though they infect a wider variety of insects, their mode of action is slow, taking several days to weeks before harming their host. Rather than killing the pest insect directly these products may frequently reduce their reproduction or feeding activity. Being a rélatively new technology, it requires more research work in order to perfect its use in the field.

12.4.1.4 Microscopic Nematodes-Based Pesticides

Nematodes are multicellular, colorless, unsegmented roundworms without any appendages. They can be parasitic, free-living or predaceous by nature (Figure 12.1). Out of these, many of the parasitic species have been beneficial in controlling the insect pests, by either sterilizing or debilitating them. Few of them have also succeeded in causing insect death but they may be difficult to handle (e.g., tetradomatids) or expensive to mass-produce (e.g. mermithids). Another drawback is their narrow host specificity against minor insect pests and modest virulence (e.g., sphaeruliids). Successful examples of nematodes with biological control characters are from the genera *Steinernema* and *Heterorhabditis*. These are called entomogenous nematodes as they are effective only against insects or related arthropods (Moscardi, 1999). These have been used to control over 400 pest species of beetles, caterpillars and fly maggots.

The third juvenile stage or the J3 stage is the infectious stage, also referred to as "dauer" larvae that can survive without food for long periods in moist soil. While *Steinernema* juveniles enter the host insects via body openings such as the anus, mouth and spiracles, *Heterorhabditis* juveniles

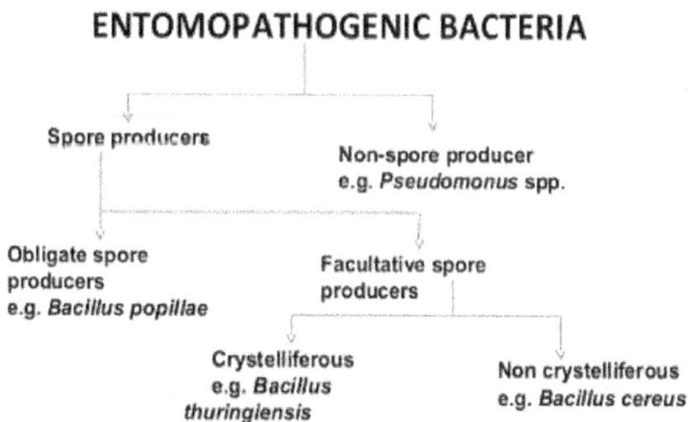

FIGURE 12.1 Classification of entomopathogenic bacteria.

may additionally penetrate through the insect's cuticle. Under favourable conditions, they molt into adults inside the infected host to complete their life cycle and start the next generation. Once the J3 stage is reached, they leave the previous dead insect to seek a new host (Ma et al., 2012).

Another commonly used nematode is *Nosema locustae*, which successfully reduces grasshopper populations in rangeland areas by infecting the young hoppers that in turn consume less forage. Though 100% mortality is not attained, the infected females produce a smaller number of viable eggs. *N. locustae* persisting on egg pods assures subsequent infection (Usta, 2013).

12.4.1.5 Bacteria-Based Bacterial Pesticides

These are the widely used cheapest form of microbial pesticides easily available for use in the field. They are highly specific to individual species of moths, butterflies, beetles, flies and mosquitoes but have to be ingested by the target pest to be effective. Only the soil-dwelling, rod-shaped Bacillus genera that produces spores are employed for this purpose. These are present all around us ranging from sea water to soil, and even in the hot springs (Harwood and Wipat, 1996). Many valuable characteristics make these organisms one of the major sources of potential microbial biopesticides (Ongena and Jacques, 2008). A well-studied example is *B. subtilis*, which are categorized as "generally regarded as safe" (GRAS) and non-pathogenic by the US Food and Drug Administration (USFDA), an essential feature of biopesticides (Monteiro et al., 2005). Also, under unfavorable conditions, they produce spores (resistant dormant form) which can easily survive in extreme temperatures, pH, even without nutrients or water (Piggot and Hilbert, 2004). Sporulation can also be induced artificially for exploitation in industrial production. Another advantage is that these bacteria are viable in the open so denaturation is delayed and the control measure is intact for a longer period.

Another important and widely used bacterium is *Bacillus thuringiensis* (Bt) which is marketed worldwide for control of many important plant pests, mainly caterpillars of the Lepidopterans, mosquito larvae, and simuliid blackflies, etc. Bt can also produce endogenous crystals protein inclusions called endotoxins or Cry proteins, with insecticidal properties (Schnepf et al., 1998) during its sporulation. These consist of one or more crystal (Cry) and cytolytic (Cyt) toxins known as δ-endotoxins or insecticidal crystal proteins that are highly toxic to certain insects but pose no harm to other organisms. The commercial Bt products are usually available in powder form containing dried spores and toxin crystals. Since they must be ingested, these are applied mostly on feeding surfaces of the insect larvae like leaves. Several crop plants have also been genetically engineered with the toxin genes. Once ingested, the endotoxin is activated by the alkalinity of the insect's gut and attaches to specific receptors present on its wall, causing gut paralysis due to breakdown of gut lining, probably by upsetting the ionic balance of the gut. The spore survives its passage through the gut, penetrates the weakened midgut wall, and multiplies in the hemolymph. Death results from either intoxication or septicemia.

Since the discovery of its insecticidal potential in 1901, Bt has been used as a potential source of eco-friendly biopesticide all over the world. It has many different strains, each producing a different concoction of proteins, specific for killing one or a few related insect species. In order to be effective against a target pest species, *Bt* must produce a protein specific for binding to the receptors present on the larval gut, causing its starvation. Studies have shown that Bt spores can survive well both on the ground and in animals. They can also be carried to neighboring areas by wind, rain and animals. They are even able to transfer their toxicity genes to other bacteria present in the region of application (Usta, 2013).

Another species, *Bacillus popilliae*, or milky spore disease, has been marketed for controlling Japanese beetle larvae (Weinzierl and Henn, 1989; US EPA, 2004). *Paenibacillus popilliae* causes a fatal illness called milky disease in Japanese beetle larvae. After ingestion by the larvae, *B. popilliae* germinates in the gut, begins to multiply, and invades the haemolymph. After about 10 days, a typical milky appearance is observed due to the massive numbers of bacteria.

Bacillus cereus strains are also often pathogenic for insects. They produce phospholipase C, an α-exotoxin that permits the bacteria to pass through the barrier of the intestinal epithelial cells. Subsequent penetration into the haemolymph followed by multiplication kills the insect.

Bacillus sphaericus is a rod-shaped, aerobic, gram-positive bacterium used against certain strains of insect vectors, such as mosquitoes and black flies. It is the most commonly used microbial bio-control agent along with *Bacillus thuringiensis* (Kevan and Shipp, 2017). Its mode of action is similar to that of Bt.

12.4.2 Advantages of Using Microbial Pesticides

- The greatest plus-point of these preparations is their assured environmental safety, being non-toxic and non-pathogenic to humans, wild animals and other unrelated organisms.
- Their toxic action is highly species-specific and does not cause any direct harm to beneficial organisms (natural enemies of pests) in the applied field. Thus, ensuring ecological balance and perseverance of biodiversity.
- They leave very little or no residue in food.
- If required, most of these products can be used in combination with chemical insecticides without being denatured or altered by the latter.
- They can even be applied at a time close to crop harvesting as their residues are harmless to other organisms.
- At times, the infectious or active microorganisms present in these products may establish themselves in a pest population or its habitat to provide persistent control during subsequent seasons.
- Additionally, these preparations also increase the crop yield by encouraging the beneficial soil microflora that in turn enhance root and plant growth (Usta, 2013).

12.4.3 Disadvantages of Microbial Insecticides

In addition to several benefits there are also a few limitations of these preparations, but these do not hamper their successful use.

- Target specificity is a great advantage of these pesticides but this narrow range is also their main limitation. Application of a single microbial pesticide may control only a small segment of the entire pest complex, while many other types of pests present in the applied field continue to survive and cause damage.
- Heat or desiccation (drying out) decreases their shelf life.
- Many of these products exhibit reduced effectiveness on exposure to UV radiation.
- Variations in lab-to-land performances also need to be accounted for their efficiency as lab conditions are homogenous. Therefore, good timing and mindful procedures are especially important for their application.
- Some of the microbial products may need special formulation and storage procedures, complicating the manufacturing and dispensing of certain products. It is important to store microbial insecticides by strictly adhering to label directions.
- Most of these pesticides are heavily priced, making them out of reach for local farmers.
- Their high pest-specificity also limits their potential market. Since the overall production cost involved in developing and registering these products cannot span over a wide range of pest control sales, availability of a few products may be affected while their cost escalates (several insect viruses, for example) (Usta, 2013; Rani et al., 2021).

12.5 CO-INOCULATION

Initially a majority of the manufactured microbial inoculant contained a single species of microorganism for a particular crop. Later, experiments that used two or more beneficial microorganisms together as inoculant for a particular crop plant started to gain prominence, being more effective than a single inoculant. One of the earliest reports of successful field application is from Brazil for the soybean crop, where the combination of two *Bradyrhizobium* strains has been successfully used by farmers since the 1950s (Santos et al., 2019). The use of two similar strains ensures that at least one of them will nodulate and bring about the desired effect. Co-inoculation of two strains of *Azotobacter chroococcum*, AC1 and AC10 also exhibited a greater beneficial effect on growth parameters of cotton plant compared to single inoculation (Romero-Perdomo et al., 2017) (Table 12.4).

Over the last two decade the use of mixed co-inoculation, i.e., inoculant with two or more different types of microbes, has gained popularity (Table 12.4). The combined use of two or more PGPMs possessing different beneficial properties produces a better overall combined effect. Most of the successful reports of mixed co-inoculation use microorganisms performing processes such as BNF (e.g., *Rhizobium* spp., *Bradyrhizobium* spp.), phosphate solubilization (e.g., *Bacillus* spp., *Pseudomonas* sp.), phytohormone production (e.g., *Azospirillum* spp., *Pseudomonas* sp.), or biological control (e.g., *Pseudomonas* sp., *Bacillus* sp.). Utkarsh Microbes marketed by Utkarshagro is one such product which is a combination of a biofertilizer and a biopesticide available for commercial use in India. It is a unique combination of nitrogen-fixing bacteria like *Azotobacter, Acetobacter, Rhizobium* and microbial pesticides such as *Trichoderma viride, Pseudomonas fluorescens, Veticillium lecanii*, etc. Its target crops include chillies, tomato, brinjal, cabbage, onion, bottle gourd, ridge gourd, sponge gourd, mango, papaya, coconut, potato, etc.

Some non-rhizobial strains have been shown to improve the nodulation and N-fixing potential of the inoculated rhizobial strains (Chibeba et al., 2015; Hungria et al., 2013; Mishra et al., 2009). *Azospirillun brasilense* co-inoculated with *Rhizobium* significantly increased nodule number and nitrogen fixation in common bean (*Phaseolus vulgaris* L.), compared to single *Rhizobium* inoculation (Burdman et al., 1997). *Azospirillun* sp. enhances root hair formation, creating more infection sites for rhizobium, and increases flavonoid exudation from roots that are crucial for the establishment of the *Rhizobium*–legume symbiosis (Volpin et al., 1996). Some specific environmental conditions and even the host genotype affect the success of some PGPR and *Rhizobium* co-inoculation in legumes (Remans et al., 2007; Remans et al., 2008). This suggests that extensive research is required to understand the mechanisms employed by various PGPRs and the factors affecting their response before preparing formulations for different crops.

Some phosphate-solubilizing bacteria, in addition to providing phosphorus to the plants, also stimulate nitrogen fixation, accelerate accessibility of other trace elements and are known to produce growth hormones exerting substantial measurable positive effects on plants. The synergistic effect of nitrogen-fixing bacterium *Phyllobacterium* sp. and PSB, *Bacillus licheniformis* was reported by Rojas et al. (2001). Similar increase in yield on dual inoculation with N2 fixing and phosphate-solubilizing bacteria has been reported in sorghum (Alagawadi & Gaur, 1992), barley (Belimov et al., 1995) and black gram (Tanwar et al., 2002).

Many legumes form tripartite symbiotic associations with nodule-inducing rhizobia and arbuscular mycorrhizal fungi that provide fitness benefits to legumes especially under some environmental stress conditions. Co-inoculation of *Rhizobium* and AMF biofertilizer was found to be more effective in promoting growth of faba bean than individual treatment under alkaline soil conditions (Abd-Alla et al., 2014). Mobilization of certain key elements such as P, Fe, K and other minerals by AMF involved in synthesis of nitrogenase and leghaemoglobin required for nitrogen fixation by *Rhizobium* may be the reason for this better performance. Improvement in symbiotic performance of *Mesorhizobium ciceri* with its host plant, chickpea, under saline soil conditions was recorded when IAA-producing *Bacillus subtilis* NUU4 was co-inoculated with *M. ciceri* IC53 (Egamberdieva et al., 2017).

TABLE 12.4
Microbial Consortium Used in Agriculture for Plants Growth

Microbial Strain	Plants	Effects	Region	References
I) **BACTERIAL CONSORTIUM**				
Azospirillun brasilense and *Rhizobium*	Common bean (*Phaseolus vulgaris* L.),	Increase in nitrogen fixation and nodule number	Lab work, Jerusalem, Israel	Burdman et al. (1997).
Azospirillum spp. + *Azoarcus* spp. + *Azorhizobium* spp.	Wheat (*Triticum aestivum)*	Enhanced plant growth and nitrogen content	Italy	Dal Cortivo et al. (2020)
Rhizobia strains (IITA-PAU 987 and IITA-PAU 983) + *Bacillus megaterium*	Common bean varieties	Enhanced growth of common bean Plants in a phosphorous deficient soil.	Greenhouse pot experiment, Kenya	Korir et al. (2017)
Bradyrhizobium and *Pseudomonas* spp.	Mung bean (*Vigna radiate* L.)	Enhanced nodulation	Pot experiment, Pakistan	Shaharoona et al. (2006)
Number of Native rhizobia strains	Common beans	Better growth improvement compared to the commercial rhizobia strain, viz. CIAT 899 and Strain 446	Kenya	Kawaka et al. (2014)
Pseudomonas Sp. DN 13–01, *Sphingobacterium suaedae* T47, *Bacillus pimilus* X22 and *Bacillus cereus* 263AG5	*Zea mays*	Improved growth parameters, viz. plant height, fresh areal and root biomass.	Greenhouse pot experiment, Morocco	Benbrik et al. (2020)
Bacillus thuringiensis-KR1 + *Rhizobium leguminosarum*-PR1	Field pea (*Pisum sativum* L.) and lentil (*Lens culinaris* L.)	Improved nodulation and growth of legumes	India	Mishra et al. (2009)
Azotobacter chroococcum and *Pseudomonas fluorescens* + Manure	Strawberries	To increase crop yield	Uttarakhand, India	Negi et al. (2021)
Paenibacillus polymyxa and *Bacillus megaterium* var. *phosphaticum* (phosphate dissolvers)	Tomato plants	Crop yield comparable to those produced on using phosphate fertilizer	Egypt	El-Yazeid and Abou-Aly (2011)
Rhizobium and *Pseudomonas* species	Maize plants	Positive adaptive responses under salinity		Bano and Fatima (2009)
B. subtilis NUU4 and *Mesorhizobium ciceri* IC53	Chickpea (*Cicer arietinum* L.)	Increased root and shoot biomass and improved nodule formation under salt stress	Greenhouse pot experiment, Country-NA	Egamberdieva et al. (2017)
PSBs, *Pantoea agglomerans* and *Eurkholderia anthina*	Mung bean (*Vigna radiata* [L.]	Seed inoculation enhanced shoot and root length & their dry matter, and P uptake.	South Korea	Walpola and Yoon (2013)
Pseudomonas chlororaphis and *Bacillus cereus* and *Pseudomonas fluorescens*	Walnut	Increase in shoot and root dry weight, & improvement in P and nitrogen (N) uptake.	NA	Yu et al. (2011)
Paenibacillus polymyxa and *Bacillus megaterium*	Squash seeds (*Cucurbita pepo* L.)	Increase in levels of growth promoters, GA, IAA and cytokinines and decrease of ABA inhibitor	Moshtohor, Egypt	El-Yazeid et al. (2007)

(Continued)

TABLE 12.4 *(Continued)*

Microbial Strain	Plants	Effects	Region	References
Paenibacillus polymyxa and *Bacillus megaterium*	Tomato	Increased growth and P uptake have	Egypt	El-Yazeid and Abou-Aly (2011)
Two Bradyrhizobial isolates (R25B and IRj 2180A)	Soyabean	symbiotic effectiveness of bacteria increased grain yield	Nigeria	Sanginga et al. (2000)
Rhizobium leguminismarum Thal-8/SK8 and PSB, *Pseudomonas* sp. strain 54RB	Wheat (*Triticum aestivum*)	Yield increase	Islamabad, Pakistan	Afzal and Bano (2008)
Bacillus megaterium var. *phosphaticum* (PSB) and *Bacillus mucilaginosus* (KSB)	Cucumber and pepper	Increased mineral availability, uptake and plant growth	Korea	Han et al. (2006)
Bacillus mucilaginosus (KSB) and *Azotobacter chroococcum* A-41 (nitrogen (N) fixing bacteria)	Sudan grass (*Sorghum vulgare* Pers.)	Significant increase in biomass and nutrient acquisition	Jharkhand, India	Basak and Biswas (2010)
N-free fixing bacteria (*Azotobacter* and *Azospirillium*)+ P-dissolving bacteria (*Bacillus megatherium*) + silicate dissolving bacteria (*Bacillus circulans*).	Potato (*Solanum tuberosum* L. cv. Sponta)	Increase in yield, nutrient uptake (N, P and K), tuber size, leaf area, and total Chlorophyll	Egypt	Abdel-Salam and Shams, (2012)
II) **BACTERIAL + FUNGAL CONSORTIUM**				
Rhizophagus irregularis + *Azotobacter vinelandii* & *R. irregularis* + *Bacillus megaterium* + *Frateuria aurantia*,	Wheat (*Triticum aestivum*)	Enhanced plant growth and nitrogen accumulation.	Italy	Dal Cortivo et al. (2020)
Rhizobium leguminosarum bv. *viciae* and arbuscular mycorrhizal fungi	Faba bean (*Vicia faba* L.)	Increase in nodules, nitrogenase activity, leghaemoglobin content, mycorrhizal colonization, root and shoot dry mass.	Assiut, Egypt	Abd-Alla et al. (2014)
Paenibacillus polymyxa & AMF, *Glomus mosseae*	Squash seeds (*Cucurbita pepo* L.)	Early production of fruits and high yield.	Moshtohor. Egypt	El-Yazeid et al. (2007)
AMF + *Diversispora versiformis*	*Chrysanthemum morifolium* (Hangbaiju)	Improved growth and root nitrogen uptake under moderate soil salinity.	Hangzhou, Zhejiang, China	Wang et al. (2018)
AMF and *Bradyrhizobium*	Green grams (*Vigna radiata* L. Wildzek)	Enhances nitrogen fixation and growth		Musyoka et al. (2020)
AMF, Glomus manihotis + rhizobacteria Entrophospora colombiana & PSB, Pseudomonas	Shorghum plant	Increasing yield under acid and low phosphate level soil	Indonesia	Widada et al. (2007)
III) **FUNGAL + FUNGAL CONSORTIUM**				
Aspergillus fumigatus & *Glomus mosseae*	Wheat	Increase in shoot and root dry weight, root length and phosphatase activity on seed inoculation.	**NA**	Tarafdar and Marschner (1995)

Li et al. (2020) designed a multispecies PGPR inoculum containing *Providencia rettgeri, Advenella incenata, Acinetobacter calcoaceticus* and *Serratia plymuthica*, which was able to fix nitrogen, solubilize phosphorus and produce auxin. The PGPR microbial inoculants had a plant growth-promoting effect on oat (*Avena sativa*), alfalfa (*Medicago sativa*) and cucumber (*Cucumis sativus*) seedlings by increasing soil enzyme activity and available nutrient content.

The two commercial biofertilizers, TripleN® composed of bacterial consortium (*Azospirillum* spp. + *Azoarcus* spp. + *Azorhizobium* spp.) and RhizosumN®, a fungal + bacterial consortium (*Rhizophagus irregularis* + *Azotobacter vinelandii*) are being successfully applied in common wheat (*Triticum aestivum* L.) in open field before stem elongation using foliar spray (Dal Cortivo et al., 2017; Dal Cortivo et al., 2018) (Table 12.5).

TABLE 12.5
Some Commercial Microbe-Based Biofertilizers

Name of the Product	Strain	Crops Suited	Country/Region	References
A) Rhizobia based biofertilizers				
Biofix	*Bradyrhizobium japonicum* strain USDA 110	Legumes	MEA, Kenya	Ulzen et al. (2016)
TripleN®,	PGPRs *Azorhizobium* spp., *Azoarcus* spp., and *Azospirillum* spp.	Wheat	Mapleton Agri Biotec, Mapleton, Australia	Dal Cortivo et al. (2020)
Rhizosum N®	AMF *Rhizophagus irregularis* and *Azotobacter vinelandii*	Wheat	Biosum Technology, Madrid, Spain	Dal Cortivo et al. (2020)
Rhizosum PK®	AMF *Rhizophagus irregularis* and *Frateuria aurantia*	Wheat	Biosum Technology, Madrid, Spain	Dal Cortivo et al. (2020)
Legumefix	*Bradyrhizobium japonicum* 532 C	Soybean and cowpea	(Becker Underwood, UK),	Ulzen et al. (2016)
Nitrofix P	*Bradyrhizobium japonicum* and *Bradyrhizobium elkanii*	Soybeans	Agro-Input Suppliers Limited (AISL) (Malawi)	Parnell et al. (2016)
SEMIA 5079	*Bradyrhizobium japonicum*	Soybean (*Glycine max*, L.)	Brazil	Santos et al. (2021)
SEMIA 5080	*Bradyrhizobium diazoefficiens*	Soybean (*Glycine max*, L.)	Brazil	Santos et al. (2021)
Vault®	*Bradyrhizobium japonicum* + dual-strain biofungicide	Soybean	BASF, USA	Parnell et al. (2016)
B) *Azotobacter*-based biofertilizers				
Azotobacterin,	*Azotobacter chroococcum*	Field pea, spyabean, chickpea, tomato pepper, brinjal, etc.	Southern and Eastern Russia Natural resources	Aasfar et al. (2021)
Nitrofix AC	*Azotobacter chroococcum*	Wheat	Agri Life, India	Aasfar et al. (2021)
Dimargon1	*Azotobacter chroococcum*	Rice, cotton	Colombia	Aasfar et al. (2021)
TwinN	*Azotobacter*	Legume & cereal crops	Australia Mapleton Agri Biotec Pty Ltd	Aasfar et al. (2021)
Nutri-Life Bio-P Nutri-Life Bio-N	*Azotobacter* ssp. and *Bacillus subtilis Azotobacter* ssp.	All crops All crops	Canada Nutri-Tech solutions	Aasfar et al. (2021)
Phylazonit-M	*Bacillus megaterium* and *Azotobacter chroococcum*	Rice, maize	Hungary PhylazonitKft	Aasfar et al. (2021)

(Continued)

TABLE 12.5 *(Continued)*

Name of the Product	Strain	Crops Suited	Country/Region	References
C) *Azospirillum*-based biofertilizers				
Bio-N	Two *Azospirillum* sp.	Rice, corn	BIOTECH-UPLB	Monsalud, (2008)
MicroAZ-ST™	*Azospirillum brasilense* & *A. lipoferum*	Wheat, Corn and Grain Sorghum	TerraMax (Minnesota, USA)	Parnell et al. (2016)
Mazospirflo-2	Azospirillum brasilense	Maize &soybean	Soilgro Ltd. South Africa	Owen et al. (2015)
Az39	*Azospirillum brasilense*	Maize and wheat	Argentina	Cassán et al. (2015)
Azo-NR and Azo-N Plus R	*Azospirillum*-based bioinoculant	Grain and cover crop legumes	South Africa, BioControl Products SA (Pty) Ltd	Raimi et al. (2017)
Ab-V5 (=CNPSo 2083) and Ab-V6 (=CNPSo 2084)	*Azospirillum brasilense*	Sugarcane, corn, wheat, rice & co-inoculation of legumes, such as soybean and common bean	Brazil	Santos et al. (2021)
D) *Pseudomonas* based phosphate solubilizing biofertilizers				
Pr70RELEASE	*Pseudomonas radicum*	Wheat	(BioCare Pty Ltd, Somersby, Australia	(Wakelin and Ryder 2004).
JumpStart®	*Pseudomonas bilaiae*	Wheat & many other crops	PhilomBios Ltd., Saskatoon, Canada	Leggett et al. (2007).
E) Mycorrhizae based biofertilizers				
AM-fungal spores	~200 spores/g mixture of *Glomus intraradices, Glomus etunicatum, Glomus mosseae, Glomus geosporum*, and *Glomus clarum*)	Tomatoes (*Solanum lycopersicum* L.)	BioSym B.V. (Hengelo, Netherlands)	Balliu et al. (2015)
Mycormax®		Cotton, roses, nuts, citrus, grapes, raspberries, vegetables, etc.	JH Biotech, USA	Parnell et al. (2016)
Micronized Endomycorrhizal Inoculant	Nine types of Endo mycorrhizal spores	Most vegetables, grapes, fruit trees, berries, turfgrass, and flowers	BioOrganics™, **California**	Parnell et al. (2016)
F) Products Containing Multiple Biofertility Microbes				
Symbion-N non-associative type	*Azospirillum, Rhizobium, Acetobacter*, and *Azotobacter*	Sugar cane, sorghum, jowar, maize, cotton, tea and coffee	India, T. Stanes & Company Limited	Aasfar et al. (2021)
BioGro	*Pseudomonas fluorescens, Bacillus subtilis, Bacillus amyloliquefaciens, Candida tropicalis*	Rice (Oryza sativa)	Nguyen Thanh Hien in Hanoi University, Vietnam	Nguyen et al. (2017)
Microbin and Azottein	*Klebsiella, Bacillus, Azotobacter, Azospirillum,*	Barley cultivar	Egyptian Ministry of Agriculture	El-Sayed et al. (2000)
QuickRoots®	*Bacillus amyloliquefaciens* and the filamentous fungus *Trichoderma virens*	Corn	Monsanto BioAg Alliance	Parnell et al. (2016)
Excalibre-SA	*Trichoderma* & *Bradyrhizobium*	Soybean	ABM, USA	Parnell et al. (2016)

12.6 CONCLUSION

The global hue and cry about the hazardous effects of conventional pesticides on human health and the environment has led to an ever-increasing demand for alternate ecologically sound management strategies in agriculture. The use of microbes in agriculture is one of the many sustainable approaches employed to fulfil this goal. This chapter summarizes the incorporation of the most commonly used microbes and their related products for sustainable agriculture. These microbes, earlier seen as a mere nuisance, are today being marketed as biofertilizers, biopesticides and stimulants of natural resistance in crop plants. Microbes employ numerous mechanisms like nitrogen fixation, potassium and phosphorus solubilization, production of phytohormone- and phytopathogen-suppressing substances and protecting plants from various abiotic and biotic stresses for promoting plant growth (Mateusz et al., 2020). These products can either be used singly or as a mixture or even in combination with each other to enhance their efficiency. But such products need to be cost-effective for farmers so that they opt for these ecologically safer methods rather than their cheaper counterparts like conventional pesticides. Environmental safety, thus, becomes a great liability for governing bodies around the globe, as it is their responsibility to set the stage for the use of these alternate products by cutting their costs via mass production and making them available to local farmers. We can thus conclude that microbial products will surely play a promising role in sustainable agricultural production in the future, especially with the changing worldwide perspectives of concerned consumers and global governing bodies.

REFERENCES

Aasfar, A., Bargaz, A., Yaakoubi, K., Hilali, A., Bennis, I., Zeroual, Y., & Meftah Kadmiri, I. (2021). Nitrogen fixing *Azotobacter* species as potential soil biological enhancers for crop nutrition and yield stability. *Front. Microbiol.*, *12*, 628379. https://doi.org/10.3389/fmicb.2021.628379

Abd-Alla, M. H., El-Enany, A. W., Nafady, N. A., Khalaf, D. M., & Morsy, F. M. (2014). Synergistic interaction of Rhizobium leguminosarum bv. viciae and arbuscular mycorrhizal fungi as a plant growth promoting biofertilizers for faba bean (Vicia faba L.) in alkaline soil. *Microbiol. Res.*, *169*(1), 49–58. https://doi.org/10.1016/j.micres.2013.07.007

Abdel-Salam, M. A. & Shams, A. S. (2012). Feldspar-K fertilization of potato (Solanum tuberosum L.) augmented by biofertilizer. *American-Eurasian J. Agric. & Environ. Sci.*, *12*(6), 694–699. https://doi.org/10.5829/idosi.aejaes.2012.12.06.1802

Abhijit, D., Kole, S. C., & Mukhim, J. (2014). Evaluation of the efficacy of different sulfur amendments and sulphur oxidizing bacteria in relation to its transformation in soil and yield of mustard (Brassica juncea). *Res. Crops*, *15*(3), 578–584. https://doi.org/10.5958/2348-7542.2014.01380.1

Abril, A., Zurdo-Piñeiro, J. L., Peix, A., Rivas, R., & Velázquez, E. (2007). Solubilization of phosphate by a strain of Rhizobium leguminosarum bv. trifolii isolated from Phaseolus vulgaris in El Chaco Arido soil (Argentina). In: Velázquez E, Rodriguez-Barrueco C, eds. *First International Meeting on Microbial Phosphate Solubilization*. Dordrecht: Springer Netherlands, 135–138.

Afzal, A., Ashraf, A., Saeed, A., & Farooq, A. M. (2005). Effect of phosphate solubilizing microorganisms on phosphorus uptake, yield and yield traits of *wheat (Triticum aestivum L.)* in rainfed area. *Int. J. Agric. Biol. 7*, 1560–8530.

Afzal, A., & Bano, A. (2008). Rhizobium and phosphate solubilizing bacteria improve the yield and phosphorus uptake in wheat (Triticum aestivum L.). *Int J Agric Biol.*, *10*, 85–88.

Alagawadi, A. R., & Gaur, A. C. (1992). Inoculation of Azospirillium brasilense and phosphate-solubilizing bacteria on yield of sorghum (Sorghum bicolor (L.) Moench) in dry land. *Trop. Agri.*, *69*, 347–350.

Alam, S., Khalil, S., Ayub, N., & Rashid, M. (2002). In vitro solubilization of inorganic phosphate by phosphate solubilizing microorganism (PSM) from maize rhizosphere. *Intl. J. Agric. Biol.*, *4*, 454–458.

Alori, E. T., Glick, B. R., & Babalola, O. O. (2017). Microbial phosphorus solubilization and its potential for use in sustainable agriculture. *Front. Microbiol.*, *8*, 971. https://doi.org/10.3389/fmicb.2017.00971

Alzate Zuluaga, M. Y., Martinez de Oliveira, A. L., Valentinuzzi, F., Tiziani, R., Pii, Y., Mimmo, T., & Cesco, S. (2021). Can inoculation with the bacterial biostimulant *Enterobacter* sp. Strain 15S be an approach for the smarter P fertilization of maize and cucumber plants? *Front. Plant Sci.*, *12*, 719873. https://doi.org/10.3389/fpls.2021.719873

Anandham, R., Gandhi, P. I., Madhaiyan, M., & Sa, T. (2008). Potential plant growth promoting traits and bioacidulation of rock phosphate by thiosulfate oxidizing bacteria isolated from crop plants. *J. Basic Microbiol.*, *48*(6), 439–447. https://doi.org/10.1002/jobm.200700380

Anandham, R., Sridar, R., Nalayini, P., Poonguzhali, S., Madhaiyan, M., & Sa, T. (2007). Potential for plant growth promotion in groundnut (Arachis hypogaea L.) cv. ALR-2 by co-inoculation of sulfur-oxidizing bacteria and Rhizobium. *Microbiol. Res.*, *162*(2), 139–153. https://doi.org/10.1016/j.micres.2006.02.005

Anjanadevi, I. P., John, N. S., John, K. S., Jeeva, M. L., & Misra, R. S. (2016). Rock inhabiting potassium solubilizing bacteria from Kerala, India: characterization and possibility in chemical K fertilizer substitution. *J. Basic Microbiol.*, *56*(1), 67–77. https://doi.org/10.1002/jobm.201500139

Argaw, A. (2016). Effectiveness of Rhizobium inoculation on common bean productivity as determined by inherent soil fertility status. *J. Crop Sci. Biotechnol.*, *19*, 311–322. https://doi.org/10.1007/s12892-016-0074-8

Atieno, M., Herrmann, L., Nguyen, H. T., Phan, H. T., Nguyen, N. K., Srean, P., Than, M. M., Zhiyong, R., Tittabutr, P., Shutsrirung, A., Bräu, L., & Lesueur, D. (2020). Assessment of biofertilizer use for sustainable agriculture in the Great Mekong Region. *J. Environ. Manag.*, *275*, 111300. https://doi.org/10.1016/jjenvman.2020.111300

Babalola, O. O. & Glick, B. R. (2012). The use of microbial inoculants in African agriculture: Current practice and future prospects. *J. Food, Agricult. Environ. 10*(3&4), 540–549.

Badr, M. A., Shafei, A. M., & Sharaf El-Deen, S. H. (2006). The dissolution of K and phosphorus bearing minerals by silicate dissolving bacteria and their effect on sorghum growth. *Res. J. Agricult. Biol. Sci.*, *2*, 5–11.

Bakhshandeh, E., Pirdashti, H., & Lendeh, K. S. (2017). Phosphate and potassium-solubilizing bacteria effect on the growth of rice. *Ecol. Eng.*, *103*(Part A), 164–169. https://doi.org/10.1016/j.ecoleng.2017.03.008

Bal, H. B. & Adhya, T. K. (2021). Alleviation of submergence stress in rice seedlings by plant growth-promoting rhizobacteria with ACC deaminase activity. *Front. Sustain. Food Syst. 5*, 606158. https://doi.org/10.3389/fsufs.2021.606158

Baldani, V. L. D., Baldani, J. I. & Dobereiner, J. (2000). Inoculation of rice plants with the endophytic diazotrophs *Herbasprillum seropedicae* and *Burkholderia* spp. *Biol. Fertil. Soils*, *30*, 485–491.

Balliu, A., Sallaku, G., & Rewald, B. (2015). AMF inoculation enhances growth and improves the nutrient uptake rates of transplanted, Salt-Stressed Tomato Seedlings. *Sustainability*, *7*(12), 15967–15981. https://doi.org/10.3390/su71215799

Bano, A., & Fatima, M. (2009). Salt tolerance in *Zea mays* (L.) following inoculation with *Rhizobium* and *Pseudomonas*. *Biol. Fertility Soils*, *45*, 405–413. https://doi.org/10.1007/s00374-008-0344-9

Basak, B. B., & Biswas, D. R. (2009). Influence of potassium solubilizing microorganism (Bacillus mucilaginosus) and waste mica on potassium uptake dynamics by sudangrass (Sorghum vulgare Pers.) grown under two Alfisols. *Plant Soil*, *317*, 235–255.

Basak, B. B., & Biswas, D. R. (2010). Coinoculation of potassium solubilizing and nitrogen fixing bacteria on solubilization of waste mica and their effect on growth promotion and nutrient acquisition by a forage crop. *Biol. Fert. Soils*, *46*, 641–648.

Belimov, A. A., Kojemiakov, A. P., & Chuvarliyeva, C. V. (1995). Interaction between barley and mixed cultures of nitrogen fixing and phosphate solubilizing bacteria. *Plant Soil 173*, 29–37.

Benbrik B., Elabed, A., El Modafar, C., Douira, A., Amir, S., Filali-Maltouf, A., El Abed, S., El Gachtouli, N., Mohammed, I., & Koraichi, S. I. (2020). Reusing phosphate sludge enriched by phosphate solubilizing bacteria as biofertilizer: Growth promotion of Zea Mays. *Biocatalysis Agricult. Biotechnol. 30*, 101825. https://doi.org/10.1016/j.bcab.2020.101825

Berruti, A., Lumini, E., Balestrini, R., & Bianciotto, V. (2016). Arbuscular mycorrhizal fungi as natural biofertilizers: Let's benefit from past successes. *Front. Microbiol.*, *6*, 1559. https://doi.org/10.3389/fmicb.2015.01559

Biari, A., Gholami, A., & Rahmani, H. A. (2008). Growth promotion and enhanced nutrient uptake of maize (Zea mays L.) by application of plant growth promoting rhizobacteria in arid region of Iran. *J. Biol. Sci.*, *8*(6), 1015–1020. https://doi.org/10.3923/jbs.2008.1015.1020

Bottini, R., Fulchieri, M., Pearce, D., & Pharis, R. P. (1989). Identification of Gibberellins A(1), A(3), and Iso-A(3) in Cultures of Azospirillum lipoferum. *Plant Physiol.*, *90*(1), 45–47. https://doi.org/10.1104/pp.90.1.45

Burdman, S., Kigel, J., & Okon, Y. (1997). Effects of Azospirillum brasilense on nodulation and growth of common bean (Phaseolus vulgaris L.). *Soil Biol. Biochem.*, *29*, 923–929.

Carareto Alves, L. M., de Souza, J. A. M., deVarani, A. M., & de Lemos, E. G. M. (2014). The family Rhizobiaceae. In: E. Rosenberg, E. F. DeLong, S. Lory, E. Stackebrandt, and F. Thompson, eds. *The Prokaryotes: Alphaproteobacteria and Betaproteobacteria*, (Berlin; Heidelberg: Springer), 419–437. https://doi.org/10.1007/978-3-642-30197-1_297

Cassán, F. D., Penna, C., Creus, C. M., Radovancich, D., Monteleone, E., Salamone, I. G., & Cáceres, E. R. (2015) Protocol for the quality control of Azospirillum spp. inoculants. In: Cassán FD, Okon Y, & Creus CM, eds. *Handbook for Azospirillum*. Dordrecht: Springer, 487–499. https://doi.org/10.1007/978-3-319-06542-7_27

Chakraborty, U., Chakraborty, B., & Basnet, M. (2006). Plant growth promotion and induction of resistance in Camellia sinensis by Bacillus megaterium. *J. Basic Microbiol.*, *46*(3), 186–195. https://doi.org/10.1002/jobm.200510050

Chibeba, A. M., de Fátima Guimarães, M., Brito, O. R., Nogueira, M. A., Araujo, R. S., & Hungria, M. (2015). Co-Inoculation of soybean with bradyrhizobium and azospirillum promotes early nodulation. *Am. J. Plant Sci.*, *6*, 1641–1649. http://dx.doi.org/10.4236/ajps.2015.610164

Chifetete, V. W., & Dames, J. F. (2020). Mycorrhizal interventions for sustainable potato production in Africa. *Front. Sustain. Food Syst.*, *4*, 593053. https://doi.org/10.3389/fsufs.2020.593053

Choudhury, A. T. M. A., & Kennedy, I. R. (2004). Prospects and potentials for systems of biological nitrogen fixation in sustainable rice production. *Biol. Fertil. Soils 39*, 219–227. https://doi.org/10.1007/s00374-003-0706-2

Dal Cortivo, C., Barion, G., Ferrari, M., Visioli, G., Dramis, L., & Panozzo, A. (2018). Effect of field inoculation with VAM and bacteria consortia on root growth and nutrients uptake in common wheat. *Sustainability*, *10*, 3286. https://doi.org/10.3390/su10093286

Dal Cortivo, C., Barion, G., Visioli, G., Mattarozzi, M., Mosca, G., & Vamerali, T. (2017). Increased root growth and nitrogen accumulation in common wheat following PGPR inoculation: Assessment of plant-microbe interactions by ESEM. *Agric. Ecosyst. Environ.*, *247*, 396–408. https://doi.org/10.1016/j.agee.2017.07.006

Dal Cortivo, C., Ferrari, M., Visioli, G., Lauro, M., Fornasier, F., Barion, G., Panozzo, A., & Vamerali, T. (2020). Effects of seed-applied biofertilizers on rhizosphere biodiversity and growth of common wheat (*Triticum aestivum* L.) in the field. *Front. Plant Sci.*, *11*, 72. https://doi.org/10.3389/fpls.2020.00072

Dey, R., Pal, K. K., Bhatt, D. M., & Chauhan, S. M. (2004). Growth promotion and yield enhancement of peanut (Arachis hypogaea L.) by application of plant growth-promoting rhizobacteria. *Microbiol. Res.*, *159*(4), 371–394. https://doi.org/10.1016/j.micres.2004.08.004

Diagne, N., Ngom, M., Djighaly, P. I., Fall, D., & Hocher, V. (2020). Roles of arbuscular mycorrhizal fungi on plant growth and performance: Importance in biotic and abiotic stressed regulation. *Diversity, MDPI*, *12*(10). https://doi.org/10.3390/d12100370-03127168

Dixit, V. K., Misra, S., Mishra, S. K., Tewari, S. K., Joshi, N., & Chauhan, P. S. (2020). Characterization of plant growth-promoting alkalotolerant Alcaligenes and Bacillus strains for mitigating the alkaline stress in Zea mays. *Antonie van Leeuwenhoek*, *113*(7), 889–905. https://doi.org/10.1007/s10482-020-01399-1

Dutta, J., Handique, P. J., & Thakur, D. (2015). Assessment of culturable tea rhizobacteria isolated from tea estates of Assam, India for growth promotion in commercial tea cultivars. *Front. Microbiol.*, *6*, 1252. https://doi.org/10.3389/fmicb.2015.01252

Egamberdieva, D., Juraeva, D., Poberejskaya, S., Myachina, O., Teryuhova, P., Seydalieva, L., & Aliev, A. (2004). Improvement of wheat and cotton growth and nutrient uptake by phosphate solubilizing bacteria. In: *Proceeding of 26th Annual Conservation Tillage Conference for Sustainable Agriculture*, Auburn, pp 58–65.

Egamberdieva, D., Kucharova, Z., Davranov, K., Berg, G., Makarova, N., & Azarova, T. (2011). Bacteria able to control foot and root rot and to promote growth of cucumber in salinated soils. *Biol. Fertiity Soils*, *47*, 197–205. https://doi.org/10.1007/s00374-010-0523-3

Egamberdieva, D., Wirth, S. J., Alqarawi, A. A., Abd Allah, E. F., & Hashem, A. (2017). Phytohormones and beneficial microbes: Essential components for plants to balance stress and fitness. *Front. Microbiol.*, *8*, 2104. https://doi.org/10.3389/fmicb.2017.02104

Egamberdieva, D., Wirth, S. J., Shurigin, V. V., Hashem, A., & Abd Allah, E. F. (2017). Endophytic bacteria improve plant growth, symbiotic performance of chickpea (Cicer arietinum L.) and induce suppression of root rot caused by Fusarium solani under salt stress. *Front. Microbiol.*, *8*, 1887. https://doi.org/10.3389/fmicb.2017.01887

Elhaissoufi, W., Khourchi, S., Ibnyasser, A., Ghoulam, C., Rchiad, Z., Zeroual, Y., Lyamlouli, K., & Bargaz, A. (2020). Phosphate solubilizing rhizobacteria could have a stronger influence on wheat root traits and aboveground physiology than rhizosphere P solubilization. *Front. Plant Sci.*, *11*, 979. https://doi.org/10.3389/fpls.2020.00979

El-Sayed, A. A., Elenein, R. A., Shalaby, E. E., Shalan, M. A., & Said, M. A. (2000). Response of barley to biofertilizer with N and P application under newly reclaimed areas in Egypt. In *Proceedings of the 3rd International Crop Science Congress (ICSC)*, Hamburg, Germany, pp. 17–22.

El-Yazeid, A. A. & Abou-Aly, H. E. (2011). Enhancing growth, productivity and quality of tomato plants using phosphate solubilizing microorganisms. *Austral. J. Basic Appl. Sci.*, *5*(7), 371–379.

El-Yazeid, A. B., Abou-Aly, H. E., Mady, M. A., & Moussa, S. A. M., (2007). Enhancing growth, productivity and quality of squash plants using phosphate dissolving microorganisms (Bio phos-phor®) combined with boron foliar spray. *Res. J. Agric. Biol. Sci.*, 3, 274286.

Etesami, H., Alikhani, H. A., Jadidi, M., & Aliakbari, A. (2009). Effect of superior IAA producing rhizobia on N, P, K uptake by wheat grown under greenhouse condition. *World J. Appl. Sci.*, *6*(Suppl 12), 1629–1633.

Farajzadeh, D., Yakhchali, B., Aliasgharzad, N., Sokhandan-Bashir, N., & Farajzadeh, M. (2012). Plant growth promoting characterization of indigenous Azotobacteria isolated from soils in Iran. *Curr. Microbiol.*, *64*(4), 397–403. https://doi.org/10.1007/s00284-012-0083-x

Gahan, J., & Schmalenberger, A. (2014). The role of bacteria and mycorrhiza in plant sulfur supply. *Front. Plant Sci.*, *5*, 723. https://doi.org/10.3389/fpls.2014.00723

Gantar, M., & Elhai, J. (1999). Colonization of wheat para nodules by the nitrogen fixing cyanobacterium Nostoc spp. strain 2S9B. *New Phytol.*, *141*, 373–379.

Glick B. R. (2012). Plant growth-promoting bacteria: mechanisms and applications. *Scientifica*, *2012*, 963401. https://doi.org/10.6064/2012/963401

Glick B. R. (2014). Bacteria with ACC deaminase can promote plant growth and help to feed the world. *Microbiol. Res.*, *169*(1), 30–39. https://doi.org/10.1016/j.micres.2013.09.009

Govindarajan, M., Balandreau, J., Muthukumarasamy, R., Revathi, G., & Lakshminarasimhan, C. (2006). Improved yield of micropropagated sugarcane following inoculation by endophytic Burkholderia vietnamiensis. *Plant Soil*, *280*, 239–252.

Gupta, S., & Dikshit, A. K. (2010). Biopesticides: An ecofriendly approach for pest control. *J. Biopesticides*, *3*(1), 186–188.

Gupta, V. (2012). Beneficial microorganisms for sustainable agriculture. *Microbiol. Australia*, *33*(3), 113–115. https://doi.org/10.1071/MA12113

Hajeck, A. E., & Leger, St. (1994). Interactions between fungal pathogens and insect hosts. *Ann. Rev. Entomol.*, *39*, 293–322.

Hamdali, H., Hafidi, M., Virolle, M. J., & Ouhdouch, Y. (2008). Rock phosphate solubilizing Actinimycetes: Screening for plant growth promoting activities. *World J. Microbiol. Biotechnol.*, *24*, 2565–2575.

Han, H. S., Supanjani, P. & Lee, K. D. (2006). Effect of co-inoculation with phosphate and potassium solubilizing bacteria on mineral uptake and growth of pepper and cucumber. *Plant. Soil. Environ.*, *52*, 130.

Harwood, C. R., & Wipat, A. (1996). Sequencing and functional analysis of the genome of Bacillus subtilis strain 168. *FEBS Lett.*, *389*(1), 84–87. https://doi.org/10.1016/0014-5793(96)00524-8

Hungria, M., Nogueira, M. A. & Araujo, R. S. (2013). Co-inoculation of soybeans and common beans with rhizobia and azospirilla: Strategies to improve sustainability. *Biol. Fertil. Soils*, *49*, 791–801. https://doi.org/10.1007/s00374-012-0771-5

Jacobs, H., Boswell, G. P., Ritz, K., Davidson, F. A., & Gadd, G. M. (2002). Solubilization of calcium phosphate as a consequence of carbon translocation by Rhizoctonia solani. *FEMS Microbiol. Ecol.*, *40*(1), 65–71. https://doi.org/10.1111/j.1574-6941.2002.tb00937.x

Jilani, G., Akram, A., Ali, R. M., Hafeez, F. Y., Shams, I. H., & Chaudhry, A. N. (2007). Enhancing crop growth, nutrients availability, economics and beneficial rhizosphere microflora through organic and biofertilizers. *Ann. Microbiol.* *57*, 177–183. https://doi.org/10.1007/BF03175204

Joseph, B., Patra, R. R., & Lawrence, R. (2007). Characterization of plant growth promoting Rhizobacteria associated with chickpea (Cicer arietinum L). *Internat. J. Plant Product.*, *1*(Suppl 2), 141–152.

Kalayu, G. (2019). Phosphate solubilizing microorganisms: Promising approach as biofertilizers. *Internat. J. Agron.*, *2019*, https://doi.org/10.1155/2019/4917256

Kannaiyan, S. (1993). Nitrogen contribution by Azolla to rice crop. *Proc. Indian Nat. Sci. Acad.*, *3*, 309–314.

Kawaka, F., Dida, M. M., Opala, P. A., Ombori, O., Maingi, J., Osoro, N., Muthini, M., Amoding, A., Mukaminega, D., & Muoma, J. (2014). Symbiotic efficiency of native rhizobia nodulating common bean (Phaseolus vulgaris L.) in soils of Western Kenya. *Internat. Scholarly Res. Not.*, *2014*, 258497. https://doi.org/10.1155/2014/258497

Kayama, M., & Yamanaka, T. (2014). Growth characteristics of ectomycorrhizal seedlings of Quercus glauca, Quercus salicina, and Castanopsis cuspidata planted on acidic soil. *Trees*, *28*, 569–583. https://doi.org/10.1007/s00468-013-0973-y

Kevan, P. G., & Shipp, L. (2017). Biological control as biotechnological amelioration and ecosystem intensification in managed ecosystem. In: Moo-Young, M., ed. *Comprehensive Biotechnology*, 757–761. Academic Press, Elsevier

Khakipour, N., Khavazi, K., Mojallali, H., Pazira, E., & Asadirahmani, H. (2008). Production of Auxin hormone by Fluorescent Pseudomonads. *American-Eurasian J. Agricult. Environ. Sci.*, *4*(Suppl 6), 687–692.

Khan, A. L., Hamayun, M., Kim, Y.-H., Kang, S. M., Lee, J. H., & Lee, I. N. (2011). Gibberellins producing endophytic Aspergillus fumigatus sp. LH02 influenced endogenous phytohormonal levels, isoflavonoids production and plant growth in salinity stress. *Process Biochem.*, *46*, 440–447. https://doi.org/10.1016/j.procbio.2010.09.013

Khan, M. R., & Khan, S. M. (2002). Effects of root-dip treatment with certain phosphate solubilizing microorganisms on the fusarial wilt of tomato. *Bioresource Technol.*, *85*(2), 213–215. https://doi.org/10.1016/s0960-8524(02)00077-9

Koppenhöfer, A. M., & Wu, S. (2017). Microbial control of insect pests of turfgrass. In: *Microbial Control of Insect and Mite Pests*, Lawrence A. Lacey, ed. Academic Press, pp. 331–341, ISBN 9780128035276.

Korir, H., Mungai, N. W., Thuita, M., Hamba, Y., & Masso, C. (2017). Co-inoculation effect of rhizobia and plant growth promoting rhizobacteria on common bean growth in a low phosphorus soil. *Front. Plant Sci.*, *8*, 141. https://doi.org/10.3389/fpls.2017.00141

Koskey, G., Mburu, S. W., Awinol, R., Njerul, E. M., & Maingi, J. M. (2021). Potential use of beneficial microorganisms for soil amelioration, phytopathogen biocontrol, and sustainable crop production in smallholder agroecosystems. *Front. Sustainable Food Syst.*, *5*, Article 606308. https://doi.org/10.3389/fsufs.2021.606308

Kour, D., Rana, K. L., Yadav, A. N., Yadav, N., Kumar, M., Kumar, V., Vyas, P., Dhaliwal, H. S., & Saxena, A. K. (2020). Microbial biofertilizers: Bioresources and eco-friendly technologies for agricultural and environmental sustainability. *Biocatal. Agric. Biotechnol.*, *23*, 101487.

Kruasuwan, W., & Thamchaipenet, A. (2016). Diversity of culturable plant growth-promoting bacterial endophytes associated with sugarcane roots and their effect of growth by co-inoculation of diazotrophs and actinomycetes. *J. Plant Growth Regul.*, *35*, 1074–1087. https://doi.org/10.1007/s00344-016-9604-3

Kucey, R. M. N. (1983). Phosphate solubilizing bacteria and fungi in various cultivated and virgin Alberta soils. *Can. J. Soil Sci.*, *63*, 671–678.

Kuenen, J. G. & Beudeker, R. F. (1982). Microbiology of bacilli and other souxtotroph, mixotrophs and heterotrophs. *Transport Res. Soc. London*, *298*, 473–497.

Kumar, V., Behl, R. K., & Narula, N. (2001). Establishment of phosphate-solubilizing strains of Azotobacter chroococcum in the rhizosphere and their effect on wheat cultivars under green house conditions. *Microbiol. Res.*, *156*(1), 87–93. https://doi.org/10.1078/0944-5013-00081

Leggett, M., Cross, J., Hnatowich, G., & Holloway, G. (2007). Challenges in commercializing a phosphate-solubilizing microorganism: Penicillium biliae, a case history. In: Velázquez E., Rodriguez-Barrueco C., eds. *First International Meeting on Microbial Phosphate Solubilization. Developments in Plants and Soil Sciences*, vol *102*. Dordrecht: Springer. https://doi.org/10.1007/978-1-4020-5765-6_32

Li, H., Qiu, Y., Yao, T., Ma, Y., Zhang, H., & Yang, X. (2020). Effects of PGPR microbial inoculants on the growth and soil properties of Avena sativa, Medicago sativa, and Cucumis sativus seedlings. *Soil Tillage Res.*, *199*, 104577.

Lobo, L. L. B., dos, R. M., & Rigobelo, E. C. (2019). Promotion of maize growth using endophytic bacteria under greenhouse and field condition. *Aust. J. Crop Sci.*, *13*, 2067–2074. https://doi.org/10.21475/ajcs.19.13.12.p2077

Luciano, A., Vetúria, L. O., & Germano, N. S. F. (2010). Utilization of rocks and ectomycorrhizal fungi to promote growth of eucalypt. *Brazilian J. Microbiol.: [publication of the Brazilian Soc. Microbiology]*, *41*(3), 676–684. https://doi.org/10.1590/S1517-83822010000300018

Ma, J., Chen, S., Li, X., Han, R., Khatri-Chhetri, B. H., De Clercq, P., & Moens, M. (2012). A new entomopathogenic nematode, Steinernema tielingense n. sp. (Rhabditida: Steinernematidae), from north China. *Nematology*, *14*(3), 321–338.

Mahanta, D., Rai, R. K., Mishra, S. D., Raja, A., Purakayastha, T. J., & Varghese, E. (2014). Influence of phosphorus and biofertilizers on soybean and wheat root growth and properties. *Field Crops Res.*, *166*, 1–9.

Malhotra, M., & Srivastava, S. (2008). An ipdC gene knock-out of Azospirillum brasilense strain SM and its implications on indole-3-acetic acid biosynthesis and plant growth promotion. *Antonie van Leeuwenhoek*, *93*(4), 425–433. https://doi.org/10.1007/s10482-007-9207-x

Mateusz, M., Gryta, A. & Frąc, M. (2020). Chapter Two - Biofertilizers in agriculture: An overview on concepts, strategies and effects on soil microorganisms. In: Donald L., ed. *Sparks, Advances in Agronomy*, Academic Press, Vol. *162*, 31–87. https://doi.org/10.1016/bs.agron.2020.02.001. https://www.sciencedirect.com/science/article/pii/S0065211320300274

Mazid, M., & Khan, T. A., (2014). Future of Bio-fertilizers in Indian Agriculture: An Overview. *Internat. J. Agricult. Food Res.*, *3*(3), 10–23. ISSN 1929-0969.

Meadows, M. P. (1993). Bacillus thuringiensis in the environment-ecology and risk as-sessment. In: Entwistle, P.F.; Cory, J.S.; Bailey, M.J. and Higgs, S., eds. *Bacillus Thuringiensis: An Environmental Biopesticide; Theory and Practice*. Chichester: John Wiley, 193–220.

Menge, E. M., Njeru, E. M., Koskey, G., & Maingi, J. (2018). Rhizobial inoculation methods affect the nodulation and plant growth traits of host plant genotypes: A case study of Common bean Phaseolus vulgaris L. germplasms cultivated by smallholder farmers in Eastern Kenya. *Advances in Agricultural Science*, *6*(03), 77–94.

Mettenmeyer, A. (2002). Viral insecticides hold promise for bio-control. *Farming Ahead*, *124*, 50–51.

Michelle, J. E. (2016). Brundtland Report. Encyclopedia Britannica, https://www.britannica.com/topic/Brundtland-Report

Mishra, P. K., Mishra, S., Selvakumar, G., Gupta, H. S., Bisht, J. K., & Kundu, S. (2009). Coinoculation of Bacillus thuringeinsis-KR1 with Rhizobium leguminosarum enhances plant growth and nodulation of pea (Pisum sativum L.) and lentil (Lens culinaris L.). *World J. Microbiol. Biotechnol.*, *25*(5), 753–761. https://doi.org/10.1007/s11274-009-9963-z

Mishra, S. K., Khan, M. H., Misra, S., Dixit, V. K., Khare, P., Srivastava, S., & Chauhan, P. S. (2017). Characterisation of Pseudomonas spp. and Ochrobactrum sp. isolated from volcanic soil. *Antonie van Leeuwenhoek*, *110*(2), 253–270. https://doi.org/10.1007/s10482-016-0796-0

Monsalud, R. G. (2008). Harnessing microbial resources for sustainable crop production: The Philippine experience. *J. Faculty Agricult. Shinshu Univer.*, *44*(1), 2.

Monteiro, S. M., Clemente, J. J., Henriques, A. O., Gomes, R. J., Carrondo, M. J., & Cunha, A. E. (2005). A procedure for high-yield spore production by Bacillus subtilis. *Biotechnol. Progr.*, *21*(4), 1026–1031. https://doi.org/10.1021/bp050062z

Moscardi, F. (1999). Assessment of the application of baculoviruses for control of Lepidoptera. *Ann. Rev. Entomol.*, *44*, 257–289. https://doi.org/10.1146/annurev.ento.44.1.257

Munda, S., Shivakumar, B. G., Rana, D. S., Gangaiah, B., Manjaiah, K. M., Dass, A., Layek, J., & Lakshman, K. (2018). Inorganic phosphorus along with biofertilizers improves profitability and sustainability in soybean (Glycine max)–potato (Solanum tuberosum) cropping system. *J. Saudi Soc. Agricult. Sci.*, *17*(2), 107–113. https://doi.org/10.1016/jjssas.2016.01.008

Murray J. D. (2011). Invasion by invitation: rhizobial infection in legumes. *Molecular Plant-Microbe Interactions: MPMI*, *24*(6), 631–639. https://doi.org/10.1094/MPMI-08-10-0181

Musyoka, D. M., Njeru, E. M., Nyamwange, M. M., & Maingi, J. M. (2020). Arbuscular mycorrhizal fungi and Bradyrhizobium co-inoculation enhances nitrogen fixation and growth of green grams (Vigna radiata L.) under water stress. *J. Plant Nutrit.*, *43*(7), 1036–1047. https://doi.org/10.1080/01904167.2020.1711940

Muthukumarasamy, R., Govindarajan, M., Vadivelu, M., & Revathi, G. (2006). N-fertilizer saving by the inoculation of Gluconacetobacter diazotrophicus and Herbaspirillum sp. in micropropagated sugarcane plants. *Microbiol. Res.*, *161*(3), 238–245. https://doi.org/10.1016/jmicres.2005.08.007

Naik, K., Mishra, S., Srichandan, H., Singh, P. K., & Sarangi, P. K. (2019). Plant growth promoting microbes: Potential link to sustainable agriculture and environment. *Biocatalysis Agricult. Biotechnol.*, *21*, 101326. https://doi.org/10.1016/jbcab.2019.101326

Negi, Y. K., Sajwan, P., Uniyal, S., & Mishra, A. C. (2021). Enhancement in yield and nutritive qualities of strawberry fruits by the application of organic manures and io fertilizers, *Scientia Horticulturae*, *283*, 110038. https://doi.org/10.1016/jscienta.2021.110038

Nguyen, T. H., Phan, T. C., Choudhury, A. T. M. A., Rose, M. T., Deaker, R. J., & Kennedy, I. R. (2017). BioGro: A plant growth-promoting biofertilizer validated by 15 years' research from laboratory selection to rice farmer's fields of the Mekong Delta. In: *Agro-Environmental Sustainability*. Cham: Springer, 237–254.

Nosheen, S., Ajmal, I., & Song, Y. (2021). Microbes as biofertilizers, a potential approach for sustainable crop production. *Sustainability*, *13*, 1868. https://doi.org/10.3390/su13041868

Obreht, Z., Kerby, N. W., Gantar, M., & Rowell, P. (1993). Effect of root-associated N_2-fixing cyanobacteria on the growth and nitrogen content of wheat (Triticum vulgare L.) seedlings. *Biol. Fertil. Soil*, *15*, 68–72.

Ojo, O. A. & Fagade, O. E. (2002). Persistence of Rhizobium inoculants originating from Leucaena leucocephala fallowed plots in Southwest Nigeria. *African J. Biotechnol.*, *1*(1), 23–27. https://www.academicjournals.org/AJB

Ongena, M., & Jacques, P. (2008). Bacillus lipopeptides: Versatile weapons for plant disease biocontrol. *Trendsin Microbiol.*, *16*(3), 115–125. https://doi.org/10.1016/jtim.2007.12.009

Owen, D., Williams, A. P., Griffith, G. W., & Withers, P. J. A. (2015). Use of commercial bioinoculants to increase agricultural production through improved phosphrous acquisition. *Appl. Soil Ecol.*, *86*, 41–54. https://doi.org/10.1016/japsoil.2014.09.012

Parmar, P., & Sindhu, S. S. (2013). Potassium solubilization by rhizosphere bacteria: Influence of nutritional and environmental conditions. *J. Microbiol. Res.*, *3*, 25–31.

Parnell, J. J., Berka, R., Young, H. A., Sturino, J. M., Kang, Y., Barnhart, D. M., & Di Leo, M. V. (2016). From the lab to the farm: An industrial perspective of plant beneficial microorganisms. *Front Plant Sci.*, *7*, 1110.

Pastor-Bueis, R., Sánchez-Cañizares, C., James, E. K., & González-Andrés, F. (2019). Formulation of a highly effective inoculant for common bean based on an autochthonous Elite Strain of *Rhizobium leguminosarum* bv. *phaseoli*, and genomic-based insights into its agronomic performance. *Front. Microbiol.*, *10*, 2724. https://doi.org/10.3389/fmicb.2019.02724

Peix, A., Rivas-Boyero, A. A., Mateos, P. F., Rodriguez-Barrueco, C., Martı´nez-Molina, E., & Velázquez, E. (2001) Growth promotion of chickpea and barley by a phosphate solubilizing strain of Mesorhizobium mediterraneum under growth. *Soil Biol. Biochem.*, *33*, 103–110.

Peoples, M. B., Herridge, D. F., & Ladha, J. K. (1995). Biological nitrogen fixation: an efficient source of nitrogen for sustainable agricultural production? *Plant Soil*, *174*, 3–28.

Piggot, P. J., & Hilbert, D. W. (2004). Sporulation of Bacillus subtilis. *Curr. Opin. Microbiol.*, *7*(6), 579–586. https://doi.org/10.1016/jmib.2004.10.001

Pourbabaee, A. A., Koohbori Dinekaboodi, S., Seyed Hosseini, H. M., Alikhani, H. A., & Emami, S. (2020). Potential application of selected sulfur-oxidizing bacteria and different sources of sulfur in plant growth promotion under different moisture conditions. *Commun. Soil Sci. Plant Anal.*, *51*, 735–745. https://doi.org/10.1080/00103624.2020.1729377

Prasad, M., Chaudhary, M., Choudhary, M., Kumar, T. K., & Jat, L. K. (2017). Rhizosphere microorganisms towards soil sustainability and nutrient acquisition. In: Meena V. P., Mishra J., Bisht A., and Pattanayak A., eds. *Agriculturally Important Microbes for Sustainable Agriculture*, Singapore: Springer, 31–49.

Raimi, A., Adeleke, R., & Roopnarain, A. (2017). Soil fertility challenges and Biofertiliser as a viable alternative for increasing smallholder farmer crop productivity in sub-Saharan Africa. *Cogent Food Agric.*, *3*, 1400933. https://doi.org/10.1080/23311932.2017.1400933

Rani, A. T., Kammar, V., Keerthi, M. C., Rani, V., Majumder, S., Pandey, K. K., & Singh, J. (2021). Biopesticides: An alternative to synthetic insecticides. In: Bhatt P., Gangola S., Udayanga D., Kumar G., eds *Microbial Technology for Sustainable Environment*. Singapore: Springer. https://doi.org/10.1007/978-981-16-3840-4_23

Raza, A., & Faisal, M. (2013). Growth promotion of maize by desiccation tolerant Micrococcus luteus-chp37 isolated from Cholistan desert. *Pakistan. Austr. J. Crop Sci.*, *7*, 1693–1698.

Remans, R., Croonenborghs, A., Torres-Gutierrez, R., Michiels, J., & Vanderleyden, J. (2007). Effects of plant growth promoting rhizobacteria on nodulation of Phaseolus vulgaris L. are dependent on plant P nutrition. *Eur. J. Plant Pathol.*, *119*, 341–351.

Remans, R., Ramaekers, L., Schalkens, S., Hernandez, G., Garcia, A., Reyes, J. L., Mendez, N., Toskano, V., Mulling, M., Galvez, L., & Vanderleyden, J. (2008). Effect of Rhizobium-Azospirillum coinoculation on nitrogen fixation and yield of two contrasting Phaseolus vulgaris L. genotypes cultivated across different environments in Cuba. *Plant Soil*, *312*, 25–37.

Rojas, A., Holguin, G., Glick, B. R., & Bashan, Y. (2001). Synergism between Phyllobacterium sp. (N(2)-fixer) and Bacillus licheniformis (P-solubilizer), both from a semiarid mangrove rhizosphere. *FEMS Microbiol. Ecol.*, *35*(2), 181–187. https://doi.org/10.1111/j1574-6941.2001.tb00802.x

Romero-Perdomo, F., Abril, J., Camelo, M., Moreno-Galván, A., Pastrana, I., Rojas-Tapias, D., & Bonilla, R. (2017). Azotobacter chroococcum as a potentially useful bacterial biofertilizer for cotton (Gossypium hirsutum): Effect in reducing N fertilization. *Revista Argentina de Microbiologia*, *49*(4), 377–383. https://doi.org/10.1016/jram.2017.04.006

Rouphael, Y., Franken, P., Schneider, C., Schwarz, D., Giovannetti, M., & Agnolucci, M. (2015). Arbuscular mycorrhizal fungi act as bio-stimulants in horticultural crops. *Sci. Hort. 196*, 91–108. https://doi.org/10.1016/jscienta.2015.09.002

Safikhani. S., Morad, M., & Reza, C. M. (2013). Effects of seed inoculation by Rhizobium strains on chlorophyll content and protein percentage in common bean cultivars (Phaseolus vulgaris L.). *International J. Biosci.*, *3*(3), 1–8. https://doi.org/10.12692/ijb/3.3.1-8

Saharan, B. S., & Nehra, V. (2011). Plant growth promoting rhizobacteria: A critical review. *Life Sci. Med. Res.*, *21*, 1–30.

Saia, S., Amato, G., Frenda, A. S., Giambalvo, D., & Ruisi, P. (2014). Influence of arbuscular mycorrhizae on biomass production and nitrogen fixation of berseem clover plants subjected to water stress. *PloS One*, *9*(3), e90738. https://doi.org/10.1371/journal.pone.0090738

Saikia, S. P., & Jain, V. (2007). Biological nitrogen fixation with non-legumes: An achievable target or a dogma. *Curr. Sci.*, *92*(3), 317–322.

Salomon, M. V., Bottini, R., de Souza Filho, G. A., Cohen, A. C., Moreno, D., Gil, M., & Piccoli, P. (2014). Bacteria isolated from roots and rhizosphere of Vitis vinifera retard water losses, induce abscisic acid accumulation and synthesis of defense-related terpenes in in vitro cultured grapevine. *Physiologia Plantarum*, *151*(4), 359–374. https://doi.org/10.1111/ppl.12117

Sanginga, N., Danso, S. K. A. & Mulongoy, K. (1994). Persistence and recovery of introduced Rhizobium ten years after inoculation on Leucaena leucocephala grown on an Alfisol in southwestern Nigeria. *Plant Soil*, *159*, 199–204. http://doi.org/10.1007/BF00009281

Sanginga, P., Thottappilly, G., & Dashiell, K. (2000). Effectiveness of rhizobia nodulating recent promiscuous soyabean selections in the moist savanna of Nigeria. *Soil Biol. Biochem.*, *32*, 127–133.

Santos, M. S., Nogueira, M. A., & Hungria, M. (2019). Microbial inoculants: reviewing the past, discussing the present and previewing an outstanding future for the use of beneficial bacteria in agriculture. *AMB Express*, *9*(1), 205. https://doi.org/10.1186/s13568-019-0932-0

Santos, M. S., Nogueira, M. A., & Hungria, M. (2021). Outstanding impact of Azospirillum brasilense strains Ab-V5 and Ab-V6 on the Brazilian agriculture: Lessons that farmers are receptive to adopt new microbial inoculants. *Rev. Bras. Cienc Solo.*, *45*, e0200128. https://doi.org/10.36783/18069657rbcs20200128

Santos, R. M. D. & Rigobelo, E. C. (2021). Growth-promoting potential of rhizobacteria isolated from sugarcane. *Front. Sustain. Food Syst.*, *5*, 596269. https://doi.org/10.3389/fsufs.2021.596269

Sarma, R. K., & Saikia, R. (2014). Alleviation of drought stress in mung bean by strain Pseudomonas aeruginosa GGRJ21. *Plant Soil*, *377*, 111–126. https://doi.org/10.1007/s11104-013-1981-9

Satyaprakash, M., Nikitha, T., Reddi, E. U. B., Sadhana, B., & Satya Vani, S. (2017). Phosphorous and phosphate solubilising bacteria and their role in plant nutrition. *Int. J. Curr. Microbiol. App. Sci.*, *6*(4), 2133–2144. https://doi.org/10.20546/ijcmas.2017.604.251

Schnepf, E., Crickmore, N., Van Rie, J., Lereclus, D., Baum, J., Feitelson, J., Zeigler, D. R., & Dean, D. H. (1998). Bacillus thuringiensis and its pesticidal crystal proteins. *Microbiol. Molecular Biol. Revi.: MMBR*, *62*(3), 775–806. https://doi.org/10.1128/MMBR.62.3.775-806.1998

Shaharoona, B., Arshad, M., & Zahir, Z. A. (2006). Effect of plant growth promoting rhizobacteria containing ACC-deaminase on maize (Zea mays L.) growth under axenic conditions and on nodulation in mung bean (Vigna radiata L.). *Letters Appl. Microbiol.*, *42*(2), 155–159. https://doi.org/10.1111/j.1472-765X.2005.01827.x

Sharma, S. B., Sayyed, R. Z., Trivedi, M. H., & Gobi, T. A. (2013). Phosphate solubilizing microbes: sustainable approach for managing phosphorus deficiency in agricultural soils. *SpringerPlus*, *2*, 587. https://doi.org/10.1186/2193-1801-2-587

Sheng, X. F. (2005). Growth promotion and increased potassium uptake of cotton and rape by a potassium releasing strain of Bacillus edaphicus. *Soil Biol. Biochem.*, 37(10), 1918–1922.

Sheng, X. F., & He, L. Y. (2006). Solubilization of potassium-bearing minerals by a wild-type strain of Bacillus edaphicus and its mutants and increased potassium uptake by wheat. *Canad. J. Microbiol.*, *52*(1), 66–72. https://doi.org/10.1139/w05-117

Shia, W. B., & Feng, M. G. (2004). Lethal effect of Beauveria bassiana, Metarhizium anisopliae, and Paecilomyces fumosoroseus on the eggs of Tetranychus cinnabarinus (Acari: Tetranychidae) with a description of a mite egg bioassay system. *Biol. Cont.*, *30*, 165–173.

Singh, G., Biswas, D. R. & Marwah, T. S. (2010). Mobilization of potassium from waste mica by plant growth promoting rhizobacteria and its assimilation by maize (Zea mays) and wheat (Triticum aestivum L.). *J. Plant Nutrit.*, *33*, 1236–1251.

Singh, R. P., & Jha, P. N. (2016). A Halotolerant Bacterium *Bacillus licheniformis* HSW-16 Augments Induced Systemic Tolerance to Salt Stress in Wheat Plant (*Triticum aestivum*). *Front. Plant Sci.*, *7*, 1890. https://doi.org/10.3389/fpls.2016.01890

Souza, R. D., Ambrosini, A., & Passaglia, L. M. (2015). Plant growth-promoting bacteria as inoculants in agricultural soils. *Genet. Molecul. Biol.*, *38*(4), 401–419. https://doi.org/10.1590/S1415-475738420150053

Stamenković, S., Beškoski, V., Karabegović, I., Lazić, M., & Nikolić, N. (2018). Microbial fertilizers: A comprehensive review of current findings and future perspectives. *Spanish J. Agricult. Res.*, *16*(1). https://doi.org/10.5424/sjar/2018161-12117

Tanwar, S. P. S., Sharma, G. L., & Chahar, M. S. (2002). Effects of phosphorus and biofertilizers on the growth and productivity of black gram. *Annals Agri. Res.*, *23*(3), 491–493.

Tarafdar, J. C. (2019). Fungal Inoculants for native Phosphorus Mobilization. In: Giri B., Prasad R., Wu Q. S., Varma A., eds. *Biofertilizers for Sustainable Agriculture and Environment: Soil Biology*, Vol. 55. Cham: Springer. https://doi.org/10.1007/978-3-030-18933-4_2

Tarafdar, J. C., & Gharu, A. (2006). Mobilization of organic and poorly soluble phosphates by Chaetomium globossum. *Appl. Soil Ecol.*, *32*, 273–283.

Tarafdar, J. C., & Marschner, H. (1995). Mobilization of organic and poorly soluble phosphates by Chaetomium globossum: Supplied with organic phosphorus as Na-phytate. *Plant Soil*, *173*, 97–102.

Thajuddin, N., & Subramanian, G. (2005). Cyanobacterial biodiversity and potential applications in biotechnology. *Curr. Sci.*, *89*, 47–57.

Thimann K. V. (1936). On the physiology of the formation of nodules on legume roots. *Proc. Nat. Acad. Sci. United States Am.*, *22*(8), 511–514. https://doi.org/10.1073/pnas.22.8.511

Ulzen, J., Abaidoo, R. C., Mensah, N. E., Masso, C., & AbdelGadir, A. H. (2016). Bradyrhizobium inoculants enhance grain yields of Soybean and Cowpea in Northern Ghana Front. *Plant Sci.*, *7*, 1770.

US EPA. (2004). Bacillus popilliae Spores (054502) Fact Sheet. https://www3.epa.gov/pesticides/chem_search/reg_actions/registration/fs_PC-054502_19-Oct-04.pdf

Usta, C. (2013). Microorganisms in biological pest control: A review (bacterial toxin application and effect of environmental factors). *Curr. Prog. Biol. Res.*, *24*, 287–317.

Venkateswarlu, B., Rao, A. V., Raina, P., & Ahmad, N. (1984). Evaluation of phosphorus solubilization by microorganisms isolated from arid soil. *J. Indian Soc. Soil Sci.*, *32*, 273–277.

Volpin, H., Burdman, S., Castro-Sowinski, S., Kapulnik, Y., & Okon, Y. (1996). Inoculation with Azospirillum increased exudation of Rhizobial nod-gene inducers by alfalfa roots. *Molecular Plant Microbe Interactions*, *1996*, 388–394.

Wakelin, S. A., & Ryder, M. H. (2004). Plant growth-promoting inoculants in Australian agriculture. *Crop Manage.* https://doi.org/10.1094/CM-2004-0301-01-RV

Walpola, B. C., & Yoon, M. (2013). Phosphate solubilizing bacteria: Assessment of their effect on growth promotion and phosphorous uptake of mung bean (Vigna radiata [L.] R. Wilczek). *Chil. J. Agric. Res. 73*, 275–281. https://doi.org/10.4067/S0718-58392013000300010

Walpola, B. C., & Yoon, M.- H. (2012). Prospectus of phosphate solubilizing microorganisms and phosphorus availability in agricultural soils: A review. *African J. Microbiol. Res.*, *6*(37), 6600–6605. https://www.academicjournals.org/AJMR. https://doi.org/10.5897/AJMR12.889

Wang, Y., Wang, M., Li, Y., Wu, A., & Huang, J. (2018). Effects of arbuscular mycorrhizal fungi on growth and nitrogen uptake of Chrysanthemum morifolium under salt stress. *PloS One*, *13*(4), e0196408. https://doi.org/10.1371/journal.pone.0196408

Wani, S. P., & Gopalakrishnan, S. (2019). Plant growth-promoting microbes for sustainable agriculture. In: *Plant Growth Promoting Rhizobacteria (PGPR): Prospect for Sustainable Agriculture*, Singapore: Springer, 19–45.

Waqas, M., Khan, A. L., Kamran, M., Hamayun, M., Kang, S. M., Kim, Y. H., & Lee, I. J. (2012). Endophytic fungi produce gibberellins and indoleacetic acid and promotes host-plant growth during stress. *Molecules (Basel, Switzerland)*, *17*(9), 10754–10773. https://doi.org/10.3390/molecules170910754

Weinzierl, R., & Henn, T. (1989). *Alternatives in Insect Management: Microbial Insecticides. Cooperative Extension*, University of Illinois, Circular 1295. 12 p.

Whitelaw, M. A., Harden, T. J., & Bender, G. L. (1997). Plant growth promotionof wheat inoculated with *Penicillium radicum sp.* nov. *Austral. J. Soil Res.*, *35*, 291–300. https://doi.org/10.1071/S96040

Widada, J., Damarjaya, D. I., & Kabirun, S. (2007).The interactive effects of arbuscular mycorrhizal fungi and rhizobacteria on the growth and nutrients uptake of sorghum in acid soil. In: Rodriguez-Barrueco C., Velázquez E, ed. *First International Meeting on Microbial Phosphate Solubilization*, Switzerland: Springer Nature, 173–177.

Xiao, Y., Wang, X., Chen, W., & Huang, Q. (2017). Isolation and identification of three potassium-solubilizing bacteria from rape rhizospheric soil and their effects on ryegrass, *Geomicrobiol. J.*, *34*(10), 873–880. https://doi.org/10.1080/01490451.2017.1286416

Yazdani, M., Bahmanyar, M. A., Pirdashti, H., & Esmaili, M. A. (2009). Effect of phosphate solubilization microorganisms (PSM) and plant growth promoting rhizobacteria (PGPR) on yield and yield components of corn (*Zea mays* L.). *World Acad. Sci. Eng. Technol.*, *49*, 90–92.

Yu, X., Liu, X., Zhu, T. H., Liu, G. H., & Mao, C. (2011). Isolation and characterization of phosphate-solubilizing bacteria from walnut and their effect on growth and phosphorus mobilization. *Biol. Fertil. Soils*, 47, 437–446. https://doi.org/10.1007/s00374-011-0548-2.

Zaidi, S., Usmani, S., Singh, B. R., & Musarrat, J. (2006). Significance of Bacillus subtilis strain SJ-101 as a bioinoculant for concurrent plant growth promotion and nickel accumulation in Brassica juncea. *Chemosphere*, *64*(6), 991–997. https://doi.org/10.1016/j.chemosphere.2005.12.057

Zeffa, D. M., Perini, L. J., Silva, M. B., de Sousa, N. V., Scapim, C. A., Oliveira, A., Amaral Júnior, A., & Azeredo Gonçalves, L. S. (2019). Azospirillum brasilense promotes increases in growth and nitrogen use efficiency of maize genotypes. *PloS One*, *14*(4), e0215332. https://doi.org/10.1371/journal.pone.0215332

Zhang, C., & Kong, F. (2014). Isolation and identification of potassium-solubilizing bacteria from tobacco rhizospheric soil and their effect on tobacco plants. *Appl. Soil Ecol.*, *82*, 18–25. http://dx.doi.org/10.1016/j.apsoil.2014.05.002

13 Microbial Biopesticides
An Eco-Friendly Approach for a Sustainable Agro-ecosystem

Nidhi Srivastava
School of Basic and Applied Sciences, Maharaja Agrasen University, Solan, India

Usha Kumari
Gargi College, University of Delhi, New Delhi, India

Jai Kumar
Gargi College, University of Delhi, New Delhi, India

CONTENTS

DOI: 10.1201/9781003306931-15

13.1 INTRODUCTION

Pesticides are chemical compounds used in the agro-ecosystem to fortify flora and fauna against various disease-causing pests that may inhibit their growth (Nicolopoulou-Stamati et al., 2016). Regardless of their benefits, pesticides have also proved dangerous for the community as well as the surrounding environment. Their widespread use has led to the emergence of many pesticide-resistant pests and pathogens (Gomiero et al., 2011). Several pesticides do not get degraded and remain bioaccumulated in the environment, and hence are considered harmful to us. These formulations contain two types of ingredients—active and inert. Of course, the inert ingredients are so called because they are considered to be a helping compound for the application of the active ingredients in the field, but sometimes they are also reported to have toxic potential and hence are not safe to use. Hence, the inert ingredients also contribute in raising the toxicity of commercial products not only in the target organisms but also the non-target ones (Bernardes et al., 2015). Pesticides have been used extensively in the agro-ecosystem for decades but the research has now proved that they are a menace to our health and the ecosystem (Attathom 2002; Karaborkl & Akbulut, 2011).

Since our agricultural system has been so dependent on chemical treatments and has suffered lots of negative impacts, now the technology is moving towards the use of innovative and environment-friendly bioactive compounds that are both cheap and reliable (Attathom, 2002). Biopesticides are derived from natural resources such as microorganisms, flora and fauna present around us (Hubbard et al., 2014). These are very specific to their targets and are ecologically safe, so they are used widely in the various pest management programs (Pathak et al., 2017). Among various classes of biopesticides, this chapter focuses on microbial biopesticides that are based on pathogenic microorganisms including bacteria, fungi, protozoans and viruses. Various reviews and research articles have reported that these pesticides are target specific and hence have the potential to effectively control the insect pest problems in our fields. As they are the natural components, they are biodegradable and hence safe for the environment as well as for human health.

13.2 FORMULATION OF MICROBIAL BIOPESTICIDES

A formulation is defined as a specific combination of ingredients, processes, and equipment to form a useful commercial product. The correct formulation has the potential to provide numerous benefits for biopesticides, such as longer storage, easier handling, and greater field efficacy (Behle & Birthisel, 2014). Biopesticides are prepared in a similar way to chemical pesticides. Since microbial biopesticides originate from living organisms, it is important to maintain their viability during formulation and storage. It is also important to make sure that the organism is properly revived so that it is active when applied in the field (Gašić & Tanović, 2013). Biopesticides can be prepared in a variety of forms based on their shelf life, stability, and the microorganisms used (Thakur et al., 2020).

Biopesticide preparations come in two forms: liquid and dry. Liquid formulations are prepared using different bases such as water, oil, polymers or different combinations. Water-based formulations require support from inert ingredients such as stabilizers, stickers, surfactants, coloring agents, antifreeze compounds, and additional nutrients. Dry formulations are produced by adding binder, dispersant, wetting agents, etc., using various technologies such as spray drying, freeze-drying or air drying (Brar et al., 2006; Knowles, 2008, Mishra et al., 2019). Dusts, seed dressing powders, granules, micro granules, water-dispersible granules and wettable powders are examples of dry formulations (Knowles, 2005, 2006). Liquid formulations have a high purity level and a two-year shelf life. They are simple to handle, store, and transport, which increases their export potential. As they

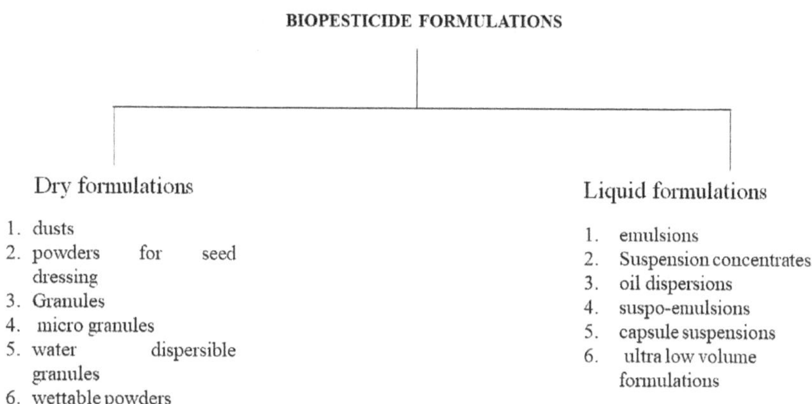

FIGURE 13.1 Types of biopesticide formulations

are compatible with machinery used in big farms, they are generally preferred by farmers and industries. Emulsions, suspension concentrates, oil dispersions, suspo-emulsions, capsule suspensions and ultra-low volume formulations are the most common liquid biopesticides (Rao et al., 2015) (Figure 13.1).

Some important points should always be kept in mind when formulating a microbial biopesticide. Production should not be very costly, it should be easy to store, and the biocontrol agent should remain viable. It must always be easy to handle and use for applications. Above all it should always be effective in the control of anticipated pests. These requirements can be met by formulating biocontrol agents in a variety of ways (Seaman 1990; Mollet & Grubenmann, 2008). Depending on the type of product, the microbial component is blended with a carrier during the formulation process. Other adjuvants are also added that can aid the microbial agent's survival, improving the efficacy of the final product (Mishra et al., 2019).

13.3 TYPES OF MICROBIAL BIOPESTICIDES

Insect pests are infected by a wide range of naturally occurring microscopic organisms, including bacteria, fungi, viruses, and others, and thus play an important role in their management (Dara, 2017). Synthetic pesticides basically exploit a single chemical associated with their mode of action, but microbial biopesticides have multifaceted modes of action (Hubbard et al., 2014). The action mechanisms of microbial biopesticides can be divided into three types: (i) biological or ecological, which includes predation or competition, (ii) physical, which involves the creation of barriers, and (iii) chemical or biochemical processes that disrupt the physiological activities of the targeted pest (Hubbard et al., 2014). In order to successfully apply these microbial biopesticides in agriculture, it is critical to understand their types, mechanisms of action, range of hosts, and the dynamics of pathogen–arthropod–plant interactions. The following sections list the various types of microbial biopesticides, along with their modes of action and genetic modifications.

13.3.1 Bacteria

Bacteria are not very good at spreading diseases when free, but once they enter the midgut of the insect and reach the haemocoel, their potential increases, they become pathogenic and are hence called 'entomopathogenic' (Park & Federici, 2009). Although gram-negative bacteria can control disease-causing vectors and insect pests, gram-positive bacteria have demonstrated their ability to control insect pests (Priest, 2000). *Bacillus* spp., *Paenibacillus* spp., and *Clostridium* spp. are among the spore-forming gram-positive bacterial pathogens. *Bacillus* and *Paenibacillus* are harmful to coleopteran, dipteran, and lepidopteran insects. The bacterial entomopathogens are largely dominated by the *Bacillus thuringiensis* (*Bt*), a useful alternative to synthetic pesticides as they are highly specific to the target insect, easy to formulate and mass-produce, and can be easily stored for many years without losing activity (Federici, 1991). *Bt* subspecies are effective for controlling various groups of target insects, including caterpillars, mosquito larvae, and coleopterans. *Bt* subspecies *aizawai* and *kurstaki* are good for controlling caterpillars, *Bt* subspecies *israelensis* and *sphaericus* are effective against mosquito larvae, and *Bt* subspecies *tenebrionis* can control some coleopterans (Dara, 2017).

The genes from *B. thuringiensis*is are also used for transgenic expression which can help provide pest resistance in plants. Since about 90% of biopesticides are derivatives of *B. thuringiensi* (Thakur et al., 2020) we will now explain the mode of action of this particular bacterium.

13.3.1.1 Mechanism of Action

The δ-endotoxin (crystalline inclusions) is the molecule produced by *Bt* that is responsible for its entomopathogenic properties. In 1989, Höfte and Whiteley classified δ-endotoxins into four major classes: Crystal (*Cry*) I, II, III and IV. In addition, cytolysins (*Cyt*), found in the crystals of the mosquitocidal strains, were also reported to have insecticidal properties (Agaisse & Lereclus, 1995; Höfte and Whiteley, 1989). These molecules have been successfully used as a toxin or bioinsecticide against caterpillars, beetles, mosquitoes, and blackflies. *Bt* also produces VIPs (vegetative insecticidal proteins) and SIPs (secreted insecticidal proteins) during the vegetative development phase, which have insecticidal effects against lepidopteran, coleopteran, homopteran, and exclusively coleopteran pests, respectively (Palma et al., 2014).

When *Cry* toxins enter insect pests, they are activated and solubilized by enzymes (proteases) in the midgut. The toxins then interact with the receptors of the epithelial cells leading to their insertion into the membrane which in turn results in the death of the cell (Soberón et al., 2009; Bravo et al., 2007, 2011).

13.3.1.2 Genetic Manipulations in Entomopathogenic Bacterial Strains

Cry toxins derived from *Bt* are successfully employed to control various insect pests and disease-causing vectors due to their target specificity and harmfulness towards various insects (Lucena et al., 2014). Due to the development of resistance in insect pests, the long-term efficacy of *Bt* toxins is being challenged (Pardo-López et al. 2009) and measures to address all these issues are required. The situation can be rectified by enhancing the activity of *Cry* toxins using additional proteins such as serine protease inhibitors, chitinases, *Cyt* toxins, or a fragment of cadherin receptor containing a toxin-binding site. Aside from that, hazardous genes can be altered utilizing a variety of techniques such as site-specific mutations, protein cleavage, and deletion of tiny fragments from the amino-terminal region (Pardo-López et al. 2009). The high toxicity and specificity of these *Bt* genes have caused them to be introduced into several other bacterial species too. The use of plasmids that are able to replicate in the host and the introduction of these genes into the host chromosomal DNA can improve the efficacy of these biopesticides.

During the vegetative development and sporulation phases, the mosquitocidal strains of *B. sphaericus* produce a variety of protein toxins, including *Mtx* toxins and *Bin* toxin (Charles et al., 1990; Federici et al., 2003). The *Mtx* proteins, when fused with glutathione S-transferase, show an enhanced toxic activity against *C. quinquefasciatus* mosquitoes. Further, these proteins, when applied in combination with a recombinant *Bt* protein (Cry11Aa), also appeared to be highly effective against *Cry*11A-resistant larvae (Wirth et al., 2007). *Bin* toxin is a crystal toxin generated by *B. sphaericus*, and mosquitoes living in fields where this bacterium is employed extensively have acquired considerable tolerance to it (Sinegre et al., 1994; Rao et al., 2015; Silva-Filha et al., 1995; Yuan et al., 2000; Su & Mulla, 2004). Bin toxins have been coupled with *Cry* and *Cyt* proteins from *B. thuringiensis sp. israelensis* to combat insect resistance (Park & Fedrice, 2009). Such initiatives aid in the prevention of mosquito-borne diseases such as malaria and filariasis (Wirth et al., 2010).

13.3.2 Fungi

As biocontrol agents, pathogenic fungi are found to be very useful against insect pest populations (Sharma & Malik, 2012). Pathogenic fungi have an advantage as their spores need not be directly ingested by the target pest population, hence they can infect a wide variety of pests, even those who only have mouthparts for sucking and piercing purposes (Dara, 2017). As direct ingestion of entomopathogenic fungi is not necessary to kill pests, these fungi can also target insects like aphids and mosquitoes (Zhao et al., 2016). These pathogenic fungi require certain ideal conditions, including mild temperatures and high relative humidity in their surroundings, for infection to take place. Under these conditions, spores of these entomopathogenic fungi come into interface with the host pest and with the help of mechanical pressure and enzymatic activity they break the cuticle and enter the host's body, following which the fungi invade tissue and multiply resulting in the pest's death and the release of new spores from its body (Dara, 2017). Biological pesticides of fungal origin are used as mycelium or conidia which may sporulate once employed in the field (Usta, 2013).

Despite the fact that there are over 700 different species of entomopathogenic fungus, some are commercially produced in greater quantities than others, e.g., *Isaria, Lecanicillium, Beauveria and Metarhizium* (Vega et al., 2009). Under field conditions, the characteristic features required for their production are indelible stability, low cost, and above all persistent efficacy. The mass production of the diaspores (dispersion entities) is achieved by promoting in-air conidiation on growth media (solid), growth of hyphal biomass on solid/liquid media or blastospores generation in liquid media by yeast-like growth (Faria & Wraight 2007). To increase the dispersal potential, hydrophobic conidia are generally developed in oil or used as a spray mixed with adjuvants (wetting agents). With hydrophilic blastospores, spray preparation is performed along with wetting agents which are used as adjuvants but customarily prepared as water-dispersible granules or wettable powders. Because on the foliar surface, solar radiation has effects on the propagule persistence of these entomopathogenic fungi, extensive efforts have been made to give these entomopathogens sun protection by incorporating sunscreens and solar blockers (Strasser, 2001).

13.3.2.1 Mechanism of Action

The infective stage of the entomopathogenic fungi is the spores that are retained on the integument surface of the insect. Adhesins are the molecules produced by the fungus that intermediates the process of spore sticking to the cuticle of the host insect (Mora et al., 2017). Optimal conditions and parameters like temperature, humidity, and the chemical, physical and nutritional state of the target insect's cuticle should be correct enough to allow the conidium to start the process of penetration with the help of structures like appressorium, germ tubes or mucilaginous substances to permeate through the outmost layer of cuticle or the pores present on the insect body (Shah and Pell, 2003). As the conidium sticks to the cuticle of the insect host, at the end of the germ tube an infection structure called appressorium forms which further leads to invasion of the host by an infection or penetration peg. Mechanical pressure is required to rupture the cuticle (Zacharuk, 1970) along with the cuticle-dissolving enzymes produced by the fungus to colonize the host body and also to procure nutrition from it (Vega et al., 2012). Several factors like the presence of nutritional and antifungal substances and other cuticle properties like sclerotization and thickness dictate the penetration process of entomopathogenic fungi into the host (Charnley, 2003). Cuticle degradation is aided by cuticle-dissolving enzymes such as lipases, proteases, and chitinases, which are required to break down lipids, proteins, and chitin, respectively (Mora et al., 2017). Upon breaching and invading the host insect, the fungi start producing hyphal bodies that propagate through the haemocoel, invading distinct muscle tissues. Malpighian tubes, fatty bodies, haemocytes, and mitochondria are all destroyed within 3–14 days of infection, resulting in the death of the host pest. After the infected insect dies, the fungi start invading the rest of the organs by producing micelles. Afterwards, the fungal hyphae grow and breach the cuticle from the inside of the insect body, emerging at the body surface, and once optimal conditions are reached, spore formation is initiated so that it can infect other pest insects (Pucheta et al., 2006).

13.3.2.2 Genetic Manipulations in Entomopathogenic Fungi

Genetic manipulation/engineering is an exciting technique to enhance the effectiveness of fungal insecticides by reforming their virulence and resilience to substantial stresses (Zhao et al., 2016). Toxins, hormones, enzymes, and other physiological regulators are targeted by manipulation techniques (Karabörklü et al., 2018). For example, when the scorpion toxin gene AaIT1 was introduced into the haemolymph of *Metarhizium anisopliae*, the fungal toxicity against tobacco hornworm (*Manduca sexta*) caterpillars rose 22-fold (Wang & St Leger, 2007). To reinforce the virulence against *Locusta migratoria manilensis, BjαIT*, an insect-selective neurotoxin isolated from *Buthotus judaicus* was infused in *Metarhizium acridum* (Peng & Xia, 2015).

13.3.3 Viruses

To become a successful viral biopesticide, a virus must be virulent and infectious enough to kill a sufficient number of pests to efficiently control pest damage. It should possess the tendency to kill insects in order to control damage to economically important plants (Erlandson, 2008). There are two types of entomogenous viruses: (i) those that secrete inclusion bodies, called inclusion viruses and (ii) those that do not, called non-inclusion viruses (Kacchawa, 2017, Thakur et al., 2020). Many viruses have been shown to infect insects, including DNA and RNA viruses. Baculoviruses are the most widely accepted viruses for the control of pests, as they are never reported to cause disease in any organism outside the phylum Arthropoda. These are also found to be responsible for several natural epizootics observed in insects (Miller, 1997; Payne, 1982) (Reid et al., 2014, Kacchawa, 2017) and hence are considered potential candidates for biopesticides. Baculoviruses are detectable by light microscopy in the form of occlusion bodies (OBs) within infected host cells (Erlandson, 2008; Hubbard et al., 2014), which adds to their advantages for field application.

13.3.3.1 Mechanism of Action

Entomopathogenic viruses are good for controlling pests that have chewing-type mouthparts (example, lepidopteran insects), as they should be primarily ingested by the insects. Occlusion-derived virions (ODVs) and budded virions (BVs) are two forms of virions that help propagate diseases. The epithelial cells that line the midgut are infected by ODVs. With the support of BVs, they later disseminate the infection to all tissues throughout the host. Thus, the ODVs are responsible for the distribution of infection among the insect colony and BVs help spread the infection inside the host body. With the progression of infection, the insects move to the top of their host plants, where they die, degrade and release the occlusion bodies. These degraded insect products serve as a source of inoculum and help in the transmission of infection to other pests in the field (Erlandson, 2008, Dara, 2017).

Baculoviruses finish off their hosts by interrupting their metabolic pathways which compels them to produce the genetic material and the proteins of the baculovirus, which may in turn increase the virus production. Though this method is very advanced, no toxins are expressed during replication. This makes the baculovirus biopesticide a slow-acting one (Hubbard et al., 2014). However, baculoviruses encode a variety of genes that could improve their ability to infect, reproduce quickly in specific insects and survive in the host's body (Hubbard et al., 2014).

13.3.3.2 Genetic Manipulations in Entomopathogenic Viral Agents

Recombination in the genes of baculovirus has led to the production of economically important biopesticides. *vEV–Tox34* is a modified baculovirus which expresses a mite gene *Tox-34* and helps in the rapid killing of the corn earworm, *Helicoverpazea* (Tomalski & Miller, 1991). Similarly,

AcMNPV (Autographa californicanuclear polyhedrosis virus), a genetically modified virus which expresses insect-specific neurotoxin genes *vAcTalTX-1* and *vAcDTX9.2* from the spiders *Diguetia canities* and *Tegenaria agrestis*, respectively, have been used commercially against lepidopteran insects.

Lymantria disparmulticapsid nucleo polyhedrovirus (*LdMNPV*) is used as a control agent for gypsy moth (L. dispar). Its efficacy was improved by preparing a recombined viral strain *vEGT*. It was observed that the median lethal time (time until death) of fifth-instar larvae infected with *vEGT* was 33% lower than that with *LdMNPV*. Neuro Bactrus is a recombinant baculovirus constructed from *Btcry1-5*, *AcMNPV*, *AaIT* and *orf603* genes from various viral sources in order to increase its efficiency in the field (Shim et al., 2013). Neuro Bactrus is used to control *Plutella xylostella* and *Spodoptera exigua* larvae, and it has been shown to have good insecticidal action when compared to wild-type *AcMNPV*. Furthermore, *Buthu seupeus* insect toxin-1, the *Manduca sexta* diuretic hormone, the *Bt. kurstaki* HD-73 delta-endotoxin, the *Heliothis virescens* juvenile hormone esterase, the *P. Tritici TxP-I* toxin, *Androctonus australis* neurotoxin, *Dol m V* gene, and *Turf 13* genes have been recombined with the baculovirus genes for the purpose of developing strong and effective viral biopesticides (Khan et al., 2009).

13.3.4 Microsporidia

Microsporidians, also known as entomopathogenic protozoans, are found all over the world and infect nearly all invertebrates and vertebrates (Han & Weiss, 2017). These organisms were formerly called "primitive" protozoans (Vossbrinck et al., 1987), but the phylogenetic analysis at the molecular level demonstrated that they are not primitive but are related to the fungi (Weiss et al., 1999). Infections in insects caused by microspores are very frequent, which may lead to their natural death. These are occasionally used as biological control agents for insect pests, but their use is limited because they are slow-growing organisms that require sophisticated culture conditions (Han & Weiss, 2017).

Microsporidian infections are chronic in nature, i.e., slow-acting and time-consuming rather than acute. Even though the effects can range from benign to lethal, the main symptoms of such infections include delayed larval development of the insect, lethargic movement, low pupal weight, and low fecundity, and once infected, the immature insects do not survive for long (Solter et al., 2012). Though these effects can reduce the number of insect pests, microsporidia are not as competitive as chemical or microbial insecticides in terms of causing quick pest clearance (Lacey et al., 2001; Solter and Becnel, 2007).

13.3.4.1 Mechanism of Action

Microsporidia infect hosts either by ingestion or through egg transmission. The spores germinate in an exclusive manner after intake. Inside the spore, a polar tube abruptly everts out to produce a hollow tube that facilitates the transport of sporoplasm into the host cell cytoplasm. It induces massive infection and demolishes organs and tissues of the host. As the microsporidia lead an intracellular parasitic life cycle, they have lost several gene contents including those of the mitochondria. This damage results in a deficiency of the energy supply and resources required for the development of the parasite. To make up for this loss, the parasite fulfills its needs from the host, causing energetic stress (Huang et al., 2021).

Only a few species of microsporidium have been successfully used as biopesticides. *Nosema pyrausta* (=*Pereziapyraustae*) infects several insect species such as the European corn borer. *Nosema locustae*, the only commercially accessible species of microsporidium, is used to control grasshoppers and crickets. Other *Nosema* species are known to be used for controlling spider mites and webworms but are still in the early stages of commercial development. Similarly, *Vairimorpha necatrix* is also used widely against caterpillars of corn earworm, European corn borer, various armyworms, fall webworms and cabbage looper (Hoffmann & Frodsham, 1993).

Microsporidia are obligate intracellular pathogens, i.e., they need the inside environment of the host body to ease the DNA incorporation effectively. This requirement makes them incompatible with development in artificial media or fermentation tanks. As this method is costly and not feasible for the mass production of biopesticides, microsporidia are not used frequently in agricultural fields and hence are not very successful.

13.4 MICROBIAL-DERIVED ANTI-INSECTAN METABOLITES AND COMPOUNDS

Microorganisms frequently create metabolites with a low molecular mass. These are secondary metabolites, which are formed during the late growth phase of microorganisms and have a significant tendency to interfere with the pathogen's disease-causing capacity (Keswani et al., 2020). Since these microbial secondary metabolites are not vital for its primary growth, they are extracted and utilized by biotechnologists in the formulation of anti-insectan compounds useful for pest management systems. The following is a brief description of some important secondary metabolites (along with their structures) derived from various microbial groups.

13.4.1 Bacteria-Isolated Anti-Insectan Compounds

13.4.1.1 Thuringiensin (Thu)

Thuringiensin is a tiny oligosaccharide molecule generated by *Bacillus thuringiensis* as a secondary metabolite (Figure 13.2a). Also called β-exotoxin, it is a thermostable toxin. Thuringiensin works against a huge variety of insects which includes Lepidoptera, Isoptera, Orthoptera, Diptera, Hymenoptera and Coleoptera (Tamez-Guerra et al., 2004; Toledo et al., 1999). It is also found to be toxic against plant nematodes (Liu et al., 2010). Its chemical formula is $C_{22}H_{32}O_{19}N_5P$. Its insecticidal mechanism involves interfering with RNA polymerase as an ATP analog (Farkaš et al., 1969; Šebesta & Horska, 1970).

13.4.1.2 Thiolutin (THL)

Thiolutin is a disulphide containing bicyclic antibiotic (Figure 13.2b) and angiogenesis inhibiting in nature, formed by several strains of *Streptomycetes luteosporeus* during submerged fermentation. It is used as a yeast inhibitor and inhibits RNA (ribonucleic acid) and protein synthesis (Jimenez et al., 1973).

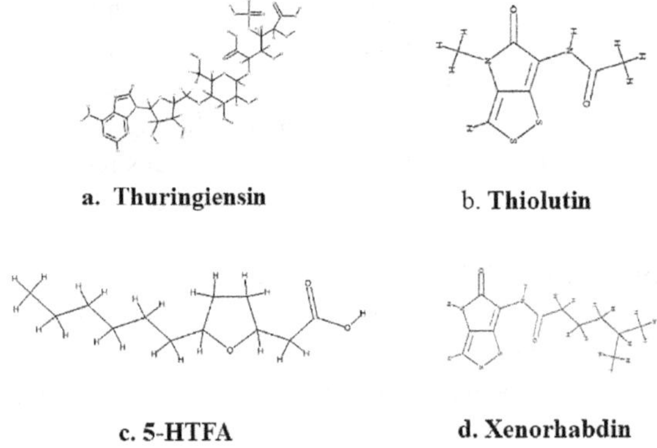

FIGURE 13.2 Structure of metabolites obtained from bacteria.

13.4.1.3 5-n-hexyl-tetrahydrofuran-2-acetic acid (5-HTFA)

Bacillus lentus produces 12-hydroxyoctadecanoic acid which bioconverts into 5-n-hexyl tetrahydrofuran-2-acetic acid (Huang et al., 1995) (Figure 13.2c).

13.4.1.4 Xenorhabdins

Xenorhabdins are produced and isolated from bacteria *Xenorhabdus spp*. They are found to be effective against *Helicoverpa punctigera*. (Figure 13.2d).

13.4.2 Fungi-Isolated Anti-Insectan Compounds

Several anti-insectan compounds are isolated from fungi and some common ones are described below.

13.4.2.1 Phenolic

Fungi can survive in different types of environments as they produce several metabolites that help them adapt to their habitats (Fox & Howlett, 2008). Some of these fungi produce phenolic compounds that are anti-insectan in nature. Ferulic acid is a phenolic compound (Figure 13.3a) produced by the *Rhizoctonia* which is effective as a pesticide against *Botrytis cinerea* (Patzke & Schieber, 2018). Another phenolic compound is benzoic acid and its derivatives produced by *Eurotium, Lentinus, Aspergillus, Lambertella, Penicillium, Hansenul and Mycotorula*, which is used as an antimicrobial additive and preservative in foods against the proliferation of bacteria and yeasts (Kalpana & Rajeswari, 2019).

13.4.2.2 Polyacetylenes

Polyacetylenes are a group of 500 hydrocarbons that are naturally occurring and have one or more acetylenic groups (Figure 13.3b). Polyacetylenes are produced by fungi *Russula, Stereum, Peniophora, Fistulina, Collybia, Agrocybe, Clitocybe, Daedalea, Hydnum, Ramaria, Polyporus, Resinicium* and *Kuehneromyces* (Turner, 1971). Most of these polyacetylenes are C-9 or C-10 photosensitizers and toxic to *Spodoptera* larvae (Arnason et al., 1983).

13.4.2.3 Aflatoxins

Some fungi also produce anti-insectan compounds like aflatoxins (Figure 13.3c). Aflatoxins are produced by *A. parasiticus* and *A. flavu* (Cole & Cox, 1981). Sterigmatocystin is an aflatoxin precursor which is toxic to *S. frugiperda, H. zea* and *Tyrophagus putrescentiae* (Dowd, 1988; Rodriguez et al., 1980). Aflatoxins cause anomalies during DNA transcription by reacting to guanidine and converting it into formamidopyrimide (Lillehoj, 1992). Aflatoxins were also found to be toxic to housefly (Al-Adil et al., 1973).

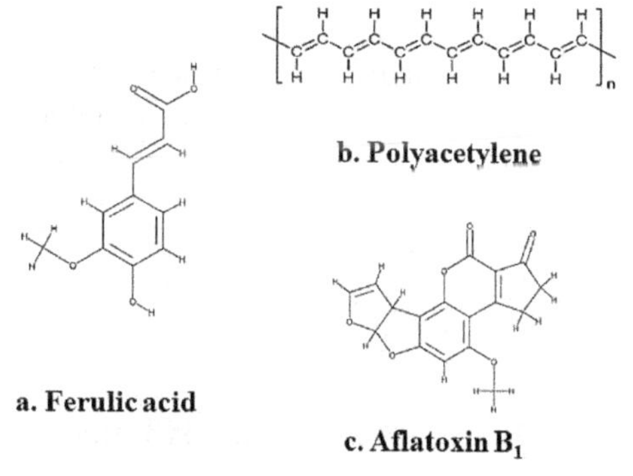

b. Polyacetylene

a. Ferulic acid

c. Aflatoxin B_1

FIGURE 13.3 Structure of metabolites obtained from fungi.

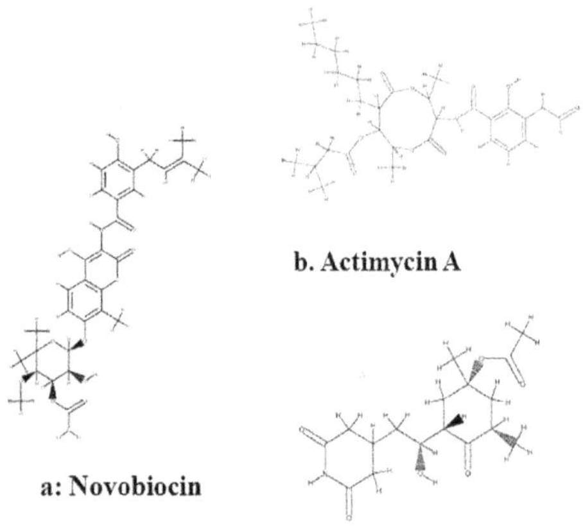

b. Actimycin A

a: Novobiocin

c. Cyclohexamide

FIGURE 13.4 Structure of metabolites obtained from actinomycetes.

13.4.3 Actinomycetes-Isolated Anti-Insectan Compounds

Many actinomycetes produce anti-insectan compounds known as antibiotics. Streptomycetes spp. are the primary producers of most antibiotics. Novobiocin (Figure 13.4a) is an antibiotic that inhibits the production of RNA in insects belonging to the Lepidoptera, Homoptera, and Mite families. Antimycin A (Figure 13.4b) is also dangerous to Lepidopterans, Homopterans, Coleopterans, Orthopterans, and Mites because it inhibits RNA synthesis. Cycloheximide (Figure 13.4c) interferes with protein synthesis and is toxic to insects belonging to *Lepidoptera*, *Homoptera*, and Mites. *Streptomyces lavendulae* produce antibiotics known as Racemomycins (Racemomycin A, B, C and D). Racemomycins majorly affects Malpighian tubules. Racemomycin A, B, and C are toxic to *B.germanica* and Racemomycin D is toxic to *Bombyx mori* (Kubo et al., 1981).

13.4.4 Anti-Insectan Protein Metabolites

13.4.4.1 Cholesterol Oxidase

Cholesterol oxidase is an anti-insectan protein (52.5 kDa) produced by *Pseudomonas fluorescens*. It was found to be toxic to *Anthonomus grandis* as it disrupts the epithelial lining of the midgut (Purcell et al., 1993). In the microvilli carrying membranes of various insects, cholesterol oxidase was found to inhibit alkaline phosphatase activity (Shen et al., 1997).

13.4.4.2 Restrictocin

Aspergillus restrictris produces restrictocin, an anti-insectan protein (Brandhorst et al., 1996). It is a phosphodiesterase that has the ability to inactivate ribosomes by cleaving rRNA linkages (Jimenez & Vazquez, 1985). Restrictocin is toxic to *S. frugiperda* and *H. zea* (Brandhorst et al., 1996).

13.4.4.3 Clostridium Toxin

Clostridium toxin is an anti-insectan protein produced by *Clostridium bifermentans*. This 66Kda protein exhibits larvicidal proteolytic activity against *A. aegypti, C. pipiens*, and *A. stephensi* (Barloy et al., 1996; Charles et al., 1990; Nicolas et al., 1993).

13.5 SIGNIFICANCE OF MICROBIAL BIOPESTICIDES

In recent years, the use of biopesticides has been gaining attention because they can be efficiently used in sustainable agricultural practices. Biopesticides are very effective in small amounts and disintegrate quickly without leaving residues, allowing them to supplement the limited use of

TABLE 13.1
A Comparison Between Conventional Pesticides and Microbial Biopesticides

Features	Conventional Pesticides	Microbial Biopesticides
Active ingredients	Synthetic chemicals	Compounds derived from living organisms (i.e., microbes)
Environment friendly	No; create pollution	Yes
Specificity	Harmful even to non-target organisms	Affects only the target pests
Resistance	The target pest usually gains resistance against this	Does not lead to resistance development in the pest
Contamination	Yes; contaminates soil and water	No; cannot create contamination
Bioaccumulation	Yes; often gets bioaccumulated	No

conventional pesticides in IPM programs (Kumar et al., 2021). Table 13.1 gives a snapshot of the significance of microbial biopesticides over conventional pesticides.

13.6 DISADVANTAGES OF MICROBIAL BIOPESTICIDES

Though biopesticides have various significant and advantageous points, they also have some limitations which sometimes keep them away from agriculture areas. The downsides to using biopesticides include the following:

1. They have a slow rate of action and less persistence than conventional pesticides.
2. They are extremely vulnerable to harsh environmental conditions such as high temperatures, UV light, and so on.
3. Microbial insecticides necessitate special formulation and storage processes.
4. The development, registration, and production costs of all microbial biopesticides are high and hence generally not the first choice for farmers.

13.7 CONCLUSION

Insect resistance to synthetic pesticides is a substantial and escalating problem in agriculture. It is critical to address this issue, and pesticides of microbial origin have emerged as a promising tool for long-term pest management programs. Microbial biopesticides should be used extensively for such programs because they are very eco-friendly, harmless to users, and specific to the host, and arthropods do not develop resistance to them. More research should be conducted in order to identify active metabolic compounds of microbial origin, and the use of microbial biopesticides must be promoted in order to develop a sustainable agro-ecosystem.

REFERENCES

Agaisse, H., & Lereclus, D. (1995). How does Bacillus thuringiensis produce so much insecticidal crystal protein? *Journal of Bacteriology*, 177(21), 6027–6032.

Al-Adil, K. M., Kilgore, W. W., & Painter, R. R. (1973). Effects of orally ingested aflatoxin B1 on nucleic acids and ribosomes of housefly ovaries. *Toxicology and Applied Pharmacology*, 26(1), 130–136. https://doi.org/10.1016/0041-008X(73)90093-8

Arnason, T., Towers, G., Philogene, B., & Lambert, J. (1983). The role of natural photosensitizers in plant resistance to insects. https://doi.org/10.1021/bk-1983-0208.ch008

Attathom, T. (2002). Biotechnology for insect pest control. *Sustainable Agricultural System in Asia*, 2, 73–84.

Barloy, F., Delécluse, A., Nicolas, L., & Lecadet, M.-M. (1996). Cloning and expression of the first anaerobic toxin gene from *Clostridium bifermentans* subsp. malaysia, encoding a new mosquitocidal protein with homologies to *Bacillus thuringiensis* delta-endotoxins. *Journal of Bacteriology*, 178(11), 3099–3105. https://doi.org/10.1128/jb.178.11.3099-3105.1996

Behle, R., & Birthisel, T. (2014). Formulations of entomopathogens as bioinsecticides. In *Mass Production of Beneficial Organisms* (pp. 483–517). Academic Press. https://doi.org/10.1016/B978-0-12-391453-8.00014-5

Bernardes, M. F. F., Pazin, M., Pereira, L. C., & Dorta, D. J. (2015). Impact of pesticides on environmental and human health. *Toxicology Studies-Cells, Drugs and Environment*, 195–233.

Brandhorst, T., Dowd, P. F., & Kenealy, W. R. (1996). The ribosome-inactivating protein restrictocin deters insect feeding on *Aspergillus restrictus*. *Microbiology*, 142(6), 1551–1556. https://doi.org/10.1099/13500872-142-6-1551

Brar, S. K., Verma, M., Tyagi, R. D., & Valero, J. R. (2006). Recent advances in downstream processing and formulations of *Bacillus thuringiensis* based biopesticides. *Process Biochemistry*, 41(2), 323–342.

Bravo, A., Gill, S., & Soberón, M. (2007). Mode of action of *Bacillus thuringiensis* Cry and Cyt and their potential for insect control. *Toxicon*, 49, 423–435. https://doi.org/10.1016/j.toxicon.2006.11.022

Bravo, A., Likitvivatanavong, S., & Gill, S., & Soberón, M. 2011. *Bacillus thuringiensis*: A story of a successful bioinsecticide. *Insect Biochemistry and Molecular Biology*, 41, 423–431. https://doi.org/10.1016/j.ibmb.2011.02.006

Charles, J.-F., Nicolas, L., Sebald, M., & de Barjac, H. (1990). *Clostridium bifermentans* serovar malaysia: Sporulation, biogenesis of inclusion bodies and larvicidal effect on mosquito. *Research in Microbiology*, 141(6), 721–733. https://doi.org/10.1016/0923-2508(90)90066-Y

Charnley, A. K. (2003). Fungal pathogens of insects: Cuticle degrading enzymes and toxins. *Advances in Botanical Research*, 40, 241–321.

Cole, R., & Cox, R. (1981). Fusarium toxins In: Cole, R.J., (ed.). *Handbook of Toxic Fungal Metabolites*. Elsevier, Amsterdam, the Netherlands.

Dara, S. K. (2017). *Entomopathogenic Microorganisms: Modes of Action and Role in IPM*. Agriculture and Natural Blogs, University of California

Dowd, P. F. (1988). Synergism of aflatoxin B1 toxicity with the co-occurring fungal metabolite kojic acid to two caterpillars. https://doi.org/10.1007/BF00186717

Erlandson, M. (2008). Insect pest control by viruses. *Encyclopedia of Virology*, Third Edition 3: 125–133.

Faria, M. R., & Wraight, S. P. (2007). Mycoinsecticides and mycoacaricides: A comprehensive list with worldwide coverage and international classification of formulation types. *Biological Control* 43, 237–256. https://doi.org/10.1016/j.biocontrol.2007.08.001

Farkaš, J., Šebesta, K., Horska, K., Samek, Z., Dolejš, L., & Šorm, F. (1969). The structure of exotoxin of Bacillus thuringiensis var. gelechiae. Preliminary communication. *Collection of Czechoslovak Chemical Communications*, 34(3), 1118–1120. https://doi.org/10.1135/cccc19691118

Federici, B. A. (1991). Microbial insecticides. *Pesticide Outlook*, 2(3), 22–28.

Federici, B. A., Park, H. W., Bideshi, D. K., Wirth, M. C., & Johnson, J. J. (2003). Recombinant bacteria for mosquito control. *Journal of Experimental Biology*, 206(21), 3877–3885.

Fox, E. M., & Howlett, B. J. (2008). Secondary metabolism: Regulation and role in fungal biology. *Current Opinion in Microbiology*, 11(6), 481–487. https://doi.org/10.1016/j.mib.2008.10.007

Gašić, S., & Tanović, B. (2013). Biopesticide formulations, possibility of application and future trends. *Pesticidiifitomedicina*, 28(2), 97–102. https://doi.org/10.2298/PIF1302097G

Gomiero, T., Pimentel, D., & Paoletti, M. G. (2011). Is there a need for a more sustainable agriculture? *Critical Reviews in Plant Sciences*, 30 (1–2), 6–23. https://doi.org/10.1080/07352689.2011.553515

Han, B., & Weiss, L. M. (2017). Microsporidia: obligate intracellular pathogens within the fungal kingdom. *Microbiology Spectrum*, 5(2), 5(2). https://doi.org/10.1128/microbiolspec.FUNK-0018-2016

Hoffmann, M. P., & Frodsham, A. C. (1993) *Natural Enemies of Vegetable Insect Pests*. Cooperative Extension, Cornell University, Ithaca, NY. 63 p.

Höfte, H., & Whiteley, H. R. (1989). Insecticidal crystal proteins of Bacillus thuringiensis. *Microbiological reviews*, 53(2), 242–255. https://doi.org/10.1128/mr.53.2.242-255.1989

Huang, J.-K., Keudell, K., Zhao, J., Klopfenstein, W., Wen, L., Bagby, M., Lanser, A., Plattner, R., Peterson, R., & Abbott, T. (1995). Microbial transformation of 12-hydroxyoctadecanoic acid to 5-n-hexyl-tetrahydrofuran-2-acetic acid. *Journal of the American Oil Chemists' Society*, 72(3), 323–326. https://doi.org/10.1007/BF02541090

Huang, Q., Wu, Z. H., Li, W. F., Guo, R., Xu, J. S., Dang, X. Q., … & Evans, J. D. (2021). Genome and evolutionary analysis of Nosema ceranae: A microsporidian parasite of honey bees. *Frontiers in Microbiology*, 12, 1303. https://doi.org/10.3389/fmicb.2021.645353

Hubbard, M., Hynes, R. K., Erlandson, M., & Bailey, K. L. (2014). The biochemistry behind biopesticide efficacy. *Sustainable Chemical Processes*, 2(1), 1–8

Jimenez, A., Tipper, D. J., & Davies, J. (1973). Mode of action of thiolutin, an inhibitor of macromolecular synthesis in Saccharomyces cerevisiae. *Antimicrobial Agents and Chemotherapy*, 3(6), 729–738. https://doi.org/10.1128/AAC.3.6.729

Jimenez, A., & Vazquez, D. (1985). Plant and fungal protein and glycoprotein toxins inhibiting eukaryote protein synthesis. *Annual Review of Microbiology*, 39(1), 649–672. https://doi.org/10.1146/annurev.mi.39.100185.003245

Kalpana, V., & Rajeswari, V. D. (2019). Preservatives in beverages: Perception and needs. In *Preservatives and Preservation Approaches in Beverages* (pp. 1–30). Elsevier. https://doi.org/10.1016/B978-0-12-816685-7.00001-X

Karaborkl, S., & Akbulut, M. (2011). Characterization of local bacillus thuringiensis isolates and their toxicity to *Ephestiakuehniella* (Zeller) and *Plodia interpunctella* (Hubner) Larvae Ugur Azizoglul*; Semih Ydmaz; Abdurrahman Ayvaz. *Egyptian Journal of Biological Pest Control*, 21(2), 143–150.

Karabörklü, S., Ayvaz, A., Yilmaz, S., Azizoglu, U., & Akbulut, M. (2015). Native entomopathogenic nematodes isolated from Turkey and their effectiveness on pine processionary moth, Thaumetopoea wilkinsoni Tams. *International Journal of Pest Management*, 61(1), 3–8.

Kachhawa, D. (2017). Microorganisms as a biopesticides. *Journal of Entomology and Zoology Studies*, 5(3), 468–473.

Keswani, C., Singh, H. B., García-Estrada, C., Caradus, J., He, Y. W., Mezaache-Aichour, S., & Sansinenea, E. (2020). Antimicrobial secondary metabolites from agriculturally important bacteria as next-generation pesticides. *Applied Microbiology and Biotechnology*, 104(3), 1013–1034. https://doi.org/10.1007/s00253-019-10300-8

Khan, M. S., Zaidi, A., & Musarrat, J. (2009). *Microbial Strategies for Crop Improvement*. Springer, UK.

Knowles, A. (2005). *New Developments in Crop Protection Product Formulation* (pp. 153–156). Agrow Reports UK: T and F Informa UK Ltd

Knowles, A. (2006). *Adjuvants and Additives* (pp. 126–129). Agrow Reports UK: T&F Informa UK Ltd

Knowles, A. (2008). Recent developments of safer formulations of agrochemicals. *Environmentalist*, 28(1), 35–44. doi:10.1007/s10669-007-9045-4

Kubo, M., Kato, Y., Morisaka, K., Inamori, Y., Nomoto, K., Takemoto, T., Sakai, M., Sawada, Y., & Taniyama, H. (1981). Insecticidal activity of streptothricin antibiotics. *Chemical and Pharmaceutical Bulletin*, 29(12), 3727–3730. https://doi.org/10.1248/cpb.29.3727

Kumar, J., Ramlal, A., Mallick, D., & Mishra, V. (2021). An overview of some biopesticides and their importance in plant protection for commercial acceptance. *Plants*, 10, 1185. https://doi.org/10.3390/plants10061185

Lacey, L. A., Frutos, R., Kaya, H. K., & Vail, P. (2001). Insect pathogens as biological control agents: Do they have a future? *Biological Control*, 21(3), 230–248.

Lillehoj, E. (1992). Aflatoxin: Genetic mobilization agent. In: *Handbook of Applied Mycology: Mycotoxins in Ecological Systems* (pp. 1–22). Marcel Dekker, New York.

Liu, X.-Y., Ruan, L.-F., Hu, Z.-F., Peng, D.-H., Cao, S.-Y., Yu, Z.-N., Liu, Y., Zheng, J.-S., & Sun, M. (2010). Genome-wide screening reveals the genetic determinants of an antibiotic insecticide in Bacillus thuringiensis. *Journal of Biological Chemistry*, 285(50), 39191–39200. https://doi.org/10.1074/jbc.M110.148387

Lucena, W. A., Pelegrini, P. B., Martins-de-Sa, D., Fonseca, F. C., Gomes Jr, J. E., De Macedo, L. L., ... & Grossi-de-Sa, M. F. (2014). Molecular approaches to improve the insecticidal activity of Bacillus thuringiensis Cry toxins. *Toxins*, 6(8), 2393–2423.

Mishra, A., Arshi, A., Mishra, S. P., & Bala, M. (2019). Microbe-based biopesticide formulation: A tool for crop protection and sustainable agriculture development. In *Microbial Technology for the Welfare of Society* (pp. 125–145). Springer, Singapore. https://doi.org/10.1007/978-981-13-8844-6_6

Miller, L. K. (1997). Introduction to the Baculoviruses. In *The Baculoviruses* (pp. 1–6). Springer.

Mollet, H., & Grubenmann, A. (2008). *Formulation Technology: Emulsions, Suspensions, Solid Forms*. John Wiley & Sons.

Mora, M. A. E., Castilho, A. M. C., & Fraga, M. E. (2017). Classification and infection mechanism of entomopathogenic fungi. *Arquivos do Instituto Biológico*, 84, 1–10.

Nicolas, L., Charles, J.-F., & de Barjac, H. (1993). Clostridium bifermentans serovar malaysia: characterization of putative mosquito larvicidal proteins. *FEMS Microbiology Letters*, 113(1), 23–28. https://doi.org/10.1111/j.1574-6968.1993.tb06482.x

Nicolopoulou-Stamati, P., Maipas, S., Kotampasi, C., Stamatis, P., & Hens, L. (2016). Chemical pesticides and human health: The urgent need for a new concept in agriculture. *Frontiers in Public Health*, 4, 148. https://doi.org/10.3389/fpubh.2016.00148

Palma, L., Muñoz, D., Berry, C., Murillo, J., & Caballero, P. (2014). Bacillus thuringiensis toxins: An overview of their biocidal activity. *Toxins*, 6(12), 3296–3325.

Pardo-Lopez, L., Munoz-Garay, C., Porta, H., Rodríguez-Almazán, C., Soberón, M., & Bravo, A. (2009). Strategies to improve the insecticidal activity of Cry toxins from Bacillus thuringiensis. *Peptides*, 30(3), 589–595.

Park, H. W., & Federici, B. A. (2009). Genetic engineering of bacteria to improve efficacy using the insecticidal proteins of Bacillus species. In *Insect Pathogens: Molecular Approaches and Techniques* (pp. 275–305). CABI International, Oxfordshire, UK.

Pathak, D. V., Yadav, R., & Kumar, M. (2017). Microbial pesticides: Development, prospects and popularization in India. In *Plant-microbe Interactions in Agro-Ecological Perspectives* (pp. 455–471). Springer, Singapore. https://doi.org/10.1007/978-981-10-6593-4_18

Patzke, H., & Schieber, A. (2018). Growth-inhibitory activity of phenolic compounds applied in an emulsifiable concentrate-ferulic acid as a natural pesticide against Botrytis cinerea. *Food Research International*, 113, 18–23. https://doi.org/10.1016/j.foodres.2018.06.062

Payne, C. C. (1982). Insect viruses as control agents. *Parasitology*, 84(4), 35–77.

Peng, G., & Xia, Y. (2015). Integration of an insecticidal scorpion toxin (BjαIT) gene into Metarhiziumacridum enhances fungal virulence towards *Locusta migratoriamanilensis*. *Pest Management Science*. Jan; 71(1): 58–64. doi:10.1002/ps.3762. PMID: 25488590

Priest, F. G. (2000). Biodiversity of the entomopathogenic, endosporeforming bacteria. In *Entomopathogenic Bacteria: From Laboratory to Field Application* (pp. 1–22). Springer, Dordrecht. https://doi.org/10.1007/978-94-017-1429-7_1

Pucheta, D. M., Macias, A. F., Navarro, S. R., & Mayra, D. L. T. (2006). Mechanism of action of entomopathogenic fungi. *Microbiology*, 156, 2164–2171.

Purcell, J. P., Greenplate, J. T., Jennings, M. G., Ryerse, J. S., Pershing, J. C., Sims, S. R., Prinsen, M. J., Corbin, D. R., Tran, M., & Sammons, R. D. (1993). Cholesterol oxidase: A potent insecticidal protein active against boll weevil larvae. *Biochemical and Biophysical Research Communications*, 196(3), 1406–1413. https://doi.org/10.1006/bbrc.1993.2409

Rao, M. S., Umamaheswari, R., Chakravarthy, A. K., Grace, G. N., Kamalnath, M., & Prabu, P. (2015). A frontier area of research on liquid biopesticides: Ahe way forward for sustainable agriculture in India. *Current Science*, 108(9), 1590–1592.

Reid, S., Chan, L., & van Oers, M. M. (2014). Production of entomopathogenic viruses. In *Mass production of Beneficial Organisms* (pp. 437–482). Academic Press. https://doi.org/10.1016/B978-0-12-391453-8.00013-3

Rodriguez, J., Potts, M., & Rodriguez, L. (1980). Mycotoxin toxicity to *Tyrophagusputrescentiae*. *Journal of Economic Entomology*, 73(2), 282–284. https://doi.org/10.1093/jee/73.2.282

Seaman, D. (1990). Trends in the formulation of pesticides—an overview. *Pesticide Science*, 29(4), 437–449. https://doi.org/10.1002/ps.2780290408

Šebesta, K., & Horska, K. (1970). Mechanism of inhibition of DNA-dependent RNA polymerase by exotoxin of Bacillus thuringiensis. *Biochimica et Biophysica Acta (BBA)-Nucleic Acids and Protein Synthesis*, 209(2), 357–367. https://doi.org/10.1016/0005-2787(70)90734-3

Shah, P. A., & Pell, J. K. (2003). Entomopathogenic fungi as biological control agents. *Applied Microbiology and Biotechnology*, 61(5), 413–423.

Shen, Z., Corbin, D. R., Greenplate, J. T., Grebenok, R. J., Galbraith, D. W., & Purcell, J. P. (1997). Studies on the mode of action of cholesterol oxidase on insect midgut membranes. *Archives of Insect Biochemistry and Physiology: Published in Collaboration with the Entomological Society of America*, 34(4), 429–442. https://doi.org/10.1002/(SICI)1520-6327(1997)34:4%3C429::AID-ARCH3%3E3.0.CO;2-N

Shim, H. J., Choi, J. Y., Wang, Y., Tao, X. Y., Liu, Q., Roh, J. Y., … & Je, Y. H. (2013). NeuroBactrus, a novel, highly effective, and environmentally friendly recombinant baculovirus insecticide. *Applied and Environmental Microbiology*, 79(1), 141–149. https://doi.org/10.1128/AEM.02781-12

Silva-Filha, M.-H., Regis, L., Nielsen-LeRoux, C., & Charles, J.-F. (1995). Low-level resistance to Bacillus sphaericus in a field-treated population of Culex quinquefasciatus (Diptera: Culicidae). *Journal of Economic Entomology*, 88, 525–530.

Sinègre, G., Babinot, M., Quermel, J.-M., & Gaven, B. (1994). First field occurrence of Culex pipiens resistance to Bacillus sphaericus in southern France, p. 17. In Proceedings, 8th European Meeting of Society for Vector Ecology, 5–8 September 1994, Barcelona, Spain. Society for Vector Ecology, Santa Ana, CA.

Soberón, M., Gill, S., & Bravo, A. (2009). Signaling *versus* punching hole: How do *Bacillus thuringiensis* toxins kill insect midgut cells? *Cellular and Molecular Life Sciences*, 66, 1337–1349. https://doi.org/10.1007/s00018-008-8330-9

Solter, L. F., Becnel, J. J., & Oi, D. H. (2012). Microsporidian entomopathogens. In *Insect Pathology* (221–263), 2nd edition. Academic Press, Elsevier Inc, San Diego.

Strasser, H. (2001). Use of hyphomycetous fungi for managing insect pests. In: *Fungi as Biocontrol Agents: Progress Problems and Potential* (p. 23). CABI, Wallingford, UK.

Su, T., & Mulla, M. S. (2004). Documentation of high-level Bacillus sphaericus 2362 resistance in field population of Culex quinquefasciatus breeding in polluted water in Thailand. *Journal of the American Mosquito Control Association*, 20, 405–411

Tamez-Guerra, P., Iracheta, M. M., Pereyra-Alférez, B., Galán-Wong, L. J., Gomez-Flores, R., Tamez-Guerra, R. S., & Rodrıguez-Padilla, C. (2004). Characterization of Mexican Bacillus thuringiensis strains toxic for lepidopteran and coleopteran larvae. *Journal of Invertebrate Pathology*, 86(1–2), 7–18. https://doi.org/10.1016/j.jip.2004.02.009

Thakur, N., Kaur, S., Tomar, P., Thakur, S., & Yadav, A. N. (2020). Microbial biopesticides: Current status and advancement for sustainable agriculture and environment. In *New and Future Developments In Microbial Biotechnology and Bioengineering* (pp. 243–282). Elsevier. https://doi.org/10.1016/B978-0-12-820526-6.00016-6

Toledo, J., Liedo, P., Williams, T., & Ibarra, J. (1999). Toxicity of Bacillus thuringiensis β-exotoxin to three species of fruit flies (Diptera: Tephritidae). *Journal of Economic Entomology*, 92(5), 1052–1056. https://doi.org/10.1093/jee/92.5.1052

Tomalski, M. D., & Miller, L. K. (1991). Insect paralysis by baculovirus mediated expression of a mite neurotoxin gene. *Nature*, 352, 82–95. https://doi.org/10.1038/352082a0

Turner, W. B. (1971). *Fungal Metabolites*. pp. xi+446, Academic Press, London; New York, USA.

Usta, C. (2013). Microorganisms in biological pest control: A review (bacterial toxin application and effect of environmental factors). *Current Progress in Biological Research*, 13, 287–317.

Vega, F. E., Goettel, M. S., Blackwell, M., Chandler, D., Jackson, M. A., Keller, S., … & Roy, H. E. (2009). Fungal entomopathogens: New insights on their ecology. *Fungal Ecology*, 2(4), 149–159. https://doi.org/10.1016/j.funeco.2009.05.001

Vega, F. E., Meyling, N. V., Luangsa-ard, J. J., & Blackwell, M. (2012). Fungal entomopathogens. In F. Vega, & H. Kaya (Eds.), *Insect pathology* (2 ed., pp. 171–220). Elsevier. https://doi.org/10.1016/B978-0-12-384984-7.00006-3

Vossbrinck, C. R., Maddox, J. V., Friedman, S., Debrunner-Vossbrinck, B. A., & Woese, C. R. (1987). Ribosomal RNA sequence suggests microsporidia are extremely ancient eukaryotes. *Nature*, 326(6111), 411–414. https://doi.org/10.1038/326411a0

Yuan, Z., Zhang, Y., Cai, Q., & Liu, E.-Y. 2000. High level field resistance to Bacillus sphaericus C3-41 in Culex quinquefasciatus from southern China. *Biocontrol Science and Technology*, 10, 41–49.

Wang, C., St Leger, R. (2007). A scorpion neurotoxin increases the potency of a fungal insecticide. *Nature Biotechnology* 25, 1455–1456. https://doi.org/10.1038/nbt1357

Weiss, L. M., Edlind, T. D., Vossbrinck, C. R., & Hashimoto, T. (1999). Microsporidian molecular phylogeny: The fungal connection. *The Journal of Eukaryotic Microbiology*, 46(5), 17S–18S. https://doi.org/10.1111/j.1550-7408.1999.tb06055.x

Wirth, M. C., Walton, W. E., & Federici, B. A. (2010). Evolution of resistance to the Bacillus sphaericus Bin toxin is phenotypically masked by combination with the mosquitocidal proteins of Bacillus thuringiensis subspecies israelensis. *Environmental Microbiology*, 12(5), 1154–1160. 10.1111/j.1462-2920.2010.02156.x

Wirth, M. C., Yang, Y., Walton, W. E., Federici, B. A., & Berry, C. (2007). Mtx toxins synergize *Bacillus sphaericus* and Cry11Aa against susceptible and insecticide-resistant *Culex quinquefasciatus* larvae. *Applied and Environmental Microbiology*, 73(19), 6066–6071. https://doi.org/10.1128/AEM.00654-07

Zacharuk, R. Y. (1970). Fine structure of the fungus Metarhizium anisopliae infecting three species of larval Elateridae (Coleoptera). III. Penetration of the host integument. *Journal Invertebrate Pathology*, 15, 372–396.

Zhao, H., Lovett, B., & Fang, W. (2016). Genetically engineering entomopathogenic fungi. *Advances in Genetics*, 94, 137–163. https://doi.org/10.1016/bs.adgen.2015.11.001

14 Application and Impact of Biofertilizers in Sustainable Agriculture

Sourabh
ICAR - Central Arid Zone Research Institute, Jodhpur, India

Nirmal Singh
CCS Haryana Agricultural University, Hisar, India

Preeti
National Bureau of Plant Genetic Resources, New Delhi, India

CONTENTS

DOI: 10.1201/9781003306931-16

14.1 INTRODUCTION

According to reports by the United Nations' Food and Agriculture Organization (FAO), the world's population is approximately 7.6 billion and is anticipated to exceed 9 billion by the year 2050. This ever-growing human population imposes pressure on agrarian lands and other resources to ensure food security, making farmers dependent on the widespread use of chemical fertilizers and pesticides in order to achieve ever-increasing productivity levels (Santos et al. 2012). From the mid-twentieth century to the present day, green revolution using chemical fertilizers has been feeding the world's population with an outstanding increase in food grain production but led to a lack of concern for the long-term viability of the ecosystem. Approximately 53 billion tonnes of nitrogen, phosphorus, and potassium fertilizers are consumed every year to augment the nutrient requirements of the plant in order for it to grow properly and produce efficiently. Chemical fertilizers are produced in factories and include a predetermined amount of nutrients such as nitrogen, phosphorous, potassium, and sulfur. However, only about 10–40% of these nutrients are utilized by plants, while the majority (60–90%) is lost through leaching, immobilization, volatilization, precipitation, or other gaseous losses. Accumulation of these agrochemicals and synthetic fertilizers has caused ground water pollution through the process of eutrophication in water bodies, and destroyed friendly insects, micro-flora and fauna, causing more susceptibility of crops to insects and disease outbreaks. Inappropriate use of agrochemicals reduces the water retention capacity of soil, raises salinity, and causes nutritional disproportionality which ultimately results in decreased soil fertility and crop production. Various studies have shown that careful use of chemical inputs is permissible, but that continuous extensive and inappropriate use of agro-chemicals will result in soil degradation, water contamination, soil ecosystem imbalance, and an unsustainable burden on the country's economic system (Osman, 2014). Thus, recent efforts have been focused on producing 'quality food' while guaranteeing agricultural sustainability and biosecurity via the use of beneficial soil microbes. Organic agriculture practices are gaining favor in this area, as they promote biodiversity, biological transformation, and soil microbiological activity, all of which contribute to the creation of a socially, environmentally, and economically viable ecosystem. Organic farming is a more sustainable approach because it ensures food security while protecting and nurturing the naturally prevailing soil biodiversity. Among the various strategies of organic food production, uses of microbial inoculants or biofertilizers are critical in nutrient management to maintain agricultural output and a healthy environment.

Biofertilizers are seen as a sustainable source of plant nutrients that are capable of mobilizing growth ingredients into readily digested forms and have developed into an integral part of an integrated crop and soil management system. Biofertilizers are compounds that are combined with microbes or living cells which accelerate microbial processes in soil, helping to nurture growth of plant and encourage soil health by escalating nutrient accessibility when used as seed or soil treatment. Biofertilizers reduce carbon emissions by cutting back on chemical fertilizers in many regions of the globe. They are products that contain naturally occurring microorganisms that have been intentionally grown in order to increase soil fertility and crop yield (Mazid & Khan, 2015). According to Mishra et al. (2013), biofertilizers are "live or latent cells preparations that contain nitrogen fixers, phosphate solubilizers, S-oxidisers, organic matter decomposers, cyanobacteria, endo and ecto mycorrhizal fungi, plant growth promoting rhizobacteria (PGPRs) etc" in order to increase mutually beneficial microorganisms. A Dutch scientist first detected biofertilizers in 1888 and their use began in 1895 with a laboratory growth of *Rhizobia* (Ghosh, 2004). In India, N.V. Joshi studied legume–Rhizobium symbiosis in 1920, laying the foundations for their commercial production in 1956 (Rana et al. 2013). Subsequently *Azotobacter, Azospirillum*, blue green algae, vesicular arbuscular mycorrhizae (VAM) etc. were discovered. In the last few years, other rhizobacteria such as *Aeromonas veronii, Azotobacter* sp., *Azoarus* sp., Cyanobacteria (*Anabaena* and *Nostoc*), *Alcaligenes, Erwinia, Comamonas acidororans, Enterobacter, Burkholderia, Flavobacterium, Rhizobia* (*Mesorhizobium, Bradyrhizoblum, Azorhizobium, Allorhizobium, Rhizobium and Sinorhizobium*), *Gluconacetobacter diazotrophicus, Herbaspirillum seroepdicae, Serratia, Variovorax paradoxus* and *Xanthomonas maltophilia* have been identified as biofertilizers (Wahane et al., 2020).

Biofertilizers boost the protein, amino acid, vitamin, nitrogen, and other vital element content of food. They help stimulate proper root architecture, improving photosynthesis and enhancing nutrient solubilization. Biofertilizers produce helpful plant hormones such as auxins, cytokinins, biotins, and vitamins in addition to providing critical micro and macro elements for growth and development of the plant. The secretion of antibiotics by biofertilizers imparts biotic and abiotic resistance in plants. Beneficial soil microorganisms improve rhizosphere management through multidimensional mechanisms including production of siderophore, lytic acid, hydrogen cyanide and indole acetic acid, and phosphate solubilization. Isolation of various strains of agriculturally important soil microorganisms, including mycorrhiza fungi and plant growth-promoting rhizobacteria (PGPR), are now being employed in biotechnology to increase food security and agricultural sustainability. Application of biofertilizers improves soil structure and water uptake, and restores soil nutrients, build-up of soil organic matter, growth and tolerance of plants to abiotic and biotic factors (Adesemoye et al., 2008). Biofertilizers are a cost-effective, environmentally friendly and natural way to keep the system alive and functioning while also preserving biodiversity by improving soil quality. Thus, encouraging sustainable agriculture will be a major focus in the coming years, with greater awareness of the application of biofertilizers as a green technology. *Rhizobium* with phosphotika (phosphate-mobilizing biofertilizer) is suggested for seed treatment of various legumes like *Cajanus cajan, Vigna radiata, Vigna mungo, Vigna unguiculata, Arachis hypogaea* and *Glycine max*. Similarly, *Azotobacter* with phosphotika as seed treatment has proved beneficial for sorghum, wheat, cotton, mustard, maize, and transplanted rice. Apart from cereal crops, *Azotobacter* and phosphate-solubilizing biofertilizers (PSBs) are being used in horticultural crops to improve quality and yield of fruit crops, in addition to reducing fertilization through chemicals (Sourabh et al., 2018). Besides fixing atmospheric nitrogen in soil, they absorb phosphate from different soil layers, generate hormones and anti-metabolites, and solubilize insoluble forms of phosphates unavailable to the plant, notably tri-calcium, iron, and aluminium phosphates which encourage root growth, mineralization and decomposition of soil organic matter, and ultimately up to one-fourth increase in production without negative effects on the soil environment. The regular application of biofertilizers builds up microbial population in soil, which aids in soil fertility maintenance and helps to achieve sustainability in agriculture (Choudhury & Kennedy, 2004; Malik et al., 2011). Overall, the protracted usage of biofertilizers is cost-effective, ecofriendly, more effective, fruitful, and accessible to marginal and small farmers, making it an essential and powerful tool for organic and sustainable agriculture (Venkataraman and Shanmugasundaram, 1992). Thus, to extend the opportunities for adoption of biofertilizers, there is an urgent need to understand the limiting variables and to use this information to inform policy decisions. Therefore, this chapter describes various kinds of biofertilizers and their mechanisms of action, improved tolerance to environmental stress, potential for increasing nutrient profiles, plant growth and productivity, with a special emphasis on their role in sustainable agriculture.

14.2 PLANT–MICROBE INTERACTIONS

The highly diverse interactions displayed by plants and microorganisms form an integral part of the terrestrial ecosystem. Microbes exist in all parts of the plants: in association with the plant rhizosphere (roots), phyllosphere (aerial parts), and within plant cells (endophytic). These interactions are not always beneficial for plant species and can be categorized into competition, commensalism, mutualism, and parasitism (Wu et al., 2009). While the microbial community is always benefitted by associations with plants, as they gain a supply of nutrients and habitat, these interactions can be classified as positive or negative in terms of the effect on host plants (Dhankhar et al., 2021). The negative interactions principally relate to the various diseases caused by phyto-pathogenic microbes. However, the present chapter focuses on the positive interactions pertaining to the role of microbes as biofertilizers. Microbes assist plant growth via various mechanisms like plant disease control, via modulation of plant hormones, by acting as biofertilizers to enhance nutrient accessibility for plants, and by carrying out carbon sequestration and phytoremidiation to provide a nutrient-rich environment.

A variety of evidence suggests that plant microflora produce different phytohormones such as auxins, abscisic acid, gibberellic acid, cytokinins, and ethylene (Wu et al. 2009). For instance, the nitrogen-fixing bacteria *Gluconacetobacter diazotrophicus* and *Bacillus amyloliquefaciens* produced significant amounts of IAA (indole acetic acid, an auxin molecule) that support the development of *Lemna minor* (Spaepen et al., 2007; Idris et al., 2007). Similarly, Arkhipova and co-workers (2005) observed improved growth in lettuce plants grown in association with *Bacillus subtilis* due to the production of cytokinins by the bacteria.

Plant-associated microbes also act as bio-control agents by inhibiting the growth of pathogenic microbes. They produce antimicrobial compounds to resist the phytopathogens. Besides providing protection from pathogens, plant-associated microbes also help provide pollutant-free soil for the growth of plants. Microbes and plants establish symbiotic relationships in order to eliminate harmful chemicals and heavy metals from the soil (phytoremediation). Many studies have proved the remediation of various harmful compounds like polychlorinated biphenyl (PCB) hexachlorocyclohexane and 4-chloronitrobenzene (4CNB) by plant-associated microbes, for example *Rhodococcus, Pseudomonas putida*, *Sphingomonas sp*. and *Comamonas sp* (Leigh et al., 2006; Liu et al., 2007; Narasimhan et al., 2003).

Microbes also show a significant role in carbon sequestration, i.e., deposition of atmospheric carbon dioxide in plant material and incorporation into plant-associated microbes for a balanced atmosphere (Wu et al., 2009). Another major contribution of microbes in plant growth is enhancing nutrient availability to plants. Microbes carry out various processes like nitrogen fixation, phosphorus solubilization, and sulfur oxidation to make insoluble soil nutrients available for plant uptake. The following sections examine this aspect of plant–microbe interaction in further depth.

14.3 BIOFERTILIZERS VS CHEMICAL FERTILIZER

Chemical fertilizers are non-organic materials that are entirely or partially synthetic in nature and are given to soil to support plant development, e.g., ammonium sulphate, ammonium phosphate, urea. They are made artificially. They primarily supply nitrogen, phosphate, and potash as main soil nutrients. Chemical fertilizers are full of nutrients that crops require and are used to provide quick nutrient supply to plants. Several chemical fertilizers have high acid content and adversely affect soil fertility. Applying chemical fertilizers has a negative impact on soil quality and fertility, which reduces the nutritional value and edible qualities of fruits and vegetables (Singh et al., 2021). Reduction in dry matter content in tomatoes due to application of chemical fertilizers was reported by Marzouk and Kassem (2011). Heavy metals accumulate in plant tissues as a result of the long-term use of chemical fertilizers, reducing the nutritional value and quality of the fruit (Shimbo et al., 2001). The negative effects of excessive chemical fertilizer inputs in traditional agriculture operations have been clearly recognized (Banerjee et al., 2011; Garai et al., 2014). Biofertilizers are ready-to-use live formulations of a specific or a group of advantageous microorganisms which mobilize the nutrient availability by their biological action and boost soil productivity. Biofertilizers contain microorganisms such as bacteria (*Azotobacter*, *Rhizobium* etc.), fungi, etc. that fix free nutrients from the atmosphere, which are then used by the crops. Biofertilizers are slow nutrient releasing as they need time to establish themselves for raising the supply of vital nutrients to the plants. Biofertilizers enhance soil texture and plant output while also being environmentally friendly and cost-effective. Biofertilizers reduce environmental pollution as they are natural fertilizers and been proven to be beneficial even under water-scarce conditions. The use of biofertilizers may help prevent or mitigate the negative consequences of increased dependence on chemical fertilizers for higher production.

14.4 TYPES OF BIOFERTILIZERS

Biofertilizers are categorized into different classes according to the microorganisms they carry and the nutrients they make available to plants. A biofertilizer containing selected appropriate living microbial cultures can colonize the rhizosphere or internal organs of the host plant when applied to seed, soil, or as a spray on plants, boosting plant growth by increasing the availability, delivery, or absorption of crucial nutrients to the host. Furthermore, small and marginal farmers can obtain biofertilizers at reasonable prices, unlike conventional fertilizers. The most important types of microbes used in the production of microbial biofertilizers are bacteria, fungi, and cyanobacteria, a large percentage of which have symbiotic relationships with plants. The different categories of biofertilizers are depicted in Figure 14.1.

14.4.1 Nitrogen-Fixing Biofertilizers

Nitrogen is a vital component of plant development and productivity. It is an essential constituent of chlorophyll, the principal pigment involved in photosynthesis, as well as amino acids, the building blocks of proteins. Although nitrogen is one of the most ample substances in nature (notably in the environment as di-nitrogen gas (N2)), this element is only used by plants in reduced forms. Plants receive these types of 'combined nitrogen' through: 1) the incorporation of manures to soil or the use of ammonia and/or nitrate fertilizer (from the Haber–Bosch process); 2) the production of these chemicals during the breakdown of organic materials; 3) natural processes, such as lightning, that convert atmospheric nitrogen into chemicals; and 4) biological nitrogen fixation (Vance, 2001). Most nitrogen fixation is carried out biologically (Figure 14.2). Microorganisms perform the first transformation of N_2 into ammonia, and then into proteins, in a process known as nitrogen fixation.

14.4.1.1 Mechanism of Biological Nitrogen Fixation

The following equation represents biological nitrogen fixation, where one mole of nitrogen gas at the expenditure of 16 moles of ATP and a supply of electrons and protons (hydrogen ions) generate two moles of ammonia:

$$N_2 + 8H^+ + 8e^- + 16\,ATP = 2NH_3 + H_2 + 16ADP + 16\,Pi$$

It is only prokaryotes (bacteria and related organisms) that can accomplish this reaction, which is done by an enzyme compound called nitrogenase. There are two proteins in this enzyme: a Fe protein and a Mo-Fe protein. When nitrogen is attached to the nitrogenase enzyme complex, reactions occur. Electrons contributed by ferredoxin reduce the Fe protein which binds ATP and decreases the molybdenum-iron protein. This molybdenum-iron protein gives electrons to nitrogen (N_2), producing HN=NH. HN=NH is converted into $H_2N\text{-}NH_2$, which is further converted into 2 molecules of ammonia in two advance phases of this procedure (each needing electrons given by ferredoxin). The reduced ferredoxin that gives electrons for this activity is produced by photosynthesis, respiration, or fermentation, depending on the nature of the microorganism.

14.4.1.2 Sub-groups of N-fixers

Nitrogen-fixing biofertilizers have living microorganisms in the form of microbial inoculants which are able to fix atmospheric nitrogen into plant-available form. Nitrogen fixers (N-fixers) are principally grouped into (i) free-living (ii) symbiotic and (iii) associative, and a comparative account of these three is presented in Table 14.1 and depicted in Figure 14.3.

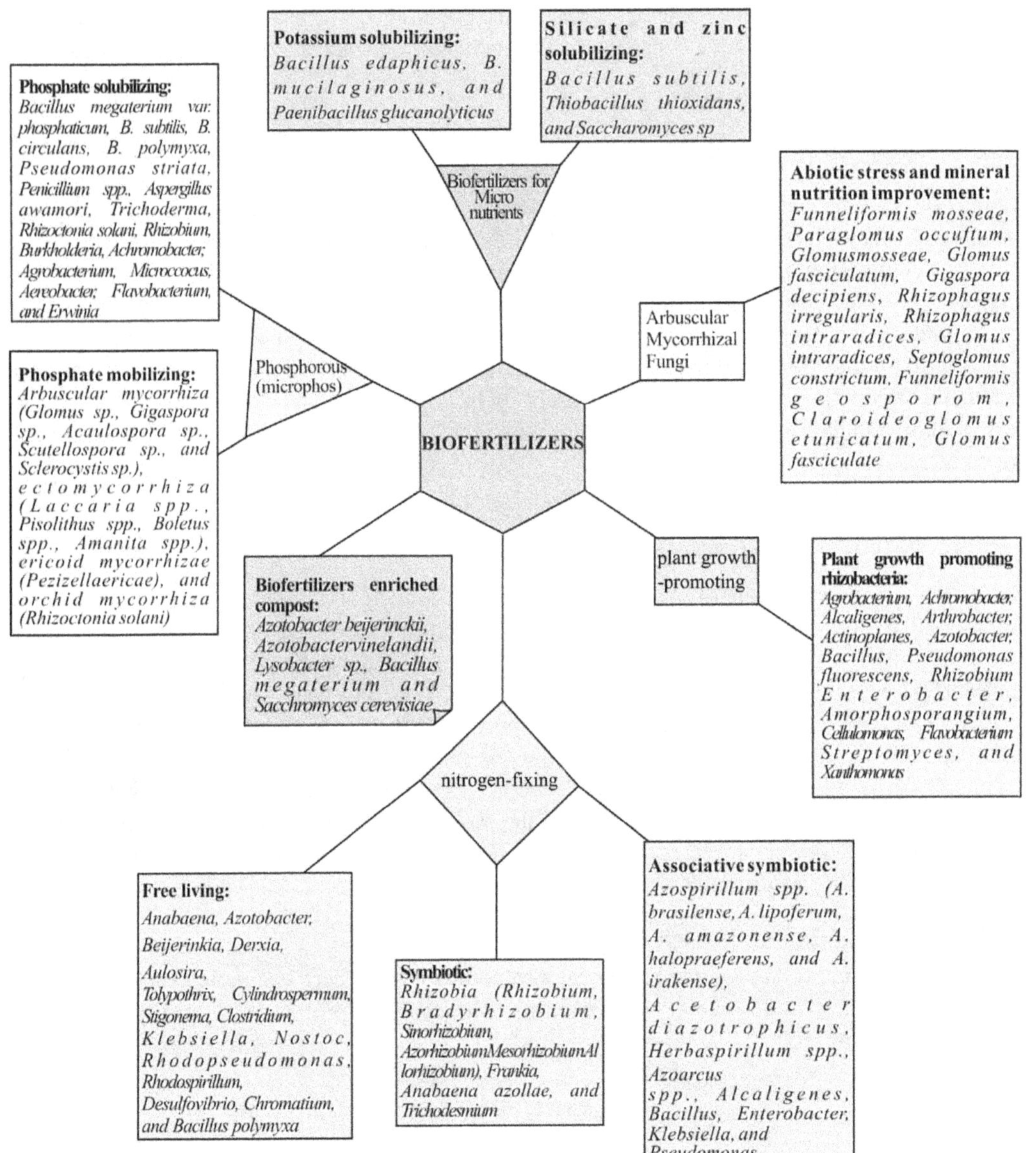

FIGURE 14.1 Different types of biofertilizers and representative microorganisms.

i. Free-living N-fixers

These are heterotrophic bacteria that fix significant quantities of nitrogen without having direct association with other organisms. *Clostridium, Bacillus, Azotobacter* and *Klebsiella* are free-living nitrogen-fixing bacteria. These organisms require their own source of energy for N-fixation, which is usually acquired through the oxidation of organic molecules produced from other organisms or through breakdown. Some free-living organisms can use inorganic molecules as a source of energy because of their chemolithotrophic capabilities. While fixing nitrogen, free-living organisms act as anaerobes because oxygen inhibits nitrogenase activity. Due to lack of sufficient carbon and energy sources, the contribution of free-living bacteria to the global nitrogen fixation rate is widely regarded as minimal.

Fixation of atmospheric nitrogen

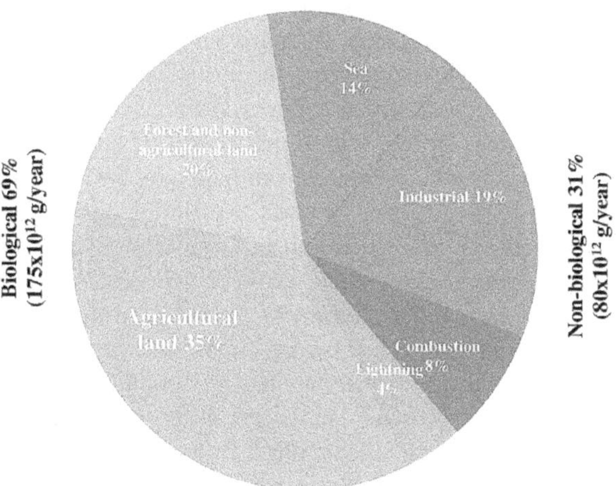

FIGURE 14.2 Annual fixation of nitrogen through various biological and non-biological processes.

TABLE 14.1
Comparative Analysis of Different Types of Nitrogen-Fixing Microorganisms

Characteristic	Symbiotic Bacteria (Root Nodules)	Associative Nitrogen-Fixing Bacteria	Free-Living Nitrogen-Fixing Bacteria
Energy source	High	Moderate	Moderate
Oxygen protection	High	Moderate	Low
Transfer of fixed N	High	Moderate	Low
Estimates of nitrogen fixation rates, kg N/ha/year	50–465	2–170	1–80

Table modified from Wagner (2011)

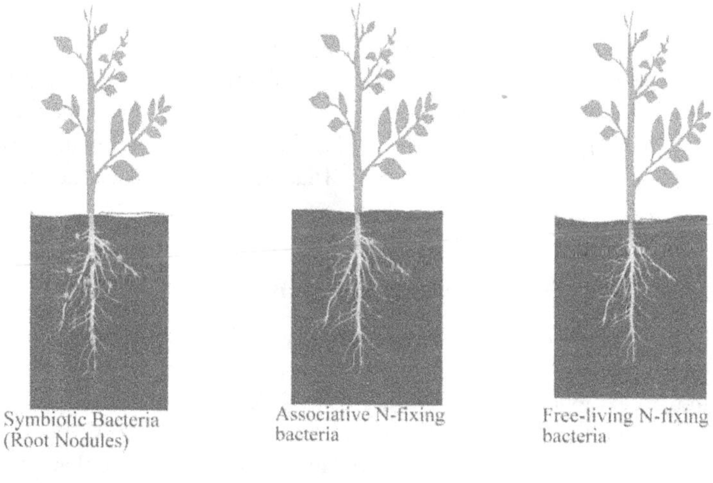

Symbiotic Bacteria (Root Nodules)
Associative N-fixing bacteria
Free-living N-fixing bacteria

FIGURE 14.3 Diagrammatic representation of different modes of nitrogen fixation by microorganisms (symbiotic, associative and free living).

ii. Symbiotic N-fixers
Symbiotic nitrogen fixation by several microorganisms occurs in conjunction with host plants. The photosynthates (sugars) provided by the host plant are used by the nitrogen-fixing bacteria to meet their energy requirements for nitrogen fixation. The microorganism supplies biologically fixed plant-accessible nitrogen to the host plant in return for these carbon sources, allowing it to grow and develop. Symbiotic nitrogen-fixing bacteria invade host plants' roots, where they multiply and foster the growth of root nodules, plant cell expansions, and bacteria closer proximity. Free nitrogen is converted to ammonia by bacteria in the nodules, which the host plant utilizes for growth. Seeds of legumes like alfalfa, beans, clover, peas, and soybean are often subcultured with commercial cultures of a suitable *Rhizobium* species to make sure there are enough nodules and that the plants grow at their best, especially in soils insufficient in the bacterium. *Rhizobium* and *Bradyrhizobium* bacteria form one of the most important symbiotic associations with legumes. Cowpea, soybean, beans, alfalfa, clover, lupines and peanut are the most commonly cultivated legumes in agricultural systems. Soybean is grown on half of the world's legume-growing land, accounting for 68% of total worldwide legume production (Vance, 2001). Other significant nitrogen-fixing symbiotic connections include the one between the water fern *Azolla* and the cyanobacterium *Anabaena azollae*. *Azolla* provides space for cavity formation by *Anabaena* and *Anabaena* fix substantial nitrogen in specialized cells present in cavities called heterocysts. This association has existed for 1000 years in the paddy fields of Asian countries. Another example is the actinorhizal symbiosis between the actinomycete *Frankia* and actinorhizal plants and shrubs like alder (*Alnus* sp.). They are by far the most abundant non-leguminous nitrogen fixers, typically appearing first in successional plant groups. Alpine, chapparal, coastal dune, glacial till, xeric, riparian, forest, and arctic tundra habitats all have actinorhizal plants (Benson & Silvester, 1993).

iii. Associative N-fixing bacteria
These are bacteria that fix di-nitrogen gas to ammonia while in casual association with plants, i.e., loose mutualism; this process is called associative symbiosis. Bacteria live in the zone between soil and roots (the rhizosphere) and may invade the roots. The plant roots are fixed with nitrogen and in return the bacteria get nourishment from the carbohydrates released by the roots. The most prominent examples of associative symbiosis are *Azospirillum brasilense* in association with cereal roots, *Beijerinckia* in association with the roots of sugarcane, *Azotobacter paspali* in association with roots of tropical grass *Paspalum notatum*. *Azospirillum sp.* makes close symbiosis with numerous members of the Poaceae (grasses), such as *Oryza sativa*, *Triticum aestivum*, *Zea mays*, *Avena sativa*, and *Hordeum vulgare*.

These bacteria generate a significant amount of nitrogen within the host plant's roots (Stephan et al., 1981). Soil temperature (*Azospirillum* species thrive more in tropical and temperate situations), the ability of the host plant to deliver a low oxygen pressure rhizosphere environment, the accessibility and availability of photosynthates by the host for the bacteria, the effectiveness of the bacteria, and the effectiveness of nitrogenase are all factors that influence the amount of nitrogen fixation (Vlassak & Reynders, 1979) (Table 14.1).

14.4.2 Phosphate-Solubilizing Biofertilizers (PSB)

Phosphorus is the most vital macronutrient for plant development after nitrogen and plays a major role in cell division and development, energy transfer, movement of genetic characteristics from one generation to the next, photosynthesis, movement of nutrients, and transformation of nutrients (Kalayu, 2019; Tairo and Ndakidemi, 2013; Bhattacharjee and Dey, 2014). Phosphorus is always in ample quantity in soil but it is found in non-available form for plants (Schachtman et al., 1998). Microorganisms are required to convert the insoluble form to a plant-adsorbable/soluble form, HPO_4^{-2} and $H_2PO_4^{-3}$. This process, called solubilization, is explained in Figure 14.4 (Nacoon et al.,

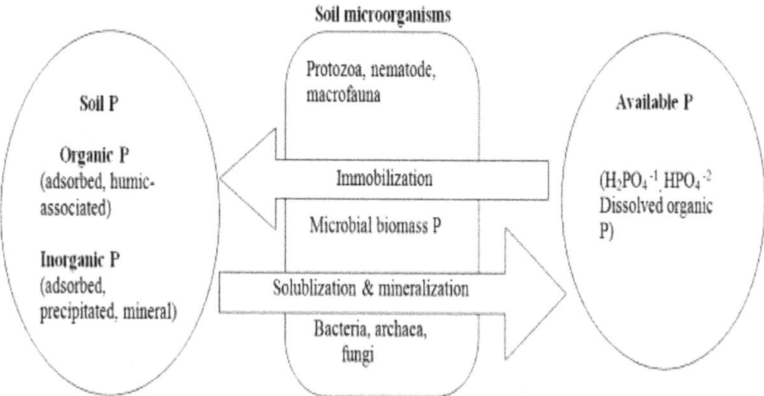

FIGURE 14.4 Schematic representation of the importance of microorganisms to phosphorus availability in soil.

(adapted from Dhankhar et al., 2013)

2020). Solubilization is done by various bacteria like *Bacillus megaterium* var. *phosphoricum*, *B. coagulans*, *B. circulans*, *Torulospora globasa*, *Pseudomonas fluorescens*, *Pseudomonas putida*, *Pseudomonas striata*, *Bacillus polymyxa*, *Bacillus pulvifaciens*, *Bacillus sircalmous*, *Thiobacillus ferrooxidans*, *Pseudomonas canescens*, *Pseudomonas calcis*, *Rhizobium meliloti*, *Mesorhizobium mediterraneum*, *Bacillus fusiformis*; fungi like *Penicillium oxalicum*, *Penicillium simplicissimum*, *Penicillium digitatum*, *Aspergillus sydawi*, *Penicillium lilacinium*, *Aspergillus awamori*; cynobacteria like *Calothrix braunii;* and actinomycetes like *Streptomyces albus*, *Streptomyces cyaneus; Streptoverticillium album*. These solubilizing microorganisms create organic acids that reduce soil pH and increase solubility (Hariprasad & Niranjana, 2009; Sharma et al., 2013). It is also stated that solubilization of P helps in uptake of N and K too.

Chemical fertilizers are toxic for the environment and cause soil pollution so PSBs are the best alternatives (Zak et al., 2018). PSBs also help in production of IAA and GA that are needed for plant development (Jiang et al., 2019; Awasthi et al., 2011; Mohite, 2013). They help in production of antifungal metabolites, hydrogen cyanate (OCN^-), and antibiotics that act against plant pathogens. PSBs have been confirmed to boost many crops' growth and productivity, including wheat (Rodŕiguez and Fraga 1999; Islam et al., 2007; Singh and Kapoor, 1999), maize (Walpola and Yoon, 2012; Bano and Fatima, 2009), soybean (Son et al., 2006), groundnut (Dey et al., 2004), and chickpea (Peix et al., 2001).

14.4.2.1 Mechanisms of Phosphorus Solubilization

PSBs combine a number of approaches to make phosphorus available to plants, including decreasing soil pH, chelation, and mineralization.

i. Lowering Soil pH
   Soil P is primarily dissolved by reducing soil pH through the production of organic acids or the discharge of protons by bacteria (Kalayu, 2019). Phosphate-containing rock phosphate (fluorapatite and francolite) can precipitate in alkaline soils to form calcium phosphates, which are not soluble in soil. Their solubility decreases with increase in soil pH. As seen in the solubilization of calcium phosphates, phosphate-solubilizing microorganisms (PSMs) are found to produce acidity by creating organic acids and CO_2 evolution (Satyaprakash et al., 2017, Walpola and Yoon, 2012). PSMs can produce organic acids like ketogluconic, lactic, 2-glyconic, butyric, glutaric, oxalic, succinic, glyoxalic, fumaric, tartaric, citric, malonic, propionic, gluconic, acetic, and adipic acid, which are microbial metabolism products, primarily

through oxidative respiration or fermentation using glucose as a source of carbon. Gram-negative bacteria are more active at dissolving mineral phosphates than gram-positive bacteria due to the release of many organic acids into the surrounding soil (Kumar et al., 2018).

ii. Chelation
PSMs produce inorganic and organic acids which chelate cations and compete with phosphate for adsorption sites in the soil, dissolving insoluble soil phosphates (Khan et al., 2009). The acids' carboxyl and hydroxyl groups chelate the phosphate-bound cations, turning them to soluble form. These acids may compete for fixation sites of insoluble iron and aluminium oxides, stabilize them by reacting with them, and are referred to as 'chelates'. 2- Ketogluconic acid is a potential calcium chelator.

iii. Mineralization
Soil P is further solubilized through mineralization. Organic phosphate is formed usefully by PSMs through the mineralization process, and it comes from animal and plant waste that contains a lot of organic P compounds like phospholipids, polyphosphates, nucleic acids, phytic acid, phosphonates, and sugar phosphates (Khan et al., 2009). PSMs produce phosphatases like phytase which hydrolyze organic P and release plant-immobilized inorganic phosphorus (Tarafdar et al., 2003, Dodor and Tabatabai, 2003).

Acid phosphatases and alkaline utilize organic phosphate as a substrate to transform them to inorganic form. Some of the frequently designated phytase-producing fungi are *Aspergillus niger*, *Aspergillus parasiticus*, *Trichoderma harzianum*, *Penicillium simplicissimum*, *Aspergillus fumigatus*, *Penicillium rubrum*, *Aspergillus rugulosus*, *Aspergillus terreus*, *Pseudeurotium zonatum*, *Aspergillus candidus*, and *Trichoderma viride* (Tarafdar et al., 2003).

14.4.3 Phosphorus-Mobilizing Biofertilizers: Arbuscular Mycorrhizal Fungi

Arbuscular mycorrhizal fungi (AMF) are a type of root-bound biotrophs that help nearly 80% of plants by symbiosis. Because they exchange photosynthetic products for water, nutrients, and pathogen protection for the host, they are termed natural biofertilizers. Arbuscular mycorrhiza, which forms a mutual association between plants and Glomeromycota fungi, has the widest distribution in nature among plant–fungal symbiotic associations. AM fungus infects the roots of the majority of plants, comprising pteridophytes, gymnosperms, angiosperms, and bryophytes in a variety of ecosystems, including forests, agricultural lands, and grasslands, as well as numerous stressful environments. Figure 14.5 depicts the well-documented beneficial effect of mycorrhizae, which is the enhancement of P-nutrition in plants. The mechanism for this mycorrhizal involvement is usually understood as mycorrhizal fungi exploring soil more than roots.

Aside from hyphae that extend beyond the root depletion zone, several secondary mechanisms have been suggested to describe P-uptake by mycorrhizal fungi: (i) the kinetics of P uptake in hyphae vary from those in root systems because hyphae have a higher affinity (lower K_m) or a lower threshold concentration where the inflow equals efflux (C_{min}); (ii) hyphae and roots explore microsites otherwise, particularly small patches of organic matter; and (iii) chemical changes and P solubility in the rhizosphere are influenced differently by mycorrhizal hyphae and plant roots. AMF forms arbuscules, hyphae, and vesicles in roots, as well as hyphae and spores in the rhizosphere, allowing roots to cover a broader portion of the soil's surface area and enhancing plant growth (Bowles et al., 2016). By enhancing the availability and transport of different nutrients, AMF increase plant nutrition (Rouphael et al., 2015). AMF improve soil quality by altering the structure and texture of the soil, and thus the health of the plant (Zou et al., 2016; Thirkell et al., 2017). Mycorrhizal hyphae promote the breakdown of organic matter present in soil (Paterson et al., 2016) and the availability of micronutrients such as copper and zinc to plants. Additionally, mycorrhizal fungi may influence ambient CO_2 fixation by host plants by boosting the 'sink effect' and photo-assimilate migration

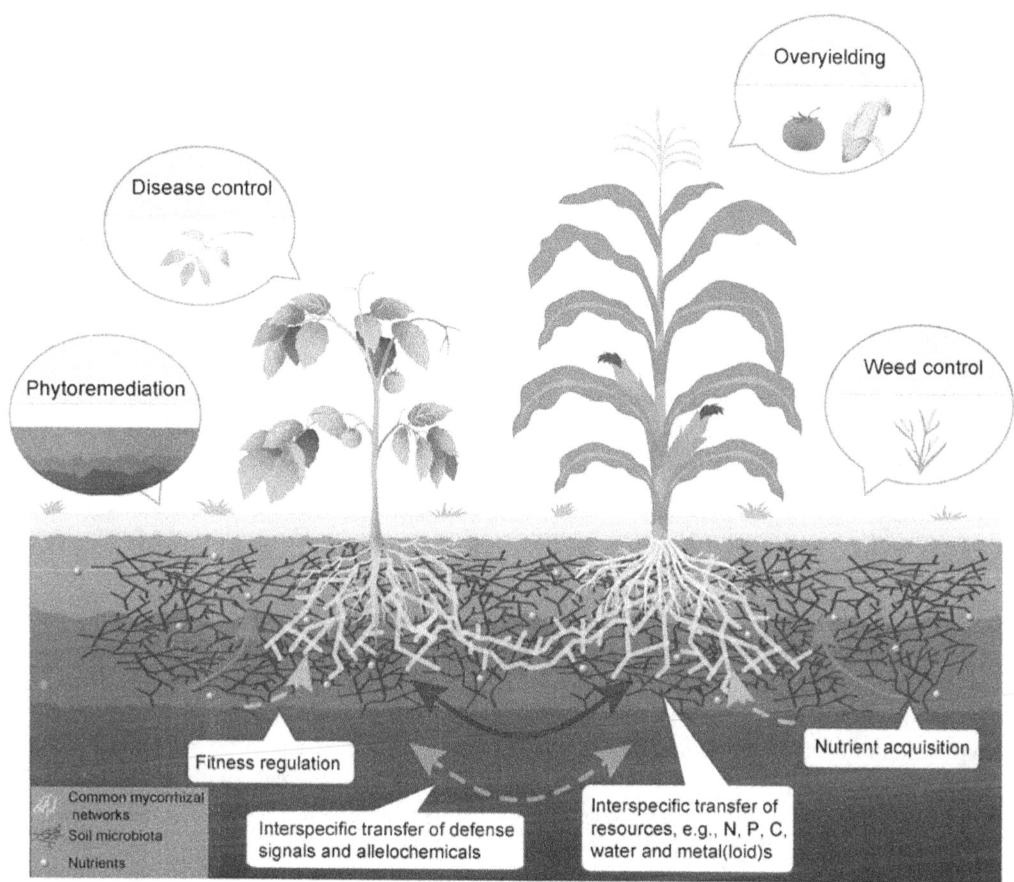

FIGURE 14.5 The potential benefits of arbuscular mycorrhizal fungi (AMF) in intercropping systems. **(adapted from Li et al., 2021)**

from above-ground portions to roots. As a result, AMF are active endosymbionts that contribute to plant productivity and ecological functioning.

14.4.3.1 Arbuscular Mycorrhizal Fungi-Facilitated Networking Beneath the Soil

The extent of fungal hyphae in soil is enormous and the mutualism between host plants and fungal species usually disperses, resulting in the establishment of mycorrhizal networks. These extensive fungal hyphae connect two or more plants of the same or different species in these networks. Thus, the network integrates numerous plant species with various fungal species, promotes their interaction, and provides reciprocal feedback, thereby forming a complex adaptive social network. Different plant species benefit differently from this network, depending on the AMF species involved, and these differences affect plant co-existence significantly. The network is regarded as evolutionarily and ecologically significant due to its constructive effects on the fitness of the member fungi and plants. Communication among the members of the network takes place through biochemical signaling (release of compounds) and resource transfers. For instance, interconnecting legumes with cereal crops through the network in mixed cropping benefits the cereals with nitrogen supply from legumes (capable of fixing atmospheric nitrogen). Nitrogen translocation mediated by the fungal network from legumes to cereals prevents the loss of nitrogen, which is generally applied to a field before cultivation of cereal crops. It is worth mentioning that this network renders the same nitrogen benefit from fewer legume plants than is the case with rotating cereals like corn and legume crops in field, thus providing maximum benefits for less time and space. Also, reduced use of nitrogen fertilizers impacts groundwater quality and soil health, providing farmers with an economic benefit that is also ecologically sound.

AM fungi are of key importance for sustainable crop production. AM fungi shape sustainable agriculture in two ways: quality of soil and productivity. Their beneficial effects on physical conditions of soil and plant performances are vital for sustainably managing agricultural ecosystems. AM fungi are considered natural biofertilizers as they provide host plants with nutrients, water, and pathogen protection. They form a key link between soil mineral nutrients and plants by allowing effective utilization of mineral elements such as phosphorus and nitrogen by plants. Plants require significant amounts of phosphorus and its deficiency leads to reduced plant development. This symbiotic relation is primarily important for plants that grow in phosphorus-deficient environments, as it increases plant growth and phosphorus levels in plants. AM fungi bulk production has been achieved with several species such as *Glomus etunicatum, G. mosseae, Glomus clarum, G. intraradices, Gigaspora ramisporophora Acaulospora laevis*, and *Gigaspora rosea*, but the most common inoculum of endomycorrhizae products is *Glomus intraradices*.

14.4.4 Potassium Solubilizers

After N and P, potassium (K) is the third-most critical element required by plants and the seventh-most prevalent element in the earth's crust. It is required in huge amounts to produce a crop's potential yield. Potassium is associated with the passage of water, minerals, and carbohydrates in plant tissue. In the case of reduced availability of K, growth as well as yield is decreased. It also helps in the regulation of osmotic pressure in plant cells as well as chemical transport.

Potassium in soil is divided into four different pools based on the ease with which plants can absorb it (Zakaria, 2019). The total K content of the top 20 cm of most agricultural soils is between 10 and 20 g per kg, whereas mineral soils have a K concentration of 0.04–3%. On the other hand, the majority of soil K (90–98%) is fused in the crystal lattice structure of minerals and so unavailable to plants directly.

14.4.4.1 Potassium-Solubilizing Microorganisms (KSMs)

Cost-effective and environmentally friendly microbial inoculants capable of dissolving K from minerals and rocks are also available. Muentz (1890) was the first to show that microbial activity is involved in rock potassium solubilization. *Bacillus mucilaginosus*, *B. edaphicus*, *B. circulans*, *Paenibacillus* spp., *Acidithiobacillus ferrooxidans*, *Pseudomonas* and *Burkholderia* (Basak and Biswas 2012; Singh et al. 2010; Sheng et al. 2008) are among the KSMs that have been recognized as providing K in an accessible form from K-bearing minerals. Microorganisms in the rhizosphere are vital in the solubilization of bound forms of soil minerals (Supanjani et al. 2006; Sindhu et al. 2014). Even on the surface of mountain rocks, fungi, bacteria, mycorrhizae, and actinomycetes colonized (Groudev, 1987; Gundala et al. 2013). According to reports, *B. mucilaginosus sub spp. siliceus* is a silicate-solubilizing bacterium that frees K from aluminosilicates and feldspar. Gram-negative bacteria have been found to be suppressed by the silicate bacterium *B. mucilaginosus* strain CS1. It's also a silicate-solubilizing bacterium that can be found in both rhizosphere and non-rhizosphere soils (Lin et al. 2002; Liu 2001).

14.4.4.2 Mechanism of Potassium Solubilization

Insoluble K and structurally inaccessible potassium compounds are mobilized and solubilized through the generation of numerous kinds of organic acids, as well as acidolysis and complexolysis exchange processes (Uroz et al. 2009). Organic and inorganic acids transform insoluble K (biotite, feldspar, muscovite, and mica) to soluble K (soil solution form). Depending on the organisms found in the microbial suspension, KSMs release different forms of organic acids (Verma et al. 2014;

Zhang and Kong 2014; Maurya et al. 2014). KSMs can weather phlogopite through acidic disintegration and aluminium chelation of the crystal network (Leyval and Berthelin 1989; Abou-elSeoud and Abdel-Megeed 2012; Meena et al. 2014). Microorganisms produce several types of organic acids to solubilize the insoluble K and convert it to a form that is easily absorbed by the plant. Plant growth is directly linked to K-solubilization and the release of organic acids by K-solubilizing strains. Under in-vitro field and greenhouse conditions, naturally abundant KSMs commonly solubilize structural K compounds (Meena et al. 2013; Maurya et al. 2014; Prajapati et al. 2013; Parmar and Sindhu 2013). KSMs are very efficient at producing K from structural K via solubilization and from interchangeable pools of total soil K via acidolysis, chelation, and solubilization by native rhizospheric bacteria (Uroz et al. 2009). Plants may use a considerable quantity of fixed K found in the biomass of rhizospheric microorganisms in the soil (Subhashini and Kumar 2014). KSMs solubilize K by reducing the pH, accelerating the chelation of the K-bound cations and acidolysis of the adjacent region of the microorganism.

14.4.5 Biofertilizers for Micronutrients

In developing countries, micronutrient deficiencies, particularly Fe and Zn, are a major concern, causing substantial health-related problems in women, teenagers, and neonates. The reduced bioavailability of Fe and Zn in tropical soils and thus in plants causes Zn and Fe deficiency in humans. In numerous fruit crops, vesicular-arbuscular mycorrhizal (VAM) fungi have demonstrated potential as mobilizers of immobile micronutrients, primarily Zn, but also Fe and Cu to a limited extent.

Researchers have discovered that by increasing solubility of micronutrients, *Oidiodendron maius* isolates and ericoid mycorrhizal fungi produce organic acids such as citrate, fumarate, and malate, which solubilize insoluble inorganic Zn compounds and thus increase their bioavailability to plants; however, bulk production of this organism as a biofertilizer has not been recognized. Some bacteria can synthesize acids and survive acidic environments, allowing micronutrient cations to solubilize and boost their bioavailability; examples include *Thiobacilus acidophilus, Thiobacilus ferrooxidans*, and *Thiobacillus thiooxidans*. Some species overcome iron's low solubility and bioavailability by producing and excreting iron-chelating molecules known as 'siderophores'. These molecules associate to Fe to produce a siderophore-iron complex that is absorbed by the plant cell, and Fe is released inside. Some commercially used siderophore producers are: *Pseudomonas putida, Pseudomonas aeruginosa* and *Pseudomonas fluorescens*.

14.4.6 Silicate-Solubilizing Bacteria (SSB)

Silicon is regarded as a stimulating element or quasi-element because its deficiency can cause abnormalities in plant growth (Lee et al., 2019). It is not important for all plants but sometimes it is a critical element for crops like rice (Ma et al., 2007). Silicon is also known to protect plants against pathogens, nematodes, and insects (Hawerroth et al., 2019; Lukacova et al., 2019). There is an abundant quantity of silicon in the earth's crust but in non-available form (silicate) (Bist et al., 2020; Lee et al., 2019) for plants like phosphorus. So, solubilization of sources like biotite is needed, and microorganisms like *Burkholderia, Dyella, Collimonas, Proteobacteria, Aminobacter, Janthinobacterium*, and *Frateuria* can make it available in adsorbable form (monosilicic acid) (Raturi et al., 2021). Silicon-based fertilizers also help to mitigate the effects of environmental stress, and improve nutrient balance and water-retaining capacity of soil (Meena et al., 2014).

14.4.7 Plant Growth-Promoting Biofertilizers

The ability of a plant's rhizosphere-derived microbial inoculants to stimulate growth of plants has received much attention in the last two decades. Plant growth-promoting rhizobacteria (PGPR) is the collective name for these bacteria. PGPR refers to a group of soil bacteria that stimulate the plant's growth when grown in association with a host plant. Fixing nitrogen, improving nutrient availability, having positive impact on root growth and shape, and encouraging additional valuable plant–microbe associations are all examples of PGPR working modes. PGPR is a special type of bacteria that colonize in the rhizosphere (a thin layer of soil around the plant root). The PGPR inoculants that are currently on the market seem to help plants grow through at least one of the following ways:

1. Restricting plant diseases (termed bio-protectants)
2. Ameliorating availability of nutrients (termed biofertilizers)
3. Phytohormone production (termed bio-stimulants). Bacillus species, *Pseudomonas fluorescens*, and pink-pigmented facultative methylotrophs (PPFM) can create as yet unidentified phytohormones or growth regulators that drive crops to have more fine roots, increasing the absorptive surface of the plant's roots for nutrient and water uptake. Cytokinins, gibberellins, Indole-acetic acid, and inhibitors of ethylene production are among the phytohormones produced by these PGPR.

14.5 PREPARATION OF BIOFERTILIZERS

Preparation of biofertilizers requires a fermentation process involving live soil microorganisms. For production of biofertilizers, solid-state fermentation and submerged fermentation are the preferred methods. In order to produce each type of biofertilizer, an efficient microbial strain is selected and cultivated in appropriate nutrient conditions, and its formulation is carried out using either a solid or liquid base. The steps involved in production of microbial inoculants is illustrated in Figure 14.6 (broth or liquid medium used is different for respective organisms).

14.5.1 Production of Starter Cultures (Mother Cultures)

Pure mother cultures of different strains are kept in labs of agricultural universities and institutions, the National Centre for Organic Farming, and regional biofertilizer labs. After determining the strains' performance in the greenhouse and in the field for a variety of nutrients and locations, the mother culture of selected strains is obtained. Different microorganisms used in biofertilizer production are

- *Rhizobium*
- *Azotobacter*
- *Azospirillum*
- Potash-mobilizing bacteria (KMB)
- Phosphate-solubilizing bacteria (PSB)
- *Trichoderma* for compost production.

The pure culture of an efficient strain of microbe is cultivated on an appropriate agar-medium slant and preserved in the laboratory for future application. A standard four mm loop of inoculum is transferred into a 250 ml conical flask containing liquid medium and placed on a rotating mixer for 3–7 days, according to the organism's growth rate. The contents of the flask is further multiplied till it attains 10^5–10^6 cells per ml in larger flasks.

14.5.2 Production of Broth Cultures

A liquid growth medium is prepared for appropriate microorganisms and distributed evenly among large conical flasks with a volume of 1000 ml. The medium is autoclaved in the flask for 30 minutes at 15 lbs of pressure to sterilize it. After sterilization, the broth is inoculated under aseptic conditions

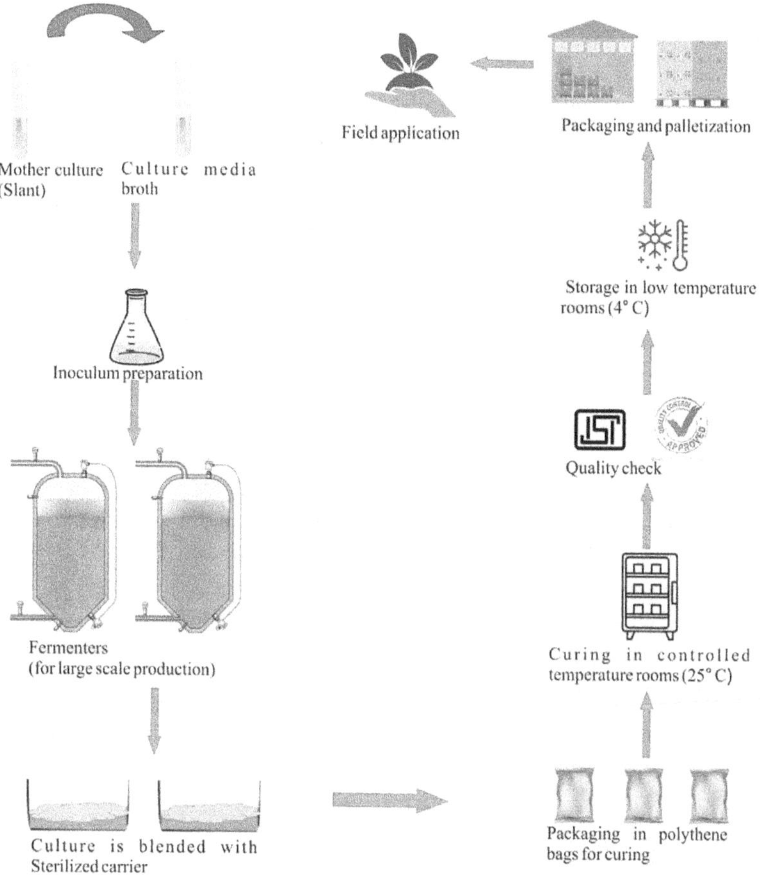

FIGURE 14.6 Flow diagram depicting various stages involved in production and commercialization of biofertilizers.

with the mother culture in a 1:5 ratio. For 96–120 hours, the flasks are kept on rotary shakers, with the broth becoming thicker in consistency, until the viable cell count per millilitre reaches 10^9–10^{10}. . This broth culture should not be kept for more than 24 hours or at a temperature lower than 4°C.

14.5.3 Production of Final Product in Fermenter

Fermenters are used to produce microbial products such as biofertilizers and biopesticides on a large scale. A fermenter is a metallic vessel that is used to moist sterilize any item. The theory of moist sterilization is based on the fact that when water is boiled in a closed system, the resulting water vapor accumulates within the vessel, increasing the internal pressure. Therefore, the boiling point of water increases above 100°C. The steam that is released from the boiling water is at a higher temperature in this circumstance. If any media are placed in this jar at this temperature, the high temperature kills any microorganisms present.

14.5.4 Preparation of Carrier Material

The carrier material for solid-state formulation is lignite or bentonite or peat of desired quality in fine powdered form. The pH of the carrier material is neutralized by adding calcium carbonate ($CaCO_3$) in 1:10 ratio. To eliminate impurities, the neutralized carrier material is sterilized in an autoclave. The solid-state formulation is temperature sensitive, and at elevated temperatures, the bacteria count falls below the threshold. Compared with solid formulation, the liquid technology has far more stability and can retain the product's shelf-life for up to 12 months. All that is needed to make bio fertilizer is water and chemicals – nothing else. The formulation is created in fermenters, then packaged and sold directly in bottles. Growth materials include Mannitol, sucrose and chemical nutrients.

The chemicals required for the production of biofertilizer are: carbon source – glucose, malic acid, sucrose; nitrogen source – peptone, ammonium sulphate yield extract; micro-nutrients – Mg sulfate, Co nitrate, Zn sulfate; stabilizers/surfactant – Polyethylene glycol.

14.5.5 Filling and Packaging

To create a solid-state bio-fertilizer, neutralized and sterilized carrier material is spread in a hygienic and dehydrated container. The bacterial culture is taken from the fermenter, mixed with carrier material manually or by a mechanical mixer with 40–50% water-holding capacity, packed in sealed bags of desirable quantity and allowed to cure at room temperature for 2–3 days. For manufacturing of liquid biofertilizer; fermenters send broth directly to an automatic filling machine, where it is packed in bottles with 2/3 space left free for aeration of the bacteria. The bottles used for pouring microbial inoculants are packed and should be labelled with the following information: a) inoculant name, b) crop name, c) instructions for use, d) manufacture date, and e) expiration date.

14.5.6 Quality Checking and Storage

It is important to ensure that the inoculants have a high microbial count at the time of manufacture in accordance with ISI criteria. It is possible to store the inoculants for up to six months in a cool place away from direct heat, at an optimum temperature of 15°C, not exceeding 30°C +/-2°C.

14.5.7 Bulk Production of Mycorrhizal Biofertilizers

Mycorrhizal fungi have been used on a small scale due to their obligate symbiotic nature, which makes laboratory culture challenging. Arbuscular mycorrhiza inoculum production has moved on from the use of field soils containing fungi to pot culture inoculum from spores of single arbuscular mycorrhiza fungus on a host plant in sterile medium. New techniques for producing arbuscular mycorrhiza fungal inoculum have been investigated, including soil-based culture and carrier-based inoculum. Pot culture is frequently used for producing fungal inoculum as a carrier-based inoculum. This inoculum is made up of roots, spores, and hyphae that have an infestation trapped in the carrier. Because the sterilizing method is time-consuming, perlite and vermiculite are useful alternatives to inoculum formation. The method of inoculum production is as follows:

- To serve as a plant-growing tub, dig a trench 1m in length, 1m width, and 0.3m depth and line it properly with black polythene sheet.
- Mix 50 kg vermiculite and 5 kg sterilized soil and fill the trench to a height of 20 cm.
- Spread 1 kg arbuscular mycorrhiza inoculum 2–5 cm below surface of the vermiculite.
- Sow sterilized maize seed (sterilized with 5% sodium hypochlorite/suitable chemical for 2 minutes).
- At the time of seeding, apply appropriate urea, super-phosphate, and muriate of potash to each trench. For each trench, urea is applied twice, 30 and 45 days after sowing.
- On the 30th and 45th days, perform quality check test on arbuscular mycorrhiza colonization in root.

- Stock plants are produced for a period of 60 days (8 weeks). Cut all the roots of stock plants to yield inoculum. The inoculum is made up of vermiculite, spores, hyphae fragments, and infected root fragments.
- From a 1 m^2 area, 55 kg of AM inoculum may be produced in 60 days.

14.6 BIOFERTILIZER APPLICATION METHODS TO AGRICULTURAL CROPS

Understanding the microbial strain's host specificity and the features of the soil and atmospheric conditions in the field are crucial to determine the effectiveness of biofertilizer application. Biofertilizers can be applied for sustainable agriculture in various ways and their effect on various crops via different application methods is shown in Table 14.3. Many researchers used biofertilizers for seed treatment of different agricultural crops, which is the most common method of biofertilizer application (Asif et al., 2018); in some cases, biofertilizers are applied directly to soil and then seeds are sown in it. Sometimes seedlings are raised in nursery conditions and then dipped in biofertilizer inoculum (Stamford et al., 1995). Each treatment has advantages and disadvantages, so it is necessary to know the proper method of biofertilizer application and appropriate storage conditions (Muraleedharan et al., 2010). Biofertilizers and seeds treated with biofertilizers should not be exposed to direct sunlight because UV rays of solar radiation reduce inoculated bacteria population. An appropriate storage temperature (0–35°C) should be ensured.

14.6.1 Seed Treatment

Seeds of agricultural crops can be treated with biofertilizers in many ways, such as dusting, slurry, and seed coating. Biofertilizers can be applied to seeds in powder as well as liquid form. This method of biofertilizer application is most common among farmers, being less costly because the biofertilizer requirement is minimal compared to other methods (Asif et al., 2018). Effectiveness of biofertilizers is least in the dusting method of seed treatment because of the low adherence of biofertilizers towards seeds. 10% jaggary solution of 5% sugar solution is added to the seed-treating solution to increase effectiveness of biofertilizers (Uddin et al., 2014). Generally, 200 ml jaggary solution is required for 200 g (1 packet) of the inoculant to be mixed. 750 g of biofertilizers is recommended for legume seeds for sowing in an area of one hectare. In the slurry method of seed treatment, wetted seeds are wrapped up in bioinoculant and left overnight to ensure proper distribution of the microorganism over seeds (Muraleedharan et al., 2010). To increase effectiveness of biofertilizers, adhesives like gum arabic, carboxy methyl cellulose, sucrose solutions, etc. can be used and in case of unavailability of adhesives, addition of 1% milk powder to the solution is recommended. According to Brahmaprakash et al., (2017), PSM inoculant is added after seed treatment with *Rhizobium, Azotobacter, Azospirillum* etc. to ensure a higher number of viable microbial cells. Nitrogen biofertilizers, viz. *Azotobacter, Azospirillum, Azolla, Acetobacter* blue-green algae, *Frankia* and *Rhizobium*, are used for seed treatment of legumes, non-legumes, sugarcane, paddy, *Casuarina* and *Alnus*, respectively (Chesti et al., 2013). Before seed treatment, seeds should be sterilized with sodium hypochloride and then dried in the shade sufficiently for seed treatment (Kumar et al., 2017). After that, seeds are placed on a petri dish and the appropriate dose of biofertilizer is applied. Seeds treated with biofertilizers should be sown within 24 hours of treatment. Seed treatment with liquid biofertilizers is done in a plastic bag (size: 21" × 10"); the bag is filled with around 2 kg of seeds and then squeezed in solution for 2–3 minutes for uniform distribution of the biofertilizer. Seed pelleting is also done for seed treatment with liquid *Rhizobium* (Singh et al., 2018). Table 14.2 shows the doses of biofertilizer for various crops.

TABLE 14.2
Recommended Liquid Biofertilizers for Seed Treatment and Their Doses for Beneficiary Crops

Recommended Biofertilizer	Quantity	Crop
Rhizobium	500 ml/hectare	Chickpea, pigeon pea, berseem, pea, black gram, soybean, cowpea, groundnut, beans, lucerne, green gram and lentil
Azotobacter/Azospirillum	500 ml/hectare	Cereals
Azospirillum	500 ml/hectare	Oat, barley, rice, wheat
Azotobacter	500 ml/hectare	Oil seeds
Azotobacter	500 ml/hectare	Millets and oil seed crops like sunflower, linseeds, mustard, castor
Azotobacter	500 ml/hectare	Grasses and forage crops

TABLE 14.3
Effect of Biofertilizers on Various Crops Via Different Application Methods

Sr. No.	Biofertilizer	Type of Application	Crop	Results	References
1	*Bacillus subtilis*	Seed treatment	Cotton	Increased bolls/plant and yield	Yao et al. (2006)
2	*Bacillus subtilis*	Seed treatment	Maize and sunflower	Increased yield and reduction in diseases	Schmiedeknecht et al. (2001)
3	*Bacillus subtilis*	Seed treatment	Tomato	Reduced disease severity and yield losses caused by soil-borne pathogens	Grosch et al. (1999)
4	*Bacillus subtilis*	Added with irrigation water	Eggplant and pepper	Better growth and yield	Bochow et al. (2001)
5	*Bacillus megaterium var. phosphaticum* and *Azotobacter chroococcum*	Seed treatments and filter paper treatments	Maize	Increased germination and weight of shoots and roots	Bakonyi et al. (2013)
6	*Azotobacter chroococcum, Bacillus megaterium* and *Bacillus mucilaginous*	Soil application	Maize	Improved soil properties and plant growth	Wu et al. (2005)
7.	Rhizobacteria	Seed treatment	Maize	Improved seed quality, growth and yield parameters	Nezarat and Gholami, (2009)
8.	*Pseudomonas, Azotobacter, Azospirillum*	Seed treatment	*Helianthus annus*	Improved seed quality, growth, yield and biochemical parameters	Kamran et al. (2010)
9.	*Azospirillum*, phosphate-solubilizing bacteria and AMF	Soil application	Lemon	Better seedling growth in nursery	Singh (2016)
10.	*Azospirillum*, phosphate solubilizing bacteria and AMF	Seed treatment	Cucumber, beet, pea and carrot	Increased germination	Siqueira et al. (1993)
11.	*Azotobacter vinelandii*	Seed treatment	Cucumber	Increased germination	Kurdish et al. (2008)

Sr. No.	Biofertilizer	Type of Application	Crop	Results	References
12.	*Azospirillum*	Seed treatment	Maize	Better germination and seedling parameters, viz. root length and shoot length, and the fresh and dry weight and vigour index of the seedlings	Pathak and Chakraborti (2014)
13.	*Azospirillum brasilense*	Seed treatment	Wheat	Increased vegetative growth and yield	Zonita et al. (2008)
14.	*Rhizobium*, VAM and PSB	Seed treatment	Green gram	Increased crop growth and seed yield parameters	Kamaraj and Padmavathi (2018)
15	*Azotobacter chroococcum*	Soil application	Wheat	Increased plant height, biomass and yield	Yadav et al. (2000)
16	Azotobacter	Soil application	*Solarium tuberosum*	Increased yield and tuber size	Singh (2001)
17	*Rhizoctonia solani, Macrophomina phaseolina, Trichoderma viride* and *Pseudomonas fluroescens*	Soil application	Cotton	Increased seed quality parameters	Shanmugaiah et al. (2009)
18	*Trichoderma harzianum*	Seed treatment	Tomato	Reduced fusarium wilt disease	Singh et al. (2015)
19	*Trichoderma harzianum*	Seed treatment and soil application	Tomato	Increased nutrient uptake	Azarmi et al. (2011)
20	*Azotobacter chroococum*	Seed treatment and soil application	Wheat	Increased germination and seed vigour indices and reduced spot blotch disease	Biswas et al. (2015)
21	*Azotobacter* + *Azospirillum*	Soil application	Wheat	Increased growth, yield, nitrogen use efficiency, nitrogen recovery and nitrogen efficiency ratio	Kachroo and Razdan (2006)
22	AMF	Soil application	Sweet potato	Increased yield	Mukhongo et al. (2017)
23	*Azospirillum, Pseudomonas*, and *Agrobacterium*	Seed treatment	Wheat	Increased root and shoot length	Mubeen et al. (2006)
24	Biofertilizer	Seed treatment	Rice	Increased biomass and harvest index	Saryoko et al. (2021)
25	Rhizobacteria	Seed treatment	Chickpea	Decreased foot and root rot disease	Khalequzzaman (2015)
26	*Azospirillum* + Phosphobacteria + VAM	Seed treatment	Rice	Increased growth and yield parameters	Anand and Kamaraj (2017)

14.6.2 Seedling Root Dip

This method is used for crops which are transplanted, such as vegetables and rice. Seedlings are raised in nurseries and before transplanting they are dipped in inoculant (Kumar et al., 2012). The appropriate period of time for dipping varies according to the crop and the nature of the biofertilizer; most commonly roots are dipped for 30 minutes in biofertilizer solution. Seedlings of *Cynodon dactylon*, sudan grass, *Pennisetum purpureum*, *Brachiaria mutica*, *Heteranthera zosterifolia* etc. are treated with liquid *Azotobacter* @500 ml/acre (Brahmaprakash et al., 2017). *Azospirillum* is used for seedling root dip, particularly for rice and vegetable crops. This method is also appropriate for crops like *Solanum lycopersicum*, *Allium cepa*, cole crops, and flowers.

14.6.3 Soil Application

Soil application of biofertilizers is done when a larger microbial population is needed (Anand et al., 2016). It is carried out with the help of biofertilizer carriers like peat, perlite, talcum powder, or soil aggregates (Jeet and Baldi, 2020). The culture is applied to soil with the addition of farmyard manure (FYM) and compost. Generally, 4–10 times the amount of biofertilizer is required in soil application as in seed treatment (Bhattacharjee and Dey, 2014). Due to the application of a large quantity of biofertilizer, each sown seed receives a greater amount quantity than during seed treatment, so the effectiveness of biofertilizers is also increased. It is more costly than the seed and seedling root dip method of application because of the higher quantity required and the consequently greater storage area (Mahanty et al., 2017). Biofertilizer application also requires special machinery. Sometimes liquid biofertilizers are sprayed directly into seed furrows and in the case of horticultural crops, 20g biofertilizer is added to the ring of a single plant.

14.6.4 Precautions to be Followed Before Biofertilizer Application

- Crop-specific biofertilizers should be used for each crop.
- Store biofertilizers in a cool, dry place out of the way of direct sunshine or any other heat source.
- Out-of-date biofertilizer should not be used for treatment.
- Other chemicals should not be added to biofertilizers.
- Biofertilizers should be mixed with FYM or compost before application to soil.
- If soil is acidic in nature, lime @ 250 kg/ha should be mixed with biofertilizer during soil application.
- Proper irrigation should be applied after biofertilizer application to ensure survival of microbial population.
- When purchasing biofertilizer, confirm that it has been prepared in accordance with the standards set by the Bureau of Indian Standards and has sufficient microbial population (10 million per gram) (see Table 14.3).

14.7 QUALITY CONTROL OF BIOFERTILIZERS

Due to the numerous harmful and adverse effects of chemical fertilizers, interest in biofertilizers for agriculture practice is growing. However, the quality of biofertilizers is a major bottleneck in their commercial success. Compared to chemical fertilizers, the composition and manufacture conditions of biofertilizers are complex, and the involvement of a live agent makes the process even more intricate. Thus, quality control and assurance are required at every step in this multi-step process of formulation development. An ideal biofertilizer should have the appropriate microbial density for the best on-field results and should be simple to use, ecofriendly and low-cost (Herrmann and Lesueur, 2013). The major parameters that can determine the quality of a biofertilizer are density and viability of microbes, the amount of contamination, pH, moisture, physical form of the formulation, type of carrier used, and additives and their amount (Malusa and Vassilev, 2014).

A common international standard for assessment of biofertilizers is currently lacking. Although some countries have their own rules and legislation, these are not properly enforced and the manufactured biofertilizers fail to give good results in the field. This leads to poor acceptance of biofertilizers by farmers. Moreover, farmers have inadequate knowledge of microorganism storage and quality checks, and have lost faith in the product after experiencing uneven outcomes and low output compared to chemical fertilizers (Husen et al., 2016). In the absence of legislation, manufactures often fail to provide the necessary details on the label (for example instructions for suitable use and storage, expiry date) and manufacturing workers are not sufficiently skilled at handling microbes to develop successful inoculation. This represents a setback both for farmers and reliable manufacturers (Malusa and Vassilev, 2014). Thus, a legal framework is required to ensure more widespread adoption of biofertilizers among farmers.

As far as Indian situation is concerned, India is one of the few countries that have defined legislation relating to biofertilizers. In 2006, the Ministry of Agriculture and Farmers' Welfare declared biofertilizers essential commodities under the Essential Commodities Act of 1955. This act defines biofertilizers as "the product containing carrier based (solid or liquid) living microorganisms which are agriculturally useful in terms of nitrogen fixation, phosphorus solubilization or nutrient mobilization, to increase the productivity of the soil and/or crop" (Ministry of Agriculture, 2009). Seven quality parameters are included (Table 14.4):

TABLE 14.4
Standards for Biofertilizers as specified in Biofertilizers and Organic Fertilizers in Fertilizer (Control) Order, 1985, India

1.	Base Material	Carrier* in form of Moist/Dry Powder or Granules, or Liquid
2.	Viable cell count	Minimum CFU 5×10^7 cell/g of powder, granules or carrier material or 1×10^8 cell/ml of liquid.
3.	Contamination level	At 10^5 dilution no contamination
4.	pH	6.5–7.5 for powder or granules carrier, 5.0–7.5 for liquid-based carrier
5.	Particle size in case of carrier-based material	Can pass through 0.15–0.212 mm IS sieve
6.	Moisture % by weight, maximum limit	30–40%
7.	Efficiency character:	
	Rhizobium	Effective nodulation on all the species listed on the packet.
	Azotobacter	Strain capable of fixing at least 10 mg of nitrogen per g of sucrose consumed.
	Azospirillum	White pellicles formation in semisolid N-free bromothymol blue media.
	Phosphate-solubilizing bacteria	Phosphate solubilizing capacity in the range of minimum 30%, when tested by spectrophotometer. For zone formation, minimum 5 mm solubilization zone in prescribed media with at least 3mm thickness.
	Potassium-mobilizing biofertilizers (KMBs)	10 mm solubilization zone in prescribed media having at least 3 mm thickness.
	Zinc-solubilizing biofertilizers (ZSBs)	10 mm solubilization zone in prescribed media having at least 3 mm thickness.

* Different types of Carrier: peat, lignite, sterile soil, humus, wood charcoal or similar material favouring growth of the organism.

> the minimum count of viable cells, the physical form, pH, the contamination level, the particle size in case of carrier-based materials, the maximum moisture percent by weight of carrier-based products, and the efficiency traits of five groups of microorganisms; Azotobacter, Azospirillum, mycorrhizal fungi phosphate-solubilizing bacteria and *Rhizobium*.

However, there is a need to widen the scope to other categories of agriculturally important microbes.

There is also a research gap in this field. In a literature survey on rhizobial research, it was found that the majority of studies were focused on bacterial genetics and physiology, with less than 1% of the research focusing on the formulation element (Xavier et al., 2004). The efficacy of biofertilizers is also evaluated in controlled conditions rather than in fields. Quality control of biofertilizers therefore needs the active participation of researchers, manufacturers, and farmers. More research is required in this field, and skills training in the proper production and application of biofertilizers should be provided for both manufacturing workers and farmers.

14.8 FUTURE PROSPECTS AND CONCLUSION

For decades, crop production was dependent on conventional cultivation practices involving the injudicious use of chemically synthesized fertilizers that amplified crop productivity to an extent but also polluted the environment and had hazardous effects on human and beneficial microbes. There is thus a need to integrate the beneficial traits of biofertilizers into soil management practices in order to maintain long-lasting soil quality, health, and sustainability. However, although biofertilizers are now attracting more attention, there remain some initial stages of adoption. The majority of currently available biofertilizers aim to improve soil macronutrients (nitrogen, potassium, and phosphorus) or particular nutrient (zinc) status. Research efforts are needed to develop multifunctional biofertilizers with improved nutrient density, which can safeguard crop productivity and soil fertility holistically. Research should be conducted to confirm the capability of biofertilizers for use alongside other growth-promoting inputs without any detrimental effects, which will help to increase their adoption. The application of molecular biotechnological tools that identify the biochemical pathways for biofertilizer preparation can help with constructing mandatory hormones and overcome environmental stress conditions. Research on ecofriendly carrier materials, shelf-life, and temperature sensitivity can assist in encapsulating active cultures to ensure targeted and sustained action that will increase the usefulness and commercial adoption of biofertilizers. To evaluate biofertilizers in a number of agricultural production systems, long-term inter-disciplinary studies involving collaboration by soil microbiologists, agronomists, plant breeders, plant pathologists, and economists, must be conducted. The enactment of a Biofertilizer Act that could implement stringent quality control regulation for the manufacture and field application of inoculants is needed to safeguard and further explore the benefits of plant–microorganism symbiosis. Scientific investigation of the performance of various efficient strains, their effect on other beneficial microbes, and the microbial persistence of biofertilizers in soil, especially under adverse conditions, is needed.

Heavy dependency on chemical fertilizers has stimulated the industrial production of lethal chemicals which has resulted in ecological imbalances. Excessive dependence on chemical fertilizers for crop production is not a viable approach for future agriculture.

The changing agricultural ecosystem, which attaches greater importance to the holistic management of crops as well as to soil health, has generated vast prospects for bio-based fertilizers. The integration of biofertilizers into agricultural production techniques to fulfil the increased food demand due to a rising global population, has gained more traction. The application of biofertilizers is ecofriendly, inexpensive, non-toxic and has significant potential to provide nutrition in the plant rhizosphere. It is concluded that the noticeable effect of biofertilizers on bio-control and bio-remediation also encourages crop productivity, soil fertility, and ecosystem sustainability. Biofertilizer use in India will have a positive impact on the economic development of agriculture and will contribute to a sustainable ecosystem.

ACKNOWLEDGMENTS

The authors are grateful to K.R. Mangalam University, Gurugram, India and Chaudhary Charan Singh Haryana Agricultural University, Hisar for their institutional support. The authors would also like to acknowledge the timely and much needed support of Rakhi Dhankhar during the preparation of this manuscript.

REFERENCES

Abou-elSeoud, I. I., & Abdel-Megeed, A. (2012). Impact of rock materials and biofertilizations on P and K availability for maize (*Zea Maize*) under calcareous soil conditions. *Saudi Journal of Biological Sciences, 19*(1), 55–63. DOI: https://doi.org/10.1016/j.sjbs.2011.09.001

Adesemoye, A. O., Torbert, H. A., & Kloepper, J. W. (2008). Enhanced plant nutrient use efficiency with PGPR and AMF in an integrated nutrient management system. *Canadian Journal of Microbiology, 54*(10), 876–886. DOI: https://doi.org/10.1139/W08-081

Anand, K., Kumari, B., & Mallick, M. A. (2016). Phosphate solubilizing microbes: An effective and alternative approach as biofertilizers. *Journal of Pharmacy and Pharmaceutical Sciences, 8*(2), 37. Retrieved from https://innovareacademics.in/journals/index.php/ijpps/article/view/9747

Anand, S., & Kamaraj, A. (2017). Influence of pre sowing biofertilizer seed treatment on growth and yield parameters in rice (*Oryza sativa* L.). *Plant Archives, 17*(2), 1377–1380. Retrieved from http://plantarchives.org/17-2/1377-1380__3846_.pdf

Arkhipova, T. N., Veselov, S. U., Melentiev, A. I., Martynenko, E. V., & Kudoyarova, G. R. (2005). Ability of bacterium *Bacillus subtilis* to produce cytokinins and to influence the growth and endogenous hormone content of lettuce plants. *Plant and Soil, 272*(1), 201–209. DOI: https://doi.org/10.1007/s11104-004-5047-x

Asif, M., Mughal, A. H., Bisma, R., Mehdi, Z., Saima, S., Ajaz, M., … & Sidique, S. (2018). Application of different strains of biofertilizers for raising quality forest nursery. *International Journal of Current Microbiology and Applied Sciences, 7*(10), 3680–3686. DOI: https://doi.org/10.20546/ijcmas.2018.710.425

Awasthi, R., Tewari, R., & Nayyar, H. (2011). Synergy between plants and P-solubilizing microbes in soils: effects on growth and physiology of crops. *International Research Journal of Microbiology, 2*(12), 484–503. Retrieved from https://www.researchgate.net/profile/Rashmi-Awasthi/publication/265209336_Synergy_between_Plants_and_P-Solubilizing_Microbes_in_soils_Effects_on_Growth_and_Physiology_of_Crops/links/540551f90cf2bba34c1d2f83/Synergy-between-Plants-and-P-Solubilizing-Microbes-in-soils-Effects-on-Growth-and-Physiology-of-Crops.pdf

Azarmi, R., Hajieghrari, B., & Giglou, A. (2011). Effect of *Trichoderma* isolates on tomato seedling growth response and nutrient uptake. *African Journal of Biotechnology, 10*(31), 5850–5855. DOI: https://doi.org/10.5897/AJB10.1600

Bakonyi, N., Bott, S., Gajdos, E., Szabó, A., Jakab, A., Tóth, B., … & Veres, S. (2013). Using biofertilizer to improve seed germination and early development of maize. *Polish Journal of Environmental Studies, 22*(6), 1595–1599. Retrieved from http://www.pjoes.com/Using-Biofertilizer-to-Improve-Seed-Germination-r-nand-Early-Development-of-Maize,89127,0,2.html

Banerjee, A., Datta, J. K., Mondal, N. K., & Chanda, T. (2011). Influence of integrated nutrient management on soil properties of old alluvial soil under mustard cropping system. *Communications in Soil Science and Plant Analysis, 42*(20), 2473–2492. DOI: https://doi.org/10.1080/00103624.2011.609256

Bano, A., & Fatima, M. (2009). Salt tolerance in *Zea mays* (L). following inoculation with *Rhizobium* and *Pseudomonas*. *Biology and Fertility of Soils, 45*(4), 405–413. DOI: https://doi.org/10.1007/s00374-008-0344-9

Basak, B., & Biswas, D. (2012). *Modification of waste mica for alternative source of potassium: evaluation of potassium release in soil from waste mica treated with potassium solubilizing bacteria (KSB)*. LAMBERT Academic Publishing, Germany. ISBN 978-3659298424

Benson, D. R., & Silvester, W. B. (1993). Biology of Frankia strains, actinomycete symbionts of actinorhizal plants. *Microbiological Reviews, 57*(2), 293–319. DOI: https://doi.org/10.1128/mr.57.2.293-319.1993

Bhattacharjee, R., & Dey, U. (2014). Biofertilizer, a way towards organic agriculture: A review. *African Journal of Microbiology Research, 8*(24), 2332–2343. DOI: https://doi.org/10.5897/AJMR2013.6374

Bist, V., Niranjan, A., Ranjan, M., Lehri, A., Seem, K., & Srivastava, S. (2020). Silicon-solubilizing media and its implication for characterization of bacteria to mitigate biotic stress. *Frontiers in Plant Science, 11*, 28. DOI: https://doi.org/10.3389/fpls.2020.00028

Biswas, S. K., Shankar, U., Kumar, S., Kumar, A., Kumar, V., & Lal, K. (2015). Impact of bio-fertilizers for the management of spot blotch disease and growth and yield contributing parameters of wheat. *Journal of Pure and Applied Microbiology*, 9(4), 3025–3031. Retrieved from https://link.gale.com/apps/doc/A481650429/AONE?u=anon~b1d575e8&sid=googleScholar&xid=6275958f

Bochow, H., El-Sayed, S. F., Junge, H., Stavropoulou, A., & Schmiedeknecht, G. (2001). Use of *Bacillus subtilis* as biocontrol agent. IV. Salt-stress tolerance induction by *Bacillus subtilis* FZB24 seed treatment in tropical vegetable field crops, and its mode of action. *Journal of Plant Diseases and Protection*, *108*(1), 21–30. Retrieved from https://www.jstor.org/stable/43215378

Bowles, T. M., Barrios-Masias, F. H., Carlisle, E. A., Cavagnaro, T. R., & Jackson, L. E. (2016). Effects of arbuscular mycorrhizae on tomato yield, nutrient uptake, water relations, and soil carbon dynamics under deficit irrigation in field conditions. *Science of the Total Environment*, *566*, 1223–1234. DOI: https://doi.org/10.1016/j.scitotenv.2016.05.178

Brahmaprakash, G. P., Sahu, P. K., Lavanya, G., Nair, S. S., Gangaraddi, V. K., & Gupta, A. (2017). Microbial functions of the rhizosphere. In *Plant-microbe interactions in agro-ecological perspectives* (pp. 177–210). Springer, Singapore. DOI: https://doi.org/10.1007/978-981-10-5813-4_10

Chesti, M. U. H., Qadri, T. N., Hamid, A., Qadri, J., Azooz, M. M., & Ahmad, P. (2013). Role of Bio-fertilizers in crop improvement. In *Crop Improvement* (pp. 189–208). Springer, Boston, MA. DOI: https://doi.org/10.1007/978-1-4614-7028-1_5

Choudhury, A. T. M. A., & Kennedy, I. R. (2004). Prospects and potentials for systems of biological nitrogen fixation in sustainable rice production. *Biology and Fertility of Soils*, *39*(4), 219–227. DOI: https://doi.org/10.1007/s00374-003-0706-2

Dey, R. K., Pal, K. K., Bhatt, D. M., & Chauhan, S. M. (2004). Growth promotion and yield enhancement of peanut (*Arachis hypogaea* L.) by application of plant growth-promoting rhizobacteria. *Microbiological Research*, *159*(4), 371–394. DOI: https://doi.org/10.1016/j.micres.2004.08.004

Dhankhar, R., Mohanty, A., & Gulati, P. (2021). Microbial diversity of phyllosphere: Exploring the unexplored. *Phytomicrobiome Interactions and Sustainable Agriculture*, 66–90. DOI: https://doi.org/10.1002/9781119644798.ch5

Dhankhar, R., Sheoran, S., Dhaka, A., & Soni, R. (2013). The role of phosphorus solubilizing bacteria (PSB) in soil management-an overview. *International Journal of Development Research*, *3*(9), 31–36.

Dodor, D. E., & Tabatabai, M. A. (2003). Effect of cropping systems on phosphatases in soils. *Journal of Plant Nutrition and Soil Science*, *166*(1), 7–13. DOI: https://doi.org/10.1002/jpln.200390016

Garai, T. K., Datta, J. K., & Mondal, N. K. (2014). Evaluation of integrated nutrient management on boro rice in alluvial soil and its impacts upon growth, yield attributes, yield and soil nutrient status. *Archives of Agronomy and Soil Science*, *60*(1), 1–14. DOI: https://doi.org/10.1080/03650340.2013.766721

Ghosh, N. (2004). Promoting biofertilizers in Indian agriculture. *Economic and Political Weekly*, *39*(52), 5617–5625. Retrieved from https://www.jstor.org/stable/4415978

Grosch, R., Junge, H., Krebs, B., & Bochow, H. (1999). Use of *Bacillus subtilis* as a biocontrol agent. III. Influence of *Bacillus subtilis* on fungal root diseases and on yield in soilless culture. *Journal of Plant Diseases and Protection*, *106*(6), 568–580. Retrieved from https://www.jstor.org/stable/43390116

Groudev, S. N. (1987). Use of heterotrophic microorganisms in mineral biotechnology. *Acta Biotechnology*, *7*, 299–306. DOI: https://doi.org/10.1002/abio.370070404

Gundala, P. B., Chinthala, P., & Sreenivasulu, B. (2013). A new facultative alkaliphilic, potassium solubilizing, *Bacillus* spp. SVUNM9 isolated from mica cores of Nellore district, Andhra Pradesh, India. *Journal of Microbiology Biotechnology*, *2*(1), 1–7. Retrieved from https://www.researchgate.net/publication/285012770_A_new_facultative_alkaliphilic_potassium_solubilizing_Bacillus_Sp_SVUNM9_isolated_from_mica_cores_of_Nellore_District_Andhra_Pradesh_India_Research_and_Reviews

Hariprasad, P., & Niranjana, S. R. (2009). Isolation and characterization of phosphate solubilizing rhizobacteria to improve plant health of tomato. *Plant and Soil*, *316*(1), 13–24. DOI: https://doi.org/10.1007/s11104-008-9754-6

Hawerroth, C., Araujo, L., Bermúdez-Cardona, M. B., Silveira, P. R., Wordell Filho, J. A., & Rodrigues, F. A. (2019). Silicon-mediated maize resistance to macrospora leaf spot. *Tropical Plant Pathology*, *44*(2), 192–196. DOI: https://doi.org/10.1007/s40858-018-0247-8

Herrmann, L., & Lesueur, D. (2013). Challenges of formulation and quality of biofertilizers for successful inoculation. *Applied Microbiology and Biotechnology*, *97*(20), 8859–8873. DOI: https://doi.org/10.1007/s00253-013-5228-8

Husen, E. H., Simanungkalit, R. D. M., Saraswati, R., & Irawan, I. (2016). Characterization and quality assessment of Indonesian commercial biofertilizers. *Indonesian Journal of Agricultural Science*, *8*(1), 31–38. DOI: http://dx.doi.org/10.21082/ijas.v8n1.2007.p31-38

Idris, E. E., Iglesias, D. J., Talon, M., & Borriss, R. (2007). Tryptophan-dependent production of indole-3-acetic acid (IAA) affects level of plant growth promotion by *Bacillus amyloliquefaciens* FZB42. *Molecular Plant-Microbe Interactions*, *20*(6), 619–626. DOI: https://doi.org/10.1094/MPMI-20-6-0619

Islam, M. T., Deora, A., Hashidoko, Y., Rahman, A., Ito, T., & Tahara, S. (2007). Isolation and identification of potential phosphate solubilizing bacteria from the rhizoplane of *Oryza sativa* L. cv. BR29 of Bangladesh. *Zeitschrift für Naturforschung C*, *62*(1–2), 103–110. DOI: https://doi.org/10.1515/znc-2007-1-218

Jeet, K., & Baldi, A. (2020). Development of inorganic carrier based bioformulation of *Sebacina Vermifera* and its evaluation on *Trigonella Foenumgraecum*. *International Journal of Pharmacy and Pharmaceutical Sciences*, *11*(2), 69–82. DOI: https://doi.org/10.22376/ijpbs.2020.11.2.p69-82

Jiang, H., Qi, P., Wang, T., Chi, X., Wang, M., Chen, M., … & Pan, L. (2019). Role of halotolerant phosphate-solubilising bacteria on growth promotion of peanut (*Arachis hypogaea*) under saline soil. *Annals of Applied Biology*, *174*(1), 20–30. DOI: https://doi.org/10.1111/aab.12473

Kachroo, D., & Razdan, R. (2006). Growth, nutrient uptake and yield of wheat (*Triticum aestivum*) as influenced by biofertilizers and nitrogen. *Indian Journal of Agronomy*, *51*(1), 37–39. Retrieved from https://www.indianjournals.com/ijor.aspx?target=ijor:ija&volume=51&issue=1&article=012

Kalayu, G. (2019). Phosphate solubilizing microorganisms: Promising approach as biofertilizers. *International Journal of Agronomy*, *2019*, 1–7. DOI: https://doi.org/10.1155/2019/4917256

Kamaraj, A., & Padmavathi, S. (2018). Alleviation of saline salt stress through pre-sowing biofertilizer seed treatment on crop growth and seed yield in green gram CV ADT3. *Journal of Pharmacognosy and Phytochemistry*, *7*(1), 2205–2209. Retrieved from https://www.phytojournal.com/archives/2018/vol7issue1S/PartAG/SP-7-1-641.pdf

Kamran, S., Shazia, A., & Shahida, H. (2010). Growth responses of *Helianthus annus* to plant growth promoting *Rhizobacteria* used as a biofertilizers. *International Journal of Agricultural Research*, *5*(11), 1048–1056. Retrieved from https://www.cabdirect.org/cabdirect/abstract/20113198204

Khalequzzaman, K. M. (2015). Seed treatment with *Rhizobium* biofertilizer for controlling foot and root rot of chickpea. *International Journal of Scientific Research in Agricultural Sciences*, *2*(6), 144–150. DOI: http://dx.doi.org/10.12983/ijsras-2015-p0144-0150

Khan, A. A., Jilani, G., Akhtar, M. S., Naqvi, S. M. S., & Rasheed, M. (2009). Phosphorus solubilizing bacteria: Occurrence, mechanisms and their role in crop production. *Journal of Agriculture and Biology Science*, *1*(1), 48–58. Retrieved from https://www.uaar.edu.pk/jabs/files/jabs_1_1_6.pdf

Kumar, A., Kumar, A., & Patel, H. (2018). Role of microbes in phosphorus availability and acquisition by plants. *International Journal of Current Microbiology and Applied Sciences*, *7*(5), 1344–1347. Retrieved from https://www.ijcmas.com/7-5-2018/Amarjeet%20Kumar,%20et%20al.pdf

Kumar, A., Usmani, Z., & Kumar, V. (2017). Biochar and flyash inoculated with plant growth promoting rhizobacteria act as potential biofertilizer for luxuriant growth and yield of tomato plant. *Journal of Environmental Management*, *190*, 20–27. DOI: https://doi.org/10.1016/j.jenvman.2016.11.060

Kumar, K. V. K., Yellareddygari, S. K., Reddy, M. S., Kloepper, J. W., Lawrence, K. S., Zhou, X. G., … & Miller, M. E. (2012). Efficacy of *Bacillus subtilis* MBI 600 against sheath blight caused by *Rhizoctonia solani* and on growth and yield of rice. *Rice Science*, *19*(1), 55–63. DOI: https://doi.org/10.1016/S1672-6308(12)60021-3

Kurdish, I. K., Bega, Z. T., Gordienko, A. S., & Dyrenko, D. I. (2008). The effect of *Azotobacter vinelandii* on plant seed germination and adhesion of these bacteria to cucumber roots. *Applied Biochemistry and Microbiology*, *44*(4), 400–404. DOI: https://doi.org/10.1134/S000368380804011X

Lee, K. E., Adhikari, A., Kang, S. M., You, Y. H., Joo, G. J., Kim, J. H., … & Lee, I. J. (2019). Isolation and characterization of the high silicate and phosphate solubilizing novel strain *Enterobacter ludwigii* GAK2 that promotes growth in rice plants. *Agronomy*, *9*(3), 144. DOI: https://doi.org/10.3390/agronomy9030144

Leigh, M.B., Prouzova, P., Mackova, M., Macek, T., Nagle, D.P., & Fletcher, J.S. (2006). Polychlorinated biphenyl (PCB)-degrading bacteria associated with trees in a PCB-contaminated site. *Appl Environ Microbiology*, *72*, 2331–2342. DOI: https://doi.org/10.1128/AEM.72.4.2331-2342.2006

Leyval, C., & Berthelin, J. (1989). Interaction between *Laccaria laccata*, *Agrobacterium radiobacter* and beech roots: influence on P, K, Mg and Fe mobilization from minerals and plant growth. *Plant and Soil*, *117*, 103–110. DOI: https://doi.org/10.1007/BF02206262

Li, M., Hu, J. & Lin, X. (2021). The roles and performance of arbuscular mycorrhizal fungi in intercropping systems. *Soil Ecology Letters*, 1–9. DOI: https://doi.org/10.1007/s42832-021-0107-1

Lin, Q., Rao, Z., Sun, Y., Yao, J., Xing, L. (2002). Identification and practical application of silicate-dissolving bacteria. *Agricultural Science in China*, *1*, 81–85. Retrieved from https://en.cnki.com.cn/Article_en/CJFDTotal-ZGNX200201015.htm

Liu, G. Y. (2001). Screening of silicate bacteria with potassium releasing and antagonistic activity. *The Chinese Journal of Applied and Environmental Biology*, *7*, 66–68

Liu, L., Jiang, C.Y., Liu, X.Y., Wu, J.F., Han, J.G., & Liu, S.J. (2007) Plant-microbe association for rhizoremediation of chloronitroaromatic pollutants with *Comamonas* sp strain CNB-1. *Environmental Microbiology*, *9*, 465–473. DOI: https://doi.org/10.1111/j.1462-2920.2006.01163.x

Lukacova, Z., Svubova, R., Janikovicova, S., Volajova, Z., & Lux, A. (2019). Tobacco plants (*Nicotiana benthamiana*) were influenced by silicon and were not infected by dodder (*Cuscuta europaea*). *Plant Physiology and Biochemistry*, *139*, 179–190. DOI: https://doi.org/10.1016/j.plaphy.2019.03.004

Ma, J. F., Yamaji, N., Mitani, N., Tamai, K., Konishi, S., Fujiwara, T., … & Yano, M. (2007). An efflux transporter of silicon in rice. *Nature*, *448*(7150), 209–212. DOI: https://doi.org/10.1038/nature05964

Mahanty, T., Bhattacharjee, S., Goswami, M., Bhattacharyya, P., Das, B., Ghosh, A., & Tribedi, P. (2017). Biofertilizers: a potential approach for sustainable agriculture development. *Environmental Science and Pollution Research*, *24*(4), 3315–3335. DOI: https://doi.org/10.1007/s11356-016-8104-0

Malik, A. A., Suryapani, S., & Ahmad, J. (2011). Chemical vs organic cultivation of medicinal and aromatic plants: the choice is clear. *International Journal of Medicinal and Aromatic Plants*, *1*(1), 5–13. Retrieved from https://www.researchgate.net/profile/Afaq-Malik/publication/233934049_Chemical_Vs_Organic_Cultivation_of_Medicinal_and_Aromatic_Plants_the_choice_is_clear/links/09e4150d1c30fd41ab000000/Chemical-Vs-Organic-Cultivation-of-Medicinal-and-Aromatic-Plants-the-choice-is-clear.pdf

Malusa, E., & Vassilev, N. (2014). A contribution to set a legal framework for biofertilizers. *Applied Microbiology and Biotechnology*, *98*(15), 6599–6607. DOI: https://doi.org/10.1007/s00253-014-5828-y

Marzouk, H. A., & Kassem, H. A. (2011). Improving fruit quality, nutritional value and yield of Zaghloul dates by the application of organic and/or mineral fertilizers. *Scientia Horticulturae*, *127*(3), 249–254. DOI: https://doi.org/10.1016/j.scienta.2010.10.005

Maurya, B. R., Meena, V. S., & Meena, O. P. (2014). Influence of Inceptisol and Alfisol's potassium solubilizing bacteria (KSB) isolates on release of K from waste mica. *Vegetos*, *27*(1), 181–187. DOI: https://doi.org/10.5958/j.2229-4473.27.1.028

Mazid, M., & Khan, T. A. (2015). Future of bio-fertilizers in Indian agriculture: An overview. *International Journal of Agricultural and Food Research*, *3*(3), 10–23. Retrieved from https://pdfs.semanticscholar.org/74f9/56da9cf13a1fb1ccfd46e9a5a25230484e8f.pdf

Meena, O. P., Maurya, B. R., & Meena, V. S. (2013). Influence of K-solubilizing bacteria on release of potassium from waste mica. *Agronomy for Sustainable Development*, *1*, 53–56. Retrieved from https://d1wqtxts1xzle7.cloudfront.net/53418472/12._Meena_et_al-with-cover-page-v2.pdf?Expires=1634730916&Signature=aoN2HbmiOXmWxqNdXMW-z2HzJnYlTguk9hYG94S4qzT-OWGs0~N7rZgd1YZUHUkrvaL2zfN-JCETfqqXP31b-N5DaDE~OWwfQ5f3YNDGVjB-EgYtIxuGYy46UiHo9PLy4n1Bw6JyP3NrTiqvJCAV3rwkCUA0uhN1OFRX6VTVNkp0NN~61~Xnw-kFm8Cu6rtmfKVyhUMw~tFXkKNJun61fkuSAvcJT7fVkarEqItffojBDJPmk1NhXD6XPQ2i5-SuPmaeSZPenD8CVPBFnxBqk8z~okQxcJxnkviNPuskHeFyHzkklGctU-pCLjLHd~5XKR1hiS26aAAA8FEyLSPeUQ__&Key-Pair-Id=APKAJLOHF5GGSLRBV4ZA

Meena, V. D., Dotaniya, M. L., Coumar, V., Rajendiran, S., Kundu, S., & Rao, A. S. (2014). A case for silicon fertilization to improve crop yields in tropical soils. *Proceedings of the National Academy of Sciences, India Section B: Biological Sciences*, *84*(3), 505–518. DOI: https://doi.org/10.1007/s40011-013-0270-y

Meena, V. S., Maurya, B. R., & Verma, J. P. (2014). Does a rhizospheric microorganism enhance K+ availability in agricultural soils? *Microbiological Research*, *169*(5–6), 337–347. DOI: https://doi.org/10.1016/j.micres.2013.09.003

Ministry of Agriculture, Government of India. Biofertilizers and organic fertilizers covered in fertilizer (Control) Order, 1985 (as amended, March 2006 and November 2009). Official Gazette 3 November, 2009. Retrieved from https://cuts-cart.org/pdf/Useful_Information-Biofertilizers_and_Organic_Fertilizers_in_Fertilizer_(Control)_Order_1985.pdf

Mishra, D., Rajvir, S., Mishra, U., & Kumar, S. S. (2013). Role of bio-fertilizer in organic agriculture: a review. *Research Journal of Recent Sciences ISSN, 2277*, 2502. Retrieved from https://citeseerx.ist.psu.edu/viewdoc/download?doi=10.1.1.1046.9269&rep=rep1&type=pdf

Mohite, B. (2013). Isolation and characterization of indole acetic acid (IAA) producing bacteria from rhizospheric soil and its effect on plant growth. *Journal of Soil Science and Plant Nutrition*, *13*(3), 638–649. DOI: http://dx.doi.org/10.4067/S0718-95162013005000051

Mubeen, F., Aslam, A, Sheikh, M. A., Iqbal, T, Hameed, S., Malik, K. A., & Hafeez, F. Y. (2006). Response of wheat yield under combine use of Fungicide and Biofertilizer. *International Journal of Agriculture and Biology*, *8*(5), 580–582. Retrieved from GATXUWWYDFFHN4SK64F6H3X6UVUCRGMR6BXJ4JAPT2MMG5QI5VRQLQNE

Muentz, A. (1890). Sur la décomposition des roches et la formation de la terre arable. *CR Academic Science*, 110, 1370–1372.

Mukhongo, R. W., Tumuhairwe, J. B., Ebanyat, P., AbdelGadir, A. H., Thuita, M., & Masso, C. (2017). Combined application of biofertilizers and inorganic nutrients improves sweet potato yields. *Frontiers in Plant Science*, *8*, 219. DOI: https://doi.org/10.3389/fpls.2017.00219

Muraleedharan, H., Seshadri, S., & Perumal, K. (2010). Biofertilizer (Phosphobacteria), Booklet published by Shri AMM Murugappa Chettiar Research Centre, Chennai. Retrieved from http://www.amm-mcrc.org/Publications/Reports.html

Nacoon, S., Jogloy, S., Riddech, N., Mongkolthanaruk, W., Kuyper, T. W., & Boonlue, S. (2020). Interaction between phosphate solubilizing bacteria and arbuscular mycorrhizal fungi on growth promotion and tuber inulin content of *Helianthus tuberosus* L. *Scientific Reports*, *10*(1), 1–10. DOI: https://doi.org/10.1038/s41598-020-61846-x

Narasimhan, K., Basheer, C., Bajic, V.B., and Swarup, S. (2003) Enhancement of plant-microbe interactions using a rhizosphere metabolomics-driven approach and its application in the removal of polychlorinated biphenyls. *Plant Physiology 132*, 146–153. DOI: https://doi.org/10.1104/pp.102.016295

Nezarat, S., & Gholami, A. (2009). Screening plant growth promoting rhizobacteria for improving seed germination, seedling growth and yield of maize. *Pakistan Journal of Biological Sciences*, *12*(1), 26. DOI:https://doi.org/10.1016/jejsobi.2008.07.001.

Osman, K. T. (2014). *Soil Degradation, Conservation and Remediation*. Dordrecht: Springer Netherlands. Retrieved from https://link.springer.com/book/10.1007%2F978-94-007-7590-9

Parmar, P., & Sindhu, S. S. (2013). Potassium solubilization by rhizosphere bacteria: Influence of nutritional and environmental conditions. *Journal of Microbiology Research*, *3*(1), 25–31. DOI: https://doi.org/10.5923/j.microbiology.20130301.04

Paterson, E., Sim, A., Davidson, J., & Daniell, T. J. (2016). Arbuscular mycorrhizal hyphae promote priming of native soil organic matter mineralisation. *Plant and Soil*, *408*(1), 243–254. DOI: https://doi.org/10.1007/s11104-016-2928-8

Pathak, A. & Chakraborti, S. K. (2014). Impact of bio-fertilizer seed treatment on seed and seedling parameters of maize (*Zea mays* L.). *The Bioscan*, *9*(1), 133–135. Retrieved from https://citeseerx.ist.psu.edu/viewdoc/download?doi=10.1.1.1066.6372&rep=rep1&type=pdf

Peix, A., Rivas-Boyero, A. A., Mateos, P. F., Rodriguez-Barrueco, C., Martınez-Molina, E., & Velazquez, E. (2001). Growth promotion of chickpea and barley by a phosphate solubilizing strain of Mesorhizobium mediterraneum under growth chamber conditions. *Soil Biology and Biochemistry*, *33*(1), 103–110. DOI: https://doi.org/10.1016/S0038-0717(00)00120-6

Prajapati, K., Sharma, M. C., & Modi, H. A. (2013). Growth promoting effect of potassium solubilizing microorganisms on *Abelmoscus esculantus*. *International Journal of Agricultural Sciences*, *3*(1), 181–188. Retrieved from https://www.researchgate.net/publication/235943493_Growth_promoting_effect_of_potassium_solubilizing_microorganisms_on_okra_Abelmoschus_esculentus

Rana, R., & Kapoor, P. (2013). Biofertilizers and their role in Agriculture. *Popular Kheti 1*(1), 56–61. Retrieved from http://popularkheti.info/documents/2013-1/PK-1-1-11-56-61.pdf

Raturi, G., Sharma, Y., Rana, V., Thakral, V., Myaka, B., Salvi, P., ... & Deshmukh, R. (2021). Exploration of silicate solubilizing bacteria for sustainable agriculture and silicon biogeochemical cycle. *Plant Physiology and Biochemistry*. DOI: https://doi.org/10.1016/j.plaphy.2021.06.039

Rodŕiguez, H. & Fraga, R. (1999). Phosphate solubilizing bacteria and their role in plant growth promotion. *Biotechnology Advances*, *17*(4-5), 319–339. DOI: https://doi.org/10.1016/S0734-9750(99)00014-2

Rouphael, Y., Franken, P., Schneider, C., Schwarz, D., Giovannetti, M., Agnolucci, M., ... & Colla, G. (2015). Arbuscular mycorrhizal fungi act as biostimulants in horticultural crops. *Scientia Horticulturae*, *196*, 91–108. DOI: https://doi.org/10.1016/j.scienta.2015.09.002

Santos V. B., Araujo S. F., Leite L. F., Nunes L. A., & Melo J. W. (2012). Soil microbial biomass and organic matter fractions during transition from conventional to organic farming systems. *Geoderma*, *170*, 227–231. DOI: https://doi.org/10.1016/j.geoderma.2011.11.007

Saryoko, A., Kusumawati, S., & Pohan, A. (2021). Seed treatment using biofertilizer to improve plant growth and yield performances of upland rice cultivars under various planting densities. In *IOP Conference Series: Earth and Environmental Science*, Surakarta, Indonesia, 648(1), 012028. IOP Publishing. Retrieved from https://iopscience.iop.org/article/10.1088/1755-1315/648/1/012028/meta

Satyaprakash, M., Nikitha, T., Reddi, E. U. B., Sadhana, B., & Vani, S. S. (2017). Phosphorous and phosphate solubilising bacteria and their role in plant nutrition. *International Journal of Current Microbiology and Applied Sciences*, *6*(4), 2133–2144. DOI: https://doi.org/10.20546/ijcmas.2017.604.251

Schachtman, D. P., Reid, R. J., & Ayling, S. M. (1998). Phosphorus uptake by plants: From soil to cell. *Plant Physiology*, *116*(2), 447–453. DOI: https://doi.org/10.1104/pp.116.2.447

Schmiedeknecht, G., Issoufou, I., Junge, H., & Bochow, H. (2001). Use of *Bacillus subtilis* as biocontrol agent. V. Biological control of diseases on maize and sunflowers. *Journal of Plant Diseases and Protection*, 500–512. Retrieved from https://www.jstor.org/stable/45154885

Shanmugaiah, V., Balasubramanian, N., Gomathinayagam, S., Manoharan, P. T., & Rajendran, A. (2009). Effect of single application of *Trichoderma viride* and *Pseudomonas fluorescens* on growth promotion in cotton plants. *African Journal of Agricultural Research*, *4*(11), 1220–1225. DOI: https://doi.org/10.5897/AJAR.9000227

Sharma, S. B., Sayyed, R. Z., Trivedi, M. H., & Gobi, T. A. (2013). Phosphate solubilizing microbes: sustainable approach for managing phosphorus deficiency in agricultural soils. *Springer Plus*, *2*(1), 1–14. DOI: https://doi.org/10.1186/2193-1801-2-587

Sheng, X. F., Zhao, F., He, L. Y., Qiu, G., & Chen, L. (2008). Isolation and characterization of silicate mineral-solubilizing *Bacillus globisporus* Q12 from the surfaces of weathered feldspar. *Canadian Journal of Microbiology*, *54*(12), 1064–1068. DOI: https://doi.org/10.1139/W08-089

Shimbo, S., Zhang, Z. W., Watanabe, T., Nakatsuka, H., Matsuda-Inoguchi, N., Higashikawa, K., & Ikeda, M. (2001). Cadmium and lead contents in rice and other cereal products in Japan in 1998–2000. *Science of the Total Environment*, *281*(1–3), 165–175. DOI: https://doi.org/10.1016/S0048-9697(01)00844-0

Sindhu, S. S., Parmar, P., & Phour, M. (2014). Nutrient cycling: Potassium solubilization by microorganisms and improvement of crop growth. In *Geomicrobiology and Biogeochemistry* (pp. 175–198). Springer, Berlin, Heidelberg. https://doi.org/10.1007/978-3-642-41837-2_10

Singh, A. (2016). Effect of biofertilizers on growth and buddability of rough lemon seedlings in containerised nursery system (Doctoral dissertation, Punjab Agricultural University, Ludhiana). Retrieved from https://krishikosh.egranth.ac.in/handle/1/92126

Singh, G., Biswas, D. R., & Marwah, T. S. (2010). Mobilization of potassium from waste mica by plant growth promoting rhizobacteria and its assimilation by maize (*Zea mays*) and wheat (*Triticum aestivum* L.). *Journal of Plant Nutrition 33*(8), 1236–1251. DOI: https://doi.org/10.1080/01904161003765760

Singh, K. (2001). Response of potato (*Solarium tuberosum*) to bio-fertilizer and nitrogen under North-Eastern hill conditions. *Indian Journal of Agronomy*, *46*(2), 375–379. Retrieved from https://www.indianjournals.com/ijor.aspx?target=ijor:ija&volume=46&issue=2&article=035

Singh, N., Bhuker, A., & Jeevanadam, J. (2021). Effects of metal nanoparticle-mediated treatment on seed quality parameters of different crops. *Naunyn-Schmiedeberg's Arch Pharmacol*, *394*, 1067–1089. DOI: https://doi.org/10.1007/s00210-021-02057-7

Singh, N., Thakur, A. K., Kaushal, R., Mehta, D. K., & Bhardwaj, R. K. (2018). Effects of seed pelleting on seed quality of cowpea (*Vigna unguiculata* L.) during storage. *International Journal of Economic Plants*, *5*(2), 76–79. DOI: https://doi.org/10.23910/IJEP/2018.5.2.0237

Singh, R., Biswas, S. K., Nagar, D., Singh, J., Singh, M., & Mishra, Y. K. (2015). Sustainable integrated approach for management of Fusarium wilt of tomato caused by *Fusarium oxysporum* f. sp. lycopersici (Sacc.) Synder and Hansen. *Sustainable Agriculture Research*, *4*(526-2016-37870). DOI: https://doi.org/10.22004/ag.econ.230412

Singh, S., & Kapoor, K. K. (1999). Inoculation with phosphate-solubilizing microorganisms and a vesicular-arbuscular mycorrhizal fungus improves dry matter yield and nutrient uptake by wheat grown in a sandy soil. *Biology and Fertility of Soils*, *28*(2), 139–144. DOI: https://doi.org/10.1007/s003740050475

Siqueira, M. F. B., Sudre, C. P., Almeida, L. H., Pegorerl, A. P. R., Akiba, F., Foundation, M. O., & de Janeiro, R. (1993). Influence of Effective Microorganisms on seed germination and plantlet vigor of selected crops. In *Proceedings of the Third International Conferences on Nature Farming*, eds. J. F. Parr, S. B. Hornick, M. E. Simpson. Washington, DC: US Department of Agriculture (pp. 22–45). Retrieved from https://www.infrc.or.jp/wxp/wp-content/uploads/KNFC/KNFC3/KNFC3-7-4-Siqueira-Sudre-Almeida.pdf

Son, H. J., Park, G. T., Cha, M. S., & Heo, M. S. (2006). Solubilization of insoluble inorganic phosphates by a novel salt-and pH-tolerant *Pantoea agglomerans* R-42 isolated from soybean rhizosphere. *Bioresource Technology*, *97*(2), 204–210. DOI: https://doi.org/10.1016/j.biortech.2005.02.021

Sourabh Sharma, J.R., Baloda, S., Kumar, R., Sheoran, V., & Saini, H. (2018) Response of organic amendments and biofertilizers on growth and yield of guava during rainy season. *Journal of Pharmacognosy and Phytochemistry*, *7*(6), 2692–2695.

Spaepen, S., Vanderleyden, J., & Remans, R. (2007). Indole-3-acetic acid in microbial and microorganism-plant signaling. *FEMS Microbiology Reviews*, *31*(4), 425–448. DOI: https://doi.org/10.1111/j.1574-6976.2007.00072.x

Stamford, N. P., Chamber-Perez, M., & Camacho-Martínez, M. (1995). Symbiotic effectiveness of several tropical *Bradyrhizobium* strains on cowpea under a long-term exposure to nitrate: relationships between nitrogen fixation and nitrate reduction activities. *Journal of Plant Physiology*, *147*(3–4), 378–382. DOI: https://doi.org/10.1016/S0176-1617(11)82171-1

Stephan, M. P., Pedrosa, F. O., & DöBEREINER, J. (1981). Physiological studies with Azospirillum spp. *Associative N*, *2*, 7–13.

Subhashini, D. V., & Kumar, A. (2014). Phosphate solubilising Streptomyces spp obtained from the rhizosphere of *Ceriops decandra* of Corangi mangroves. *Indian Journal of Agricultural Sciences 84*(5), 560–564. Retrieved from https://krishi.icar.gov.in/jspui/bitstream/123456789/9605/1/corangi.pdf

Supanjani, H. H., Jung, J. S., & Lee, K. D. (2006). Rock phosphate-potassium and rock-solubilising bacteria as alternative, sustainable fertilizers. *Agronomy for Sustainable Development*, *26*(4), 233–240.

Tairo, E. V., & Ndakidemi, P. A. (2013). Possible benefits of rhizobial inoculation and phosphorus supplementation on nutrition, growth and economic sustainability in grain legumes. *American Journal of Research Communication*, *1*(12), 532–556. Retrieved from http://www.usa-journals.com/wp-content/uploads/2013/11/Tairo_Vol112.pdf

Tarafdar, J. C. (2003). Efficiency of some phosphatase producing soil-fungi. *Indian Journal of Microbiology Research*, *43*, 27–32. Retrieved from https://ci.nii.ac.jp/naid/10026513906/

Thirkell, T. J., Charters, M. D., Elliott, A. J., Sait, S. M., & Field, K. J. (2017). Are mycorrhizal fungi our sustainable saviours? Considerations for achieving food security. *Journal of Ecology*, *105*(4), 921–929. DOI: https://doi.org/10.1111/1365-2745.12788

Uddin, M., Hussain, S., Khan, M. M. A., Hashmi, N., Idrees, M., Naeem, M., & Dar, T. A. (2014). Use of N and P biofertilizers reduces inorganic phosphorus application and increases nutrient uptake, yield, and seed quality of chickpea. *Turkish Journal of Agriculture and Forestry*, *38*(1), 47–54. Retrieved from https://journals.tubitak.gov.tr/agriculture/abstract.htm?id=14423

Uroz, S., Calvaruso, C., Turpault, M. P., & Frey-Klett, P. (2009). Mineral weathering by bacteria: ecology, actors and mechanisms. *Trends in Microbiology*, *17*(8), 378–387. DOI: https://doi.org/10.1016/j.tim.2009.05.004

Vance, C. P. (2001). Symbiotic nitrogen fixation and phosphorus acquisition: Plant nutrition in a world of declining renewable resources. *Plant Physiology*, *127*(2), 390–397. DOI: https://doi.org/10.1104/pp.010331

Venkataraman, G. S., & Shanmugasundaram, S. (1992). Algal biofertilizers technology for rice. *DBT Centre for BGA. Bio-fertilizer, Madurai Kamraj University, Madurai*, *625021*, 1–24.

Verma, J. P., Yadav, J., Tiwari, K. N., & Jaiswal, D. K. (2014). Evaluation of plant growth promoting activities of microbial strains and their effect on growth and yield of chickpea (*Cicer arietinum* L.) in India. *Soil Biology and Biochemistry*, *70*, 33–37. DOI: https://doi.org/10.1016/j.soilbio.2013.12.001

Vlassak, K., & Reynders, R. (1979). Agronomic aspects of biological dinitrogen fixation by *Azospirillum* spp. *Associative N*, *2*, 93–102.

Wagner, S. C. (2011). Biological nitrogen fixation. *Nature Education Knowledge*, *3*(10), 15. Retrieved from https://www.nature.com/scitable/knowledge/library/biological-nitrogen-fixation-23570419/

Wahane, M. R., Meshram, N. A., More, S. S., & Khobragade, N. H. (2020). Biofertilizer and their role in sustainable agriculture-A review. *The Pharma Innovation Journal*, *9*(7), 127–130. Retrieved from https://www.thepharmajournal.com/archives/2020/vol9issue7/PartB/9-6-73-209.pdf

Walpola, B. C., & Yoon, M. H. (2012). Prospectus of phosphate solubilizing microorganisms and phosphorus availability in agricultural soils: A review. *African Journal of Microbiology Research*, *6*(37), 6600–6605. Retrieved from https://academicjournals.org/journal/AJMR/article-abstract/F5B8B1D29561

Wu, C. H., Bernard, S. M., Andersen, G. L., & Chen, W. (2009). Developing microbe–plant interactions for applications in plant-growth promotion and disease control, production of useful compounds, remediation and carbon sequestration. *Microbial Biotechnology*, *2*(4), 428–440. DOI: https://doi.org/10.1111/j.1751-7915.2009.00109.x

Wu, S. C., Cao, Z. H., Li, Z. G., Cheung, K. C., & Wong, M. H. (2005). Effects of biofertilizer containing N-fixer, P and K solubilizers and AM fungi on maize growth: A greenhouse trial. *Geoderma*, *125*(1–2), 155–166. DOI: https://doi.org/10.1016/j.geoderma.2004.07.003

Xavier, I. J., Holloway, G., & Leggett, M. (2004). Development of rhizobial inoculant formulations. *Crop Management*, *3*(1), 1–6. DOI: https://doi.org/10.1094/CM-2004-0301-06-RV

Yadav, K. S., Singh, D. P., Sunita, S., Neeru, N., & Lakshminarayana, K. (2000). Effect of *Azotobacter chroococcum* on yield and nitrogen economy in wheat (*Triticum aestivum*) under field conditions. *Environment and Ecology*, *18*(1), 109–113. Retrieved from https://www.cabdirect.org/cabdirect/abstract/20000708560

Yao, A. V., Bochow, H., Karimov, S., Boturov, U., Sanginboy, S., & Sharipov, A. K. (2006). Effect of FZB 24® *Bacillus subtilis* as a biofertilizer on cotton yields in field tests. *Archives of Phytopathology and Plant Protection*, *39*(4), 323–328. DOI: https://doi.org/10.1080/03235400600655347

Zak, D., Goldhammer, T., Cabezas, A., Gelbrecht, J., Gurke, R., Wagner, C., … & McInnes, R. (2018). Top soil removal reduces water pollution from phosphorus and dissolved organic matter and lowers methane emissions from rewetted peatlands. *Journal of Applied Ecology*, *55*(1), 311–320. DOI: https://doi.org/10.1111/1365-2664.12931

Zakaria, A. A. B. (2019). Growth optimization of potassium solubilizing bacteria isolated from biofertilizer. *Faculty of Chemical & Natural Resources Engineering, University Malaysia Pahang*. Retrieved from http://umpir.ump.edu.my/id/eprint/753/1/AHMAD_AZINUDDIN_BIN_ZAKARIA.pdf

Zhang, C., & Kong, F. (2014). Isolation and identification of potassium-solubilizing bacteria from tobacco rhizospheric soil and their effect on tobacco plants. *Applied Soil Ecology*, *82*, 18–25. DOI: https://doi.org/10.1016/j.apsoil.2014.05.002

Zonita, M.D., Virginia, M., & Canigia, F. (2008). Field performance of a liquid formulation of *Azospirillum basilense* on dry land wheat productivity. *The European Journal of Soil Biology*, *45*(1), 3–11 DOI:https://doi.org/10.1016/jejsobi.2008.07.001.

Zou, Y. N., Srivastava, A. K., & Wu, Q. S. (2016). Glomalin: a potential soil conditioner for perennial fruits. *International Journal of Agriculture and Biology*, *18*, 293–297. DOI: https://doi.org/10.17957/IJAB/15.0085

15 Microbial Endophytes of Medicinal Plants as an Emerging Bioresource for Novel Therapeutic Compounds

Ashish Kumar
Sant Gahira Guru Vishwavidyalaya, Sarguja Ambikapur, India

Rameshwari A. Banjara
Rajeev Gandhi Government Post Graduate College, Ambikapur, India

Roman Kumar Aneshwari
Institute of Pharmacy, Pt. Ravishankar Shukla University, Raipur, India

Nagendra Kumar Chandrawanshi
School of Studies in Biotechnology, Pt. Ravishankar Shukla University, Raipur, India

CONTENTS

15.1 INTRODUCTION

Since ancient times, medicinal plants have been widely being used for human welfare, and their metabolites act as a most important stockpile for plant-based therapeutic drugs (Farnsworth and Soejarto, 1991). Due to structural variability and low production costs, renewed interest has arisen in the wide range of applications of active ingredients to treat various human diseases, and in natural product drug research. Alkaloids, steroids, flavonoids, terpenoids, and other secondary metabolites of plants are known to have a variety of biological actions. Furthermore, metabolites generated from plants and microbes are increasingly sought for medicinal applications around the world due to their

DOI: 10.1201/9781003306931-17

natural origin. Total herbal raw medicine use in the country for the year 2014–15 has been estimated at 5, 12,000 MT, with corresponding trade value of Rs. 5,500 crore. A total of 1178 species were used in the industry, of which 242 species traded more than 100 MT/year (National Medicinal Plants Board, Ministry of AYUSH, Govt. of India). The over-harvesting of medicinal plants for their bioactive substances is rapidly reducing this future valuable resource, as well as negatively affecting forest ecosystems and biodiversity.

Therefore, it remains very important to explore rarely encountered microbial resources such as the endophytes of medicinal plants, for their novel bioactive compounds as a source of therapeutic agents. Researchers have concluded that a plant can have multiple types of endophytes (Strobel and Daisy, 2003). Extensive testing and initiatives to introduce a large number of novel endophytes are ongoing. There is a positive linear correlation between medicinal plants and endophytes, in metabolite production, due to genetic recombination with the host over evolutionary time (Cai et al., 2004). Endophytes which indwell inhabit healthy living tissues without causing side-effects also yield certain biologically active secondary metabolites (Schulz et al., 2002). Endophytes are important sources of structurally distinct, natural bioactive metabolites like alkaloids, benzoquinones, benzopyranones, phenols, flavonoids, steroids, tetralones, tetralones, terpenoids, and xanthones, all of which have much potential for innovative therapeutic development (Tan and Zou, 2001). Endophytes have been discovered to be highly effective producers of immunosuppressive, cytotoxic, antiviral, antifungal, and antibacterial compounds (Wiyakrutta et al., 2004), which are vital in the management of various infectious diseases.

15.2 MICROBIAL ENDOPHYTES

The endophytes are a group of microorganisms that include fungi as well as bacteria, are gathered together in plants, and may be repeatedly isolated by any plant growth media (Singh and Dubey, 2015). Normally endophytic fungi and bacteria develop in plant internal tissues and resemble the host plant's chemistry. Endophytes can colonize plant leaf segments, stems, petioles, roots, weed inflorescences, buds, fruit, and seeds, as well as dead and hollowed hyaline cells (Specian et al., 2012; Stepniewska and Kuzniar, 2013). The endophyte population inside a plant relies on several factors, including the host species, inoculum density, environmental conditions, and the developmental stage of the host (Dudeja and Giri, 2014). Many endophytes have been found to be promising sources of novel, potent and active metabolites; as a result, they have attracted much attention from scientists all over the world. Endophytes have been found to produce a wide range of biologically active compounds in a particular plant or microorganism; consequently, they are a good source of medication for treating a wide range of illnesses, as well as having potential applications in medicine, agriculture, the food business, and cosmetics (Jalgaonwala et al., 2011). The study of endophytes from medicinal plants is therefore, important because metabolites in plants and endophytes may overlap.

15.3 ENDOPHYTES AS SOURCES OF ANTIMICROBIAL AGENTS

Approximately half of all deaths globally are caused by parasitic and infectious diseases (Menpara and Chanda, 2013). The global human population is experiencing exponential growth, and a slew of new health problems are emerging. For example, the rise in drug-resistant microorganisms is a matter of concern. Throughout the global fight against resistance to antibiotics, research on pharmaceuticals and other microbiological natural remedies has been critical. To address this challenge, new antibiotics are required. Natural sources have already been demonstrated to be the most reliable source for drug development. Medicinal plants, including accompanying endophytes, are a major source of valuable secondary metabolites and bioactive chemicals, accounting for almost 80% of all natural drugs on the market (Singh and Dubey, 2015). The use of therapeutic plant species in traditional medicine dates back to the beginning of time, and it is now widely assumed that all species of vascular plants harbor microbial endophytes. The natural medicinal substances produced by endophytic bacteria and fungi can be used in the pharmaceutical sector in a variety of ways. Ecomycins, xiamycinspseudomycins, and munumbicins are antibacterial, antiparasitic, and antimycotic natural compounds produced from endophytic bacteria. A few of these natural substances have antiviral properties against human immunodeficiency virus (HIV). Endophytes may be a source of novel antibiotics and antimicrobial agents to fight the rising number of microorganisms that are resistant to drugs. The potential of endophytes to manufacture diverse natural compounds is of key importance in the biopharmaceutical industry, as they are known to boost host plant tolerance to infections (Strobel and Daisy, 2003).

Coronamycins and rapamycin peptide antibiotics generated by a verticillate *Streptomyces lygroscopicus* (MSU-2110) endophytic on *Monstera* sp. were found by Ezra et al. (2004) to be effective inhibitors of *Saccharomyces cerevisiae* growth in fermented and bakery products. Terephthalic acid phomodione, a furandione from endophytes *Streptomyces* sp. *Achyranthes bidentata*, of *Saurauiascaberrinae* has been found by Hoffman et al. (2008) to have strong antibacterial action against *Staphylococcus aureus* grown in egg, meat, and dairy products. A number of antibacterial compounds produced by endophytic actinomycetes have been discovered,*Streptomyces* being the most common genera (Zhao et al., 2011). Antifungal and antibacterial substances such as jesterone (*Pestalotiopsisjesteri*), sordaricin (*Fusarium* sp.), and javanicin (*Chloridium*sp.) have been discovered to be effective against a number of foodborne bacterial infections (Jalgaonwala et al., 2011). Phytohormones and other bioactive substances of biotechnological significance (enzymes and pharmaceutical drugs) are produced by endophytes. *Nothapodytesfoetida* produces camptothecin, which has cytotoxic and antifungal effects (Joseph and Priya, 2011). Kaul et al. (2012) identified munumbicins, saadamycin from endophytes *Phomopsis* sp. *Cinnamomum mollissimum* which have antinfungal activity against fungal disease caused by *Aspergillus niger* in maize, groundnuts, cereals, and tree nuts. Fabatin, tyrosol from endophytes *Diaporthehelianthi*, was reported by Specian et al., 2012; Godstime et al. (2014) to cure nosocomial infection caused by *Enterococcus hirae* in hospitalized patients. Alvin et al. (2014) investigated the antibacterial chemical valinomycin and its potential in medicinal plant endophytes *Streptomyces tsusimaensis*. Cytonic acids A and B were discovered in *Cytonaema* sp. by Bhardwaj and Agrawal (2014). These are effective against CMV and hepatitis virus in humans. Endophytes *Cryptosporiopsisquercina* and *Nigrospora* sp. produce the bioactive agent saadamycin; in maize, cereals, groundnuts, and tree nuts, it serves as an antipathogen against disease caused by *Campylobacter jejuniand Fusarium oxysporum* plants (El-Gendy and El-Bondkly, 2010; Dutta et al., 2014). Cardiac glycosides and phenolic compounds from *Cladosporium* sp. and *Xylaria* sp. endophytes of medicinal plants have been found to exhibit substantial antibacterial activity against *Klebsiella pneumoniae* and *Streptococcus pyogenes*, according to Selvi and Balagengatharathilagam (2014). Munumbicins and the triterpenoid helvolic acid are produced by endophytic fungi *Streptomyces* sp., *Cytonaema* sp., and *Kennedia nigricans*, which have significant antibacterial effects against *Vibrio cholerae* (Kumar et al., 2014). Beauvericin from *Fusarium proliferatum* is effective against *Clostridium botulinum* developed in poorly processed canned food (Meca et al., 2010; Golinska et al., 2015). Endophytic actinomycetes, such as *Boesenbergia rotunda* and *Streptomyces coelicolor*, are reported to produce a variety of bioactive compounds with different structures, such as munumbicins, which have great medicinal usefulness against *Escherichia coli* (Singh and Dubey, 2015). Investigations have shown that these substances are isomers of numerous different phytohormones, essential oils, and other endophyte compounds (Molina et al., 2012; Nicoletti and Fiorentino, 2015).

15.4 MICROBIAL ENDOPHYTES AS SOURCES OF THERAPEUTIC DRUGS

The metabolites of microorganisms, plants, and animals are known as natural products. These natural products are significant because they have traditionally served as therapeutic sources. Natural compounds have often been used as sources of lead molecules, resulting in the development of several synthetic medications. Endophytes produce a variety of natural compounds that are antifungal, antibacterial, antidiabetic, immunosuppressive, and antioxidants. Paclitaxel (Taxol), the first billion-dollar anticancer drug in the world, is an amazing example of a primary substance derived from the yew tree (*Taxus wallachiana*) (Wani et al., 1971). Strobel et al. (1996) have shown that an endophytic fungus found in *P. microspora*, may likewise manufacture Taxol. The immunosuppressive cyclosporine, which was discovered in an endophytic fungus, *Tolypocladium inflatum*, greatly improved the importance and usefulness of endophytes (Toofanee and Dulymamode, 2002). Like endophytic fungi, endophytic bacteria offer tremendous potential for producing innovative organic products. In reality, scientists are investigating endophytes for innovative and unique natural compounds with commercial value. As a result, endophytes could be a major source of naturally occurring physiologically active chemicals. The majority of endophytic bacteria produce a wide variety of antibiotics, and they are one of the few undiscovered promising sources of new antibiotics. Microbial endophytes are a rich source of novel secondary metabolites that can be used as anti-arthritic, antibiotic, anticancer, antidiabetic, anti-insect, and immunosuppressive medicines (Godstime et al., 2014). To date, several medicinal plants have been studied for endophytic bacteria and fungi, and their potential to produce biologically active secondary compounds. Endophytes and associated plant hosts are thought to produce a greater amount and variety of biomolecules than epiphytes or soil-associated microorganisms due to their tight biological relationship. Furthermore, because this interaction is symbiotic, endophytic bioactive substances are predicted to have lower cell toxicity because they do not destroy the eukaryotic host. This is important for the health sector since potential medications may not harm human cells. It has been established that novel therapeutic substances for drug development exist in natural sources. Medicinal herbs and associated endophytes are indeed a major source of valuable phytochemical substances and secondary metabolites, contributing to more than 80% of all natural medications on the market (Singh and Dubey, 2015). Because of their enormous ability to contribute to the identification of new bioactive chemicals, endophytes have recently sparked the interest of the microbial biochemistry community. Endophytic variability and their capacity to create biologically active compounds have to date only been investigated in a few plants. The development of novel antibacterial secondary metabolites and bioactive substances from types of various endophytic microbes isolated from medicinal plants may be a significant

TABLE 15.1
Details of Some Bacterial Endophytes Having Therapeutic Properties

S. No.	Microbial Endophytes	Therapeutic Properties	Name of Medicinal Plants	Medicinal Plant Family	Reference
1.	*LactobacilliusRhamnosus*	Antidiabetic	*Euterpe oleraceae*	Arecaceae	Bichara and Rogez, (2011)
2.	*LactobacilliusCasei*	Antidiabetic	*Malusdomestica*	Rosaceae	Savino et al. (2012)
3.	*Exigobacterium Indicum*	Antidiabetic	*Solanum tuberosum*	Solanaceae	Collins et al. (1983)
4.	*Pseudumonarprotegens*	Antidiabetic	*Piper auritum*	Piperaceae	Kidarsa et al. (2011)
5.	*PeniBacillusPolympeal*	Anticancerous	*Lilium lancifolium*	Liliaceae	Weselowski et al. (2016)
6.	*Bacillius*	Anticancerous	*Malus domestica*	Rosaceae	Xiao et al. (2017)
7.	*B.Licheniformis*	Anticancerous	*Prosecco*	Vitaceae	Baldan et al. (2015)
8.	*B.Pseudomycoides*	Anticancerous	*Perennial ryegrass*	Poaceae	Berdy (2012)
9	*PaeriBacilliusDentriformis*	Anticancerous	*Aquilaria Malaccensis*	Thymelaeaceae	Ryu et al. (2013)
10	*Methylo Bacterium Radiotolerens*	Antioxidant	*Arachis hypogaea*	Fabaceae	Dourado and Cesar (2015)
11	*Macrococus*	Antioxidant	*Cyperus conglomerats*	Cyperaceae	Lafi et al. (2016)
12	*Aeromonar*	Antioxidant	*Gyycine max*	Fabaceae	Akinsanya et al. (2015)
13	*Cedecea*	Antioxidant	*Arugula*	Brassicaceae	Akinsanya et al. (2015)
14	*Streptimyces Sp.*	Antimalerial	*Monstera deliciosa*	Araceae	Ezra et al. (2004)

TABLE 15.2
Detail of Some Fungal Endophytes Having Therapeutic Properties

S. No.	Microbial Endophytes	Therapeutic Properties	Name of Medicinal Plants	Medicinal Plant Family	Reference
1.	*Aspergillus awamori*	Antidiabetic	*Acacia nilotica*	Fabaceae	Singh et al. (2015)
2.	*Syncephalastrum*	Antidiabetic	*Adhatodabeddomei*	Asteraceae	Ushasri and Anusha (2015)
3.	*Dendryphionnanum*	Antidiabetic	*Ficus religiosa*	Moraceae	Mishra et al. (2013)
4.	*Phomopsis sp*	Antidiabetic	*Paeonia delavayi*	Paeoniaceae	Huang et al. (2018)
5.	*Pseudumonarprotegns*	Antidiabetic	*Piper auritum*	Piperaceae	Gutierrez et al. (2012)
6.	*Phomasp*	Antidiabetic	*Salvadoraoleoides*	Salvadoraceae	Dhankhar et al. (2013)
7.	*A.niger*	Antidiabetic	*Tabebuia aurea*	Bignoniaceae	Kumar et al. (2017)
8.	*Phomabetae*	Anticancerous	*Costusspicatus*	Costaceae	Palanichamy et al. (2018)
9.	*Aternariasolani*	Anticancerous	*Ocimum sanctum*	Lamiaceae	Chandra (2012)
10	*Bacillius SP*	Anticancerous	*Withaniasomnifera*	Solanaceae	Ingle (2012)
11	*Aspergillus parasiticus*	Anticancerous	*Sequoia sempervirens*	Cupressaceae	Stierle et al. (1999)
12	*Periconiaatropurpurea*	Anticancerous	*Xylopiaaromatica*	Annonaceae	Teles et al. (2007)
13	*Phomabetae*	Antioxidant	*Costusspicatus*	Costaceae	Palanichamy et al. (2018)
14	*Cladosporium*	Antioxidant	*Moringa peregrina*	Moringaceae	Khan et al. (2017)
15	*Aspergillus versicolor*	Antioxidant	*Centella asiatica*	Apiaceae	Netala et al. (2016)
16	*Diaporthesp*	Antioxidant	*Emblica officinalis*	Phyllanthaceae	Ascencio et al. (2014)
17	*Chaetomium sp*	Antioxidant	*Eugenia jambalana*	Mrytaceae	Nath et al. (2012)
18	*Fusarium redolens*	Antioxidant	*Fritillaria unibracteata*	Liliaceae	Yadav et al. (2016)
19	*Phyllostictasp*	Antioxidant	*Gauzuma tomentosa*	Malvaceae	Srinivasan et al. (2010)
20	*Acremonium*	Antioxidant	*Kandis gajah*	Clusiaceae	Elfita and Munawar (2012)
21	*Phomabetae*	Antimalarial	*Costaceae*	Costaceae	Palanichamy et al. (2018)
22	*Fusarium sp*	Antimalarial	*Mentha longifolia*	Lamiaceae	Ibrahim et al. (2018)
23	*Purpureocilliumlilacinum*	Antimalarial	*Rauvolfia macrophylla*	Apocynaceae	Lentaet al. (2016)
24	*Preussiasp*	Antimalarial	*Emantiachloranthiaoliv*		Lentaet al. (2016)
25	*Diaporthesp*	Antimalerial	*Guapirastandleyana*	Araceae	Calcul et al. (2013)
26	*Verticillium sp*	Antimalarial	*Kandeliaobovata*	Rhizophoraceae	Calcul et al. (2013)
27	*Aspergillus sp*	Antimalarial	*Guapirastandleyana*	Araceae	Calcul et al. (2013)
28	*Phomopsis archeri*	Antimalerial	*Vanilla albidia*	Orchidaceae	Hemtasin et al. (2011)

alternative for combating drug resistance in many deadly pathogens, and for treating various human diseases (Godstime et al., 2014). Several bioactive chemicals that have been commercialized, such as camptothecin, diosgenin, paclitaxel, hypericin, vinblastine, and podophyllotoxin, are created by diverse endophytic fungi found in various plants, and they are relevant to the pharmaceutical and biotechnology industries (Zhao et al., 2011; Joseph and Priya, 2011). Tables 15.1 and 15.2 contain a comprehensive list of bacterial and fungal endophytes isolated from diverse medicinal plants, bioactive compounds, and their application as therapeutic drugs for the treatment of different diseases.

15.5 MICROBIAL ENDOPHYTES AS SOURCE OF ANTIDIABETIC DRUGS

Diabetes mellitus is a metabolic disorder caused by improper glucose metabolism, which is connected to low blood insulin levels or insulin sensitivity in target organs. This disease affects many individuals around the world; the number of patients has increased dramatically from 108 million in 1980 to 422 million in 2014, as well as the incidence rising from 171 million in 2000 to an expected 366 million in 2030 (WHO, 2016). In both industrialized and developing countries, diabetes is one of the main causes of death. Synthetic antidiabetic medications and allopathic hypoglycemic treatments, which reduce the function of some metabolic enzymes, notably amylases, can control diabetes. However, existing synthetic antidiabetic medicines and allopathic hypoglycemic medications are thought to have several drawbacks and adverse reactions. Endophytic fungus has recently been identified as a source of secondary bioactive compounds such as distinct and novel antibiotics, antioxidant anticancer and antidiabetic molecules.

To date over 2,000 naturally derived bioactive compounds have been identified from endophytes of medicinal plants, including amines and amides, chlorinated metabolites, indole and isocoumarin derivatives, lactone, polysaccharides, phenolic acids, phenol, phenylpropanoids, and xanthones (Strobel, 2003). Zhang et al. (1999) reported that the fungal endophytes fungus *pseudomassaria* sp. found in African rainforests generates L-783,281, an antidiabetic chemical that is an insulin mimic, and does not degrade in the digestive system as insulin does. Artanti et al. (2011) identified endophytic fungi *Taxus sumatrana* with antioxidative and alpha-glucosidase inhibitory properties. Recently, Khan et al. (2019) observed that fungal endophytes isolated from plant leaves such as *Azadirachta indica* and *Mangifera indica* exhibited significant amylase inhibition potential, implying that endophytic fungi isolated from these plants' leaves could be a good source of natural antidiabetic molecules. Many studies examining the potential impacts of microbial endophytes as antiviral, anticancer, and antibacterial agents have been undertaken in India, but only a few scientists have looked into the antidiabetic and hypolipidemic activities of fungal endophytes identified from antidiabetic botanicals. Endophytic fungi with antidiabetic activities were isolated from *Salvadoraoleoides* and *Adathodabeddomei* by Dhankhar et al. (2013). Singh and Kaur (2015) established that endophytic fungus *Aspergillus awamori* isolated from *Acacia nilotica* has antidiabetic properties. Singh et al. (2015) discovered an endophytic fungus *Cladosporuim* sp. from *Tinospora cordifolia* that inhibited alpha-glucosidase with high potency. Govindappa et al. (2015) extracted lectin (N-acetylgalactosamine, 64 kDa) from endophytic fungi, *Alternaria* species found on plant *Viscum* album. *Penicillium* species isolated from *Tabebuia argentea* were recently described as a source of strong bioactive chemicals with significant *in silico* and *in vitro* antidiabetic potential.

15.6 MICROBIAL ENDOPHYTES AS SOURCE OF ANTICANCER DRUGS

Cancer is a serious health problem in both industrialized and developing countries, with a greater fatality rate in the latter. Despite breakthroughs in treatment, cancer is still one of the leading causes of death worldwide (Chen et al., 2013). Drug resistance and substantial side-effects are significant drawbacks of traditional chemotherapy and radiotherapy, prompting the discovery of novel anticancer medicines with high therapeutic efficacy and minimal or no side-effects. The potential of plant-derived chemicals as inhibitors of various phases of carcinogenesis and associated inflammatory processes is becoming increasingly clear, highlighting the significance of these substances in anticancer treatment. Many endophytic actinomycete substances have been identified, and they have been used as antimicrobials and toxic metabolites against cancerous cells. Because they are less harmful to healthy cells while also being more effective against drug-resistant bacteria, endophyte extracts have been shown to be a better alternative than chemotherapeutic medicines in terms of antitumor activity efficacy and adverse effects. As a result, natural endophyte-derived metabolites have received a lot of attention for their potential as human cancer chemoprevention and anticancer chemotherapeutic drugs (Cardoso-Filho, 2018). Natural compounds from endophytic gram-positive bacteria, such as anthracyclines, anthraquinones, aureolic acids, coumarins, glucans, glycopeptides, carzinophilin, flavonoids, mitomycins, macrotetrolides, naphthoquinones, quinoxalines, and polysaccharides have proved themselves one of the best sources of alternative medicine (Igarashi et al., 2007).

Exopolysaccharides from *Ophiopogon japonicas* endophytic bacteria *amyloliquefaciens* sp. were found to have anticancer action against the human gastric carcinoma cell lines SGC-7901 and MC-4 (Chen et al., 2013). The endophytic *Streptomyces alnumycin* inhibited the proliferation of human leukemia cells K562 (Bieber et al., 1998), while salaceyins-producing *Streptomyces laceyi* strain MS53 of plant *Ricinus communis* was found to be cytotoxic to a cell line of human breast cancer SKBR3 (Bieber et al., 1998; Kim et al., 2006). Furthermore, *Streptomyces hygroscopicus* strain TP-A0451, a pterocidin-producing endophyte, was found to inhibit the growth of human cancer cell lines LOX-IMVI, NCI-H522, SF539, and OVCAR-3 (Igarashi et al., 2006; Qin et al., 2011), *Streptomyces aureofaci*, an endophyte that produces 4-arylcoumarin (Taechowisan et al., 2007). Several plant growth promoters have been demonstrated to be cytotoxic to tumor cells. Invasion of mouse colon carcinoma cells 26-L5 was considerably reduced by *Lupinus angustifolius* endophytic anthraquinones generating *Micromonospora* sp. actinomycete nodules (Igarashi et al., 2007). Taechowisan et al. (2007) revealed that the Boesenbergia rotunda-endophytic biphenyls-producing *Streptomyces* sp. strain BO-07 has recently been found to have anticancer potential against the human HepG2 and Huh7 liver cell lines, as well as the HeLa cervical tumor cell line. Moreover, *Cinnamomum cassiaendophytic*, *Streptomyces cavourensis* strain YBQ59 was found to suppress the growth of EGFR-TKI-resistant human lung cancer cells H1299, and A549 (Vu et al., 2018).

Guo et al. (1998) reported that the endophytic fungus *Alternaria* sp. produces anticancerous compound vinblastine living in *Catharanthus roseus*. Both vinblastine and vincristine were synthesized by the endophytic fungus *Fusarium oxysporum*, *Talaromycesradicus*, and *Eutypella* spp. from *C. roseus* (Zhang et al., 2000; Palem et al., 2015; Kuriakose et al., 2016). Podophyllotoxin (podofilox) is an anticancer and antiviral aryl tetralin lactone lignan found in the *Podophyllum* sp. *Diphylleia*, *Dysosma*, and *Juniperus* plant that has therapeutic properties. Yang et al. described the anticancerous podophyllotoxin-producing endophytic fungus from *P. hexandrum*, *Dysosmaveitchii*, and *Diphylleia sinensis* for the first time in 2003 (Yang et al., 2003). Puri et al. (2006) discovered a fungal endophyte,*Trameteshirsuta*, in *P. hexandrum* rhizomes podophyllotoxin and other aryl tetralin lignans are produced by this organism with significant anticancer activities. There are a number of other endophytic fungi, including *Phialocephalafortinii* isolated from *Podophyllum peltatum* rhizomes, *F. oxysporum* isolated from *Juniperus recurva*, *Phialocephalahexandrum*, and *A. fumigatus* isolated from *J. communisAlternaria* sp. isolated from *J vulgaris* have since been reported as alternative sources of pod (Lu et al., 2006; Eyberger et al., 2006; Cao et al., 2007; Kusari et al., 2009). Recently, podophyllotoxin was discovered in the secondary metabolites of *A. tenuissima*, *Fusarium* sp. and a fungal endophyte from the roots of *Podophyllum emodi* from *Dysosmaversipellis* (Liang et al., 2016; Tan et al., 2018). Toxin podophyllotoxin has been produced from 17 different endophytic fungi and ten different host plant types were used to gather the species. Diosgenin, an anticancer and anti-inflammatory compound, is derived mostly from *Dioscoreazingiberensis*. Endophytes may be appropriate alternatives to generate diosgenin, given the depletion of populations in their natural habitat and the requirement of a long period of root development for its principal source *Dioscorea zingiberensis*. *Paecilomyces* sp., found in *Paris polyphylla*, was first described as an endophytic fungus that produces adiosgenin by Zhou et al. (2004). Later, when complemented using the extract of the root of its host plant, in liquid cultures, an endophytic strain of *Fusarium* sp. from *Dioscoreanipponica* produced more diosgenin (Ding et al., 2014). Hypericin (naphthodianthrone) is an antidepressant, antineoplastic, anticancer, antiviral, and photosensitizer chemical produced from *Hypericum perforatum*. Hypericin was later discovered to be produced by endophytic fungus *Thielaviasubthermophila*, *C. globosum*, and *Epicoccum nigrum* isolated from *H. perforatum* (Kusari et al., 2008; Vigneshwari et al., 2019). Tanshinones (tanshinone I, tanshinone IIA, tanshinone IIB, cryptotanshinone, and isotanshinone I) are diterpenoid quinine metabolites discovered in the roots of *Salvia* spp. and are thought to be effective anticancer, antiatherosclerosis, antihypertensive, and neuroprotective drugs. Tanshinones were created by endophytic fungus *Phoma glomerata* and *Alternaria* sp. living in the roots of *Salvia miltiorrhiza* (Li et al., 2016; Lou et al., 2016). Forskolin, a physiologically active labdane Indian *Coleus forskohlii* whose roots contain a diterpene compound has antiglaucoma, anti-HIV, and anticancer properties. *Rhizoctonia bataticola*, an endophytic fungus isolated from *C. forskohlii*, was discovered to produce forskolin (Mir et al., 2015). Rohitukine is largely extracted from the bark of *Dysoxylumbinectariferum* and is a lead for the semisynthetic potential anticancer medicines flavopiridol and P-276-00. Kumara and colleagues revealed an endophyte *Fusarium proliferatum* from *D. binectariferum* that manufactures rohitukine from the host in 2012 (Kumara et al., 2012). Other *Fusarium* species that produce rohitukine were later discovered in *D. binectariferum* and *Amoorarohituka* (Kumara et al., 2014). Sanguinarine is a lethal benzophenanthridine alkaloid discovered in the roots of *Sanguinaria canadensis* and *Macleaya cordata* leaves that has recently attracted interest for its cytotoxicity and anticancer properties (Wang et al., 2014). In cancer cells, it also suppresses microtubule synthesis and particularly causes DNA damage. Endophytic fungi *F. proliferatum* isolated from *M. cordata* leaves has been found to produce SA (Wang et al., 2014). Vinblastine and vincristine are anticancer terpenoid indole alkaloids are found in abundance in Madagascar periwinkle (*Catharanthus roseus*). Vincristine, an alkaloid, inhibits malignant cells spindle assembly, ion channels, and revascularization while not affecting normal cells. Guo et al. (1998) isolated and identified vinblastine from *Alternaria* sp., an endophytic fungus living in *C. roseus*. Both vincristine and vinblastine were generated by the endophytic fungus *Talaromycesradicus*, *Fusarium oxysporum*, and *Eutypella*spp. from *C. roseus* (Palem et al., 2015; Kuriakose et al., 2016).

15.7 MICROBIAL ENDOPHYTES AS SOURCE OF ANTIOXIDANTS

Sanguinarine, a poisonous benzophenanthridine alkaloid reported in *Sanguinaria canadensis* roots and *Macleaya cordata* leaves, has recently attracted interest for its cytotoxic and anticancer properties. Endophytic fungi *F. proliferatum* isolated from *M. cordata* leaves has also generated SA (Wang et al., 2014). Due to its flavonoid and steroidal alkaloid content the medicinal plant *Solanum nigrum* has anticancer, antibacterial, antioxidant, antipyretic, anti-inflammatory, diuretic, and hepatoprotective activities. When an endophytic fungus *A. flavus* was separated, more solamargine was formed from its stem than from the host callus culture, which was surprising (El-Hawary et al., 2016). Curcumin has already been extracted from *Chaetomium globosum* and an unnamed fungal endophyte (Wang et al., 2011; Yan et al., 2014).Quercetin is a reddish pigment found in plants that have antioxidant, anticancer, and anti-inflammatory, antidiabetic, antiviral, neuroprotective, and cardiovascular activities. In cancer cells, quercetin promotes cell cycle arrest in the S phase and initiates apoptosis (Srivastava et al., 2016). The neuroprotective action of quercetin may be due to the stimulation of nuclear erythroid 2-related factor 2 (Nrf2)/in aspect of antioxidant response (ARE) and antioxidant paraoxonase 2 (PON_2), which may help to reduce oxidative stress (Costa et al., 2016). Three quercetin derivatives were recently derived from *Nigrosporaoryzae*an endophytic fungus isolated from the leaves of *Nigerian mistletoe*, and *Loranthus micranthus* (Ebada et al., 2016). Silymarin is a natural bioactive chemical found in milk thistle (*Silybum marianum*) fruits that exhibits antihepatitic, antioxidant, anti-inflammatory, cardioprotective, hepatoprotective, and immunomodulatory properties (Ramasamy and Agarwal, 2008). Silymarin's hepatoprotective properties are achieved through a rise in glutathione levels, suppression of lipid peroxidation, antioxidant defense activation, and translational actions in hepatic cells (Vargas-Mendoza et al., 2014). Endophytic silymarin has been discovered for the first time in *Aspergillus iizukae* strains isolated from *S. marianum* leaves and stems (El-Elimat et al., 2014).

15.8 MICROBIAL ENDOPHYTES AS SOURCE OF ANTIMALARIAL DRUGS

In tropical countries, malaria (a protozoal infection) is extremely frequent. Chloroquin, the gold standard antimalarial drug, has several fatal adverse effects, including hemolysis. In the treatment of malaria, a bioactive natural product or its derivatives that are free of such side-effects could be beneficial. Antibiotics derived from endophytic bacteria, such as *Munumbicin* and *Kakadumycins*, are effective against *P. falciparum*, one of the malaria-causing organisms. Fixing these as lead compounds in the hunt for an anti-plasmodial medication could result in a drug that has fewer adverse effects. *Cinchona* spp. root and stem bark have long been utilized as quinine sources. Meanwhile, until the 1940s, when synthetic antimalarial drugs were developed, it remained the only effective therapy for malaria for decades. Quinine is an antimalarial drug that works as an intraerythrocytic schizonticide as well as a gametocytocidal against *Plasmodium vivax* and *P. malariae*, but not *P. falciparum* (Achan et al., 2011). Simanjuntak et al. (2002) derived quinine from endophytic fungus. Maehara and colleagues discovered 21 quinine-producing strains of endophytic fungi *Cinchona ledgeriana*, and identified them as *Diaporthe*, *Arthrinium*, *Fomitopsis*, *Penicillium*, *Phomopsis*, and *Schizophyllum*(Maehara et al., 2013). Hidayat et al. found seven different *Fusarium* strains from three different *Fusarium* species that can produce quinine (Hidayat et al., 2016).

15.9 CONCLUSION

Endophytes are a type of endosymbiotic microorganism that colonizes plants and any microbial or plant growth medium can be used to isolate it. They serve as repositories for unique secondary metabolites with bioactive properties such as alkaloids, steroids, phenolic acids, saponins, quinones, terpenoids, and tannins, which have anticancer, antibacterial, anti-insect, and other characteristics. While plant sources of novel drugs are being researched extensively for medicinal purposes, endophytic microbes are also a key source for drug development. Endophytes have been shown to be reliable sources of bioactive unique molecules and have been functional in the development of new drugs. Since the first reports of endophytic bacteria producing secondary metabolites with industrial potential, more evidence has emerged that endophytic bacteria and fungi isolated from medicinal plants are one of the most important sources of new chemicals with verified potential for agriculture, medical, and pharmaceutical applications, owing to their plant growth promotion, anticancer, and antimicrobial activities. From therapeutic plants all over the world, several endophytic bacteria and fungi have been discovered, and this is not by chance. It has been discovered that medicinal plants' endophytes generate plant hormones, numerous enzymes, aromatic compounds, lipopeptides, and polysaccharides, indicating that they have a lot of potential for the discovery of novel therapeutic bioactive substances.

REFERENCES

Achan, J., Talisuna, A. O., Erhart, A., Yeka, A., Tibenderana, J. K., Baliraine, F. N., Rosenthal, P. J., & D'Alessandro, U. (2011). Quinine, an old anti-malarial drug in a modern world: role in the treatment of malaria. *Malaria Journal, 10*, 144.

Akinsanya, M. A., Goh, J. K., Lim, S. P., & Ting, A. S. (2015). Diversity, antimicrobial and antioxidant activities of culturable bacterial endophyte communities in *Aloe vera. FEMS Microbiol Lett.*, 362 (23), 184.

Alvin, A., Kristin, I., Miller, B., & Neilan, A. (2014). Exploring the potential of endophytes from medicinal plants as sources of antimycobacterial compounds. *Microbiol Res.*,169, 483–495.

Artanti, N., Tachibana, S., Leonardus, K., & Sukiman, H. (2011). Screening of endophytic fungi having the ability for antioxidative and α-glucosidase inhibitor activities isolated from *Taxus sumatrana. Pakistan J Biol Sci.*,14, 1019–1023.

Ascencio, P. G. M., Ascencio, S. D., Aguiar, A. A., Fiorini, A., & Pimenta, R. Z. (2014). Chemical assessment and antimicrobial and antioxidant activities of endophytic fungi extract isolated from *Costus spiralis* (Jacq.) Roscoe (Costaceae). *Evid Based Comp. Alternat Med.*, 190543. doi:10.1155/2014/190543

Baldan, E., Nigris, S., Romualdi, C., D'Alessandro, S., Clocchiatti, A., & Zottini, M. (2015). Beneficial bacteria isolated from grapevine inner tissues shape *Arabidopsis thaliana* roots. *PLoS One*, 10, 10.

Berdy, J. (2012). Thoughts and facts about antibiotics: where we are now and where we are heading. *J.Antibiot.*, 65 (8), 385–395.

Bhardwaj, A., & Agrawal, P. (2014). A review fungal endophytes: asastorehouse of bioactive compound. *World J.Pharm.Pharm.Sci.*, 3, 228–237.

Bichara, C. M. G., & Rogez, H. (2011). Acai (*Euterpe oleracea* Martius). In: YahiaE M (Ed.) *Postharvest biology and technology of tropical and subtropical fruits*, Vol.2: Açaí to citrus, 1st edn. Woodhead Publishing Series in Food Science, Technology and Nutrition, Cambridge, UK pp. 1–27.

Bieber, B., Nuske, J., Ritzau, M., & Grafe, U. (1998). Alnumycin a new naphthoquinone antibiotic produced by an endophytic *Streptomyces* sp. *J. Antibiot.*51, 381–382.

Cai, Y., Luo, Q., Sun, M., & Corke, H. (2004). Antioxidant activity and phenolic compounds of 112 traditional Chinese medicinal plants associated with anticancer. *Life Sci.*, 74 (17), 2157–2184.

Calcul, L., Waterman, C., Ma, W. S., Lebar, M. D., Harter, C., Mutka, T., Morton, L., Maignan, P., Van Olphen, A., Kyle, D. E., Vrijmoed, L., Pang, K. L., Pearce, C., & Baker, B. J. (2013). Screening mangrove endophytic fungi for antimalarial natural products. *Marine Drugs*, *11*(12), 5036–5050.

Cao, L., Huang, J., & Li, J. (2007). Fermentation conditions of *Sinopodophyllumhexandrum* endophytic fungus on the production of podophyllotoxin. *Food Ferment Ind.*, *33*, 28–32.

Cardoso-Filho, J. A. (2018). Endophytic microbes as a novel source for producing anticancer compounds as multidrug resistance modulators, in *Anticancer Plants: Natural Products and Biotechnological Implements*, M. Akhtar and M. Swamy(Eds) (Singapore: Springer), pp 343–381.

Chandra, S. (2012). Endophytic fungi: Novel sources of anticancer lead molecules. *Appl. Microbiol Biotechnol.*, 95, 47–59.

Chen, Y. T., Yuan, Q., Shan, L. T., Lin, M. A., Cheng, D. Q., & Li, C. Y. (2013). Antitumor activity of bacterial exopolysaccharides from the endophyte *Bacillus amyloliquefaciens* sp. isolated from *Ophiopogon japonicus*. *Oncol. Lett.*, 5, 1787–1792.

Collins, M. D., Lund, B. M., Farrow, J., & Schleifer, K. H. (1983). Chemotaxonomic study of an alkalophilic bacterium, *Exiguobacteriumaurantiacum* gen. *Microbiol*, 129 (7), 2037–2042.

Costa, L. G., Garrick, J. M., Roque, P. J., & Pellacani, C. (2016). Mechanisms of neuroprotection by quercetin: Counteracting oxidative stress and more. *Oxid Med Cell Longev.*, 1–10. doi:10.1155/2016/2986796

Dhankhar, S., Dhankhar, S., & Yadav, J. P. (2013). Investigations towards new antidiabetic drugs from fungal endophytes associated with *Salvadoraoleoides* Decne. *Med Chem.*,9 (4), 624–632.

Ding, C. H., Du, X. W., Xu, Y., Xu, X. M., Mou, J. C., Yu, D., Wu, J. K., Meng, F. J. Liu, Y., Wang, W. L., & Wang, L. J. (2014). Screening for differentially expressed genes in endophytic fungus strain 39 during co-culture with herbal extract of its host *Dioscorea nipponica* Makino. *Curr. Microbial*. 69, 517–524.

Dourado, G. K., & Cesar, T. B. (2015). Investigation of cytokines, oxidative stress, metabolic, and inflammatory biomarkers after orange juice consumption by normal and overweight subjects. *FoodNut Res.*, 59 (1), 28147.

Dudeja, S. S., & Giri, R. (2014). Beneficial properties, colonization, establishment, and molecular diversity of endophytic bacteria in legume and non-legume. *Afr. J.Microbiol.Res.* 8, 1562–1572.

Dutta, D., Puzari, K. C., & Gogoi, R., & Dutta, P. (2014). Endophytes: Exploitation as a tool in plant protection. *Braz Arch Biol Technol* 57, 621–629.

Ebada, S. S., Eze, P., Okoye, F. B., Esimone, C. O., & Proksch, P. (2016). The fungal endophyte *Nigrosporaoryzae* produces quercetin monoglycosides previously known only from plants. *Chem. Select*, 16, 2767–2771.

El-Elimat, T., Huzefa, R., Tyler, G., Stanley, F., Nadja, C., & Nicholas, O. (2014). Flavonolignans from *Aspergillus iizukae*, a fungal endophyte of milk thistle (*Silybum marianum*). *J. Nat. Prod.*, 77 (2), 193–199.

Elfita, M., & Munawar, R. (2012). Isolation of antioxidant compound from endophytic fungi *Acremonium* sp. from the Twigs of Kandis Gajah. *Makara J. Sci.*, 16, 46–50.

El-Gendy, M. M. A., & El-Bondkly, A. M. A. (2010). Production and genetic improvement of a novel antimycotic agent, saadamycin, against dermatophytes and other clinical fungi from endophytic *Streptomyces* sp. Hedaya 48. *J. Ind. Microbiol. Biotechnol.*, 37, 831–841.

El-Hawary, S., Mohammad, R., Sameh, A., Walid, B., Rainer, E., Ahmed, S., & Mostafa, R. (2016). Solamargine production by a fungal endophyte of *Solanum nigrum*. *J. App.Microbio.*, 120. n/a–n/a. 10.1111/jam.13077.

Eyberger, A. L., Dondapati, R., & Porter, J. R. (2006). Endophyte fungal isolates from *Podophyllum peltatum* produce podophyllotoxin. *J Nat Prod.*, 69, 1121–1124.

Ezra, D., Castillo, U. F., Strobel, G. A., Hess, W. M., Porter, H., Jensen, J. B., Condron, M. A.M., & Teplow, D. B. (2004). Coronamycins, peptide antibiotics produced by a verticillate *Streptomyces sp.* (MSU-2110) endophytic on *Monstera sp. Microbiology (Reading)*, 4, 785–793.

Farnsworth, N. R., & Soejarto, D. D. (1991). Global importance of medicinal plants. In: Akerele, O., HeywoodV, Synge H. (Eds). *The conservation of medicinal plants.* Cambridge University Press, Cambridge, UK. pp. 25–51.

Godstime, O. C., Enwa, F. O., Augustina, J. O., & Christopher, E. O. (2014). Mechanisms of antimicrobial actions of phytochemicals against enteric pathogens: A review. *J Pharm Chem Biol Sci.*, 2, 77–85.

Golinska, P., Wypij, M., Agarkar, G., Rathod, D., Dahm, H., & Rai, M. (2015). Endophytic actinobacteria of medicinal plants: diversity and bioactivity. *Antonie Van Leeuwenhoek*, 108, 267–289.

Govindappa, M., Sadananda, T. S., & Channabasava, R., Ramachandra, Y. L., Chandrappa, C. P., Padmalatha, R. S., & Prasad, S. K. (2015). Integrative obesity and diabetes *in vitro* and *in vivo* antidiabetic activity of lectin (N-acetyl-galactosamine, 64 kDa) isolated from endophytic fungi, *Alternaria* species from *Viscum album* on alloxan induced diabetic rats. doi:10.15761/IOD.1000104.

Guo, B., Li, H., & Zhang, L. (1998). Isolation of a fungus producingvinbrastine. *J. Yunnan Univ. Nat. Sci.*, *20*, 214–215.

Gutierrez, R. M., Gonzalez, A. M., & Ramirez, A. M. (2012). Compounds derived from endophytes: A review of phytochemistry and pharmacology. *Curr. Med. Chem.*, 19, 2992–3030.

Hemtasin, C., Kanokmedhakul, S., Kanokmedhakul, K., Hahnvajanawong, C., Soytong, K., Prabpai, S., & Kongsaeree, P. (2011). Cytotoxic pentacyclic and tetracyclic aromatic sesquiterpenes from *Phomopsis archeri*. *J Nat Prod.*, 74, 609–613.

Hidayat, I., Radiastuti, N., Rahayu, G., Achmadi, S., & Okane, I. (2016). Three quinine and cinchonidine producing *Fusarium* species from Indonesia. *Curr Res Environ Appl Microbiol.*,*6*, 20–34.

Hoffman, A. M., Mayer, S. G., Strobel, G. A., Hess, W. M., SovocoolG, W., & Grange, A. H. (2008). Purification, identification and activity of phomodione, a furandione from an endophytic *Phoma* species. *Phytochemistry*, 69, 1049–1056.

Huang, R., Jiang, B. G., Li, X. N., Wang, Y. T., Liu, S. S., Zheng, K. X., He, J., & Wu, S. H. (2018). Polyoxygenated cyclohexenoids with promising α-glycosidase inhibitory activity produced by *Phomopsis* sp. YE3250, an endophytic fungus derived from *Paeonia delavayi*. *J Agric Food Chem.*, 66, 1140–1146.

Ibrahim, S. R. M., Abdallah, H. M., Elkhayat, E. S., Al Musayeib, N. M., Asfour, H. Z., Zayed, M. F., Mohamed, G. A., & Fusaripeptide, A. (2018). New antifungal and anti-malarial cyclodepsipeptide from the endophytic fungus *Fusarium* sp. *J Asian Nat Prod Res.*, 20 (1), 75–85.

Igarashi, Y., Miura, S. S., Fujita, T., & Furumai, T. (2006). Pterocidin, a cytotoxic compound from the endophytic *Streptomyces hygroscopicus*. *J Antibiot.*, 59, 193–195.

Igarashi, Y., Trujillo, M. E., Martínez-Molina, E., Yanase, S., Miyanaga, S., & Obata, T. (2007). Antitumor anthraquinones from an endophytic actinomycete *Micromonospora lupine*sp, nov. *Bioorg Med Chem Lett.*17, 3702–3705.

Ingle, S. (2012). Isolation and characterization of endophytes from medicinal plant, *Withania Somnifera* (Ashwagandha). *IUP J. Biotechnol.* 5, 7–13.

Jalgaonwala, R. E., Mohite, B. V., & Mahajan, R. T. (2011). Natural products from plant-associated endophytic fungi. *J MicrobiolBiotechnol Res.*, 1, 21–32.

Joseph, B., & Priya, R. M. (2011). Bioactive compounds from endophytes and their potential in pharmaceutical effect: A review. *Am J Biochem Mol Bio.*, 1, 291–309.

Kaul, S., Gupta, S., Ahmed, M., & Dhar, M. K. (2012). Endophytic fungi from medicinal plants: A treasure hunt for bioactive metabolites. *Phytochem Rev.*, 11, 487–505.

Khan, A. L., Gilani, S. A., Waqas, M., Al-Hosni, K., Al-Khiziri, S., Kim, Y., Ali, L., Kang, S., Asaf, S., Shahzad, R., Hussain, J., Lee, I., & Al-Harrasi, A. (2017). Endophytes from medicinal plants and their potential for producing indole acetic acid, improving seed germination, and mitigating oxidative stress. *J Zhejiang Univ-Sci B (Biomed Biotechnol)*, 18, 125–137.

Khan, R., Naqvi, S.T.Q., Fatima, N., & Muhammad, S.A. (2019) Study of antidiabetic activities of endophytic fungi isolated from plants. *Pure Appl Biol.*, 8(2), 1287–1295.

Kidarsa, T., Neal, G., Zabriskie, T., & Joyce, L. (2011). Phloroglucinol mediates cross-talk between the pyoluteorin and 2,4-diacetylphloroglucinol biosynthetic pathways in *Pseudomonas fluorescens* Pf-5. *Molecular Microbiol.*, 81 (2), 395–414.

Kim, N., Shin, J. C., Kim, W., Hwang, B. Y., Kim, B. S., & Hong, Y. S. (2006). Cytotoxic 6-alkylsalicylic acids from the endophytic *Streptomyces laceyi*. *J Antibiot.*, 59, 797–800.

Kumar, K. M., Poojari, C. C., Ryavalad, C., Lakshmikantha, R. Y., Satwadi, P. R., Vittal, R. R., & Melappa, G. (2017). Anti-diabetic activity of endophytic fungi, *Penicillium species* of *Tabebuia argentea*; in silico and experimental analysis. *Res J Phytochem.*, 11, 90–110.

Kumar, S., Aharwal, R. P., Shukla, H., Rajak, R. C., &Sandhu, S. S. (2014). Endophytic fungi: As a source of antimicrobial bioactive compounds. *World J Pharm Pharm Sci.*, 3, 1179–1197.

Kumara, P. M., Soujanya, K. N., Ravikanth, G., Vasudeva, R., Ganeshaiah, K. N., & Shaanker, R. U. R. (2014). A chromone alkaloid and a precursor of flavopiridol is produced by endophytic fungi isolated from *Dysoxylum binectariferum* Hook.f and *Amoorarohituka* (Roxb). Wight &Arn. *Phytomedicine*, 21, 541–546.

Kumara, P. M., Zuehlke, S., Priti, V., Ramesha, B. T., Shweta, S., Ravikanth, G., Vasudeva, R., Santhoshkumar, T. R., Spiteller, M., & Umashaanker, R. (2012). *Fusarium proliferatum*, an endophytic fungus from *Dysoxylumbinectariferum* Hook.f, produces rohitukine, a chromane alkaloid possessing anti-cancer activity. *Antonie Van Leeuwenhoek*, 101, 323–329.

Kuriakose, G. C., Palem, P. P., & Jayabaskaran, C. (2016). Fungal vincristine from *Eutypella* spp-CrP14 isolated from *Catharanthus roseus* induces apoptosis in human squamous carcinoma cell line-A431. *BMC Complement Alternat Med.*, 16, 302.

Kusari, S., Lamshoft, M., & Spiteller, M. (2009). *Aspergillus fumigatus* Fresenius, an endophytic fungus from *Juniperus communis* L. Horstmann as a novel source of the anticancer pro-drug deoxypodophyllotoxin. *J Appl Mmicrobiol.*, 107, 1019–1030.

Kusari, S., Lamshoft, M., Zuhlke, S., & Spiteller, M. (2008). An endophytic fungus from *Hypericum perforatum* that produces hypericin. *J Nat Prod.*, 71, 159–162.

Lafi, F. F., Bokhari, A., Alam, I., Bajic, V. B., Hirt, H., & Saad, M. M. (2016). Draft genome sequence of the plant growth-promoting *Cupriavidusgilardii* strain JZ4 isolated from the desert plant *Tribulus terrestris*. *Genome Announc*, 4 (4), e00678–e00716

Li, X., Zhai, X., Shu, Z., Dong, R., Ming, Q., & Qin, L. (2016). *Phoma glomerata* D14: An endophytic fungus from *Salvia miltiorrhiza* that produces Salvianolic acid C. *CurrMicrobiol.*, 73, 31–37.

Liang, Z., Zhang, J., Zhang, X., Li, J., Zhang, X., & Zhao, C. (2016). Endophytic fungus from *Sinopodophyllumemodi* (Wall.) Ying that produces Podophyllotoxin. *J Chromatogr Sci.*, 54, 175–178.

Lou, J., Yu, R., Wang, X., Mao, Z., Fu, L., Liu, Y., & Zhou, L. (2016). Alternariol 9-methyl ether from the endophytic fungus *Alternaria* sp. Samif01 and its bioactivities. *Braz J Microbiol.*, 47, 96–101.

Lu, L., He, J., Yu, X., Li, G., & Zhang, X. (2006). Studies on isolation and identification of endophytic fungi strain SC13 from pharmaceutical plant *Sabina vulgaris* Ant. and metabolites. *Acta Agric Boreali-Occident Sin.*,15, 85–89.

Maehara, S., Simanjuntak, P., Maetani, Y., Kitamura, C., Ohashi, K., & Shibuya, H. (2013). Ability of endophytic filamentous fungi associated with *Cinchona ledgeriana* to produce *Cinchona* alkaloids. *J Nat Med.*, 67, 421–423.

Meca, G., Sospedra, I., Soriano, J. M., Ritieni, A., Moretti, A., & Manes, J. (2010). Antibacterial effect of the bioactive compound beauvericin produced by *Fusarium proliferatum* on the solid medium of wheat. *Toxicon*, 56, 349–354.

Menpara, D., & Chanda, S. (2013). Endophytic bacteria - the unexplored reservoir of antimicrobials for combating microbial pathogens, in *Microbial Pathogens and Strategies for Combating them: Science, Technology and Education*, ed. A. Méndez-Vilas (Badajoz: Formatex Research Center), 1095–1103.

Mir, R. A., Kaushik, S. P., Chowdery, R. A., & Anuradha, M. (2015). Elicitation of forskolin in cultures of *Rhizactoniabataticola*-a phytochemical synthesizing endophytic fungi. *Int J Pharm Pharma Sci.*, 7, 185–189.

Mishra, P. D., Verekar, S. A., Kulkarni-Almeida, A., Roy, S. K., Jain, S., Balakrishnan, A., Vishwakarma, R., & Deshmuk, S. K. (2013). Anti-inflammatory and antidiabetic naphthaquinones from an endophytic fungus *Dendryphionnanum* (Nees) S. Hughes. *Indian J Chem.*, 52, B, 556–565.

Molina, G., Pimentel, M. R., Bertucci, T. C. P., & Pastore, G. M. (2012). Application of fungal endophytes in biotechnological processes. *Chem Eng Trans.*, 27, 289–294

Nath, A., Raghunatha, P., & Joshi, S. R. (2012). Diversity and biological activities of endophytic fungi of *Emblica officinalis*, an ethnomedicinal plant of India. *Mycobiology*, 40 (1), 8–13.

Netala, V. R., Kotakadi, V. S., Bobbu, P., Gaddam, S. A., & Tartte, V. (2016) Endophytic fungal isolate mediated biosynthesis of silver nanoparticles and their free radical scavenging activity and antimicrobial studies. *3Biotech*, 6, 132.

Nicoletti, R., & Fiorentino, A. (2015). Plant bioactive metabolites and drugs produced by endophytic fungi of Spermatophyta. *Agriculture*, 5, 918–970.

Palanichamy, P., Krishnamoorthy, G., Kannan, S., & Murugan, M. (2018). Bioactive potential of secondary metabolites derived from medicinal plant endophytes. *Egyptian J Basic Appl Sci.*, 5 (4), 303–312.

Palem, P. P., Kuriakose, G. C., & Jayabaskaran, C. (2015). An endophytic fungus, *Talaromycesradicus*, isolated from *Catharanthus roseus*, produces vincristine and vinblastine, which induce apoptotic cell death. *PLoS ONE*, 10, 12.

Puri, S. C., Nazir, A., Chawla, R., Arora, R., Riyaz-ul-Hasan, S., Amna, T., Ahmed, B.Verma, V., Singh, S., & Sagar, R. (2006). The endophytic fungus *Trameteshirsuta* as a novel alternative source of podophyllotoxin and related aryl tetralin lignans. *J Biotechnol.*, 122, 494–510.

Qin, S., Xing, K., Jiang, J. H., Xu, L. H., & Li, W. J. (2011). Biodiversity, bioactive natural products, and biotechnological potential of plant-associated endophytic actinobacteria. *Appl MicrobiolBiotechnol.*, 89, 457–473.

Ramasamy, K., & Agarwal, R. (2008). Multitargeted therapy of cancer by silymarin. *Cancer Lett.*, 269, 352–362.

Ryu, C. M., Farag, M. A., Hu, C. H., Reddy, M. S., Wei, H. X., Pare, P. W., & Kloepper, J. W. (2013). Bacterial volatiles promotes growth in Arabidopsis. *Proc Natl Acad Sci USA*, 15; 100(8), 4927–4932.

Savino, M. J., Sanchez, L. A., Saguir, F. M., & Nadra, M. C. M. (2012). Lactic acid bacteria isolated from apples are able to catabolise arginine. *World J MicrobiolBiotechnol.*, 28, 1003–1012.

Schulz, B., Boyle, C., Draeger, S., Rommert, A. K., & Krohn, K. (2002). Endophytic fungi: A source of novel biologically active secondary metabolites. *Mycol Res.*, 106 (9), 996–1004.

Selvi, B. K., & Balagengatharathilagam, P. (2014). Isolation and screening of endophytic fungi from medicinal plants of Virudhunagar district for antimicrobial activity. *Int J Sci Nat.*, 5, 147–155.

Simanjuntak, P., Parwati, T., Bustanussalam, P., Wibowo, S., & Shibuya, H. (2002). Isolasidankultivasimikrobaed of it penghasilsenyawa alkaloid kinkonadari*Cinchona* spp. *J MikrobiolIndones.*, 7, 27–30.

Singh, B., & Kaur, A. (2015). Antidiabetic potential of a peptide isolated from an endophytic Aspergillus awamori. *J Appl Microbiol.*, 120, 301–311.

Singh, B., Kaur, T., Kaur, S., Manhas, R. K., & Kaur, A. (2015). An alpha-glucosidase inhibitor from an endophytic *Cladosporium* sp. with potential as a biocontrol agent. *Appl BiochemBiotechnol.*, 175, 2020–2034.

Singh, R., & Dubey, A. K. (2015). Endophytic actinomycetes as the emerging source for therapeutic compounds. *Indo Global J Pharm Sci.*, 5, 106–116.

Specian, V., Sarragiotto, M. H., Pamphile, J. A., & Clemente, E. (2012). Chemical characterization of bioactive compounds from the endophytic fungus *Diaporthehelianthi* isolated from *Lueheadivaricata. Braz J Microbiol.*, 43, 1174–1182.

Srinivasan, K., Jagadish, L. K., Shenbhagaraman, R., & Muthumary, J. (2010). Antioxidant activity of endophytic fungus *Phyllosticta* sp. isolated from *Guazuma tomentosa. J Phytology.*, 2, 37–41.

Srivastava, S., Somasagara, R. R., Hegde, M., Nishana, M., Tadi, S. K., Srivastava, M., Choudhary, B., & Raghavan, S. C. (2016). Quercetin, a natural flavonoid interacts with DNA, arrests the cell cycle, and causes tumor regression by activating mitochondrial pathway of apoptosis. *Sci Rep.*, 6, 240–249.

Stepniewska, Z., & Kuzniar, A. (2013). Endophytic microorganisms promising applications in bioremediation of greenhouse gases. *Appl MicrobiolBiotechnol.*, 97, 9589–9596.

Stierle, A. A., Stierle, D. B., & Bugni, T. (1999). Sequoiatones, A. and B: Novel Antitumor metabolites isolated from a redwood endophyte. *J Org Chem.*, 23, 64 (15), 5479–5484.

Strobel, G., Yang, X., Sears, J., Robert, K., Sidhu, R. S., & Hess, W. M. (1996). 5 Taxol from *Pestalotiopsismicrospora*, an endophytic fungus of *Taxus wallachiana. Microbiology* (Reading), 142(2), 435–440.

Strobel, G. A. (2003). Endophytes as sources of bioactive products. *Microb Infect.*, 5(6), 535–544.

Strobel, G. A., & Daisy, B. (2003). Bioprospecting for microbial endophytes and their natural products. *Microbiol Mol Biol Rev*, 67, 491–502.

Taechowisan, T. C., Lu, C. H., Shen, Y. M., & Lumyong, S. (2007). Antitumor activity of 4-arylcoumarins from endophytic *Streptomyces aureofaciens* CMUAc130. *J Cancer Res Ther.*, 3, 86–91.

Tan, R. X., & Zou, W. X. (2001). Endophytes: A rich source of functional metabolites. *Nat Prod Rep.*, 18(4), 448–459.

Tan, X., Zhou, Y., Zhou, X., Xia, X., Wei, Y., He, L., Tang, H., & Yu, L. (2018). Diversity and bioactive potential of culturable fungal endophytes of *Dysosma versipellis*; a rare medicinal plant endemic to China. *Sci. Rep.*, 8, 1–9. doi:10.1038/s41598-018-31009-0.

Teles, H., Sordi, R., Silva, G., Castro-Gamboa, I., Bolzani, V., Pfenning, L., Abreu, L., Costa-Neto, C., Marx, Y., Maria, C., & Angela, A. (2007). Aromatic compounds produced by *Periconiaatropurpurea*, an endophytic fungus associated with *Xylopiaaromatica. Phytochemistry.*, 67, 2686–2690.

Toofanee, S. B., & Dulymamode, R. (2002). Fungal endophytes associated with *Cordemoyaintegrifolia. Fungal Divers.*, 11, 169–175.

Ushasri, R., & Anusha, R. (2015). In vitro anti-diabetic activity of ethanolic and acetone extracts of endophytic fungi *Syncephalastrumracemosum* isolated from the seaweed *Gracilariacorticata* by alpha-amylase inhibition assay method. *Int J CurrMicrobiol App Sci.*, 4(1), 254–259.

Vargas-Mendoza, N., Madrigal-Santillán, E., Morales-Gonzalez, A., Esquivel-Soto, J., Esquivel-Chirino, C., Gonzalez-Rubio, M. G. L., Gayosso-de-Lucio, J. A., & Morales-Gonzalez, J. A. (2014). Hepatoprotective effect of silymarin. *World J Hepatol.*, *6*, 144–149.

Vigneshwari, A., Rakk, D., Nemeth, A., Kocsube, S., Kiss, N., & Csupor, D. (2019). Host metabolite producing endophytic fungi isolated from *Hypericum perforatum. PLoS ONE*, 14, e0217060.

Vu, H. N. T., Nguyen, D. T., Nguyen, H. Q., Chu, H. H., Chu, S. K., & Van Chau, M. (2018). Antimicrobial and cytotoxic properties of bioactive metabolites produced by *Streptomyces cavourensis* YBQ59 isolated from *Cinnamomum cassia*prels in Yen Bai province of Vietnam. *CurrMicrobiol.*, 75, 1247–1255.

Wang, X. J., Min, C. L., Ge, M., & Zuo, R. H. (2014). An endophytic sanguinarine-producing fungus from *Macleaya cordata, Fusarium proliferatum* BLH51. *CurrMicrobiol.*, 68, 336–341.

Wang, Y., Xu, L., Ren, W., Zhao, D., Zhu, Y., & Wu, X. (2011). Bioactive metabolites from *Chaetomium globosum* L18, an endophytic fungus in the medicinal plant *Curcuma wenyujin* Phytomedicine. *Int J Phytotherapy Phytopharmacol.*, 19, 364–368.

Wani, M. C., Taylor, H. L., Wall, M. E., Goggon, P., & McPhail, A. T. (1971). Plant antitumor agents VI. The isolation and structure of Taxol, a novel antileukemic and antitumor agent from *Taxus brevifolia. J Am Chem Soc.*, 93, 2325–2327.

Weselowski, B., Nathoo, N., Eastman, A. W., MacDonald, J., & Yuan, Z. C. (2016). Isolation, identification, and characterization of *Paenibacilluspolymyxa* CR1 with potentials for biopesticide, bio fertilization, biomass degradation and biofuel production. *BMC Microbiology*, 16, 244.

WHO (2016) *Global report on diabetes*, World Health Organization.

Wiyakrutta, S., Sriubolmas, N., Panphut, W., Thongon, N., Danwisetkanjana, K., & Ruangrungsi, N. (2004). Endophytic fungi with anti-microbial, anti-cancer, and anti-malarial activities isolated from Thai medicinal plants. *World J MicrobiolBiotechnol.*, 20 (3), 265–272.

Xiao, Z., Zhang, Y., Chen, X., Wang, Y., Chen, W., Xu, Q., Li, P., & Ma, F. (2017). Extraction, identification, and antioxidant and anticancer tests of seven dihydrochalcones from Malus 'Red Splendor' fruit. *Food Chem.*, 15(231), 324–331.

Yadav, M., Yadav, A., Kumar, S., & Yadav, J. P. (2016). Spatial and seasonal influences on culturable endophytic mycobiota associated with different tissues of *Eugenia jambolana* Lam. and their antibacterial activity against MDR strains. *BMC Microbiol.*, 16, 44.

Yan, J., Qi, N., Wang, S., Gadhave, K., & Yang, S. (2014). Characterization of secondary metabolites of an endophytic fungus from *Curcuma wenyujin. CurrMicrobiol.*69, 740–744.

Yang, X., Guo, S., Zhang, L., & Shao, H. (2003). Select of producing podophyllotoxin endophytic fungi from podophyllin plant. *Nat Prod Res Dev.*, 15, 419–422.

Zhang, B., Salituro, G., Szalkowski, D., Li, Z., Zhang, Y., Royo, I., Vilella, D., Dez, M., Pelaez, F., Ruby, C., Kendall, R. L., Mao, X., Griffin, P., Calaycay, J., Zierath, J. R., Heck, J. V., Smith, R. G., & Moller, D. E. (1999). Discovery of small molecule insulin mimetic with antidiabetic activity in mice. *Science*, 284, 974–981.

Zhang, L., Guo, B., Li, H., Zeng, S., Shao, H., Gu, S., & Wei, R. (2000). Preliminary study on the isolation of endophytic fungus of *Catharanthus roseus* and its fermentation to produce products of therapeutic value. *Chin Tradit Herb Drugs*, 31, 805–807.

Zhao, J., Shan, T., Mou, Y., & Zhou, L. (2011). Plant-derived bioactive compounds produced by endophytic fungi. *Mini Rev Med Chem.*, 11, 159–168.

Zhou, L., Cao, X., Yang, C., Wu, X., & Zhang, L. (2004). Endophytic fungi of *Paris polyphylla* var. *yunnanensis* and steroid analysis in the fungi. *Nat Prod Res Dev.*, 16, 198–200.

16 Utility of Probiotics in Aquaculture

Reema Mishra, Renu Soni, Preeti Agarwal and Aparajita Mohanty
Gargi College, University of Delhi, New Delhi, India

CONTENTS

16.1 INTRODUCTION

Aquaculture, often known as aquafarming or fish farming, is the practice of growing aquatic organisms in a controlled environment, primarily for human consumption. Aquaculture is a significant and quickly developing area as it plays a significant part in fulfilling the worldwide need for protein

DOI: 10.1201/9781003306931-18

supplements. According to FAO (2014), aquaculture products along with fish provide a very important supply of the micronutrients and proteins required for balanced and healthy diets across the globe. The world population is growing at an annual rate of 1.6%. The lack of consistent food supply will require searching for protein sources in particular in future. Aquaculture is one of the most economical and reliable option to address the growing demand for fish proteins (Eyo and Ivon, 2017; Eyo and Akanse, 2018).

Aquaculture not only plays an important role in improving the socio-economic condition of an area by acting as the source of important nutrients, but is also significant in creating employment (Araújo et al., 2015). Aquaculture plays a dynamic role in providing sustainable livelihoods and food security for a growing population. Therefore, the expansion, divergence, and amplification of aquaculture methods are needed to obtain higher yields. This will increase pressure on the environment and aquatic animals, resulting in disease outbreaks, reduced production, and threats to the sustainability of aquaculture (Martínez Cruz et al., 2012). Although the trend is encouraging, some constraints have negatively affected aquaculture growth, disease being the main limiting factor. Bacterial diseases cause a massive number of deaths amongst wild and farmed fish, restricting aquaculture production, which has affected the social and economic situation of the native population, as well as the national economy, in various countries (Bondad-Reantaso et al., 2005; Hai, 2015).

Excessive amounts of chemotherapeutic agents like antibiotics and other compounds are used extensively in the regulation of diseases. The use of expensive chemotherapy drugs to control diseases has been widely criticized because of its negative effects, including accumulation of residues, resistance to drugs and immunosuppression. The use of antibiotics causes many problems and also results in the evolution and spread of resistant pathogens which ultimately affect human health (Martínez Cruz et al., 2012).

Therefore, it is necessary to develop alternative environment-friendly techniques for creating healthy microbial surroundings in the aquaculture system to maintain the health of aquatic organisms. The use of probiotics to control potential pathogens and achieve sustainable yields of aquatic produce is becoming increasingly important (Chauhan and Singh, 2019). Probiotics are live organisms (typically bacteria or yeast, or a combination of both) that are taken with food and have a variety of beneficial effects on the host (Fuller, 1989). It is now widely accepted that probiotics are efficient new agents that play a significant part in gut wellbeing by modulating the structure of the gut microbiota (microbial community) (Nayak, 2010). It has been shown that probiotics perform a wide range of functions in the host such as improving health, reducing illness and stress, improving immunity and digestion, regulating intestinal flora, aiding nutrition, improving water quality, preventing infection and bioremediation (Verschuere et al., 2000; Gobi et al., 2018; Nandi et al., 2018; Ramesh and Souissi, 2018; Helmy et al., 2019). Kozasa (1986) conducted the first experimental attempt at probiotic application in aquaculture, based on the benefits of probiotics to poultry and humans.

Probiotics, which have antibacterial, antiviral, and antifungal effects, aid aquatic animals, particularly fish, in fighting infections and improving their health. The beneficial effects of probiotics also help increase the nutritional value and growth of the animals, and increase the spawning and hatching rate of the aquaculture system (Chauhan and Singh, 2019; Hasan and Banerjee, 2020).

This chapter aims to provide useful information on the importance of probiotics, their selection criteria, widely accepted probiotic organisms, their probable mechanism of action, administrative procedures in aquaculture, the importance of the aquaculture industry, and its future prospects.

16.2 PROBIOTICS AND THEIR TYPES

Dr Elie Metchnikoff (1907) was the first to identify the beneficial effect of bacteria among farmers who ingested bacteria-containing milk. He observed that "reliance on bacteria" was a key factor. Lilly and Stillwell (1965) coined the term 'probiotic' as a variation of the original word *probiotika* from the Greek *pro* (meaning "favor") and *bios* (meaning "life") (Gismondo et al., 1999). They defined probiotics as "substances secreted by one microbe that boost the growth of other bacteria." Parker (1974) used the word probiotics to refer to "microbial feed," defining probiotics as "organisms and chemicals that contribute to the intestinal microbial balance." Probiotics are "organisms and substances that contribute to the balance of intestinal microbes," according to their original definition. "Live microbial food supplements that benefit the host animals by improving their intestinal microbial balance," as modified by Fuller (1989). In general, the term probiotics is used to refer to bacteria which are beneficial in promoting the health of host organisms (Balcázar et al., 2006). According to Schrezenmeir and de Vrese (2001), a probiotic is "a preparation of or a product containing viable, specified microorganisms in sufficient numbers, which affect the microflora in a compartment of the host via implantation or colonisation, and thereby exert favorable effects on host health." According to the WHO, probiotics are live microbes that, when administered in appropriate amounts, can improve the host's health (FAO 2001, FAO/WHO 2002). Merrifield et al. (2010) offered a classical and wider description of probiotics as "any microbial cell provided via the diet or rearing water that benefits the host fish, fish farmer, or fish consumer, which is achieved, at least in part, by improving the fish's microbial balance." Probiotics have been shown to provide a wide range of health benefits for both humans and animals as functional foods and therapeutic as well as prophylactic supplements (Ouwehand et al., 1999; Mombelli and Gismondo, 2000; Vijayaram and Kannan, 2018). There are mainly two types of probiotics, as discussed below.

16.2.1 Feed or Gut Probiotics

Gut probiotics enhance the gut's beneficial microbial flora, and can be combined with diet or given orally. Some bacterial and fungal strains can be mixed with feeding pellets, encapsulated into live feed stock, or fed orally to raising animals to avoid illness and enhance essential gut microbial ecology (Prasad et al., 2003; Nageswara and Babu, 2006). Before feeding animals, strains should be checked for viability. The inclusion of probiotics in Atlantic cod fry feed, such as lactic acid bacteria, resulted in appropriate growth, survival and immunity (Gildberg et al., 1997). Regular usage of probiotics in fish feed will provide various health benefits, including improved digestion, effective nutrient integration into body parts, improved development, increased immunity and increased survival (Bazar et al., 2021).

16.2.2 Water Probiotics

Water probiotics can be used in two different forms: dry and liquid. Because water probiotics have a lower density, they can grow in water medium and devour all the existing resources, thereby excluding the harmful bacteria. The dangerous bacteria are eliminated due to starvation (Nageswara and Babu, 2006). Bacteria in liquid form provide favorable outcomes in a shorter time than bacteria in dry and spore form. Water probiotics are used to minimize organic pollutants and different toxins in water by simply administering them to the rearing medium (Prasad et al., 2003). These have a significant impact on the water quality of the culture pond. They break down organic debris into tiny units, which helps to improve water quality (Bazar et al., 2021). Organic waste in aquaculture systems can be reduced by using probiotic bacteria like *Bacillus* sp. *Nitrosomonas*, *Nitrobacter* and sulfur-reducing bacteria. These bacteria lead to purification of the water in the hatchery, resulting in increased larval survival and growth (Lipton, 1997; Moriarty, 1998).

Some of the candidate microorganisms used as probiotics are (Hai, 2015, Zorriehzahra et al., 2016; Shefat, 2018; Hoseinifar et al., 2018; Hasan and Banerjee, 2020):

- Gram-positive bacteria: e.g., *Bacillus, Lactobacillus, Lactococcus, Streptomyces*
- Gram-negative bacteria: e.g., *Alteromonas, Enterobacter, Neptunomonas, Nitrobacter, Nitrosomonas, Phaeobacter, Pseudomonas*
- Bacteriophage: e.g., *Podoviridae, Myoviridae*

- Microalgae: e.g., *Dunaliella, Isochrysis, Navicula, Phaedactylum*
- Yeast: e.g., *Debaryomyces, Saccharomyces, Yarrowia*

16.3 SELECTION CRITERIA FOR PROBIOTICS

Probable candidates isolated from various sources are put through a series of tests in order to determine their suitability as optimal probiotics. Their safety and lack of pathogenicity must be established first (Verschuere et al., 2000; Chythanya et al., 2002). After that, a successful probiotic candidate must fulfill a series of criteria. The main objective of using probiotics is to restore a healthy balance between beneficial and harmful bacteria in the fish gut microbiota. A well-known probiotic should possess a few essential characteristics in order to have a positive impact. When choosing microorganisms for use as probiotics in aquaculture, Merrifield et al. (2010) identified two types of criteria: a) essential and b) favorable. Over the last few decades, new investigations have led to the addition of new criteria which are discussed in the following section (Banerjee and Ray, 2017; Chauhan and Singh, 2019).

16.3.1 Essential Criteria

It should not be pathogenic in terms of the host, other aquatic creatures and humans.

It should be devoid of plasmid-encoded antibiotic resistance genes.

It must be able to withstand high bile salt concentrations and a wide pH range.

16.3.2 Favorable Criteria

A probiotic should have the ability to grow well within the host intestinal secretion.

It must have the ability to colonize the epithelial surface of the host intestines.

It must have the ability to be used as feed supplement.

It must exhibit favorable growth features (such as short lag time, rapid growth and replication at optimum temperature required for host nurturing).

It must possess a broad range of antagonistic action against important pathogens.

It must secrete appropriate enzymes important for digestion (like chitinase, cellulases) or vitamins.

It must be able to endure the native host environment, should have long-term viability under normal storage surroundings and should be strong enough to survive different manufacturing processes.

It must possess good sensory characteristics, fermentative activity, freeze-drying resistance, and feed sustainability during the packing and storage procedure.

It should have immune system stimulation and enzyme and vitamin production properties.

Probiotics should have the following characteristics in order to be used as an effective feed: acid and bile tolerance; gastric juice resistance; antagonism towards pathogens; adhesion to the digestive system's surface; increased gastrointestinal motility; and mucus survival.

16.4 PROBIOTICS ADMINISTRATION METHODS

To use probiotics in aquaculture, it is essential to understand the best way to administer them. The administration of probiotics depends on a number of factors like the probionts, vector required for administration, dosage, duration, and supplementation form. Probiotics widely utilized in aquaculture are *Bacillus* sp., *Lactobacillus* sp., or *Saccharomyces cerevisiae* (Robertson et al., 2000). Probiotics can be taken as dietary supplements (in the form of live food or pellet food) or simply added to the water (Moriarty, 1998; Skjermo and Vadstein, 1999). In most cases, probiotics are added directly to the feed components or sprayed straight into the prepared meal. Numerous alternative administration methods have been employed (Jahangiri and Esteban, 2018) (Figure 16.1).

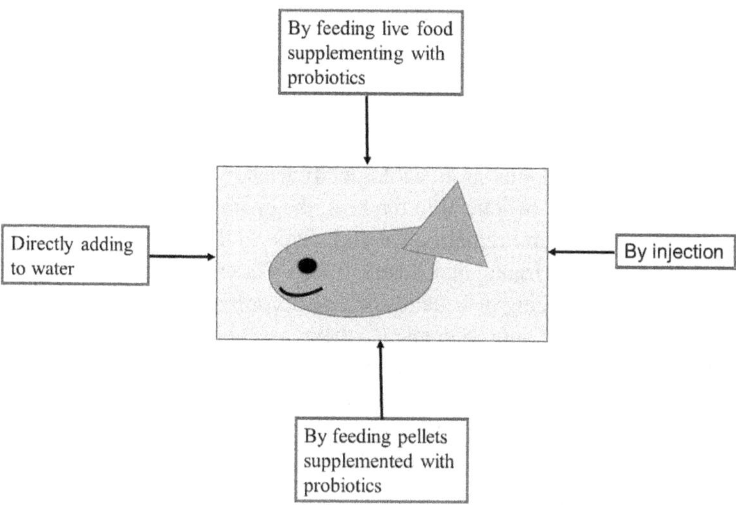

FIGURE 16.1 Modes of administration of probiotics in aquaculture.

Several probiotics are known to be added singly or administered in combinations. Mixed culture of probiotic strains has been found to be more efficient than probiotics based on a single strain. The advantage of using multiple strains is that they are active against a broad range of species and conditions. It was found that Nile tilapia when administered with mixed strain of *Bacillus subtilis* and *Lactobacillus acidophilus* showed higher bactericidal activity than when a single strain was used (Aly et al., 2008a, 2008b). Probiotics are sometimes added to feed as a supplement in the form of freeze-dried culture, occasionally combined with lipids (Ringø et al., 2020). In some cases, heat-inactivated bacteria are even given orally. Examples are *L. delbrueckii* ssp. *lactis* and *B. subtilis* (Salinas et al., 2005). In some studies, live feed was enriched with encapsulated probiotics. This method has a major advantage as the probiotic bacteria remain viable and even have the ability to multiply on the live feed. Therefore, probiotics delivery via live feed to the host is one of the efficient modes of administration of probiotics (Hai, 2015). There are several examples of live feed in which encapsulated probiotics have been used, like copepods, rotifer and *Artemia* species (Chauhan and Singh, 2019). In some cases, probiotic bacteria that produce spores are also being used as the spores help the bacteria to survive harsh conditions like extreme temperature, UV, drought, low nutrition, etc. (Elisashvili et al., 2019).

Probiotic bacteria can be applied to culture water directly, which can modify the microbial composition and water quality of water and sediments. Probiotic treatment directly in pond and tank water also had a positive effect on fish health. *Bacillus* spp. and other probiotic bacteria such as *Nitrobacter* sp., *Aerobacter* sp., and *S. cerevisiae* (yeast) significantly improved the quality of water (Venkateswara, 2007). Probiotic administration by injection has been documented by LaPatra et al. (2014). Delivery of the well-known probiotic *Micrococcus luteus* via injection to *Oreochromis niloticus* resulted in mortality of just 25%, compared to 90% mortality with *Pseudomonas* (Yassir et al., 2002). The use of an appropriate administration technique adds to probiotics' positive performance, and understanding the action modes, as well as appropriate administration strategies, may be crucial to their use in aquaculture (Dawood and Koshio, 2016).

The duration of administration also plays a very important role. It has been reported that the duration can be small (6 days) or extensive (5 months or even 8 months). It has been observed that the prolonged administration of probiotics can induce repression of the immune system (Aly et al., 2008a, 2008b; Chauhan and Singh, 2019). Supplementation for a short period has been found to be more useful than long-term administration of probiotics (Skjermo et al., 2015). In addition to duration, the number of times (frequency) that the probiotics are administered also plays a critical role in maintaining their efficiency. It has been found that daily application of probiotics is much better than application three times a week during the culture period (Guo et al., 2009).

16.5 COMMERCIAL PROBIOTIC PREPARATIONS

Probiotics are becoming more popular as an ecologically friendly option, and their use is both scientific and empirical. In order to be beneficial to the host, the enclosed microorganisms must be able to withstand storage environments and remain alive and stable in the digestive system of aquatic creatures, ultimately boosting output (Irianto and Austin, 2002). These products must be both harmless and helpful in safeguarding the wellbeing of aquatic species, according to the makers (Wang et al., 2008). Aside from laboratory-prepared microorganisms, there are now several commercially accessible probiotic items. One of the earliest commercial product assessments focused on Biostart, a bacterial preparation produced from *Bacillus* strains. It was used to test the influence of inoculum concentration on the generation of cultured catfish (Queiroz and Boyd, 1998). There are numerous commercial probiotic formulations that comprise one or more live microbes that have been developed to promote aquatic organism growth (Martínez Cruz et al., 2012; Akhter et al., 2015; Rahman et al., 2017; Hasan and Banerjee, 2020). Some of the commercially available probiotics are listed in Table 16.1.

TABLE 16.1
Some Commercially Available Probiotics for Aquaculture

Commercial Name with Trademark Sign	Manufacturing Company, Their Location and Logo	Probiotic Mixture Constitution	Uses
AquaStar® GUT PERFORMANCE MANAGEMENT	Biomin Chennai, Tamil Nadu, India Biomin	*Bacillus* sp., *Enterococcus* sp., *Lactobacillus* sp., *Pediococcus* sp.,	Effective against (pathogen inhibition) *Aeromonas hydrophila, A. salmonicida, Edwardsiella. tarda, Flavobacterium. indologenes, Streptococcus agalactiae, S. iniae, Vibrio alginolyticus, V. diazotrophicus, V. harveyi, V. parahaemolyticus, Yersinia ruckeri*
Aqua gold 	CENZONE TECH-EUROPE, Ltd. California, USA 	*Paecilomyces sp., Ganoderma lucidum, Saccharomyces cerevisiae, Schizophyllum commune Rhodopseudomonas sp., Chlorobium limicola. Bacillus subtilis, Bacillus licheniformis, Aspergillus oryzae and Aspergillus niger*	Increase the pace of growth and the ability to prevent diseases.
Aqua photo	ACI Animal Health Dhaka, Bangladesh 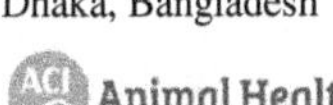	*Bacillus subtilis* and *Rhodopseudomonas*	Control undesired gas and sediment while boosting plankton growth.
ECO MARINE™ 	Virbac Animal Health India Pvt Ltd. Mumbai, Maharashtra, India 	*Bacillus subtilis, B. pumilis, B. amyloliquefaciens, B. megaterium*	Control vibriosis and luminescent bacteria and also reduces toxic gases

Commercial Name with Trademark Sign	Manufacturing Company, Their Location and Logo	Probiotic Mixture Constitution	Uses
Hydroyeast Aquaculture® 	Agranco Corp. Florida, USA 	*Saccharomyces sp., Bifidobacterium spp., Lactobacillus acidophilus, Streptococcus faecium, Oligosaccharide and Enzymes*	Act as a growth promoter by improving feed efficacy in shrimps and fish (*Oreochromis niloticus*).
Procon-PS 	PVS Laboratories Limited, Andhra Pradesh, India  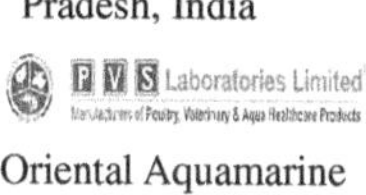	*Bacillus sp. Rhodococcus, and Rhodobacter*	Effective against pathogenic microorganisms especially bacterial species and also helps control unwanted gas and sediment
Enterotrophotic	Oriental Aquamarine Biotech India (P) Ltd Coimbatore, India	*Bacillus cereus* and *Arthrobacter nicotianae*	Effective against *Vibrio* spp.
Super PS 	CP Aquaculture Tamil Nadu, India 	*Rhodobacter sp. Rhodococcus sp.*	Improve quality of soil and minimize hazardous gas emissions at the bottom

16.6 PROBIOTICS ACTION MECHANISMS

The effectiveness of probiotics depends to a great extent on their mode of action. Probiotics exhibit specific mechanisms of action for the protection of the host in the aquatic environment (reviewed by Oelschlaeger, 2010; Chauhan and Singh, 2019; Simón et al., 2021). Oelschlaeger (2010) categorized those mechanisms into three main types: (i) those that protect the host by modulating its immune system; (ii) those that can prevent the growth of disease-causing pathogenic microorganisms by secreting various inhibitory antimicrobial compounds; and (iii) those that can inactivate the pathogen's toxic products, thereby detoxifying the host (Oelschlaeger, 2010). Different specific modes of action of probiotics for improving the host growth (Figure 16.2) are discussed in the next section.

16.6.1 Competitive Exclusion of Opportunistic Pathogens by Probiotics

In order to develop a disease, pathogenic microorganisms mostly colonize and spread inside the host after binding to its intestinal mucosa layer (Adams, 2010). One probable way of averting the colonization of pathogenic microorganisms is competition by probiotics for those adhesion sites. This is known as competitive inhibition/exclusion. Once probiotics bind to the adhesion locations they form a physical barricade to prevent the binding of pathogens to the gastrointestinal tract of the host. Isolates (potential probiotics) from *Sparus aurata* were shown to interfere with the binding of its pathogen, *Listonella anguillarum*, to the mucosal layer. The researchers observed a decline in the mortality of fish fed with a diet augmented with probiotics (Chabrillón et al., 2006). The fish, *Puntius conchonius*, supplemented with diet containing the probiotic bacteria *B. coagulans*, *Bifidobacterium infantis* and *B. mesentericus* showed a reduction in the gut pathogenic bacteria (Divya et al., 2012). *Lactobacillus* secretes mucus-binding proteins, MUB, which facilitate the binding of the bacteria to the mucous layer. Bermudez-Brito et al. (2012) have stated that electrostatic, hydrophobic interactions

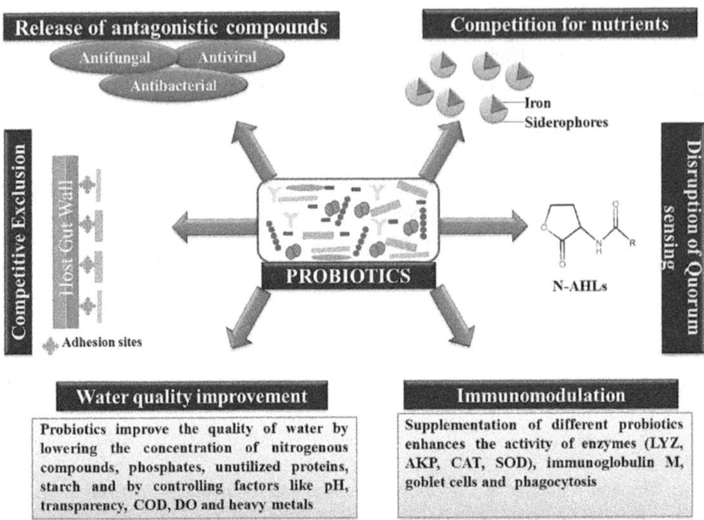

FIGURE 16.2 Mode of action of probiotics in aquaculture. (COD, chemical oxygen demand; DO dissolved oxygen; LYZ, lysozyme; AKP, alkaline phosphatase; CAT, catalase; SOD, superoxide dismutase; N-AHLs, N-acyl homoserine lactones).

and lipoteichoic acids are among the factors that might affect the binding of probiotics to the adhesion sites. In a recent report, different strains of *Lactobacillus* were shown to avert the binding of pathogenic bacteria *Escherichia coli* and *Salmonella* to Caco-2 and HT-29 cells (Fonseca et al., 2021).

16.6.2 Mitigation of Pathogens by Production of Antimicrobial Materials

Probiotics may hinder both colonization and proliferation of pathogenic bacteria by releasing antagonistic bactericidal compounds. Among the antimicrobial compounds produced by probiotic bacteria are siderophores, antibiotics, enzymes (proteases, lysozymes), organic acids (which change the pH), bacteriocin, and volatile fatty acids (Tinh et al., 2008; Chauhan and Singh, 2019).

A bacteriocin isolated from *L. murinus* (AU06) was found to inhibit fish pathogens *S. aureus*, *Micrococcus* sp., and *E. coli* (Elayaraja et al., 2014). Another report suggested that different strains of LAB extracted from finfish exhibited antibacterial activity against finfish bacterial pathogens (Ringø et al., 2018). *B. velezensis* strain JW displayed bacteriocin activity against a wide spectrum of pathogenic bacteria (*Aeromonas hydrophila*, *Lactococcus garvieae*, *V. parahemolyticus* etc.) infecting fish (Yi et al., 2018). Similarly, *B. velezensis* (LF01) was found to inhibit the streptococcosis disease-causing pathogen, *Streptococcus agalactiae* in *O. niloticus*. The antagonistic compounds released were shown to be thermostable, UV and pH stable, and resilient to proteolytic enzymes, had a long shelf life and exhibited antibacterial activity against many other fish pathogens (Zhang et al., 2019). Kuebutornye et al. (2020) discussed different mechanisms (bacteriocin, antibiotic, lytic enzyme secretion, competitive inhibition, etc.) exhibited by *Bacillus* probiotic for conferring disease resistance against fish pathogens.

There are very limited investigations reporting the antiviral and antifungal mechanism of probiotics. In *Paralichthys olivaceus*, Lactobacil singly or mixed with Sporolac has been shown to stimulate the immune response and conferred resistance against lymphocystis disease virus (LCDV) (Harikrishnan et al., 2010). *B. subtilis* E20 fed to *Epinephelus coioides* led to increased disease resistance and immunity against an iridovirus (Liu et al., 2012). Lakshmi et al. (2013) also suggested that probiotics act by conferring disease resistance and stimulate the shrimp immune system. Administration of *Bacillus* sp. improved the immune response as well as resistance of *Oreochromis* sp. (tilapia) to Tilapia lake viral disease (Waiyamitra et al., 2020). *L. plantarum* FNCC 226 has been shown to mitigate *Saprolegnia parasitica* A3 (oomycete pathogen) infection in *Pangasius hypophthalamus* Sauvage (Nurhajati et al.,

2012). *Bacillus* sp secretes lytic enzymes and synthesizes antibiotics which have antifungal activities and can be used in aquaculture to inhibit fungal fish pathogens (Kuebutornye et al., 2020).

16.6.3 Inhibition of Pathogens by Competing for Nutrients

Constant supply of nutrients is the basic need for the survival, development and spread of both fish pathogens and probiotics. Probiotics have been shown to compete with these pathogens by consuming the nutrients and energy available in the aquaculture environment. Ferric-iron chelators, also called siderophores, function by dissolving or extracting iron (an important nutrient) from different complexes, thereby contributing to microbial nutrition (Wilson et al., 2016; Khan et al., 2018). Therefore probiotic bacteria that synthesize siderophores can be employed as probiotics for sequestering iron. *Pseudomonas fluorescens* repressed the growth of *A. salmonicida* because of iron competition (Smith and Davey, 1993). In an iron-limited environment *P. fluorescens* has also been shown to competitively repress the growth of *V. anguillarum* (Gram et al., 1999). Vazquez-Gutierrez et al. (2016) also suggested that *Bifidobacteria pseudolongum* and *Bifidobacteria kashiwanohense* can be used as potential probiotics as they can deter the growth of enteric pathogens (*Salmonella typhimurium* and *E. coli*) by competing for iron.

16.6.4 Probiotics Act by Improving the Quality of Water

Effluents present in aquaculture consist of suspended items, organic material, nitrogenous compounds, phosphates, heavy metals, etc. and are hazardous to the water ecosystem (Jegatheesan et al., 2011; Hlordzi et al., 2020). Application of probiotics in aquaculture improves the quality of water. *B. licheniformis* (X3914) was shown to reduce the concentration of ammonia as well as unutilized proteins and starch present in waste water (Qing Hua et al., 2011). Decline in levels of nitrogenous compounds and phosphates was observed in fish ponds supplemented with *Nitrosomonas* and *Nitrobacter* species (Padmavathi et al., 2012). Sanolife PRO-F probiotic (a commercial mixture of different *Bacillus* strains) improved water quality by lowering ammonia levels and enhancing pH in tilapia (*O. niloticus*) ponds (Elsabagh et al., 2018). When applied to water, *B. subtilis* AHAHBS001 led to a decline in ammonia levels (Kewcharoen and Srisapoome, 2019). Different species of *Bacillus* have been reported to maintain the quality of water in aquaculture by controlling many factors including transparency, pH, dissolved oxygen, alkalinity, chemical oxygen demand and heavy metals (reviewed by Hlordzi et al., 2020).

16.6.5 Interruption of Quorum Signaling

Quorum sensing or quorum signaling (QS) is a phenomenon that allows bacteria to regulate the expression of genes in cell density-dependent mode. Through this mechanism bacteria communicate with other bacteria by producing and releasing chemical signaling molecules called auto-inducers (Miller and Bassler, 2001). Secretion of virulence factors, biofilm formation and sporulation are important factors contributing to pathogenicity of aquaculture pathogens and are also regulated by QS. Therefore, QS disruption can be used as an effective strategy for averting infection by bacteria in aquaculture (Zhao et al., 2015). Auto-inducers such as N-acyl homoserine lactones (AHLs) (the most common type) are released by gram-negative bacteria. Degradation of AHLs has been shown to disrupt QS and therefore can be employed as an anti-infective approach to controlling fish diseases (Cao et al., 2014; Chu et al., 2014). QSI-1 quorum quencher extracted from *Bacillus* spp prevented formation of biofilm and decreased the pathogenicity of zebrafish (*Danio rerio*) pathogen *A. hydrophila* (Chu et al., 2014). Zhou et al. (2016) also reported that quorum quenching (QQ) enzyme QSI-1 resulted in a decline in *A. hydrophila* due to degradation of AHLs, and positively altered the gut microbiota of goldfish (*Carassius auratus*). QQ enzymes from other strains of *Bacillus* have also been characterized and were found to be effective against different bacterial pathogens in aquaculture (Chen et al., 2020; Ghanei-Motlagh et al., 2020). In another report, *Enterococcus faecium* QQ12 extracted from *O. niloticus* degraded AHLs and disrupted the pathogenicity of *A. hydrophila* (Vadassery and Pillai, 2020).

16.6.6 Immunomodulation/Augmentation of Host Immune System

Probiotics stimulate the immune system (innate and adaptive) of fish by releasing cytokines, interleukins, interferons, chemokines and tumor necrosis factors from different immune cells. It is also reported that probiotics can enhance macrophage phagocytosis, natural killer cell toxicity, and can interact with Th1, Treg, enterocytes, and dendritic Th2 cells to improve the host immune response (Azad et al., 2018; Simón et al., 2021). This is significant for protecting the host from different infectious gut diseases caused by pathogens in aquaculture.

Fish immune systems (both innate and adaptive immunity) have been shown to be stimulated by probiotics. *O. niloticus* (tilapia) fish tanks when subjected to probiotic microorganisms *B. coagulans* strain B16 and *Rhodopseudomonas palustris* G06 resulted in enhanced respiratory burst immune response and better growth (Zhou et al., 2010a). Combined mixture of *Lactococcus lactis* and *L. plantarum* enhanced the innate immune response of *Paralichthys olivaceus* (Beck et al., 2015). *Puntius gonionotus*, commonly known as Javanese carp, when fed with food enriched with *E. faecalis* and challenged with pathogen *A. hydrophila*, exhibited enhanced antibody concentration (Allameh et al., 2017). *L. acidophilus*-rich diet fed to *Cyprinos carpio* (juvenile common carp) led to a significant increase in immunoglobulin and protection against pathogens *P. aeruginosa* and *A. hydrophila* (Adeshina, 2018). Red tilapia fish fed with a diet supplemented with *Helianthus tuberosus* and *L. rhamnosus* (synbiotics) displayed an increased number of goblet cells when exposed to pathogen *A. veronii*, suggesting the role of probiotics in enhancing mucosa immunity (Sewaka et al., 2019). Pulse supplementation of *E. faecium* probiotic every 14 days to tilapia resulted in improved immune system responses (Tachibana et al., 2020).

Other recently investigated probiotics that can be used as immunomodulating feed supplements for host protection in aquaculture are tabulated in Table 16.2.

TABLE 16.2
Probiotics Used as Immunomodulating Feed Supplements in Aquaculture

	Host Organism	Probiotic Administered	Immune Response	Reference
1	*Oreochromis niloticus*	*Bacillus pumilus* and protease	Enhanced goblet cells, enterocytes, lysozyme (LYZ), immunoglobulin M (IgM) and phagocytic parameters	Hassaan et al. (2021)
2	*Trachinotus ovatus*	*B. pumilus* A97	Enhanced non-specific immune responses	Liu et al. (2020)
3	*Larimichthys crocea*	*Clostridium butyricum*	Enhanced immune enzymes [LYZ, alkaline phosphatase (AKP), catalase (CAT), superoxide dismutase (SOD)] activity	Yin et al. (2021)
4	*Litopenaeus vannamei*	*Monascus purpureus*	Enhanced AKP and LYZ activities	Wang et al. (2021)
5	*Paralichthys olivaceus*	*Lactobacillus plantarum* and *B. subtilis*	Increased LYZ, IgM and complement protein 4 levels	Li et al. (2021)
6	*Oncorhynchus mykiss*	*Lactococcus lactis* subsp. *lactis* PTCC 1403	Increased LYZ and IgM levels	Yeganeh et al. (2021)
7	*Cherax cainii*	*L. acidophilus* and *L. plantarum*	Enhanced expression of genes involved in innate immune response	Foysal et al. (2020)
8	*Oreochromis niloticus*	*Spirulina platensis* and *B. amyloliquefaciens*	Increased IgM levels	Al-Deriny et al. (2020)
9	*Litopenaeus vannamei*	*B. subtilis* WB60, *Lactococcus lactis* and *Pediococcus pentosaceus*	Better LYZ and SOD activities	Won et al. (2020)
10	*Litopenaeus vannamei*	*B. pumilus* SE5, *E. faecium* MM4, *Psychrobacter* sp. SE6 or *B. claussi* DE5	Better SOD and AKP activities	Zhang et al. (2020)

16.7 THE IMPORTANCE OF PROBIOTICS IN AQUACULTURE

Aquaculture is one of the most important food industries and is growing rapidly; however, aquaculture production is limited as it is susceptible to various types of disease. To control these diseases, various kinds of chemicals and antibiotics are used, which is an increasing concern as they are harmful to both aquatic animals and their environment. This in turn has led to the rapid growth of the use of probiotics in aquaculture as an alternative strategy for the prevention of diseases. They also improve health, and enhance growth and performance of fish and other aquatic animals. There are numerous benefits associated with the use of probiotics in aquaculture, which are discussed in the following sections.

16.7.1 Maintenance and Preservation of Water Quality

It is very important to improve the quality of water in aquaculture, which is a prerequisite for fish and shrimp farming. The high organic matter in the water derived from feed, phytoplankton and feces induces the growth and proliferation of aquatic pathogens. The decomposition of organic matter is a natural process; however, the amount of organic matter in the aquaculture production system is sometimes too high to be removed by the natural decomposition process. Probiotics are well known to improve water quality as they play a critical role in the turnover of organic nutrients in aquaculture. Probiotics are highly efficient in the conversion of organic matter to CO_2 and thus are maintained at high level in production ponds. This reduces the load of organic carbon, enhances water quality and improves the health of fish and other aquatic animals. For details of probiotics used in maintenance of water quality, see Section 6.4 (Wang and Wang, 2008; Martínez Cruz et al., 2012; Hasan and Banerjee, 2020).

16.7.2 Augmentation of Growth and Survival Rate

There are numerous reports on the use of probiotics in aquaculture to enhance the growth of diverse species. Planas et al. (2004) observed that addition of LAB increases the growth of the rotifer *Brachionus plicatilis* and it was further observed that the addition of *L. casei* ssp. *casei*, *Lactococcus lactis* ssp. *lactis* and *Pediococcus acidilactici* gives the best result. The use of *E. faecalis* was found to increase the weight of Javanese carp over that of carp under control conditions (Allameh et al., 2016). A study found that Nile tilapia fed with *Lactobacillus* sp. increased body weight as well as crude lipid and total protein content (Hamdan et al., 2016). There was increase in growth as well as survival rate of *Xiphophorus helleri*, *X. maculates* and *Poecilia reticulata* when feed was supplemented with *B. subtilis* and *Streptomyces* sp. (Tan et al., 2016; Banerjee and Ray, 2017).

16.7.3 Improves Nutrient Utilization, Digestion and Feed Efficiency

It is well established that probiotics are beneficial for the digestive processes in aquatic animals as they can synthesize extracellular enzymes like amylases, proteases and lipases. They also provide energy and growth factors (Hasan and Banerjee, 2020). LAB are most frequently used in probiotic preparations (Ringø et al., 2018). They are present in huge numbers in the gut of healthy animals. They are considered safe and have been given GRAS status by Food and Drug Administration (Giri et al., 2013).

16.7.4 Effects on Phytoplankton

Probiotics have been found to have a considerable effect on phytoplankton (various species of microalgae), mainly red tide phytoplankton. Bacteria have an adverse effect on algae which is undesirable in hatchery. However, this is beneficial when unwanted algal species develop in the culture pond (Sahu et al., 2008). There are numerous reports of the effect of probiotics on phytoplankton (Bonnet et al., 2010; de Paiva-Maia et al., 2013). Martínez Cruz et al. (2012) examined the effect of probiotics on microalgae like diatoms (*Chaetoceros* species), which serves as a healthier live food, although its production is limited due to nutritional requirements. Gomez-Gil et al. (2002) observed that when probiotic *V. alginolyticus* C7b and *Chaetoceros muelleri* are co-cultured, they achieve high density growth which can be used for feeding shrimps.

16.7.5 Bacteriostatic Effects of Probiotics

Probiotics are reported to release a wide range of chemicals like bacteriocin, lysozyme, proteases, hydrogen peroxides, siderophores, ammonia, diacetyl and antibiotics. These possess bacteriostatic and bactericidal properties. Some bacteria also produce volatile fatty acids and organic acid. These substances reduce the pH in the lumen of the gastrointestinal tract, and thus restrict the growth and proliferation of infectious and pathogenic microbes (Giri et al., 2013). A new compound that has been identified in bacteria, indole (s,3-benzopyrrole), has antifungal and antibacterial properties and a suppressive effect against pathogens (Zorriehzahra et al., 2016). For details of the bacteriostatic effects of probiotic strains, see Section 16.6.

16.7.6 Strengthening of the Immune System

Probiotics are reported to induce both local and systematic immunity of the host. This is attributed to the various cytokines like TNF (Tumor Necrosis Factors), ILS (interleukin), IFS (interferon) and chemokines (Simón et al., 2021). Rainbow trout when co-administered with *L. sakei*, *Lactococcus lactis* subsp. *lactis* and *Leuconostoc mesenteroides* showed a mark increase in the phagocytic activity of its gut leukocytes (Balcázar et al., 2007a, 2007b). The expression of TNFα, TIL-1b and lysozyme-C increased when fish were fed a diet supplemented with *V. lentus, A. veronii* and *Flavobacterium sasangense* (Dawood & Koshio, 2016). There are many examples of immunomodulatory effects of probiotics; for example, the administration of *L. rhamnosus* to Nile tilapia led to increase in number of lymphocytes and acidophilic granulocytes (Pirarat et al., 2011). In another study, fish fed with LAB showed an increased production of inflammatory cytokines like IL-1, IL-2, IL-6, IFN-g and TNF-a, I as well as increased production of anti-inflammatory cytokines like IL-10 and TGF-b. There was an increase in the number of mucin-secreting goblet cells and a decrease in the concentration of total Ig (Mohammadian et al., 2019; Soltani et al., 2019; Shabirah et al., 2019; Ringø et al., 2020). Immunomodulatory effects are discussed in more detail in Section 16.6.

16.7.7 Antibacterial Properties of Probiotics

Many studies document the antibacterial properties of probiotics. It was found that *Lactococcus lactis* has an antibacterial effect against *Yersinia rukeri* and *A. salmonicida* (Balcázar et al., 2007b). In another study by Dhanasekaran et al. (2010), species of *Lactobacilli* displayed an antimicrobial effect against *Aeromonas* and *Vibrio* species. It was also found that *L. plantarum* (LP1, LP2), *L. lactis* subsp. *lactis* (LL2), *S. cerevisiae* (SC3), *Staphylococcus arlettae* and *Candida glabrata* (CG2) display high inhibitory activity to *Listeria monocytogenes* and *Staphylococcus aureus*. There are reports in which various species of LAB like *L. acidophilus*, *L. fermentum*, *L. buchneri*, *Lactococcus lactis*, and *Streptococcus salivarius* were found to suppress growth of *Listeria innocua* (Moosavi-Nasab et al., 2014). Later, Ghosh et al. (2008) and Newaj-Fyzul and Austin (2015) observed that use of *Bacillus subtilis* appreciably decreased the population of total coliforms, pseudomonads and aeromonads in fish. There are examples of *Lactobacillus* species like *L. acidophilus, L. pentosus*, *L. fermentum* and *L. casei* that are used as candidates for probiotics for enhancing protection from bacterial pathogens. It was observed that when carp feed was supplemented with species of *Lactococcus*, its resistance to *A. hydrophila* was enhanced. Other examples are improved resistance of olive flounder to *Edwardsiella tarda*; rainbow trout from *A. salmonicida*; Nile tilapia from *Staphylococcus aureus* and tilapia (*O. niloticus*) to *A. hydrophila* (Zhou et al., 2010b; Beck et al., 2017; Abdelfatah and Mahboub 2018; Sun et al., 2018; Feng et al., 2019; Simón et al., 2021).

16.7.8 Antifungal Properties of Probiotics

Limited studies are available on the antifungal properties of probiotics. In a study by Lategan et al. (2004), *Aeromonas* strain (A199) from *Anguilla australis* (eel) displayed a repressive effect to species of *Saprolegnia*. In catfish, use of *L. plantarum* (FNCC 226) as probiotic suppresses *S. parasitica* (A3) (Nurhajati et al., 2012). Zorriehzahra et al. (2016) found that in *Oncorhynchus mykiss*, the use of *Pseudomonas* M162, M174 and species of *Janthinobacterium* (M169) improved their resistance to saprolegniasis.

16.7.9 Antiviral Activity of Probiotics

Probiotics have been found to exert an antiviral effect on certain viral pathogens. Kamei et al. (1988) reported the antiviral property of *Pseudomonas*, *Vibrio*, *Coryneforms* and *Aeromonas* spp. against hematopoietic necrosis virus. In shrimp (*Litopenaeus vannamei*), species of *Bacillus* and *Vibrio* improved their resistance to WSSV (white spot syndrome virus) (Balcázar, 2003). Li et al. (2009) observed that shrimp (*L. vannamei*) fed with *B. megaterium* improved their resistance to disease caused by WSSV. It was found that *Lactobacillus*, used either alone or combined with tablets of Sporolac, improves *Paralichthys olivaceus* resistance to lymphocystis (Harikrishnan et al., 2010). *L. plantarum* fed to fish enhanced both growth and innate immune responses (Son et al., 2009). *S. cerevisiae* P13 improved the resistance of *Epinephelus coioides* to GIV (grouper iridovirus). It was found that the mortality rate of fish fed with *B. subtilis* E20 reduced when treated with GIV; survival rate and innate immune responses were also enhanced (Chiu et al., 2007). Although the effect of probiotics on viral pathogens has received much attention, the mode of action by which they achieve antiviral effects is unknown.

16.7.10 Effects on Reproduction

Probiotics are known to affect reproduction of aquatic animals in various ways including fertilization, fecundity, gonadosomatic index and fry production of female species. However, there are limited studies on these topics and more experimental data is needed (Hasan & Banerjee, 2020).

Various strains like *B. subtilis, L. acidophilus* and *L. casei* have been used to analyze the effect of probiotics on the rate of reproduction of aquatic animals (Abasali & Mohammad, 2011). In one study, *B. subtilis* was supplemented in feed at different concentrations and fed to fish like *Poecilia sphenops*, *Poecilia reticulata*, *X. maculates* and *X. helleri*. It was found to improve the viability rate, fecundity, gonadosomatic index and fry production of the females of all the species (Ghosh et al., 2007). In zebra fish, probiotics were found to increase the production of ovulated eggs in comparison to control, having a high hatching rate and quicker embryonic development (Gioacchini et al., 2013).

16.7.11 Improvement in Stress Tolerance

Various kinds of stresses like thermal, hypoxic, anoxic, oxidative, chemical, toxic, nutritional, and high density are known to have a deleterious impact on fish and other aquatic animals (Martínez Cruz et al., 2012; Zorriehzahra et al., 2016). The use of probiotics either in the form of feed supplement or water can help overcome the deleterious effects of stress as well as improving host immunity (Zorriehzahra et al., 2016). Early evidence that probiotics helps to overcome stress came from the study in which *Dicentrarchus labrax* feed was supplemented with *L. delbrueckii ssp. Delbrueckii.*

The cortisol level in the treated fish was considerably lower than that of the control (Carnevali et al., 2006). In another study, flounder (*Paralichthys olivaceus*) was subjected to stress conditions. The activity of plasma lysozyme in the group fed with probiotic diet was considerably higher than in the control. When subjected to heat shock, flounder fed with probiotics displayed increased tolerance to heat stress (Taoka et al., 2006). Varela et al. (2010) observed that administration of probiotic enhanced the resistance of *Sparus aurata* (gilthead sea bream) to stress under conditions of high stocking density. In shrimp (*Litopenaeus vannamei*), addition of *L. plantarum* enhanced antioxidant state and subsequently improved the resistance of shrimp against *V. alginolyticus* (Chiu et al., 2007).

16.8 LIMITATIONS OF PROBIOTICS AND SAFETY CONSIDERATIONS

The administration of feed additives (probiotics) in aquaculture has widely been accepted as an important disease control tool, but it is associated with certain limitations and challenges (Wang et al., 2008; Martínez Cruz et al., 2012; Amenyogbe et al., 2020). Probiotics are principally used for the prevention rather than cure of diseases. The biggest limitation with probiotics is that they are easily impaired by some of the chemicals or drugs that are used for the treatment of fungal or bacterial diseases in aquaculture or are important for the establishment of beneficial microorganisms. These probiotics should therefore be applied in sterile water (Sahu et al., 2008). Prior to their supplementation in feed, careful evaluation of these probiotic strains in terms of their efficiency, dosage, administration and effect on the environment is required, which incurs extra costs to the aquaculture industry. Industries that commercially manufacture these probiotics also employ advanced technologies for quality assurance and safety, further increasing production costs (Wang et al., 2008). Storage and viability are also important challenges to be addressed during probiotic manufacturing (Wang et al., 2008). Martínez Cruz et al. (2012) have provided a theoretical proposal on the side-effects of probiotics on susceptible organisms, according to which probiotics can be harmful to the metabolism of vulnerable individuals, can cause extreme stimulation of immune system, can lead to systemic infections and can contribute to the transfer of drug-resistance genes in susceptible organisms. These hypothetical prerogatives need strong research evidence. Some aquaculture produce is eaten raw or partially cooked, therefore the effect of residual probiotics on human health also needs to be evaluated (Martínez Cruz et al., 2012).

The FAO and WHO have formulated guidelines for methodical and scientific assessment of probiotics in edible items to determine their suitability for human consumption. In order to regulate the probiotic industry in India, the Indian Council of Medical Research in association with the Department of Biotechnology (DBT), has also drafted guidelines for safeguarding the effectiveness and safety of probiotics (Ganguly et al., 2011). Careful evaluation of probiotics is required to avert or overcome their adverse effects on beneficial organisms. Enforceable regulations are also a prerequisite for marketing aquaculture probiotics.

16.9 CONCLUSION AND FUTURE PROSPECTS

Aquaculture is one of the fastest-growing industries and accounting for over 50% of fish produced for human consumption throughout the world.; however, frequent eruption of disease in aquaculture caused by pathogenic microbes limits both growth and production. Initially antibiotics proved a valuable strategy, but they have been found to have various harmful effects on fish and other aquatic organisms as well as the environment. Drawbacks included increased antibiotic resistance in bacteria and fish pathogens; transfer of resistant genes; reduced immunity, increased toxicity as a result of accumulation of antibiotic residues; and reduced demand for drug-treated aquatic products. Probiotics is one of the alternatives that have decreased the dependence of aquaculture on antibiotics and helped overcome several drawbacks associated with them. Probiotics serve as a valuable resource for beneficial microbes as they have antifungal, antibacterial, antiviral, bactericidal or bacteriostatic properties against pathogenic bacteria. They improve water quality and the health and performance of fish, enhance their immunity, increase their resistance to pathogens, thereby decreasing the spread of disease, and also help overcome the harmful and undesirable effects of stress.

However, much work is still required to streamline the available probiotics based on their mode of action. There is also a need to develop specific strains for precise targeting of fish species, as well as a demand to develop probiotics which are more efficient, cost-effective and eco-friendly. In vivo studies are crucial for optimizing these probiotic organisms in terms of colonization, survival and proliferation rate. Effective modes of administration also need to be developed. It is very important to understand and identify the fate of these probiotic microbes. Other aspects related to probiotics that need to be addressed include longevity of their effect on health and vigor of the host; viability, usefulness and mode of action; resistance to antibiotics; and optimal environment for probiotic–host interactions.

Advances in molecular tools can be used to understand and unravel the microbial diversity of aquaculture as well as deciphering the molecular mechanisms of the mode of action of these probiotic organisms, which would help with identification of their genes as well as new applications.

REFERENCES

Abasali, H., & Mohammad, S. (2011). Dietary prebiotic immunogen supplementation in reproductive performance of platy (*Xiphophorus maculatus*). *Veterinary Research (Pakistan)*, *4*(3), 66–70. https://doi.org/10.3923/vr.2011.66.70

Abdelfatah, E. N., & Mahboub, H. H. H. (2018). Studies on the effect of *Lactococcus garvieae* of dairy origin on both cheese and Nile tilapia (*O. niloticus*). *International journal of Veterinary. Science and Medicine*, *6*(2), 201–207. https://doi.org/10.1016/j.ijvsm.2018.11.002

Adams, C. A. (2010). The probiotic paradox: Live and dead cells are biological response modifiers. *Nutrition Research Reviews*, *23*(1), 37–46. https://doi.org/10.1017/S0954422410000090

Adeshina, I. (2018). The effect of *Lactobacillus acidophilus* as a dietary supplement on nonspecific immune response and disease resistance in juvenile common carp, *Cyprinos carpio*. *International Food Research Journal*, *25*(6), 2345–2351

Akhter, N., Wu, B., Memon, A. M., & Mohsin, M. (2015). Probiotics and prebiotics associated with aquaculture: A review. *Fish & Shellfish Immunology*, *45*(2), 733–741. https://doi.org/10.1016/j.fsi.2015.05.038.

Al-Deriny, S. H., Dawood, M. A., Abou Zaid, A. A., Wael, F., Paray, B. A., Van Doan, H., & Mohamed, R. A. (2020). The synergistic effects of *Spirulina platensis* and *Bacillus amyloliquefaciens* on the growth performance, intestinal histomorphology, and immune response of Nile tilapia (*Oreochromis niloticus*). *Aquaculture Reports*, *17*, 100390. https://doi.org/10.1016/j.aqrep.2020.100390

Allameh, S. K., Ringø, E., Yusoff, F. M., Daud, H. M., & Ideris, A. (2017). Dietary supplement of *Enterococcus faecalis* on digestive enzyme activities, short-chain fatty acid production, immune system response and disease resistance of Javanese carp (*Puntius gonionotus*, Bleeker 1850). *Aquaculture Nutrition*, *23*(2), 331–338. https://doi.org/10.1111/anu.12397

Allameh, S. K., Yusoff, F. M., Ringø, E., Daud, H. M., Saad, C. R., & Ideris, A. (2016). Effects of dietary mono and multiprobiotic strains on growth performance, gut bacteria and body composition of Javanese carp (*Puntius gonionotus*, Bleeker 1850). *Aquaculture Nutrition*, *22*(2), 367–373. https://doi.org/10.1111/anu.12265

Aly, S. M., Ahmed, Y. A. G., Ghareeb, A. A. A., & Mohamed, M. F. (2008b). Studies on *Bacillus subtilis* and *Lactobacillus acidophilus*, as potential probiotics, on the immune response and resistance of *Tilapia nilotica* (*Oreochromis niloticus*) to challenge infections. *Fish & Shellfish Immunology*, *25*(1–2), 128–136. https://doi.org/10.1016/j.fsi.2008.03.013

Aly, S. M., Mohamed, M. F., & John, G. (2008a). Effect of probiotics on the survival, growth and challenge infection in *Tilapia nilotica* (*Oreochromis niloticus*). *Aquaculture Research*, *39*(6), 647–656. https://doi.org/10.1111/j.1365-2109.2008.01932.x

Amenyogbe, E., Chen, G., Wang, Z., Huang, J., Huang, B., & Li, H. (2020). The exploitation of probiotics, prebiotics and synbiotics in aquaculture: Present study, limitations and future directions: A review. *Aquaculture International*, *28*(3), 1017–1041. https://doi.org/10.1007/s10499-020-00509-0

Araújo, C., Muñoz-Atienza, E., Nahuelquín, Y., Poeta, P., Igrejas, G., Hernández, P. E., & Cintas, L. M. (2015). Inhibition of fish pathogens by the microbiota from rainbow trout (*Oncorhynchus mykiss*, Walbaum) and rearing environment. *Anaerobe*, *32*, 7–14. https://doi.org/10.1016/j.anaerobe.2014.11.001

Azad, M. A. K., Sarker, M., & Wan, D. (2018). Immunomodulatory effects of probiotics on cytokine profiles. *BioMed Research International, 2018*, 8063647. https://doi.org/10.1155/2018/8063647

Balcázar, J. L. (2003). Evaluation of probiotic bacterial strains in *Litopenaeus vannamei*. Final Report. *National Center for Marine and Aquaculture Research, Guayaquil, Ecuador.*

Balcázar, J. L., De Blas, I., Ruiz-Zarzuela, I., Cunningham, D., Vendrell, D., & Múzquiz, J. L. (2006). The role of probiotics in aquaculture. *Veterinary Microbiology, 114*(3–4), 173–186. https://doi.org/10.1016/j.vetmic.2006.01.009

Balcázar, J. L., De Blas, I., Ruiz-Zarzuela, I., Vendrell, D., Gironés, O., & Muzquiz, J. L. (2007a). Enhancement of the immune response and protection induced by probiotic lactic acid bacteria against furunculosis in rainbow trout (Oncorhynchus mykiss). FEMS Immunology & Medical Microbiology, 51(1), 185–193. https://doi.org/10.1111/j.1574-695X.2007.00294.x

Balcázar, J. L., Vendrell, D., de Blas, I., Ruiz-Zarzuela, I., Gironés, O., & Múzquiz, J. L. (2007b). *In vitro* competitive adhesion and production of antagonistic compounds by lactic acid bacteria against fish pathogens. *Veterinary Microbiology, 122*(3–4), 373–380. https://doi.org/10.1016/j.vetmic.2007.01.023.

Banerjee, G., & Ray, A. K. (2017). The advancement of probiotics research and its application in fish farming industries. *Research in Veterinary Science, 115*, 66–77. https://doi.org/10.1016/j.rvsc.2017.01.016

Bazar, K. K., Pemmineti, N. J., & Mohammad, S. A. (2021). Effect of soil probiotic on water quality and soil quality maintenance and growth of freshwater fish *Pangasius hypophthalmus*. 11(1) 3291–3304. https://doi.org/10.33263/LIANBS111.32913304

Beck, B. R., Kim, D., Jeon, J., Lee, S.-M., Kim, H. K., Kim, O.-J., Lee, J. I., Suh, B. S., Do, H. K., Lee, K. H., Holzapfel, W. H., Hwang, J. Y., Kwon, M. G., & Song, S. K. (2015). The effects of combined dietary probiotics *Lactococcus lactis* BFE920 and *Lactobacillus plantarum* FGL0001 on innate immunity and disease resistance in olive flounder (*Paralichthys olivaceus*). *Fish & Shellfish Immunology, 42*(1), 177–183. https://doi.org/10.1016/j.fsi.2014.10.035

Beck, B. R., Lee, S. H., Kim, D., Park, J. H., Lee, H. K., Kwon, S. S., Lee, K.H., Lee, J.I., & Song, S. K. (2017). A *Lactococcus lactis* BFE920 feed vaccine expressing a fusion protein composed of the OmpA and FlgD antigens from *Edwardsiella tarda* was significantly better at protecting olive flounder (*Paralichthys olivaceus*) from edwardsiellosis than single antigen vaccines. *Fish & Shellfish Immunology, 68*, 19–28. https://doi.org/10.1016/j.fsi.2017.07.004

Bermudez-Brito, M., Plaza-Díaz, J., Muñoz-Quezada, S., Gómez-Llorente, C., & Gil, A. (2012). Probiotic Mechanisms of Action. *Annals of Nutrition and Metabolism, 61*(2), 160–174. https://doi.org/10.1159/000342079

Bondad-Reantaso, M. G., Subasinghe, R. P., Arthur, J. R., Ogawa, K., Chinabut, S., Adlard, R., & Shariff, M. (2005). Disease and health management in Asian aquaculture. *Veterinary Parasitology, 132*(3–4), 249–272. https://doi.org/10.1016/j.vetpar.2005.07.005

Bonnet, S., Webb, E. A., Panzeca, C., Karl, D. M., Capone, D. G., & Wilhelmy, S. A. S. (2010). Vitamin B12 excretion by cultures of the marine cyanobacteria *Crocosphaera* and *Synechococcus. Limnology and Oceanography, 55*(5), 1959–1964. https://doi.org/10.4319/lo.2010.55.5.1959

Cao, Y., Liu, Y., Mao, W., Chen, R., He, S., Gao, X., Zhou, Z., & Yao, B. (2014). Effect of dietary N-acyl homoserin lactonase on the immune response and the gut microbiota of zebrafish, *Danio rerio*, infected with *Aeromonas hydrophila. Journal of the World Aquaculture Society, 45*(2), 149–162. https://doi.org/10.1111/jwas.12105

Carnevali, O., de Vivo, L., Sulpizio, R., Gioacchini, G., Olivotto, I., Silvi, S., & Cresci, A. (2006). Growth improvement by probiotic in European sea bass juveniles (*Dicentrarchus labrax*, L.), with particular attention to IGF-1, myostatin and cortisol gene expression. *Aquaculture, 258*(1–4), 430–438. https://doi.org/10.1016/j.aquaculture.2006.04.025

Chabrillón, M., Arijo, S., Díaz-Rosales, P., Balebona, M. C., & Moriñigo, M. A. (2006). Interference of *Listonella anguillarum* with potential probiotic microorganisms isolated from farmed gilthead seabream (*Sparus aurata*, L.). *Aquaculture Research, 37*(1), 78–86. https://doi.org/10.1111/j.1365-2109.2005.01400.x

Chauhan, A., & Singh, R. (2019). Probiotics in aquaculture: A promising emerging alternative approach. *Symbiosis, 77*(2), 99–113. https://doi.org/10.1007/s13199-018-0580-1

Chen, B., Peng, M., Tong, W., Zhang, Q., & Song, Z. (2020). The Quorum Quenching Bacterium *Bacillus licheniformis* T-1 Protects Zebrafish against *Aeromonas hydrophila* Infection. *Probiotics and Antimicrobial Proteins, 12*(1), 160–171. https://doi.org/10.1007/s12602-018-9495-7

Chiu, C. H., Guu, Y. K., Liu, C. H., Pan, T. M., & Cheng, W. (2007). Immune responses and gene expression in white shrimp, *Litopenaeus vannamei*, induced by *Lactobacillus plantarum. Fish & Shellfish Immunology, 23*(2), 364–377. https://doi.org/10.1016/j.fsi.2006.11.010

Chu, W., Zhou, S., Zhu, W., & Zhuang, X. (2014). Quorum quenching bacteria Bacillus sp. QSI-1 protect zebrafish (*Danio rerio*) from *Aeromonas hydrophila* infection. *Scientific Reports*, *4*(1), 5446. https://doi.org/10.1038/srep05446

Chythanya, R., Karunasagar, I., & Karunasagar, I. (2002). Inhibition of shrimp pathogenic vibrios by a marine *Pseudomonas* I-2 strain. *Aquaculture*, *208*(1–2), 1–10. https://doi.org/10.1016/S0044-8486(01)00714-1

Dawood, M. A., & Koshio, S. (2016). Recent advances in the role of probiotics and prebiotics in carp aquaculture: A review. *Aquaculture*, *454*, 243–251. https://doi.org/10.1016/j.aquaculture.2015.12.033

De Paiva-Maia, E., Alves-Modesto, G., Otavio-Brito, L., Vasconcelos-Gesteira, T. C., & Olivera, A. (2013). Effect of a commercial probiotic on bacterial and phytoplankton concentration in intensive shrimp farming (*Litopenaeus vannamei*) recirculation systems. *Latin American Journal of Aquatic Research*, *41*(1), 126–137.

Dhanasekaran, D., Saha, S., Thajuddin, N., Rajalakshmi, M., & Panneerselvam, A. (2010). Probiotic effect of *Lactobacillus* isolates against bacterial pathogens in fresh water fish. *Journal of Coastal Development*, *13*(2), 103–112.

Divya, K. R., Isamma, A., Ramasubramanian, V., Sureshkumar, S., & Arunjith, T. S. (2012). Colonization of probiotic bacteria and its impact on ornamental fish *Puntius conchonius*. *Journal of Environmental Biology*, *33*(3), 551–555.

Elayaraja, S., Annamalai, N., Mayavu, P., & Balasubramanian, T. (2014). Production, purification and characterization of bacteriocin from *Lactobacillus murinus* AU06 and its broad antibacterial spectrum. *Asian Pacific Journal of Tropical Biomedicine*, *4*, S305–S311. https://doi.org/10.12980/APJTB.4.2014C537

Elisashvili, V., Kachlishvili, E., & Chikindas, M. L. (2019). Recent advances in the physiology of spore formation for *Bacillus* probiotic production. *Probiotics and Antimicrobial Proteins*, 11(3), 731–747. https://doi.org/10.1007/s12602-018-9492-x

Elsabagh, M., Mohamed, R., Moustafa, E. M., Hamza, A., Farrag, F., Decamp, O., Dawood, M. A. O., & Eltholth, M. (2018). Assessing the impact of *Bacillus* strains mixture probiotic on water quality, growth performance, blood profile and intestinal morphology of Nile tilapia, *Oreochromis Niloticus*. *Aquaculture Nutrition*, *24*(6), 1613–1622. https://doi.org/10.1111/anu.12797

Eyo V.O., and Akanse N.N. (2018) Comparative study on the condition factor, hematological and serum biochemical parameters of wild and hatchery collected broodfish of the african catfish *Heterobranchus longifilis* (Valenciennes 1840). *Asian Journal of Advances in Agricultural Research* *5*(4), 1–8. https://doi.org/10.9734/AJAAR/2018/39238

Eyo, V. O., & Ivon, E. A. (2017). Growth performance, survival and feed utilization of the African Catfish *Heterobranchus longifilis* (Valenciennes, 1840) fed diets with varying inclusion levels of *Moringa oleifera* leaf meal (MLM). *Asian Journal of Biology*, *4*(1), 1–10. https://doi.org/10.9734/AJOB/2017/35614

FAO (2001). *Probiotics in Food, Health and Nutritional Properties and Guidelines for Evaluation. Joint FAO/WHO Consultation on Evaluation of Health and Properties of Probiotics in Food Including Powder Milk with Live Lactic Acid Bacteria*, Cordoba, Argentina: FAO.

FAO (2014) *The State of the World Fisheries and Aquaculture*. Rome: FAO Fisheries and Aquaculture Department, p. 106.

FAO/WHO (2002) *Joint FAO/WHO Working Group Report on Drafting Guidelines for the Evaluation of Probiotics in Food*, London, Ontario, Canada.

Feng, J., Chang, X., Zhang, Y., Yan, X., Zhang, J., & Nie, G. (2019). Effects of *Lactococcus lactis* from *Cyprinus carpio* L. as probiotics on growth performance, innate immune response and disease resistance against *Aeromonas hydrophila*. *Fish & Shellfish Immunology*, *93*, 73–81. https://doi.org/10.1016/j.fsi.2019.07.028

Fonseca, H. C., de Sousa Melo, D., Ramos, C. L., Dias, D. R., & Schwan, R. F. (2021). Probiotic properties of lactobacilli and their ability to inhibit the adhesion of enteropathogenic bacteria to Caco-2 and HT-29 cells. *Probiotics and Antimicrobial Proteins*, *13*(1), 102–112. https://doi.org/10.1007/s12602-020-09659-2

Foysal, M. J., Fotedar, R., Siddik, M. A. B., & Tay, A. (2020). *Lactobacillus acidophilus* and *L. plantarum* improve health status, modulate gut microbiota and innate immune response of marron (*Cherax cainii*). *Scientific Reports*, *10*(1), 5916. https://doi.org/10.1038/s41598-020-62655-y

Fuller, R. (1989). A review: Probiotics in man and animals. *Journal of Applied Bacteriology*, *66*(5), 365–378.

Ganguly, N. K., Bhattacharya, S. K., Sesikeran, B., Nair, G. B., Ramakrishna, B. S., Sachdev, H. P. S., Batish, V. K., Kanagasabapathy, A. S., Muthuswamy, V., Kathuria, S. C., Katoch, V. M., Satyanarayana, K., Toteja, G. S., Rahi, M., Rao, S., Bhan, M. K., Kapur, R., & Hemalatha, R. (2011). ICMR-DBT guidelines for evaluation of probiotics in food. *The Indian Journal of Medical Research*, *134*(1), 22–25. https://www.ncbi.nlm.nih.gov/pmc/articles/PMC3171912/

Ghanei-Motlagh, R., Mohammadian, T., Gharibi, D., Menanteau-Ledouble, S., Mahmoudi, E., Khosravi, M., Zarea, M., & El-Matbouli, M. (2020). Quorum quenching properties and probiotic potentials of intestinal associated bacteria in Asian Sea Bass *Lates calcarifer*. *Marine Drugs*, *18*(1), 23. https://doi.org/10.3390/md18010023

Ghosh, S., Sinha, A., & Sahu, C. (2007). Effect of probiotic on reproductive performance in female live bearing ornamental fish. *Aquaculture Research*, *38*(5), 518–526. https://doi.org//10.1111/j.1365-2109.2007.01696.x

Ghosh, S., Sinha, A., & Sahu, C. (2008). Dietary probiotic supplementation in growth and health of live-bearing ornamental fishes. *Aquaculture Nutrition*, *14*(4), 289–299. https://doi.org/10.1111/j.1365-2095.2007.00529.x

Gildberg, A., Mikkelsen, H., Sandaker, E., & Ringø, E. (1997). Probiotic effect of lactic acid bacteria in the feed on growth and survival of fry of Atlantic cod (*Gadus morhua*). *Hydrobiologia*, *352*(1), 279–285. https://doi.org/10.1023/A:1003052111938

Gioacchini, G., Dalla Valle, L., Benato, F., Fimia, G. M., Nardacci, R., Ciccosanti, F., Piacentini, M., Borini, A., & Carnevali, O. (2013). Interplay between autophagy and apoptosis in the development of *Danio rerio* follicles and the effects of a probiotic. *Reproduction, Fertility and Development*, *25*(8), 1115–1125. https://doi.org/10.1071/RD12187

Giri, S. S., Sukumaran, V., & Oviya, M. (2013). Potential probiotic *Lactobacillus plantarum* VSG3 improves the growth, immunity, and disease resistance of tropical freshwater fish, *Labeo Rohita*. *Fish and Shellfish Immunology*, *34*(2), 660–666. https://doi.org/10.1016/j.fsi.2012.12.008

Gismondo, M. R., Drago, L., & Lombardi, A. (1999). Review of probiotics available to modify gastrointestinal flora. *International Journal of Antimicrobial Agents*, *12*(4), 287–292. https://doi.org/10.1016/s0924-8579(99)00050-3

Gobi, N., Vaseeharan, B., Chen, J. C., Rekha, R., Vijayakumar, S., Anjugam, M., & Iswarya, A. (2018). Dietary supplementation of probiotic *Bacillus licheniformis* Dahb1 improves growth performance, mucus and serum immune parameters, antioxidant enzyme activity as well as resistance against *Aeromonas hydrophila* in tilapia *Oreochromis mossambicus*. *Fish & Shellfish Immunology*, *74*, 501–508. https://doi.org/10.1016/j.fsi.2017.12.066

Gomez-Gil, B., Roque, A., & Velasco-Blanco, G. (2002). Culture of *Vibrio alginolyticus* C7b, a potential probiotic bacterium, with the microalga *Chaetoceros muelleri*. *Aquaculture*, *211*(1–4), 43–48. https://doi.org/10.1016/S0044-8486(02)00004-2

Gram, L., Melchiorsen, J., Spanggaard, B., Huber, I., & Nielsen, T. F. (1999). Inhibition of *Vibrio anguillarum* by *Pseudomonas fluorescens* AH2, a Possible Probiotic Treatment of Fish. *Applied and Environmental Microbiology*, *65*(3), 969–973. https://doi.org/10.1128/AEM.65.3.969-973.1999

Guo, J. J., Liu, K. F., Cheng, S. H., Chang, C. I., Lay, J. J., Hsu, Y. O., Yang, J.Y., & Chen, T. I. (2009). Selection of probiotic bacteria for use in shrimp larviculture. *Aquaculture Research*, *40*(5), 609–618. https://doi.org/10.1111/j.1365-2109.2008.02140.x

Hai, N. V. (2015). The use of probiotics in aquaculture. *Journal of Applied Microbiology*, *119*(4), 917–935. https://doi.org/10.1111/jam.12886

Hamdan, A. M., El-Sayed, A. F. M., & Mahmoud, M. M. (2016). Effects of a novel marine probiotic, *Lactobacillus plantarum* AH 78, on growth performance and immune response of Nile tilapia (*Oreochromis niloticus*). *Journal of Applied Microbiology*, *120*(4), 1061–1073. https://doi.org/10.1111/jam.13081.

Harikrishnan, R., Balasundaram, C., & Heo, M. S. (2010). Effect of probiotics enriched diet on *Paralichthys olivaceus* infected with lymphocystis disease virus (LCDV). *Fish & Shellfish Immunology*, *29*(5), 868–874. https://doi.org/10.1016/j.fsi.2010.07.031

Hasan, K. N., & Banerjee, G. (2020). Recent studies on probiotics as beneficial mediator in aquaculture: A review. *The Journal of Basic and Applied Zoology*, *81*(1), 1–16. https://doi.org/10.1186/s41936-020-00190-y

Hassaan, M. S., Mohammady, E. Y., Soaudy, M. R., Elashry, M. A., Moustafa, M. M. A., Wassel, M. A., El-Garhy, H. A. S., El-Haroun, E. R., & Elsaied, H. E. (2021). Synergistic effects of *Bacillus pumilus* and exogenous protease on Nile tilapia (*Oreochromis niloticus*) growth, gut microbes, immune response and gene expression fed plant protein diet. *Animal Feed Science and Technology*, *275*, 114892. https://doi.org/10.1016/j.anifeedsci.2021.114892

Helmy, Q., Kardena, E., & Gustiani, S. (2019). Probiotics and bioremediation. In *Microorganisms*. IntechOpen. https://doi.org/10.5772/intechopen.90093

Hlordzi, V., Kuebutornye, F. K. A., Afriyie, G., Abarike, E. D., Lu, Y., Chi, S., & Anokyewaa, M. A. (2020). The use of *Bacillus* species in maintenance of water quality in aquaculture: A review. *Aquaculture Reports*, *18*, 100503. https://doi.org/10.1016/j.aqrep.2020.100503

Hoseinifar, S. H., Sun, Y. Z., Wang, A., & Zhou, Z. (2018). Probiotics as means of diseases control in aquaculture, a review of current knowledge and future perspectives. *Frontiers in Microbiology*, *9*, 2429. https://doi.org/10.3389/fmicb.2018.02429

Irianto, A., & Austin, B. (2002). Probiotics in aquaculture. *Journal of Fish Diseases*, *25*(11), 633–642.

Jahangiri, L., & Esteban, M. Á. (2018). Administration of probiotics in the water in finfish aquaculture systems: A review. *Fishes*, *3*(3), 33. https://doi.org/10.3390/fishes3030033

Jegatheesan, V., Shu, L., & Visvanathan, C. (2011). Aquaculture effluent: Impacts and remedies for protecting the environment and human health. *Encyclopedia of Environmental Health*, 123–135. https://dro.deakin.edu.au/view/DU:30039572

Kamei, Y., Yoshimizu, M., Ezura, Y., & Kimura, T. (1988). Screening of bacteria with antiviral activity from fresh water salmonid hatcheries. *Microbiology and Immunology*, *32*(1), 67–73. https://doi.org/10.1111/j.1348-0421.1988.tb01366.x.

Kewcharoen, W., & Srisapoome, P. (2019). Probiotic effects of *Bacillus* spp. From Pacific white shrimp (*Litopenaeus vannamei*) on water quality and shrimp growth, immune responses, and resistance to *Vibrio parahaemolyticus* (AHPND strains). *Fish & Shellfish Immunology*, *94*, 175–189. https://doi.org/10.1016/j.fsi.2019.09.013

Khan, A., Singh, P., & Srivastava, A. (2018). Synthesis, nature and utility of universal iron chelator – Siderophore: A review. *Microbiological Research*, *212–213*, 103–111. https://doi.org/10.1016/j.micres.2017.10.012

Kozasa, M. (1986). Toyocerin (*Bacillus toyoi*) as growth promotor for animal feeding. *Microbiologie Aliments Nutrition*, *4*, 121–135.

Kuebutornye, F. K. A., Abarike, E. D., Lu, Y., Hlordzi, V., Sakyi, M. E., Afriyie, G., Wang, Z., Li, Y., & Xie, C. X. (2020). Mechanisms and the role of probiotic Bacillus in mitigating fish pathogens in aquaculture. *Fish Physiology and Biochemistry*, *46*(3), 819–841. https://doi.org/10.1007/s10695-019-00754-y

Lakshmi, B., Viswanath, B., & Sai Gopal, D. V. R. (2013). Probiotics as antiviral agents in shrimp aquaculture. *Journal of Pathogens*, *2013*, e424123. https://doi.org/10.1155/2013/424123

LaPatra, S. E., Fehringer, T. R., & Cain, K. D. (2014). A probiotic *Enterobacter* sp. provides significant protection against *Flavobacterium psychrophilum* in rainbow trout (*Oncorhynchus mykiss*) after injection by two different routes. *Aquaculture*, *433*, 361–366.

Lategan, M. J., Torpy, F. R., & Gibson, L. F. (2004). Control of saprolegniosis in the eel *Anguilla australis* Richardson, by *Aeromonas* media strain A199. *Aquaculture*, *240*(1–4), 19–27. https://doi.org/10.1016/j.aquaculture.2004.04.009

Li, J., Tan, B., & Mai, K. (2009). Dietary probiotic *Bacillus* OJ and isomaltooligosaccharides influence the intestine microbial populations, immune responses and resistance to white spot syndrome virus in shrimp (*Litopenaeus vannamei*). *Aquaculture*, *291*(1–2), 35–40. https://doi.org/10.1016/j.aquaculture.2009.03.005

Li, Y., Yang, Y., Song, L., Wang, J., Hu, Y., Yang, Q., Cheng, P., & Li, J. (2021). Effects of dietary supplementation of *Lactobacillus plantarum* and *Bacillus subtilis* on growth performance, survival, immune response, antioxidant capacity and digestive enzyme activity in olive flounder (*Paralichthys olivaceus*). *Aquaculture and Fisheries*, *6*(3), 283–288. https://doi.org/10.1016/j.aaf.2020.10.006

Lilly, D. M., & Stillwell, R. H. (1965). Probiotics: Growth-promoting factors produced by microorganisms. *Science*, *147*(3659), 747–748.

Lipton, A. P. (1997, January). Disease management in shrimp culture with special reference to probionts and additives. In *Proc Workshop Natl Aquac Week. The Aquaculture Foundation of India, Chennai* (205–209).

Liu, C.-H., Chiu, C.-H., Wang, S.-W., & Cheng, W. (2012). Dietary administration of the probiotic, *Bacillus subtilis* E20, enhances the growth, innate immune responses, and disease resistance of the grouper, *Epinephelus coioides. Fish & Shellfish Immunology*, *33*(4), 699–706. https://doi.org/10.1016/j.fsi.2012.06.012

Liu, S., Wang, S., Cai, Y., Li, E., Ren, Z., Wu, Y., Guo, W., Sun, Y., & Zhou, Y. (2020). Beneficial effects of a host gut-derived probiotic, *Bacillus pumilus*, on the growth, non-specific immune response and disease resistance of juvenile golden pompano, *Trachinotus Ovatus*: *Aquaculture*, *514*, 734446. https://doi.org/10.1016/j.aquaculture.2019.734446

Martínez Cruz, P., Ibáñez, A. L., Monroy Hermosillo, O. A., & Ramírez Saad, H. C. (2012). Use of probiotics in aquaculture. *ISRN Microbiology*, *2012*, 916845. https://doi.org/10.5402/2012/916845

Merrifield, D. L., Dimitroglou, A., Foey, A., Davies, S. J., Baker, R. T., Bøgwald, J., Castex, M., & Ringø, E. (2010). The current status and future focus of probiotic and prebiotic applications for salmonids. *Aquaculture*, *302*(1–2), 1–18. https://doi.org/10.1016/j.aquaculture.2010.02.007

Metchnikoff, E. (1907). Lactic acid as inhibiting intestinal putrefaction. *The Prolongation of Life: Optimistic Studies*. London: W. Heinemann, pp. 161–183.

Miller, M. B., & Bassler, B. L. (2001). Quorum sensing in bacteria. *Annual Review of Microbiology*, *55*, 165–199. https://doi.org/10.1146/annurev.micro.55.1.165

Mohammadian, T., Nasirpour, M., Tabandeh, M. R., Heidary, A. A., Ghanei-Motlagh, R., & Hosseini, S. S. (2019). Administrations of autochthonous probiotics altered juvenile rainbow trout *Oncorhynchus mykiss* health status, growth performance and resistance to *Lactococcus garvieae*, an experimental infection. *Fish & Shellfish Immunology*, 86, 269–279. https://doi.org/10.1016/j.fsi.2018.11.052

Mombelli, B., Gismondo, M. R. (2000). The use of probiotics in medical practice. *The International Journal* of *Antimicrobial Agents* 16, 531–536. https://doi.org/10.1016/s0924-8579(00)00322-8

Moosavi-Nasab, M., Abedi, E., Moosavi-Nasab, S., & Eskandari, M. H. (2014). Inhibitory effect of isolated lactic acid bacteria from *Scomberomorus commerson* intestines and their bacteriocin on *Listeria innocua. Iran Agricultural Research*, *33*(1), 43–52. https://doi.org/10.22099/iar.2014.2380

Moriarty, D. J. W. (1998). Control of luminous *Vibrio* species in penaeid aquaculture ponds. *Aquaculture*, *164*(1–4), 351–358. https://doi.org/10.1016/S0044-8486(98)00199-9

Nageswara, P. V., & Babu, D. E. (2006). Probiotics as an alternative therapy to minimize or avoid antibiotics use in aquaculture. *Fishing Chimes*, *26*(1), 112–114.

Nandi, A., Banerjee, G., Dan, S. K., Ghosh, K., & Ray, A. K. (2018). Evaluation of in vivo probiotic efficiency of *Bacillus amyloliquefaciens* in *Labeo rohita* challenged by pathogenic strain of *Aeromonas hydrophila* MTCC 1739. *Probiotics and Antimicrobial Proteins*, *10*(2), 391–398. https://doi.org/10.1007/s12602-017-9310-x

Nayak, S. K. (2010). Role of gastrointestinal microbiota in fish. *Aquaculture Research*, *41*(11), 1553–1573. https://doi.org/10.1111/j.1365-2109.2010.02546.x

Newaj-Fyzul, A., & Austin, B. (2015). Probiotics, immunostimulants, plant products and oral vaccines, and their role as feed supplements in the control of bacterial fish diseases. *Journal of Fish Diseases*, *38*(11), 937–955. https://doi.org/10.1111/jfd.12313

Nurhajati, J., Aryantha, I. N. P., & Indah, D. G. (2012). The curative action of *Lactobacillus plantarum* FNCC 226 to *Saprolegnia parasitica* A3 on catfish (*Pangasius hypophthalamus* Sauvage). *International Food Research Journal*, *19*(4). 1723–1727

Oelschlaeger, T. A. (2010). Mechanisms of probiotic actions: A review. *International Journal of Medical Microbiology*, *300*(1), 57–62. https://doi.org/10.1016/j.ijmm.2009.08.005

Ouwehand, A. C., Kirjavainen, P. V., Shortt, C., & Salminen, S. (1999). Probiotics: Mechanisms and established effects. *International Dairy Journal*, *9*(1), 43–52.

Padmavathi, P., Sunitha, K., & Veeraiah, K. (2012). Efficacy of probiotics in improving water quality and bacterial flora in fish ponds. *African Journal of Microbiology Research*, *6*(49), 7471–7478. https://doi.org/10.5897/AJMR12.496

Parker, R. B. (1974). Probiotics, the other half of the antibiotic story. *Animal Nutrition and Health*, *29*, 4–8.

Pirarat, N., Pinpimai, K., Endo, M., Katagiri, T., Ponpornpisit, A., Chansue, N., & Maita, M. (2011). Modulation of intestinal morphology and immunity in nile tilapia (*Oreochromis niloticus*) by *Lactobacillus rhamnosus* GG. *Research in Veterinary Science*, *91*(3), e92–e97. https://doi.org/10.1016/j.rvsc.2011.02.014

Planas, M., Vázquez, J. A., Marqués, J., Pérez-Lomba, R., González, M. P., & Murado, M. (2004). Enhancement of rotifer (*Brachionus plicatilis*) growth by using terrestrial lactic acid bacteria. *Aquaculture*, *240*(1-4), 313–329. https://doi.org/10.1016/j.aquaculture.2004.07.016

Prasad, L., Baghel, D. S., & Kumar, V. (2003). Role and prospects of probiotics use in aquaculture. *Aquacult*, *4*(2), 247–251.

QingHua, Z., YongHui, F., Juan, W., Jing, G., YongHua, Z., JianZhong, G., & ZengFu, S. (2011). Study on the characteristics of the ammonia-nitrogen and residual feeds degradation in aquatic water by *Bacillus licheniformis. Acta Hydrobiologica Sinica*, *35*(3), 498–503. https://www.cabdirect.org/cabdirect/abstract/20113346628

Queiroz, J. F., & Boyd, C. E. (1998). Effects of a bacterial inoculum in channel catfish ponds. *Journal of the World Aquaculture Society*, *29*(1), 67–73. https://doi.org/10.1111/j.1749-7345.1998.tb00300.x

Rahman, M. Z., Khatun, A., Kholil, M. I., & Hossain, M. M. (2017). Aqua drugs and chemicals used in fish farms of Comilla regions. *Journal of Entomology and Zoology Studies*, *5*(6), 2462–2473. https://doi.org/10.22271/j.ento.2017.v5.i6ah.2893

Ramesh, D., & Souissi, S. (2018). Effects of potential probiotic *Bacillus subtilis* KADR1 and its subcellular components on immune responses and disease resistance in *Labeo rohita. Aquaculture Research*, *49*(1), 367–377. https://doi.org/10.1111/are.13467

Ringø, E., Hoseinifar, S. H., Ghosh, K., Doan, H. V., Beck, B. R., & Song, S. K. (2018). Lactic acid bacteria in finfish: An update. *Frontiers in Microbiology*, 9, 1818. https://doi.org/10.3389/fmicb.2018.01818

Ringø, E., Van Doan, H., Lee, S. H., Soltani, M., Hoseinifar, S. H., Harikrishnan, R., & Song, S. K. (2020). Probiotics, lactic acid bacteria and bacilli: Interesting supplementation for aquaculture. *Journal of Applied Microbiology*, *129*(1), 116–136. https://doi.org/10.1111/jam.14628

Robertson, P. A. W., O'Dowd, C., Burrells, C., Williams, P., & Austin, B. (2000). Use of *Carnobacterium* sp. as a probiotic for Atlantic salmon (*Salmo salar* L.) and rainbow trout (*Oncorhynchus mykiss*, Walbaum). *Aquaculture*, *185*(3–4), 235–243. https://doi.org/10.1016/S0044-8486(99)00349-X

Sahu, M. K., Swarnakumar, N. S., Sivakumar, K., Thangaradjou, T., & Kannan, L. (2008). Probiotics in aquaculture: Importance and future perspectives. *Indian Journal of Microbiology*, *48*(3), 299–308. https://doi.org/10.1007%2Fs12088-008-0024-3

Salinas, I., Cuesta, A., Esteban, M. Á., & Meseguer, J. (2005). Dietary administration of *Lactobacillus delbrüeckii* and *Bacillus subtilis*, single or combined, on gilthead seabream cellular innate immune responses. *Fish & Shellfish Immunology*, *19*(1), 67–77. https://doi.org/10.1016/j.fsi.2004.11.007

Schrezenmeir, J., & de Vrese, M. (2001). Probiotics, prebiotics, and synbiotics: Approaching a definition. *The American Journal of Clinical Nutrition*, *73*(2), 361s–364s. https://doi.org/10.1093/ajcn/73.2.361s

Sewaka, M., Trullas, C., Chotiko, A., Rodkhum, C., Chansue, N., Boonanuntanasarn, S., & Pirarat, N. (2019). Efficacy of synbiotic Jerusalem artichoke and *Lactobacillus rhamnosus* GG-supplemented diets on growth performance, serum biochemical parameters, intestinal morphology, immune parameters and protection against *Aeromonas veronii* in juvenile red tilapia (*Oreochromis* spp.). *Fish & Shellfish Immunology*, *86*, 260–268. https://doi.org/10.1016/j.fsi.2018.11.026

Shabirah, A., Mulyani, Y., & Lili, W. (2019). Effect of types isolated lactic acid bacteria on hematocrit and differential leukocytes fingerling common carp (*Cyprinus carpio* L.) infected with *Aeromonas hydrophila* bacteria. *World News of Natural Sciences*, 24, 22–35.

Shefat, S. H. T. (2018). Probiotic strains used in aquaculture. *International Research Journal of Microbiology*, *7*(2), 43–55. https://doi.org/10.14303/irjm.2018.023

Simón, R., Docando, F., Nuñez-Ortiz, N., Tafalla, C., & Díaz-Rosales, P. (2021). Mechanisms used by probiotics to confer pathogen resistance to teleost fish. *Frontiers in Immunology*, *12*. https://doi.org/10.3389/fimmu.2021.653025

Skjermo, J., Bakke, I., Dahle, S. W., & Vadstein, O. (2015). Probiotic strains introduced through live feed and rearing water have low colonizing success in developing Atlantic cod larvae. *Aquaculture*, *438*, 17–23. https://doi.org/10.1016/j.aquaculture.2014.12.027

Skjermo, J., & Vadstein, O. (1999). Techniques for microbial control in the intensive rearing of marine larvae. *Aquaculture*, *177*(1–4), 333–343.

Smith, P. R., & Davey, S. (1993). Evidence for the competitive exclusion of *Aeromonas salmonicida* from fish with stress-inducible furunculosis by a fluorescent pseudomonad. *Journal of Fish Diseases*, *16*(5), 521–524. https://www.cabdirect.org/cabdirect/abstract/19942202150

Soltani, M., Ghosh, K., Hoseinifar, S. H., Kumar, V., Lymbery, A. J., Roy, S., & Ringø, E. (2019). Genus Bacillus, promising probiotics in aquaculture: Aquatic animal origin, bio-active components, bioremediation and efficacy in fish and shellfish. *Reviews in Fisheries Science & Aquaculture*, *27*(3), 331–379. https://doi.org/10.1080/23308249.2019.1597010

Son, V. M., Chang, C. C., Wu, M. C., Guu, Y. K., Chiu, C. H., & Cheng, W. (2009). Dietary administration of the probiotic, *Lactobacillus plantarum*, enhanced the growth, innate immune responses, and disease resistance of the grouper *Epinephelus coioides*. *Fish & Shellfish Immunology*, 26(5), 691–698. https://doi.org/10.1016/j.fsi.2009.02.018

Sun, Y., He, M., Cao, Z., Xie, Z., Liu, C., Wang, S., Guo, W., Zhang, X., & Zhou, Y. (2018). Effects of dietary administration of *Lactococcus lactis* HNL12 on growth, innate immune response, and disease resistance of humpback grouper (*Cromileptes altivelis*). *Fish & Shellfish Immunology*, *82*, 296–303. https://doi.org/10.1016/j.fsi.2018.08.039

Tachibana, L., Telli, G. S., de Carla Dias, D., Gonçalves, G. S., Ishikawa, C. M., Cavalcante, R. B., Natori, M. M., Hamed, S. B., & Ranzani-Paiva, M. J. T. (2020). Effect of feeding strategy of probiotic *Enterococcus faecium* on growth performance, hematologic, biochemical parameters and non-specific immune response of Nile tilapia. *Aquaculture Reports*, *16*, 100277. https://doi.org/10.1016/j.aqrep.2020.100277

Tan, L. T. H., Chan, K. G., Lee, L. H., & Goh, B. H. (2016). Streptomyces bacteria as potential probiotics in aquaculture. *Frontiers in Microbiology*, *7*, 79. https://doi.org/10.3389/fmicb.2016.00079

Taoka, Y., Maeda, H., Jo, J. Y., Jeon, M. J., Bai, S. C., Lee, W. J., Yuge, K., & Koshio, S. (2006). Growth, stress tolerance and non-specific immune response of Japanese flounder *Paralichthys olivaceus* to probiotics in a closed recirculating system. *Fisheries Science*, *72*(2), 310–321. https://doi.org/10.1111/j.1444-2906.2006.01152.x

Tinh, N. T. N., Dierckens, K., Sorgeloos, P., & Bossier, P. (2008). A review of the functionality of probiotics in the larviculture food chain. *Marine Biotechnology (New York, N.Y.)*, *10*(1), 1–12. https://doi.org/10.1007/s10126-007-9054-9

Vadassery, D. H., & Pillai, D. (2020). Quorum quenching potential of *Enterococcus faecium* QQ12 isolated from gastrointestinal tract of *Oreochromis niloticus* and its application as a probiotic for the control of *Aeromonas hydrophila* infection in goldfish *Carassius auratus* (Linnaeus 1758). *Brazilian Journal of Microbiology*, *51*(3), 1333–1343. https://doi.org/10.1007/s42770-020-00230-3

Varela, J. L., Ruiz-Jarabo, I., Vargas-Chacoff, L., Arijo, S., León-Rubio, J. M., García-Millán, I., Martin del Rio, M.P., Morinigo, M.A., & Mancera, J. M. (2010). Dietary administration of probiotic Pdp11 promotes growth and improves stress tolerance to high stocking density in gilthead seabream *Sparus auratus*. *Aquaculture*, *309*(1–4), 265–271. https://doi.org/10.1016/j.aquaculture.2010.09.029

Vazquez-Gutierrez, P., de Wouters, T., Werder, J., Chassard, C., & Lacroix, C. (2016). High iron-sequestrating *Bifidobacteria* inhibit enteropathogen growth and adhesion to intestinal epithelial cells *In vitro*. *Frontiers in Microbiology*, *7*, 1480. https://doi.org/10.3389/fmicb.2016.01480

Venkateswara, A. R. (2007). Bioremediation to restore the health of aquaculture. *Pond Ecosystem Hyderabad*, *500*(082), 1–12.

Verschuere, L., Rombaut, G., Sorgeloos, P., & Verstraete, W. (2000). Probiotic bacteria as biological control agents in aquaculture. *Microbiology and Molecular Biology Reviews*, *64*(4), 655–671. https://doi.org/10.1128/mmbr.64.4.655-671.2000

Vijayaram, S., & Kannan, S. (2018). Probiotics: The marvelous factor and health benefits. *Biomedical and Biotechnology Research Journal (BBRJ)*, *2*(1), 1. https://doi.org/10.4103/bbrj.bbrj_87_17

Waiyamitra, P., Zoral, M. A., Saengtienchai, A., Luengnaruemitchai, A., Decamp, O., Gorgoglione, B., & Surachetpong, W. (2020). Probiotics Modulate Tilapia Resistance and Immune response against Tilapia Lake Virus Infection. *Pathogens*, *9*(11), 919. https://doi.org/10.3390/pathogens9110919

Wang, P., Chen, S., Wei, C., Yan, Q., Sun, Y.-Z., Yi, G., Li, D., & Fu, W. (2021). *Monascus purpureus* M-32 improves growth performance, immune response, intestinal morphology, microbiota and disease resistance in *Litopenaeus vannamei*. *Aquaculture*, *530*, 735947. https://doi.org/10.1016/j.aquaculture.2020.735947

Wang, Y. B., Li, J. R., & Lin, J. (2008). Probiotics in aquaculture: Challenges and outlook. *Aquaculture*, *281*(1), 1–4. https://doi.org/10.1016/j.aquaculture.2008.06.002

Wang, Y. M., & Wang, Y. G. (2008). Advance in the mechanisms and application of microecologics in aquaculture. *Progress in Veterinary Medicine*, 29, 72–75

Wilson, B. R., Bogdan, A. R., Miyazawa, M., Hashimoto, K., & Tsuji, Y. (2016). Siderophores in iron metabolism: From mechanism to therapy potential. *Trends in Molecular Medicine*, *22*(12), 1077–1090. https://doi.org/10.1016/j.molmed.2016.10.005

Won, S., Hamidoghli, A., Choi, W., Bae, J., Jang, W. J., Lee, S., & Bai, S. C. (2020). Evaluation of potential probiotics *Bacillus subtilis* WB60, *Pediococcus pentosaceus*, and *Lactococcus lactis* on growth performance, immune response, gut histology and immune-related genes in whiteleg shrimp, *Litopenaeus vannamei*. *Microorganisms*, *8*(2), 281. https://doi.org/10.3390/microorganisms8020281

Yassir, R. Y., Adel, M. E., & Azze, A. (2002). Use of probiotic bacteria as growth promoters, antibacterial and the effect on physiological parameters of *Orechromis niloticus*. *Journal of Fish Diseases*, *22*, 633–642.

Yeganeh, S., Adel, M., Nosratimovafagh, A., & Dawood, M. A. O. (2021). The Effect of *Lactococcus lactis* subsp. Lactis PTCC 1403 on the Growth Performance, Digestive Enzymes Activity, Antioxidative Status, Immune Response, and Disease Resistance of Rainbow Trout (*Oncorhynchus mykiss*). *Probiotics and Antimicrobial Proteins*. https://doi.org/10.1007/s12602-021-09787-3

Yi, Y., Zhang, Z., Zhao, F., Liu, H., Yu, L., Zha, J., & Wang, G. (2018). Probiotic potential of *Bacillus velezensis* JW: Antimicrobial activity against fish pathogenic bacteria and immune enhancement effects on *Carassius auratus*. *Fish & Shellfish Immunology*, *78*, 322–330. https://doi.org/10.1016/j.fsi.2018.04.055

Yin, Z., Liu, Q., Liu, Y., Gao, S., He, Y., Yao, C., Huang, W., Gong, Y., Mai, K., & Ai, Q. (2021). Early life intervention using probiotic *Clostridium butyricum* improves intestinal development, immune response, and gut microbiota in large yellow croaker (*Larimichthys crocea*) Larvae. *Frontiers in Immunology*, *12*, 640767. https://doi.org/10.3389/fimmu.2021.640767

Zhang, D., Gao, Y., Ke, X., Yi, M., Liu, Z., Han, X., Shi, C., & Lu, M. (2019). *Bacillus velezensis* LF01: *In vitro* antimicrobial activity against fish pathogens, growth performance enhancement, and disease resistance against streptococcosis in Nile tilapia (*Oreochromis niloticus*). *Applied Microbiology and Biotechnology*, *103*(21), 9023–9035. https://doi.org/10.1007/s00253-019-10176-8

Zhang, J.-J., Yang, H.-L., Yan, Y.-Y., Zhang, C.-X., Ye, J., & Sun, Y.-Z. (2020). Effects of fish origin probiotics on growth performance, immune response and intestinal health of shrimp (*Litopenaeus vannamei*) fed diets with fish meal partially replaced by soybean meal. *Aquaculture Nutrition*, *26*(4), 1255–1265. https://doi.org/10.1111/anu.13081

Zhao, J., Chen, M., Quan, C. S., & Fan, S. D. (2015). Mechanisms of quorum sensing and strategies for quorum sensing disruption in aquaculture pathogens. *Journal of Fish Diseases*, *38*(9), 771–786. https://doi.org/10.1111/jfd.12299

Zhou, S., Zhang, A., Yin, H., & Chu, W. (2016). Bacillus sp. QSI-1 modulate quorum sensing signals reduce *Aeromonas hydrophila* level and alter gut microbial community structure in fish. *Frontiers in Cellular and Infection Microbiology*, *6*, 184. https://doi.org/10.3389/fcimb.2016.00184

Zhou, X., Tian, Z., Wang, Y., & Li, W. (2010a). Effect of treatment with probiotics as water additives on tilapia (*Oreochromis niloticus*) growth performance and immune response. *Fish Physiology and Biochemistry*, *36*(3), 501–509. https://doi.org/10.1007/s10695-009-9320-z

Zhou, X., Wang, Y., Yao, J., & Li, W. (2010b). Inhibition ability of probiotic, *Lactococcus lactis*, against *A. hydrophila* and study of its immunostimulatory effect in tilapia (*Oreochromis niloticus*). *International Journal of Engineering, Science and Technology*, *2*(7). http://doi.org/10.4314/ijest.v2i7.63743

Zorriehzahra, M. J., Delshad, S. T., Adel, M., Tiwari, R., Karthik, K., Dhama, K., & Lazado, C. C. (2016). Probiotics as beneficial microbes in aquaculture: An update on their multiple modes of action: A review. *Veterinary Quarterly*, *36*(4), 228–241. https://doi.org/10.1080/01652176.2016.1172132

17 The Potential Role of Microbes in the Sustainable Growth of Ornamental Fish Culture

Archana Aggarwal and Rakhi Gupta
Maitreyi College, University of Delhi, New Delhi, India

CONTENTS

17.1 INTRODUCTION: BACKGROUND AND DRIVING FORCES

Ornamental fish, also known as living jewels, are peaceful, bright, tiny, and colorful creatures, reared in aquariums or in garden pools for aesthetic and recreational purposes. Not only the culture and trade of ornamental fish are cost-effective, but they have also evolved as an important contributor to the pet industry. Their popularity can be assessed by the fact that in Brazil, fish are the fourth most common pet after dogs, cats, and birds (Cardoso et al., 2019). This is an industry requiring very little space and attention compared to other pets, and it plays a role in poverty alleviation, marine conservation, and provision of a sustainable source of income in developing countries like India. India, despite having a vast sea coastline and perennial rivers, contributes less than 1% to the global trade in ornamental fish. India possesses a rich diversity of ornamental fish, with approximately 1074 indigenous freshwater and marine species including rosy barb, zebra fish, giant danio, goldfish, gourami, silver dollar, guppy, molly, swordtail, rasboras, killifish, hill trout, damsel fish, parrot fish, and marine angels, located in varied biodiversity hotspots such as the north-eastern states,

DOI: 10.1201/9781003306931-19

Western ghats, the Gulfs of Kutch and Mannar, Andaman and Nicobar Islands, Lakshadweep, Coast of Kerala, and Palk Bay (Government of India, 2017). Various exotic species bred in captivity are also exported from India (Pandey & Mandal, 2017; Raja et al., 2019). India offers a great opportunity in the ornamental fish trade due to its rich biodiversity, favorable climatic conditions, and the availability of cheap labor. Ornamental aquaculture provides valuable foreign exchange, and generates job opportunities and self-employment in rural areas to millions of people, making an immense contribution to the economy of the country.

In 2017, India's ornamental fish exports were only US$1.4 million, 0.4% of total world exports, the top exporter countries being Singapore (43%), Japan (14%), Malaysia (10%), and China (7%) (Government of India, 2017). A possible explanation for this low export level could be the trade in ornamental fish that fails to meet technical standards, lack of quality brooders, unavailability of species-specific feed, and safety procedures which may enable the propagation and growth of microbes causing disease and stress, a leading cause of high mortality. Disease outbreaks restrict the output of ornamental fish aquaculture because of productivity losses and trade restrictions (Cardoso et al., 2019; Raja et al., 2019). Ornamental fish are highly susceptible to many microbial pathogens, some of which are readily identifiable. Poor culture conditions, overstocking, and improper handling are triggers for the growth of latent pathogens and the onset of disease (Bernoth & Crane, 1995). Thus, it is imperative to maintain water quality, oxygenation, low levels of organic matter, and good quality feed to enhance disease-free production of ornamental fish. Feed plays an indispensable role and can be modified and supplemented with probiotics to achieve the desired results. It is important to explore the opportunities and role of probiotics in supporting ornamental fish culture by modifying the culture environment, reducing growth of pathogens and ultimately supporting growth of cultured fish.

17.2 MICROBES AND HEALTH MANAGEMENT STRATEGIES IN AQUARIUM FISH

Ornamental fish culture requires a very small area to set up culture and the initial cost incurred is also very low, making it a very profitable sector, and the market is growing exponentially. It is important to maintain standard culture conditions to ensure healthy growth of fish as the presence of a variety of pathogenic organisms or parasitic etiology in the water body are the primary cause of fish diseases (Table 17.1). They may also have zoonotic features, a potential risk for workers handling cultures (Beran et al., 2006; Lowry and Smith, 2007; Fulde & Valentin-Weigand, 2012; Gauthier, 2015).

Fish are continuously exposed through their gut to a diverse variety of pathogens in the external environment, including bacteria, fungi, and parasites. The majority of them are natural inhabitants of the aquatic environment and causative agents for various diseases. Some of them, including bacterial species *Acinetobacter*, *Aeromonas, Citrobacter* and *Shewanella putrefaciens*, *Plesiomonas shigelloides*, are zoonotic in nature. It is important to identify these bacteria in ornamental fish to avoid further contamination. Routine methods using antibiotics and chemotherapeutics for prevention/treatment of various diseases (Magada, 2019) may offer instant relief but do not provide a long-term benefit and have many negative impacts. Studies have found that treating fish with antibiotic drugs not only leads to increased antibiotic resistance in bacteria but also interferes with the normal flora of culture (Weir et al., 2012), which is an alarming situation for the ornamental fish trade. It has been observed that in natural conditions fish are less crowded, stress is minimal, and diseased animals are removed by predators. Parasitic and infectious agents become a major limiting factor for the growth of fish under cultured conditions due to crowding and stress. There is also a significant increase in production costs and further losses to stakeholders (Francis-Floyd, 2013; Magada, 2019). It is indispensable to monitor the negative effect of pathogens, and of overcrowding and stressed conditions on growth, and to identify alternative preventive methods (Rao et al., 2013).

To ensure sustainable growth of aquarium fish and beneficial outcomes, health management strategies that go beyond prophylactics should be planned from the outset to prevent various

TABLE 17.1
Diseases Caused by Various Pathogens in Ornamental Fish and Probiotics Used for Their Treatment

Disease	Pathogen	Affected Fish	Symptoms	Probiotic Strain	References
Protozoan Disease					
Velvet or rust disease	*Oödinium (Piscinoodinium)*	*Carassius carassius, Danio rerio, Betta splendens*	Clamped fins, respiratory distress and yellow-brown dust on body	*Lactobacillus, Bifidobacterium, Vibrio* and *Aeromonas*	Lieke et al. (2020)
Marine velvet disease	*Amyloodinium ocellatum*	*Sciaenops ocellatus, Amphiprion ocellaris*	Respiratory disease, loss of appetite, rubbing or scratching against décor	*Lactobacillus, Bifidobacterium, Vibrio* and *Aeromonas*	Lieke et al. (2020)
White spot/Ich disease	*Ichthyophthirius multifiliis*	*Carassius carassius, Danio rerio, Poecilia sphenops, Betta splendens* (Tropical fishes)	Hyperplasia of mucus releasing cells, shortening and adhesion of secondary gill lamellae, water channels proliferate and become narrow	*Brochothrix thermophasta BA211, Aeromonas sobria*	Jørgensen (2016)
Epistylis (red sore disease)	*Heteropolaria colisarum*	*Hypostomus plecostomus*	Show significantly slow swimming, low appetite, vent is swollen and red, hemorrhages on the head, mouth and at base of fins, yellow to pink ascites	*Bacillus cereus* var. *Toyoi* and *Bacillus subtilis* C-3201 (No significant changes reported)	Khoi (2011)
Chilodonellosis	*Chilodonella*	*Carassius carassius*	Degeneration, necrosis and consequent degradation of the branchial epithelium and occlusion of the capillaries	*Pseudomonas* sp. strain MT5	Suomalainen et al. (2005)
Guppy killer disease/ Tetrahymenosis/Tet disease	*Tetrahymena*	*Poecilia reticulata, Poecilia sphenops*	Inflammatory responses	*Bacillus subtilis*	Sharon et al. (2014)
Trichodiniasis	*Trichodina*	*Oreochromis niloticus*	Skin and gill damage, respiratory discomfort, low appetite and loss of scales	*Bacillus cereus* var. *Toyoi* and *Bacillus subtilis* C-3201 (No significant change)	Martins et al. (2012); EL-Sayed (2020)
Scuticociliatosis	*Uronema*	*Apolemichthys trimaculatus*	Degeneration of muscle fiber and hepatocytes, atrophized intestinal villi, inflammation and granuloma in liver and kidney	*Lactobacillus*	Harikrishnan et al. (2010)
Viral Disease					
Lymphocystis	Iridovirus	*Trichogaster lalius, Aplocheilichthys centralis*	Hypertrophy of the connective tissue cells	Sporolac (*Lactobacillus* sp.)	Sudthongkong et al. (2002)
Viral nervous necrosis (VNN), or viral encephalopathy and retinopathy (VER)	Betanodaviruses	*Carassius auratus auratus, Epalzeorhynchos frenatum*	Mild to moderate vacuolation in brain	Artemia-encapsulated recombinant *Escherichia coli*	Shetty et al. (2012)

(Continued)

TABLE 17.1 *(Continued)*

Disease	Pathogen	Affected Fish	Symptoms	Probiotic Strain	References
Vibrosis	*Vibrio vulnificus*	*Carassius auratus auratus*	Dropsy, exophthalmia, detachment scale, and hemorrhage on body surface	*Saccharomyces cerevisiae*	El-Deen and Elkamel (2015)
Bacterial Disease					
Mycobacteriosis	*Mycobacterium*	*Danio rerio*	Emaciation, poor growth, retarded sexual maturation or decreased reproductive performance	*Aeromonas veronii and Pseudomonas entomophila*	Pacheco et al. (2021)
Columnaris infection	*Flavobacterium columnare*	*Carassius auratus, Cyprinus carpio, Danio rerio, Xiphophorus maculatus, X. helleri, Poecilia latipina, P. reticulata and P. sphenops*	Pale or colorless patches on skin surrounded by reddish zones, skin and gill lesions, fin necrosis	*Enterobacter* sp., *Ent. amnigenus, Pseudomonas fluorescens*	Seghouani et al. (2017)
Edwardsiellosis	*Edwardsiella,*	*Eigenmannia, Puntius, Danio, Betta*	Loss of pigmentation in skin, exophtalmia and spiral movement	*E. faecium* (from Cernivet®LBC)	Vandepitte et al. (1983)
Fish tank granuloma	*Mycobacterium marinum*	*Carassius, Xiphophorus, Poecilia & Gambusia*	Excessive mucus secretion, ulcerations, detachment of the scales, exophthalmia and body deformity with external hemorrhages	*Bacillus subtilis*	Marzouk et al. (2009)
Tail rot, fin rot	*Pseudomonas, Vibrio*/Listonella	*Carassius, Xiphophorus*	Red sores, rotten tails with inflamed margins, swelling on tail rays, hemorrhages at base of fins, pale gill, bulging of eyes and corneal opacity, ulcerated skin, splenomegaly and enteritis	*B. subtilis, Lactobacilli*	Marudhupandi et al. (2017)
Streptococcosis	*Streptococcus*	*Epalzeorhynchos*	Darkening of the skin, lethargy, hemorrhages, brain and kidney tissues	*Bacillus sp.*	Russo et al. (2006)
Fungal Disease					
Saprolegniasis	*Saprolegnia*	*Pterophyllum scalare*	Mild inflammatory response, Lesions in superficial musculature, disruption to osmoregulation	*L. plantarum* FNCC 226	Nurhajati et al. (2012)
Epizootic ulcerative syndrome	*Aphanomyces invadans*	All ornamental fishes	Hemorrhages, ascites and scale loss, with infection spreading to anterior and posterior kidneys and spleen	*Saccharomyces cerevisiae*	Devi et al. (2019)
Aspergillomycosis	*Aspergillus flavus* and *A. parasiticus*	*Oreochromis niloticus*	Infected fish inactive, dark, edematous, exophalmia with corneal changes	Lactic acid bacteria (LAB) and Bifidobacterium bifidum	Abdeltawab et al. (2020)

pathogen-borne diseases and attain optimum growth. Preventive measures can be practiced using beneficial microorganisms or probiotics to create a culture environment (no toxin, side effects, residue, and resistance) appropriate for the growth of fish. This would essentially be achieved by a modified culture environment, leading to increased immunity to fight diseases and creating a sustainable ecosystem (Li, 2008) (Table 17.2).

TABLE 17.2
Use of Probiotics in the Health Management of Ornamental Fish

Probiotics Strains	Ornamental Fish Where Probiotic Strain is Used (Genus)	Health Management	References
Bacillus subtilis	*Poecilia, Xiphophorus, Carassius*	Improvement in reproductive parameters, intestinal microflora and resistance to pathogen. Improvement in growth. Decreased number of motile aeromonads and total coliforms in culture	Ahmadifard et al. (2019)
Bacillus subtilis C-3102 (Calsporin®)	*Cyprinus carpio* (Koi Carp)	Enhanced growth and feed utilization. Intestinal microbiota population regulated	He et al. (2011)
Bacillus subtilis	*Cyprinus carpio* (Koi Carp)	Inhibited growth of pathogenic bacteria *Aeromonas hydrophila*, reduced concentrations of ammonium, nitrite, nitrate and phosphate ions in the body fluid	Lalloo et al. (2007)
Bacillus subtilis	*Carassius auratus gibelio*	Attenuated Pb-induced toxicity, Pb bioaccumulation and enhanced antioxidant responses and immune response	Yin et al. (2018)
Lactic acid bacteria (LAB) strain (*Enterococcus faecium)*	*Pterophyllum scalare*	Pathogen *Pterophyllum scalare* inhibition, improving growth and viability of fish	Dias et al. (2019)
L. rhamnosus PTCC 1637	*Carassius auratus*	Increased reproductive efficiency	Ahmadnia-Motlagh et al. (2017)
Pediococcus acidilactici (PA)	*Danio rerio, Carassius auratus*	Improved health, mucosal immune responses, growth and reproductive production	Arani et al. (2021), Mehdinejad et al. (2019)
Bacillus Subtilis PXN 21	*Astronotus ocellatus*	Improved health, mucosal immune responses, growth and reproductive performance	Firouzbakhsh et al. (2011)
Lactobacillus acidophilus	*Xiphophorus helleri* (Black swordtail) *Carassius auratus* (Gold fish)	Increased skin mucus immune parameters, modulated intestinal microbiota, improved growth performance, elevated stress resistance	Hoseinifar et al. (2015)
Lactobacillus sp.	*Carassius auratus and Xiphophorus helleri*	Significant improvement in wet weight gain, food conversion ratio and specific growth rate	Nandi et al. (2009)
Lactobacillus sporogens, L. acidophilus, B. licheniformis, Streptococcus faecium, Saccharomyces cerevisiae	*Carassius auratus, Xiphophorus helleri*	Activated gut immune system, enhanced growth	Abraham et al. (2008)

(Continued)

TABLE 17.2 *(Continued)*

Probiotics Strains	Ornamental Fish Where Probiotic Strain is Used (Genus)	Health Management	References
Micrococcus luteus	*Oreochromis niloticus*	Improvement in fish growth and health	Abd El-Rhman et al. (2009)
Pseudomonas	*Oreochromis niloticus*	Enhanced fish growth and health	Abd El-Rhman et al. (2009)
Enterococcus faecium	*Pterophyllum scalare, Oreochromis niloticus*	Improved growth and viability of fish	Dias et al. (2019), Wang et al. (2008)
Bacillus subtillis, B. Licheniformes, Trichoderma viride, Nitrosomona europaea, Nitrobacter winogradskyi, Aspergillus oryzae, Rhodococcus, Rhodospirillum ubrum, Cyanobacteria, Pseudomonas denitrificans, Pseudomonas oxalacticus	*Pangasius hypophthalmus,*	Enhanced disease resistance against *P. fluorescens*	Biswas et al. (2016)
Pseudomonas fluorscens	*Oreochromis niloticus*	Protection against *Pseudomonas angulliseptica* and *Streptococcus faecium*	Eissa et al. (2014)
Pseudomonas aeruginosa	*Pseudotropheus lombardoi*	Enhanced color, growth and disease resistance	Ezhil and Narayanan (2013)
Citrobacter sp, Exiguobacterium sp and Enterobacter sp,	*Puntius conchonius*	Biological control of *Aeromonas salmonicida*	Núñez-Cardona (2016)
Saccharomyces cerevisiae	*Oreochromis niloticus, Carassius auratus*	Growth and immunity promoter, better resistance against *Aeromonas hydrophila*	Abdel-Tawwab et al. (2008), Kumari, (2020)
Pediococcus acidilactici	*Aequidens rivulatus, Carassius carassius*	Positive response on growth indices and innate immune system	Neissi et al. (2013), Mehdinejad et al. (2019)
Lactobacillus rhamnosus, Bacillus coagulans, B. mesentericus, Bifidobacterium infantis and B. longum	*Brachydanio rerio*	Significant effect on the pathogenic gut inhabitants	Akbar et al. (2014)
Bacillus coagulans, B. mesentericus, and Bifidobacterium infantis	*Puntius conchonius*	Gut pathogens significantly inhibited	Divya et al. (2012)
Chromobacterium aquaticum	*Danio rerio*	Enhanced hepatic expression of genes related to metabolism and growth, resistance against *A. hydrophila* and *S. iniae*, modulated innate immunity and enhanced defense mechanism against bacterial infection	Yi et al. (2019)
Lactobacillus delbrueckii delbruekii	*Dicentrarchus labrax*	Significantly lowered cortisol level	Carnevali et al. (2006)

17.3 TRADITIONAL ORNAMENTAL FISH FOOD

Before planning any supplement feed for healthy enhanced growth, it is important to know the basic nutritional requirements for ornamental fish species. Most of the successfully cultured ornamental fish are omnivores, and have adapted to captivity, including available feed which may contain essential nutrients (Ray et al., 2012). Conventional fish feed is composed of 40% protein (either animal protein, plant protein or both), 10–30% fats, minerals, vitamin mixture, beta-glucan, pectin, yeast, brewers' dried grain, poultry by-product meal, and a source of starch (wheat) that serves as a binder for the pellets (Pandey, 2013). Inert food with various combinations of live food containing zooplanktons (rotifers, copepods, cladocerans, larval and adult invertebrates) have been extensively used in ornamental fish feed considering their nutritional and productive advantages. Ornamental fish are especially attractive due to their unique pigmentation, and therefore the use of carotenoids as dietary supplement is also recommended (Yohana & Wilson, 2011). The role of dietary nutrients or additives must be considered to ensure proper growth, health and protection from infectious diseases by strengthening the immune system. The correct formulation of the feed improves the digestibility, absorption, and utilization of food, and also fulfills metabolic needs. Additionally, the correct formulation of fish feed further reduces maintenance costs and water pollution. However, prepared fish feed is not always sufficient to develop immunity against various pathogens in ornamental fish culture. Thus, it is essential to enhance immunity with alternative methods. The role of probiotics in overcoming constraints on improving the economical and productive potential of ornamental fish due to various diseases should be emphasized.

17.4 PROBIOTICS: A HEALTHY ALTERNATIVE FOOD FOR ORNAMENTAL FISH

In 1908, Russian scientist and Nobel laureate Eli Metchnikoff studied the positive effects of selected bacteria on humans and proposed that the gut flora can be modified by inclusion of useful microbes. The word "probiotics" was first used by Parker in 1974 for such beneficial microorganisms (BM) (Parker, 1974). In the late 20th century, Fuller (1989) defined probiotics as live microbial feed additives that are beneficial for the health of the host by improving microbial balance in the intestine. The use of probiotics as growth promoters through dietary supplements dates back to 1970; they have been used in the feed to enhance innate immunity and promote growth by resisting the growth of pathogens. The positive impact of selected bacteria has been reported in ruminants, pig, and humans (Vila et al., 2010; Allameh et al., 2017). The implementation of probiotics in aquaculture is gaining momentum and has great future prospects as an alternative to antibiotics (Pandiyan et al., 2013; Banerjee et al., 2017) (Figure 17.1).

FIGURE 17.1 The role of probiotics in health management of ornamental fish

17.5 SELECTION OF PROBIOTICS

Selection of appropriate probiotic bacteria is important for achieving higher productivity and avoiding many possible adverse effects on host organisms. Limited understanding with respect to the gut flora of aquatic organisms and their role poses a constraint in the selection of desired microorganisms as probiotics. There are various selection criteria and it is crucial to study the mechanisms of action and adaptability of potential probiotics (Hosain & Liangyi, 2020). General selection criteria as guided by biosafety considerations are:

- Processing methods
- Administration route
- Site of activity of microorganism in host species

Probiotics used in fish are essentially different from those in terrestrial species because of close association with the aquatic environment. The gut microbiota of fish is transient due to continuous exposure of the digestive tract to cultured water and thus its dependence on the external environment. Lactic acid bacteria isolated from animal sources are popular as probiotics due to their ability to tolerate the low pH of the gut (Martínez Cruz et al., 2012). Some of the commonly used probiotics include gram-positive bacteria belonging to *Lactobacillus*, *Bifidobacterium, Enterococcus, Streptococcus, Pediococcus, Leuconostoc, Bacillus* etc. species. Gram-negative bacteria like *Escherichia coli* and many other organisms including *Saccharomyces*, bacteriophage, and single-celled algae also possess many health benefits. They manifest beneficial physiological responses by improving the host gut microbiota and thus ameliorating feed utilization efficiency, immune response, and quality of culture environment (Verschuere et al., 2000; Yohana & Wilson, 2011). The rate of rejection of foreign microorganisms can be minimized by isolating the probiotic from the host fish itself. The microflora colonized in a species depends largely on its feeding habits. It has been reported that carnivore fish largely harbor species of *Benecka vulnifica, Enterococcus*, and *Vibrio* in their digestive tract while gut microflora of herbivore fish are colonized by *Bifidobacteria* and *Lactobacillus acidophilus* (Hengping and Yanqing, 1997)

Studies have shown that bacteria of the genus *Bacillus* have potential to convert organic matter into carbon dioxide and also reduce nitrite, nitrate and ammonium concentration from the cultured water. They are most popular as probiotics due to their spore-forming capacity and long shelf life. Probiotic bacteria also decrease the load of harmful agents and heavy metals. Besides this, *Bacillus Nitrosomonas, Nitrobacter* sp. *Rhodopseudomonas, Lactobacillus*, and *Saccharomyces* help in reducing the pathogen load from the culture ponds and improve water quality (Sayes et al., 2018). Gram-negative facultative bacteria applied in the digestive system of crustacean and herbivore fish have been shown to lead to better productivity. *Aeromonas*, *Enterobacteriaceae*, and *Plesiomonas* are the preferred microbes used in fresh water fish, while *Vibrio* and *Pseudomonas* are used in marine fish, crustaceans and bivalves (Moriarty, 1990; Hosain & Liangyi, 2020)

17.6 PROBIOTICS: MECHANISM OF ACTION

The ultimate role of selected probiotics is to create a hostile/antagonist environment for the growth of pathogens in the host and reduce stress for the host species. This is achieved by the following mechanisms and depends on the choice of probiotic (Mohapatra et al., 2013; Zorriehzahra et al., 2016; Loh, 2017) (Figure 17.2):

- Creating a security guard around gut, larvae and eggs.
- Facilitating increased secretion of inhibitory chemicals like bacteriocins, lysozymes, proteases and hydrogen peroxide to inhibit growth of pathogens.
- Competition with pathogens for binding sites, essential nutrients, vitamins, enzymes and chemicals which leads to decreased colonization of pathogens.

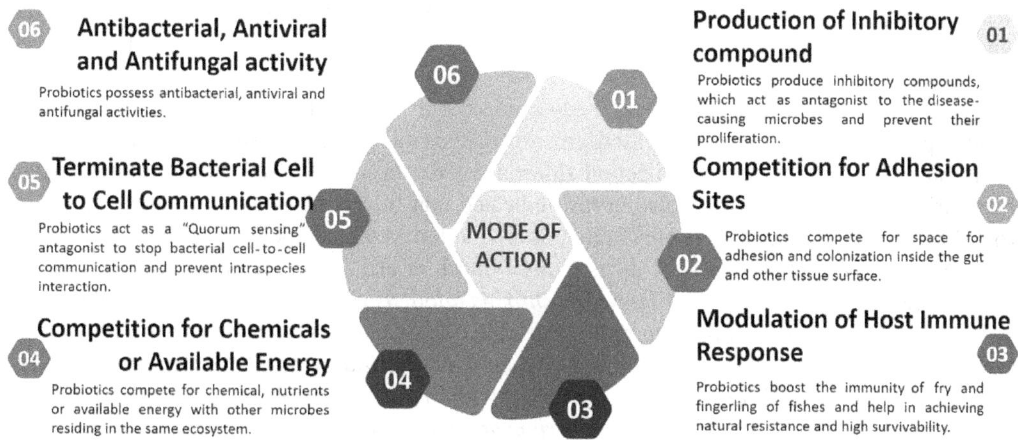

FIGURE 17.2 Mechanism of action of probiotics

- Producing acetic acid, lactic acid, and propionic acid *via* metabolism resulting in low pH of the gut which further facilitates dominance of beneficial microorganisms.
- Modulating non-specific immune systems which further strengthen antibody level and macrophage activity.
- Enhancing disease resistance.
- Acting as an antiviral agent.
- Modulating interaction with the environment.

The mode of delivery of probiotics is also a key to determining outcomes. It is therefore of utmost importance that the correct mode of delivery is chosen, depending on the host species and issues faced. There are various means of probiotic application and selected bacteria/probiotics can be mixed with feed or applied in water in the case of filter feeders. Probiotic mixed with the inoculum of microalgae facilitates early hatching if used as a food supplement in larval culture. The introduction of probiotics 25 days after fish hatching through live feed (rotifers and *Artemia*) is also a common practice. Probiotic inoculum can also be applied in culture via skin to provide microflora with a surface layer to colonize and then enter through it (Sayes et al., 2018). Once the probiotic is administered, various biotic and abiotic factors further determine and stimulate the growth of microorganisms/probiotics.

17.7 APPLICATIONS OF PROBIOTICS IN ORNAMENTAL FISH CULTURE

To meet the increasing demand for ornamental fish both as pets and for photography, various culture farms for breeding and rearing are being maintained all over the world. However, the growth of a variety of pathogens in captive environments leads to disease outbreaks and consequent economic losses. To manage these diseases as well as to enhance growth, antibiotics are generally used in aquaculture. To avoid the negative impact of antibiotics, a plausible solution is the use of probiotics for their potential to ameliorate gut physiology and function.

17.7.1 Probiotics and Fish Metabolism

Studies show that giving supplement microorganisms/probiotics as feed has many beneficial effects on gut physiology, primarily because of the presence of endogenous enzymes as well as their ability to synthesize extracellular enzymes such as proteases, amylases, and lipases. Probiotics also activate the digestive enzymes of aquatic organisms and serve as a source of growth factors such as vitamins, fatty acids, and amino acids. Therefore, they improve the ecosystem of the fish gut and enhance the digestion and nutritional value of food, also improving efficient absorption of nutrients (Martínez Cruz et al., 2012). Probiotics have also shown positive effects on feed conversion and protein efficiency ratio, thus impacting body composition (Allameh et al., 2017). The first probiotics applied in 1986 led to increased growth of hydrobionts, essentially by improving water quality and decreasing bacterial infection (Martínez Cruz et al., 2012). Research carried out in live bearing fish, *Poecilia* and *Xiphophorus*, indicated that the probiont *Bacillus subtilis*, isolated from the intestine of *Cirrhinus mrigala* (Hamilton) and fed as diet supplementation to live bearers, leads to positive effects on nutrient digestibility (Ghosh et al., 2008b). Similar research has also shown increased nutrient digestibility in *Oreochromis niloticus* after using the commercial probiotics Premalac (a combination of skim milk, vegetable oil and calcium carbonate with Torula yeast, *Aspergillus oryzae* extract, *Bifedobacterium bifidum, Lactobacillus acidophilus, Streptococcus faecium*) and Biogen (*Bacillus subtilis*, Allicin mixed with high unit hydrolytic enzymes, and ginseng extract) (Ghazalah et al., 2010). *Chromobacterium aquaticum* significantly enhanced the expression of hepatic genes responsible for carbohydrate metabolism in the ornamental fish *Danio rerio* (Yi et al., 2019). The primary remedial capabilities of *Lactobacillus acidophilus* and *Bifidobacteria* in the digestive tract are improvement in inflammation of the intestines, irritable bowel syndrome, and lactose tolerance, prevention of colon cancer, prevention of high cholesterol levels, gastrointestinal tract diseases and diarrheal diseases, and stabilization of the gut mucosal barrier (Chauhan et al., 2018). Thus, probiotics positively influence the metabolism by increasing the population of beneficial microorganisms and therefore microbial enzyme activity, which is ultimately manifested as improved digestion, absorption and utilization of feed.

17.7.2 Role of Probiotics in the Growth of Fish Body

The ultimate goal of ornamental fish culture is to maintain the healthy growth of cultured fish, although the mechanisms for achieving this are not fully understood. Some reports believe that probiotics enhance the appetite of fish and others believe that they improve the absorbability of nutrients. A few studies have considered both factors in determining outcomes (Martínez Cruz et al., 2012). It is believed that probiotic microorganisms such as *Streptomyces, Bacillus subtilis* etc. have a higher multiplication rate as opposed to expulsion rate, enabling their colonization in the host digestive tract. Thus, when included in fish culture, they associate with the mucosa of the host organism's digestive tract, and start synthesizing macromolecules and vitamins (Dharmaraj & Dhevendaran, 2010). The survival and acceptance of probiotic is primarily determined by genetic make-up and enzyme composition of the host species along with culture conditions. Administration of *B. subtilis* resulted in an increase in body ash and protein content along with increased protease and amylase enzyme activity in all the fish. *Bacillus* is a popular probiotic for water treatment due to its ability to sequester matter into CO_2. Probiotic-fed fish are also shown to have a lower count of various pathogens (motile aeromonads, presumptive pseudomonads and total coliforms). It is of utmost importance to standardize the concentration of probiotic for adequate growth and performance to avoid the negative impact of sub- and supra-threshold concentrations (Anuar et al., 2017). Bo (2006) reported enhanced enzymatic activity and growth in Allogygenetic crucian carp after ingestion of 0.02–0.3% *Bacillus licheniformis* per kilogram body weight.

Zebra fish (*Danio rerio*) fed with *Chromobacterium aquaticum* showed increased expression of genes responsible for growth including growth hormone receptor and insulin-like growth factor-1 (Yi et al., 2019). Lara-Flores et al. (2003) reported that all the probiotic-containing diets resulted in higher growth for tilapia (*Oreochromis niloticus L.*) effectively by mitigating the effects of the stress factors. It has been verified that bacteria of genus *Bacillus* decrease nitrite, nitrate and ammonium concentrations in ornamental fish water (Lalloo et al., 2007; Chauhan et al., 2018). In ornamental fish, growth improvement in terms of increase in length and weight of body was observed in guppy

(*Poecilia reticulate, P. sphenops*) and swordtail (*Xiphophorus helleri, X. maculatus*) after *Bacillus subtilis* and *Streptomyces* supplementation respectively (Ghosh et al., 2008a).

17.7.3 Probiotics and Stress Management

Aquaculture demands high production levels on a tight timescale with minimum costing, leading to general increase in stress for the crop species. Slow movement, lower intake of food and lethargy are general indicators of stress or disease in fish species. Oxidative or pathogenic stress is due to the captive environment, overcrowding and culture conditions. Increased oxidative stress further increases protein loss and apoptosis in cultured fish. General depression of synthesis of muscle proteins was studied in zebrafish, *Danio rerio*, due to chronic stress. The hormone cortisol is a known stress marker and its increased level represents the stress level of fish (Vianello et al., 2003). To assess the increase in stress tolerance after probiotic supplementation, the level of cortisol was compared with a control, and cortisol level was reported to be lower in European sea bass (*Dicentrarchus labrax*) when provided with *Lactobacillus delbrueckii* (Carnevali et al., 2006). Some of the Amazonian ornamental fish, such as cardinal tetra (*Paracheirodon axelrodi*), when fed with *Lactobacillus plantarum* and *Lactobacillus fructivorans*, displayed decreased cortisol level when subjected to acute transportation stress condition (Gomes et al., 2009). Irianto et al. (2003) reported a positive impact of formalin-inactivated cells of *Aeromonas hydrophila* A3-51 as feed supplement in containing *A. salmonicida* infection in goldfish. Similarly, *Bacillus subtilis* and *B. licheniformis* (BioPlus2B) have been beneficial in increasing resistance to infection with *Yersinia ruckeri* in normal trout (Raida et al., 2003).

17.7.4 The Role of Probiotics in the Reproductive Cycle

The nutritional value of feed directly affects various aspects of the reproductive cycle including fertility, fertilization, birth rate and larvae development. Thus, it is essential to maintain brood stock quality and nutritious feed should be provided containing appropriate amounts of macromolecules (lipids, proteins, fats), vitamins and minerals. Popular brood stock diets including fish by-products and live organisms do not provide adequate nutrients and generally increase the possibilities of pathogenic growth and transmission to both parents and offspring (Ghosh et al., 2007; Abasali & Mohamad, 2010). By improving the food value and reducing the risk of pathogenic infection, probiotics improve reproductive outcomes. Four species of ornamental fish, *Poecilia sphenops, P. reticulata, Xiphophorus maculates* and *X. helleri*, showed increased reproductive performance (viability, gonadosomatic index, fecundity and production of fry) when supplemented with the probiotic bacterial strain *Bacillus subtilis* isolated from *Cirrhinus mrigala*. The number of dead or deformed fry also reduced (Chitra and Krishnaveni, 2013). Swordtail (*Xiphophorus helleri*) supplemented with primalaca, a commercial probiotic containing *Bifidobacterium thermophilum, Enterococcus faecium, Lactobacillus acidophilus* and *L. casei*, showed increased reproductive performance of female broodstock in terms of number of offspring and fecundity. *Lactobacillus rhamnosus*-supplemented feed also leads to increase in number of eggs ovulated by zebrafish in vivo and a positive effect on reproductive system physiology in the long term (Carnevali et al., 2013; Aydin and Şehriban, 2019).

17.7.5 Probiotics and Immune System Enhancement

Preventive health care involving dietary considerations is one of the best approaches in sustainable management of aquaculture practices. The role of nutrition in modulating fish immune systems is substantiated by various studies (Caipang and Lazado, 2015). The roles of immunostimulants, prebiotics, and probiotics as health promoters have been studied in aquaculture species. The role of feed supplement in modulating the immune system of fish is achieved via different pathways/modes. It has been reported that they play pivotal roles in boosting specific and non-specific immune systems and mucosal immunity, increasing disease resistance during pathogenic infection in fish (Ibrahem, 2015). Gatesoupe (2008) reported that probiotics increase the parameters for cellular and humoral immune systems as well as stimulating the piscine complement system. Probiotics have also been shown to enhance gut immunity with a marked increase in proliferation of acidophilic granulocytes and immunoglobulin (Vázquez, et al., 2005). It has been reported that fish feed supplemented with probiotic *Pseudomonas aeruginosa* VSG-2 leads to significant increase in level of lysozymes, antibodies (IgM) and alternative complement pathway (ACP) activities as well as macrophage activity. It also leads to increased survival against the pathogen *Aeromonas hydrophila* (Chauhan et al., 2018). In *Scophthalmus maximus*, probiotic bacteria *Lactococcus lactis* increased immune functions (Villamil, et al., 2002). *L. delbrueckii* sp. *Delbrueckii* supplemented through live carriers like artemia and rotifers resulted in an increase in level of T cells in the gut of sea bass (Picchietti et al., 2009). In rainbow trout *(Oncorhynchus mykiss)*, enhanced activity of gut mucosal lysozymes and phagocytic activity of mucosal leukocytes was observed with *Carnobacterium maltaromaticum* and *C. divergens*, and *Lactococcus lactis* spp. *L. mesenteroides* and *L. sakei* respectively (Ibrahem, 2015).

17.7.6 Use of Microbes and Water Quality

Aquatic pathogens flourish in environments loaded with organic matter derived from fish feed and feces in cultured water. Decomposition of organic matter leads to accumulation of nitrogenous waste (ammonia, nitrite and nitrate) formed as part of the nitrogen cycle which supports the growth of pathogens (adapted from Hagenbuch, 2007). Since fish are continuously exposed to aquatic environments, it is of utmost importance to maintain the quality of farming water to avoid growth of pathogens. Controlling the quality of farming water with probiotics is an eco-friendly option and a feasible alternative to chemical treatment. Incubation with an augmenter was found to lead to significant reduction in concentration of dissolved organic matter as well as impacting the nitrification and denitrification processes in the culture tanks of live bearing fish. This process lowered the motile aeromonad and total coliform counts (Ghosh et al., 2008a). The addition of photosynthetic bacteria results in elimination of toxic metabolites and by-products, leading to water quality improvement. Heterotrophic probiotic bacteria help in decomposition of food by-products, dead plankton or organic material to inorganic salts like carbon dioxide, phosphates and nitrates. This further supports the sustainable growth of microalgae, photosynthetic bacteria and cultured fish, and inhibits growth of pathogens (Ibrahem, 2015). Supplementation of Nile tilapia food with commercially available probiotics prepared from *Bacillus subtilis* and *B. licheniformis* leads to improvement in dissolved oxygen, ammonia concentration and level of pH (El-Haroun, et al., 2006). Various strains of *Bacillus* isolated from *Cyprinus carpio* inhibited the growth of pathogens (*Aeromonas hydrophila*) and improved the water quality (downregulation of ammonia, phosphate and nitrite levels) in ornamental fish culture (Lalloo et al., 2007). Wang et al. (2007) reported a total of 27 bacterial strains by 16S rDNA analysis capable of nitrate and nitrite reduction, improving fish pond water quality. Thus, probiotics play a pivotal role in maintaining water quality by essentially removing waste and pathogens.

17.7.7 Probiotics and Inhibition of Pathogens

Water provides an appropriate medium for the rapid growth of pathogenic microorganisms in aquatic animals/fish. High culture density and uncontrolled feeding adds to the problem. Common fish diseases like hemorrhagic septicemia, Edwardsiellosis/Edwardsiella septicemia, gill disease, kidney disease, dropsy, pop eye, vibriosis, tail rot and fin rot are caused by pathogenic bacteria. It has been reported that fish pathogenic gram-negative bacteria include *Aeromonas* spp. and *Vibrio* spp., causing *vibriosis* and *furunculosis* and resulting in high mortality in cultured and wild fish species globally. Mortality due to disease is a limiting factor for the growth of ornamental fish culture

(Allameh et al., 2017). Probiotics offer an excellent opportunity of a safe, effective and viable source without adverse effects on host species. Probiotic microorganisms isolated from different kinds of freshwater cultured fish have played an important role in the antagonism of pathogenic bacteria in the gut of fish, essentially by producing bactericidal or bacteriostatic effects and thus stopping the growth of pathogens (Sihag and Sharma, 2012). They are also known to produce hydrogen peroxide, proteases, lysozymes and bacteriocins. In addition, many probiotics inhibit the multiplication of pathogens by increasing the acidity of the gut through secretion of organic acid, volatile fatty acids like lactic acid, butyric acid and propionic acid (Tinh et al., 2008; Zorriehzahra et al., 2016). *Lactobacillus lactis* and *Leuconostoc mesenteroides* have been shown to exhibit inhibitory activities against pathogenic bacteria in tilapia (*Oreochromis niloticus*) and also to inhibit the growth of *Aeromonas salmonicida* and *Yersinia rukeri* (Zhou et al., 2010). The decline in population of total coliforms, motile aeromonads and presumptive pseudomonads was observed following *Bacillus subtilis* administration (Newaj-Fyzul and Austin 2015). Various species of *Lactobacillus* isolated from *Scomberomorus commerson* (Spanish mackerel) have been reported to down-regulate the growth of *Listeria innocua* (Moosavi-Nasab et al., 2014). The culture of swordtail (*Xiphophorus helleri*) incubated with *Lactobacillus* P21 had a lower count of motile aeromonads, presumptive pseudomonads, lactose fermenters and lactose non-fermenters, and total plate counts, total MRS agar counts in the gut, compared to control (Abraham et al., 2007).

17.8 SAFETY CONCERNS WITH THE USE OF PROBIOTICS

The use of probiotics in aquaculture is gaining popularity as an eco-friendly, affordable and sustainable alternative to control growth of pathogens. It is important to consider and check the safety profile and mechanisms of action of a potential probiotic in order to avoid any adverse effects on cultured species. A potential probiotic must be checked for resistance to multiple antibiotics and the risk of developing antibiotic resistance should be evaluated critically, as incorrect probiotic selection may lead to adverse effects, and unwanted pathogens may flourish and become more resistant. Probiotics work more slowly than antibiotics, and studies have mostly focused on the role of probiotics in the gut. More information and research should be available with respect to targeted delivery of probiotics for maximum output. Quality control guidelines should be followed in the manufacture and application of probiotics, as a quality probiotic helps disease control and sustainable growth of ornamental fish culture. Advanced molecular technologies may be used to identify beneficial microorganisms, their gene structure and molecular functioning, and to ensure safety.

17.9 CONCLUSIONS

Ornamental fish represent a rapidly growing and economically significant field. It is therefore important to develop a sustainable approach to growth increase and disease containment in these species. India has rich marine and freshwater ornamental fish resources with the potential to contribute to the international market. This potential has not been fully exploited for a number of reasons. The use of probiotics is essentially based on the principle of competitive exclusion and requires the living beneficial microorganisms to be used in the diet or culture water to ensure that the gut of the host species is colonized with beneficial microorganisms to improve digestion. Various commercially available products have been promoted to increase output. Immobilization of probiotics, especially using microencapsulation, is also being carried out. Microbial cells at high density are encapsulated in a suitable colloidal matrix to physically and chemically protect the microorganisms during administration. While this shows promising outcomes, active research and study are needed with respect to selection mechanisms, application and target delivery. Farmers should be provided with risk factors with respect to host/microbe interaction in culture conditions. The use of probiotics should be promoted and various advanced technologies should be utilized to increase their efficiency in cultured conditions.

ACKNOWLEDGMENT

We are thankful to the Principal of Maitreyi College, University of Delhi for her constant support and encouragement.

REFERENCES

Abasali H., & Mohamad S. (2010). Effect of dietary supplementation with probiotic on reproductive performance of female live bearing ornamental fish. *Research Journal of Animal Sciences*, *4* (4), 103–107.

Abd El-Rhman, A. M., Khattab, Y. A., & Shalaby, A. M. (2009). *Micrococcus luteus* and *Pseudomonas* species as probiotics for promoting the growth performance and health of Nile tilapia, *Oreochromis niloticus*. *Fish & Shellfish Immunology*, *27*(2), 175–180.

Abdeltawab, A., Tawab, E. L., Hofy, F., Moustafa, E., & Halawa, M. (2020). Isolation and Molecular Identification of *Aspergillus* species from Cultured Nile Tilapia (*Oreochromis niloticus*). *Benha Veterinary Medical Journal*, *38*, 136–140. https://doi.org/10.21608/bvmj.2020.27514.1198

Abdel-Tawwab, M., Abdel-Rahman, A. M., & Ismael, N. E. (2008). Evaluation of commercial live bakers' yeast, Saccharomyces cerevisiae as a growth and immunity promoter for Fry Nile tilapia, *Oreochromis niloticus* (L.) challenged in situ with *Aeromonas hydrophila*. *Aquaculture*, *280*(1–4), 185–189.

Abraham, T. J., Babu, S., & Banerjee, T. (2007). Influence of a fish bacterium *Lactobacillus* sp. on the production of swordtail *Xiphophorus helleri* (Heckel 1848). *Bangladesh Journal of Fisheries Research*, *11*(1), 65–74.

Abraham, T. J., Mondal, S., & Babu, C. S. (2008). Effect of commercial aquaculture probiotic and fish gut antagonistic bacterial flora on the growth and disease resistance of ornamental fishes *Carassius auratus* and *Xiphophorus helleri*. *Su Ürünleri Dergisi*, *25*(1), 27–30.

Ahmadifard, N., Aminlooi, V. R., Tukmechi, A., & Agh, N. (2019). Evaluation of the impacts of long-term enriched Artemia with *Bacillus subtilis* on growth performance, reproduction, intestinal microflora, and resistance to Aeromonas hydrophila of ornamental fish *Poecilia latipinna*. *Probiotics and Antimicrobial Proteins*, *11*(3), 957–965. https://doi.org/10.1007/s12602-018-9453-4

Ahmadnia-Motlagh, H., Hajimoradlo, A., Gorbani, R., Naser, A. G. H., Safari, O., & Lashkarizadeh-Bami, M. (2017). Reproductive performance and intestinal bacterial changes of *Carassius auratus* fed supplemented lactoferrin and *Lactobacillus rhamnosus* PTCC 1637 diet. *Iranian Journal of Ichthyology*, *4*(2), 150–161.

Akbar, I., Radhakrishnan, D. K., Venkatachalam, R., Sathrajith, A. T., Velayudhannair, K., & Sureshkumar, S. (2014). Studies on probiotics administration and its influence on gut microflora of ornamental fish *Brachydanio rerio* larvae. *International Journal of Current Microbiology and Applied Sciences*, *3*(8), 336–344.

Allameh, S. K., Noaman, V., & Nahavandi, R. (2017). Effects of probiotic bacteria on fish performance. *Advanced Techniques in Clinical Microbiology*, *1*(2), 11.

Anuar, N. S., Omar, N. S., Noordiyana, M. N., & Sharifah, N. E. (2017). Effect of commercial probiotics on the survival and growth performance of goldfish *Carassius auratus*. *Aquaculture, Aquarium, Conservation & Legislation*, *10*(6), 1663–1670.

Arani, M. M., Salati, A. P., Keyvanshokooh, S., & Safari, O. (2021). The effect of *Pediococcus acidilactici* on mucosal immune responses, growth, and reproductive performance in zebrafish (*Danio rerio*). *Fish Physiology and Biochemistry*, *47*(1), 153–162.

Aydin, F., & Şehriban, Ç. Y. (2019). Effect of probiotics on reproductive performance of fish. *Natural and Engineering Sciences*, *4*(2), 153–162.

Banerjee, G., Nandi, A., & Ray, A. K. (2017). Assessment of hemolytic activity, enzyme production and bacteriocin characterization of *Bacillus subtilis* LR1 isolated from the gastrointestinal tract of fish. *Archives of Microbiology*, *199*(1), 115–124.

Beran, V., Matlova, L., Dvorska, L., Svastova, P., & Pavlik, I. (2006). Distribution of mycobacteria in clinically healthy ornamental fish and their aquarium environment. *Journal of Fish Diseases*, *29*(7), 383–393. http://dx.doi.org/10.1111/j.1365-2761.2006.00729.x

Bernoth, E. M., & Crane, M. S. J. (1995). Viral diseases of aquarium fish. In *Seminars in Avian and Exotic Pet Medicine* (Vol. 4, No. 2, pp. 103–110). WB Saunders, Elsevier Science.

Biswas, C., Hossain, M. M. M., Hasan-Uj-Jaman, M., Roy, H. S., Alam, M. E., & Banerjee, S. (2016). Dietary probiotics enhance the immunity of Thai pangas (*Pangasius hypophthalmus*) against *Pseudomonas fluorescens*. *International Journal of Fisheries and Aquatic Studies*, *4*(4), 173–178.

Bo, L., Jun, X., Wenbin, L., Tian, W., Huifen, W., & Wei, D. (2006). Effects of *Bacillus licheniformis* on digestive performance and growth of Allogynogenetic crucian carp. *Journal of Dalian Fisheries University, 21*, 336–340.

Caipang, C. M. A., & Lazado, C. C. (2015). Nutritional impacts on fish mucosa: Immunostimulants, pre-and probiotics. In *Mucosal Health in Aquaculture* (pp. 211–272). Academic Press, London

Cardoso, P. H. M., Moreno, A. M., Moreno, L. Z., de Oliveira, C. H., Baroni, F., de Maganha, S. R., de Souza, R. L. M., & Balian, S. (2019). Infectious diseases in aquarium ornamental pet fish: Prevention and control measures. *Brazilian Journal of Veterinary Research and Animal Science, 56*(2), e151697. https://doi.org/10.11606/issn.1678-4456.bjvras.2019.151697

Carnevali, O., Avella, M.A., & Gioacchini, G. (2013). Effect of probiotic administration on zebrafish development and reproduction. *General and Comparative Endocrinology, 188*, 297–302.

Carnevali, O., de Vivo, L., Sulpizio, R., Gioacchini, G., Olivotto, I., Silvi, S., & Cresci, A. (2006). Growth improvement by probiotic in European sea bass juveniles (*Dicentrarchus labrax*, L.), with particular attention to IGF-1, myostatin and cortisol gene expression. *Aquaculture, 258*(1–4), 430–438.

Chauhan, A., Kaur, S., & Singh, R. (2018). A review on probiotics and fish farming. *Research Journal of Pharmacy and Technology, 11*(11), 5143–5146.

Chitra, G., & Krishnaveni, N. (2013). Effect of probiotics on reproductive performance in female livebearing ornamental fish *Poecilia sphenops. International Journal of Pure and Applied Zoology*, 1(3), 249–254.

Devi, G., Harikrishnan, R., Paray, B. A., Al-Sadoon, M. K., Hoseinifar, S. H., & Balasundaram, C. (2019). Comparative immunostimulatory effect of probiotics and prebiotics in *Channa punctatus* against Aphanomyces invadans. *Fish & Shellfish Immunology, 86*, 965–973. https://doi.org/10.1016/j.fsi.2018.12.051

Dharmaraj, S., & Dhevendaran, K. (2010). Evaluation of Streptomyces as a probiotic feed for the growth of ornamental fish *Xiphophorus helleri. Food Technology and Biotechnology, 48*(4), 497–504.

Dias, J. A. R., Abe, H. A., Sousa, N. C., Silva, R. D. F., Cordeiro, C. A. M., Gomes, G. F. E., Ready, J.S., Mouriño, J.L.P., Martins, M.L., Carneiro, P.C.F., Maria, A.N., & Fujimoto, R. Y. (2019). *Enterococcus faecium* as potential probiotic for ornamental neotropical cichlid fish, *Pterophyllum scalare* (Schultze, 1823). *Aquaculture International, 27*(2), 463–474.

Divya, K. R., Isamma, A., Ramasubramanian, V., Sureshkumar, S., & Arunjith, T. S. (2012). Colonization of probiotic bacteria and its impact on ornamental fish *Puntius conchonius. Journal of Environmental Biology, 33*(3), 551.

Eissa, N., Abou El-Gheit, N., & Shaheen, A. A. (2014). Protective effect of *Pseudomonas fluorescens* as a probiotic in controlling fish pathogens. *American Journal of BioScience, 2*(5), 175–181.

El-Deen, A. G. S., & Elkamel, A. A. (2015). Clinical and experimental study on vibriosis in ornamental fish. *Assiut Veterinary Medical Journal (AVMJ)*, 61(146), 147–153.

El-Haroun, E. R., Goda, A. S., & Kabir Chowdhury, M. A. (2006). Effect of dietary probiotic Biogen® supplementation as a growth promoter on growth performance and feed utilization of Nile tilapia *Oreochromis niloticus* (L.). *Aquaculture Research, 37*(14), 1473–1480.

El-Sayed, A. F. M. (2020) Chapter 9 - Stress and diseases. A. F. M. El-Sayed ed., *Tilapia Culture* (2nd ed.), Academic Press, pp. 205–243, ISBN 9780128165096, https://doi.org/10.1016/B978-0-12-816509-6.00009-4

Ezhil, J., & Narayanan, M. (2013). *Pseudomonas aeruginosa* as a Potential Probiont and Pigment Enhancer in the Ornamental Cichlid, *Pseudotropheus lombardoi. Proceedings of the Zoological Society*, 66(2), 154–158).

Firouzbakhsh, F., Noori, F., Khalesi, M. K., & Jani-Khalili, K. (2011). Effects of a probiotic, protexin, on the growth performance and hematological parameters in the Oscar (*Astronotus ocellatus*) fingerlings. *Fish Physiology and Biochemistry, 37*(4), 833–842.

Francis-Floyd, R. (2013). *Introduction to fish health management.* Florida Cooperative Extension Service, Institute of Food and Agricultural Sciences, University of Florida.

Fulde, M., & Valentin-Weigand, P. (2012). Epidemiology and pathogenicity of zoonotic *Streptococci. Host-Pathogen Interactions in Streptococcal Diseases*, 49–81. http://dx.doi.org/10.1007/82_2012_277

Fuller, R. (1989). Probiotic in man and animals. *Journal of Applied Microbiology, 66*, 131–139.

Gatesoupe, F. J. (2008). Updating the importance of lactic acid bacteria in fish farming: natural occurrence and probiotic treatments. *Journal of Molecular Microbiology and Biotechnology, 14*(1–3), 107–114.

Gauthier, D. T. (2015). Bacterial zoonoses of fishes: a review and appraisal of evidence for linkages between fish and human infections. *The Veterinary Journal, 203*(1), 27–35. http://dx.doi.org/10.1016/j.tvjl.2014.10.028

Ghazalah, A. A., Ali, H. M., Gehad, E. A., Hammouda, Y. A., & Abo-State, H. A. (2010). Effect of probiotic on performance and nutrients digestibility of Nile tilapia (*Oreochromis niloticus*) fed low protein diets. *Nature and Science*, *8*(5), 46–53.

Ghosh, S., Sinha, A. and Sahu, C. (2007), Effect of probiotic on reproductive performance in female livebearing ornamental fish. *Aquaculture Research*, *38*(5), 518–526. https://doi.org/10.1111/j.1365-2109.2007.01696.x

Ghosh, S., Sinha, A., & Sahu, C. (2008a). Bioaugmentation in the growth and water quality of livebearing ornamental fishes. *Aquaculture International*, *16*(5), 393–403. https://doi.org/10.1007/s10499-007-9152-8

Ghosh, S., Sinha, A., & Sahu, C. (2008b). Dietary probiotic supplementation in growth and health of live-bearing ornamental fishes. *Aquaculture Nutrition*, *14*(4), 289–299. https://doi.org/10.1111/j.1365-2095.2007.00529.x

Gomes, L. C., Brinn, R. P., Marcon, J. L., Dantas, L. A., Brandão, F. R., De Abreu, J. S., Lemos, P. E. M., McComb, D. M., & Baldisserotto, B. (2009). Benefits of using the probiotic Efinol® L during transportation of cardinal tetra, *Paracheirodon axelrodi* (Schultz), in the Amazon. *Aquaculture Research*, *40*(2), 157–165.

Government of India (2017). Promotion of ornamental fisheries under PMMSY. Department of Fisheries, Ministry of Fishesries, *Animal Husbandry & Dairying*, 1–21. http://www.dof.gov.in/sites/default/files/2020-07/Ornamental_fisheries_development_under_PMMSY.pdf

Hagenbuch, B. (2007). Cellular entry of thyroid hormones by organic anion transporting polypeptides. *Best Practice & Research Clinical Endocrinology & Metabolism*, *21*(2), 209–221.

Harikrishnan, R., Balasundaram, C., & Heo, M. S. (2010). Scuticociliatosis and its recent prophylactic measures in aquaculture with special reference to South Korea: Taxonomy, diversity and diagnosis of scuticociliatosis: Part I Control strategies of scuticociliatosis: Part II. *Fish & Shellfish Immunology*, *29*(1), 15–31.

He, S. Liu, W., Zhou, Z., Mao, W., Ren, P., Marubashi, T., & Ring, E. (2011). Evaluation of probiotic strain *Bacillus subtilis* C-3102 as a feed supplement for koi carp (*Cyprinus carpio*). *Journal of Aquaculture Research and Development*, S1–S7. http://dx.doi.org/10.4172/2155-9546.S1-005

Hengping, X., & Yanqing, X. (1997). Preliminary study on the relationship between aquaculture and micro ecology and ecology of microbes. *Journal of Feed Industries* 18(2), 25.

Hosain, M. A. & Liangyi, X. (2020). Impacts of probiotics on feeding technology and its application in aquaculture. *Journal of Aquaculture, Fisheries & Fish Science*, *3*(1), 174–185.

Hoseinifar, S. H., Roosta, Z., Hajimoradloo, A., & Vakili, F. (2015). The effects of *Lactobacillus acidophilus* as feed supplement on skin mucosal immune parameters, intestinal microbiota, stress resistance and growth performance of black swordtail (*Xiphophorus helleri*). *Fish & Shellfish Immunology*, *42*(2), 533–538.

Ibrahem, M. D. (2015). Evolution of probiotics in aquatic world: Potential effects, the current status in Egypt and recent prospectives. *Journal of Advanced Research*, *6*(6), 765–791.

Irianto, A., Robertson, P. A., & Austin, B. (2003). Oral administration of formalin-inactivated cells of *Aeromonas hydrophila* A3-51 controls infection by atypical *A. salmonicida* in goldfish, *Carassius auratus* (L.). *Journal of Fish Diseases*, *26*(2), 117–120.

Jørgensen, V. L. G. (2016). Infection and immunity against Ichthyophthirius multifiliis in zebrafish (Danio rerio). *Fish & Shellfish Immunology*, 57, 335–339, ISSN 1050-4648, https://doi.org/10.1016/j.fsi.2016.08.042

Khoi, L. N. D. (2011). *Quality management in the Pangasius export supply chain in Vietnam: The case of small-scale Pangasius farming in the Mekong River Delta*. University of Groningen, SOM Research School, Netherlands.

Kumari, D.V. (2020). Effect of preferred dietary probiotics on the growth execution of *Carassius auratus*. *Test Engineering Management*, *83*, 18010–18013.

Lalloo, R., Ramchuran, S., Ramduth, D., Görgens, J., & Gardiner, N. (2007). Isolation and selection of *Bacillus* spp. as potential biological agents for enhancement of water quality in culture of ornamental fish. *Journal of Applied Microbiology*, *103*(5), 1471–1479. https://doi.org/10.1111/j.1365-2672.2007.03360.x

Lara-Flores, M., Olvera-Novoa, M. A., Guzmán-Méndez, B. E., & López-Madrid, W. (2003). Use of the bacteria *Streptococcus faecium* and *Lactobacillus acidophilus*, and the yeast *Saccharomyces cerevisiae* as growth promoters in Nile tilapia (*Oreochromis niloticus*). *Aquaculture*, *216*(1–4), 193–201.

Li, K. (2008). Extending Integrated Fish Farming Technology to Mariculture in China. In: UNESCO/EOLSS Certification No. 6-58-08-10.

Lieke, T., Meinelt, T., Hoseinifar, S. H., Pan, B., Straus, D. L., & Steinberg, C. E. (2020). Sustainable aquaculture requires environmental-friendly treatment strategies for fish diseases. *Reviews in Aquaculture*, *12*(2), 943–965.

Loh, J. Y. (2017). The role of probiotics and their mechanisms of action: an aquaculture perspective. *JWAS*, *18*(1), 19–23.

Lowry, T., & Smith, S. A. (2007). Aquatic zoonoses associated with food, bait, ornamental, and tropical fish. *Journal of the American Veterinary Medical Association*, *231*(6), 876–880. http://dx.doi.org/10.2460/javma.231.6.876

Magada, S. (2019). Health Management in Ornamental Fish Farming. In *BMP Guidelines for Ornamental Fish*. National Fisheries Development Board, Hyderabad, pp. 93–106.

Martínez Cruz, P., Ibáñez, A. L., Monroy Hermosillo, O. A., & Ramírez Saad, H. C. (2012). Use of probiotics in aquaculture. *International Scholarly Research Notices*, *2012*, 916845–916857.

Martins, M. L., Marchiori, N., Roumbedakis, K., & Lami, F. (2012). *Trichodina nobilis* Chen, 1963 and *Trichodina reticulata* Hirschmann et Partsch, 1955 from ornamental freshwater fishes in Brazil. *Brazilian Journal of Biology*, *72*(2), 281–286. https://dx.doi.org/10.1590/S1519-69842012000200008

Marudhupandi, T., Kumar, T. T. A., Prakash, S., Balamurugan, J., & Dhayanithi, N. B. (2017). *Vibrio parahaemolyticus* a causative bacterium for tail rot disease in ornamental fish, *Amphiprion sebae*. *Aquaculture Reports*, *8*, 39–44. https://doi.org/10.1016/j.aqrep.2017.09.004

Marzouk, M. S. M., Essa, M. A., El-Seedy, F. R., Kenawy, A. M., & Abd El-Gawad, D. M. (2009). Epizootiological and histopathological studies on mycobacteriosis in some ornamental fishes. *Global Veterinaria*, *3*(2), 137–143.

Mehdinejad, N., Imanpour, M. R., & Jafari, V. (2019). Combined or individual effects of dietary probiotic, *Pediococcus acidilactici* and nucleotide on reproductive performance in goldfish (*Carassius auratus*). *Probiotics and Antimicrobial Proteins*, *11*(1), 233–238.

Mohapatra, S., Chakraborty, T., Kumar, V., DeBoeck, G., & Mohanta, K.N. (2013), Aquaculture and stress management: A review of probiotic intervention. *Journal of Animal Physiology and Animal Nutrition*, *97*, 405–430. https://doi.org/10.1111/j.1439-0396.2012.01301.x

Moosavi-Nasab, M., Abedi, E., Moosavi-Nasab, S., & Eskandari, M. H. (2014). Inhibitory effect of isolated lactic acid bacteria from *Scomberomorus commerson* intestines and their bacteriocin on *Listeria innocua*. *Iran Agricultural Research*, *33*(1), 43–52.

Moriarty, D. J. W. (1990). Interactions of microorganisms and aquatic animals, particularly the nutritional role of the gut flora. In: Lesel R, Ed. *Microbiology in Poecilotherms*. Elsevier Amsterdam, 217–222.

Nandi, A., Rout, S. K., Dasgupta, A., & Abraham, T. J. (2009). Water quality characteristics in controlled production of ornamental fishes as influenced by feeding a probiotic bacterium, *Lactobacillus* sp. bioencapsulated in *Artemia* sp. *The Indian Journal of Fisheries*, *56*(4), 283–286.

Neissi, A., Rafiee, G., Nematollahi, M., & Safari, O. (2013). The effect of *Pediococcus acidilactici* bacteria used as probiotic supplement on the growth and non-specific immune responses of green terror, *Aequidens rivulatus: Fish & Shellfish Immunology*, *35*(6), 1976–1980.

Newaj-Fyzul, A., & Austin, B. (2015). Probiotics, immunostimulants, plant products and oral vaccines, and their role as feed supplements in the control of bacterial fish diseases. *Journal of Fish Diseases*, *38*(11), 937–955.

Núñez-Cardona, M. T. (2016). Biological control of *Aeromonas salmonicida* in *Puntius conchonius* culture using probiotics under laboratory and fish farm conditions. *International Journal of Fisheries and Aquatic Studies*, *4* (4), 440–443.

Nurhajati, J., Aryantha, I. N. P., & Indah, D. G. (2012). The curative action of *Lactobacillus plantarum* FNCC 226 to *Saprolegnia parasitica* A3 on catfish (*Pangasius hypophthalamus* Sauvage). *International Food Research Journal*, *19*(4), 1723–1727.

Pacheco, I., Díaz-Sánchez, S., Contreras, M., Villar, M., Cabezas-Cruz, A., Gortázar, C., & de la Fuente, J. (2021). Probiotic bacteria with high alpha-gal content protect zebrafish against mycobacteriosis. *Pharmaceuticals*, *14*(7), 635.

Pandey, G. (2013). Feed formulation and feeding technology for fishes. *International Research Journal of Pharmacy*, *4*(3), 23–30. https://doi.org/10.7897/2230-8407.04306

Pandey, P. K., & Mandal, S. C. (2017). Present status, challenges and scope of ornamental fish trade in India. In *Conference: Aqua Aquaria India*, Mangalore.

Pandiyan, P., Balaraman, D., Thirunavukkarasu, R., George, E. G. J., Subaramaniyan, K., Manikkam, S., & Sadayappan, B. (2013). Probiotics in aquaculture. *Drug Invention Today*, *5*(1), 55–59. https://doi.org/10.1016/j.dit.2013.03.003

Parker, R. B. (1974). Probiotics, the other half of the antibiotics story. *Animal Nutrition & Health*, 29, 4–8.

Picchietti, S., Fausto, A. M., Randelli, E., Carnevali, O., Taddei, A. R., Buonocore, F., Scapiglati, G. & Abelli, L. (2009). Early treatment with *Lactobacillus delbrueckii* strain induces an increase in intestinal T-cells and granulocytes and modulates immune-related genes of larval *Dicentrarchus labrax* (L.). *Fish & Shellfish Immunology*, *26*(3), 368–376.

Raida, M. K., Larsen, J. L., Nielsen, M. E., & Buchmann, K. (2003). Enhanced resistance of rainbow trout, *Oncorhynchus mykiss* (Walbaum), against *Yersinia ruckeri* challenge following oral administration of *Bacillus subtilis* and *B. licheniformis* (BioPlus2B). *Journal of Fish Diseases*, *26*(8), 495–498.

Raja, K., Aanand, P., Padmavathy, S., & Sampathkumar, J. S. (2019). Present and future market trends of Indian ornamental fish sector. *International Journal of Fisheries and Aquatic Studies*, *7*(2), 6–15.

Rao, M. V., Kumar, T. A., & Haq, M. B. (2013). Diseases in the Aquarium fishes: Challenges and areas of concern: An overview. *International Journal of Environment*, *2*(1), 127–146.

Ray, A. K. Ghosh, K., & Ringo, E. (2012). Enzyme-producing bacteria isolated from fish gut: A review. *Aquaculture Nutrition*, *18*(5), 465–492. https://doi.org/10.1111/j.1365-2095.2012.00943.x

Russo, R., Mitchell, H., & Yanong, R. P. (2006). Characterization of *Streptococcus iniae* isolated from ornamental cyprinid fishes and development of challenge models. *Aquaculture*, *256*(1–4), 105–110. https://doi.org/10.1016/j.aquaculture.2006.02.046

Sayes, C., Leyton, Y., & Riquelme, C. (2018). Probiotic bacteria as a healthy alternative for fish aquaculture. *Antibiotics use in animals*, Savic, S, editor. Rijeka, Croatia: InTech Publishers, 115–132. https://doi.org/10.5772/intechopen.71206

Seghouani, H., Garcia-Rangel, C. E., Füller, J., Gauthier, J., & Derome, N. (2017). *Walleye autochthonous* bacteria as promising probiotic candidates against *Flavobacterium columnare*. *Frontiers in Microbiology*, *8*, 1349.

Sharon, G., Leibowitz, M. P., Chettri, J. K., Isakov, N., & Zilberg, D. (2014). Comparative study of infection with *Tetrahymena* of different ornamental fish species. *Journal of Comparative Pathology*, *150*(2–3), 316–324. https://doi.org/10.1016/j.jcpa.2013.08.005

Shetty, M., Maiti, B., Santhosh, K. S., Venugopal, M. N., & Karunasagar, I. (2012). *Betanodavirus* of marine and freshwater fish: Distribution, genomic organization, diagnosis and control measures. *Indian Journal of Virology*, *23*(2), 114–123. https://doi.org/10.1007/s13337-012-0088-x

Sihag, R. C., & Sharma, P. (2012). Probiotics: The new eco-friendly alternative measures of disease control for sustainable aquaculture. *Journal of Fisheries and Aquatic Science*, *7*(2), 72–103.

Sudthongkong, C., Miyata, M., & Miyazaki, T. (2002). *Iridovirus* disease in two ornamental tropical freshwater fishes: African lampeye and dwarf gourami. *Diseases of Aquatic Organisms*, *48*(3), 163–173.

Suomalainen, L. R., Tiirola, M. A., & Valtonen, E. T. (2005). Effect of *Pseudomonas* sp. MT5 baths on *Flavobacterium columnare* infection of rainbow trout and on microbial diversity on fish skin and gills. *Diseases of Aquatic Organisms*, *63*(1), 61–68.

Tinh, N. T. N., Dierckens, K., Sorgeloos, P., & Bossier, P. (2008). A review of the functionality of probiotics in the larviculture food chain. *Marine Biotechnology*, *10*(1), 1–12.

Vandepitte, J., Lemmens, P., & De Swert, L. (1983). Human edwardsiellosis traced to ornamental fish. *Journal of Clinical Microbiology*, *17*(1), 165–167.

Vázquez, J. A., González, M., & Murado, M. A. (2005). Effects of lactic acid bacteria cultures on pathogenic microbiota from fish. *Aquaculture*, *245*(1–4), 149–161.

Verschuere, L., Rombaut, G., Sorgeloos, P., & Verstraete, W. (2000). Probiotic bacteria as biological control agents in aquaculture. *Microbiology and Molecular Biology Reviews*, *64*(4), 655–671.

Vianello, S., Brazzoduro, L., Dalla Valle, L., Belvedere, P., & Colombo, L. (2003). Myostatin expression during development and chronic stress in zebrafish (*Danio rerio*). *Journal of Endocrinology*, *176*(1), 47–60.

Vila, B., Esteve-Garcia, E., & Brufau, J. (2010). Probiotic micro-organisms: 100 years of innovation and efficacy; modes of action. *World's Poultry Science Journal*, *66*(3), 369–380.

Villamil, L. A., Tafalla, C., Figueras, A., & Novoa, B. (2002). Evaluation of immunomodulatory effects of lactic acid bacteria in turbot (*Scophthalmus maximus*). *Clinical and Diagnostic Laboratory Immunology*, *9*(6), 1318–1323.

Wang, A., Zheng, G., Liao, S., Huang, H., & Sun, R. (2007). Diversity analysis of bacteria capable of removing nitrate/nitrite in a shrimp pond. *Acta Ecologica Sinica*, *27*(5), 1937–1943.

Wang, Y. B., Tian, Z. Q., Yao, J. T., & Li, W. F. (2008). Effect of probiotics, *Enteroccus faecium*, on tilapia (*Oreochromis niloticus*) growth performance and immune response. *Aquaculture*, *277*(3-4), 203–207.

Weir, M., Rajić, A., Dutil, L., Cernicchiaro, N., Uhland, F. C., Mercier, B., & Tuševljak, N. (2012). Zoonotic bacteria, antimicrobial use and antimicrobial resistance in ornamental fish: a systematic review of the existing research and survey of aquaculture-allied professionals. *Epidemiology & Infection*, *140*(2), 192–206. http://dx.doi.org/10.1017/S0950268811001798

Yi, C. C., Liu, C. H., Chuang, K. P., Chang, Y. T., & Hu, S. Y. (2019). A potential probiotic *Chromobacterium aquaticum* with bacteriocin-like activity enhances the expression of indicator genes associated with nutrient metabolism, growth performance and innate immunity against pathogen infections in zebrafish (*Danio rerio*). *Fish & Shellfish Immunology*, *93*, 124–134.

Yin, Y., Zhang, P., Yue, X., Du, X., Li, W., Yin, Y., … & Li, Y. (2018). Effect of sub-chronic exposure to lead (Pb) and *Bacillus subtilis* on *Carassius auratus gibelio*: Bioaccumulation, antioxidant responses and immune responses. *Ecotoxicology and Environmental Safety*, *161*, 755–762. https://doi.org/10.1016/j.ecoenv.2018.06.056

Yohana, V., & Wilson, C. (2011). Nutritional requirements of freshwater Ornamental Fish. *La Revista MVZ Cordoba*, *16*(2), 2458–2469.

Zhou, X., Wang, Y., Yao, J., & Li, W. (2010). Inhibition ability of probiotic, *Lactococcus lactis*, against *A. hydrophila* and study of its immunostimulatory effect in tilapia (*Oreochromis niloticus*). *International Journal of Engineering, Science and Technology*, 2(7), 73–80.

Zorriehzahra, M. J., Delshad, S. T., Adel, M., Tiwari, R., Karthik, K., Dhama, K., & Lazado, C. C. (2016). Probiotics as beneficial microbes in aquaculture: An update on their multiple modes of action: A review. *Veterinary Quarterly*, *36*(4), 228–241. https://doi.org/10.1080/01652176.2016.1172132

18 Microbial Metabolites as Fish Immunostimulants

Implications for Aquaculture and Fish Vaccines

M. Divya Gnaneswari
Gargi College, New Delhi, India

Parasuraman Aiya Subramani
Vels Institute of Science, Technology, and Advanced Studies, Chennai, India

Chandrasekaran Binuramesh
Thiagarajar College, Madurai, India

CONTENTS

DOI: 10.1201/9781003306931-20

18.1 INTRODUCTION

Aquaculture is a key, rapidly growing food sector that contributed 96.4 million tonnes (MT) of production to total global fisheries production of 178.5 MT in the year 2018. Asia is the largest producer with 72.8 million tonnes (value: 210.19 billion USD) in 2018 (Figures 18.1 and 18.2). Almost 87% of global aquaculture production is used for human consumption. Globally, India ranks fourth, with total aquaculture production increasing steadily since 2000 to reach 7 MT in 2018, a rise of nearly 263.75%. The per-capita fish supply in India is 6.9 kg, with consumption of 2 g/capita/day, which constitutes 13.8% of all animal proteins and 3.1% of total protein in the year 2017. The per-capita supply was 5.1 kg and 4.5 kg in 2007 and 1997 respectively (FAO, 2020).

The demand for and increased consumption rate of aquaculture products has resulted in the expansion of aquaculture practices in pond or net cages. The most viable practice to boost production is intensification, which often leads to deterioration of water quality and spatial constraint stress. This combination quite often results in sudden and frequent outbreak of disease, and a drastic increase in the number of disease outbreaks has been reported in many countries. Motile aeromonad septicemia, furunculosis, enteric septicemia, columnaris, vibriosis, bacterial gill disease, yersiniosis, bacterial kidney disease, infectious hematopoietic necrosis, viral haemorrhagic septicemia, infectious pancreatic necrosis, branchial amoebiasis, edwardsiellosis, streptococcosis, tail rot and fin rot are some of the major diseases reported in fish caused by bacteria, viruses and parasites. Though there are more than 20, 000 known fish species, the innate immune system in general has the upper hand in protecting fish from disease-causing agents. The fish immune system has close

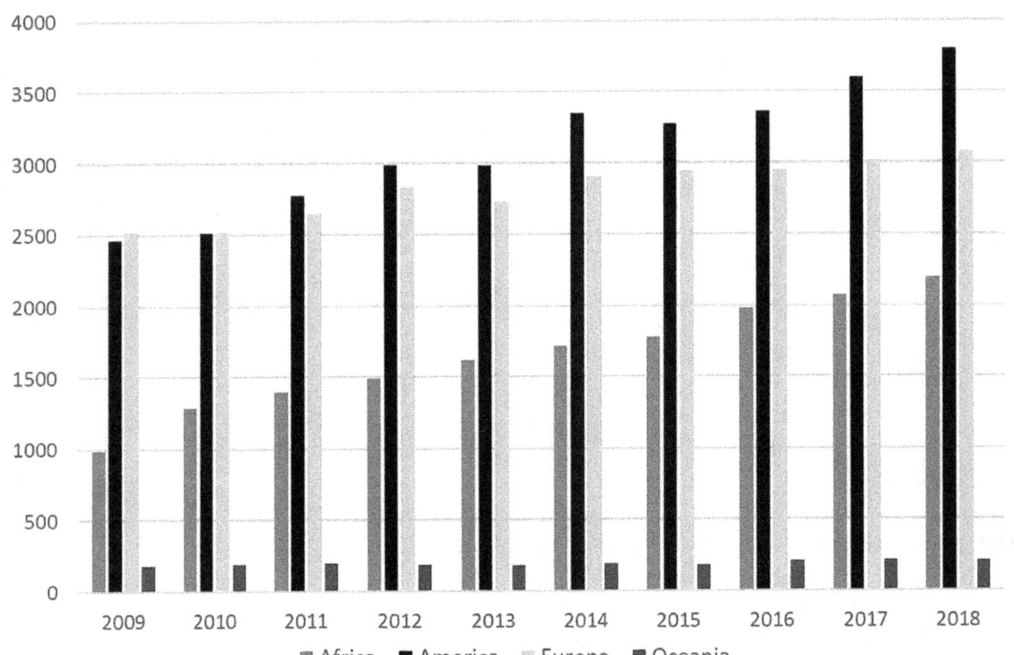

FIGURE 18.1 World aquaculture production in quantity by continent from 2009 to 2018

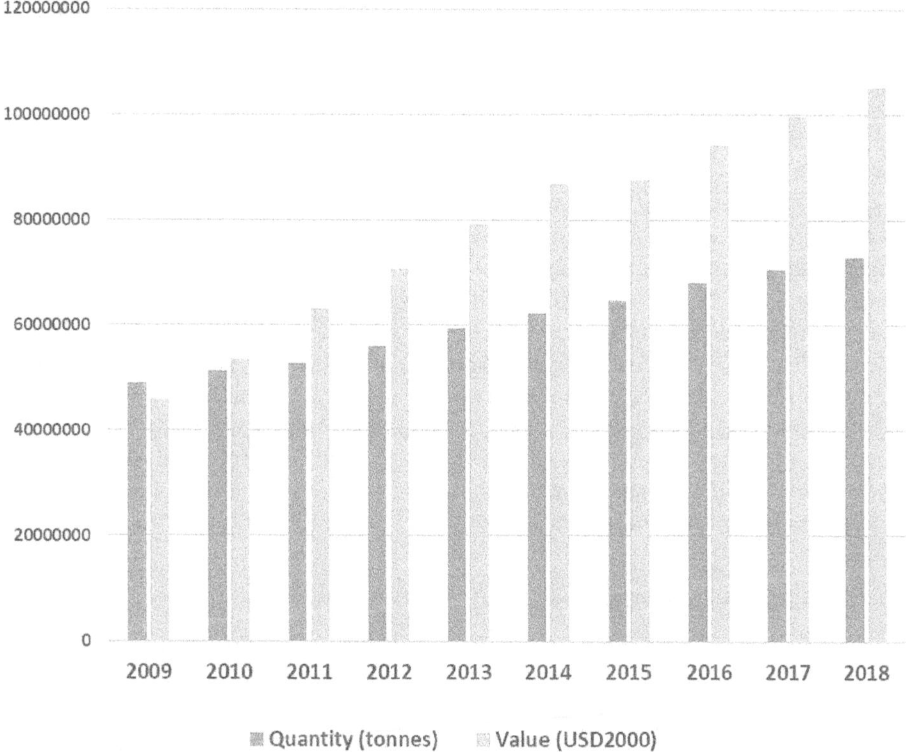

*Includes, fish, crustaceans and molluscs excluding aquatic plants and all other aquatic organisms (FAO, 2020)

FIGURE 18.2 Increase in aquaculture production in Asia from 2009 to 2018.

similarity with mammals in terms of antigen recognition, antibody production, memory response, and mounting a strong immune response against these pathogens. Figure 18.3 summarizes various immune organs, innate and adaptive mechanisms reported in fish. Fish have both specific and non-specific/innate immunity consisting of both humoral and cellular components while mounting an immune response.

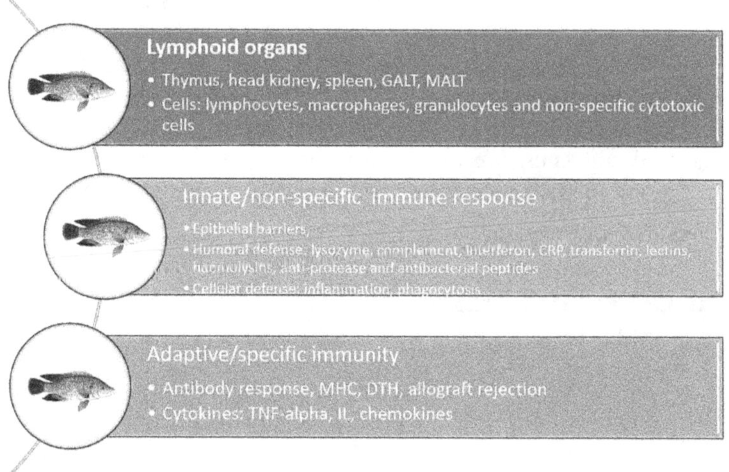

FIGURE 18.3 Lymphoid organs and mechanisms in fish

Maintenance of fish health is the prime concern of the farmer in order to increase production and profit. Chemotherapeutics are successfully used for the treatment of many diseases in farmed animals. The emergence of disease-resistant bacteria, immunosuppression and bioaccumulation of chemical residue in fish and fish products are major concerns for both scientists and consumers. Vaccines and immunostimulants are used as immune-prophylactic measures to prevent infection in fish. Though vaccines are effective in preventing fish diseases, the lack of vaccines against many diseases, the frequent emergence of new strains, difficulties with administration and increased costs make vaccination an unattractive choice as an immunoprohylactic. The development of a vaccine for intracellular pathogens such as *Renibacterium salmoninarum* poses a much greater challenge than for extracellular ones. There is always a need for other promising alternatives such as immunostimulants to prevent diseases in fish (Sakai, 1999).

Immunostimulants are defined as agents that stimulate either non-specific immune mechanisms on their own or specific immune mechanisms when coupled with an antigen (Anderson, 1992). The prophylactic activation of immune mechanisms in animals helps them to fight against bacterial, fungal, viral and parasitic infections even in stressful environments. The wide variety of agents that are reported for their immunostimulant activity are derivatives of microbes, plant extracts, animal extracts, vitamins, hormones and synthetic chemicals.

18.1.1 Advantages and Disadvantages of Immunostimulants

An immunostimulant enhances the non-specific immune response to produce immunocompetency and disease resistance in fish. It can also act as an adjuvant and enhances specific immune response when administered together with a vaccine or an antigen. It also supports the fish to fight against opportunistic pathogens, so it can be used when the animal is subjected to stressful conditions such as handling, transportation or even in a diseased state. It is reported to be non-toxic and safe, eco-friendly and biodegradable (Jadhav et al., 2006; Barman et al., 2013). Immunostimulation during the larval stage is an important strategy to stimulate non-specific immunity and thereby increase survival rate and overall production, providing economic relief to farmers (Bricknell and Dalmo, 2005). Boonyaratpalin et al. (1995) reported the added benefits of combining growth stimulation and immune enhancement, an additional advantage of using immunostimulants to augment production of commercially viable species. There are reports of immunostimulants failing to provide protection against certain bacteria that are resistant to phagocytosis for e.g., *Renibacterium salmoninarum*, *Pseudomonas piscicida* and *Edwardsiella ictaluri* infection. Since most immunostimulants activate non-specific immune mechanisms by enhancing the activity of macrophages, these bacteria probably do evade macrophage defense mechanisms and thus can cause diseases (Sakai, 1999).

In this chapter we focus on reviewing the present state of knowledge of microbial derivatives tested for their immunomodulatory activity, adjuvant properties and disease resistance in fish for application in aquaculture. Four categories are considered in detail: bacterial, fungal and algal immunostimulants, and adjuvants. We have also incorporated brief information on the mode of administration tested and the fish species, realizing the importance in deciding the magnitude of immune activation.

18.2 FUNGAL DERIVATIVES TESTED FOR THEIR IMMUNOSTIMULANT ACTIVITY IN FISH

The most promising fungi-derived immunostimulant reported to protect fish and shellfish is β-glucan, a mushroom and yeast (fungal) cell wall component. It is a poly glucose molecule joined by β- 1, 3 bonds in a linear chain with β-1, 6 branches with single or multiple glucose molecules. The particulate form of glucan has been reported to be more effective in activating fish macrophages than the soluble form. The number of β-1, 6 branches attached with the backbone is an important factor deciding the activity of glucan in fish (Gannam and Schrock, 2001).

The mode of action of β–glucan as described by Joyce Czop (1986) exhibits binding to specific receptors present on the surface of macrophages, stimulates a cascade of events and activates the cell. It can also bind specifically with the receptors present on the monocytes, T cells, granulocytes (Rodrigues et al., 2020) and NK cells (Engstad, 1994). It has been suggested that the immunomodulatory effect of glucan could be due to its binding with C-type lectin receptor family based on in vitro macrophage activation studies conducted in *Cyprinus carpio* (Petit et al., 2019). It has been shown to increase macrophage activity (Jørgensen & Robertsen, 1995), lysozyme activity and complement activity. Esteban et al. (2004) also suggested that the glucan receptors present on the blood leucocytes of seabream (*Sparus aurata* L.) are involved in the recognition and phagocytosis of heat-killed yeast cells (*S. cerevisiae*). In crustaceans, β–glucan binds with specific proteins on the hemocytes and activates them resulting in degranulation and release of prophenol oxidase. This pro-form is converted into an active enzyme by serine proteases. This activated enzyme oxidizes a phenolic group containing amino acids (Tyrosine) into semiquinones which are reported to have antibacterial activity (Philip et al., 2003). Though there are many reports on the immunostimulatory activity of yeast glucan in fish, it did not provide protection against *V. anguillarum* infection in rainbow trout upon injection (Thompson et al., 1995). However, oral administration of the same in Atlantic salmon resulted in increased protection against *V. anguillarum* and *V. salmonicida* (Raa et al., 1992). As listed in Table 18.1, there are several reports on the improvement of growth in several fish and shrimps which might be due to the enhancement of non-specific immunity and reduced bacterial stress on the animals. Cornet et al. (2021) carried out comparative analysis of β-glucan derived from yeast and null mutant, Gas 1, where they found increased lysozyme activity, immunoglobulin and immune gene expression (MCSFR-a hepcidin) and protection against *A. salmonicida* infection in juvenile rainbow trout fed with Gas 1 as compared to wild-type glucan and MacroGard™. The cell wall composition of *S. cerevisiae* has been modified in such a way as to have low glucan and high chitin and this has been tested for its immunostimulatory activity. The duration of immunostimulant effect generated in Brook trout (*Salvelinus fontinalis*) upon single injection or 30 min immersion of either glucan or chitosan was studied by Anderson and Siwicki (1994). They found a higher degree of protection on challenge *A. salmonicida* culture after 1, 2 and 3 days of immunostimulant administration. The reduction in the protective activity was observed 14 days post treatment with these immunostimulants, which suggests short-term elevation of the immune system upon treatment with such agents. There are several reviews available on fungal immunostimulants tested in fish and shrimp (Mohan et al., 2019a; Meena et al., 2013; Ringo et al., 2010; Song et al., 2014)

18.2.1 Other Fungal Derivatives

1. *Lentinan*. A polysaccharide isolated from basidiomycete *Lentinus edodes* and a soluble glucose polymer with β- 1, 3 linked backbone with two glucose branches for every five β- 1, 3 linkages. All these can form a triple-helix structure which is very important for their biological activity as immunostimulants (Saito et al., 1991).
2. *Schizophyllan*. This is extracted from the basidiomycete fungus *Schizophyllum commune* fermented broth. It is structurally more or less similar to lentinan with one additional glucose branch for every third glucose unit in the β- 1, 3 backbone.
3. *Scleroglucan*. Produced by the basidiomycete *Sclerotium glucanum* & *S. rolfsii*, it is structurally similar to Schizophyllan, having 1, 6 β-D-glucopyranosyl side chain.
4. *SSG*. A highly branched β- 1, 3 glucan preparation derived from the culture of *Sclerotina sclerotiorum* and reported to have immunostimulant activity by activating phagocytic cells.
5. *Chitin*. Chitin (poly [1-4]-β-N-acetyl-D-glucosamine) is an insoluble polysaccharide found in fungi cell walls and also a major component of crustaceans and insect exoskeletons.

TABLE 18.1
Immunostimulatory Effect of Fungal Derivatives Reported in Fish

S. No	Fish	Mode of Administration	Reported Activity	References
***S. cerevisiae* – Glucan**				
1	Atlantic salmon, *Salmo salar* L.	IP	Increased resistance to *Vibrio anguillarum, V. salmonicida* and *Yersinia ruckeri;* increased lysozyme and complement activity	Robertsen et al. (1990), Engstad et al. (1992)
2	Atlantic salmon	Oral	Increased resistance to *Vibrio anguillarum, V. salmonicida*	Raa et al. (1992)
3	Atlantic salmon	IP	Increased resistance to *A. salmonicida*, as an adjuvant	Aakre et al. (1994)
4	Atlantic salmon	Feed	Increased expression of genes involved in innate and adaptive immune response	Rodriguez et al. (2016)
5	Atlantic salmon	In vitro – glucan + LPS	Enhanced lysozyme production, transcript accumulation.	Paulsen et al. (2001)
6	Atlantic salmon	IP- glucan, glycogen, FIA	Accumulation of macrophages, neutrophils and thrombocytes, head kidney macrophages from glucan-treated fish showed enhanced ability to kill *A. salmonicida*	Jørgensen, et al. (2006)
7	Turbot	Feed	Increased lysozyme after two weeks	De Baulny et al. (1996)
8	Rainbow trout, *Oncorhynchus mykiss*	IP/ oral	Increased lysozyme activity and phagocytic activity	Jørgensen et al. (1993), Thompson et al. (1995)
9	Rainbow trout	IP and immersion – glucan+ bacterin	Enhanced neutrophil activity and phagocytic activity	Jeney and Anderson (1993)
10	Rainbow trout	Feed	Enhanced immune response and reduced mortality on *Y. ruckeri* infection	Tukmechi, et al. (2011)
11	Rainbow trout	Feed	Enhanced activity of total superoxide dismutase (T-SOD), peroxidase (POD) and catalase (CAT) in serum and their mRNA expressions in the head kidney; increased serum lysozyme and its gene expression in the head kidney	Ji et al. (2017)
12	Channel cat fish, *Ictalurus punctatus*	IP	Increased phagocytosis, antibody and increased resistance *to E. ictaluri*	Chen & Ainsworth (1992)
13	Channel catfish	Feed	Increased NBT; enhanced cell migration and phagocytosis, no resistance to *E. ictaluri* infection	Yoshida et al. (1995), Duncan and Klesius (1996a)
14	Cat fish, *Pangasianodon hypophthalmus*	Feed	Increased respiratory burst activity, lysozyme activity, anti-protease activity, antibody titre and complement activity	Sirimanapong et al. (2015)
15	Nile tilapia, *Oreochromis niloticus*	Feed - β-glucan	Increased serum lysozyme, bactericidal activity, serum nitric oxide and the lymphocyte transformation index after 21 days of diet	El-Boshy et al. (2010)
16	Nile tilapia	Feed	Increased gene expression – HSP70, CXC chemokine, MHC IIβ, MX, VTG, TNF-α, CAS, IgM-H, GST, Il-8; protection against *Streptococcus iniae*	Salah et al. (2017)

17	Pacu, *Piaractus mesopotamicus*	Feed	Increased resistance to *A. hydrophila;* leucocyte production, lysozyme activity	Biller-Takahashi et al. (2014)
18	Pacu	Feed	Enhanced complement activity and stress protection	De Mello et al. (2019)
19	Common carp, *Cyprinus carpio*	IP	Increased SOD and catalase activity	Kim et al. (2009)
20	Common carp	Feed	Increased neutrophil activity, serum lysozyme activity and enhanced protection against *A. hydrophila* infection	Gopalakannan, and Arul (2010)
21	Common carp	*In vitro*	Activated macrophages, ROS, RNS production,	Pietretti, et al. (2013)
22	Indian Major Carp, *Labeo rohita*	Feed	Enhanced specific immunity against *E. tarda* infection and reduced mortality	Sahoo and Mukherjee (2002)
23	*L. bata*	Feed	Enhanced expression of I-IFN	Sahoo et al. (2020)
24	Juvenile pompano, *Trachinotus ovatus*	Feed	Enhanced non-specific immunity, survival rate and tolerance to *S. iniae*	Do-Huu, et al. (2019)
25	Sea bass, *Dicentrarchus labrax*	Feed	Enhanced complement activity	Bagni et al. (2000)
27	Sea bass	Feed	Increased respiratory burst activity in head kidney macrophages	Bonaldo et al. (2007)
28	Sea bass	Diet - *MOS* (extracted from *S. cerevisiae*)	Increased phagocytic macrophage HK-activity; significant reduction in infection with *V. alginolyticus*	Torrecillas et al. (2007)
29	Japanese flounder, *Paralichthys olivaceus*	Feed	Increased respiratory burst activity	Galindo-Villegas et al. (2002)
30	Climbing perch, *Anabas testudineus*	immersion	Enhanced lysozyme and bactericidal activities and survival of spawns when challenged with *A. hydrophila* and fungal pathogen *Saprolegnia parasitica*	Das et al. (2009, 2013)
31	Zebra fish, *Danio rerio*	IP	Increase in the myelomonocytic cell population and enhanced survival upon *A. hydrophila* challenge	Rodríguez et al. (2009)
32	Yellow croaker, *Pseudosciaena crocea*	Feed - β-glucan extract	Enhanced serum lysozyme content and phagocytic, respiratory burst activity in head kidney macrophages; decreased mortality rate upon *V. harveyi* challenge	Ai et al. (2007)
33	Japanese sea bass, *Lateolabrax japonicus*	YCW – feed – 72 days	No significant effect on complement, myeloperoxidase activity, antibody response and respiratory burst activity; increased survival when challenged with *A. veronii.*	Yu et al. (2014)
34	*Catla catla*	Feed - β-glucan from Mushroom, *Pleurotus florida*	Enhanced superoxide anion production, phagocytosis, bactericidal activity, lysozyme activity and enhanced survival against *A. hydrophila* challenge	Kamilya et al. (2008)

(Continued)

TABLE 18.1 *(Continued)*

S. No	Fish	Mode of Administration	Reported Activity	References
Lentinan				
35	*Cyprinus carpio*	Injection	Increased phagocytic activity of kidney leucocyte and disease resistance to *E. tarda*	Yano, et al. (1989)
Schizophyllan				
36	*Cyprinus carpio* Yellow tail, *Seriola quinqueradiata*	Injection	Increased phagocytic activity of kidney leucocytes, complement activity and lysozyme levels	Yano et al. (1991); Matsuyama et al. (1992)
37	*Cyprinus carpio*	Feed	Increased disease protection against *E. tarda* and *A. hydrophila*, increase in the number of peripheral macrophages & neutrophils; enhanced phagocytic and lysozyme activity	Kwak et al. (2003)
Scleroglucan				
38	Yellow tail, *Seriola quinqueradiata*	Injection	Increased disease resistance	Matsuyama, Mangindaan, & Yano (1992)
39	*Cyprinus carpio*	Injection	Upregulation of genes – complement factor B and C2 in kidney and spleen	Nakao et al. (2002)
40	Seabream, *Sparus aurata L.*	Feed – 4 weeks	Enhancement in respiratory burst activity, phagocytic ability, natural cytotoxic activity and MPO content in the head kidney leukocytes; increase in IgM level	Ortuno et al. (2002), Cuesta et al. (2004)
41	Seabream, *Sparus aurata L.*	Feed	Enhanced serum lysozyme activity, leucocyte phagocytic ability; decreased serum complement activity and peroxidase content after 6 weeks of the diet	Rodriguez et al. (2003)
42	*Tilapia aureus* P., and grass carp, *Ctenopharyngodon idellus*	Bar, krestin, scleroglucan, and zymosan	Increased protection against *A. hydrophila* and *E. tarda* infection; increase in number of NBT-positive staining cells	Wang and Wang (1997)
43	Hybrid tilapia and Japanese eels Anguilla japonica	Barley, krestin, MacroGard, scleroglucan, and zymosan	Enhanced lysozyme activity, phagocytic activity, in head kidney and PBL and complement activity	Wang et al. (2007)

Chitin				
44	Rainbow trout	IP	Enhanced macrophage activity and resistance to *V.anguillarum* infection	Sakai et al. (1992)
45	Rainbow trout	IP/Immersion	Increased protection against *A. salmonicida;* increased phagocytic activity, MPO and total Ig concentration	Siwicki et al. (1994), Anderson et al. (1995)
46	Rainbow trout	Feed – *Lentinula edodes*	Increase in the number of total leukocytes, phagocytic activity, lysozyme activity and serum IgM levels; higher survival on *Lactococcus garvieae* infection	Baba et al. (2015)
47	Brook trout, *S. fontinalis*	IP/Immersion	Increased protection against *A. salmonicida*	Anderson and Siwicki (1994)
48	Yellowtail, *Seriola quinqueradiata*	IP	Increased resistance to *P. piscicida*	Kawakami et al. (1998)
49	*Roach, Rutilus rutilus*	β-1,3/1,6-glucan, *Lactobacillus plantarum* bacteria – feed	Increased respiratory burst activity and killing activity of head kidney phagocytes against *A. hydrophila*	Kazun et al. (2020)
50	Prussian carp, *Carassius auratus gibelio*	*Coriolus versciclor* polysaccharides – feed	Increase rate of survival after infection with *A. hydrophila*	Wu et al. (2013)
51	Grass carp, *Ctenopharyngodon idella*	*Agaricus bisporus* polysaccharides (ABPs) – feed	Increased ROS, SOD activity, enhanced gene expression - β-defensin, LEAP-2A, IL-6, NF-κB P65, IFN-γ2, IL-10, and TNFα upon *A. hydrophila* infection	Harikrishnan et al. (2021)
52	Spotted grouper, *Epinephelus coioides*	Feed	Enhanced lysozyme, alternate complement activity, phagocytic activity and respiratory burst activity, protection against *Vibrio alginolyticus*	Chang et al. (2013)
53	Cat fish, *Pangasius bocourti*	*Eryngii mushroom, Pleurotus eryngii* (PE), and L. *plantarum* - feed	Enhanced lysozyme, phagocytic activity and respiratory burst activity, protection against *A. hydrophila* challenge	Van Doan et al. (2016a)

6. *Peptidoglucan*. Derived from *Brevibacterium lactofermentum*, it enhanced phagocytosis in yellowtail and increased disease resistance upon infection with *Enterococcus seriola* (Itami et al., 1996). Matsuo and Miyazano (1993) also reported its effect in rendering protection against vibriosis in rainbow trout. The stimulatory activity of peptidoglucan has also been tested in black tiger shrimp against yellow head baculovirus infection (Boonyaratpalin et al. 1995).
7. *Ganoderma lucidum (GL)*. Polysaccharides isolated from this mushroom have been reported to be efficient in modulating immune responses, inhibiting tumor growth, preventing oxidative damage and stimulating B lymphocytes (Bai et al., 2015). Mushroom β-glucan mixture which is composed of 34.06% of macro molecular polymers with bioactive linkages (3, (3, 4), (4, 6) - glucopyranosyl and 6-linked galactopyranosyl residues) has also been reported to have immunostimulatory activity in fish (Chang et al., 2013). There are recent reports of symbiotic stimulatory effects of mushroom-derived glucans and Lactobacillus in catfish and *Rautilus*. Of two different glucans (PS-I and PS-II) isolated from alkaline extract of *Pleurotus florida*, PS-I was found to be responsible for macrophage, splenocyte and thymocyte activations (Dey et al., 2012).

18.3 COMMERCIAL PRODUCTS TESTED FOR THEIR IMMUNOSTIMULATORY ACTIVITY IN FISH

Some of the commercial products using fungal constituents tested for their activity include (Table 18.2):

- Biosaf (concentrated live yeast strain Sc 47)
- DVAQUA (feed additive with fermented product of yeast)
- Ecoactiva (β-glucan feed additive)
- Ergosan (extract of *Laminaraia digitata*)
- Fibosel (inactive yeast cell fraction rich in β-glucan)
- MacroGard™ (derived from yeast cell wall)
- VitaStim (probiotic with beneficial bacteria)
- Others: chitin, LPS, mannan-oligosaccharides (MOS), and PGN

MacroGard™. This is derived from the cell wall of *Saccharomyces cerevisiae*. It is mainly made up of poly glucose with β 1-3 linkage and few β 1, 6 linkages. In Atlantic salmon, it has been reported that it binds with specific receptors on macrophages and elevates their phagocytic and microbicidal activity, induces higher production of lysozyme, IL-1 and TNF, enhances production of complements and antibodies when administered along with antigen. Increased disease resistance against *A. salmonicidae*, *Vibrio anguillarum*, and *Yersinia ruckeri* has also been reported (Siwicki et al., 2004). However, Douxfils et al. (2017) reported that MacroGard™ did not induce gene expression of TNFα-1, IL-1β, COX-2, TGF-β, and IL-10 in head kidney and gills in *O. mykiss* upon oral administration of MacroGard™ or on challenge with *A. hydrophila*. But in the same study, they observed dose-dependent immune modulation in spleen. Though certain doses generated random effects too difficult to interpret, feeding of high doses of glucan resulted in stress, immune exhaustion, or feedback regulation.

Betafectin. This is derived from genetically modified yeast, *S. cerevisiae*. It has β- 1, 3 glucan with β- 1, 6 tri-glucosyl side chains (Onderdonk, et al., 1992).

Vitastim-Taito. A commercial preparation of Schizophyllan. It has been reported to stimulate the immune response of Coho salmon and catfish (Ainsworth et al., 1994) by enhancing disease resistance upon administration through feed. Nikl et al. (1993) reported no enhancement upon immersion of *Oncorhynchus tshawytscha* in VST.*Ergosan*. Rich in alginate and polysaccharides.

TABLE 18.2
Immunostimulatory Effect of Commercial Microbial Products Tested in Fish

S. No	Commercial Product	Fish	Effect	References
1	Biosaf	Tilapia – feed	Increase in serum lysozyme activity, bactericidal activity, phagocytic activity; reduction in mortality when challenged with *A. hydrophila*	El-Boshy et al. (2010)
2	Chitin(Sigma)	Sea bream – feed	Increase in natural haemolytic complement activity and natural cytotoxic activity of HK leukocyte after 2 weeks; enhanced respiratory burst activity in head kidney leukocyte after 4 weeks; an increase in serum IgM	Cuesta et al. (2004), Esteban et al. (2001)
3	DVAQUA	Hybrid tilapia – Feed(*Oreochromis niloticus* ♀ *x O. aureus* ♂)	Enhanced serum lysozyme activity, serum C3 & C4 content and phagocytic activity of head kidney macrophages	He et al. (2009)
4	EcoActiva	Snapper –feed	Enhanced respiratory burst activity of head kidney macrophages even after 56 days of administration	Cook et al. (2003)
5	Ergosan	*Oncorhynchus mykiss* – IP	Increase in leucocyte migration, phagocytosis, respiratory burst activity, expression of IL-β, IL-8, TNF α in peritoneal leucocytes. Slight increase in complement activity.	Peddie et al. (2002)
6	Ergosan and Salar-bec, a vitamin premix,	*Channa striata* – IP	Increased survival upon *Aphanomyces invadans* infection and bactericidal activity	Miles et al. (2001)
7	Ergosan and Vitacel	*Huso huso* – feed	Enhanced lymphocyte count, neutrophil profile and lysozyme activity	Heidarieh et al. (2011)
8	Fibosel and VitaStim	Seabream - feed	Enhanced spleen macrophage respiratory burst activity; reduced mortality when challenged with *P. damselae*	Couso et al. (2003)
9	Glucan- MOS	Juvenile pacu, *Piaractus mesopotamicus* - feed	Enhanced respiratory burst activity, lysozyme activity, number of thrombocytes, neutrophils and monocytes and reduced stress response	Soares et al. (2018)
10	β-glucan - (Sigma)	Indian major carp, *Labeo rohita* - feed	Increased bacterial agglutination, haemagglutination and haemolysin titre, bactericidal activity, serum phagocytic ratio, serum phagocytic index and serum leukocrit l and enhanced survival to *A. hydrophila* challenge	Sahoo and Mukherjee (2001)
11	β-glucan (Sigma)	Carp, *Cyprinus carpio*, feed	No significant difference in bacterial killing, O_2production	Selvaraj et al. (2005)
12	β-glucan (Sigma)	Rohu fingerlings – feed	Enhanced leukocytes count, phagocytic ratio, phagocytic index, lysozyme activity, complement activity and serum bactericidal activity; significant reduction in mortality upon challenge with *A. hydrophila* & *E. tarda*	Misra et al. (2006)

(Continued)

TABLE 18.2 *(Continued)*

S. No	Commercial Product	Fish	Effect	References
13	Nucleotides and β-glucan (*S. cerevisiae*)	Carp – feed	Increase in expression of inflammatory cytokines, IL-1β, TNF-α and IL-12p35, IL-12p40 and IFN-γ2 in the head kidney; expression CXC-chemokines; reduction of IL-10 gene expression; Increased O_2– production and phagocytic activity in head kidney leukocytes; enhanced disease resistance upon *A. hydrophila* challenge	Biswas et al. (2012)
14	β-glucan and LPS	Carp, *Cyprinus carpio*, feed	Increase in bactericidal activity; an increase in survival against *A. hydrophila*	Selvaraj et al. (2006)
15	β-glucan and synthetic levamisole (Sigma)	Asian catfish, *Clarias batrachus* – feed	Enhanced phagocyte respiratory burst activity and leucocyte myeloperoxidase content	Kumari and Sahoo (2006)
16	Grobiotic ™ and Brewers yeast	Hybrid striped bass (*Morone chrysops*×*M. saxatilis*) - Feed	Enhanced ROS production in neutrophil and intracellular superoxide anion production of head kidney macrophages; enhanced survival upon *S. iniae* infection	Li and Gatlin (2004, 2005)
17	LPS extracted from *A. salmonicida*	Atlantic salmon – feed	No increase in plasma immunoglobulin level	Guttvik et al. (2002)
18	LPS	Carp, *Cyprinus carpio* – feed	No immunostimulatory effect; survival and enhanced antibody titre upon *A. hydrophila* challenge	Selvaraj et al. (2009)
19	LPS extracted from *P. agglomerans*	Rainbow trout – feed	Increased bactericidal activity, lysozyme activity, blood hemolytic activity and NBT	Skalli et al. (2013)
20	MacroGard, *C. utilis and S. cerevisiae*, deacylated chitin	Rainbow trout - feed	Increased respiratory burst activity, phagocytic index, MPO activity, serum total Ig and blood bactericidal killing activity after 1 week	Siwicki, et al. (1994)
21	MacroGard	Seabream – feed	Increase in head kidney macrophage phagocytic activity and spleen macrophage respiratory burst and phagocytic activity; reduction in the mortality rate when challenged with *P. damselae*	Couso et al. (2003)
22	MacroGard	Rainbow trout – feed	Increase in number of antibody-secreting cells, level of serum immunoglobulin and increased the effectiveness of *Yersinia* vaccine	Siwicki et al. (2004)
23	MacroGard	*O. mykiss* – feed	Enhanced respiratory burst activity after 21 days	Kunttu et al. 2009
24	MacroGard	Tench, *Tinca tinca* (L.) – feed	Increased phagocytic activity of macrophages, proliferative response of mitogen stimulated lymphocytes, serum lysozyme activity, Ig level, reduction in mortality rate against *A. hydrophila*.	Siwicki et al. (2010)

25	MacroGard - diet	*Cyprinus carpio* – feed	Downregulation of TNF-α2 (gut, HK) and IL-10 (gut). Increased production of TNF-α1 and TNF-α2 upon challenge with *A. salmonicida*	Falco et al. (2012)
26	MacroGard	*Cyprinus carpio* - feed	Increase in complement activity	Sych et al. (2013)
27	Macrogard	*Cyprinus carpio* - feed	Downregulation of (IL-1b, IL-10, TNF-α1, TNF-α2 and CXCa) and increased expression of Mx in liver and mid-gut	Falco et al. (2014)
28	Macrogard	*Cyprinus carpio* – feed	Downregulation of complement-related mRNA transcripts at 7 and 25 days in liver and head kidney; high serum CRP level after 7 days increase in alternative complement activity after 25 days of feeding	Pionnier et al. (2014)
29	MacroGard	*Cyprinus carpio* – feed	Increased leucocyte infiltration in the epithelial layer of fish and blood monocyte count	Kühlwein, et al. (2014)
30	MacroGard	*O. mykiss* – feed	Enhanced ACH50 and lysozyme activity; higher total serum Ig and IgM content; increase in survival rate	Ghaedi et al. (2015)
31	MacroGard	Juvenile Persian sturgeon, *Acipenser persicus* – feed	Increased lysozyme activity and ACH50; Increased growth parameters	Aramli et al. (2015)
32	MacroGard	Rainbow trout juveniles – feed	No change in lysozyme activity and ACH50, induced gene expression of TNFα-1, IL-1β, COX-2, TGF-β, & IL-10 in spleen	Douxfils et al. (2017)
33	MacroGard	Nile Tilapia – feed	Increased lysozyme activity in plasma, liver and intestine, increased respiratory burst activity and disease resistance	Koch et al. (2021)
34	MacroGard and Ergosan	Sea bass – feed	Increased serum complement activity and plasma lysozyme activity; gills HSP70	Bagni et al. (2005)
35	MacroGard and β-glucan extracted from *H. vulgare*	Rainbow trout – feed	No immunostimulatory effect	Sealey et al. (2008)
36	Mannan Oligosaccharide	*Cyprinus carpio* – feed	Increased alternate complement activity, lysozyme activity and Ig level in serum	Momeni-Moghaddam et al. (2015)
37	Microbial levan	Juvenile rohu – feed	Increase in hemoglobin content, total leucocyte count and serum total protein (1%); enhanced serum lysozyme activity and respiratory burst activity; reduction in mortality upon challenge with *A. hydrophila*	Gupta et al. (2008)
38	Nutriferm ™	*Labeo rohita* – feed	Enhanced *in vitro* ROS, RNI production, phagocytosis of leukocytes, and proliferation of lymphocytes	Pal et al. (2007, 2021)
39	peptidoglycan (PG)	Rainbow trout – feed	Early expression of defensins and delay in the expression of cathelicidins and LEAPs	Casadei et al. (2013), Casadei et al. (2015)

18.4 BACTERIAL PRODUCTS TESTED FOR THEIR IMMUNOSTIMULATORY ACTIVITY IN FISH

18.4.1 Microbial Surfactants

These are molecules with hydrophobic and hydrophilic moieties produced by bacteria. Among several other important biological activities, immunostimulant activity is reported for microbial surfactants. The surfactants reported to have immunomodulating activity in fish are derived from *Bacillus licheniformis*, *Bacillus thuringiensis*, *B. subtilis*, *Pseudomonas* and *Staphylococcus hominis*. The survival of Nile tilapia larvae is reported to increase by feeding them with poly-β-hydroxybutyrate (PHB)-enriched *Artemia nauplii* (Situmorang, et al., 2016). Similarly, in European sea bass (*Dicentrarchus labrax*) post larvae, the expression of MHC Class II and antimicrobial peptides such as dicentracin and hepcidin was found to increase after feeding them with PHB-enriched *A. nauplii* (Franke et al., 2017). The major classes of microbial surfactants, sources, modes of action and effect on immune system were reviewed by Giri et al. (2020).

18.4.2 Freund's (complete) Adjuvant (FCA)

This is the preparation or *Mycobacterium* cell wall suspended in mineral oil. Other related bacteria used in this preparation include *Nocardia* and *Corynebacteria*. The oil in the adjuvant is responsible for holding the cell wall component within oil globules and releasing it slowly into the tissues after injection. Among the disadvantages of using FCA are that it can cause granulomas, lesions at the injection site and 30–40% reduction in fish growth (Table 18.3).

18.4.3 Muramyl Dipeptide (MDP)

Muramyl dipeptide (N-acetyl-muramyl-L-alanyl-D-isoglutamine) is a derivative of *Mycobacterium* that stimulates immune mechanism by activating macrophages and lymphocyte, alternate pathway of complement (Olivier et al., 1985). It has been reported that when fish are injected with MDP-Lys, it can stimulate colony formation of kidney cells (Kodama et al., 1994).

18.4.4 Lipopolysaccharide (LPS)

It is derived from the outer cell wall of gram-negative bacteria, also known as endotoxins. Though it is toxic to higher animals, it is a potent immunostimulant, since it is recognized by the fish immune system, possibly through toll-like receptor -4 (TLR) and is responsible for activating non-specific immune mechanisms. It is known to act as B cell mitogen, and also activates macrophages and T lymphocytes which results in enhanced secretion of interferons, which in turn activates other leukocytes. At low doses it can enhance disease resistance and is less toxic to fish and shrimps (Song & Sung, 1990). The toxic nature of this compound may prove a limitation to its use as an immunostimulatory agent. LPS interacts with β2-integrins found on the surface of mononuclear phagocytes in rainbow trout. Injecting LPS did not induce inducible nitric oxide synthase (iNOS) in *Cyprinus carpio* as reported by Miest et al. (2012). Similarly, a suppressive effect or no effect was observed in *C. carpio* juveniles upon oral administration of *A. salmonicida* LPS, yeast DNA (CpG motifs), or high M-alginate when assessed in terms of kinetics of B cells present in head kidney and peripheral blood leucocytes on the plasma IgM level, on cytokine and iNOS expression and acute-phase protein expression in the liver (Huttenhuis et al., 2006). The biological effect of LPS in fish has been extensively reviewed by Swain et al. (2008).

TABLE 18.3
Immunostimulatory Effect of Bacterial Derivatives Reported in Fish

S. No	Fish	Mode of Administration	Reported Activity	References
FCA				
1	Coho salmon – *Oncorhynchus kisutch*	Injection	Increased resistance to *A. Salmonicida, A. hydrophila* and *Vibrio ordalii*	Olivier, et al. (1985)
2	Brook trout – *Salvelnus fontalis*	IP	Increased phagocytic and antibacterial activity against virulent and avirulent strain of *A. salmonicida*	Olivier et al. (1986)
3	Rainbow trout *Oncorhynchus mykiss*	Injection	Enhanced respiratory burst, phagocytic and NK cell activity of leucocytes & resistance to *V. anguillarum*	(Kajita, Sakai, Atsuta, & Kobayashi, 1992)
4	Nile tilapia, *Oreochromis nilotica*	IP	Enhanced non-specific response, neutrophil activity, oxygen radical production & serum lysozyme activity	Chen et al. (1998)
LPS				
5	Channel catfish	NA	Production of IL-1 from monocytes	(Clem, Sizemore, Ellsaesser, & Miller, 1985)
6	*Pleuronectes platessa*	Injection	Increased macrophage migration	(MacArthur, Thomson, & Fletcher, 1985)
7	Red sea bream, *Pagrus major*	Injection	Enhanced phagocytic activity of macrophages	(Salati, Hamaguchi, & Kusuda, 1987)
8	Atlantic salmon	*In vitro*	Enhanced phagocytosis and production of superoxide anions	Solem, Jørgensen, & Robertsen, (1995)
9	Goldfish	In vitro	Increased production of macrophage activating factor	(Neumann, Fagan & Belosevic, 1995)
10	Common carp, *Cyprinus carpio*	Oral	Enhanced serum lysozyme activity, phagocytic activity, enhanced expression of IL-1β, TNFα and reduced expression of IL-6 in head kidney	Kadowaki et al. (2013)
MDP				
11	Coho salmon, *O. kisutch*	IP – mixture of MDp and FCA	Increased resistance to *A. salmonicida*	Olivier et al. (1985)
12	Rainbow trout	IP	Enhanced phagocytic activities, respiratory burst and migration activities of head kidney leucocytes; disease resistance to *A. salmonicida* challenge	Kodama et al. (1993)

(Continued)

TABLE 18.3 *(Continued)*

S. No	Fish	Mode of Administration	Reported Activity	References
Bacterins				
13	Rainbow trout	Feed	*Clostridium butyricum* – Increased phagocytic activity and superoxide anion production; NK cell activity and improved protection against *Candida* infection	Sakai et al. (1995a)
14	Rainbow trout	NA	*Aeromonas hydrophila* A3-51, *Vibrio fluvialis* A3-47S, *Carnobacterium* sp. BA211 or *Micro-coccus luteus* A1-6, enhanced serum lysozyme activity	Irianto & Austin (2002)
15	European sea bass, *Dicentrarchus labrax*	IP and immersion	FKC bacterins of *Listonella anguillarum* or *Vibrio alginolyticus*, enhanced protection against vibriosis	Diab et al. (2021)
Microbial surfactants				
16	*Oreochromis mossambicus*	Feed	Enhanced lysozyme level, bactericidal activity, anti-protease and serum peroxidase activity, specific immune response and disease resistance	Suguna et al. (2014)
17	*Labeo rohita*	IP	Enhanced lysozyme level, alternate complement activity, phagocytic activity, bactericidal activity and enhanced resistance to challenge with *A. hydrophila*.	Giri et al. (2016)
18	*Oreochromis mossambicus*	IP	Enhanced lysozyme level, bactericidal activity, anti-protease and serum peroxidase activity, specific immune response and disease resistance	Rajeswari, et al. (2016)
19	Juvenile grass carp, *Ctenopharyngodon idella*	Feed	Enhanced lysozyme activity, acid phosphatase activity, complement content and antimicrobial peptide, upregulation of IL-10, TGF- β and IKB-α and TOR	Feng et al. (2016)
20	*Labeo rohita*	IP	Enhanced lysozyme level, alternate complement activity and phagocytic activity, downregulation of IL-1β, TNF-α, NK-κB, p65 and IKK- β and upregulation of IL-10, TGF- β and IKB-α.	Giri et al. (2017)
21	Soiny mullet *Liza haematocheila*	Feed	Upregulation of MHC Class II and immune related genes	Qiao et al. (2019)

18.4.5 Curdlan

This is derived from bacterium *Alcaligenes faecalis*, made up of β- 1, 3 glucan. Though it is structurally similar to yeast-derived glucan, the difference exists in the low number of branches in curdlan. It has anti-tumor activity and is used as a food additive. It was found to interact with various cells through TLR-2 (Aizawa et al., 2018)

18.4.6 Flagellin

This is present in the flagella of bacteria and a potent PAMP (pathogen-associated molecular pattern) by activating toll-like receptor 5 (TLR5) in fish. The full-length recombinant flagellin (FlaB) and amino terminus of the D1 domain has been tested in head kidney macrophages of gilthead seabream. These protein genes are isolated from *V. anguillarum*, cloned, expressed and tested for their immunostimulatory activity. In this in vitro study, both the proteins induced the expression of pro-inflammatory cytokines IL-1β, tumor necrosis factor (TNF-α) and chemokine IL-8. In rainbow trout, FlaB induced the expression of IL-8 (González-Stegmaie et al., 2015). The recombinant proteins of three flagellin genes found in *Y. ruckeri* (Fla A, B, C) were tested for their activity in channel catfish cultured head kidney cells in vitro. All three recombinant proteins were reported to enhance the expression of pro-inflammatory cytokines Il1-β1, TNF-α, IL-8, iNOS1 and antibacterial protein hepcidin. They also enhanced the expression of TLR5M, TLR5S, NF-κB and MHC II β in head kidney macrophages and monocytes. Fla C was found to be more active in enhancing these gene expression than Fla A and Fla B (Jiang et al., 2017).

18.4.7 Bacterins

The increased chemiluminescent response of kidney cells, complement activation and enhanced protection against *A. salmonicida* challenge was observed in fish white spotted char, *Salvelinus leucomaenis* upon injecting with inactivated *A. stenohalis* (Kawahara et al., 1994). The bacterin solution prepared from *V. anguillarum* was also shown to increase protection against *Streptococcus* infection in rainbow trout upon immersion in bacterin solution (Sakai, et al., 1995a). Jeney and Anderson (1993) also reported enhanced non-specific immunity provided by bacterins. In Atlantic salmon, a mixture of *A. salmonicida* bacterin and yeast glucan elicited antibody response (Aakre et al., 1994). Similar enhancement of antibody response was reported when *A. salmonicida* bacterin was administered along with vitamin C (Thompson et al., 1993) and FK-565 (Kitao et al., 1987). The elevated resistance to vibriosis in rainbow trout was demonstrated by Sakai et al. (1995b) and they also reported enhancement of leucocyte activation, phagocytosis and increased superoxide anion production. The injection of *Vibrio* bacterin induced Mx gene expression (antiviral protein) in Atlantic salmon that was higher than LPS, but lower than the induction by synthetic ds-poly ribonucleotide (Salinas et al., 2004).

18.4.8 FK-565

This is a peptide (Heptanoyl-R-D-glutamyl-(L)-meso-diaminopimelyl-D-alanine) related to tetrapeptide (FK-156, D-Lactoyl-L-alanyl-7-D-glutamyl-L-mesodiaminopimelyl-(L)glycine) isolated from cultures of *Streptomyces olivaceogriseus*. It has been shown to be active in rendering protection against *A. salmonicida* infections in rainbow trout (Kitao and Yoshida, 1986). They also reported increased phagocytic activity in peritoneal and pronephric phagocytic cells and restoration of immunosuppressed conditions induced by cyclophosphamide in trout.

18.5 ALGAL DERIVATIVES

Marine-derived polysaccharides (MDP) are the major groups of bioactive substances used as prebiotic substances to improve growth, immune status and disease resistance of aquatic animals. They have been broadly reviewed by Mohan et al. (2019b) and Thepot et al. (2021). MDPs are either homo- or hetero-polysaccharides consisting of fucose, glucose, galactose, mannose, rhamnose, xylose, glucuronic acid and mannuronic acid. Alginate, fucoidan, carrageenan, laminarin, ulvan, galactan and agar are some of the MPDs tested in fish for their immunostimulatory effect, measured in terms of lysozyme activity, alternate complement activity, anti-protease activity and phagocytosis. Lorenz & Cysewski (2000) also reviewed the potential of Astaxanthin derived from freshwater algae, *Haematococcus* and its application in Salmon, trout and Sea bream farming. Enhancement of specific immune response and disease resistance has also been tested upon oral administration of alginate, fucoidan and agar. The sulfated galactan and carrageenan isolated from red algae (Rhodophyta); alginate, laminarin, fucoidan isolated from brown algae (Phaeophyta); ulvan isolated from green algae (chlorophyta) are some of the polysaccharides tested (Mohan et al., 2019b). They act as prebiotics which means they assist the growth of beneficial bacteria in the gastrointestinal tract of the host upon consumption and can improve overall health of the animal. Also, they can work as PAMP which is recognized by intestinal epithelial cells that mount an immune response (Vallejos-Vidal et al. 2016). Laminarin is derived from Laminarae, *Laminaria hyperborea*, which is β- 1, 3 D glucan with β- 1, 6 branch. It is more soluble than fungal-derived glucan and can be considered a suitable candidate for administration as immunostimulant through diet. The increase in complement activity measured in terms of ACH50 was reported upon dietary inclusion of *Ulva* spp. in Nile tilapia with no beneficial effect on lysozyme and peroxidase activity (Valente et al., 2016). In contrast, in rainbow trout, increased complement activity, lysozyme and peroxidase activity was observed after feeding them with a red algae (5% of *Gracilaria vermiculophylla*)-incorporated diet (Araujo et al., 2016). Alginates isolated from brown algae have been explored for their immunostimulant activity. They are polysaccharide having β-1,4-D-mannuronic acid (M) and C5-epimer α-L-glucuronic acid (G) (Remminghorst & Rehm, 2006). Alginates are extracted from *Laminaria hyperborean, Ascophyllum nodosum* and *Macrocystis pyrifera*, where the active component comprises up to 40% of the dry weight (Smidsrød & Skja, 1990). Another algal product is Ergosan, which has 1% alginic acid isolated from *Laminaria digitata*. Thepot et al. (2021), have noted in their review that a combination of prebiotic and probiotic synergistically enhances serum lysozyme activity, complement activity, phagocytic activity and respiratory burst activity. As given in Table 18.4, these four assays are the major techniques used in most studies to assess the immune status of the animal after vaccination.

18.6 MECHANISMS OF ACTION

The fish immune system recognizes specific conserved structures of the pathogen through pathogen/pattern recognition receptors (PRRs) and is thus activated. These conserved structures include carbohydrates (LPS, mannose, fructose, sucrose), peptidoglycans, lipoproteins, flagellin, nucleic acids (CpG DNA) and Muramyl dipeptide (MDP) from bacteria, β-glucans, zymosan and ds RNA of viruses. They are collectively known as PAMPs (pathogen-associated molecular patterns). One such receptor is the toll-like receptors (TLRs) with which immunostimulants bind, inducing signaling pathways responsible for mounting immune response in several leucocytes (Bricknell and Dalmo, 2005). It has been identified in zebrafish (Jault et al., 2004; Meijer et al., 2004), puffer fish (Oshiumi et al., 2003; Jaillon et al., 2004), Japanese flounder, goldfish (Hirono et al., 2004; Stafford et al., 2003), *Labeo rohita* (Samanta et al., 2014), *Cyprinus carpio* (Kongchum et al., 2011), rainbow trout, Atlantic salmon and channel catfish (Bilodeau & Waldbieser, 2005). Binding of immunostimulants with TLR may result in the activation of adaptor proteins and kinases and thereby a cascade of activation as demonstrated in mitogen-activated protein kinase pathway (MAP kinase) and NF-kB

TABLE 18.4
Immunostimulatory Effect of Algal Derivatives Reported in Fish

S.No	Name	Fish	Effect	References
1	*Ulva pertusa*	Red seabream, *Pagrus major* – feed	Enhanced phagocytosis, haemolytic activity, bactericidal activity, disease resistance and no change in granulocyte number and agglutinin titre	Satoh et al. (1987)
2	*U. clathrate*	Nile tilapia, *O. niloticus – feed*	Enhanced phagocytic activity and WBC count	del Rocio Quezada-Rodriguez, & Fajer-Avila (2017)
3	*U. prolifera, Gracilaria lemaneiformis, U. pertusa*	*Siganus canaliculatus* – feed	Enhanced lysozyme activity, SOD, acid phosphatase and disease resistance to *V. parahaemolyticus*	Xie et al. (2019)
4	*Ulva rigida* – water-soluble polysaccharides extract	Juvenile gray mullet, *Mugil cephalus* – feed	Enhanced lysozyme, respiratory burst activity, phagocytic activity and decreased mortality	Akbary, & Aminikhoei (2018)
5	*Sargassum wightii*	*Mugil cephalus* – feed	Enhanced WBC count, lysozyme activity, respiratory burst activity and relative percent survival on challenge with *P. fluorescence*	Kanimozhi et al. (2013)
6	*Sargassum wightii* – fucoidan	*Pangasianodon hypophthalmus* fingerlings – feed	Enhanced respiratory activity, lysozyme activity, phagocytic activity, TLC and IFNγ and higher survival upon challenge with *A. hydrophila*	Prabhu et al. (2016)
7	*Sargassum sp.* – hot water extract	Asian seabass – *Lates calcarifer*	Enhanced total Ig, lysozyme and highest survival.	Yangthong et al. (2016)
8	*Sargassum wightii* and fucoidan rich extract	*Labeo rohita* fingerlings – feed	Enhanced expression of antimicrobial peptides in liver, skin and intestinal tissues, higher survival; enhanced NBT reduction, lysozyme activity and phagocytic activity	Gora et al. (2018)
9	*Ecklonia cava*	Olive flounder, *Paralichthys olivaceus* – feed	Enhanced serum lysozyme activity, myeloperoxidase activity and respiratory burst activity, survival upon infection with *E. tarda, S. iniae, & Vibrio harveyi*	Lee et al. (2016)
10	*Padina gymnospora* – polysaccharide fraction	*Cyprinus carpio* – feed	Enhanced serum lysozyme, myeloperoxidase activity, antibody response and survival upon challenge with *A. hydrophila* and *E. tarda*	Rajendran et al. (2016)
11	*Spirulina platensis*	Channel catfish	Enhanced antibody response to thymus dependent antigen	Duncan and Klesius, 1996b
12	*Spirulina platensis*	Carp	Upregulation of IL-1β and TNF-α mRNAs, increased phagocytic activity and superoxide anion production in leukocytes	Watanuki et al. (2006)

(Continued)

TABLE 18.4 *(Continued)*

S.No	Name	Fish	Effect	References
13	*Spirulina platensis*	Rainbow trout	Enhanced haemato-immune parameters (RBC, WBC & total protein);reduction in plasma cortisol and glucose	Yeganeh et al. (2015)
14	*Navicula* sp	Gilthead seabream –feed Pacific red snapper, *Lutjanus peru*	Enhanced growth rate, humoral immune responses and antioxidant ability	Reyes-Becerril et al. (2013), Reyes-Becerril et al. (2014)
15	Alginate	Wolfish, *Anarhichas minor, Atlantic cod, Gadus morhua* – feed	Increase in SGR	Vollstad et al. (2006)
16	Alginate from *Ascophyllum nodosum* L.	*Salmo salar* L. – feed	Enhanced lysozyme activity	Gabrielsen, & Austreng, 1998
17	Alginate	Clarias sp. – feed	Increased NBT activity, phagocytic activity and phagocytic index	Isnansetyo et al. (2014)
18	Alginate	*Sparus aurata L.*	Enhanced IgM level, serum peroxidase activity, and gene expression of MHCIIα and TCRβ in head kidney leucocytes	Cordero et al. (2015)
19	Alginic acid	*Oncorhynchus mykiss* – Juveniles,	Enhanced immune related genes – Il-1β, IL-8 and TLR3	Gioacchini, et al. (2010)
20	High M-alginate	Halibut (*Hippoglossus hippoglossus* L.) larvae – rotifers	Increased survival upon challenge with *V. anguillarum*	Skjermo and Bergh, 2004
21	Sodium alginate	*Cyprinus* *Carpio* - IP	Enhanced non-specific immune response; increase in disease resistance against *E. tarda*	Fujiki et al. (1994), Fujiki and Yano (1997)
22	Sodium alginate	Juvenile grouper *Epinephelus fuscoguttatus* – feed	Higher survival upon challenge with *Streptococcus* sp. and an *iridovirus*, increased ACH50 level, lysozyme activity, respiratory burst activity, phagocytic activity and SOD	Chiu et al. (2008)
23	Sodium alginate	*Epinephelus coioides* – feed	Higher survival upon challenge with *Streptococcus* sp. and an *iridovirus*, increased ACH50 level, lysozyme activity, respiratory burst activity, phagocytic activity and SOD	Yeh et al. (2008)
24	Sodium alginate	*Epinephelus brneus* – feed	Higher survival upon challenge with *Streptococcus* sp. increased ACH50 level, lysozyme activity, respiratory bursts, phagocytic activity, and SOD	Harikrishnan et al. (2011)
25	Sodium alginate	*O. niloticus* – feed	Enhanced lysozyme, phagocytic activity and respiratory burst activity, complement activity, protection against A. *hydrophila* challenge	Van Doan et al. (2016b)
26	Sodium alginate	Malaysian mahseer – diet	Improved hematocrit value and respiratory burst activity along with other growth parameters	Asaduzzaman et al. (2019)
27	Sodium alginate and l- Carrageenan	Orange – spotted grouper *E. coicoides* – IP	Enhanced ACH50 activity, respiratory burst, SOD and phagocytic activity and increased against *V. alginolyticus* infection	Cheng et al. (2007)

28	Sodium alginate and K-Carrageenan	*Epinephelus fuscoguttatus* – feed	Enhanced lysozyme, complement, phagocytic and respiratory burst activity, enhanced survival	Cheng et al. (2008)
29	K-Carrageenan – Red seaweed	*Cyprinus Carpio* – IP	Increase in macrophage phagocytic activity and resistance against bacterial infections	Fujiki et al. (1997a, 1997b)
30	Astaxanthin, a high-value carotenoid produced from microalgae	Carp – feed	Enhanced production of IL-1 and TNF-α; increase in RBC and WBC, hemoglobin, hematocrit, and enhanced survival against *A. hydrophila*	Jagruthi et al. (2014)
31	Laminaran	Atlantic salmon	Activate macrophages, increase in respiratory burst activity of head kidney leucocytes	Dalmo and Seljelid (1995), Dalmo et al. (1996)
32	Laminaran	Blue gourami, *Trichogaster trichopterus – IP*	Enhanced CL response of head kidney phagocytic cells, enhanced protection against *A. hydrophila* when injected along with bacterin	Samuel et al. (1996)
33	Laminarin	*Epinephelus coioides* – feed	Enhanced lysozyme, SOD, TP, CAT and enhanced expression of IL-1β, IL-8 and TLR-2	Yin et al. (2014)
34	Laminarin	*O. mykiss* – feed and IP	Enhanced phagocytic activity and increased production of TNFα and IL-8 in gills.	Morales-Lange et al. (2015)
35	Lentinan	Rainbow trout – feed	Reduction in the gene expression which is involved in acute inflammatory reactions; MHC class I antigen presentation, IFN and TNF in spleen	Djordjevic et al. (2009)
36	*Ascophyllum nodosum* brown seaweed meal (FAM)	Nile tilapia – feed	Reduction in the lesions caused by *A. hydrophila*	Oliveira et al. (2014)
37	*Laminaria digitata* (SW1) and a commercial blend of seaweeds (Oceanfeed®)	Atlantic salmon	Upregulation in the expression of MHC class I, related gene protein; receptor-mediated endocytosis and cell adhesion; immune receptors involved in response to LPS and inflammatory response	Palstra et al. (2018)
38	β-(1 → 3, 1 → 6)-glucan derived from the marine diatom *Chaetoceros mülleri*	Atlantic cod larvae –rotifers	Increased survival	Skjermo et al. (2006)
39	β-1,3/1,6-glucan (*Laminarina digitata*) and Pdp 11 (*Shewanella putrefaciens*) single or combined	Gilthead seabream – *Sparus aurata*	Enhanced anti-protease activity, phagocytic activity, IL-1β, INFγ in head kindney, downregulation of IgM gene expression	Guzmán-Villanueva et al. (2014)
40	Agar and *Lactobacillus plantarum*	*Pangasius bocourti* – feed	Enhanced respiratory burst activity, phagocytosis, alternate complement activity, lysozyme activity and enhanced survival against *A. hydrophila* challenge	Van Doan et al. (2014)

pathway in humans. The effect of this signal transduction pathway includes expression of genes which are responsible for activating antigen-presenting cells, leucocytes, and increased phagocytic activity of macrophages and neutrophils (Vallejos-Vidal et al., 2016). These activated cells also secrete chemokines, cytokines, antimicrobial peptides and interferons which are responsible for heightened immune response. Other receptors that can also recognize specific molecular patterns include C-type lectin receptors (CLR), NOD-like receptors (NOR) and RIG-1 like receptors (RLR) (Kawai and Akira, 2010).

18.7 MODE OF ADMINISTRATION

The efficiency of any immunostimulant depends on the method of preparation, and timing, dose and route of administration. The enhanced immune system helps the animal to fight well against infection, so as suggested by Anderson (1992), it should be administered before the outbreak of disease. It can also be used during such situations as handling, grading, transportation (especially in the juvenile stage), acclimatization and high stocking density. These are times when the fish experiences stress, causing immunosuppression which makes them vulnerable to disease (Raa, 2000). High doses of immunostimulant may not enhance immunity, rather they can inhibit the immune status of the animal. The maximum respiratory activity in macrophages was observed with a glucan concentration of 0.1–1 µg/mL, which is not observed with increased concentration of 10 µg/ml; rather, it was inhibited at higher concentration i.e., 50 µg/ml (Robertsen et al., 1994). It is therefore very important to explore the dose/response relationship and the duration of the stimulatory effect (Sohn et al., 2000). The other factors which can have an impact on the efficiency of immunostimulants are the site of pathogen entry, different strains of pathogen, environmental factors, stress conditions and the developmental stage of the animal. In the field, due to the high degree of variation, there is no guarantee that the same stimulatory effect will be reported based on in vitro/in vivo studies carried out in controlled lab conditions. The dose, duration, and method of administration are important limiting factors in deciding the rate of absorption, assimilation, distribution and persistence in the tissue and thereby the efficacy of any immunostimulant.

18.7.1 Injection

A strong immune response can be induced by delivering the desired dosage. It can be used in adults and in larger juveniles intended for high-value purposes such as broodstock and genetic stock (Jadhav et al., 2006). But it is time-consuming, labor-intensive and not feasible for small fish and on huge farms.

18.7.2 Feed

This is the best way of administration, as it is non-stressful and cost-effective. This method is really helpful in the field where animals are of various sizes. Either continuous feeding or pulse feeding can be done depending on the culture system, whether a flow-through system or a recirculation system. It can induce a non-specific immune system but either there is no control or it is not possible to monitor the amount of immunostimulant entering the animal. Due to the existence of hierarchies in many fish, immunostimulation by feed may severely interfere with the amount of food intake and thereby the concentration of immunostimulant taken by dominant and subordinate fish. The efficiency of this method also depends on the extent to which the immunostimulant adheres to feed. The high molecular weight of some substances that are poorly digested and absorbed by the animal will be another limiting factor. Since there are few reports on the inactivity or immunosuppression upon oral administration for a longer period of time than shorter duration (Matsuo and Miyazano, 1993 and Yoshida et al., 1995), it is important to investigate for the selected compound in selected fish over a longer period before implementing it in the field.

Bio-encapsulation is another method of administering immunostimulants to aquatic animals that prefer live animal feed. Immunostimulants can be administered to other animals that can be used as fish feed. Artemia and rotifers have been used for the administration of algal derivatives.

18.7.3 Immersion

Immersion is not as effective as injection, though it does enhance a non-specific immune system, as reported in many studies. This method is cost-effective and feasible in large fish farms. Its disadvantage is the stress experienced by fish during the treatment, and the need for higher volumes and dosage of the immunostimulant. Two hours' immersion in the immunostimulant is recommended for an effective response (Jadhav et al., 2006).

18.8 MICROBIAL DERIVATIVES AS ADJUVANTS IN FISH VACCINES

There are 24 commercial vaccines available for fish which can be used for various species, such as Atlantic salmon, rainbow trout, sea bass, sea bream, tilapia, amberjack, yellowtail, catfish and Vietnamese catfish. The category of vaccine used for the prevention of disease includes live attenuated vaccine, formalin-killed whole-cell vaccine, subunit vaccine, multivalent vaccine, recombinant vaccine and DNA vaccine. These commercial vaccines are administered intraperitoneally and include adjuvants as one of the components (Adams, 2019). Adjuvants are crucial for enhancing the immunogenicity of the antigen(s) present in vaccines.

Adjuvants (from the Latin, meaning 'to help') are used for the slow release of active components such as antigens to achieve long-lasting effect during vaccination. Adjuvants enhance the potency and efficacy of fish vaccines by increasing disease resistance, as well as providing protection against the specific antigen used in vaccines. In general, adjuvants can enhance adaptive immune responses which exhibit a memory response that is faster and stronger on subsequent encounters with the same antigen. It is advantageous to use adjuvants in fish vaccines since it elevates immunogenicity of an antigen with few or no side-effects (Guy, 2007). Immunostimulants are known to activate non-specific immune mechanisms, whereas adjuvants can activate specific immune responses that can be measured in terms of antibody titre. They can be administered before or together with vaccine either by immersion or injection (Anderson, 1992). Identifying the potential adjuvant, and optimizing the dose and mode of administration to elevate the beneficial effect of any vaccine preparation in fish are important aspects of vaccine research. Microbial products are a major group of adjuvants that that have been tested for their properties by administering it together with several fish vaccine preparations included in this review.

The activation of adaptive immune response requires activation of both T and B lymphocytes by antigen presentation (signal 1) along with additional secondary signals (signal 2). The presentation of antigen present in the vaccine is a crucial step in generating an adaptive immune response, which also requires the presence of co-stimulatory signals. Adjuvants can be grouped into two categories based on their ability to either enhance antigen presentation or to facilitate secondary signals (Schijns, 2001; Ribeiro & Schijns, 2010). Oil emulsions, FCA, FIA and flagellin are reported to facilitate the presentation of antigen, (signal 1). Signal 2 adjuvants include β-glucan, oligonucleotides and flagellin.

The advantages of adjuvants are that they act as immunostimulants and are involved in lengthening the duration of the immune response generated against the antigen administered along with it, by augmenting its affinity and specificity, and extending the antibody response, enhancing cell-mediated immune response and enhancing the immunogenicity of the antigen, especially when the antigen is poorly immunogenic. The adjuvant is also responsible for inducing mucosal immunity. Since the fish has more mucosal surface, development of mucosal immunity or development of effective mucosal vaccine will be a vital tool to counter the antigen. But in most of the studies conducted, vaccine and adjuvant are administered intraperitoneally and very few reports are available on oral administration. Lack of understanding of and techniques for mucosal immunity is a major shortcoming with these studies. The adjuvant helps to reduce the burden of booster vaccines and also facilitates the reduction of antigen dose (Tafalla et al., 2014). Altogether, adjuvants have immense potential in increasing the survival rate and relative percentage survival (RPS) of fish when challenged with virulent strains of bacterial pathogens, as evident in most of the studies conducted by various research groups in different fish (Table 18.5).

18.8.1 FCA

This is composed of heat-killed mycobacteria, mineral oil and a surfactant. When antigen is administered to an animal together with FCA, Th1 and Th7 responses increase through the MyD88 pathway. The disadvantage of using FCA with fish vaccine is the formation of granulomas at the site of injection, which may interfere with the growth and health of the fish. Though there are several reports on the protective effect of FCA when administered with vaccines (Table 18.5), there are few cases where it did not give protection. The administration of LPS mixed chloroform-killed bacterin against pasteurellosis together with FCA in *Seriola quinqueradiata* did not significantly enhance protection when challenged with virulent bacteria (Kawakami et al., 1998). Similarly, FCA did not enhance protection against *S. iniae* when vaccinated with formalin-killed culture of this bacteria with FCA in rainbow trout (Soltani et al., 2007).

18.8.2 B-GLUCAN

The adjuvant property of this product has also been tested in many species. Nikl et al. (1991) reported enhanced protection in salmon when administered with β-glucan after vaccination with formalin-treated *A. salmonicida* bacterin. However, no significant increase in antibody level was reported for this combination. In contrast, significant reduction in mortality was not observed in *Catla* when administered with β-glucan after vaccinating with formalin-inactivated *A. hydrophila* (Kamilya et al., 2006). The mode of administration affects the adjuvant property of β-glucan as reported in the following studies. Enhancement in RPS was reported in carp vaccinated with LPS of *A. hydrophila* in the presence of β-glucan when administered through IP rather than oral or bath (Selvaraj et al., 2006). The administration of adjuvant together with vaccine through immersion increased neither survival nor antibody level in carp. In blue gourami, laminaran administered through IP failed to have an adjuvant effect against *A. hydrophila*, though it enhanced the phagocytic activity of head kidney phagocytes and rendered protection (Samuel et al., 1996).

18.8.3 LIPOPEPTIDES

There are few reports on the adjuvant property of these peptides which are isolated from mycobacteria and mycoplasmas, known for their adjuvant effect, and enhance innate and adaptive immune response in mammals (Tafalla et al., 2013).

18.8.4 FLAGELLIN

The molecular weight of monomeric flagellin is 30–60 kDa and it has N-terminal, middle and C-terminal domains. The middle region is highly variable and is composed of β-sheets exhibiting adhesion-like property that is essential for immunogenic properties of flagellin. It is reported to have an adjuvant effect and also to enhance cytokine production. These cytokines can induce the expression of several acute-phase proteins, antimicrobial proteins and complement genes which are very important innate defense components involved in protecting fish from pathogens. The proteinaceous nature of flagellin is a major drawback to its use as immunostimulant, since it undergoes degradation in the gastric and intestinal tracts upon administration along with feed. It enhances the production of cytokines involved in recruiting T, B cells and APC to the site of immunization, thereby acting as

TABLE 18.5
Microbial Derivatives Tested for Their Adjuvant Property in Fish

S. No	Name of the Adjuvant	Fish	Effect	References
1	FCA- IP	Coho salmon, *Oncorhynchus kisutch*	Formalin-killed *A. salmonicida* – enhanced protection	Olivier et al. (1985)
	FCA-IP/bath	Rainbow trout, *Salmo gairdneri* Richardson; bath – Atlantic salmon	Enhanced protection against furunculosis, vibriosis & redmouth disease	Adams et al. (1988)
2	FCA- IP	Tilapia, *Oreochromis niloticus*	Enhanced antibody response – *F. columnare*	Grabowski et al. (2004)
3	FCA- IP	Gourami, *Trichogaster trichopterus*	Adhesin of *A. hydrophila* – enhanced protection	Fang et al. (2004)
4	FCA- IP	Rainbow trout, *O. mykiss*	Enhanced protection – *Flavobacterium*	Högfors et al. (2008)
5	FCA- IP	Rainbow trout, *Oreochromis niloticus*	*A. hydrophila* – enhanced protection	LaPatra et al. (2010)
6	FCA- IP	Japanese flounder, *Paralichthys olivaceus*	Inactivated *M. bovis* – protection	Kato et al. (2012)
7	FCA - IP	Turbot	Pentavalent vaccine – *V. anguillarum, V. scophtalmi, E. tarda, V. harveyi & V. alginolyticus* – enhanced antibody titre, lysozyme activity, SOD activity and RPS	Zheng et al. (2012)
7	β-glucan – IP	Channel catfish	*E. ictaluri* – higher serum antibody level	Chen & Ainsworth (1992)
8	β-glucan – IP	Atlantic Salmon	Furunculosis vaccine – enhanced protection	Rørstad et al. (1993)
9	β-glucan	Turbot	Oral vaccine – Higher protection against vibriosis – not significant	De Baulney et al. (1996)
10	β-glucan – IP	Atlantic Salmon	Furunculosis vaccine – enhanced protection	Midtlyng et al. (1996)
11	β- glucan – IP	Turbot	Enhanced phagocytic index and antibody titre	Figueras et al. (1998)
12	Glycopeptidolipids isolated from *Mycobacterium chelonae*	Atlantic salmon	Enhanced antibody response when vaccinated with *A. salmonicida*	Hole & Lillehaug (1997)
13	FliC– flagellin of *E. tarda*	Japanese flounder	Vaccine – Eta6 – enhanced specific immune response against *E. tarda* measured interms of serum antibody level and IL-1ß, IFN, Mx, CD8α, MHC-Iα, MHC-IIα, IgM	Jiao et al. (2009, 2010)
14	Flagellin	Atlantic salmon	Enhanced production of TNF-α, IL-6, IL-8 and IL-1ß and no elevated antibody response	Hynes et al. (2011)
15	FlgD and FliD – Recombinant	Zebra fish and turbot	High RPS against *E. tarda*	Zhang et al. (2012)
16	Fusion protein – FlaA and FlaB	Red Snapper – *Lutjanus sanguineus* -IP	Enhanced antibody level and survival upon challenge with *V. alginolyticus*	Liang et al. (2012)

(Continued)

TABLE 18.5 *(Continued)*

S. No	Name of the Adjuvant	Fish	Effect	References
17	FlaA, FlaB, FlaD, FlaE - Recombinant	Japanese Flounder - *Paralichthys olivaceus - IP*	Fla E - Enhanced RPS of rEsa1 (subunit vaccine), highest serum antibody level	Jia et al. (2013)
18	FlgD	Turbot – IP	Enhanced RPS, serum antibody level (*E. tarda*) and enhanced expression of MHC-I, IgM, IL-1β, TCR, and TNFα	Liu et al. (2017)
19	FlgD and ompA	Japanese flounder – *Paralichthys olivaceus – oral*	Enhanced T cell gene expression *(CD4-1, CD4-2 & CD8α)*, Th1 subset indicator genes (*T-bet & IFN-γ*) & increase in antigen-specific antibodies and enhanced survival	Beck et al. (2017)
20	FlaC – recombinant	Japanese flounder – *Paralichthys olivaceus* – IP	Three bivalent vaccine – induced the proliferation of sIg^+ B lymphocytes & $CD3^+$ T lymphocytes in PBLs, total and specific antibody, upregulation of CD3, IgM, CD4-1, CD4-2, CD8α and CD8β, high RPS	Xing et al. (2018)
21	Flagellin – recombinant from the salmonid pathogen *Yersinia ruckeri*	Rainbow trout – IP	Enhanced expression of IL-1β, TNFα, IL-6, IL-11 and chemokines (CXCL_F4 & CXCL-8), acute-phase proteins, antimicrobial peptides, complement genes in liver and mucosal tissues, secreted form of TLR5 induction; enhanced antibody production	Wangkahart et al. (2019)
22	CpG ODN	Chinook salmon – IP	Protection against challenge with *R. salmoninarum* and reduction in the bacterial antigens in the kidney of naturally infected fish	Rhodes et al. (2004)
23	CpG ODN	Rainbow trout – IP	Furovac 5 and CpG ODN 2143 combination achieved significant reduction in mortality	Carrington & Secombes, 2007
24	CpG ODN	Turbot, *Scophthalmus maximus*, Japanese flounder – IP	Enhanced protection when vaccinated with *V. harveyi* recombinant subunit vaccine and challenged with virulent strain of the same	Liu et al. (2010a, 2010b)
25	CpG ODN	Rainbow trout – IP	Increased antibody titre and enhanced expression of MHC-I when administered as part of VHSV DNA vaccine	Martinez-Alonso et al. (2011)
26	CpG and Poly I:C	Atlantic salmon – IP	Enhanced protection, production of antibodies against *Salmonid* alpha virus and IFN type I expression	Thim et al. (2012)
27	LPS	Rainbow trout, *O. mykiss* – oral, immersion and IP	Enhanced specific and non-specific immunity	Tulaby Dezfuly, et al. (2020)

signal 1 facilitator. Wangkahart et al. (2019) suggested that it may also act as signal 2 (enhance co-stimulatory signal) and signal 3 facilitator (enhance cytokine production which are responsible for lymphocyte differentiation and expansion). They also stated that flagellin can be easily incorporated with DNA vaccines which induce both early and late immune responses in fish. Non-adjuvanted flagellin has also been explored by Scott et al. (2013) for its potential as a vaccine to protect from enteric redmouth disease caused by *Y. ruckeri* in rainbow trout.

18.8.5 CpG Oligodeoxynucleotides

Either synthetic or bacterial DNA-expressing unmethylated CpG motifs can act as immunostimulant and adjuvant and are involved in the maturation, differentiation and proliferation of immune cells such as B cells, T lymphocytes, NK cells, monocytes, macrophages and dendritic cells. It can bind with TLR 9 and act as an adjuvant when it is administered along with a vaccine by enhancing the specific immune response (Bode et al., 2011). Since DNA vaccines are able to stimulate both cell-mediated and humoral immunity, including CpG ODN along with DNA vaccine is an attractive strategy to achieve enhanced immunity in fish. Synthetic CpG ODN were screened for their inhibitory activity and selected for cloning. The cloned plasmids were injected intramuscularly into Japanese flounder and challenged with *A. hydrophila* and *E. tarda*, where a non-specific protective effect measured in terms of reduction in mortality was observed (Liu et al., 2010a, 2010b). CpG ODN 1668 and CpG ODN 2359 have also been reported to protect fish from parasitic infection such as *Miamiensis avidus* (Kang and Kim, 2012).

18.8.6 Algal Derivatives

Alginate is a polysaccharide derived from brown algae cell wall and composed of M- and G-blocks (Haug et al., 1967). The alginate-coated chitosan microspheres have been tested for their ability to work as antigen carriers for oral vaccination in fish. The combinations tried for testing include antigen-loaded alginate-coated chitosan microsphere without alginate, alginate-coated chitosan without antigen and free whole-cell antigen. Out of these different combinations tested, alginate-coated chitosan microsphere-enhanced innate immune response measured in terms of lysozyme activity and respiratory burst activity and no significant enhancement was observed in alternate complement activity (ACH50). It also enhanced the adaptive immune response measured by antibody response in *Labeo rohita*. Though no significant protection against *A. hydrophila* infection was observed, the enhancement in immune status is very promising for use as antigen delivery system in fish (Behera and Swain, 2014).

18.9 CONCLUSION

There are several reports on the nil effect of immunostimulants in some fish and there is no linear relationship between dose and effect. The discrepancies in the results may be due to variations in the immune status of the animal, method of preparation, mode of administration and dose. However, the majority of the studies are in favor of its stimulatory effect; it is a wise strategy for use in the field to protect fish from diseases. This is especially feasible for farmers involved in large-scale aquaculture practices where the treatment of disease or mitigation is really challenging. Though immunostimulants are not as effective as antibiotics and other chemotherapeutics, they can be used to compensate for the impact of using such chemicals on the environment and consumers. Another problem can be using immunostimulants for larva or juvenile form, which can cause immune tolerance/immunosuppression/undesirable effects since the immune system of such an animal is not mature enough to handle immunostimulants. More progressive research is needed to recognize and address this issue and also to understand the effect of immunostimulants with reference to their absorption, structure–activity relationship, signaling event which results in the activation of immune responses.

REFERENCES

Aakre, R., Wergeland, H.I., Aasjord, P.M., & Endersen, C. (1994). Enhanced antibody response in Atlantic salmon (*Salmo salar* L.) to *Aeromonas salmonicida* cell wall antigens using a bacterin containing β-1, 3-M-glucan as adjuvant. *Fish & Shellfish Immunology*, 4, 47–61. https://doi.org/10.1006/fsim.1994.1005

Adams A. (2019). Progress, challenges and opportunities in fish vaccine development. *Fish & Shellfish Immunology*, 90, 210–214. https://doi.org/10.1016/j.fsi.2019.04.066

Adams, A., Auchinachie, N., Bundy, A., Tatner, M.F., & Horne, M.T. (1988). The potency of adjuvanted injected vaccines in rainbow trout (*Salmo gairdneri* Richardson) and bath vaccines in Atlantic salmon (*Salmo salar* L.) against furunculosis. *Aquaculture*, 69, 15–26.

Ai, Q., Mai, K., Zhang, L., Tan, B., Zhang, W., Xu, W., & Li, H. (2007). Effects of dietary β-1, 3 glucan on innate immune response of large yellow croaker, *Pseudosciaena Crocea. Fish & Shellfish Immunology*, 22, 394–402. https://doi.org/10.1016/j.fsi.2006.06.011

Ainsworth, A., Mao, C., & Boyle, C. (1994). Immune response enhancement in channel catfish Ictalurus punctatus using β-glucan from Schizcophyllum commune. In J. Stolen, & T. Fletcher (Eds.), *Modulators of Fish Immune Response* (pp. 67–81). Fair Haven, NJ: SOS Publications Fair Haven.

Aizawa, M., Watanabe, K., Tominari, T., Matsumoto, C., Hirata, M., Grundler, F., Inada, M., & Miyaura, C. (2018). Low Molecular-Weight Curdlan, (1→3)-β-Glucan Suppresses TLR2-Induced RANKL-Dependent Bone Resorption. *Biological & Pharmaceutical Bulletin*, 41(8), 1282–1285. https://doi.org/10.1248/bpb.b18-00057

Akbary, P., & Aminikhoei, Z. (2018). Effect of water-soluble polysaccharide extract from the green alga *Ulva rigida* on growth performance, antioxidant enzyme activity and immune stimulation of grey mullet *Mugil cephalus*. *Journal of Applied Phycology*, 30, 1345–1353. https://doi.org/10.1007/s10811-017-1299-8

Anderson, D. (1992). Immunostimulants, adjuvants and vaccine carriers in fish: Application to aquaculture. *Annual REview of Fish Diseases*, 2, 281–307.

Anderson, D.P. & Siwicki, A.K. (1994). Duration of Protection against *Aeromonas salmonicida* in Brook trout immunostimulated with glucan or chitosan by injection or immersion. *The Progressive Fish-Culturist*, 56(4), 258–261. https://doi.org/10.1577/1548-8640(1994)056<0258:DOPAAS>2.3.CO;2

Anderson, D.P., Siwicki, A.K., & Rumsey, G.L. (1995). Injection or immersion delivery of selected immunostimulants to trout demonstrate enhancement of nonspecific defense mechanisms and protective immunity. In Shariff, M., Subasinghe, R.P., Arthur, J.R. (Eds), *Diseases in Asian Aquaculture* Vol. 11. Fish Health Section, Asian Fisheries Society, Manila, Philippines, pp. 413–426.

Aramli, M.S., Kamangar, B., & Nazari, R.M. (2015). Effects of dietary β-glucan on the growth and innate immune response of juvenile Persian sturgeon. *Acipenser Persicus, Fish & Shellfish Immunology*, 47(1), 606–610, https://doi.org/10.1016/j.fsi.2015.10.004

Araujo, M., Rema, P., Sousa-Pinto, I., Cunha, L.M., Peixoto, M.J., Pires, M.A., Seixas, F., Brotas, V., Beltran, C., & Valente, L.M.P. (2016). Dietary inclusion of IMTA- cultivated *Gracilaria vermiculophylla* in rainbow trout (*Oncorhynchus mykiss*) diets: Effect on growth, intestinal morphology, tissue pigmentation and immunological response. *Journal of Applied Phycology*, 28, 679–689. https://doi.org/10.1007/s10811-015-0591-8

Asaduzzaman, M., Iehata, S., Islam, M.M., Kader, M.A., Bolong A.A., Ikeda, D., & Kinoshita, S. (2019). Sodium alginate supplementation modulates gut microbiota, health parameters, growth performance and growth-related gene expression in Malaysian Mahseer *Tor tambroides*. *Aquaculture Nutrition*, 25(6), 1300–1317. https://doi.org/10.1111/anu.12950

Baba, E., Uluköy, G., & Öntaş, C. (2015). Effects of feed supplemented with *Lentinula edodes* mushroom extract on the immune response of rainbow trout, *Oncorhynchus mykiss*, and disease resistance against *Lactococcus garvieae*. *Aquaculture*, 448, 476–482. https://doi.org/10.1016/j.aquaculture.2015.04.031

Bagni, M., Archetti, L., Amadori, M., & Marino, G. (2000). Effect of long-term oral administration of an immunostimulant diet on innate immunity in sea bass (*Dicentrarchus labrax*). *Journal of Veterinary Medicine. B, Infectious Diseases and Veterinary Public Health*, 47(10), 745–751. https://doi.org/10.1046/j.1439-0450.2000.00412.x

Bagni, M., Romano, N., Finoia, M. G., Abelli, L., Scapigliati, G., Tiscar, P.G., Sarti, M., & Marino, G. (2005). Short- and long-term effects of a dietary yeast β-glucan (Macrogard) and alginic acid (Ergosan) preparation on immune response in sea bass (*Dicentrarchus labrax*). *Fish & Shellfish Immunology*, 18(4), 311–325. https://doi.org/10.1016/j.fsi.2004.08.003

Bai, D., Xu, H., Wu, X., Zhai, S., Yang, G., Qiao, X., & Guo, Y. (2015). Effect of Dietary *Ganoderma lucidum* Polysaccharides (GLP) on Cellular Immune Responses and Disease Resistance of Yellow Catfish (*Pelteobagrus fulvidraco*). *The Israeli Journal of Aquaculture - Bamidgeh*, 67, 1200–1210. https://doi.org/10.46989/001c.20691

Barman, D., Nen, P., Mandal, S.C., & Kumar, V. (2013). Immunostimulants for Aquaculture Health Management. *Journal of Marine Science: Research & Development*, 3, 134. https://doi.org/10.4172/2155-9910.1000134

Beck, B.R., Lee, S.H., Kim, D., Park, J.H., Lee, H.K., Kwon, S.S., Lee, K.H., Lee, J.I., & Song, S.K. (2017). A *Lactococcus lactis* BFE920 feed vaccine expressing a fusion protein composed of the OmpA and FlgD antigens from *Edwardsiella tarda* was significantly better at protecting olive flounder (*Paralichthys olivaceus*) from *edwardsiellosis* than single antigen vaccines. *Fish & Shellfish Immunology*, 68, 19–28. https://doi.org/10.1016/j.fsi.2017.07.004

Behera, T. & Swain, P. (2014). Antigen encapsulated alginate-coated chitosan microspheres stimulate both innate and adaptive immune responses in fish through oral immunization. *Aquaculture International*, 22, 673–688. https://doi.org/10.1007/s10499-013-9696-8

Biller-Takahashi, J.D., Takahashi, L.S., Marzocchi-Machado, C.M., Zanuzzo, F.S., & Urbinati, E.C. (2014). Disease resistance of pacu, *Piaractus mesopotamicus* (Holmberg, 1887) fed with β-glucan. *Brazilian Journal of Biology*, 74, 698–703.

Bilodeau, A.L. & Waldbieser, G.C. (2005). Activation of TLR3 and TLR5 in channel catfish exposed to virulent *Edwardsiella ictaluri. Developmental and Comparative Immunology*, 29(8), 713–721, https://doi.org/10.1016/j.dci.2004.12.002

Biswas, G., Korenaga, H., Takayama, H., Kono, T., Shimokawa, H., & Sakai, M. (2012). Cytokine responses in the common carp, *Cyprinus carpio* L. treated with baker's yeast extract. *Aquaculture*, 356–357, 169–175. https://doi.org/10.1016/j.aquaculture.2012.05.019

Bode, C., Zhao, G., Steinhagen, F., Kinjo, T., & Klinman, D.M. (2011). CpG DNA as a vaccine adjuvant. *Expert Review of Vaccines*, 10(4), 499–511. https://doi.org/10.1586/erv.10.174

Bonaldo, A., Thompson, K.D., Manfrin, A., Adams, A., Murano, E., Mordenti, A.L., & Gatta, P.P. (2007). The influence of dietary β-glucans on the adaptive and innate immune responses of European sea bass (*Dicentrarchus labrax*) vaccinated against vibriosis. *Italian Journal of Animal Science*, 6(2), 151–164.

Boonyaratpalin, S., Boonyaratpalin, M., Supamattaya, K., & Toride, Y. (1995). Effects of peptidoglucan (PG) on growth, survival, immune responses, and tolerance to stress in black tiger shrimp, *Penaeus monodon*. In Shariff, M., Subasighe, R.P., Arthur, J.R. (Eds.), Diseases in Asian Aquaculture Vol. 11. Fish Health Section, Asian Fisheries Society, Manila, Philippines, pp. 469–477.

Bricknell, I., & Dalmo, R.A. (2005). The use of immunostimulants in fish larval aquaculture. *Fish & Shellfish Immunology*, 19(5), 457–472. https://doi.org/10.1016/j.fsi.2005.03.008

Carrington, A.C., & Secombes, C.J. (2007). CpG oligodeoxynucleotides up-regulate antibacterial systems and induce protection against bacterial challenge in rainbow trout (*Oncorhynchus mykiss*). *Fish & Shellfish Immunology*, 23(4), 781–792. https://doi.org/10.1016/j.fsi.2007.02.006

Casadei, E., Bird, S., Vecino, J.L.G., Wadsworth, S., & Secombes, C.J. (2013). The effect of peptidoglycan enriched diets on antimicrobial peptide gene expression in rainbow trout (*Oncorhynchus mykiss*). *Fish & Shellfish Immunology*, 34, 529–537. https://doi.org/10.1016/j.fsi.2012.11.027

Casadei, E., Bird, S., Wadsworth, S., González Vecino, J.L., & Secombes, C.J. (2015). The longevity of the antimicrobial response in rainbow trout (*Oncorhynchus mykiss*) fed a peptidoglycan (PG) supplemented diet. *Fish & Shellfish Immunology*, 44, 316–320. https://doi.org/10.1016/j.fsi.2015.02.039

Chang, C.S., Huang, S.L., Chen, S., & Chen, S.N. (2013). Innate immune responses and efficacy of using mushroom β-glucan mixture (MBG) on orange-spotted grouper, *Epinephelus coioides*, aquaculture. *Fish & Shellfish Immunology*, 35(1), 115–125. https://doi.org/10.1016/j.fsi.2013.04.004

Chen, D., & Ainsworth, A. (1992). Glucan administration potentiates immune defense mechanisms of channel catfish Ictalurus punctatus Rafineque. *Journal of Fish Diseases*, 15, 295–304. https://doi.org/10.1111/j.1365-2761.1992.tb00667.x

Chen, S.C., Yoshida, T., Adams, A., Thompson, K.D., Richards, R.H. (1998). Non-specific immune response of Nile tilapia, *Oreochromis nilotica*, to the extracellular products of *Mycobacterium* spp. and to various adjuvants. *Journal of Fish Diseases*, 21(1), 39–46. https://doi.org/10.1046/j.1365-2761.1998.00075.x

Cheng, A., Chen, Y., & Chen, J. (2008). Dietary administration of sodium alginate and κ-carrageenan enhances the innate immune response of brown-marbled grouper *Epinephelus fuscoguttatus* and its resistance against *Vibrio alginolyticus*. *Veterinary Immunology and Immunopathology*, 121(3–4), 206–215. https://doi.org/10.1016/j.vetimm.2007.09.011

Cheng, A., Tu, C., Chen, Y., Nan, F., & Chen, J. (2007). The immunostimulatory effects of sodium alginate and ι-carrageenan on orange-spotted grouper *Epinephelus coicoides* and its resistance against *Vibrio alginolyticus*. *Fish & Shellfish Immunology*, 22(3), 197–205. https://doi.org/10.1016/j.fsi.2006.04.009

Chiu, S., Tsai, R., Hsu, J., Liu, C., & Cheng, W. (2008). Dietary sodium alginate administration to enhance the non-specific immune responses, and disease resistance of the juvenile grouper *Epinephelus fuscoguttatus*. *Aquaculture*, 277, (1–2), 66–72

Clem, L., Sizemore, R., Ellsaesser, C., & Miller, N. (1985). Monocytes as accessory cells in fish immune responses. *Developmental and Comparative Immunology*, 9, 803–809. https://doi.org/10.1016/0145-305X(85)90046-1

Cook, M.T., Hayball, P.J., Hutchinson, W., Nowak, B.F., & Hayball, J.D. (2003). Administration of a commercial immunostimulant preparation, EcoActiva (TM) as a feed supplement enhances macrophage respiratory burst and the growth rate of snapper (*Pagrus auratus*, Sparidae (Bloch and Schneider)) in winter. *Fish & Shellfish Immunology*, 14, 333–345. https://doi.org/10.1006/fsim.2002.0441

Cordero, H., Guardiola, F.A., Tapia-Paniagua, S.T., Cuesta, A., Meseguer, J., Balebona, M.C., Moriñigo, M.Á., & Esteban, M.Á. (2015). Modulation of immunity and gut microbiota after dietary administration of alginate encapsulated *Shewanella putrefaciens* Pdp11 to gilthead seabream (*Sparus aurata* L.). *Fish & Shellfish Immunology*, 45(2), 608–618. https://doi.org/10.1016/j.fsi.2015.05.010

Cornet, V., Khuyen, T.D., Mandiki, S., Betoulle, S., Bossier, P., Reyes-López, F.E., Tort, L., & Kestemont, P. (2021). GAS1: A New β-Glucan Immunostimulant Candidate to tncrease Rainbow trout (*Oncorhynchus mykiss*) Resistance to bacterial infections with *Aeromonas salmonicida achromogenes. Frontiers in Immunology*, 12, 693613. https://doi.org/10.3389/fimmu.2021.693613

Couso, N., Castro, R., Magariños, B., Obach, A., & Lamas, J. (2003). Effect of oral administration of glucans on the resistance of gilthead seabream to pasteurellosi. *Aquaculture*, 219, 99–109. https://doi.org/10.1016/S0044-8486(03)00019-X

Cuesta, A., Meseguer, J., & Esteban, M.A. (2004). Total serum immunoglobulin M levels are affected by immunomodulators in seabream (*Sparus aurata* L.). *Veterinary Immunology and Immunopathology*, 101, 203–210. https://doi.org/10.1016/j.vetimm.2004.04.021

Czop J.K. (1986). The role of beta-glucan receptors on blood and tissue leukocytes in phagocytosis and metabolic activation. *Pathology and Immunopathology Research*, 5(3–5), 286–296. https://doi.org/10.1159/000157022

Dalmo, R.A., Bøgwald, J., Ingebrigtsen, K., & Seljelid, R. (1996). The immunomodulatory effect of laminaran [ß(1,3)-D-glucan] on Atlantic salmon, *Salmo salar* L., anterior kidney leucocytes after intraperitoneal, peroral and peranal administration. *Journal of Fish Diseases* 19, 449–457.

Dalmo, R.A. & Seljelid, R. (1995). The immunomodulatory effect of LPS, laminaran and sulphated laminaran [ß(1, 3)-D-glucan] on Atlantic salmon, *Salmo salar* L., macrophages in vitro. *Journal of Fish Disease*, 18, 175–185.

Das, B.K., Debnath, C., Patnaik, P., Swain, D.K., Kumar, K., & Misrhra, B.K. (2009). Effect of beta-glucan on immunity and survival of early stage of Anabas testudineus (Bloch). *Fish & Shellfish Immunology*, 27(6), 678–683. https://doi.org/10.1016/j.fsi.2009.08.002

Das, B.K., Pattnaik, P., Debnath, C., Swain, D.K., & Pradhan, J. (2013). Effect of β-glucan on the immune response of early stage of Anabas testudineus (Bloch) challenged with fungus Saprolegnia parasitica. *SpringerPlus*, 2(1), 197. https://doi.org/10.1186/2193-1801-2-197

De Baulney, M., Quentel, C., Fournier, V., Lamour, F., & Le Gouvello, R. (1996). Effect of long term oral administration of β-glucan as an immunostimulant or an adjuvant on some non-specific parameters of the immune response of turbot *Scophthalmus maximus. Diseases of Aquatic Organisms*, 26, 139–147. https://doi.org/10.3354/dao026139

de Mello, M.M.M., de Faria, C.F.P., Zanuzzo, F.S., & Urbinati, E.C. (2019). β-glucan modulates cortisol levels in stressed pacu (*Piaractus mesopotamicus*) inoculated with heat-killed *Aeromonas hydrophila. Fish & Shellfish Immunology*, 93, 1076–1108. https://doi.org/10.1016/j.fsi.2019.07.068

del Rocio Quezada-Rodriguez, P., & Fajer-Avila, E.J. (2017) The dietary effect of ulvan from Ulva clathrata on hematological-immunological parameters and growth of tilapia (*Oreochromis niloticus*). *Journal of Applied Phycology*, 29, 423–431. https://doi.org/10.1007/s10811-016-0903-7

Dey, B., Bhunia, S.K., Maity, K.K., Patra, S., Mandal, S., Maiti, S., Maiti, T.K., Sikdar, S.R., & Islam, S.S. (2012). Glucans of *Pleurotus florida* blue variant: Isolation, purification, characterization and immunological studies. *International Journal of Biological Macromolecules*, 50(3), 591–597. https://doi.org/10.1016/j.ijbiomac.2012.01.031

Diab, A.M., Khalil, R.H., Leila, R.H.M.A., Abotaleb, M.M., Khallaf, M.A., & Dawood, M.A.O. (2021). Cross-protection of *Listonella anguillarum* and *Vibrio alginolyticus* FKC bacterins to control vibriosis in European sea bass (*Dicentrarchus labrax*). *Aquaculture*, 535, 736379, https://doi.org/10.1016/j.aquaculture.2021.736379

Djordjevic, B., Skugor, S., Jørgensen, S.M., Overland, M., Mydland, L.T., & Krasnov, A. (2009). Modulation of splenic immune responses to bacterial lipopolysaccharide in rainbow trout (*Oncorhynchus mykiss*) fed lentinan, a β-glucan from mushroom *Lentinula edodes. Fish & Shellfish Immunology*, 26(2), 201–209. https://doi.org/10.1016/j.fsi.2008.10.012

Do-Huu, H., Nguyen, T.H.N., & Tran, V.H. (2019). Effects of dietary β-glucan supplementation of growth, innate immune, and capacity against pathogen *Streptococcus iniae* of juvenile pompano (*Trachinotus ovatus*). *Israeli Journal Aquac*, 71, 1622–1632.

Douxfils, J., Fierro-Castro, C., Mandiki, S.N.M., Emile, W., Tort, L., & Kestemont, P. (2017). Dietary β-glucans differentially modulate immune and stress-related gene expression in lymphoid organs from healthy and *Aeromonas hydrophila*-infected rainbow trout (*Oncorhynchus mykiss*). *Fish & Shellfish Immunology*, 63, 285–296, https://doi.org/10.1016/j.fsi.2017.02.027

Duncan, P.L., & Klesius, P.H. (1996a). Dietary immunostimulants enhance nonspecific immune responses in channel catfish but not resistance to *Edwardsiella ictaluri*. *Journal of Aquatic Animal Health* 8, 241–248.

Duncan, P.L., & Klesius, P.H. (1996b). Effects of feeding *Spirulina* on specific and nonspecific immune responses of channel catfish. *Journal of Aquatic Animal Health*, 8, 308–313.

El-Boshy, M.E., El-Ashram, A.M., Abdelhamid, F.M., & Gadalla, H.A. (2010). Immunomodulatory effect of dietary *Saccharomyces cerevisiae*, β-glucan and laminaran in mercuric chloride treated Nile tilapia (*Oreochromis niloticus*) and experimentally infected with *Aeromonas hydrophila*. *Fish & Shellfish Immunology*, 28, 802–808. https://doi.org/10.1016/j.fsi.2010.01.017

Engstad, R. (1994). Yease β glucan as an immunostimulant in Atlantic salmon (*Salmon salar* L.): Biological effects, recognition and structural aspects. *Dr. Scient Thesis*. University of Tromso.

Engstad, R.E., Robertsen, B., & Frivold, E. (1992). Yeast glucan induces increase in lysozyme and complement-mediated haemolytic activity in Atlantic salmon blood. *Fish & Shellfish Immunology*, 2, 287–297.

Esteban, M.A., Cuesta, A., Ortuno, J., & Meseguer, J. (2001). Immunomodulatory effects of dietary intake of chitin on gilthead seabream (*Sparus aurata* L.) innate immune system. *Fish & Shellfish Immunology*, 11, 303–315. https://doi.org/10.1006/fsim.2000.0315

Esteban, M.A., Rodriguez, A., & Meseguer, J. (2004). Glucan receptor but not mannose receptor is involved in the phagocytosis of *Saccharomyces cerevisiae* by seabream (*Sparus aurata* L) blood leucocytes. *Fish & Shellfish Immunology*, 16, 447–451.

Falco, A., Frost, P., Miest, J., Pionnier, N., Irnazarow, I., & Hoole, D. (2012). Reduced inflammatory response to *Aeromonas salmonicida* infection in common carp (*Cyprinus carpio* L.) fed with β-glucan supplements. *Fish & Shellfish Immunology*, 32, 1051–1057. https://doi.org/10.1016/j.fsi.2012.02.028

Falco, A., Miest, J.J., Pionnier, N., Pietretti, D., Forlenza, M., Wiegertjes, G.F., & Hoole, D. (2014). β-Glucan supplemented diets increase poly(I:C)-induced gene expression of Mx, possibly via Tlr3-mediated recognition mechanism in common carp (*Cyprinus carpio*). *Fish & Shellfish Immunology*, 36, 494–502. https://doi.org/10.1016/j.fsi.2013.12.005

Fang, H.M., Ge, R., & Sin, Y. M. (2004). Cloning, characterisation and expression of *Aeromonas hydrophila* major adhesin. *Fish & Shellfish Immunology*, 16(5), 645–658. https://doi.org/10.1016/j.fsi.2003.10.003

FAO. (2020). FAO yearbook. *Fishery and Aquaculture Statistics*. https://www.fao.org/fishery/static/Yearbook/YB2019_USBcard/index.htm

Feng, L., Chen, Y.P., Jiang, W.D., Liu, Y., Jiang, J., Wu, P., Zhao, J., Kuang, S.Y., Tang, L., Tang, W.N., Zhang, Y.A., & Zhou, X.Q. (2016). Modulation of immune response, physical barrier and related signalling factors in the gills of juvenile grass carp (*Ctenopharyngodon idella*) fed supplemented diet with phospholipids. *Fish & Shellfish Immunology*, 48, 79–93. https://doi.org/10.1016/j.fsi.2015.11.020

Figueras, A., Santarém, M.M., & Novoa, B. (1998). Influence of the sequence of administration of β-glucans and a *Vibrio damsela* vaccine on the immune response of turbot (*Scophthalmus maximus* L.). *Veterinary Immunology and Immunopathology*, 64(1), 59–68. https://doi.org/10.1016/s0165-2427(98)00114-7

Franke, A., Clemmesen, C., De Schryver, P., Garcia-Gonzalez, L., Miest, J.J., & Roth, O. (2017). Immunostimulatory effects of dietary poly-β-hydroxybutyrate in European sea bass postlarvae. *Aquaculture Research*, 48(12), 5707–5717. https://doi.org/10.1111/are.13393

Fujiki, K., Matsuyama, H., & Yano, T., (1994). Protective effect of sodium alginates against bacterial infection in common carp, *Cyprinus carpio* L. *Journal of Fish Diseases*, 17, 349–355. https://doi.org/10.1111/j.1365-2761.1994.tb00230.x

Fujiki, K., Shin, D., Nakao, M., & Yano, T., (1997a). Effects of κ-carrageenan on the non-specific defense system of carp *Cyprinus carpio*. *Fisheries Science*, 63(6), 934–938. https://doi.org/10.2331/fishsci.63.934

Fujiki, K., Shin, D., Nakao, M., & Yano, T., (1997b). Protective effect of κ-carrageenan against bacterial infections in carp *Cyprinus carpio*. *Journal – Faculty of Agriculture Kyushu University*, 42, 113–119. https://doi.org/10.5109/24198

Fujiki, K., & Yano, T. (1997). Effects of sodium alginate on the non-specific defence system of the common carp (*Cyprinus carpio* L.). *Fish & Shellfish Immunology*, 7(6), 417–427. https://doi.org/10.1006/fsim.1997.0095

Gabrielsen, B.O., & Austreng, E. (1998). Growth, product quality and immune status of Atlantic salmon, *Salmo salar* L., fed wet feed with alginate. *Aquaculture Research*, 29(6), 397–401. https://doi.org/10.1046/j.1365-2109.1998.00215.x

Galindo-Villegas, J., Masumoto, T., & Hosokawa, H. (2002) Immunostimulants and disease resistance in Japanese flounder, *Paralichthys olivaceus*. In: *Proceedings of the 5th Korean-Japan Biannual Joint Meeting*, Kunsan University, Kunsan, Korea, p. 26.

Gannam, A.L., & Schrock, R.M. (2001). Immunostimulants in fish diets, In. Lim, C., Webster, C.D. (Eds), *Nutrition and fish health* Food Products Press, The Haworth Press, Inc., New York

Ghaedi, G., Keyvanshokooh, S., Mohammadi Azarm, H., & Akhlaghi, M. (2015). Effects of dietary β-glucan on maternal immunity and fry quality of rainbow trout (*Oncorhynchus mykiss*). *Aquaculture*, 441, 78–83. https://doi.org/10.1016/j.aquaculture.2015.02.023

Gioacchini, G., Lombardo, F., Avella, M.A., Olivotto, I., & Carnevali, O. (2010). Welfare improvement using alginic acid in rainbow trout (*Oncorhynchus mykiss*) juveniles. *Chemistry and Ecology*, 26(2), 111–121. https://doi.org/10.1080/02757541003627738

Giri, S.S., Kim, H.J., Kim, S.G., Kim, S.W., Kwon, J., Lee, S.B., & Park, S.C. (2020). Immunomodulatory Role of Microbial Surfactants, with Special Emphasis on Fish. *International Journal of Molecular Sciences*, 21(19), 7004. https://doi.org/10.3390/ijms21197004

Giri, S.S., Sen, S.S., Jun, J.W., Sukumaran, V., & Park, S.C. (2016). Role of *Bacillus subtilis* VSG4-derived biosurfactant in mediating immune responses in *Labeo rohita*. *Fish & Shellfish Immunology*, 54, 220–229. https://doi.org/10.1016/j.fsi.2016.04.004

Giri, S.S., Sen, S.S., Jun, J.W., Sukumaran, V., & Park, S.C. (2017). Role of *Bacillus licheniformis* VS16-Derived Biosurfactant in Mediating Immune Responses in Carp, Rohu and its Application to the Food Industry. *Frontiers in Microbiology*, 8, 514. https://doi.org/10.3389/fmicb.2017.00514

González-Stegmaie, R., Romero, A., Estepa, A., Montero, J., Mulero, V., & Mercado, L. (2015). Effects of recombinant flagellin B and its ND1 domain from *Vibrio anguillarum* on macrophages from gilthead seabream (*Sparus aurata* L.) and rainbow trout (*Oncorhynchus mykiss*, W.). *Fish & Shellfish Immunology*, 42(1), 144–152. https://doi.org/10.1016/j.fsi.2014.10.034

Gopalakannan, A., & Arul, V. (2010). Enhancement of the innate immune system and disease-resistant activity in *Cyprinus carpio* by oral administration of β-glucan and whole cell yeast. *Aquacultural Research*, 41(6), 884–892. https://doi.org/10.1111/j.1365-2109.2009.02368.x

Gora, A.H., Sahu, N.P., Sahoo, S., Rehman, S., Dar, S.A., Ahmad, I. & Agarwal, D. (2018). Effect of dietary *Sargassum weightii* and its fucoidan rich extract on growth, immunity, disease resistance and antimicrobial peptide gene expression in *Labeo rohita*. *International Aquatic Research*, 10, 115–131. https://doi.org/10.1007/s40071-018-0193-6

Grabowski, L.D., LaPatra, S.E., & Cain, K.D. (2004). Systemic and mucosal antibody response in tilapia, *Oreochromis niloticus* (L.), following immunization with *Flavobacterium columnare*. *Journal of Fish Diseases*, 27(10), 573–581. https://doi.org/10.1111/j.1365-2761.2004.00576.x

Gupta, S.K., Pal, A.K., Sahu, N.P., Dalvi, R., Kumar, V., & Mukherjee, S.C. (2008). Microbial levan in the diet of *Labeo rohita* Hamilton juveniles: Effect on non-specific immunity and histopathological changes after challenge with *Aeromonas hydrophila*. *Journal of Fish Diseases*, 31, 649–657. https://doi.org/10.1111/j.1365-2761.2008.00939.x

Guttvik, A., Paulsen, B., Dalmo, R.A., Espelid, S., Lund, V., & Bøgwald, J. (2002). Oral administration of lipopolysaccharide to Atlantic salmon (*Salmo salar* L.) fry. *Uptake, Distribution, Influence on Growth and Immune Stimulation, Aquaculture*, 214, 35–53. https://doi.org/10.1016/S0044-8486(02)00358-7

Guy B. (2007). The perfect mix: recent progress in adjuvant research. *Nature Reviews Microbiology*, 5, 396–397. https://doi.org/10.1038/nrmicro1681

Guzmán-Villanueva, L.T., Tovar-Ramírez, D., Gisbert, E., Cordero, H., Guardiola, F.A., Cuesta, A., Meseguer, J., Ascencio-Valle, F., & Esteban, M.A. (2014). Dietary administration of β-1,3/1,6-glucan and probiotic strain *Shewanella putrefaciens*, single or combined, on gilthead seabream growth, immune responses and gene expression. *Fish & Shellfish Immunology*, 39(1), 34–41. https://doi.org/10.1016/j.fsi.2014.04.024

Harikrishnan, R., Devi, G., Van Doan, G., Balasundaram, C., Thamizharasan, S., Hoseinifar, S.H., & Abdel-Tawwab, M. (2021). Effect of diet enriched with *Agaricus bisporus* polysaccharides (ABPs) on antioxidant property, innate-adaptive immune response and pro-anti inflammatory genes expression in *Ctenopharyngodon idella* against *Aeromonas hydrophila*. *Fish & Shellfish Immunology*, 114, 238–252. https://doi.org/10.1016/j.fsi.2021.04.025

Harikrishnan, R., Kim, M.C., Kim, J.S., Han, Y.J., Jang, I.S., Balasundaram, C., & Heo, M.S. (2011). Immunomodulatory effect of sodium alginate enriched diet in kelp grouper *Epinephelus brneus* against *Streptococcus iniae*. *Fish & Shellfish Immunology*, 30(2), 543–549. https://doi.org/10.1016/j.fsi.2010.11.023

Haug, A., Myklestad, S., Larsen, B., Smidsrød, O., Eriksson, G., Blinc, R., Paušak, S., Ehrenberg, L., & Dumanović, J. (1967). Correlation between chemical structure and physical properties of alginates. *Acta Chemica Scandinavica*, 21, 768–778. https://doi.org/10.3891/acta.chem.scand.21-0768

He, S., Zhou, Z., Liu, Y., Shi, P., Yao, B., Ringø, E., & Yoon, I. (2009). Effects of dietary *Saccharomyces cerevisiae* fermentation product (DVAQUA®) on growth performance, intestinal autochthonous bacterial community and non-specific immunity of hybrid tilapia (*Oreochromis niloticus* ♀×*O. aureus* ♂) cultured. *Aquaculture*, 294, 99–107. https://doi.org/10.1016/j.aquaculture.2009.04.043

Heidarieh M., Soltani, M., Tamimi, A.H., & Toluei, M.H. (2011). Comparative effect of raw fiber (Vitacel) and alginic acid (Ergosan) on growth performance, immunocompetent cell population and plasma lysozyme content of giant sturgeon (*Huso Huso*). *Turkish Journal of Fisheries and Aquatic Sciences*, 11, 445–450. https://doi.org/10.4194/1303-2712-v11_3_15

Hirono, I., Takami, M., Miyata, M., Miyazaki, T., Han, H., Takano, T., Endo, M., & Aoki, T. (2004). Characterization of gene structure and expression of two toll-like receptors from Japanese flounder, *Paralichthys olivaceus*. *Immunogenetics*, 56, 38–46. https://doi.org/10.1007/s00251-004-0657-2

Högfors, E., Pullinen, K.R., Madetoja, J., & Wiklund, T. (2008). Immunization of rainbow trout, *Oncorhynchus mykiss* (Walbaum), with a low molecular mass fraction isolated from *Flavobacterium psychrophilum*. *Journal of Fish Diseases*, 31(12), 899–911. https://doi.org/10.1111/j.1365-2761.2008.00956.x

Hole, K., & Lillehaug, A. (1997). Adjuvant activity of polar glycopeptidolipids from *Mycobacterium chelonae* in experimental vaccines against *Aeromonas salmonicida* in salmonid fish. *Fish & Shellfish Immunology*, 7(6), 365–376. https://doi.org/10.1006/fsim.1997.0091

Huttenhuis, H.B., Ribeiro, A.S., Bowden, T.J., Van Bavel, C., Taverne-Thiele, A.J., & Rombout, J.H. (2006). The effect of oral immuno-stimulation in juvenile carp (*Cyprinus carpio* L.). *Fish & Shellfish immunology*, 21(3), 261–271. https://doi.org/10.1016/j.fsi.2005.12.002

Hynes, N.A., Furnes, C., Fredriksen, B.N., Winther, T., Bøgwald, J., Larsen, A.N., & Dalmo, R.A. (2011). Immune response of Atlantic salmon to recombinant flagellin. *Vaccine*, 29(44), 7678–7687. https://doi.org/10.1016/j.vaccine.2011.07.138

Irianto, A., & Austin, B. (2002). Use of probiotics to control furunculosis in rainbow trout, *Oncorhynchus mykiss* (Walbaum). *Journal of Fish Diseases*, 25, 333–342.

Isnansetyo, A., Irpani, H.M., Wulansari, T.A., & Kasanah, N. (2014). Oral administration of alginate from a tropical brown seaweed, *Sargassum* sp. to enhance non-specific defense in walking catfish (*Clarias* sp.). *Aquacultura Indonesiana*, 15(1), 14–20. https://doi.org/10.21534/ai.v15i1.29

Itami, T., Kondo, M., Uozu, M., Suganuma, A., Abe, T., Nakagawa, A., Suzuki, N., & Takahashi, Y. (1996). Enhancement of resistance against *Enterococcus seriolicida* infection in yellowtail, *Seriola quinqueradiata* (Temminck and Schlegel), by oral administration of peptidoglucan derived from *Bifidobacterium thermophilum*. *Journal of Fish Diseases*, 19, 185–187.

Jadhav V.S., Khan, S.I., Girkar, M.M., & Gitte, M.J. (2006). The role of immunostimulants in fish and shrimp aquaculture. *Aquaculture Asia*, 11(3), 24–27

Jagruthi, C., Yogeshwari, G., Anbazahan, S.M., Mari, L.S.S., Arockiaraj, J., Mariappan, P., Sudhakar, G.R.L., Balasundaram, C., & Harikrishnan, R. (2014). Effect of dietary astaxanthin against *Aeromonas hydrophila* infection in common carp, *Cyprinus carpio*. *Fish & Shellfish Immunology*, 41, 674–680. https://doi.org/10.1016/j.fsi.2014.10.010

Jaillon, O., Aury, J.M., Brunet, F. et al. (2004). Genome duplication in the teleost fish *Tetraodon nigroviridis* reveals the early vertebrate proto-karyotype. *Nature*, 431, 946–957. https://doi.org/10.1038/nature03025

Jault, C., Pichon, L., & Chluba, J. (2004). Toll-like receptor gene family and TIR-domain adapters in *Danio rerio*. *Molecular Immunology*, 40(11), 759–771. https://doi.org/10.1016/j.molimm.2003.10.001

Jeney, G., & Anderson, D.P. (1993). Glucan injection or bath exposure given alone or in combination with bacterin enhance the non-specific defence mechanism in rainbow trout (*Oncorhynchus mykiss*). *Aquaculture*, 116, 315–329.

Ji, L., Sun, G., Li, J., Wang, Y., Du, Y., Li, X., & Liu, Y. (2017). Effect of dietary β-glucan on growth, survival and regulation of immune processes in rainbow trout (*Oncorhynchus mykiss*) infected by *Aeromonas salmonicida*. *Fish & Shellfish Immunology*, 64, 56–67. https://doi.org/10.1016/j.fsi.2017.03.015

Jia, P.P., Hu, Y.H., Chi, H., Sun, B.G., Yu, W.G., & Sun, L. (2013). Comparative study of four flagellins of *Vibrio anguillarum*: vaccine potential and adjuvanticity. *Fish & Shellfish Immunology*, 34, 514–520. https://doi.org/10.1016/j.fsi.2012.11.039

Jiang, J., Zhao, W., Xiong, Q., Wang, K., He, Y., Wang, J., Chen, D., Geng, Y., Huang, X., Ouyang, P. & Lai, W. (2017). Immune responses of channel catfish following the stimulation of three recombinant flagellins of *Yersinia ruckeri in vitro* and *in vivo*. *Developmental & Comparative Immunology*, 73, 61–71. https://doi.org/10.1016/j.dci.2017.02.015

Jiao, X., Hu, Y., & Sun, L. (2010). Dissection and localization of the immunostimulating domain of *Edwardsiella tarda* FliC. *Vaccine*, 28(34), 5635–5640, https://doi.org/10.1016/j.vaccine.2010.06.022

Jiao, X., Zhang, M., Hu, Y., & Sun, L. (2009). Construction and evaluation of DNA vaccines encoding *Edwardsiella tarda* antigens. *Vaccine*, 27(38), 5195–5202, https://doi.org/10.1016/j.vaccine.2009.06.071

Jørgensen, J.B., Lunde, H., & Robertsen, B. (2006). Peritoneal and head kidney cell response to intraperitoneally injected yeast glucan in Atlantic salmon, *Salmo salar* L. *Journal of Fish Diseases*, 16, 313. https://doi.org/10.1111/j.1365-2761.1993.tb00865.x

Jørgensen, J.B., & Robertsen, B. (1995). Yeast ß-glucan stimulates respiratory burst activity of Atlantic salmon (*Salmo salar* L.) macrophages. *Developmental and Comparative Immunology*, 19, 43–57.

Jørgensen, J.B., Sharp, G.J.E., Secombes, C.J., & Robertsen, B. (1993). Effect of yeast-cell glucan in the bactericidal activity of rainbow trout macrophages. *Fish & Shellfish Immunology*, 3, 267–277.

Kadowaki, T., Yasui, Y., Nishimiya, O., Takahashi, Y., Kohchi, C., Soma, G., & Inagawa, H. (2013). Orally administered LPS enhances head kidney macrophage activation with down-regulation of IL-6 in common carp (*Cyprinus carpio*). *Fish & Shellfish Immunology*, 34(6), 1569–1575. https://doi.org/10.1016/j.fsi.2013.03.372

Kajita, Y., Sakai, M., Atsuta, S., & Kobayashi, M. (1992). Immunopotentiation activity of freund's complete adjuvant in rainbow trout *Oncorhynchus mykiss*. *Nippon Suisan Gakkaishi.*, 58, 433–437. https://doi.org/10.2331/suisan.58.433

Kamilya, D., Joardar, S.N., Mal, B.C. & Maiti, T. (2008). Effects of a glucan from the edible mushroom (*Pleurotus florida*) as an immunostimulant in farmed Indian Major Carp (*Catla catla*). *The Israeli journal of aquaculture – Bamidgeh*, 60(1),37–45. https://doi.org/10.46989/001c.20471

Kamilya, D., Maiti, T.K., Joardar, S.N., & Mal, B.C. (2006). Adjuvant effect of mushroom glucan and bovine lactoferrin upon *Aeromonas hydrophila* vaccination in catla, *Catla catla* (Hamilton). *Journal of Fish Diseases*, 29(6), 331–337. https://doi.org/10.1111/j.1365-2761.2006.00722.x

Kang, Y., & Kim, K. (2012). Effect of CpG-ODNs belonging to different classes on resistance of olive flounder (*Paralichthys olivaceus*) against viral hemorrhagic septicemia virus (VHSV) and *Miamiensis avidus* (Ciliata; Scuticociliatia) infections. *Aquaculture*, 324–325, 39–43. https://doi.org/10.1016/j.aquaculture.2011.11.008

Kanimozhi, S., Krishnaveni, M., Deivasigmani, B., Rajasekar, T., & Priyadarshni, P. (2013). Immunomostimulation effects of *Sargassum whitti* on *Mugil cephalus* against *Pseudomonas fluorescence*. *International Journal of Current Microbiology and Applied Sciences*, 2(7), 93–103.

Kato, G., Kondo, H., Aoki, T., & Hirono, I. (2012). *Mycobacterium bovis* BCG vaccine induces non-specific immune responses in Japanese flounder against *Nocardia seriolae*. *Fish & Shellfish Immunology*, 33(2), 243–250. https://doi.org/10.1016/j.fsi.2012.05.002

Kawahara, E., Sakai, M., Nomura, S., Chang, K., & Muraki, K. (1994). Immunomodulatory effects on white-spotted char, *SalÕelinus leucomaenis*, injected with *Achromobacter stenohalis*. In L. E. Chou, *The Third Asian Fisheries Forum* (pp. 390–393). Manila, Philippines: The Third Asian Fisheries Forum. Asian Fisheries Society.

Kawai, T., & Akira, S. (2010). The role of pattern-recognition receptors in innate immunity: update on Toll-like receptors. *Nature Iimmunology*, 11(5), 373–384. https://doi.org/10.1038/ni.1863

Kawakami, H., Shinohara, N., & Sakai, M. (1998). The non-specific immunostimulation and adjuvant effects of *Vibrio anguillarum* bacterin, M-glucan, chitin or Freund's complete adjuvant in yellowtail *Seriola quinqueradiata* to *Pasteurella piscicida* infection. *Fish Pathology*, 33(4), 287–292. https://doi.org/10.3147/jsfp.33.287

Kazuń, B., Małaczewska, J., Kazuń, K., Kamiński, R., Adamek-Urbańska, D., & Żylińska-Urban, J. (2020). Dietary administration of β-1,3/1,6-glucan and *Lactobacillus plantarum* improves innate immune response and increases the number of intestine immune cells in roach (*Rutilus rutilus*). *BMC Veterinary Research*, 16(1), 216. https://doi.org/10.1186/s12917-020-02432-1

Kim, Y., Ke, F., & Zhang, Q.Y. (2009). Effect of β-glucan on activity of antioxidant enzymes and Mx gene expression in virus infected grass carp. *Fish & Shellfish Immunology*, 27(2), 336–340. https://doi.org/10.1016/j.fsi.2009.06.006

Kitao, T., Yoshida, T., Anderson, D.P., Dixon, O.W., & Blanch, A. (1987). Immunostimulation of antibody-producing cells and humoral antibody to fish bacterins by a biological response modifier. *Journal of Fish Biology*, 31, 87–91.

Kitao, T., & Yoshida, Y. (1986). Effect of an immunopotentiator on *Aeromonas salmonicida* infection in rainbow trout (*Salmo gairdneri*). *Veterinary Immunology and Immunopathology*, 12(1–4), 287–296. https://doi.org/10.1016/0165-2427(86)90132-7

Koch, J.F.A., de Oliveira, C.A.F., & Zanuzzo, F.S. (2021). Dietary β-glucan (MacroGard®) improves innate immune responses and disease resistance in Nile tilapia regardless of the administration period. *Fish & Shellfish Immunology*, 112, 56–63. https://doi.org/10.1016/j.fsi.2021.02.014

Kodama, H., Hirota, Y., Mukamoto, M., Baba, T., & Azuma, I. (1993). Activation of rainbow trout (*Oncorhynchus mykiss*) phagocytes by muramyl dipeptide. *Developmental and Comparative Immunology*, 17(2), 129–140. https://doi.org/10.1016/0145-305X(93)90023-J

Kodama, H., Mukamoto, M., Baba, T., & Mule, D. (1994). Macrophage-colony stimulating activity in rainbow trout (*Oncorhynchus mykiss*) serum. *Modulators of Fish Immune Responses*, 1, 59–66.

Kongchum P., Hallerman E.M., Hulata G., David L., & Palti Y. (2011). Molecular cloning, characterization and expression analysis of TLR9, MyD88 and TRAF6 genes in common carp (*Cyprinus carpio*). *Fish & Shellfish Immunology*, 30,361–371. https://doi.org/10.1016/j.fsi.2010.11.012

Kühlwein, H., Merrifield, D.L., Rawling, M.D., Foey, A.D., & Davies, S.J. (2014). Effects of dietary β-(1,3)(1,6)-D-glucan supplementation on growth performance, intestinal morphology and haemato-immunological profile of mirror carp (*Cyprinus carpio L.*). *Journal of Animal Physiology and Animal Nutrition*, 98(2), 279–289. https://doi.org/10.1111/jpn.12078

Kumari, J., & Sahoo, P.K. (2006). Non-specific immune response of healthy and immunocompromised Asian catfish (*Clarias batrachus*) to several immunostimulants. *Aquaculture*, 255, 133–141. https://doi.org/10.1016/j.aquaculture.2005.12.012

Kunttu, H.M.T., Valtonen, E.T., Suomalainen, L.R., Vielma, J., & Jokinen, I.E. (2009). The efficacy of two immunostimulants against *Flavobacterium columnare* infection in juvenile rainbow trout (*Oncorhynchus mykiss*). *Fish & Shellfish Immunology*, 26, 850–857. https://doi.org/10.1016/j.fsi.2009.03.013

Kwak, J.-K., Park, S.W., Koo, J.-G., Cho, M.-G., Buchholz, R., & Goetz, P. (2003). Enhancement of the Non-Specific Defence Activities in Carp (*Cyprinus carpio*) and Flounder (*Paralichthys olivcaces*) by Oral Administration of Schizophyllan. *Engineering in Life Sciences*, 23(4), 359–371. https://doi.org/10.1002/abio.200390046

LaPatra, S.E., Plant, K.P., Alcorn, S., Ostland, V., & Winton, J. (2010). An experimental vaccine against *Aeromonas hydrophila* can induce protection in rainbow trout, *Oncorhynchus mykiss* (Walbaum). *Journal of Fish Diseases*, 33(2), 143–151. https://doi.org/10.1111/j.1365-2761.2009.01098.x

Lee, W., Ahn, G., Oh, J.Y., Kim, S.M., Kang, N., Kim, E. A., Kim, K.N., Jeong, J.B., & Jeon, Y.J. (2016). A prebiotic effect of *Ecklonia cava* on the growth and mortality of olive flounder infected with pathogenic bacteria. *Fish & Shellfish Immunology*, 51, 313–320. https://doi.org/10.1016/j.fsi.2016.02.030

Li, P., & Gatlin, D.M. (2004). Dietary brewer's yeast and the prebiotic GroBiotic™ AE influence growth performance, immune responses and resistance of hybrid striped bass (*Morone chrysops×M. saxatilis*) to *Streptococcus iniae* infection. *Aquaculture*, 231, 445–456.

Li, P., & Gatlin, D.M. (2005). Evaluation of the prebiotic GroBiotic®-A and brewer's yeast as dietary supplements for sub-adult hybrid striped bass (*Morone chrysops×M. saxatilis*) challenged *in situ* with *Mycobacterium marinum*. *Aquaculture*, 248, 197–2005.

Liang, H.Y., Wu, Z.H., Jian, J.C., & Liu, Z.H. (2012). Construction of a fusion flagellin complex and evaluation of the protective immunity of it in red snapper (*Lutjanus sanguineus*). *Letters in Applied Microbiology*, 55(2), 115–121. https://doi.org/10.1111/j.1472-765X.2012.03267.x

Liu, C.S., Sun, Y., Hu, Y.H., & Sun, L. (2010a). Identification and analysis of a CpG motif that protects turbot (*Scophthalmus maximus*) against bacterial challenge and enhances vaccine-induced specific immunity. *Vaccine*, 28(25), 4153–4161. https://doi.org/10.1016/j.vaccine.2010.04.016

Liu, C.S., Sun, Y., Hu, Y.H., & Sun, L. (2010b). Identification and analysis of the immune effects of CpG motifs that protect Japanese flounder (*Paralichthys olivaceus*) against bacterial infection. *Fish & Shellfish Immunology*, 29(2), 279–285. https://doi.org/10.1016/j.fsi.2010.04.012

Liu, X., Zhang, H., Jiao, C., Liu, Q., Zhang, Y., & Xiao, J. (2017). Flagellin enhances the immunoprotection of formalin-inactivated *Edwardsiella tarda* vaccine in turbot. *Vaccine*, 35, 369–374. https://doi.org/10.1016/j.vaccine.2016.11.031

Lorenz, R.T., & Cysewski, G.R. (2000). Commercial potential for *Haematococcus* microalgae as a natural source of astaxanthin. *Trends Biotechnology* 18, 160–167. https://doi.org/10.1016/S0167-7799(00)01433-5

MacArthur, J., Thomson, A., & Fletcher, T. (1985). Aspects of leucocyte migration in the plaice, Pleuronectes platessa L. *Journal of Fish Biology*, 27, 667–676.

Martinez-Alonso, S., Martinez-Lopez, A., Estepa, A., Cuesta, A., & Tafalla, C. (2011). The introduction of multi-copy CpG motifs into an antiviral DNA vaccine strongly up-regulates its immunogenicity in fish. *Vaccine*, 29(6), 1289–1296. https://doi.org/10.1016/j.vaccine.2010.11.073

Matsuo, K., & Miyazano, I. (1993). The influence of long-term administration of peptidoglucan on disease resistance and growth of juvenile rainbow trout. *Nippon Suisan Gakkaishi*, 59, 1377–1379.

Matsuyama, H., Mangindaan, R., & Yano, T. (1992). Protective effect of schizophyllan and scleroglucan against *Streptococcus* sp. infection in yellowtail (*Seriola quinqueradiata*). *Aquaculture*, 101(3-4), 197–203. https://doi.org/10.1016/0044-8486(92)90023-E

Meena, D.K., Das, P., Kumar, S., Mandal, S.C., Prusty, A.K., Singh, S.K., Akhtar, M.S., Behera, B.K., Kumar, K., Pal, A.K., & Mukherjee, S.C. (2013). β-glucan: an ideal immunostimulant in aquaculture (a review). *Fish Physiology and Biochemistry*, 39(3), 431–457. https://doi.org/10.1007/s10695-012-9710-5

Meijer, A.H., Krens, S.F.G., Rodriguez, I.A.M., He, S., Bitter, W., Snaar-Jagalska, B.E., & Spaink, H.P. (2004). Expression analysis of the Toll-like receptor and TIR domain adaptor families of zebrafish. *Molecular Immunology*, 40(11), 773–783, https://doi.org/10.1016/j.molimm.2003.10.003

Midtlyng, P.J., Reitan, L.J., & Speilberg, L. (1996). Experimental studies on the efficacy and side-effects of intraperitoneal vaccination of Atlantic salmon (*Salmo salar L.*) against furunculosis. *Fish & Shellfish Immunology*, 6(5), 335–350. https://doi.org/10.1006/fsim.1996.0034

Miest, J.J., Falco, A., Pionnier, N.P., Frost, P., Irnazarow, I., Williams, G.T., & Hoole, D. (2012). The influence of dietary β-glucan, PAMP exposure and *Aeromonas salmonicida* on apoptosis modulation in common carp (*Cyprinus carpio*). *Fish & Shellfish Immunology*, 33(4), 846–856. https://doi.org/10.1016/j.fsi.2012.07.014

Miles, D.J.C., Polchana, J., Lilley, J.H., Kanchanakhan, S., Thompson, K.D., & Adams, A. (2001). Immunostimulation of striped snakehead *Channa striata* against epizootic ulcerative syndrome. *Aquaculture*, 195(1–2), 1–15. https://doi.org/10.1016/S0044-8486(00)00529-9

Misra, C., Das, B., Mukherjee, S., & Pattnaik, P. (2006). Effect of long term administration of dietary β-glucan on immunity, growth and survival of *Labeo rohita* fingerlings. *Aquaculture*, 255, 82–94.

Mohan, K., Ravichandran, S., Muralisankar, T., Uthayakumar, V., Chandirasekar, R., Sreedevi, P., Abirami, R.G., & Rajan, D.K. (2019b). Application of marine-derived polysaccharides as immunostimulants in aquaculture: A review of current knowledge and further perspectives. *Fish & Shellfish Immunology*, 86, 1177–1193. https://doi.org/10.1016/j.fsi.2018.12.072

Mohan, K., Ravichandran, S., Muralisankar, T., Uthayakumar, V., Chandirasekar, R., Sreedevi, P., & Rajan, D.K. (2019a). Potential uses of fungal polysaccharides as immunostimulants in fish and shrimp aquaculture: A review. *Aquaculture*, 500, 250–263. https://doi.org/10.1016/j.aquaculture.2018.10.023

Momeni-Moghaddam, P., Keyvanshokooh, S., Ziaei-Nejad, S., Parviz Salati, A., & Pasha-Zanoosi, H. (2015). Effects of mannan oligosaccharide supplementation on growth, some immune responses and gut lactic acid bacteria of common carp (*Cyprinus Carpio*) fingerlings. *Veterinary Research Forum: An International Quarterly Journal*, 6(3), 239–244.

Morales-Lange, B., Bethke, J., Schmitt, P., & Mercado, L. (2015). Phenotypical parameters as a tool to evaluate the immunostimulatory effects of laminarin in *Oncorhynchus mykiss*. *Aquaculture Research*, 46(11), 2707–2715. https://doi.org/10.1111/are.12426

Nakao, M., Matsumoto, M., Nakazawa, M., Fujiki, K., & Yano, T. (2002). Diversity of complement factor B/C2 in the common carp (*Cyprinus carpio*): three isotypes of B/C2-A expressed in different tissues. *Developmental and Comparative Immunology*, 26(6), 533–541. https://doi.org/10.1016/s0145-305x(01)00083-0

Neumann, N., Fagan, D., & Belosevic, M. (1995). Macrophage activating factor(s) secreted by mitogen stimulated goldfish kidney leukocytes synergize with bacterial lipopolysaccharide to induce nitric oxide production in teleost macrophages. *Developmental and Comparative Immunology*, 19, 473–482. https://doi.org/10.1016/0145-305X(95)00032-O

Nikl, L., Albright, L., & Evelyn, T. (1991). Influence of seven immunostimulants on the immune response of coho salmon to *Aeromonas salmonicida*. *Diseases of Aquatic Organisms*, 12, 7–12. https://doi.org/10.3354/dao012007

Nikl, L., Evelyn, T.P.T., & Albright, L.J. (1993). Trials with an orally and immersion-administered ß-1,3 glucan as an immunoprophylactic against *Aeromonas salmonicida* in juvenile chinook salmon *Oncorhynchus tshawytscha*. *Diseases of Aquatic Organisms* 17, 191–196.

Oliveira, S.T.L., Veneroni-Gouveia, G., Santos, A.C., Sousa, S.M.N., Veiga, M.L., Krewer, C.C., & Costa, M.M. (2014). *Ascophyllum nodosum* in the diet of tilapia (*Oreochromis niloticus*) and its effect after inoculation of *Aeromonas hydrophila*. *Livestock Diseases*, 34(5), 403–408. https://doi.org/10.1590/S0100-736X2014000500003

Olivier, G., Eaton, C.A., & Campbell, N. (1986). Interaction between *Aeromonas salmonicida* and peritoneal macrophages of brook trout (*Salvelinus fontinalis*). *Veterinary Immunology and Immunopathology*, 12(1–4), 223–234. https://doi.org/10.1016/0165-2427(86)90126-1

Olivier, G., Evelyn, T.P.T., & Lallier, R. (1985). Immunity to *Aeromonas salmonicida* in coho salmon (*Oncorhynchus kisutch*) induced by modified Freund's complete adjuvant: its non-specific nature and the probable role of macrophages in the phenomenon. *Developmental and Comparative Immunology*, 9, 419–432.

Onderdonk, A., Cisneros, R., Hinkson, P., & Ostroff, G. (1992). Anti-infective effect of poly-β 1-6-glucotriosyl-β 1-3-glucopyranose glucan *in vivo*. *Infectious Immunology*, 60, 1642–1647. https://journals.asm.org/doi/10.1128/iai.60.4.1642-1647.1992

Ortuño, J., Cuesta, A., Rodríguez, A., Esteban, M.A., & Meseguer, J. (2002). Oral administration of yeast, *Saccharomyces cerevisiae*, enhances the cellular innate immune response of gilthead seabream (*Sparus aurata* L.), *Veterinary Immunology and Immunopathology*, 85, 41–50.

Oshiumi, H., Tsujita, T., Shida, K., Matsumoto, M., Ikeo, K., & Seya, T. (2003). Prediction of the prototype of the human Toll-like receptor gene family from the pufferfish, *Fugu rubripes*, genome. *Immunogenetics*, 54(11), 791–800. https://doi.org/10.1007/s00251-002-0519-8

Pal, D., Joardar, S.N., Abraham, T.J. & Roy, B. (2021). Immunostimulatory role of yeast cell wall preparation from *Saccharomyces cerevisiae* in Indian major carp, *Labeo rohita*. *Indian Journal of Animal Health*, 48(2), 105–112.

Pal, D., Joardar, S.N., & Roy, B. (2007). Immunostimulatory effects of a yeast (*Saccharomyces cerevisiae*) cell wall feed supplement on rohu (*Labeo rohita*), an Indian major carp. *The Israeli Journal of Aquaculture–Bamidgeh*, 59, 175–181.

Palstra, A.P., Kals, J., Blanco Garcia, A., Dirks, R.P., & Poelman, M. (2018). Immunomodulatory effects of dietary seaweeds in LPS challenged Atlantic Salmon *Salmo salar* as determined by deep RNA sequencing of the head kidney transcriptome. *Frontiers in Physiology*, 9, 625. https://doi.org/10.3389/fphys.2018.00625

Paulsen, S.M., Engstad, R.E., & Robertsen, B. (2001). Enhanced lysozyme production in Atlantic salmon (*Salmo salar* L.) macrophages treated with yeast β-glucan and bacterial lipopolysaccharide. *Fish & Shellfish Immunology*, 11(1), 23–37. https://doi.org/10.1006/fsim.2000.0291

Peddie, S., Zou, J., & Secombes, C.J. (2002). Immunostimulation in the rainbow trout (*Oncorhynchus mykiss*) following intraperitoneal administration of Ergosan. *Veterinary Immunology and Immunopathology*, 86(1–2), 101–113. https://doi.org/10.1016/s0165-2427(02)00019-3

Petit, J., Bailey, E.C., Wheeler, R.T., de Oliveira, C., Forlenza, M., & Wiegertjes, G.F. (2019). Studies into β-glucan recognition in fish suggests a key role for the C-Type Lectin pathway. *Frontiers in Immunology*, 10, 280. https://doi.org/10.3389/fimmu.2019.00280

Philip, R., Sahoo, P., Sreekumar, A., Anas, A., Mukherjee, S., & Ayyappan, S. (2003). Immunostimulants-Source, diversity and mode of application. In I. Bright Singh, S. Somnath Pai, R. Philip, & A. Mohandas, *Aquaculture Medicine* (pp. 135–147). Kochi: Centre for Fish Disease Diagnosis and Mangaement CUSAT, Kochi, India.

Pietretti, D., Vera-Jimenez, N.I., Hoole, D., & Wiegertjes, G.F. (2013). Oxidative burst and nitric oxide responses in carp macrophages induced by zymosan, MacroGard(®) and selective dectin-1 agonists suggest recognition by multiple pattern recognition receptors. *Fish & Shellfish Immunology*, 35, 847–857.

Pionnier, N., Falco, A., Miest, J.J., Shrive, A.K., & Hoole, D. (2014). Feeding common carp *Cyprinus carpio* with β-glucan supplemented diet stimulates C-reactive protein and complement immune acute phase responses following PAMPs injection. *Fish & Shellfish Immunology*, 39, 285–295. https://doi.org/10.1016/j.fsi.2014.05.008

Prabhu, D.L., Sahu, N.P., Pal, A.K., Dasgupta, S., & Narendra, A. (2016). Immunomodulation and interferon gamma gene expression in sutchi cat fish, *Pangasianodon hypophthalmus*: Effect of dietary fucoidan rich seaweed extract (FRSE) on pre and post challenge period. *Aquaculture Research*, 47(1), 199–218. https://doi.org/10.1111/are.12482

Qiao, G., Xu, C., Sun, Q., Xu, D.H., Zhang, M., Chen, P., & Li, Q. (2019). Effects of dietary poly-β-hydroxybutyrate supplementation on the growth, immune response and intestinal microbiota of soiny mullet (*Liza haematocheila*). *Fish & Shellfish Immunology*, 91, 251–263. https://doi.org/10.1016/j.fsi.2019.05.038

Raa, J. (2000). The use of immune-stimulants in fish and shellfish feeds. In Cruz-Suzrez, L. E., Ricque-Marie, D., Tapia-Salazar, M., Olvera-Novoa, M. A.Y., Civera-Cerecedo, R., (Eds.). *Avances en Nutricion Acuicola V. Memorias del V Simposium Internacional de Nutricion Acuicola*. 19–22 November, 2000. Merida, Yucatan, Mexico.

Raa, R., Rørstad, G., Engstad, R., & Robertsen, B. (1992). The use of immunostimulants to increase resistance of aquatic organisms to microbial infections. In Shariff, M., Subasighe, R.P., Arthur, J.R. (Eds.), *Diseases in Asian Aquaculture* Vol. 1. Fish Health Section, Asian Fisheries Society, Manila, Philippines, pp. 39–50.

Rajendran, P., Subramani, P.A., & Michael, D. (2016). Polysaccharides from marine macroalga, *Padina gymnospora* improve the nonspecific and specific immune responses of *Cyprinus carpio* and protect it from different pathogens. *Fish & Shellfish Immunology*, 58, 220–228. https://doi.org/10.1016/j.fsi.2016.09.016

Rajeswari, V., Kalaivani Priyadarshini, S., Saranya, V., Suguna, P., & Shenbagarathai, R. (2016). Immunostimulation by phospholipopeptide biosurfactant from *Staphylococcus hominis* in *Oreochromis mossambicus*. *Fish & Shellfish Immunology*, 48, 244–253. https://doi.org/10.1016/j.fsi.2015.11.006

Remminghorst, U., & Rehm, B.H. (2006). Bacterial alginates: From biosynthesis to applications. *Biotechnology Letters*, 28(21), 1701–1712. https://doi.org/10.1007/s10529-006-9156-x

Reyes-Becerril, M., Angulo, C., Estrada, N., Murillo, Y., & Ascencio-Valle, F. (2014). Dietary administration of microalgae alone or supplemented with *Lactobacillus sakei* affects immune response and intestinal morphology of Pacific red snapper (*Lutjanus peru*). *Fish & Shellfish Immunology*, 40, 208–216. https://doi.org/10.1016/j.fsi.2014.06.032

Reyes-Becerril, M., Guardiola, F., Rojas, M., Ascencio-Valle, F., & Esteban, M.A. (2013). Dietary administration of microalgae *Navicula* sp. affects immune status and gene expression of gilthead seabream (*Sparus aurata*). *Fish & Shellfish Immunology*, 35, 883–889. https://doi.org/10.1016/j.fsi.2013.06.026

Rhodes, L. D., Rathbone, C.K., Corbett, S.C., Harrell, L.W., & Strom, M.S. (2004). Efficacy of cellular vaccines and genetic adjuvants against bacterial kidney disease in Chinook salmon (*Oncorhynchus tshawytscha*). *Fish & Shellfish Immunology*, 16(4), 461–474. https://doi.org/10.1016/j.fsi.2003.08.004

Ribeiro, C., & Schijns, V.E.J.C. (2010). Immunology of vaccine adjuvants. In: Davies, G., ed. *Vaccine Adjuvants, Methods in Molecular Biology*, Humana Press, Totowa, NJ. pp. 1–14.

Ringo, E., Olsen, R.E., Gifstad, T.O., Dalmo, R.A., Amlund, H., Hemre, G.I., & Bakke, A.M. (2010). Prebiotics in aquaculture: A review. *Aquaculture Nutrition*, 16(2), 117–136. https://doi.org/10.1111/j.1365-2095.2009.00731.x

Robertsen, B., Engstad, R.E., & Jørgensen, J.B. (1994). β-glucan as immunostimulants in fish. In Stolen, J.S., Fletcher, T.C. (Eds), *Modulators of Fish Immune Responses* Vol. 1, SOS Publications, Fair Haven, NJ, pp. 83–99.

Robertsen, B., Rorstad, G., Engstad, R., & Raa, J. (1990). Enhancement of non-specific disease resistance in Atlantic salmon, *Salmo salar* L., by a glucan from Saccharomyces cerevisiae cell walls. *Journal of Fish Diseases.*, 13, 391–400. https://doi.org/10.1111/j.1

Rodrigues, M.V., Zanuzzo, F.S., Koch, J.F.A., de Oliveira, C.A.F., Sima, P., & Vetvicka, V. (2020). Development of Fish immunity and the role of β-Glucan in immune responses. *Molecules*, 25, 5378. https://doi.org/10.3390/molecules25225378

Rodríguez, A., Cuesta, A., Ortuño, J., Esteban, M.A., & Meseguer, J. (2003). Immunostimulant properties of a cell wall modified whole *Saccharomyces cerevisiae* strain administered by diet to seabream (*Sparus aurata* L.). *Veterinary Immunology and Immunopathology*, 96, 183–192. https://doi.org/10.1016/S0165-2427(03)00164-8

Rodriguez, F.E., Valenzuela, B., Farias, A., Sandino, A.M., & Imarai, M. (2016). β-1,3/1,6-glucan-supplemented diets antagonize immune inhibitory effects of hypoxia and enhance the immune response to a model vaccine. *Fish & Shellfish Immunology*, 59, 36–45.

Rodríguez, I., Chamorro, R., Novoa, B., & Figueras, A. (2009). β-Glucan administration enhances disease resistance and some innate immune responses in zebrafish (*Danio rerio*). *Fish & Shellfish Immunology*, 27(2), 369–373. https://doi.org/10.1016/j.fsi.2009.02.007

Rørstad, G., Aasjord, P.M., & Robertsen, B. (1993). Adjuvant effect of a yeast glucan in vaccines against furunculosis in Atlantic salmon (*Salmo salar* L.) *Fish & Shellfish Immunology*, 3, 179–190.

Sahoo, L., Debnath, C., Parhi, J., Choudhury, J., Choudhury, T., PaniPrasad, K., & Kandpal, B.K. (2020). Molecular characterization and immunostimulant-induced expression analysis of type I interferon gene in *Labeo bata* (Ham.) *Aquaculture Reports*, 18, 100490. https://doi.org/10.1016/j.aqrep.2020.100490

Sahoo, P.K., & Mukherjee, S.C. (2001). Effect of dietary β-1,3 glucan on immune responses and disease resistance of healthy and aflatoxin B1-induced immunocompromised rohu (*Labeo rohita* Hamilton). *Fish & Shellfish Immunology*, 11, 683–695. https://doi.org/10.1006/fsim.2001.0345

Sahoo, P.K. & Mukherjee, S.C. (2002). The effect of dietary immunomodulation upon *Edwardsiella tarda* vaccination in healthy and immunocompromised Indian major carp (*Labeo rohita*). *Fish & Shellfish Immunology*, 12(1), 1–16, https://doi.org/10.1006/fsim.2001.0349

Saito, H., Yoshioka, Y., & Uehara, N. (1991). Relationship between conformation and biological response of 1, 3 β – D glucans in the activation of coagulation factor from limulus amoebocyte lysate and host mediated anti-tumouractivity; demonstration of single helix conformation as a stimulant. *Carbohydrate Research*, 217, 181–190.

Sakai, M. (1999). Current research status of fish Immunostimulants. *Aquaculture*, 172(1–2), 63–92. https://doi.org/10.1016/S0044-8486(98)00436-0

Sakai, M., Atsuta, S., & Kobayashi, M. (1995b). Efficacies of combined vaccine for *Vibrio anguillarum* and *Streptococcus* sp. *Fisheries Science*, 61(2), 359–360.

Sakai, M., Kamiya, H., Ishii, S., Atsuta, S., & Kobayashi, M. (1992). The immunostimulating effects of chitin in rainbow trout, *Oncorhynchus mykiss*. In Shariff, M., Subasighe, R.P., Arthur, J.R. (Eds.), *Diseases in Asian Aquaculture* Vol. 1. Fish Health Section, Asian Fisheries Society, Manila, Philippines, pp. 413–417.

Sakai, M., Yoshida, T., Atsuta, S., & Kobayashi, M. (1995a). Enhancement of resistance to vibriosis in rainbow trout, *Oncorhynchus mykiss* (Walbaum), by oral administration *Clostridium butyricum* bacterin. *Journal of Fish Diseases*, 18, 187–190.

Salah, A.S., El Nahas, A.F., & Mahmoud, S. (2017). Modulatory effect of different doses of β-1,3/1,6-glucan on the expression of antioxidant, inflammatory, stress and immune-related genes of *Oreochromis niloticus* challenged with *Streptococcus iniae*. *Fish & Shellfish Immunology*, 70, 204–213. https://doi.org/10.1016/j.fsi.2017.09.008

Salati, F., Hamaguchi, M., & Kusuda, R. (1987). Immune response of Red Sea bream to *Edwardsiella tarda* antigens. *Fish Pathology*, 22, 93–98. https://doi.org/10.3147/jsfp.22.93

Salinas, I., Lockhart, K., Bowden, T.J., Collet, B., Secombes, C., & Ellis, A.E. (2004). Assessment of immunostimulants as Mx inducers in Atlantic salmon (*Salmo salar* L.) parr and the effect of temperature on the kinetics of Mx responses. *Fish & Shellfish Immunology*, 17, 159–170. https://doi.org/10.1016/j.fsi.2004.01.003

Samanta, M., Swain, B., Basu, M., Mahapatra, G., Sahoo, B.R., Paichha, M., Lenka, S.S., & Jayasankar, P. (2014). Toll-like receptor 22 in Labeo rohita: molecular cloning, characterization, 3D modeling, and expression analysis following ligands stimulation and bacterial infection. *Applied Biochemistry and Biotechnology*, 174(1), 309–327. https://doi.org/10.1007/s12010-014-1058-0

Samuel, M., Lam, T.J., & Sin, Y.M. (1996). Effect of Laminaran [β(1,3)-D-Glucan] on the protective immunity of blue gourami, *Trichogaster trichopterus* against *Aeromonas hydrophila*. *Fish & Shellfish Immunology*, 6(6), 443–454. https://doi.org/10.1006/fsim.1996.0042

Satoh, K., Nakagawa, H., & Kasahara, S. (1987). Effect of *Ulva* meal supplementation on disease resistance of red sea bream. *Nippon Suisan Gakkaishi*, 53, 1115–1120.

Schijns V.E. (2001). Induction and direction of immune responses by vaccine adjuvants. *Critical Reviews in Immunology*, 21(1–3), 75–85.

Scott, C.J.W., Austin, B., Austin, D.A., & Morris, P.C. (2013). Non-adjuvanted flagellin elicits a non-specific protective immune response in rainbow trout (*Oncorhynchus mykiss*, Walbaum) towards bacterial infections. *Vaccine*, 31, 3262–3267. https://doi.org/10.1016/j.vaccine.2013.05.025

Sealey, W.M., Barrows, F.T., Hang, A., Johansen, K.A., Overturf, K., LaPatra, S.E., & Hardy, R.W. (2008). Evaluation of the ability of barley genotypes containing different amounts of β-glucan to alter growth and disease resistance of rainbow trout *Oncorhynchus mykiss*. *Animal Feed Science & Technology*, 141(1–2), 115–128. https://doi.org/10.1016/j.anifeedsci.2007.05.022

Selvaraj, V., Sampath, K., & Sekar, V. (2005). Administration of yeast glucan enhances survival and some non-specific and specific immune parameters in carp (*Cyprinus carpio*) infected with *Aeromonas hydrophila*. *Fish & Shellfish Immunology*, 19, 293–306. https://doi.org/10.1016/j.fsi.2005.01.001

Selvaraj, V., Sampath, K., & Sekar, V. (2006). Adjuvant and immunostimulatory effects of β-glucan administration in combination with lipopolysaccharide enhances survival and some immune parameters in carp challenged with *Aeromonas hydrophila*. *Veterinary Immunology and Immunopathology*, 114(1–2), 15–24. https://doi.org/10.1016/j.vetimm.2006.06.011

Selvaraj, V., Sampath, K., & Sekar, V. (2009). Administration of lipopolysaccharide increases specific and non-specific immune parameters and survival in carp (*Cyprinus carpio*) infected with *Aeromonas hydrophila*. *Aquaculture*, 286, 176–183. https://doi.org/10.1016/j.aquaculture.2008.09.017

Sirimanapong, W., Adams, A., Ooi, E.L., Green, D.M., Nguyen, D.K., Browdy, C.L., Collet, B., & Thompson, K.D. (2015). The effects of feeding immunostimulant β-glucan on the immune response of *Pangasianodon hypophthalmus*. *Fish & Shellfish Immunology*, 45(2), 357–366. https://doi.org/10.1016/j.fsi.2015.04.025

Situmorang, M.L., De Schryver, P., Dierckens, K., & Bossier, P. (2016). Effect of poly-β-hydroxybutyrate on growth and disease resistance of Nile tilapia Oreochromis niloticus juveniles. *Veterinary Microbiology*, 182, 44–49. https://doi.org/10.1016/j.vetmic.2015.10.024

Siwicki, A.K., Anderson, D.P., & Rumsey, G.L. (1994). Dietary intake of immunostimulants by rainbow trout affects non-specific immunity and protection against furunculosis. *Veterinary Immunology and Immunopathology*, 41(1-2), 125–139. https://doi.org/10.1016/0165-2427(94)90062-0

Siwicki, A.K., Kazuń, K., Głąbski, E., Terech-Majewska, E., Baranowski, P., & Trapkowska, S. (2004). The effect of β-1.3/1.6 – glucan in diets on the effectiveness of anti-*Yersinia ruckeri* vaccine – an experimental study in rainbow trout (*Oncorhynchus mykiss*). *Polish Journal of Food and Nutrition Sciences*, 54(2s), 59–61.

Siwicki, A.K., Zakęś, Z., Terech-Majewska, E., Kazun, K., Lepa, A., & Glabski, E. (2010). Dietary Macrogard reduces *Aeromonas hydrophila* mortality in tench (*Tinca tinca*) through the activation of cellular and humoral defence mechanisms. *Reviews in Fish Biology and Fisheries*, 20(3), 435–439. https://doi.org/10.1007/s11160-009-9133-2

Skalli, A., Castillo, M., Andree, K.B., Tort, L., Furones, D., & Gisbert, E. (2013). The LPS derived from the cell walls of the Gram-negative bacteria *Pantoea agglomerans* stimulates growth and immune status of rainbow trout (*Oncorhynchus mykiss*) juveniles. *Aquaculture*, 416–417, 272–279. https://doi.org/10.1016/j.aquaculture.2013.09.037

Skjermo, J., & Bergh, O. (2004). High-M alginate immunostimulation of Atlantic halibut (*Hippoglossus hippoglossus* L.) larvae using Artemia for delivery, increases resistance against vibriosis. *Aquaculture*, 238 (1–4), 107–113, https://doi.org/10.1016/j.aquaculture.2004.05.038

Skjermo, J., Størseth, T.R., Hansen, K., Handå, A., & Øie, G. (2006). Evaluation of β-(1→3, 1→6)-glucans and High-M alginate used as immunostimulatory dietary supplement during first feeding and weaning of Atlantic cod (*Gadus morhua* L.). *Aquaculture*, 261(3), 1088–1101. https://doi.org/10.1016/j.aquaculture.2006.07.035

Smidsrød, O. & Skja, G. (1990). Alginate as immobilization matrix for cells. *Trends in Biotechnology*, 8, 71–78. https://doi.org/10.1016/0167-7799(90)90139-O

Soares, M.P., Oliveira, F.C., Cardoso, I.L., Urbinati, E.C., de Campos, C.M., & Hisano, H. (2018). Glucan-MOS® improved growth and innate immunity in pacu stressed and experimentally infected with *Aeromonas hydrophila*. *Fish & Shellfish Immunology*, 73, 133–140. https://doi.org/10.1016/j.fsi.2017.11.046

Sohn, K., Kim, M., Kim, J., & Han, I. (2000). The role of immunostimulants in monogastric animal and Fish-Review. *Asian Australasian Journal of Animal Sciences*, 13(8), 1178–1187. https://doi.org/10.5713/ajas.2000.1178

Solem, S., Jørgensen, J., & Robertsen, B. (1995). Stimulation of respiratory burst and phagocytic activity in Atlantic salmon Ž. *Salmo salar* L. macrophages by lipopolysaccharide. *Fish & Shellfish Immunology*, 5, 475–491.

Soltani, M., Alishahi, M., Mirzargar, S., & Nikbakht, G. (2007). Vaccination of rainbow trout against *Streptococcus iniae* infection: comparison of different routes of administration and different vaccines. *Iranian Journal of Fisheries Sciences*, 7, 129–140. http://hdl.handle.net/1834/11186

Song, S.K., Beck, B.R., Kim, D., Park, J., Kim, J., Kim, H.D., & Ringø, E. (2014). Prebiotics as immunostimulants in aquaculture: A review. *Fish & Shellfish Immunology*, 40(1), 40–48, https://doi.org/10.1016/j.fsi.2014.06.016

Song, Y., & Sung, H. (1990). Enhancement of growth in Tiger shrimp (*Penaeus monodon*) by bacterin prepared from *Vibrio vulnificus*. *Bulletin of the European Association of Fish Pathologists*, 10(4), 98–99

Stafford, J.L., Ellestad, K.K., Magor, K.E., Belosevic, M., & Magor, B.G. (2003). A toll-like receptor (TLR) gene that is up-regulated in activated goldfish macrophages. *Developmental & Comparative Immunology*, 27(8), 685–698. https://doi.org/10.1016/S0145-305X(03)00041-7

Suguna, P., Binuramesh, C., Abirami, P., Saranya, V., Poornima, K., Rajeswari, V., & Shenbagarathai, R. (2014). Immunostimulation by poly-β hydroxybutyrate-hydroxyvalerate (PHB-HV) from *Bacillus thuringiensis* in *Oreochromis mossambicus*. *Fish & Shellfish Immunology*, 36(1), 90–97. https://doi.org/10.1016/j.fsi.2013.10.012

Swain, P., Nayak, S.K., Nanda, P.K., & Dash, S. (2008). Biological effects of bacterial lipopolysaccharide (endotoxin) in fish: A review. *Fish & Shellfish immunology*, 25(3), 191–201. https://doi.org/10.1016/j.fsi.2008.04.009

Sych, G., Frost, P., & Irnazarow, I. (2013). Influence of β-glucan (Macrogard®) on innate immunity of carp fry. *Journal of Veterinary Research*, 57(2), 219–223.

Tafalla, C., Bøgwald, J., & Dalmo, R.A. (2013). Adjuvants and immunostimulants in fish vaccines: Current knowledge and future perspectives. *Fish & Shellfish Immunology*, 35(6), 1740–1750. https://doi.org/10.1016/j.fsi.2013.02.029

Tafalla, C., Bogwald, J., Dalmo, R.A., Munand, H.M. & Evensen, O. (2014). Adjuvants in fish vaccines, In *Fish Vaccination*, Roar Gudding, Atle Lillehaug and Øystein Evensen (eds.), I edition, John Wiley & Sons.

Thepot, V., Campbell, A.H., Rimmer, M.A., & Paul, N.A. (2021). Meta-analysis of the use of seaweeds and their extracts as immunostimulants for fish: A systematic review. *Reviews in Aquaculture*, 13, 907–933. https://doi.org/10.1111/raq.12504

Thim, H. L., Iliev, D.B., Christie, K.E., Villoing, S., McLoughlin, M.F., Strandskog, G., & Jørgensen, J.B. (2012). Immunoprotective activity of a Salmonid Alphavirus Vaccine: Comparison of the immune responses induced by inactivated whole virus antigen formulations based on CpG class B oligonucleotides and poly I:C alone or combined with an oil adjuvant. *Vaccine*, 30(32), 4828–4834. https://doi.org/10.1016/j.vaccine.2012.05.010

Thompson, I., White, A., Fletcher, T.C., Houlihan, D.F., & Secombers, C.J. (1993). The effect of stress on the immune responses of Atlantic salmon (*Salmo salar* L.) fed diets containing different amounts of vitamin C. *Aquaculture*, 114, 1–18.

Thompson, K.D., Cachos, A., & Inglis, V. (1995). Immunomodulating effects of glucans and oxytetracycline in rainbow trout, *Oncorhynchus mykiss*, on serum lysozyme and protection. In Shariff, M., Subasighe, R.P., Arthur, J.R. (Eds.), *Diseases in Asian Aquaculture* Vol. 11. Fish Health Section, Asian Fisheries Society, Manila, Philippines, pp. 433–439.

Torrecillas, S., Makol, A., Caballero, M.J., Montero, D., Robaina, L., Real, F., Sweetman, J., Tort, L., & Izquierdo, M.S. (2007). Immune stimulation and improved infection resistance in European sea bass (*Dicentrarchus labrax*) fed mannan oligosaccharides. *Fish & Shellfish Immunology*, 23, 969–981. https://doi.org/10.1016/j.fsi.2007.03.007

Tukmechi, A., Andani, H.R.R., Manaffar, R., & Sheikhzadeh, N. (2011). Dietary administration of β-mercapto-ethanol treated *Saccharomyces cerevisiae* enhanced the growth, innate immune response and disease resistance of the rainbow trout. *Oncorhynchus mykiss*, *Fish & Shellfish Immunology*, 30(3), 923–928. https://doi.org/10.1016/j.fsi.2011.01.016

Tulaby Dezfuly, Z., Alishahi, M., Ghorbanpour, M., Masbah, M., & Tabandeh, M. (2020). Survey on Immunogenicity and protective efficacy of *Yersinia ruckeri* lipopolysaccharide (LPS) against yersiniosis disease in Rainbow trout (*Oncorhyncus mykiss*). *Journal of Animal Environment*, 12(3), 293–304. https://doi.org/10.22034/aej.2020.119247

Valente, L.M.P., Araujo, M., Batista, S et al., (2016). Carotenoid deposition, flesh quality and immunological response on Nile tilapia fed increasing levels of IMTA-cultivated *Ulva* spp. *Journal of Applied Phycology*, 28, 691–701. https://doi.org/10.1007/s10811-015-0590-9

Vallejos-Vidal, E., Reyes-López, F., Teles, M., & MacKenzie, S. (2016). The response of fish to immunostimulant diets. *Fish & Shellfish Immunology*, 56, 34–69, https://doi.org/10.1016/j.fsi.2016.06.028

Van Doan, H., Doolgindachbaporn, S., & Suksri, A. (2014). Effects of low molecular weight agar and *Lactobacillus plantarum* on growth performance, immunity, and disease resistance of basa fish (*Pangasius bocourti*, Sauvage 1880). *Fish & Shellfish Immunology*, 41(2), 340–345. https://doi.org/10.1016/j.fsi.2014.09.015

Van Doan, H., Doolgindachbaporn, S., & Suksri, A. (2016b). Effects of Eryngii mushroom (*Pleurotus eryngii*) and *Lactobacillus plantarum* on growth performance, immunity and disease resistance of Pangasius catfish (*Pangasius bocourti*, Sauvage 1880). *Fish Physiology and Biochemistry*, 42(5), 1427–1440. https://doi.org/10.1007/s10695-016-0230-6

Van Doan, H., Tapingkae, W., Moonmanee, T., & Seepai, A. (2016a). Effects of low molecular weight sodium alginate on growth performance, immunity, and disease resistance of tilapia, *Oreochromis niloticus*. *Fish & Shellfish Immunology*, 55, 186–194. https://doi.org/10.1016/j.fsi.2016.05.034

Vollstad, D., Bøgwald, J., Gåserød, O., & Dalmo, R.A. (2006). Influence of high-M alginate on the growth and survival of Atlantic cod (*Gadus morhua* L.) and spotted wolffish (*Anarhichas minor* Olafsen) fry. *Fish & Shellfish Immunology*, 20, 548–561. https://doi.org/10.1016/j.fsi.2005.07.004

Wang, W.S., Hung, S.W., Lin, Y.H., Tu, C.Y., Wong, M.L., Chiou, S.H., & Shieh, M.T. (2007). The effects of five different glycans on innate immune responses by phagocytes of hybrid tilapia and Japanese eels *Anguilla japonica*. *Journal of Aquatic Animal Health*, 19(1), 49–59. https://doi.org/10.1577/H06-020.1

Wang, W.S., & Wang, D.H. (1997). Enhancement of the resistance of tilapia and grass carp to experimental *Aeromonas hydrophila* and *Edwardsiella tarda* infections by several polysaccharides. *Comparative Immunology, Microbiology and Infectious Diseases*, 20(3), 261–270. https://doi.org/10.1016/s0147-9571(96)00035-5

Wangkahart, E., Secombes, C.J., & Wang, T. (2019). Studies on the use of flagellin as an immunostimulant and vaccine adjuvant in fish aquaculture. *Frontiers in Immunology*, 9, 3054. https://doi.org/10.3389/fimmu.2018.03054

Watanuki, H., Ota, K., Tassakka, A.C.M.A.R., Kato, T., & Sakai, M. (2006). Immunostimulant effects of dietary *Spirulina platensis* on carp. *Cyprinus carpio*, *Aquaculture*, 258, 157–163. https://doi.org/10.1016/j.aquaculture.2006.05.003

Wu, Z.X., Pang, S.F., Chen, X.X., Yu, Y.M., Zhou, J.M., Chen, X., & Pang, L.J. (2013). Effect of *Coriolus versicolor* polysaccharides on the hematological and biochemical parameters and protection against *Aeromonas hydrophila* in allogynogenetic crucian carp (*Carassius auratus gibelio*). *Fish Physiology and Biochemistry*, 39(2), 181–190. https://doi.org/10.1007/s10695-012-9689-y

Xie, D., Li, X., You, C., Wang, S. & Li, Y. (2019). Supplementation of macroalgae together with non-starch polysaccharide-degrading enzymes in diets enhanced growth performance, innate immune indexes, and disease resistance against *Vibrio parahaemolyticus* in rabbitfish *Siganus canaliculatus*. *Journal of Applied Phycology*, 31, 2073–2083. https://doi.org/10.1007/s10811-018-1662-4

Xing, J., Zhou, X., Tang, X., Sheng, X., & Zhan, W. (2018). FlaC supplemented with VAA, OmpK or OmpR as bivalent subunit vaccine candidates induce immune responses against *Vibrio anguillarum* in flounder (*Paralichthys olivaceus*). *Vaccine*, 36, 1316–1322. https://doi.org/10.1016/j.vaccine.2017.11.074

Yangthong, M., Hutadilok-Towatana, N., Thawonsuwan, J., & Phromkunthong, W. (2016). An aqueous extract from *Sargassum* sp. enhances the immune response and resistance against *Stretococcus iniae* in the Asian seabass (*Lates calcarifer* Bloch). *Journal of Applied Phycology*, 28, 3587–3598. https://doi.org/10.1007/s10811-016-0859-7

Yano, T., Mangindaan, R., & Matsuyama, H. (1989). Enhancement of the Resistance of Carp, *Cyprinus carpio* to experimental *Edwardsiella tarda* Infection, by Some β-1, 3-Glucans. *Nippon Suisan Gakkaishi*, 55(10). https://doi.org/10.2331/suisan.55.1815

Yano, T., Matsuyama, H., & Mangindaan, R. (1991). Polysaccharide-induced protection of carp, *Cyprinus carpio* L., against bacterial infection. *Journal of Fish Diseases*, 14(5), 577–582. https://doi.org/10.1111/j.1365-2761.1991.tb00613.x

Yeganeh, S., Teimouri, M., & Amirkolaie, A.K. (2015). Dietary effects of *Spirulina platensis* on hematological and serum biochemical parameters of rainbow trout (*Oncorhynchus mykiss*). *Research in Veterinary Science*, 101, 84–88. https://doi.org/10.1016/j.rvsc.2015.06.002

Yeh, S.P., Chang, C.A., Chang, C.Y., Liu, C.H., & Cheng, W. (2008). Dietary sodium alginate administration affects fingerling growth and resistance to *Streptococcus* sp. and iridovirus, and juvenile non-specific immune responses of the orange-spotted grouper, *Epinephelus coioides*. *Fish & Shellfish Immunology*, 25(1–2), 19–27. https://doi.org/10.1016/j.fsi.2007.11.011

Yin, G., Li, W., Lin, Q., Lin, X., Lin, J., Zhu, Q., Jiang, H., & Huang, Z. (2014). Dietary administration of laminarin improves the growth performance and immune responses in *Epinephelus coioides*. *Fish & Shellfish Immunology*, 41(2), 402–406. https://doi.org/10.1016/j.fsi.2014.09.027

Yoshida, T., Kruger, R., & Inglis, V. (1995). Augmentation of non-specific protection in African catfish, *Clarias gariepinus* (Burchell), by the long-term oral administration of immunostimulants. *Journal of Fish Diseases*, 18(2), 195–198. https://doi.org/10.1111/j.1365-2761.1995.tb00278.x

Yu, H.H., Han, F., Xue, M., Wang, J., Tacon, P., Zheng, Y.H., Wu, X.F., & Zhang, Y.J. (2014). Efficacy and tolerance of yeast cell wall as an immunostimulant in the diet of Japanese seabass (*Lateolabrax japonicus*). *Aquaculture*, 432, 217–224. https://doi.org/10.1016/j.aquaculture.2014.04.043

Zhang, M., Wu, H., Li, X., Yang, M., Chen, T., Wang, Q., Liu, Q., & Zhang, Y. (2012). *Edwardsiella tarda* flagellar protein FlgD: A protective immunogen against edwardsiellosis. *Vaccine*, 30(26), 3849–3856, https://doi.org/10.1016/j.vaccine.2012.04.008

Zheng, Z., Yingeng, W., Qingyin, W., Nannan, D., Meijie, L., Jiangbo, Q., Bin, L., & Lan, W. (2012). Study on the immune enhancement of different immunoadjuvants used in the pentavalent vaccine for turbots. *Fish & Shellfish Immunology*, 32(3), 391–395. https://doi.org/10.1016/j.fsi.2011.11.014

Part III

Medicine

19 Probiotics for Improving COVID-19 Infection-Linked Microbiome Disparities in Gut

Shivani Tyagi, Divyanshi Chauhan, Aarti Venkatesan and Mamtesh Singh
Gargi College, University of Delhi, New Delhi, India

CONTENTS

19.1 INTRODUCTION

The variety of microorganisms that live in and on our bodies is fascinating. The human body is home to millions of microbes from all over the world. Understanding this diversity of more than 10,000 species of microorganisms and their interactions within the human body serves as a foundation for further research, diagnosis, and treatment of various illnesses. The term microbiota refers to the complete range of microbial populations that live within the host.

The host microbiota symbiosis is sensitive to genetic make-up, antibiotic use, dietary pattern, allergens and infective agents. Nutritional adequacy enhances overall health and immunity, which helps to prevent illnesses. The majority of COVID-19 cases are mild to moderate respiratory illnesses that are self-limiting. Alteration in gut microbiome after COVID-19 infection has been evidenced in many clinical studies.

The gut microbiome of healthy individuals plays a major role in various physiological processes such as immunological modulation and development, host metabolism, cell signaling, pathogen colonization resistance, homeostasis, and mucosal regeneration.

Obesity, diabetes, inflammatory bowel disease, and other disorders can all be detected using microbiota as a biomarker. Evidence suggests that an individual's intestinal microbiome is linked to their overall health.

The composition of the human gut microbiome does not remain constant and can be changed by a variety of dietary factors. It is therefore possible that alteration of the gut microbiota could help the recovery of individuals affected by COVID-19. A better understanding and investigation of the

DOI: 10.1201/9781003306931-22

diverse microbes that make up the human microbiome could lead to new treatments for bacterial or viral infections.

Probiotics are "live bacteria that bestow a health benefit on the host when administered in suitable doses" (Hill et al., 2014). The impact of microbes on the regulation of the host immune system is not restricted to the intestinal tract, but targets many different cell types within peripheral tissues. The gut microbiome produces large numbers of microbial metabolites as acetate, propionate, butyrate and secondary bile acids which contribute to the regulation of host immunity. There continues to be a demand for more extensive research to be performed to recognize the useful associations between the microbiome and disease. In this chapter we aim to present a cohesive overview of the microbial composition in a healthy individual, the microbe–microbe association and its functionality in the human gut. We aim to also present an overview of the potential challenges of analyzing the human microbiome and the importance of studying the microbiome for gaining a better knowledge of human health. When probiotics are included in the diet, they aid the host's immunity and intestinal health. Gut microbiota and oral microbiota are discussed in the light of their relation to COVID-19 disease.

19.2 HEALTHY HUMAN MICROBIOME

The microbiome demonstrates a high degree of interpersonal diversity; even in the absence of disease, just a third of its constituent genes are found in the majority of healthy people (Lloyd-Price et al., 2016). This complicates the identification of imbalances in the microbiome that cause disease.

To identify and rectify the microbial configurations implicated in disease, it is crucial to first understand the features of a healthy microbiome. However, being able to define a core set of microbial taxa that contribute to an individual's health without any illness phenotype seems unlikely because there is a large variation in the taxonomic composition of the microbiome, as studies of ecological diversity among healthy individuals revealed (Lloyd-Price et al., 2016). Figure 19.1 demonstrates microbial diversity, showing that the colon has the most abundant microbial diversity, followed by oral microbiota. A complement of the metabolic and molecular functions carried out by the microbiome does not have to be performed by the same organism in different people.

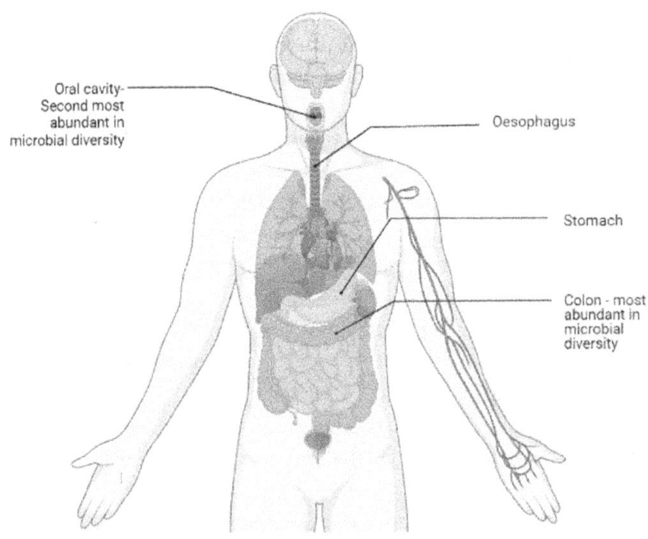

FIGURE 19.1 Microbial diversity abundant areas across the body.

The microbiome plays a critical role in function, maturation and control of the host's immune system; aging can cause significant changes in the composition and function of the microbiome associated with a decline in diversity (dysbiosis) and compromising the host's immunity (Bosco and Noti, 2021).

19.3 GUT MICROBIOTA

The human gastrointestinal tract is home to a vast and complicated diversity of bacteria which help maintain the human gut's equilibrium. The existence of disease is indicated by any changes in the microbiome. The gut microbiota is a group of bacteria that live in the gastrointestinal tract (Thursby and Juge, 2017). Multiple variables, one of the most important being diet, play a role in the development of the human gut microbiota during childhood. Over time, this microbiota co-evolves with the host, forming a mutually beneficial connection.

Understanding of human-associated microbiota has, to date, been provided by the combined data from the Meta Hit and the Human Microbiome projects. These studies identified 2172 species which were isolated from human beings. They are divided into 12 phyla, with *Proteobacteria, Firmicutes, Actinobacteria*, and *Bacteroidetes* accounting for 93.5% of the total (Thursby and Juge, 2017). 386 species out of 2172 are anaerobic and are generally found in mucosal regions such as the oral cavity and the GI tract (Thursby and Juge, 2017).

The gut microbiota is less diverse than the microbial communities found in other parts of the body. The microbiota in the intestine is usually stratified rather than continuous (Arumugam et al., 2011). Diet and drug intake may affect the well-balanced host–microbe symbiotic states in diverse ways. The diversity and composition of the gut microbiota are influenced by chemical, nutritional, and immunological gradients. *Lactobacillaceae* bacteria predominate in the small intestine of mice (Thursby and Juge, 2017).

Diet and metabolism are intricately linked, and each person's situation is different (Asnicar et al., 2021). The relative distribution of gut bacteria and archaea is unique to each individual due to strain level diversities and differences in microbial growth rate and microbial genes, as well as the influence of interpersonal variations in environmental exposure (Fan and Pedersen, 2021). Availability and manner of delivery of breastfeeding have a big impact on the microbiota of an infant's gut. The gut microbiota matures gradually during childhood as a result of environmental influences and food (Fan and Pedersen, 2021). Genes-encoding glycosaminoglycan degradation, production of short-chain fatty acids (SCFAs) via fermentation of complex polysaccharides and synthesis of specific lipopolysaccharide (LPS), and biosynthesis of some essential amino acids and vitamins are all associated with mature and healthy microbial core functions (Fan and Pedersen, 2021).

19.4 THE CONTRIBUTION OF THE GUT MICROBIOTA TO LOW IMMUNITY

Some individuals are highly susceptible to a more severe form of COVID-19 because of the presence of a low immunity state. Poor gut microbiota contributes to this low immunity state and is a key feature associated with a dysregulated metabolic state (Daoust et al., 2021). Changes in the content and features of the human gut microbiome are associated with obesity and other weight-related and metabolic disorders. The composition of the human gut microbiome does not remain constant and can be changed by a variety of dietary factors.

19.5 THE ROLE OF PROBIOTICS IN VARIOUS DISEASES

For the treatment of a large number of chronic diseases and improvement of human health, probiotics have been considered the safe and economical alternative against microbial infections (Yadav et al., 2022). In view of the importance of maintenance of microbiota and their crucial role in human health, some probiotics have been listed as nutribiotics due to their potential to improve nutritional status (Chávarri et al., 2022). Probiotics have long been used to treat a variety of infectious and non-infectious illnesses. They also have a role in promoting health by stimulation of native gut microbiota and host immunity, cholesterol reduction and several other functions. In recent reviews, their role in boosting the host's immunity and reducing pathogen interaction has been highlighted in the fight against COVID-19 (Yadav et al., 2022).

Probiotics have been around since humans began eating fermented foods. The first probiotic was created as a medicine to treat an ailment caused by bacteria *E. coli* in pigs. Mann and Spoerig (1974) found low blood cholesterol levels in people consuming fermented yogurts with *Lactobacillus* spp. Today, probiotics have been approved by the World Health Organization (WHO) as the next most significant immune defense system (Yadav et al., 2022). Probiotics have proved beneficial to human health in a variety of ways including improvement of immunity, food preservation, as antimicrobials, as ulcer treatment, and for lactose intolerance, diarrhea, colon cancer, and other ailments (Masood et al., 2011).

The way probiotics act against intestinal problems can be described in a variety of ways. Colonization is one of the most effective ways to make these bacteria start producing metabolites. Organic acids, short-chain fatty acids, carbon dioxide, hydrogen peroxide, acetaldehyde, acetoin, diacetyl, bacteriocins, and bacteriocin-like inhibitory compounds are among the inhibitory chemicals produced by probiotics. Bacteriocins produced by probiotic bacteria have become increasingly popular due to their safety in foods, and in veterinary and human medicine. These reduce cell viability and disrupt bacterial cell metabolism or toxin generation, making them inhibitory to a wide range of bacteria and viruses. Competitive inhibition on the intestinal epithelial surface by blocking the adhering sites of the intestinal epithelial surface, and a reduction in pathogen contact, are two more ways of probiotic action. Nutritional competition is also thought to be a mechanism for probiotic activity (Yadav et al., 2022).

19.6 MICROBIOTA RELATED TO OBESITY

In recent decades, obesity has become the commonest and most rapidly growing lifestyle-associated disorder experienced by individuals. Several studies have indicated that apart from genetics and lifestyle, gut microbiome plays a crucial role in the onset and progression of obesity and thus scientists have shown keen interest in analyzing gut and oral microbiota with future therapeutic implementation. Obesity is caused by an energy imbalance between calories ingested and calories spent, which can have a number of causes, including easy availability of energy-dense foods, lack of physical activity, and the putative influence of the gut microbiota on host metabolism. It has been observed that the gut of an obese individual show less diversity and richness in microbial populations than that of a healthy individual (Cunningham et al., 2021).

Metabolic function has been directly linked with microbial diversity, and lower microbial richness has been observed as a potential risk for obesity. A number of studies on the gut microbiota of obese individuals have found a considerable reduction in butyrate-producing microbes, especially microbes belonging to families of *Rikenellaceae* and *Christensenellaceae*. Obese people's microbiota have been found to have a higher energy-harvesting capacity.

A recent study examined how the gut microbiome of obese people changed before and after bariatric surgery. A detailed cladogram predicted the main differences in the gut microbiota community before and after bariatric surgery, and it was observed that after surgery, obese patients were enriched principally from the *Coriobacteriia* class, whereas after surgery enriched phylotypes from obese patients were predominantly from *Cyanobacterial* phyla and *Clostridia* class (Juárez-Fernández et al., 2021).

Several other research studies show relative abundance of families of *Prevotellaceae, Coriobacteriaceae, Erysipelotrichaceae*, and *Alcaligenaceae* along with the genus *Roseburia* in individuals with elevated BMI (Juárez-Fernández et al., 2021).

Weight problems and obesity-related metabolic illnesses are associated with specific alterations in the human intestine microbiota function. Data from observational studies allow us to hypothesize that adjustments within the composition of the microbiota may be the cause. Changes in human metabolism are the cause, not the result. If these premises are accurate, then bacterial flora might be a perfect target for feasible interventions in the prevention or treatment of diseases related to weight problems.

Although several dietary manipulation studies have suggested that modulation of the intestine microbiota may enhance metabolic variables, no evidence of the gut microbiota playing a causative role has been presented in these studies. There are many variables that could profoundly modify the outcome of customized nutritional techniques: the true correlation between nutritional intervention and a metabolic trade is yet to be found. Clinical trials are required to study the impact on the host metabolism of modulation of the gut microbiota induced by weight-reduction plans.

19.7 EFFECT OF SARS-COV-2 ON GUT MICROBIOTA

The microbiome in the gastrointestinal tract, consisting of the bacteriome and virome, responds to contamination and may also affect disease development and treatment outcome. Gut microbiome dysbiosis can also cause persistent signs and symptoms in survivors of COVID-19 for a long time after infection. These may also be regulated by production of anti-inflammatory metabolites consisting of SCFA and/or by facilitating the enrichment of opportunistic pathogens in the patients. Data suggest that depletion of salutary species can also additionally persist in maximum recovered patients in spite of SARS-CoV-2 virus clearance, suggesting exposure to SARS-CoV-2 contamination can be related to a greater long-lasting detrimental impact on the gut microbiota (Zhou et al., 2021). In comparison to healthy controls, it was discovered that COVID-19 patients had significantly greater relative abundances of *Streptococcus, Rothia, Veillonella, Erysipelatoclostridium*, and *Actinomyces*. Variations in the intestinal microbiome may also limit the efficiency of vaccination antigens as a result of prolonged intestinal inflammation (Gu et al. 2020)

19.8 BACTERIOCIN COMPOUNDS/METABOLITES RELEASED BY GUT MICROBIOME

Antimicrobial chemicals produced by probiotics include acids, peptides, and bacteriocin molecules. Lactic acid is produced by lactic acid bacteria and it assists the host cells in preventing viral replication. The expression and secretion of human angiotensin-converting enzyme (ACE-2) in *Lactobacillus paracasei* was studied by Verma et al. 2019. Binding of this secreted ACE-2 with COVID-19 binding protein can prevent its entry into the host cell and thus reduce the chances of infection (Rizzo et al. 2020).

Several bacteriocin compounds have been evaluated for their activity against bacteria, viruses, and particularly COVID-19. Bacteriocin is known to have a preservative role and act as an antimicrobial substance. Similarly, there are secondary metabolites such as non-ribosomal peptides, enterocin, etc. produced by probiotic microbes that have very broad therapeutic applications (Table 19.1).

19.9 CONCLUSION

With the increased mutation rate of SARS-CoV-2, every country across the world has been affected by the various waves of the pandemic. Different types of immunity-favored natural foods such as fruits and probiotics are considered to have beneficial effects on the host. The body benefits from probiotic microorganisms by enhancing gut health and immune system function. Probiotic strains such as *Lactobacillus acidophilus, Lactobacillus rhamnosus, Lactobacillus plantarum, Lactobacillus*

TABLE 19.1
Bacteriocin Compounds/Metabolites Released by Gut Microbiome

S. N.	Bacteriocin Compounds/Metabolites	Clinical Application	References
1	Staphylococcin 188	antiviral activity against HIV, influenza	Klebanoff and Coombs (1991)
2	Enterocin AAR-74	antiviral activity against HSV	Quereshi et al. (2006)
3	Erwiniocin NA4	antiviral activity Coliphage H1N1 virus	Conti et al. (2009), Lange-Starke et al. (2014)
4	Non-ribosomal peptides (NRPs)	antibiotics (daptomycin), anti-tumor drugs (bleomycin), antifungal drugs, and immunosuppressants (cyclosporin)	Walsh (2008).

casei, Bifidobacterium, their surface components and their metabolites such as bacteriocins have been studied as important antiviral agents as well as for their antitumor activities.

We believe that providing probiotics will help enhance host immunity and overall health. Healthy individuals are at lower risk of severe infection. With increasing awareness of healthy food choices, probiotics have become a natural and promising immunity booster which may improve the microbiome disparity of gut created by SARS Covid-19.

REFERENCES

Arumugam, M., Raes, J., Pelletier, E. et al. (2011). Enterotypes of the human gut microbiome. *Nature* 473, 174–180. https://doi.org/10.1038/nature09944

Asnicar, F., Berry, S.E., Valdes, A.M. et al. (2021). Microbiome connections with host metabolism and habitual diet from 1,098 deeply phenotyped individuals. *Nat Med* 27, 321–332. https://doi.org/10.1038/s41591-020-01183-8

Bosco, N., & Noti, M. (2021). The aging gut microbiome and its impact on host immunity. *Genes Immunity*, 22(1), 1–15.

Chávarri, M., Diez-Gutiérrez, L., Marañón, I., del Carmen Villarán, M., & Barrón, L.J.R. (2022). The role of probiotics in nutritional health: Probiotics as nutribiotics. In Dwivedi, M., Amaresan, N., Sankaranaryanan, A., Kemp, A. *Probiotics in the Prevention and Management of Human Diseases* (397–415). Academic Press, Elsevier

Conti, C., Malacrino, C., & Mastromarino, P. (2009). Inhibition of herpes simplex virus type 2 by vaginal lactobacilli. *J Physiol Pharmacol.* 60, 19–26.

Cunningham, A.L., Stephens, J.W., & Harris, D.A. (2021). A review on gut microbiota: A central factor in the pathophysiology of obesity. *Lipids Health Dis.* 20(1), 65. Published 2021 Jul 7. doi: https://doi.org/10.1186/s12944-021-01491-z

Daoust, L. et al. (2021). Perspective: Nutritional strategies targeting the gut microbiome to mitigate COVID-19 outcomes. *Adv. Nutrit. (Bethesda, Md.)* 12(4), 1074–1086. doi: 10.1093/advances/nmab031

Fan, Y., & Pedersen, O. (2021). Gut microbiota in human metabolic health and disease. *Nat Rev Microbiol.* 19, 55–71. https://doi.org/10.1038/s41579-020-0433-9

Gu, S., Chen, Y., Wu, Z. et al. (2020). Alterations of the gut microbiota in patients with coronavirus disease 2019 or H1N1 influenza. *Clin. Infect. Dis.* 71, 2669–2678.

Hill, C., Guarner, F., Reid, G. et al. (2014). Expert consensus document: The International Scientific Association for Probiotics and Prebiotics consensus statement on the scope and appropriate use of the term probiotic. *Nat Rev Gastroenterol Hepatol.*, 11(8), 506–514.

Juárez-Fernández, M., Román-Sagüillo, S., Porras, D. et al. (2021). Long-term effects of bariatric surgery on gut microbiota composition and faecal metabolome related to obesity remission. *Nutrients.*, 13(8), 2519.

Klebanoff, S.J., & Coombs, R.W. (1991). Viricidal effect of Lactobacillus acidophilus on human immunodeficiency virus type 1: Possible role in heterosexual transmission. *The Journal of Experimental Medicine*, *174*(1), 289–292.

Lange-Starke, A., Petereit, A., Truyen, U. et al. (2014). Antiviral potential of selected starter cultures, bacteriocins and d, l-lactic acid. *Food Environ Virol.*, 6, 42–47. doi: https://doi.org/10.1007/s12560-013-9135-z

Lloyd-Price, J., Abu-Ali, G., & Huttenhower, C. (2016). The healthy human microbiome. *Genome Med.*, *8*(1), 1–11.

Mann, G.V., & Spoerig, A. (1974). Studies of a surfactant and cholesterolemia in the Masai. *Am J Clin Nutr 27*, 464–469.

Masood, M.I., Qadir, M.I., Shirazi, J. H., & Khan, I. U. (2011). Beneficial effects of lactic acid bacteria on human beings. *Critical Reviews in Microbiology*, *37*(1), 91–98.

Quereshi, H., Saeed, S., Ahmed, S., & Rasool, S.A. (2006) Coliphage has as a model for antiviral studies/spectrum by some indigenous bacteriocin like inhibitory substances (BLIS) *Pak J Pharm Sci.*; 19, 182–185

Rizzo, P., Dalla Sega, F. V., Fortini, F., Marracino, L., Rapezzi, C., & Ferrari, R. (2020). COVID-19 in the heart and the lungs: Could we "Notch" the inflammatory storm? *Basic Res. Cardiol.*, 115(3), 1–8.

Thursby, E., & Juge, N. (2017). Introduction to the human gut microbiota. *Biochem. J.*, *474*(11), 1823–1836.

Verma, A., Xu, K., Du, T., Zhu, P., Liang, Z., Liao, S., … & Li, Q. (2019). Expression of human ACE2 in Lactobacillus and beneficial effects in diabetic retinopathy in mice. *Mole. Therapy-Methods & Clin. Develop.*, *14*, 161–170.

Walsh, C.T. (2008). The chemical versatility of natural-product assembly lines. *Acc Chem Res.*, 41: 4–10. doi: https://doi.org10.1021/ar7000414

Yadav, M.K., Kumari, I., Singh, B., Sharma, K.K., & Tiwari, S.K. (2022). Probiotics, prebiotics and synbiotics: Safe options for next-generation therapeutics. *Applied Microbiol. Biotechnol.*, 1–17. Advance online publication. https://doi.org/10.1007/s00253-021-11646-8

Zhou, Y., Zhang, J., Zhang, D. et al. (2021). Linking the gut microbiota to persistent symptoms in survivors of COVID-19 after discharge. *J Microbiol.*, 59, 941–948. https://doi.org/10.1007/s12275-021-1206-5

20 Algal Products in Medicine

Neelam Gandhi
Hansraj College, University of Delhi, New Delhi, India

CONTENTS

20.1 INTRODUCTION

The emergence of new strains of microbes resistant to most of the existing drugs, coupled with a rise in other ailments like cardiovascular problems, diabetes, cancer, etc., indicates an urgent need for both vaccines and effective drugs to address various health issues. Researchers have been able to obtain a large number of products of medicinal value from various species of algae. The use of natural products rather than synthetic molecules has the advantage of being not only more effective (due to the fact that these metabolites themselves have arisen in microbes by the process of natural selection operating for millions of years) but also cost-effective, and with low cytotoxicity. The study of natural products may therefore enable pharmacologists to modify/improve them in order to impart the desired drug-like properties. Algae have not only been the source of a large number of antimicrobials but also a very useful expression system for the production of heterologous proteins. Bioprospecting using the tools of genomics, transcriptomics, proteomics and metabolomics, has enabled scientists to discover different species that have the potential to produce substances which may have medicinal value (Maghembe et al., 2020). Some algae-derived products are proving useful in regenerative medicine, while others are vehicles for targeted delivery of drugs/genes/molecules. This chapter reviews some of the medical applications of various products obtained from different species of algae.

20.2 NATURAL PRODUCTS OBTAINED FROM ALGAE

Algae are a rich source of natural substances which have been found to have potential uses in medicine due to their antibacterial, antiviral, antifungal, anticancer, antiinflammatory, antioxidant, anticoagulant, hypoglycemic and antiprotozoal properties. Algae are eukaryotic, unicellular or multicellular, photosynthetic organisms occurring in freshwater and marine environments as well as on moist soil. Algae can be grouped into green algae, brown algae and red algae, and can also be

DOI: 10.1201/9781003306931-23

classified as microalgae and macroalgae. A number of substances/metabolites such as lectins, polysaccharides, peptides, fatty acids, terpenes and phlorotannins are reported to be present in algae (Shannon and Abu-Ghannam, 2016), which are pharmacologically active and help algae cope with the challenges present in their surroundings.

20.2.1 Antibacterial Properties of Algal Products

A substance may inhibit one or more of the following processes in bacteria to achieve antibacterial activity: a) cell wall synthesis; b) protein synthesis; c) nucleic acid synthesis; d) a metabolic pathway; e) membrane function; and f) ATP synthase activity.

Table 20.1 lists some bioactive compounds obtained from algae, with antibacterial activity against some medically important bacterial species.

NIH estimates suggest that up to 80% of the bacterial infections affecting humans are mediated by biofilm (self-secreted matrix)-associated microorganisms (Römling and Balsalobre, 2012). As these microbes have been found to be resistant to commonly used antibiotics, biofilm-forming microbes generally lead to persistent infections, as seen in recurrent urinary tract infections, chronic rhino-sinusitis and lung infection that is associated with cystic fibrosis. Biofilm formation may also be encountered on devices such as catheters, stents, orthopedic implants, contact lenses, etc. (Lynch and Robertson, 2008).

Low molecular weight alginate oligosaccharides (for instance OligoG), derived from alginate found in the cell walls of brown algae, have been reported to possess mucolytic activity as well as

TABLE 20.1
Bioactive Compounds of Some Species of Algae with Antibacterial Activity

Compound/Metabolite	Producer Species	Mechanism of Action/Effects	References
Lectins	*Solieria filiformis*	Inhibit the growth of *Salmonella typhi, Klebsiella pneumoniae, Pseudomonas aeruginosa* by interacting with the glycoconjugate present in bacterial cell wall	Holanda et al. (2005)
2-(20,40-dibromophenoxy)-4,6-dibromoanisol	*Cladophora fascicularis*	Inhibits growth of *E. coli, Bacillus subtilis* and *S. aureus*	Kuniyoshi et al. (1985)
Phlorotannins	*Sargassum thunbergii*	Cause cell death by damaging cell membrane and cell wall of *Vibrio parahaemolyticus*	Wei et al. (2016)
Phenols Terpenes, acetogenins, indoles, fatty acids in the algal extract	*Sargassum vulgare Sargassum fusiforme*	Cause damage to the cell walls and cytoplasmic leakage of *Staphylococcus aureus* and *Klebsiella pneumoniae*	El Shafay et al. (2016)
Long-chain fatty acids	*Planktochlorella nurekis*	Inhibitors of *Campylobacter jejuni, E. coli, Salmonella enterica var. Infantis, Arcobacter butzleri*	Cermak et al. (2015)
Sulfated polysaccharides	*Sargassum swartzii*	Inhibit both Gram-positive and Gram-negative bacteria pathogenic to humans	Vijayabaskar et al. (2012)
Ethanolic extract	*Laurencia catarinensis, and Padina pavonica*	Inhibit growth of *Klebsiella pneumonia*	Al-Enazi et al. (2018)
	Padina pavonica	Inhibits growth of *Streptococcus pyogenes, Bacillus subtilis, Staphylococcus aureus* and *Acinetobacter baumannii*	Al-Enazi et al. (2018)
Bromophycolides	*Callophycus serratus*	Inhibit methicillin-resistant *Staphylococcus aureus* and vancomycin-resistant *Enterococcus faecium*	Lane et al. (2009)

the ability to cause damage to the pathogen cell membrane. The efficacy of antibiotics against resistant pathogens such as *Pseudomonas, Acinetobacter* has been reported to be improved by OligoG.

20.2.2 Antiviral Properties of Algal Products

The human population is very adversely affected by viruses like SARS-CoV-2, HIV, hepatitis virus, dengue virus, influenza virus and many more. Vaccines are not available against several viral diseases and antiviral drugs currently in use are either not very effective due to the rapid emergence of drug-resistant strains or have serious side-effects. Many studies involve screening microbial natural products for antiviral activity.

The replication cycle of viruses involves several steps: attachment, penetration, uncoating, replication, assembly and release (Seo and Choi, 2021). Agents with antiviral activity may exert their action at any of the steps.

Many marine organisms including seaweeds contain sulfated polysaccharides in their cell walls as an adaptation to the high ionic concentration found in the marine habitat. Algal-sulfated polysaccharides show great diversity and complexity. Those well documented include galactans (agarans and carrageenans) from red algae, ulvans from green algae, and fucans and fucoidans from brown algae (Cunha and Grenha, 2016).

It was shown by Gerber et al. (1958) that polysaccharides from marine algae had inhibitory effects on mumps and influenza B virus. Studies conducted since then have reported antiviral activity of algae-derived products against several viruses (Ahmadi et al., 2015; Geahchan et al., 2021; Ghareeb et al., 2020), as briefly listed in Table 20.2.

TABLE 20.2
Bioactive Compounds of Some Species of Algae Showing Antiviral Activity

Compound	Producer Species	Mechanism of Action/Effect	References
Sulfated galactofucan	*Undaria pinnatifida*	Inhibits host cell binding and entry of viruses HSV-1, HSV-2 and HCMV	Hemmingson et al. (2006)
Fucoidan: alpha-(1-3)- linked	*Padina boryana*	Inhibits entry of SARS-CoV-2 into host cell	Song et al. (2020)
Fucoidan	*Cladosiphon okamuranus*	Inhibits viral entry into host cells by directly binding with envelope glycoprotein of DEN-2	Hidari et al. (2008)
Iota-carrageenan	Red algae	Inhibits H1N1 influenza A virus	Leibbrandt et al. (2010)
Carrageenan	Red algae	Inhibits infection by human papillomavirus (HPV)	Buck et al. (2006)
911(heparinoid alginate derivative)		Blocks the binding of HIV-1 to MT cells and inhibits the activity of reverse transcriptase	Xin et al. (2000)
Fucoidan	*Kjellmaniella crassifolia*	Inhibits viral neuraminidase and interferes with cellular EGFR pathway in influenza A virus	Wang et al. (2017)
Sulfated galactans	*Gymnogongrus griffithsiae, Cryptonemia crenulata*	Inhibits adsorption of herpes simplex virus type 1 (HSV-1) and HSV-2	Talarico et al. (2004)
Sulfated fucans	*Lobophora variegata*	Inhibits reverse transcriptase activity of HIV in in-vitro system	Queiroz et al. (2008)
Griffithsin	*Griffithsia sp.*	Blocks entry of viruses such as HIV-1, HPV, HSV-2, HCV, MERS-CoV, SARS-CoV into their host cells	Lee (2019)
Sulfated polysaccharide p-KG03	*Gyrodinium impudium*	Inhibits entry of influenza A virus into host cells	Kim et al. (2012)

According to several reports, carrageenans exert antiviral effect by directly binding to the virus, preventing its entry into the host cell. Nasal sprays containing iota-carrageenan have been developed and reported to reduce viral load in the nasal wash fluids of patients having common cold virus infections (Boots Dual Defence in the UK market) (Salih et al., 2021) and lozenges based in carrageenans for the treatment of sore throat are also available (Morokutti-Kurz et al., 2017).

Carrageenans, due to their ability to block infection of cells by virions, are finding an application in the management of sexually transmitted infections by their incorporation into lubricant gels for topical application, as reported for HPV infection (Magnan et al., 2019), and HSV-1 and 2 (Maguire et al.,1998; Talarico et al., 2004).

Several red algae species have been shown to possess lectins, the proteins capable of recognizing specific carbohydrate moiety and causing agglutination of cells or precipitation of glyco-conjugates and are being assessed for their application as antiviral, antibacterial, anticancer and anti-inflammatory agents (Singh and Walia, 2018). Griffithsin, lectin obtained from *Griffithsia* species has been reported to bind with high affinity to the mannose-rich N-linked glycans present on enveloped viruses and alpha6 integrin of non-enveloped human papillomavirus. This leads to the blocking of viral entry into the host cells. Antiviral activity of griffithsin has been reported against a number of viruses such as HIV-1, herpes simplex virus-2 (HSV-2), human papillomavirus (HPV), hepatitis C virus (HCV), Middle East respiratory syndrome coronavirus (MERS-CoV) and SARS-CoV (Decker et al., 2020), and there are reports of two Phase I clinical studies to determine the efficacy and toxicity of griffithsin for prevention of HIV infection (Lee, 2019). Due to the rising demand for griffithsin because of its broad-spectrum antiviral activity, recombinant griffithsin is being produced in engineered *E.coli* (Decker et al., 2020).

The synergistic antiviral effect of a combination of griffithsin and carrageenan against SARS-CoV-1 and SARS-CoV-2 in in-vitro studies as reported by Alsaidi et al. (2021) suggests the potential application of this combination of algal products for the treatment of these infections.

20.2.3 Antifungal Activity of Algal Products

Individuals with compromised immunity such as HIV and cancer patients, and individuals receiving immunosuppressants post transplantation surgery are more prone to opportunistic fungal infections including invasive aspergillosis caused by *Aspergillus fumigatus* responsible for very high mortality rates (Ahamefule et al., 2020). In order to meet the rising demand for effective antifungal therapy, researchers have explored and reported several new antifungal agents from a number of algal species.

Fungal specimens collected from the sputum of chronic asthmatic patients showed inhibition by the ethanolic extracts of algae *Laurencia paniculata* and *Ulva prolifera* (Mickymaray and Alturaiki, 2018), suggesting the possible use of these algal compounds for the recovery of patients from chronic asthmatic states. Alginate oligosaccharides (OligoG) have been reported to inhibit hyphal growth in germ tube assays and disrupt formation of fungal biofilm besides being able to improve the antifungal activity of commonly used antifungal drugs against *Aspergillus and Candida* spp. (Tondervik et al., 2014). Table 20.3 lists some algae-derived agents with antifungal activity against some medically important species.

20.2.4 Anticancer Properties of Algal Products

Cancer develops when abnormal or damaged cells grow in an uncontrolled manner and spread to other parts of the body. Researchers have been seeking to develop anticancer agents to curtail the active proliferation of cancer cells by targeting their increased requirements such as metabolites, various enzymes, hormones (in some cancers), and blood supply.

TABLE 20.3
Bioactive Compounds of Some Species of Algae Showing Antifungal Activity

Compound	Producer Species	Mechanism of Action/Effect	References
Algal extracts containing terpene alcohol, diterpene steroids, sesquiterpene, and sesquiterpene alcohol	*Laurencia. paniculata, Ulva prolifera*	Inhibit a variety of fungal pathogens (*Candida albicans, Aspergillus niger, Mucor sp*)	Mickymaray and Alturaiki (2018)
Bromophenol	*Odonthalia corymbifera*	Inhibits growth of *Aspergillus fumigatus, Candida albicans, Trichophyton rubrum* and *Trichophyton mentagrophytes*	Oh et al. (2008)
Ethanolic extract	*Padina pavonica*	Inhibits growth of *Candida albicans, C tropicalis, Aspergillus fumigatus*	Al-Enazi et al. (2018)
Lobophorolide	*Lobophora variegata*	Inhibits growth of *Dendryphiella salina, Lindra thalassiae*	Kubanek et al. (2003)

Phlorotannins, which are antioxidants derived from a number of species of brown algae, have been shown to exhibit activities against angiogenesis, proliferation and metastasis, and to promote apoptosis, as reviewed extensively by Catarino et al. (2021). These represent a very important category of algal-derived compounds for the treatment of cancers, which is attested by the large number of patents granted for phlorotannins.

There are several reports that lectins from a number of species of red algae exhibit anticancer properties, such as *Acrocystis nana, Bryothamnion seaforthii, B. Triquetrum, Eucheuma serra, Halosaccion glandiforme, Hypnea cervicornis* and *Solieria filiformis*. Algal lectins have been shown to interact with unique sugar chains found on carcinoma cell surfaces, inducing them to undergo apoptotic cell death (Alves et al., 2018) and algal fucoidans have been reported to exhibit strong anti-cancer and immunomodulatory activity (Atashrazm et al., 2015; Han et al., 2015)

20.2.5 Anti-inflammatory Activity of Algal Products

Inflammatory response is a normal immune defense mechanism against tissue injury or infection and stops after the resolution of the damage. However, when inflammation persists, it may contribute to progression of many chronic illnesses like autoimmune diseases, cancer, cardiovascular diseases and neurodegenerative diseases such as Alzheimer's and Parkinson's diseases (Barbalace et al., 2019).

Several species of marine algae have been reported to be the source of many compounds with potent anti-inflammatory activities (Table 20.4).

Phlorotannins obtained from algal species *Fucus vesiculosus, F. serratum, Ascophyllum nodosum*, effectively inhibit oxidative stress-induced inflammations, as reported by Catarino et al. (2021).

Zhou et al. (2015) reported that oligosaccharides obtained by enzymatic digestion of alginate were effective in inhibiting neuroinflammation, suggesting its potential in the treatment of neurodegenerative diseases.

20.2.6 Antioxidant Activity of Algal Products

Oxidative stress, due to the accumulation of reactive oxygen species, has been implicated in several diseases such as cancer, atherosclerosis, rheumatoid arthritis, cardiovascular and neurodegenerative diseases and diseases associated with the aging process (Cornish and Garbary, 2010).

A number of studies have revealed that several algal species are a very rich source of compounds such as carotenoids, vitamins and several others with antioxidant activity (Munir et al., 2013). (Table 20.4). Free radical scavenging activity of natural antioxidant, astaxanthin from alga *Haematococcus pluvialis* is 65 times more powerful than vitamin C and 100 times more effective than alpha-tocopherol.

20.2.7 Antiprotozoal Activity of Algal Products

According to WHO estimates, more than 12 million people worldwide are infected by *Leishmania sp.*, and 10 million people are currently infected by *Trypanosoma cruzi*, the causative agent of Chagas disease (Varikuti et al., 2018). *Trypanosoma brucei gambiense*, the causative agent of sleeping sickness, infects nearly 30,000 people annually. Malaria is responsible for a great deal of morbidity and mortality, especially in the developing countries.

The search for antiprotozoal agents has yielded several algae-derived compounds with potent inhibitory activity against various protozoan pathogens (Lategan et al., 2009; Torres et al., 2014) (Table 20.4).

20.2.8 Other Medically Relevant Algal Compounds

Several algae-derived compounds have been shown to exhibit anticoagulant (Li et al., 2008), anthelmintic (Gerwick, 2013) and hypoglycemic activities (Akbarzadeh et al., 2018) and could potentially be developed for the management of cardiovascular disease, worm infections and diabetes respectively (Table 20.4).

20.3 PRODUCTS OBTAINED FROM ALGAE USING BIOTECHNOLOGICAL TOOLS

Demand for biopharmaceuticals has increased considerably since the approval and use of the first recombinant protein, insulin (Leader et al., 2008). Among the requirements for the efficient production of recombinant proteins are the ease and availability of genetic manipulation tools, the rapid growth rate of the host cells, consistent expression levels of the transgene, accumulation and proper folding of the expressed protein, and scalability to obtain the desired yield. Microalgae meet these criteria (Ahmad et al., 2020; Rasala and Mayfield, 2011; Specht and Mayfield, 2014; Taunt et al., 2018) and have an advantage over bacterial expression systems as they show the post-translational modifications necessary for the protein to be functional. They are also advantageous over mammalian cell cultures because of their fast growth rate on a very simple culture medium and are thus very cost-effective. Extensive investigation of the alga *Chlamydomonas reinhardtii* (*C. reinhardtii*) has revealed that though ribosomes and translation factors within the chloroplast are similar to those found in photosynthetic prokaryotes, chloroplasts, unlike bacteria, also show the presence of the several chaperones and enzymes necessary for proper folding of the proteins, such as disulfide isomerases and peptidylprolyl isomerases (Tran et al., 2013).

A monoclonal antibody against glycoprotein D of herpes simplex virus (HSV) was the first mammalian protein to be produced in the chloroplast of *C. reinhardtii* (Mayfield et al., 2003), in which the antibodies accumulated as soluble proteins within the chloroplast and, when tested, were found to bind to HSV proteins. Using microalgae as the expression system, Hempel et al. (2011) reported the successful production of fully assembled and functional monoclonal human IgG antibodies against hepatitis surface antigen. Monoclonal antibodies provide targeted therapy and are generally free of the side-effects associated with chemotherapy.

The highly infectious nature of SARS-CoV-2 which resulted in the recent pandemic led to an urgent need to develop not only an effective vaccine but also reagents to identify antibodies in the

TABLE 20.4
Bioactive Compounds from Species of Algae Showing Various Biological Activities

Activity	Compound	Producer Species	Mechanism of Action/Effect	References
Anticancer	Heterofucan SF-1.5v	*Sargassum filipendula*	Exhibits antiproliferative effect on HeLa cells by stimulating the release of apoptosis-inducing factor	Costa et al. (2011)
	Fucoxanthin	*Ishige okamurae*	Induces apoptosis in human leukemia HL- 60 cells by generating reactive oxygen species	Kim et al. (2010)
	Fucoidan	*Undaria pinnatifida, Laminaria angustata, Fucus vesiculosus, Fucus evanescens*	Suppresses proliferation and induces apoptosis of prostate cancer cells DU 145	Choo et al. (2016)
	Ethanolic extract	*Padina pavonica, Laurencia catarinensis and Laurencia majuscula*	Shows in-vitro antitumor activity against a number of tumor cell lines such as intestinal carcinoma, colon carcinoma, cervical carcinoma, hepatocellular carcinoma, breast carcinoma	Al-Enazi et al. (2018)
	Hydrolyzed crude extract	*Sargassum hemiphyllum*	Inhibits angiogenesis by suppressing HIF-1/VEGF-regulated signaling in bladder cancer cells	Chen et al. (2015)
	Fucans	*Lobophora variegata*	Decreases viability of hepatocellular cancer cell line in in-vitro assay	Castro et al. (2015)
Anti-Inflammatory	Sulfated polysaccharides	Red microalgae	Inhibit mobility and adhesion of neutrophils, and are used for the treatment of skin inflammation	Matsui et al. (2003)
	Fucoidan	Brown alga	Inhibits excessive production of nitric oxide and prostaglandin E2 in BV2 microglia	Park et al. (2011)
	Alkaloids (Indole-4-carboxaldehyde)	*Sargassum thunbergii*	Suppress methylglyoxal-induced inflammation, a human hepatocytes, by preventing expression of genes coding for inflammatory cytokines	Cha et al. (2019)
	Lectin	*Solieria filiformis*	Exhibits reduced neutrophil migration and paw edema in a mouse model	Abreu et al. (2016)
	Fucoidan	*Lobophora variegata*	Inhibits leukocyte migration to the inflammation site	Medeiros et al. (2008)
Antioxidant	Sesquiterpenoids	*Ulva fasciata Delile*	Show good free radical scavenging activity	Chakraborty and Paulraj (2010)
	Flavonoids	*Ulva lactuca*	Show good free radical scavenging activity	Meenakshi et al. (2009)
	Aqueous extract	*Gracilaria tenuistipitata*	H1299 cell line showed recovery from H_2O_2- induced DNA damage	Yang et al. (2012)

(Continued)

TABLE 20.4 *(Continued)*

Activity	Compound	Producer Species	Mechanism of Action/Effect	References
	Sulfated polysaccharides	*Sargassum swartzii*	Show good free radical scavenging activity	Vijayabaskar et al. (2012)
	Ethanolic extract	*Padina pavonica and Laurencia majuscula*	Shows DPPH (2,2-diphenyl-1-picrylhydrazyl) radical scavenging activity	Al-Enazi et al. (2018)
	Beta-carotene and ascorbate	*Chondrus crispus, Mastocarpus stellatus*	Contain high content of antioxidant enzymes	Lohrmann et al. (2004)
	Phlorotannins	*Ascophyllum nodosum, Fucus vesiculosus*	Show high DPPH scavenging activity	Liu et al. (2017)
	Astaxanthin	*Haematococcus pluvialis*	Exhibits extremely high free radical scavenging activity	Shah et al. (2016)
Anticoagulant	Sulfated polysaccharides	*Ulva fasciata Agardhiella subulata*	Prolong prothrombin time (PT)	Faggio et al. (2016)
Hypoglycemic	Extracts	*Sargassum oligocystum*	Reduce insulin resistance, decrease glucose concentration and help in the regeneration of damaged beta-cells of pancreas	Akbarzadeh et al. (2018)
	Fucosterol	*Pelvetia siliquosa*	Decreases serum glucose levels and inhibits glycogen degradation in rats	Lee et al. (2004)
	Fucoxanthin	*Hijikia fusiformis, Sargassum fulvellum, Laminaria japonica and Undaria pinnatifida*	Prevents insulin resistance by decreasing expression of plasminogen activator inhibitor-1 and promotes lipid metabolism in mice model	D'Orazio et al. (2012)
	Fucoidan	*Fucus vesiculosus*	Inhibits alpha-glucosidase activity and lowers blood glucose levels in mice	Shan et al. (2016)
Anti-helminth	Kainic acid	*Digenea simplex*	Effectively removes Ascaris, Oxyuris, Taenia and Trichuris	Gerwick (2013)
Anti-protozoal	Fucoidan	*Fucus vesiculosus*	Activates pathways leading to production of nitric oxide and pro-inflammatory cytokines showing curative effect against leishmaniasis	Sharma et al. (2014)
	Fucoidan	*Undaria pinnatifida*	Inhibits invasion of erythrocytes by Plasmodium merozoites in in-vitro system and in-vivo system (mice)	Chen et al. (2009)
	Cold and hot water extracts	*Caulerpa sertularioides, Gracilaria corticata, G.salicornia, Sargassum oligosystum*	Leishmanicidal activity in in-vitro system	Fouladvand et al. (2011)
	Extracts (mixture of sulfoquinovosyldiacylglycerol)	*Lobophora variegata*	Show inhibitory activities against Trichomonas vaginalis, Entamoeba histolytica, Leishmania mexicana, Trypanosoma cruzi, Giardia intestinalis	Vieira et al. (2017)
	Bromophycolides J-Q	*Callophycus serratus*	Show inhibitory activity against Plasmodium falciparum in in-vitro system	Lane et al. (2009)

patient's serum. Berndt et al. (2021) have reported successful production of a correctly folded, functional receptor-binding domain of SARS-CoV-2 protein in *C. reinhardtii*. This protein can be used as a vaccine antigen as well as to detect serum antibodies to assess vaccine efficacy and formulate booster dose policy.

Tran et al. (2013) have shown that fully functional immunotoxin proteins (consisting of two parts, an antibody-binding domain which binds to B-cell surface antigen CD22, and the PE40 toxin domain of exotoxin A) are produced and accumulated in soluble form within the chloroplast of *C. reinhardtii*. The immunotoxin thus produced was found to be fully functional in inhibiting proliferation of Burkitt lymphoma cell lines. This kind of targeted delivery of drugs/toxins to specifically cancerous cells is not only more effective but also spares the adjacent healthy cells.

FDA-approved cancer vaccines Gardasil and Cervarix for protection against human papillomavirus types 16 and 18 have been produced in *C. reinhardtii* by Demurtas et al. (2013).

Production of alpha-defensin with generalized antimicrobial activity against several bacteria, viruses and pathogenic fungi in transformed *Chlorella ellipsoidea* by Bai et al. (2013) offers a promising lead towards treating infections caused by multidrug-resistant bacteria.

Antigen-expressing *Chlamydomonas* cells preserved by freeze-drying (which remain stable for up to 20 months) have been reported to be effective as orally administered vaccines in mice for protection against infection by *S. aureus* (Dreesen et al., 2010) and malarial parasites (Gregory et al., 2013), and hence can be developed for mass immunization programs.

Other medically important recombinant proteins produced successfully in microalgae include vascular endothelial growth factor (Rasala and Mayfield, 2011), Hepatitis B virus surface antigen for vaccine against hepatitis (Hempel et al., 2011) and HIV-1 antigen P24 for AIDS vaccine (Barahimipour et al., 2016).

20.4 USE OF ALGAL PRODUCTS IN TISSUE ENGINEERING AND OTHER BIOMEDICAL APPLICATIONS

Delivery of drugs to specific targets plays a very important role in the efficacy (by promoting greater bioavailability) and safety (by restricting the site of action) of the treatment of several diseases, including those requiring tissue regeneration.

The cell walls of several species of algae such as *Ascophyllum nodosum*, *Laminaria* species and many others yield a polymer alginate which forms a gel by ionic gelation (Abasalizadeh et al., 2020). It is biocompatible and is associated with low toxicity. Efficacy of the enzyme L-asparaginase, used for the treatment of acute lymphocytic leukemia (El-Gendy et al., 2021), was reported to greatly improve when introduced in alginate-based nanomaterial, due to decreased proteolytic degradation rate of the enzyme (Nunes et al., 2020). Gombotz and Wee (1998) reported that in the alginate-based hydrogels, proteins can be included under conditions conducive to preserving their native structure to a large extent, and thus alginate-based hydrogels can work as vehicles for delivery of protein drugs.

Choi et al. (2019) reported that nanoparticles constructed from chitosan (a polymer derived from fungal cell walls) and fucoidan (an algae-derived sulfated polysaccharide) could not only effectively encapsulate the anticancer drug piperlongumine but also increase its water solubility and bioavailability so as to efficiently kill human prostate cancer cells. In another study, fucoidan was used as a nanocarrier along with PLGA (poly-lactic-co-glycolic acid) to encapsulate the hydrophobic anticancer drug docetaxel, which not only functioned as a successful drug-delivery system but also greatly improved the efficacy of the treatment due to the immunostimulatory properties of fucoidan (Lai et al., 2020).

Vascular endothelial growth factor (VEGF), when enclosed in microspheres and administered in alginate-based hydrogel, has been reported to enhance vasculogenesis in vivo at the injection site (Badali et al., 2021), an important requirement in wound healing.

Protozoan *Trypanosoma brucei* causes human African trypanosomiasis (or sleeping sickness) which affects people in tropical and subtropical countries. Quinapyramine sulfate, a drug used for the treatment of trypanosomiasis, has been reported to be quite effective even in lower doses (thereby reducing toxicity) when included in nanoparticles composed of sodium alginate (due to sustained release) (Manuja et al., 2014).

A novel approach to tackle bacterial infections associated with orthopedic implants has been utilized by Barros et al. (2020), where bacteriophages encapsulated in alginate hydrogels were delivered along with the implant and showed antimicrobial activity, leaving the implant site infection free.

Many studies involved in developing therapies for diabetes and neurodegenerative disorders have employed cells encapsulated in alginate gel. DIABECELL and NTCELL are two such products that are currently being tested in clinical trials (Szekalska et al., 2016). The DIABECELL implant consists of insulin-secreting islet cells (derived from pig neonates) encapsulated in alginate gel. This implant, when introduced into the abdomen of a patient suffering from type 1 diabetes, has been reported to show encouraging results in clinical evaluation while protecting the living cells from the immune system of the host.

Researchers have developed NTCELL, an alginate-based implant consisting of encapsulated neonatal porcine cells of the choroid plexus and introduced this into the damaged site in the brain of animals which have been chemically treated to simulate the brain lesions associated with Parkinson's, the neurogenerative disease. The implant was reported to enhance release of nerve growth factors and cerebrospinal fluid. Clinical trials were undertaken to establish whether this treatment can benefit patients with Parkinson's, and clinical assessment of patients receiving the therapy reported improvement (Szekalska et al., 2016).

Alginate-based hydrogels showing such properties as absorbent capacity, mechanical stability and viscoelasticity are finding an application in wound dressing (Szekalska et al., 2016) as well as vehicles for delivery of chemotherapeutic drugs (Abasalizadeh et al., 2020)

As the structure of alginate hydrogels shows similarity to the extracellular matrix of living tissues, these gels have been used in biomedical studies that require cell cultures in which the gels form a suitable 2-D or 3-D matrix (Bidarra et al., 2019).

20.5 SUMMARY

Algae are the source of a large number of biomolecules with activities that are antibacterial, antiviral, anticancer, antifungal or antiprotozoal in nature, as well as many other medically useful substances such as anti-inflammatory, antioxidant and hypoglycemic agents, etc. Therefore, they contribute greatly in the management of several health issues faced by humans. The antimicrobial activity of compounds of some algal species are being explored for use as food additives in order to tackle issues of food-borne diseases and food poisoning. The antiviral effects of the metabolites of some algal species have been exploited to develop nasal sprays for relief against sinusitis and lubricants for topical application to prevent sexually transmitted diseases.

Algae have been used by researchers as expression systems for the production of several recombinant proteins for medical use such as hormones, the SARS-CoV antigen for vaccines and diagnostics, and the FDA-approved cancer vaccines Gardasil and Cervarix for protection against human papillomavirus types 16 and 18. Several preparations of phlorotannins showing high degree of efficacy against cancers have been granted patents.

Production of fully functional molecules of immunotoxins in the algal expression system has paved the way for targeted killing of cancer cells, sparing adjacent healthy cells.

Antigen-expressing *Chlamydomonas* cells preserved by freeze-drying have been reported to be effective in mice as orally administered vaccines obtained at low cost and have the potential to be developed for mass vaccination programs.

Algal cell wall polysaccharides with unique properties are being exploited in regenerative medicine, such as in treatment of burns, and for various biomedical applications such as targeted drug/gene/cell delivery needed for the treatment of cancer/genetic deficiencies/tissue injury. Scientists are continuing to pursue the mining of more and more species of algae to obtain natural products for their biosafety and biodegradability propertiues, as well as to overcome the problem of microbial resistance to many currently used drugs.

REFERENCES

Abasalizadeh, F., Moghaddam, S. V., Alizadeh, E., … & Akbarzadeh, A. (2020). Alginate-based hydrogels as drug delivery vehicles in cancer treatment and their applications in wound dressing and 3D bioprinting. *Journal of Biological Engineering*, 14, 8. https://doi.org/10.1186/s13036-020-0227-7

Abreu, T. M., Ribeiro, N. A., Chaves, H. V., … & Benevides, N. M. (2016). Antinociceptive and Anti-inflammatory activities of the lectin from Marine Red Alga *Solieria filiformis*. *Planta Medica*, *82*(7), 596–605. https://doi.org/10.1055/s-0042-101762

Ahamefule, C. S., Ezeuduji, B. C., Ogbonna, J. C. … & Fang, W. (2020). Marine bioactive compounds against *Aspergillus fumigatus*: Challenges and future prospects. *Antibiotics (Basel, Switzerland)*, 9(11), 813. https://doi.org/10.3390/antibiotics9110813

Ahmad, N., Mehmood, M. A., & Malik, S. (2020). Recombinant protein production in microalgae: Emerging trends. *Protein and Peptide Letters*, 27(2), 105–110. https://doi.org/10.2174/0929866526666191014124855

Ahmadi, A., Zorofchian Moghadamtousi, S., Abubakar, S., & Zandi, K. (2015). Antiviral potential of algae polysaccharides isolated from marine sources: A review. *BioMed Research International*, 2015, https://doi.org/10.1155/2015/825203

Akbarzadeh, S., Gholampour, H., Farzadinia, P., Daneshi, A., Ramavandi, B., Moazzeni, A., Keshavarz, M., & Bargahi, A. (2018). Anti-diabetic effects of Sargassum oligocystum on Streptozotocin-induced diabetic rat. *Iranian Journal of Basic Medical Sciences*, 21(3), 342–346. https://doi.org/10.22038/IJBMS.2018.25654.6329

Al-Enazi, N. M., Awaad, A. S., Zain, M. E., & Alqasoumi, S. I. (2018). Antimicrobial, antioxidant and anticancer activities of Laurencia catarinensis, Laurencia majuscula and Padina pavonica extracts. *Saudi Pharmaceutical Journal: SPJ: The Official Publication of the Saudi Pharmaceutical Society*, 26(1), 44–52. https://doi.org/10.1016/j.jsps.2017.11.001

Alsaidi, S., Cornejal, N., …& Romero J. A. F. (2021). Griffithsin and carrageenan combination results in antiviral synergy against SARS-CoV-1 and 2 in a pseudoviral model. *Marine Drugs* 19, 418. https://doi.org/10.3390/md19080418

Alves, C., Silva, J., Pinteus, S., Gaspar, H., Alpoim, M. C., Botana, L. M., & Pedrosa, R. (2018). From marine origin to therapeutics: The antitumor potential of marine algae-derived compounds front. *Pharmacology*. https://doi.org/10.3389/fphar.2018.00777

Atashrazm, F., Lowenthal, R. M., Woods, G. M., Holloway, A. F., & Dickinson, J. L. (2015). Fucoidan and cancer: A multifunctional molecule with anti-tumor potential. *Marine Drugs*, 13(4), 2327–2346. https://doi.org/10.3390/md13042327

Badali, E., Mohajer, M., Hassanzadeh, S., Saghati, S., & Khanmohammadi, M. (2021). Enzymatic crosslinked hydrogels for biomedical application. *Preprints*. https://doi.org/10.20944/preprints202104.0652.v1

Bai, L. L., Yin, W. B., Chen, Y. H., … & Hu, Z. M. (2013). A new strategy to produce a defensin: Stable production of mutated NP-1 in nitrate reductase-deficient Chlorella ellipsoidea. *PloS One*, 8(1). https://doi.org/10.1371/journal.pone.0054966

Barahimipour, R., Neupert, J., & Bock, R. (2016). Efficient expression of nuclear transgenes in the green alga *Chlamydomonas*: Synthesis of an HIV antigen and development of a new selectable marker. *Plant Molecular Biology*, 90, 403–418. https://doi.org/10.1007/s11103-015-0425-8. Epub 2016 Jan 8

Barbalace, M. C., Malaguti, M., Giusti, L., Lucacchini, A., Hrelia, S., & Angeloni, C. (2019). Anti-inflammatory activities of marine algae in neurodegenerative diseases. *International Journal of Molecular Sciences*, *20*(12), 3061. https://doi.org/10.3390/ijms20123061

Barros, J., Melo, L., Silva, R., … & Monteiro, F. J. (2020). Encapsulated bacteriophages in alginate-nanohydroxyapatite hydrogel as a novel delivery system to prevent orthopedic implant-associated infections. *Nanomedicine: Nanotechnology, Biology, and Medicine*, 24, 102145. https://doi.org/10.1016/j.nano.2019.102145

Berndt, A., Smalley, T., Ren, B., …& Mayfield, S. (2021). Recombinant production of a functional SARS-CoV-2 spike receptor binding domain in the green algae *Chlamydomonas reinhardtii* bioRxiv. https://doi.org/10.1101/2021.01.29.428890

Bidarra, S. J., & Barrias, C. C. (2019). 3D culture of mesenchymal stem cells in alginate hydrogels. *Methods in Molecular Biology* (Clifton, N.J.), 2002, 165–180. https://doi.org/10.1007/7651_2018_185

Buck, C. B., Thompson, C. D., Roberts, J. N., Müller, M., Lowy, D. R., & Schiller, J. T. (2006). Carrageenan is a potent inhibitor of papillomavirus infection. *PLoS Pathogens*, 2(7), e69. https://doi.org/10.1371/journal.ppat.0020069

Castro, L. S. E. P. W., de Sousa Pinheiro, T., Castro, A. J. G. et al. (2015). Potential anti-angiogenic, antiproliferative, antioxidant, and anticoagulant activity of anionic polysaccharides, fucans, extracted from brown algae *Lobophora variegata*. *Journal of Applied Phycology* 27, 1315–1325. https://doi.org/10.1007/s10811-014-0424-1

Catarino, M. D., Amarante, S. J., Mateus, N., Silva, A. M. S., Cardoso, S.M. (2021). Brown Algae phlorotannins: A marine alternative to break the oxidative stress, inflammation and cancer network. *Foods*, 10, 1478. https://doi.org/10.3390/foods10071478

Čermák, L., Pražáková, Š., Marounek, M., Skřivan, M., & Skřivanová, E. (2015). Effect of green alga *Planktochlorella nurekis* on selected bacteria revealed antibacterial activity *in vitro*. *Czech Journal of Animal Science* 60, 427–435. [Google Scholar]

Cha, S. H., Hwang, Y., Heo, S. J., & Jun, H. S. (2019). Indole-4-carboxaldehyde Isolated from Seaweed, *Sargassum thunbergii*, Attenuates Methylglyoxal-Induced Hepatic Inflammation. *Marine Drugs*, 17(9), 486. https://doi.org/10.3390/md17090486

Chakraborty, K., & Paulraj, R. (2010). Sesquiterpenoids with free-radical-scavenging properties from marine macroalga Ulva fasciata Delile. *Food Chemistry*, 122, 31–41

Chen, J. H., Lim, J. D., Sohn, E. H., Choi, Y. S., & Han, E. T. (2009). Growth-inhibitory effect of a fucoidan from brown seaweed *Undaria pinnatifida* on Plasmodium parasites. *Parasitology Research*, 104(2), 245–250. https://doi.org/10.1007/s00436-008-1182-2

Chen, M. C., Hsu, W. L., Hwang, P. A., & Chou, T. C. (2015). Low molecular weight fucoidan inhibits tumor angiogenesis through downregulation of HIF-1/VEGF signaling under hypoxia. *Marine Drugs*, 13(7), 4436–4451. https://doi.org/10.3390/md13074436

Choi, D. G., Venkatesan, J., & Shim, M. S. (2019). Selective anticancer therapy using pro-oxidant drug-loaded chitosan-fucoidan nanoparticles. *International Journal of Molecular Sciences*, 20(13), 3220. https://doi.org/10.3390/ijms20133220

Choo, G.-S., Lee, H.-N., Shin, S.-A., Kim, H.-J., & Jung, J.-Y. (2016) Anticancer effect of fucoidan on DU-145 prostate cancer cells through inhibition of PI3K/Akt and MAPK pathway expression. *Marine Drugs*, 14(7), 126. https://doi.org/10.3390/md14070126

Cornish, M. L., & Garbary, D. J. (2010). Antioxidants from macroalgae: Potential applications in human health and nutrition. *Algae*, 25(4), 155–171. https://doi.org/10.4490/algae.2010.25.4.155

Costa, L. S., Telles, C. B., Oliveira, R. M., Nobre, L. T., Dantas-Santos, N., Camara, R. B., Costa, M. S., Almeida-Lima, J., Melo-Silveira, R. F., Albuquerque, I. R., Leite, E. L., & Rocha, H. A. (2011). Heterofucan from *Sargassum filipendula* induces apoptosis in HeLa cells. *Marine Drugs*, 9(4), 603–614. https://doi.org/10.3390/md9040603

Cunha, L., & Grenha, A. (2016). Sulfated seaweed polysaccharides as multifunctional materials in drug delivery applications. *Marine Drugs*, *14*(3), 42. https://doi.org/10.3390/md14030042

Decker, J. S., Menacho-Melgar, R., & Lynch, M. D. (2020). Low-cost, large-scale production of the antiviral lectin griffithsin. *Frontiers in Bioengineering and Biotechnology*, 8, 1020. https://doi.org/10.3389/fbioe.2020.01020

Demurtas, O. C., Massa, S., Ferrante, P., Venuti, A., Franconi, R., & Giuliano, G. (2013). A *Chlamydomonas*-derived Human Papillomavirus 16 E7 vaccine induces specific tumor protection. *PloS One*, 8(4), e61473. https://doi.org/10.1371/journal.pone.0061473

D'Orazio, N., Gemello, E., Gammone, M. A., de Girolamo, M., Ficoneri, C., & Riccioni, G. (2012). Fucoxantin: A treasure from the sea. *Marine Drugs*, 10(3), 604–616. https://doi.org/10.3390/md10030604

Dreesen, I. A., Charpin-El Hamri, G., & Fussenegger, M. (2010). Heat-stable oral alga-based vaccine protects mice from *Staphylococcus aureus* infection. *Journal of Biotechnology*, 145(3), 273–280. https://doi.org/10.1016/j.jbiotec.2009.12.006

El Shafay, S. M., Ali, S. S., & El-Sheekh, M. M. (2016). Antimicrobial activity of some seaweeds species from Red sea, against multidrug resistant bacteria. *Egyptian Journal of Aquatic Research* 2016, 42, 65–74. [Google Scholar] https://doi.org/10.1016/j.ejar.2015.11.006

El-Gendy, M. M. A. A., Awad, M. F., El-Shenawy, F. S., & El-Bondkly, A. M. A. (2021). Production, purification, characterization, antioxidant and antiproliferative activities of extracellular L-asparaginase produced by Fusarium equiseti AHMF4 - PubMed (nih.gov) *Saudi Journal of Biological Sciences*, 28(4), 2540–2548. https://doi.org/10.1016/j.sjbs.2021.01.058. Epub 2021 Feb 10.

Faggio, C., Pagano, M., Dottore, A., Genovese, G., & Morabito, M. (2016). Evaluation of anticoagulant activity of two algal polysaccharides. *Natural Product Research, 30*(17), 1934–1937. https://doi.org/10.1080/14786419.2015.1086347

Fouladvand, M., Barazesh, A., Farokhzad, F., Malekizadeh, H., & Sartavi, K. (2011). Evaluation of in vitro anti-Leishmanial activity of some brown, green and red algae from the Persian Gulf. *European Review for Medical and Pharmacological Sciences*, 15, 597–600

Geahchan, S., Ehrlich, H., & Rahman, M. A. (2021). The anti-viral applications of marine resources for COVID-19 treatment: An overview. *Marine Drugs*, 19(8), 409. https://doi.org/10.3390/md19080409

Gerber, P., Dutcher, J. D., Adams, E. V., & Sherman, J. H. (1958). Protective effect of seaweed extracts for chicken embryos infected with influenza B or mumps virus. *Proceedings of the Society for Experimental Biology and Medicine*, *99*(3), 590–593. https://doi.org/10.3181/00379727-99-24429

Gerwick, W. H. (2013). Plant sources of drugs and chemicals: *ScienceDirect Encyclopedia of Biodiversity*, 129–139. https://doi.org/10.1016/B978-0-12-809633-8.02306-2

Ghareeb, M. A., Tammam, M. A., El-Demerdash, A., Atanas, G., & Atanasovfghi, A. G. (2020).Insights about clinically approved and preclinically investigated marine natural products. *Current Research in Biotechnology* 2, 88–102. https://doi.org/10.1016/j.crbiot.2020.09.001

Gombotz, W. R., & Wee, S. F. (1998). Protein release from alginate matrices: Science Direct. *Advanced Drug Delivery Reviews*, 31(3), 267–285. https://doi.org/10.1016/S0169-409X(97)00124-5

Gregory, J. A., Topol, A. B., Doerner, D. Z., & Mayfield, S. (2013). Alga-produced cholera toxin-Pfs25 fusion proteins as oral vaccines. *Applied and Environmental Microbiology*, 79(13), 3917–3925. https://doi.org/10.1128/AEM.00714-13

Han, Y. S., Lee, J. H., & Lee, S. H. (2015). Antitumor effects of fucoidan on human colon cancer cells via activation of Akt signaling. *Biomolecules & Therapeutics*, 23(3), 225–232. https://doi.org/10.4062/biomolther.2014.136

Hemmingson, J. A., Falshaw, R., Furneaux, R. H., & Thompson, K. (2006). Structure and antiviral activity of the galactofucan sulfates extracted from *Undaria pinnatifida (Phaeophyta), Journal of Applied Phycology* 2006, 18, 185. https://doi.org/10.1007/s10811-006-9096-9

Hempel, F., Lau, J., Klingl, A., & Maier, U. G. (2011). Algae as protein factories: Expression of a human antibody and the respective antigen in the diatom *Phaeodactylum tricornutum. PloS One*, 6(12), e28424. https://doi.org/10.1371/journal.pone.0028424

Hidari, K. I., Takahashi, N., Arihara, M., Nagaoka, M., Morita, K., & Suzuki, T. (2008). Structure and anti-dengue virus activity of sulfated polysaccharide from a marine alga. *Biochemical and Biophysical Research Communications*, 376(1), 91–95. https://doi.org/10.1016/j.bbrc.2008.08.100

Holanda, M. L., Melo, V. M., Silva, L. M., Amorim, R. C., Pereira, M. G., & Benevides, N. M. (2005). Differential activity of a lectin from *Solieria filiformis* against human pathogenic bacteria. *Brazilian Journal of Medical and Biological Research: Revista brasileira de pesquisas medicas e biologicas*, *38*(12), 1769–1773. https://doi.org/10.1590/s0100-879x2005001200005

Kim, K. N., Heo, S. J., Kang, S. M., Ahn, G., & Jeon, Y. J. (2010). Fucoxanthin induces apoptosis in human leukemia HL-60 cells through a ROS-mediated Bcl-xL pathway. *Toxicology in Vitro: An International Journal Published in Association with BIBRA*, *24*(6), 1648–1654. https://doi.org/10.1016/j.tiv.2010.05.023

Kim, M., Yim, J. H., Kim, S. Y., … & Lee, C. K. (2012). In vitro inhibition of influenza A virus infection by marine microalga-derived sulfated polysaccharide p-KG03. *Antiviral Research*, 93(2), 253–259. https://doi.org/10.1016/j.antiviral.2011.12.006

Kubanek, J., Jensen, P. R., Keifer, P. A., Sullards, M. C., Collins, D. O., & Fenical, W. (2003). Seaweed resistance to microbial attack: A targeted chemical defense against marine fungi. *Proceedings of the National Academy of Sciences* 100(12), 6916–6921. https://doi.org/10.1073/pnas.1131855100

Kuniyoshi, M., Yamada, K., & Higa, T. (1985). A biologically active diphenyl ether from the green alga *Cladophora fascicularis. Experientia* 41, 523–524. https://doi.org/10.1007/BF01966182

Lai, Y. H., Chiang, C. S., Hsu, C. H., Cheng, H. W., & Chen, S. Y. (2020). Development and characterization of a fucoidan-based drug delivery system by using hydrophilic anticancer polysaccharides to simultaneously deliver hydrophobic anticancer drugs. *Biomolecules*, 10(7), 970. https://doi.org/10.3390/biom10070970

Lane, A. L., Stout, E. P., Lin, A. S., … & Kubanek, J. (2009). Antimalarial bromophycolides J-Q from the Fijian red alga *Callophycus serratus. The Journal of Organic Chemistry*, 74(7), 2736–2742. https://doi.org/10.1021/jo900008w

Lategan, C., Kellerman, T., Afolayan, A. F., Mann, M. G., … & Beukes, D. R. (2009) Antiplasmodial and antimicrobial activities of South African marine algal extracts, *Pharmaceutical Biology* 47(5), 408–413. https://doi.org/10.1080/13880200902758832

Leader, B., Baca, Q. J., & Golan, D. E. (2008). Protein therapeutics: A summary and pharmacological classification. *Nature Reviews Drug Discovery*, 7(1), 21–39. https://doi.org/10.1038/nrd2399

Lee, C. (2019). Griffithsin, a highly potent broad-spectrum antiviral lectin from red algae: From discovery to clinical application. *Marine Drugs*, *17*(10), 567. https://doi.org/10.3390/md17100567

Lee, Y. S., Shin, K. H., Kim, B. K., & Lee, S. (2004). Anti-diabetic activities of fucosterol from *Pelvetia siliquosa*. *Archives of Pharmacal Research*, 27(11), 1120–1122. https://doi.org/10.1007/BF02975115

Leibbrandt, A., Meier, C., König-Schuster, M., … & Grassauer, A. (2010). Iota-carrageenan is a potent inhibitor of influenza A virus infection. *PloS One*, 5(12), e14320. https://doi.org/10.1371/journal.pone.0014320

Li, B., Lu, F., Wei, X., & Zhao, R. (2008). Fucoidan: Structure and bioactivity. *Molecules (Basel, Switzerland)*, 13(8), 1671–1695. https://doi.org/10.3390/molecules13081671

Liu, X., Yuan, W. Q., Sharma-Shivappa, R., & van Zanten, J. (2017). Antioxidant activity of phlorotannins from brown algae. *International Journal of Agricultural and Biological Engineering*, 10(6), 184–191.

Lohrmann, N. L., Logan, B. A., & Johnson, A. S. (2004). Seasonal acclimatization of antioxidants and photosynthesis in *Chondrus cris*pus and *Mastocarpus stellatus*, two co-occurring red algae with differing stress tolerances. *The Biological Bulletin*, 207(3), 225–232. https://doi.org/10.2307/1543211

Lynch, A. S., & Robertson, G. T. (2008). Bacterial and fungal biofilm infections. *Annual Review of Medicine*, 59, 415–428. https://doi.org/10.1146/annurev.med.59.110106.132000

Maghembe, R., Damian, D., Makaranga, A., Nyandoro, S. S., Lyantagaye, S. L., Kusari, S., & Hatti-Kaul, R. (2020) Omics for bioprospecting and drug discovery from bacteria and microalgae. *Antibiotics (Basel)*, May 4, 9(5), 229. https://doi.org/10.3390/antibiotics9050229. PMID: 32375367; PMCID: PMC7277505.

Magnan, S., Tota, J. E., El-Zein, M., …&, Franco, E. L., & CATCH Study Group (2019). Efficacy of a Carrageenan gel Against Transmission of Cervical HPV (CATCH): Interim analysis of a randomized, double-blind, placebo-controlled, phase 2B trial. *Clinical Microbiology and Infection: The Official Publication of the European Society of Clinical Microbiology and Infectious Diseases*, 25(2), 210–216. https://doi.org/10.1016/j.cmi.2018.04.012

Maguire, R. A., Zacharopoulos, V. R., & Phillips, D. M. (1998). Carrageenan-based nonoxynol-9 spermicides for prevention of sexually transmitted infections. *Sexually Transmitted Diseases*, 25(9), 494–500.

Manuja, A., Kumar, S., Dilbaghi, N., Bhanjana, G., Chopra, M., Kaur, H., Kumar, R., Manuja, B. K., Singh, S. K., & Yadav, S. C. (2014). Quinapyramine sulfate-loaded sodium alginate nanoparticles show enhanced trypanocidal activity. *Nanomedicine (London, England)*, *9*(11), 1625–1634. https://doi.org/10.2217/nnm.13.148

Matsui, M. S., Muizzuddin, N., Arad, S. et al. (2003). Sulfated polysaccharides from red microalgae have antiinflammatory properties in vitro and in vivo. *Applied Biochemistry and Biotechnology* 104, 13–22. https://doi.org/10.1385/ABAB:104:1:13

Mayfield, S. P., Franklin, S. E., & Lerner, R. A. (2003). Expression and assembly of a fully active antibody in algae. *Proceedings of the National Academy of Sciences of the United States of America*, *100*(2), 438–442.https://doi.org/10.1073/pnas.0237108100

Medeiros, V. P., Queiroz, K. C., Cardoso, M. L., … & Leite, E. L. (2008). Sulfated galactofucan from *Lobophora variegata*: Anticoagulant and anti-inflammatory properties. *Biochemistry Biokhimiia*, 73(9), 1018–1024. https://doi.org/10.1134/s0006297908090095

Meenakshi, S., Gnanambigai, D. M., Mozhi, S. T., Arumugam, M., & Balasubramanian, T. (2009). Total flavonoid and in vitro antioxidant activity of two seaweeds of Rameshwaram coast. *The Global Journal of Pharmacy and Pharmacology*, 3,59–62.

Mickymaray, S., & Alturaiki, W. (2018). Antifungal efficacy of marine macroalgae against fungal isolates from bronchial asthmatic cases. *Molecules (Basel, Switzerland)*, 23(11), 3032. https://doi.org/10.3390/molecules23113032

Morokutti-Kurz, M., Graf, C., & Prieschl-Grassauer, E. (2017). Amylmetacresol/2, 4-dichlorobenzyl alcohol, hexylresorcinol, or carrageenan lozenges as active treatments for sore throat. *International Journal of General Medicine* 2017, 10, 53.

Munir, N., Sharif, N., Naz, S., & Manzoor, F. (2013) Algae: A potent antioxidant source. *Sky Journal of Microbiology Research*, 1(3), 22–31

Nunes, J. C. F., Cristóvão, R. O., Freire, M. G., Santos-Ebinuma, V. C., Faria, J. L., Silva, C. G., & Tavares, A. P. M. (2020). Recent strategies and applications for l-Asparaginase confinement. *Molecules*, 25(24), 5827. https://doi.org/10.3390/molecules25245827

Oh, K. B., Lee, J. H., Chung, S. C., … & Lee, H. S. (2008). Antimicrobial activities of the bromophenols from the red alga Odonthalia corymbifera and some synthetic derivatives. *Bioorganic & Medicinal Chemistry Letters*, 18(1), 104–108. https://doi.org/10.1016/j.bmcl.2007.11.003

Park, H. Y., Han, M. H., Park, C., … & Choi, Y. H. (2011). Anti-inflammatory effects of fucoidan through inhibition of NF-κB, MAPK and Akt activation in lipopolysaccharide-induced BV2 microglia cells. *Food and Chemical Toxicology: An International Journal Published for the British Industrial Biological Research Association*, 49(8), 1745–1752. https://doi.org/10.1016/j.fct.2011.04.020

Queiroz, K. C., Medeiros, V. P., Queiroz, L. S., … & Leite, E. L. (2008). Inhibition of reverse transcriptase activity of HIV by polysaccharides of brown algae. *Biomedicine & Pharmacotherapy: Biomedecine & Pharmacotherapie*, 62(5), 303–307. https://doi.org/10.1016/j.biopha.2008.03.006

Rasala, B. A., & Mayfield, S. P. (2011). The microalga *Chlamydomonas reinha*rdtii as a platform for the production of human protein therapeutics. *Bioengineered Bugs*, 2(1), 50–54. https://doi.org/10.4161/bbug.2.1.13423

Römling, U., & Balsalobre, C. (2012). Biofilm infections, their resilience to therapy and innovative treatment strategies. *Journal of Internal Medicine*, *272*(6), 541–561. https://doi.org/10.1111/joim.12004

Salih, A. E. M., Thissera, B., Yaseen, M., … & Rateb, M. E. (2021). Marine sulfated polysaccharides as promising antiviral agents: A comprehensive report and modeling study focusing on SARS CoV-2. *Marine Drugs*, 19(8), 406. https://doi.org/10.3390/md19080406

Seo, D. J., & Choi, C. (2021). Antiviral bioactive compounds of mushrooms and their antiviral mechanisms: A review. *Viruses*, 13(2), 350. https://doi.org/10.3390/v13020350

Shah Md, M. R., Yuanmei Liang, Y., Cheng, J. J., & Daroch, M. (2016). Astaxanthin-producing green microalga *Haematococcus pluvialis*: From single cell to high value commercial products *Frontiers in Plant Science* 7, 531. https://doi.org/10.3389/fpls.2016.00531

Shan, X., Liu, X., Hao, J., … & Yu, G. (2016). In vitro and in vivo hypoglycemic effects of brown algal fucoidans. *International Journal of Biological Macromolecules*, 82, 249–255. https://doi.org/10.1016/j.ijbiomac.2015.11.036

Shannon, E., & Abu-Ghannam, N. (2016). Antibacterial derivatives of marine algae: An overview of pharmacological mechanisms and applications. *Marine Drugs* 14(4), 81. https://doi.org/10.3390/md14040081

Sharma, G., Kar, S., Basu Ball, W., Ghosh, K., & Das, P. K. (2014). The curative effect of fucoidan on visceral leishmaniasis is mediated by activation of MAP kinases through specific protein kinase C isoforms. *Cellular & Molecular Immunology*, 11(3), 263–274. https://doi.org/10.1038/cmi.2013.68

Singh, R. S., & Walia, A. K. (2018). Lectins from red algae and their biomedical potential. *Journal of Applied Phycology*, 30(3), 1833–1858. https://doi.org/10.1007/s10811-017-1338-5

Song, S., Peng, H., Wang, Q., Liu, Z., Dong, X., Wen, C., Ai, C., Zhang, Y., Wang, Z., & Zhu, B. (2020). Inhibitory activities of marine sulfated polysaccharides against SARS-CoV-2. *Food & Function*, 11, 7415–7420. [Google Scholar] [CrossRef]

Specht, E. A., & Mayfield, S. P. (2014). Algae-based oral recombinant vaccines. *Frontiers in Microbiology*, 17 Feb https://doi.org/10.3389/fmicb.2014.00060

Szekalska, M., Puciłowska, A., Szymańska, E., Ciosek, P., & Winnicka, K. (2016). Alginate: Current use and future perspectives in pharmaceutical and biomedical applications. *International Journal of Polymer Science*, 2016, Article ID 7697031, 17. https://doi.org/10.1155/2016/7697031

Talarico, L. B., Zibetti, R. G., Faria, P. C., Scolaro, L. A., Duarte, M. E., Noseda, M. D., Pujol, C. A., & Damonte, E. B. (2004). Anti-herpes simplex virus activity of sulfated galactans from the red seaweeds *Gymnogongrus griffithsiae* and *Cryptonemia crenulata. International Journal of Biological Macromolecules*, 34(1–2), 63–71. https://doi.org/10.1016/j.ijbiomac.2004.03.002

Taunt, H. N., Stoffels, L., & Purton, S. (2018). Green biologics: The algal chloroplast as a platform for making biopharmaceuticals. *Bioengineered* 9, 48–54. https://doi.org/10.1080/21655979.2017.1377867

Tøndervik, A., Sletta, H., Klinkenberg, G., … & Hill, K. E. (2014). Alginate oligosaccharides inhibit fungal cell growth and potentiate the activity of antifungals against *Candida* and *Aspergillus spp. PloS One*, 9(11), e112518. https://doi.org/10.1371/journal.pone.0112518

Torres, F. A. E., Passalacqua, T. G. …, & Graminha, M. A. S. (2014). New drugs with antiprotozoal activity from marine algae: A review. *ScienceDirect, Revista Brasileira de Farmacognosia*, 24(3), 265–276. https://doi.org/10.1016/j.bjp.2014.07.001

Tran, M., Van, C., Barrera, D. J., Pettersson, P. L., Peinado, C. D., Bui, J., & Mayfield, S. P. (2013). Production of unique immunotoxin cancer therapeutics in algal chloroplasts. *Proceedings of the National Academy of Sciences of the United States of America*, 110(1), E15–E22. https://doi.org/10.1073/pnas.1214638110

Varikuti, S., Jha, B. K., Volpedo, G., … & Satoskar, A. R. (2018). Host-directed drug therapies for neglected tropical diseases caused by protozoan parasites. *Frontiers in Microbiology*, 9, 2655. https://doi.org/10.3389/fmicb.2018.02655

Vieira, C., Gaubert, J., De Clerck, O., …& Thomas, O. P. (2017). Biological activities associated to the chemodiversity of the brown algae belonging to genus *Lobophora (Dictyotales, Phaeophyceae)* (researchgate.net). *Phytochemistry Reviews*. https://doi.org/10.1007/s11101-015-9445-x

Vijayabaskar, P., Vaseela, N., & Thirumaran, G. (2012). Potential antibacterial and antioxidant properties of a sulfated polysaccharide from the brown marine algae *Sargassum swartzii. The Chinese Journal of Natural Medicines* 10, 421–428. [Google Scholar] [CrossRef]

Wang, W., Wu, J., Zhang, X. et al. (2017). Inhibition of influenza a virus infection by fucoidan targeting viral neuraminidase and cellular EGFR pathway. *Scientific Reports* 7, 40760. https://doi.org/10.1038/srep40760

Wei, Y., Liu, Q., Xu, C., Yu, J., Zhao, L., & Guo, Q. (2016) Damage to the membrane permeability and cell death of vibrio parahaemolyticus caused by phlorotannins with low molecular weight from *Sargassum thunbergii, Journal of Aquatic Food Product Technology*., 25, 3, 323–333. https://doi.org/10.1080/10498850.2013.851757

Xin, X., Geng, M., Guan, H., & Li, Z. (2000). Study on the mechanism of inhibitory action of 911 on replication of HIV-1 in vitro. *Chinese Journal of Marine Drugs*, 19, 15–18.

Yang, J. I., Yeh, C. C., Lee, J. C., Yi, S. C., Huang, H. W., Tseng, C. N., & Chang, H. W. (2012). Aqueous extracts of the edible *Gracilaria tenuistipitata* are protective against H_2O_2-induced DNA damage, growth inhibition, and cell cycle arrest. *Molecules (Basel, Switzerland)*, 17(6), 7241–7254. https://doi.org/10.3390/molecules17067241

Zhou, R., Shi, X. Y., Bi, D. C., Fang, W. S., Wei, G. B., & Xu, X. (2015). Alginate-derived oligosaccharide inhibits neuroinflammation and promotes microglial phagocytosis of β-Amyloid. *Marine Drugs*, 13(9), 5828–5846. https://doi.org/10.3390/md13095828

21 Microbial Production and Emerging Applications of Lantibiotics

Indra Mani
Department of Microbiology, Gargi College, University of Delhi, New Delhi, India

Vijai Singh
Indrashil University, Rajpur, India

Sunil Kumar
Department of Biological Sciences, Sungkyunkwan University, Suwon, South Korea

CONTENTS

21.1 INTRODUCTION

Bacteriocins are ribosomally synthesized and are generally referred to as bacterial peptides with antimicrobial properties (Rea et al., 2011). They are mainly produced by numerous bacteria and by certain archaea. The first bacteriocin-producing bacterial strain was reported in 1925 (Gratia 1925; Tagg et al., 1976; Rodriguez-Valera et al., 1982). Klaenhammer (1988) proposed that at least one bacteriocin is produced by 99% of bacteria. Microbial strains which produce bacteriocins are generally protected against them through immunological systems due to the presence of related genes in the bacteriocin gene cluster (Deegan et al., 2006). Generally, the size of bacteriocins peptide is less than 10 kDa. It is in cationic and amphipathic nature (van Belkum et al., 2011; Nissen-Meyer et al., 2009). According to the Alvarez Siciro et al. (2016) classification, bacteriocins are divided into three classes. Class 1 includes peptides that are less than 10 kDa in size and heat-stable, and in which post-translational modification occurs in few amino acid residues. Class II bacteriocins are less than 10 kDa, heat-stable but unmodified peptides, while in Class III, bacteriocins are larger than 10 kDa, thermo-labile and unmodified. Class I bacteriocins are considered lantibiotics, while the other two classes are non-lantibiotics (Cintas et al., 2001).

The term "lantibiotic" was coined by Schnell et al. (1988), and it describes a wider class of lanthipeptides. Its name was devised from thioether ring comprising amino acids lanthionine and/or 3-methyl-lanthionine-containing antibiotics, subsequently identification of the first lantibiotic structural gene (Schnell et al., 1988; Kers et al., 2018). After biosynthesis of lanthipeptides, these

DOI: 10.1201/9781003306931-24

compounds normally include different post-translationally modified amino acids at their C-terminus, such as 2, 3-didehydrobutyrine (Dhb), 2, 3-didehydroalanine (Dha) and the unsaturated lanthionine derivatives aminovinyl-D-cysteine (AviCys) (Smith and Hillman 2008; Repka et al., 2017). On the basis of involvement of the post-translationally modified enzymes during maturation, lanthipeptides can be categorized into four different types. However, only types I (LanBC-modified) and II (LanM-modified) are included in lantibiotics (Knerr and van der Donk 2012), while types III and IV have not shown any antimicrobial activities (Alvarez-Sieiro et al., 2016).

The structures of lantibiotics may be either linear or globular, and various groups of lantibiotics have been identified. On the basis of their structures, they have been categorized as type A and type B (Rea et al., 2011). For different structures of lantibiotics see Figure 21.1 (Dischinger et al., 2014). Lantibiotics have different modes of action. Type A has nisin, Pep5, and epidermin, which cause pore formation in the bacterial cell membrane (Bierbaum and Sahl 2009). However, mersacidin, which comes under type B, prevents the synthesis of a peptidoglycan and forms complexes with membrane-bound substrates (Hsu et al., 2003; Böttiger et al., 2009). Genes are involved in the biosynthesis of lanthipepetide distributed across the entire bacterial genome, providing a vast repository of novel antimicrobial peptides (van Staden et al., 2021). Biotechnological techniques are being used to mine novel antimicrobial lanthipeptides, which may be useful to combat pathogenic multi-drug resistant (MDR) and extensively drug resistant (XDR) strains. This chapter discusses the production of lantibiotics from different microorganisms, including actinobacteria and lactic acid bacteria, and their applications in various fields.

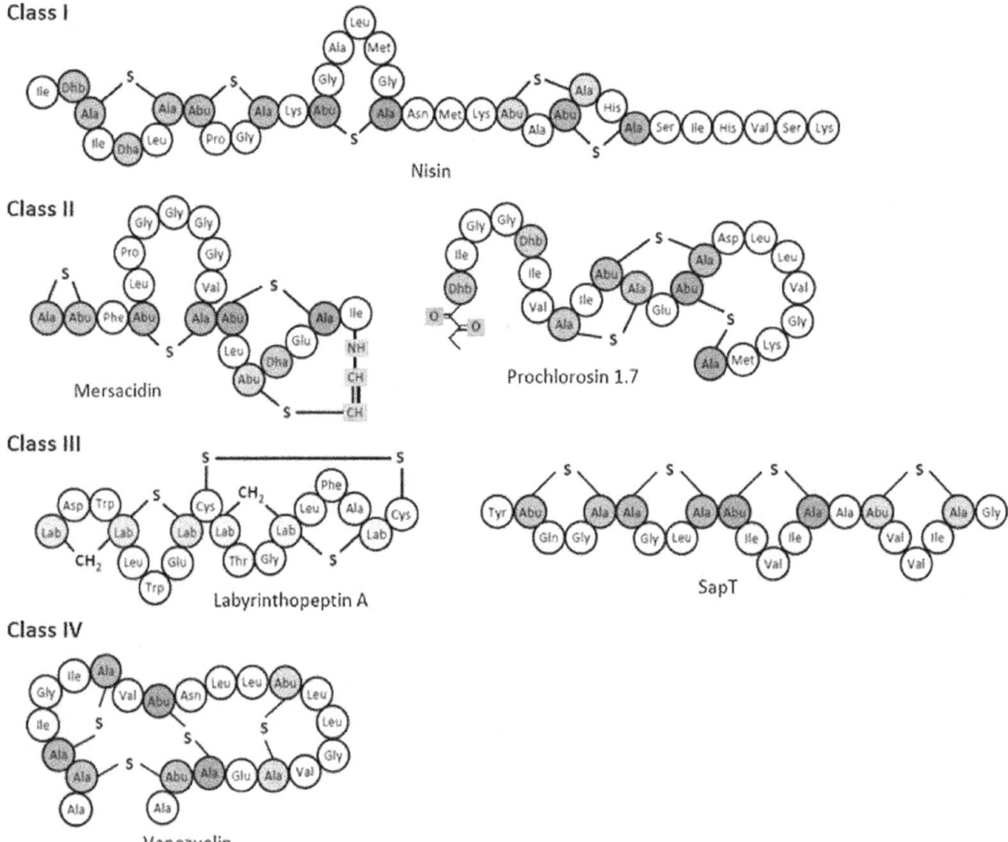

FIGURE 21.1 Structure of different classes of lantibiotics.

(Dischinger et al., 2014. Adapted with permission).

21.2 MICROBIAL PRODUCTION OF LANTIBIOTICS

Generally, lantibiotics are synthesized as an inactive precursor peptide, and their production depends upon the existence of the gene clusters present in one or more operons. Lantibiotics-encoded genes are present on bacterial genomes and plasmids (Heng et al., 2007; Bierbaum and Sahl 2009). Subtilin (Banerjee and Hansen 1988) and SA-FF22 (McLaughlin et al., 1999) gene clusters are located on the genome while in most other cases, lantibiotics-encoded genes are present in plasmids such as lacticin 3147 (Dougherty et al., 1998), epidermin (Schnell et al., 1988), Pep5 (Kaletta et al., 1989), cytolysin (Ike et al., 1990), etc. The biosynthesis of the four different classes of lanthipeptides is illustrated in Figure 21.2. For detailed illustration of genetics, biosynthesis, and regulation of lantibiotics, see excellent reviews by Entian and de Vos (1996), Field et al. (2015) and Repka et al. (2017).

Advances in DNA sequencing technology have shown that microorganisms can produce several more natural products than was earlier assumed. A list of selected lantibiotics produced by various microorganisms is given in Table 21.1. LeBel et al. (2014) identified suicin 90-1330 from *Streptococcus suis* serotype 2 (nonvirulent strain): a nisin-related lantibiotic that was effective against Gram-positive *Streptococcus suis* serotype 2 swine pathogen. Interestingly, it was inactive against all Gram-negative bacteria tested. Kers et al. (2018) reported that the lantibiotic mutacin

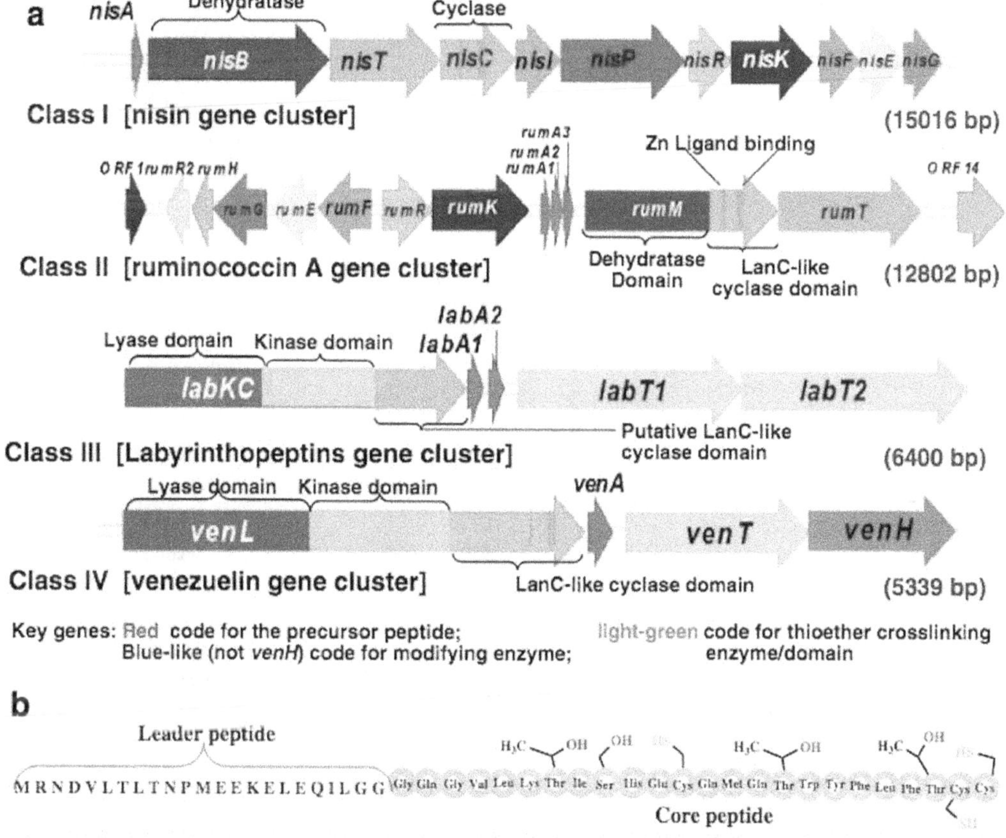

FIGURE 21.2 Schematic representation of biosynthesis of different classes of lanthipeptides. (a) Biosynthetic gene clusters showing conserved motif between the classes. Well characterized domain designated as Class I, II, III, and IV. (b) Leader and core peptide along with side chains of amino acids of precursor peptide of ruminococcin A (LanA).

(Ongey and Neubauer 2016. Adapted with permission).

TABLE 21.1
List of Selected Lantibiotics and Their Producer

Lantibiotic	Producer	References
Type A Lantibiotics		
Type AI		
Nisin group		
Nisin A	*Lactococcus lactis*	Gross and Morell (1971)
Nisin F	*Lactococcus lactis*	de Kwaadsteniet et al. (2008)
Nisin Q	*Lactococcus lactis*	Zendo et al. (2003)
Nisin U	*Streptococcus uberis*	Wirawan et al. (2006)
Nisin Z	*Lactococcus lactis*	Mulders et al. (1991)
Ericin A	*Bacillus subtilis*	Stein et al. (2002)
Ericin S	*Bacillus subtilis*	Stein et al. (2002)
Subtilin	*Bacillus subtilis*	Gross et al. (1973)
Microbisporicin	*Microbiospora sp.*	Castiglione et al. (2008)
Epidermin group		
Epidermin	*Staphylococcus epidermidis*	Allgaier et al. (1986)
Streptin	*Streptococcus pyogenes*	Wescombe and Tagg (2003)
Gallidermin	*Staphylococcus gallinarum*	Kellner et al. (1988)
Staphylococcin T	*Staphylococcus cohnii*	Furmanek et al. (1999)
Pep 5 group		
Pep5	*Staphylococcus epidermidis*	Kaletta et al. (1989)
Epicidin 280	*Staphylococcus epidermidis*	Heidrich et al. (1998)
Epilancin K7	*Staphylococcus epidermidis*	van de Kamp et al. (1995)
Epilancin 15X	*Staphylococcus epidermidis*	Ekkelenkamp et al. (2005)
Type AII		
Lacticin 481 group		
Lacticin 481	*Lactococcus lactis*	Piard et al. (1993)
Salivaricin A	*Streptococcus salivarius*	Ross et al. (1993)
Mutacin II	*Streptococcus mutans*	Novák et al. (1994)
Variacin	*Micrococcus variants*	Pridmore et al. (1996)
Plantaricin C	*Lactobacillus plantarum*	Turner et al. (1999)
Type B lantibiotics		
Mersacidin group		
Mersacidin	*Bacillus* ssp.	Chatterjee et al. (1992)
Michiganin A	*Clavibacter michiganensis*	Holtsmark et al. (2006)
Actagardine	*Actinoplanes liguriae*	Zimmermann et al. (1995)
Cinnamycin group		
Cinnamycin	*Streptomyces cinnamoneus*	Fredenhagen et al. (1990)
Ancovenin	*Streptomyces* sp.	Kido et al. (1983)
Duramycin	*Streptomyces cinnamoneus*	Fredenhagen et al. (1990)
Others		
Two peptide lantibiotics		
Haloduracin	*Bacillus halodurans*	McClerren et al. (2006)
Lacticin 3147	*Lactococcus lactis* sp.	Martin et al. (2004)
Plantaricin W	*Lactobacillus plantarum*	Holo et al. (2001)
Cytolysin LL/LS	*Enterococcus faecalis*	Booth et al. (1996)

1140 (MU1140), which is produced by *Streptococcus mutans*, limits the development of drug resistance. MU1140 has been shown to be effective in treatment of *Clostridium difficile*-associated diarrhea (CDAD) and could prevent recurrence of the infection. Recently Garcia-Gutierrez et al. (2020) characterized a new nisin variant as nisin P, which also has antimicrobial properties. However, it has weaker antimicrobial properties than nisin A and H.

21.3 LANTIBIOTICS PRODUCED BY ACTINOBACTERIA

A typical description of actinobacteria is Gram-positive filamentous bacteria, which have high GC content. They are major phyla of the domain bacteria and their features include physiological, metabolic, and morphological characteristics (Gao et al., 2012; Barka et al., 2015). Actinobacteria produce bacteriocins classes I, II, and III (Gomes et al., 2017), as well as all three types or subclasses of lantibiotics. For example, type I: Planosporicin produced by *Planomonospora alba* (Castiglione et al., 2007) and Microbisporicin A1 and A2 by *Microbispora corallina* (Castiglione et al., 2008). Type II: Cinnamycin produced by *Clavibacter michiganensis subsp. michiganensis* (Holtsmark et al., 2006), Deoxyactagardine B by *Actinoplanes liguriae* (Boakes et al., 2010), Variacin by *Kocuria varians* (Pridmore et al., 1996), Ancovenin by *Streptomyces cinnamoneus* (Pridham et al., 1956). Type III: NAI-112 produced by *Actinoplanes* species (Iorio et al., 2014) and Labyrinthopeptins A1, A2 and A3 by *Actinomadura namibiensis* (Seibert et al., 2008), etc. In addition, Maffioli et al. (2015) isolated a number of class I-related lantibiotics from various genera of actinomycetes. Various analytical techniques were used to determine the structures of the lantibiotic compounds. Analysis has shown that a new compound shares 24 amino acids with an existing lantibiotic, planosporicin (aka 97518), divergent at locations 4, 6, and 14. Isolated lantibiotics compounds have significant antimicrobial properties, including NAI-107. Moreover, NAI-107/microbisporcin is considered a promising lantibiotic because it has strong antimicrobial activity against emerging infections caused by Gram-positive pathogens (Castiglione et al., 2008). Outcomes suggest that isolated compounds can be used to treat drug-resistant microbial pathogens.

21.4 LANTIBIOTICS PRODUCED BY LACTIC ACID BACTERIA

Lactic acid bacteria (LAB) are used to increase the shelf-life of food and are generally recognized as safe (GRAS) (Deak 2014). They are a group of heterogeneous Gram-positive bacteria belonging to the Firmicutes type and having different genera (Carr et al., 2002). Several lantibiotics were produced by various LAB strains, such as Nisin A by *Lactococcus lactis* NIZOR5 (Gross and Morell 1971), Nisin Z by *Lactobacillus lactis* N8 (Mulders et al., 1991), Lactocin S by *Lactobacillus sake* L45 (Mortvedt et al., 1991), Lactcin 481 by *Lactococcus lactis* ADRIA85LO30 (van den Hooven et al., 1996), Lacticin 3147 by *Lactococcus lactis* DPC3147 (Ryan et al., 1996), Cytolysin by *Enterococcus faecalis* FA22 (Booth et al., 1996), Plantaricin W by *Lactobacillus plantarum* LMG 2379 (Holo et al., 2001), etc. Recent study has shown that *Lactococcus lactis* subsp. lactis bv. diacetylactis strain, SBR4 which produced Nisin Z has significant bio-preservative potential (Fusieger et al., 2020). This study suggests that SBR4 strain can be used as an adjunct or starter culture in the dairy industry.

21.5 APPLICATIONS OF LANTIBIOTICS

Uses of lantibiotics are emerging in a variety of areas, including medicine, the food industry, and biotechnological research. Lantibiotics exhibit antibacterial properties against Gram-positive bacteria. Remarkably, they are also effective against drug-resistance bacteria, such as vancomycin-resistant *Enterococcus* (VRE), methicillin-resistant *Staphylococcus aureus* (MRSA), *Clostridium difficile*, vancomycin intermediate *S. aureus* (VISA), *Streptococcus pneumoniae, Propionibacterium acnes*, and different mycobacteria (Cotter et al., 2013). Vancomycin-resistant *Enterococcus* (VRE) is the major reason for hospital-acquired infections (Lebreton et al., 2017). Nisin has been used therapeutically for treatment of MRSA and enterococcal infections, and bacterial mastitis, and it also has anticancer properties (Mota-Meira et al., 2000; Brumfitt et al., 2002; Cotter et al., 2005; Kamarajan et al., 2015). Various other lantibiotics have been used for the treatment of MRSA, such as microbisporcin, actagardine, planosporcin, mersacidin, and mutacin B-Ny266 (Castiglione et al., 2008; Hoffmann et al., 2002; Castiglione et al., 2007; Mota-Meira et al., 2000). Duramycin lantibiotic has been reported as a potential therapeutic agent to cure cystic fibrosis and ocular diseases (Grasemann et al., 2007; Oliynyk et al., 2010). It also has several other properties such as immunomodulatory, anticancer, antiviral, and ion channel regulation (Märki et al., 1991; Zebedin 2008; Tabata et al., 2016; Broughton et al., 2016). Recently, Kim et al. (2019) have studied the role of microbiota-derived lantibiotics against VRE. They used faecal transplantation therapy (FTT) and transplanted patient-derived feces (resistance to VRE colonization) into germ-free mice. In this consortium of commensal strains, *Blautia producta* BP_{SCSK} produced a lantibiotic, which has similar potential to nisin A. More interestingly, lantibiotic-producing commensal strains of the gut reduced VRE colonization. Findings suggest that lantibiotics producing microorganisms can be used as potential probiotics.

Apart from antimicrobial activity against multi-drug-resistant clinical isolates, lantibiotics have many significant characteristics, such as stability against heat, proteases, and oxidation, which make it an interesting candidate for innovative antimicrobial applications in the clinical sector, other food industries, and in research development (Dischinger et al., 2014; Juturuand and Wu 2018). Nowadays people demand food products that are healthy, safe, tasty, less processed, and have an extended shelf-life. Nisin has been extensively used as a food preservative in the US and Europe. It prevents microbial spoilage and inhibits the growth of food-borne pathogens, such as clostridia spores. It is added to packaged foods such as fermented vegetables, processed cheese, salad dressings, etc. (Delves-Broughton et al. 1996). Lacticin 3147 (Ross et al., 2002), lacticin 481 (Dufour et al., 2007), and variacin (O'Mahony et al., 2001) also have exhibited good food preservative qualities. Nisin and other lantibiotics are used with rapid chilling (Cao-Hoang et al., 2008), high pressure (Black et al., 2008), high-intensity pulsed electric fields (Sobrino-Lopez et al., 2006), heat and high pressure (Gao et al., 2008), and have been evaluated in combination with lysozyme (Mangalassary et al., 2007).

More recently, Yonezawa et al. (2021) have demonstrated that lantibiotics produced by oral microbiota trouble the gut microbiota dysbiosis. To understand this relationship, they took saliva and feacal samples of 69 mice and determined the antibacterial activity of Mutacin I/III and Smb produced by *Streptococcus mutans*. Both lantibiotics have significant antimicrobial activities against the Gram-positive bacteria that constitute oral and gut microbiota. Another recent study reported the use of lactic acid bacteria to control the *Listeria* (L.) *monocytogenes* in ready-to-eat (RTE) sliced emulsion-type sausages. It not only helps to enhance the shelf-life of food but is also beneficial to consumer safety (Bungenstock et al., 2021). Readers are referred to a review by van Staden et al. (2021) for an extensive summary of various lantibiotics and lanthipeptides. There is a need to explore microbes producing different varieties of antibacterial molecules with the help of "multi-omics" approaches such as metagenomics, metatranscriptomics, metaproteomics, metabolomics, fluxomics, etc.

21.6 CONCLUSIONS AND FUTURE PERSPECTIVES

Lantibiotic synthesis occurs through the ribosomes as an inactive peptide precursor, and its encoded genes are located on bacterial genome or plasmids. Recombinant DNA technology (RDT), systems and synthetic biology can be used to explore lantibiotic biosynthetic genes to develop different varieties of chemical compounds with great potential. Moreover, metagenomics (culture-independent) and in-silico approaches can be used to isolate and identify novel gene clusters of lantibiotics. Lantibiotics can be used as therapeutic peptides to treat multi-drug-resistant (MDR) and extensively drug-resistant (XDR) microorganisms. In addition, enzymes involved in the biosynthesis of lantibiotics may be used for other applications outside therapeutic purposes. Further extensive study is required to see the effects of lantibiotics on human health.

REFERENCES

Allgaier H, Jung G, Werner RG, Schneider U, Zähner H (1986) Epidermin: Sequencing of a heterodetic tetracyclic 21-peptide amide antibiotic. *Eur J Biochem*. 160(1): 9–22.

Alvarez-Sieiro P, Montalbán-López M, Mu D, Kuipers OP (2016) Bacteriocins of lactic acid bacteria: Extending the family. *Appl Microbiol Biotechnol*. 100(7): 2939–2951.

Banerjee S, Hansen JN (1988) Structure and expression of a gene encoding the precursor of subtilin, a small protein antibiotic. *J Biol Chem*. 263(19): 9508–9514.

Barka EA, Vatsa P, Sanchez L, Gaveau-Vaillant N, Jacquard C, Meier-Kolthoff JP, Klenk HP, Clément C, Ouhdouch Y, van Wezel GP (2015) Taxonomy, Physiology, and Natural Products of Actinobacteria. *Microbiol Mol Biol Rev*. 80(1): 1–43.

Bierbaum G and Sahl HG (2009) Lantibiotics: Mode of action, biosynthesis and bioengineering. *Curr Pharm Biotechnol*. 10(1): 2–18.

Black EP, Linton M, McCall RD, Curran W, Fitzgerald GF, Kelly AL, Patterson MF (2008) The combined effects of high pressure and nisin on germination and inactivation of Bacillus spores in milk. *J Appl Microbiol*. 105(1): 78–87.

Boakes S, Appleyard AN, Cortés J, Dawson MJ (2010) Organization of the biosynthetic genes encoding deoxyactagardine B (DAB), a new lantibiotic produced by Actinoplanes liguriae NCIMB41362. *J Antibiot (Tokyo)*. 63(7): 351–358.

Booth MC, Bogie CP, Sahl HG, Siezen RJ, Hatter KL, Gilmore MS (1996) Structural analysis and proteolytic activation of Enterococcus faecalis cytolysin, a novel lantibiotic. *Mol Microbiol*. 21(6): 1175–1184.

Böttiger T, Schneider T, Martínez B, Sahl HG, Wiedemann I (2009) Influence of Ca(2+) ions on the activity of lantibiotics containing a mersacidin-like lipid II binding motif. *Appl Environ Microbiol*. 75(13): 4427–4434.

Broughton LJ, Crow C, Maraveyas A, Madden LA (2016) Duramycin-induced calcium release in cancer cells. *Anticancer Drugs*. 27(3): 173–182.

Brumfitt W, Salton MR, Hamilton-Miller JM (2002) Nisin, alone and combined with peptidoglycan-modulating antibiotics: activity against methicillin-resistant Staphylococcus aureus and vancomycin-resistant enterococci. *J Antimicrob Chemother*. 50(5): 731–734.

Bungenstock L, Abdulmawjood A, Reich F (2021) Suitability of lactic acid bacteria and deriving antibacterial preparations to enhance shelf-life and consumer safety of emulsion type sausages. *Food Microbiol*. 94: 103673.

Cao-Hoang L, Marechal PA, Le-Thanh M, Gervais P (2008) Synergistic action of rapid chilling and nisin on the inactivation of Escherichia coli. *Appl Microbiol Biotechnol*. 79(1): 105–109.

Carr FJ, Chill D, Maida N (2002) The lactic acid bacteria: A literature survey. *Crit Rev Microbiol*. 28(4): 281–370.

Castiglione F, Cavaletti L, Losi D, Lazzarini A, Carrano L, Feroggio M, Ciciliato I, Corti E, Candiani G, Marinelli F, Selva E (2007) A novel lantibiotic acting on bacterial cell wall synthesis produced by the uncommon actinomycete Planomonospora sp. *Biochemistry*. 46(20): 5884–5895.

Castiglione F, Lazzarini A, Carrano L, Corti E, Ciciliato I, Gastaldo L, Candiani P, Losi D, Marinelli F, Selva E, Parenti F (2008) Determining the structure and mode of action of microbisporicin, a potent lantibiotic active against multiresistant pathogens. *Chem Biol*. 15(1): 22–31.

Chatterjee S, Chatterjee S, Lad SJ, Phansalkar MS, Rupp RH, Ganguli BN, Fehlhaber HW, Kogler H (1992) Mersacidin, a new antibiotic from Bacillus. Fermentation, isolation, purification and chemical characterization. *J Antibiot (Tokyo)*. 45(6): 832–838.

Cintas LM, Casaus MP, Herranz C, Nes IF, Hernández PE (2001) Review: Bacteriocins of lactic acid bacteria. *Food Sci. Technol. Int*. 7(4): 281–305.

Cotter PD, Hill C, Ross RP (2005) Bacterial lantibiotics: Strategies to improve therapeutic potential. *Curr Protein Pept Sci*. 6(1): 61–75.

Cotter PD, Ross RP, Hill C (2013) Bacteriocins: A viable alternative to antibiotics? *Nat Rev Microbiol*. 11(2): 95–105.

de Kwaadsteniet M, Ten Doeschate K, Dicks LM (2008) Characterization of the structural gene encoding nisin F, a new lantibiotic produced by a Lactococcus lactis subsp. lactis isolate from freshwater catfish (Clarias gariepinus). *Appl Environ Microbiol.* 74(2): 547–549.

Deak T (2014 Thermal treatment. In *Food Safety Management* (Motarjemi, Y., ed.), pp. 423–442, Elsevier Science Publishers, New York.

Deegan LH, Cotter PD, Hill C, Ross, P (2006) Bacteriocins: Biological tools for bio-preservation and shelf-life extension. *Int. Dairy J.* 16(9): 1058–1071.

Delves-Broughton J, Blackburn P, Evans RJ, Hugenholtz J (1996) Applications of the bacteriocin, nisin. *Antonie Van Leeuwenhoek.* Feb; 69(2): 193–202.

Dischinger J, Basi Chipalu S, Bierbaum G (2014) Lantibiotics: Promising candidates for future applications in health care. *Int J Med Microbiol.* 304(1): 51–62.

Dougherty BA, Hill C, Weidman JF, Richardson DR, Venter JC, Ross RP (1998). Sequence and analysis of the 60 kb conjugative, bacteriocin-producing plasmid pMRC01 from Lactococcus lactis DPC3147. *Mol Microbiol.* 29(4): 1029–1038.

Dufour A, Hindré T, Haras D, Le Pennec JP (2007) The biology of lantibiotics from the lacticin 481 group is coming of age. *FEMS Microbiol Rev.* 31(2): 134–167.

Ekkelenkamp MB, Hanssen M, Danny Hsu ST, de Jong A, Milatovic D, Verhoef J, van Nuland NA (2005) Isolation and structural characterization of epilancin 15X, a novel lantibiotic from a clinical strain of Staphylococcus epidermidis. *FEBS Lett.* 579(9): 1917–1922.

Entian KD, de Vos WM (1996) Genetics of subtilin and nisin biosyntheses: biosynthesis of lantibiotics. *Antonie Van Leeuwenhoek.* 69(2): 109–117.

Field D, Cotter PD, Hill C, Ross RP (2015). Bioengineering lantibiotics for therapeutic success. *Front Microbiol.* 6: 1363.

Fredenhagen A, Fendrich G, Märki F, Märki W, Gruner J, Raschdorf F, Peter HH (1990) Duramycins B and C, two new lanthionine containing antibiotics as inhibitors of phospholipase A2. Structural revision of duramycin and cinnamycin. *J Antibiot (Tokyo).* 43(11): 1403–1412.

Furmanek B, Kaczorowski T, Bugalski R, Bielawski K, Bohdanowicz J, Podhajska AJ (1999) Identification, characterization and purification of the lantibiotic staphylococcin T, a natural gallidermin variant. *J Appl Microbiol.* 87(6): 856–866.

Fusieger A, Perin LM, Teixeira CG, de Carvalho AF, Nero LA (2020) The ability of Lactococcus lactis subsp. lactis bv. diacetylactis strains in producing nisin. *Antonie Van Leeuwenhoek.* 113(5): 651–662.

Gao B, Gupta RS (2012) Phylogenetic framework and molecular signatures for the main clades of the phylum Actinobacteria. *Microbiol Mol Biol Rev.* 76(1): 66–112.

Gao YL, Ju XR (2008) Exploiting the combined effects of high pressure and moderate heat with nisin on inactivation of Clostridium botulinum spores. *J Microbiol Methods.* 72(1): 20–28.

Garcia-Gutierrez E, O'Connor PM, Saalbach G, Walsh CJ, Hegarty JW, Guinane CM, Mayer MJ, Narbad A, Cotter PD (2020) First evidence of production of the lantibiotic nisin P. *Sci Rep.* 10(1): 3738.

Gomes KM, Duarte RS, de Freire Bastos MDC (2017) Lantibiotics produced by Actinobacteria and their potential applications (a review). *Microbiology (Reading).* 163(2): 109–121.

Grasemann H, Stehling F, Brunar H, Widmann R, Laliberte TW, Molina L, Döring G, Ratjen F (2007) Inhalation of Moli1901 in patients with cystic fibrosis. *Chest.* 131(5): 1461–1466.

Gratia A (1925) Sur un remarquable example d'antagonisme entre deux souches de colibacille. *Compt. Rend. Soc. Biol.* 93: 1040–1042.

Gross E, Kiltz HH, Nebelin E (1973) Subtilin, VI. Die Struktur des Subtilins [Subtilin, VI: the structure of subtilin (author's transl)]. *Hoppe Seylers Z Physiol Chem.* Jul; 354(7): 810–812.

Gross E, Morell JL (1971) The structure of nisin. *J Am Chem Soc.* 93(18): 4634–4635.

Heidrich C, Pag U, Josten M, Metzger J, Jack RW, Bierbaum G, Jung G, Sahl HG (1998) Isolation, characterization, and heterologous expression of the novel lantibiotic epicidin 280 and analysis of its biosynthetic gene cluster. *Appl Environ Microbiol.* 64(9): 3140–3146.

Heng NCK, Wescombe PA, Burton JP, Jack RW, Tagg JR (2007) The diversity of bacteriocins in gram-positive bacteria. In: Riley M.A., Chavan M.A. (eds) *Bacteriocins.* Springer, Berlin, Heidelberg. https://doi.org/10.1007/978-3-540-36604-1_4

Hoffmann A, Pag U, Wiedemann I, Sahl HG (2002) Combination of antibiotic mechanisms in lantibiotics. *Farmaco.* 57(8): 685–691.

Holo H, Jeknic Z, Daeschel M, Stevanovic S, Nes IF (2001) Plantaricin W from Lactobacillus plantarum belongs to a new family of two-peptide lantibiotics. *Microbiology (Reading).* 147(Pt 3): 643–651.

Holtsmark I, Mantzilas D, Eijsink VG, Brurberg MB (2006) Purification, characterization, and gene sequence of michiganin A, an actagardine-like lantibiotic produced by the tomato pathogen Clavibacter michiganensis subsp. michiganensis. *Appl Environ Microbiol.* 72(9): 5814–5821.

Hsu ST, Breukink E, Bierbaum G, Sahl HG, de Kruijff B, Kaptein R, van Nuland NA, Bonvin AM (2003) NMR study of mersacidin and lipid II interaction in dodecylphosphocholine micelles. Conformational changes are a key to antimicrobial activity. *J Biol Chem.* 278(15): 13110–13117.

Ike Y, Clewell DB, Segarra RA, Gilmore MS (1990). Genetic analysis of the pAD1 hemolysin/bacteriocin determinant in Enterococcus faecalis: Tn917 insertional mutagenesis and cloning. *J Bacteriol.* 172(1): 155–163.

Iorio M, Sasso O, Maffioli SI, Bertorelli R, Monciardini P, Sosio M, Bonezzi F, Summa M, Brunati C, Bordoni R, Corti G, Tarozzo G, Piomelli D, Reggiani A, Donadio S (2014) A glycosylated, labionin-containing lanthipeptide with marked antinociceptive activity. *ACS Chem Biol.* 9(2): 398–404.

Juturu V, Wu JC (2018) Microbial production of bacteriocins: Latest research development and applications. *Biotechnol Adv.* 36(8): 2187–2200.

Kaletta C, Entian KD, Kellner R, Jung G, Reis M, Sahl HG (1989) Pep5, a new lantibiotic: Structural gene isolation and prepeptide sequence. *Arch Microbiol.* 152(1): 16–19.

Kamarajan P, Hayami T, Matte B, Liu Y, Danciu T, Ramamoorthy A, Worden F, Kapila S, Kapila Y, Nisin ZP (2015) A bacteriocin and food preservative, inhibits head and neck cancer tumorigenesis and prolongs survival. *PLoS One.* 10(7): e0131008.

Kellner R, Jung G, Hörner T, Zähner H, Schnell N, Entian KD, Götz F (1988) Gallidermin: A new lanthionine-containing polypeptide antibiotic. *Eur J Biochem.* 177(1): 53–59.

Kers JA, Sharp RE, Defusco AW, Park JH, Xu J, Pulse ME, Weiss WJ, Handfield M (2018) Mutacin 1140 lantibiotic variants are efficacious against *Clostridium difficile* Infection. *Front Microbiol.* 16(9): 415.

Kido Y, Hamakado T, Yoshida T, Anno M, Motoki Y, Wakamiya T, Shiba T (1983) Isolation and characterization of ancovenin, a new inhibitor of angiotensin I converting enzyme, produced by actinomycetes. *J Antibiot (Tokyo).* 36(10): 1295–1299.

Kim SG, Becattini S, Moody TU, Shliaha PV, Littmann ER, et al., (2019) Microbiota-derived lantibiotic restores resistance against vancomycin-resistant *Enterococcus. Nature.* 572(7771): 665–669.

Klaenhammer TR (1988). Bacteriocins of lactic acid bacteria. *Biochimie.* 70(3): 337–349.

Knerr PJ, van der Donk WA (2012) Discovery, biosynthesis, and engineering of lantipeptides. *Annu Rev Biochem.* 81: 479–505.

LeBel G, Vaillancourt K, Frenette M, Gottschalk M, Grenier D (2014) Suicin 90-1330 from a nonvirulent strain of Streptococcus suis: A nisin-related lantibiotic active on gram-positive swine pathogens. *Appl Environ Microbiol.* 80(17): 5484–5492.

Lebreton F, Manson AL, Saavedra JT, Straub TJ, Earl AM, Gilmore MS (2017) Tracing the enterococci from paleozoic origins to the hospital. *Cell.* 169(5): 849–861.e13.

Maffioli SI, Monciardini P, Catacchio B, Mazzetti C, Münch D, Brunati C, Sahl HG, Donadio S (2015) Family of class I lantibiotics from actinomycetes and improvement of their antibacterial activities. *ACS Chem Biol.* 10(4): 1034–1042.

Mangalassary S, Han I, Rieck J, Acton J, Jiang X, Sheldon B, Dawson P (2007) Effect of combining nisin and/or lysozyme with in-package pasteurization on thermal inactivation of *Listeria monocytogenes* in ready-to-eat turkey bologna. *J Food Prot.* 70(11): 2503–2511.

Märki F, Hänni E, Fredenhagen A, van Oostrum J (1991) Mode of action of the lanthionine-containing peptide antibiotics duramycin, duramycin B and C, and cinnamycin as indirect inhibitors of phospholipase A2. *Biochem Pharmacol.* 42(10): 2027–2035.

Martin NI, Sprules T, Carpenter MR, Cotter PD, Hill C, Ross RP, Vederas JC (2004) Structural characterization of lacticin 3147, a two-peptide lantibiotic with synergistic activity. *Biochemistry.* 43(11): 3049–3056.

McClerren AL, Cooper LE, Quan C, Thomas PM, Kelleher NL, van der Donk WA (2006) Discovery and in vitro biosynthesis of haloduracin, a two-component lantibiotic. *Proc Natl Acad Sci U S A.* 103(46): 17243–17248.

McLaughlin RE, Ferretti JJ, Hynes WL (1999) Nucleotide sequence of the streptococcin A-FF22 lantibiotic regulon: model for production of the lantibiotic SA-FF22 by strains of Streptococcus pyogenes. *FEMS Microbiol Lett.* 175(2): 171–177.

Mørtvedt CI, Nissen-Meyer J, Sletten K, Nes IF (1991) Purification and amino acid sequence of lactocin S, a bacteriocin produced by Lactobacillus sake L45. *Appl Environ Microbiol.* 57(6): 1829–1834.

Mota-Meira M, LaPointe G, Lacroix C, Lavoie MC (2000) MICs of mutacin B-Ny266, nisin A, vancomycin, and oxacillin against bacterial pathogens. *Antimicrob Agents Chemother.* 44(1): 24–29.

Mulders JW, Boerrigter IJ, Rollema HS, Siezen RJ, de Vos WM (1991) Identification and characterization of the lantibiotic nisin Z, a natural nisin variant. *Eur J Biochem.* 201(3): 581–584.

Nissen-Meyer J, Rogne P, Oppegård C, Haugen HS, Kristiansen PE (2009) Structure-function relationships of the non-lanthionine-containing peptide (class II) bacteriocins produced by gram-positive bacteria. *Curr Pharm Biotechnol.* 10(1): 19–37.

Novák J, Caufield PW, Miller EJ (1994) Isolation and biochemical characterization of a novel lantibiotic mutacin from Streptococcus mutans. *J Bacteriol.* 176(14): 4316–4320.

Oliynyk I, Varelogianni G, Roomans GM, Johannesson M (2010) Effect of duramycin on chloride transport and intracellular calcium concentration in cystic fibrosis and non-cystic fibrosis epithelia. *APMIS.* 118(12): 982–990.

O'Mahony T, Rekhif N, Cavadini C, Fitzgerald GF (2001) The application of a fermented food ingredient containing 'variacin', a novel antimicrobial produced by Kocuria varians, to control the growth of Bacillus cereus in chilled dairy products. *J Appl Microbiol.* 90(1): 106–114.

Ongey EL, Neubauer P (2016) Lanthipeptides: Chemical synthesis versus in vivo biosynthesis as tools for pharmaceutical production. *Microb Cell Fact.* 15: 97.

Piard JC, Kuipers OP, Rollema HS, Desmazeaud MJ, de Vos WM (1993) Structure, organization, and expression of the lct gene for lacticin 481, a novel lantibiotic produced by Lactococcus lactis. *J Biol Chem.* 268(22): 16361–16368.

Pridham TG, Shotwell OL, Stodola FH, Lindenfelser LA, Benedict RG et al. (1956) Antibiotics against plant disease. 2. Effective agents produced by Streptomyces cinnamomeus forma azacoluta nov. *Phytopathology.* 46: 575–581.

Pridmore D, Rekhif N, Pittet AC, Suri B, Mollet B (1996) Variacin, a new lanthionine-containing bacteriocin produced by Micrococcus varians: Comparison to lacticin 481 of Lactococcus lactis. *Appl Environ Microbiol.* 62(5): 1799–1802.

Rea MC, Ross RP, Cotter PD, Hill C (2011) Classification of bacteriocins from Gram-positive bacteria, in *Prokaryotic Antimicrobial Peptides: From genes to Applications*, D. Drider, S. Rebuffat, Eds (Springer, New York, NY), 29–53.

Repka LM, Chekan JR, Nair SK, van der Donk WA (2017) Mechanistic understanding of lanthipeptide biosynthetic enzymes. *Chem Rev.* 117(8): 5457–5520.

Rodriguez-Valera F, Juez G, Kushner DJ (1982) Halocins: Salt-dependent bacteriocins produced by extremely halophilic rods. *Can. J. Microbiol.* 28: 151–154.

Ross KF, Ronson CW, Tagg JR (1993) Isolation and characterization of the lantibiotic salivaricin A and its structural gene salA from Streptococcus salivarius 20P3. *Appl Environ Microbiol.* 59(7): 2014–2021.

Ross RP, Morgan S, Hill C (2002) Preservation and fermentation: Past, present and future. *Int J Food Microbiol.* 79(1–2): 3–16.

Ryan MP, Rea MC, Hill C, Ross RP (1996) An application in cheddar cheese manufacture for a strain of Lactococcus lactis producing a novel broad-spectrum bacteriocin, lacticin 3147. *Appl Environ Microbiol.* 62(2): 612–619.

Schnell N, Entian KD, Schneider U, Götz F, Zähner H, Kellner R, Jung G (1988) Prepeptide sequence of epidermin, a ribosomally synthesized antibiotic with four sulphide-rings. *Nature.* 333(6170): 6276–6278.

Seibert G, Vértesy L, Wink J, Winkler I, Süßmuth R et al. (2008) Antibacterial and antiviral peptides from Actinomadura namibiensis. Google Patents 2008WO2008/040469

Smith L, Hillman J (2008) Therapeutic potential of type A (I) lantibiotics, a group of cationic peptide antibiotics. *Curr Opin Microbiol.* 11(5): 401–408.

Sobrino-López A, Martín-Belloso O (2006) Enhancing inactivation of Staphylococcus aureus in skim milk by combining high-intensity pulsed electric fields and nisin. *J Food Prot.* 69(2): 345–353.

Stein T, Borchert S, Conrad B, Feesche J, Hofemeister B, Hofemeister J, Entian KD (2002). Two different lantibiotic-like peptides originate from the ericin gene cluster of Bacillus subtilis A1/3. *J Bacteriol.* 184(6): 1703–1711.

Tabata T, Petitt M, Puerta-Guardo H, Michlmayr D, Wang C, Fang-Hoover J, Harris E, Pereira L (2016) Zika virus targets different primary human placental cells, suggesting two routes for vertical transmission. *Cell Host Microbe.* 20(2): 155–166.

Tagg JR, Dajani AS, Wannamaker LW (1976) Bacteriocins of gram-positive bacteria. *Bacteriol Rev.* 40(3): 722–756.

Turner DL, Brennan L, Meyer HE, Lohaus C, Siethoff C, Costa HS, Gonzalez B, Santos H, Suárez JE (1999) Solution structure of plantaricin C, a novel lantibiotic. *Eur J Biochem.* 264(3): 833–839.

van Belkum MJ, Martin-Visscher LA, Vederas JC (2011) Structure and genetics of circular bacteriocins. *Trends Microbiol.* 19(8): 411–418.

van de Kamp M, van den Hooven HW, Konings RN, Bierbaum G, Sahl HG, Kuipers OP, Siezen RJ, de Vos WM, Hilbers CW, van de Ven FJ (1995) Elucidation of the primary structure of the lantibiotic epilancin K7 from Staphylococcus epidermidis K7. Cloning and characterisation of the epilancin-K7-encoding gene and NMR analysis of mature epilancin K7. *Eur J Biochem.* 230(2): 587–600.

van den Hooven HW, Lagerwerf FM, Heerma W, Haverkamp J, Piard JC, Hilbers CW, Siezen RJ, Kuipers OP, Rollema HS (1996). The structure of the lantibiotic lacticin 481 produced by *Lactococcus lactis*: location of the thioether bridges. *FEBS Lett.* 391(3): 317–322.

van Staden ADP, van Zyl WF, Trindade M, Dicks LMT, Smith C (2021) Therapeutic application of lantibiotics and other lanthipeptides: Old and new findings. *Appl Environ Microbiol.* 87(14): e0018621.

Wescombe PA, Tagg JR (2003) Purification and characterization of streptin, a type A1 lantibiotic produced by *Streptococcus pyogenes. Appl Environ Microbiol.* 69(5): 2737–2747.

Wirawan RE, Klesse NA, Jack RW, Tagg JR (2006) Molecular and genetic characterization of a novel nisin variant produced by *Streptococcus uberis. Appl Environ Microbiol.* 72(2): 1148–1156.

Yonezawa H, Motegi M, Oishi A, Hojo F, Higashi S, Nozaki E, Oka K, Takahashi M, Osaki T, Kamiya S (2021) Lantibiotics produced by oral inhabitants as a trigger for dysbiosis of human intestinal microbiota. *Int J Mol Sci.* 22(7): 3343.

Zebedin E, Koenig X, Radenkovic M, Pankevych H, Todt H, Freissmuth M, Hilber K (2008) Effects of duramycin on cardiac voltage-gated ion channels. *Naunyn Schmiedebergs Arch Pharmacol.* 377(1): 87–100.

Zendo T, Fukao M, Ueda K, Higuchi T, Nakayama J, Sonomoto K (2003) Identification of the lantibiotic nisin Q, a new natural nisin variant produced by Lactococcus lactis 61–14 isolated from a river in Japan. *Biosci Biotechnol Biochem.* 67(7): 1616–1619.

Zimmermann N, Metzger JW, Jung G (1995) The tetracyclic lantibiotic actagardine. 1H-NMR and 13C-NMR assignments and revised primary structure. *Eur J Biochem.* 228(3): 786–797.

22 Bacterial Drug Delivery Vehicles for Targeted Treatment of Tumors

Pooja Gulati, Anubhuti Kawatra, Sonika Dhillon and Rakhi Dhankhar
Maharshi Dayanand University, Rohtak, India

CONTENTS

Cancer poses a significant threat to global public health. According to WHO reports, cancers alone accounted for about 10 million deaths in 2020, which is expected to increase to over 17 million deaths by 2030 (https://www.who.int/news-room/fact-sheets/detail/cancer). Conventional treatments – chemotherapy, radiotherapy, and immunotherapy – have aided greatly in treating cancers, but their long-term sequela, resistance, and side-effects outweigh their merits (Yaghoubi et al., 2020), leading researchers worldwide to seek novel anticancer armamentarium with high therapeutic efficiency. Bacteriotherapy has emerged as a plausible option to precisely target tumors in vivo. The clinical history of bacteria to treat cancers dates back to 1813, when Coley's toxin (culture supernatants of *Serratia marcescens* and *Streptococcus pyogenes*) was employed to treat co-morbidities with unresected tumors (Wiemann and Starnes, 1994). The breakthrough with bacteriotherapy was achieved in 1955 by Malmgren and Flanigan (1955). Their investigation elucidated the exclusive localization of anaerobic bacteria into the hypoxic areas of cancerous tissues. Several subsequent studies focused on exploring various aspects of bacteriotherapy to treat tumors effectively.

Bacteria possess several inherent properties that can be harnessed to develop biological drug delivery systems for regressing tumors: self-propulsion/taxis, bactofection, stimulus responsiveness, on-site delivery, and colonization ability in hypoxic tumor niches (Yazawa et al., 2000; Castagliuolo et al., 2005; Pálffy et al., 2005; Storz and Hengge, 2010). Bacteria-derived delivery vehicles like minicells, bacterial spores, bacteriobots, bacteriosomes, and genetically modified biohybrid systems have been thoroughly assessed as oncolytic agents for the past few decades. A number of anaerobic and facultative bacteria and their derivative therapies – *Clostridium novyi,*

DOI: 10.1201/9781003306931-25

Lactobacillus lactis, E. coli, Listeria sp. etc. – are being examined clinically for treating cancers including melanomas, colon cancer, breast cancer, and ovarian cancer (Chen et al., 2021). The advent of nanotechnology has opened up further prospects for treatment of tumors with better pharmacokinetics and pharmacodynamics. These biological carriers also have the potency to evade immune responses and systemic toxicity, as well as to maximize therapeutic concentration in vivo (Kuzajewska et al., 2020).

This chapter provides an up-to-date insight into the advances of bacteria-derived delivery systems to treat human malignancies. It also highlights various aspects including suitable bacterial candidature for anticancer drug design, mechanistic aspects of delivery systems, and challenges associated with bacteriotherapy.

22.1 SUITABILITY OF BACTERIA FOR FABRICATING DRUG DELIVERY VEHICLES

Cancer remains a deadly disease despite all the efforts made to fight it. Although there have been many advances in cancer treatment regimes, the specificity and sensitivity of treatment procedures remains a major concern. Off-target toxicity and side-effects are major issues in current cancer treatments. Scientists have recently been working towards targeted drug delivery, and various synthetic agents like liposomes and nanoparticles have been designed. These synthetic drug delivery vehicles suffer various disadvantages including instability, leakage, and limited action in deep tumor sites (Cao and Liu, 2020). To overcome these hurdles, biological components have been employed for efficient drug delivery, with advantageous properties including biocompatibility, efficient transport across physical and biological barriers, and high concentration at cancerous sites.

Bacteria and bacteria-based drug delivery have an edge over other systems because bacteria can be grown in large amounts using straightforward conditions and can be easily modified genetically (Cao and Liu, 2020). Apart from this, a plethora of bacterial species have remained as commensals inside the human body, hence they have an inherent potential to interact with human cells via their surface appendages. For example, the affinity between lectin of *E. coli* pili and the mannose molecule of epithelial cells makes the bacteria bioadhesive. This property has been harnessed to develop bacteriobots for targeted drug delivery in the urinary and gastrointestinal tracts (Mostaghaci et al., 2017).

Bacteria possess various sensory systems that enable them to respond to such physical gradients as pH, temperature, oxygen, or some other chemical concentration. Hence, they have the property of self-propulsion in response to various stimuli. This property of bacteria taxis is crucial in making them efficient drug delivery vehicles (Hosseinidoust et al., 2016). Some anaerobic bacteria like *Clostridium* can colonize deep sites in tumors as they provide a hypoxic environment which is ideal for their growth. These bacteria or their spores can be used to deliver drugs at regions that other delivery agents fail to access (Lambin et al., 1998).

The other remarkable property of bacteria that makes them a delivery vector of choice is bactofection. Bactofection is on-site gene delivery in human cells by bacteria-mediated transfer of plasmid DNA. This is a robust approach for the targeted delivery of anticancerous agents. The innate properties of various pathogenic bacteria to escape from host defense mechanisms are often harnessed for bactofection. For example, *L. monocytogenes* possesses actin-based motility and expresses protein listeriolysin O (LLO) that enables the bacteria to get away from endosomes and phagosomes, a property that makes it an ideal gene delivery vector in cancer cells (Tangney et al., 2010).

Thus, the various properties of bacteria, including low production cost, easy manipulation, self-propulsion and taxis towards environmental stimulus, interaction with mammalian cells and bactofection ability, make them candidates of choice for efficient drug delivery in tumor cells (Figure 22.1).

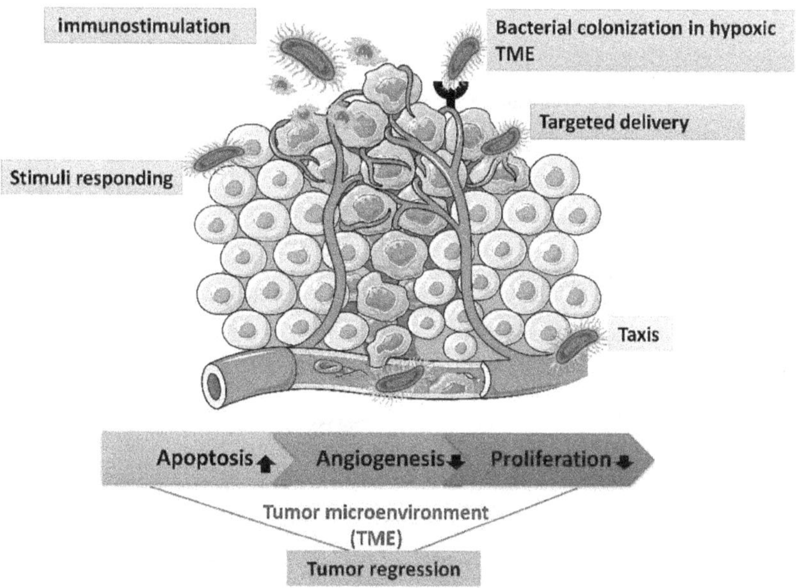

FIGURE 22.1 Multiple inherent tumor-targeting properties of bacteria – selective colonization, responsiveness to environment, motility, immune modulation ability, specific delivery via its components (vectors) – that are important for fabricating novel anticancer medicines.

22.2 BACTERIAL DRUG DELIVERY SYSTEMS FOR CANCER THERAPY

Bacterial drug delivery systems possess many features of the parental oncolytic bacterial cells, and are therefore expected to significantly contribute to the suppression of cancers with a high biocompatibility and therapeutic index. To date, several studies have been conducted to elucidate the potency of these vectors, i.e., bacteriosomes, spores, minicells, genetically modified bacteria, bacteriobots, and nano-carriers for the treatment of oncology (Table 22.1). The most promising applications of these bacterial drug delivery systems in cancer treatment are shown in Figure 22.2 and described below.

22.2.1 Bacteriosomes

Bacteriosomes, also called bacterial ghosts or bacterial vesicles, are non-infective cellular envelopes produced by the manipulation of a bacterium via gene-mediated lytic processes or chemical methods. The most common form of bacteriosomes are bacterial ghosts (BGs). Their genetic fabrication involves expression of gene E from fX174 phage which causes cell lysis in bacteria like *E. coli* K12, *Mannheimia haemolytica, S. typhimurium, Nisseria meningitidis, Helicobacter pylori, Klebsiella pneumoniae, Vibrio cholerae*, etc., leading to the formation of a bacterial envelope lacking cytoplasm and genetic content (Kudela et al., 2010; Harisa et al., 2020). In the chemical-mediated generation, agents like sodium dodecyl sulfate, hydrogen peroxide, sodium hydroxide, calcium carbonate, etc. are employed to make the bacterial cell wall porous. The cytoplasmic content is evacuated, forming acellular bacterial ghosts (Harisa et al., 2020). These bacterial ghosts can be further processed to produce bacterial vesicles by extruding them from a 100 nm membrane filter. These bacterial derivatives, however, possess instinct tissues/cells and subcellular tropism, and thus retain all the structural, bioadhesive, and immunogenic attributes of the proteins present on their surface. Bacteriosomes have attained immense importance as advanced drug delivery vehicles in

TABLE 22.1
Prospective Bacterial Drug Delivery Systems in Cancer Therapy

Parent Bacterium	Delivery Vehicle	Anticancer Agent	Therapeutic Effect	References
M. haemolytica	Bacterial ghost	Doxorubicin	Targeted Caco-2 cells effectively	Hosseinidoust et al. (2016)
Salmonella enterica	Bacterial ghost	Doxorubicin	Enhanced cellular and humoral immune response in hepatocellular carcinoma	Harisa et al. (2020)
E. coli	Bacterial ghost	Resveratrol	Decreased production of nitric oxide and promoted cell death	Harisa et al. (2020)
	Minicells	Doxorubicin	Invaded the necrotic/hypoxic regions of orthotopic breast tumors	Ali et al. (2020)
	Minicells	PFO toxin	Induced pore formation in the lipid bilayer of tumor cells	Ali et al. (2020)
M. gryphiswaldense	Magnetosomes	Anti-4-1BB antibody	Reduced tumor volume and infiltrated tumor tissue	Kuzajewska et al. (2020)
		Cytosine arabinoside	Induce apoptosis by modulating the signaling pathways associated with the activity of nuclear factor κB	Sreenivasan et al. (2003)
S. typhimurium	Genetically modified strain	FasL	Induced the apoptosis of Fas-expressing cells after binding to its receptor Fas	Chen et al. (2021)
E. coli	Genetically modified strain	Antibody against HER2/neu	Promoted bacterial infiltration in tumor tissue	Chang et al. (2011)
C. acetobutylicum	Genetically modified strain	Interleukin-2	Induced anticancer immune response and overcame immunosuppressi-on	Barbé et al. (2005)
C. novyi -NT	Genetically modified strain	Antibody against human HIF-1	Enhanced bacterial infiltration into tumors and targeted hypoxia region	Groot et al. (2007)

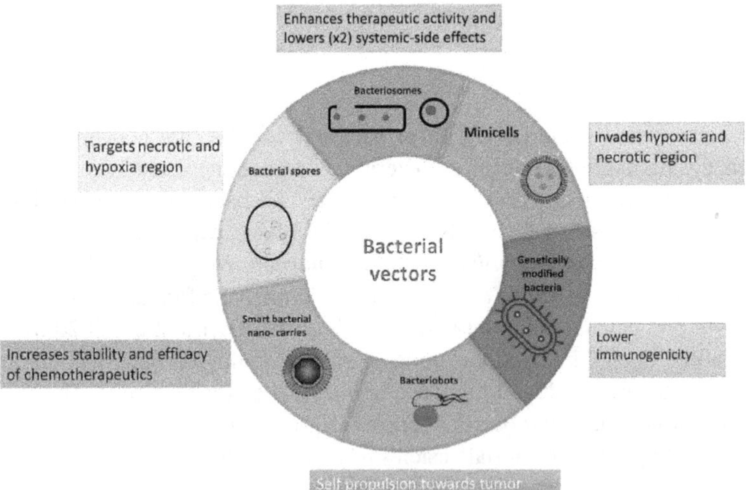

FIGURE 22.2 Various bacterial vectors and their distinctive pharmacological features for targeted cancer therapy.

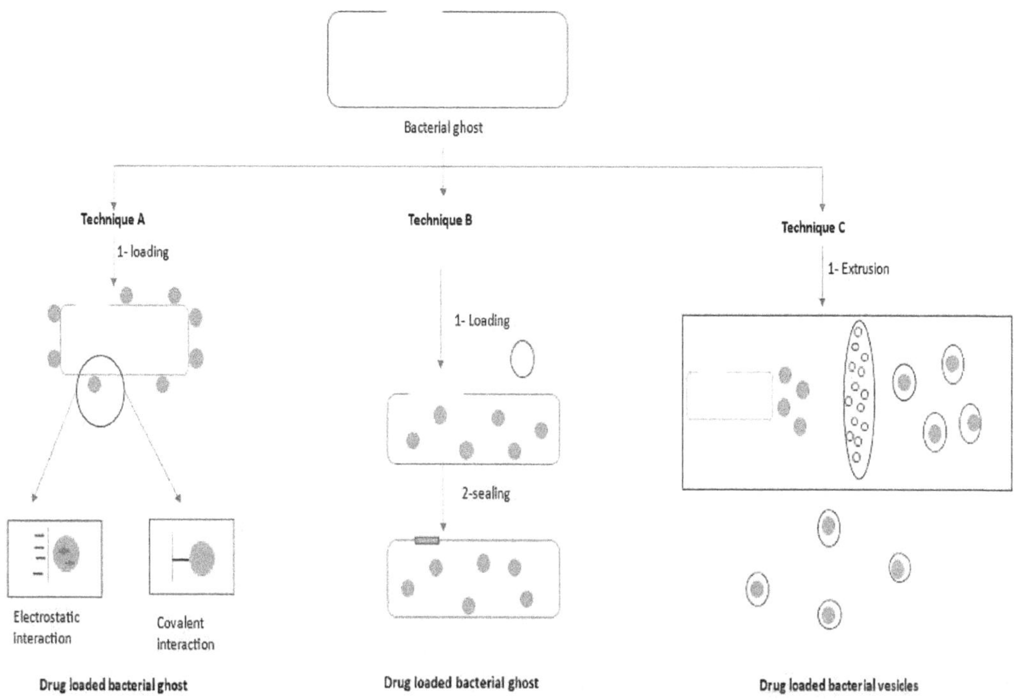

FIGURE 22.3 Methods for loading anticancer drugs into bacteriosomes,

recent times due to their ease of production, higher specificity, and versatility in entrapping various therapeutic payloads in their empty lumen (Table 22.1).

The techniques employed to load drugs into bacteriosomes are represented in Figure 22.3. BGs derived from *M. haemolytica* have been used for the delivery of doxorubicin (DOX), a cytotoxic drug used to treat human colorectal adenocarcinoma (Caco-2) (Paukner et al., 2004). Cytotoxicity analysis showed that DOX-BG formulation suppressed the proliferative activity of cancer cells two-fold compared to the free drug. In similar attempts, DOX-BG formulation developed using *S. enterica* has been evaluated against hepatocellular carcinoma (Rabea et al., 2020). The results demonstrated the enhanced apoptotic activity and immune stimulation of bacterial ghost-encapsulated DOX over free DOX after administration in animal models. A recent study has further employed *E. coli*-derived BG for targeted delivery of 5-fluorouracil against colorectal carcinoma (Abdullah et al., 2019). The results showed that BG released approximately 69.2% of the drug sustainably with promising cytotoxicity in vitro. Moreover, considering the distinctive structure of the BG with pathogen-associated molecular patterns (PAMPs), it could be effectively employed as an adjuvant, or as a transporting vehicle to stimulate cross-presentation by antigen-presenting cells on their surface including tumor-associated antigen (TAA), resulting in immune stimulation towards TAAs (Warrier et al., 2019).

22.2.2 Minicells

Minicells are small, enucleated, metabolically active, non-dividing bacterial cells produced by abnormal cellular division. The Min system associated with normal cell division of bacteria consists mainly of three genes: *min*C, *min*D, and *min*E. Mutations in the genes *min*C and *minD* frequently septates the cells towards the cell pole resulting in the formation of a small acellular spherical minicell (Ali et al., 2020). Thus, minicells have the same peptidoglycan, RNA, protein, ribosomes, and plasmids as their parent bacterial cell, but they lack chromosomal DNA. Also, they retain the ability to maintain most biological functions such as ATP synthesis, mRNA translation, transcription, and plasmid DNA

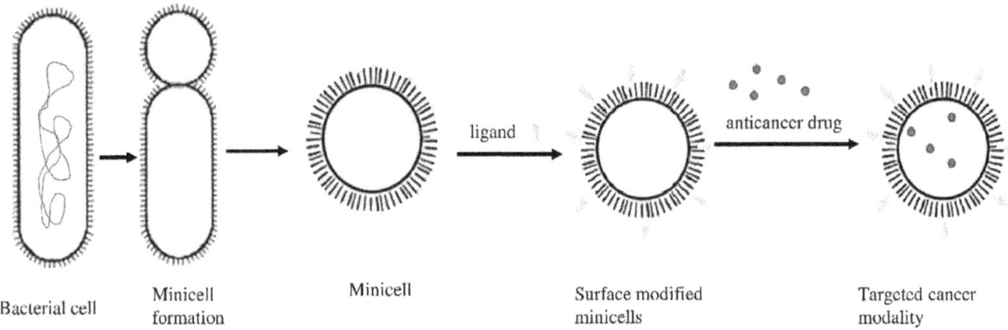

FIGURE 22.4 Schematic illustration of tumor-responsive minicells for targeted cancer therapy.

translation. These bacterial vectors have been widely used in cancer therapy to deliver drugs with low toxicity (Table 22.1). In minicells, the packing of drugs depends on the concentration of the therapeutic payload in the loading solution and the time of incubation of the drug with intact minicells. The drug penetrates the minicells through porin channels that are present in the outer membrane of the minicells. The non-specific transport of hydrophobic solutes across the outer membrane may be linked to the FadL family outer membrane proteins and the OmpW protein. The targeted efficacy of minicells has also been demonstrated with a study on dogs suffering from non-Hodgkin's lymphoma (MacDiarmid et al., 2007). Dox-loaded minicells were shown to effectively target the canine CD3 on T-cells. Further, several studies have explored the surface modification of the minicells to achieve drug delivery at a specific site. In surface modification via a bispecific antibody process, one arm of the bispecific antibodies (BsAbs) is bound with the O-polysaccharide component of the minicell's surface, while the other arm interacts with the cellular surface receptor, such as epidermal growth factor receptor (EGFR) or human epidermal growth factor receptor 2 (HER2), that is found on breast and ovarian cancer cells (El-Rayes and LoRusso, 2004). Once the therapeutic reaches the microenvironment of tumors, the BsAb-tagged minicells use the tumor receptor-specific antibody to bind to the receptors on the cellular surface of the cancer, actively targeting the cancer cells. These minicells become endocytosed and degraded in the late lysosome/endosome before being released into the nuclei of cancer cells with the payload (sh/siRNA, medicines, or toxin). Minicells conjugated with folic acid have further displayed great ability to deliver shRNAs into cancer cells, successfully downregulating the VEGFA mRNA (vascular endothelial growth factor A mRNA) both in cell lines and tumor xenografts (MacDiarmid et al., 2007). The specificity and efficacy of minicells could be further improved by decorating them with nanobodies (Nbs), binding selectively to the receptor expressing on the surface of cancerous cells. The employment of modified minicells as drug delivery vehicles to target cancers is illustrated in Figure 22.4. The minicells-Nbs conjugates have also shown the ability to pack various chemotherapeutics with a wide range of structures and properties like solubility and hydrophobicity (MacDiarmid et al., 2007; 2009).

Peptide-modified minicells have also shown high potential to deliver chemotherapeutic formulations right into the center of the cancerous cells. For example, *E. coli* Nissle 1917 (EcN), pH low insertion peptides (pHLIP)-linked minicells have been used to invade the hypoxic and necrotic regions where drug release is not possible due to the higher pressure of interstitial fluid and inadequate vasculature (Stritzker et al., 2007).

22.2.3 Bacterial Spores

Spores are dormant (metabolically inactive) and highly resistant structures formed by some bacteria to circumvent adverse environmental conditions. These rigid structures regenerate into vegetative cells once favorable conditions are provided (Basta and Annamaraju, 2021). Certain medically significant bacteria like *C. perfringens*, *C. botulinum*, *C. tetani*, *Bacillus anthracis* and *B. cereus* are known to produce endospores. Bacterial spores have often been used as vectors for gene therapy and as delivery vehicles for various anticancer agents and cytotoxic peptides as they can effectively germinate, multiply, and replicate in these regions, resulting in the targeted delivery of carriage anticancer agents (Patyar et al., 2010).

Various studies have demonstrated the effective use of bacterial spores in anticancer drug delivery. For example, in combination bacteriolytic therapy (COBALT), the spores of *C. novyi*-NT strain (*C. novyi* devoid of its lethal toxin) were administered with chemotherapeutic drugs like dolastatin-10, docetaxel, mitomycin C and vinorelbine (Dang et al., 2001). This combination resulted in a significant and prolonged anticancer effect. This enhanced anti-tumor action is attributed to the ability of anaerobic bacteria *C. novyi* to grow in the vascular regions of tumors that are inaccessible to chemotherapeutic drugs. In another study, *Bacillus* spores were used as carriers for anticancer agents (like curcumin) in the treatment of colon cancer (Yin et al., 2018). The spores of various Bacilli like *B. subtilis, B. pumilus* and *B. coagulans* are commonly used as probiotics. These can be covalently linked with anticancer drugs and act as oral drug delivery agents. The spores effectively deliver the drug in the colon area as they can tolerate the harsh conditions and pass the gastric barrier. The outer coat of the spore (where the drug is attached) disintegrates in the colon after germination of the spore.

Due to their high tumor colonization ability, spores of *C. sporogenes* are used as delivery vectors to express *E. coli* cytosine deaminase (CD) gene in solid tumors (Shende and Basarkar, 2019). CD is a prodrug-activating enzyme that carries out systemic conversion of nontoxic prodrug 5-fluorocytosine (5-FC) to the anticancer drug 5-fluorouracil (5-FU) (S. Liu et al., 2002).

22.2.4 Genetically Modified Bacteria

In addition to bacterial-derived vectors, attenuated therapeutic bacteria have attracted considerable attention in cancer therapy as they cause stronger activation of the immune system. These bacteria are genetically modified to synthesize and secrete anti-tumor agents and modify metabolic pathways of tumor cells. Furthermore, bacteria such as *Salmonella, Clostridia*, and *Bifidobacteria* show inherent oncolytic properties and are hence considered ideal tumor-targeting vectors for tumor drug delivery (Table 22.1) (Nallar et al., 2017). The details of clinical investigations of these biohybrid systems against cancers are discussed in Table 22.2. Genetic modification of these tumor-colonizing bacteria gives them the opportunity to express anti-tumor enzymes like phenylalanine ammonia lyase, arginine deiminase, methioninase, cytotoxic proteins, tumor-specific antigens, and reporter genes, along with the targeted delivery of these therapeutic agents into tumor cells (Kawatra et al., 2020; Dhankhar et al., 2020).

S. typhimurium has been the most intensively researched and evaluated anti-tumor agent in cancer treatment. Attenuated *Salmonella* has been used to deliver anti-tumor agents such as cytokines, cytotoxic proteins, regulatory factors, prodrug-converting enzymes, and small interfering RNAs (siRNAs). When *Salmonella* colonizes tumor tissue, anti-tumor agents can either be directly expressed by a prokaryotic expression system or are expressed after the corresponding encoding DNA is transferred into host cells such as cancer cells and macrophages.

Tumor-targeting *Salmonella* strain ΔppGpp has been genetically modified to produce cytolysin A, a protein that inhibits tumor growth in cancer models. ClyA shows cytotoxic effects by forming pores in the cell membrane when cultured in mammalian cells, which induce cell apoptosis. ΔppGpp *Salmonella* expressing ClyA has also been used in the treatment of tumors in mice, exhibiting a significant effect on tumor shrinkage and, in some cases, tumor elimination (Hee Sam et al., 2006). Additional studies have shown that attenuated *Salmonella* expressing cytokines such as IL-2 can reduce angiogenesis, increase necrosis within tumor tissues, and reduce metastatic osteosarcoma (Al-Ramadi et al., 2009). Attenuated *Salmonella*, engineered with the cytokine IL-18, increased the cytolytic activity of T-cells and NK cells and inhibited the proliferation of endothelial

TABLE 22.2
Clinical Studies Involving Biohybrid Bacterial Systems Against Cancers

Bacterial Strain	Strain	Target Tumor	Study	Phase	Status	NCT-Study no
Salmonella typhimurium	*VNP20009*	Metastatic melanoma, carcinomas, osteosarcoma, hemangiosarcoma, lymphoma or mast cell tumor	Formulation induced a dose-related increase in circulation of cytokines such as IL-18 and IL-12	1	Completed	NCT00006254
	VNP20009 with HSV-TK	B16F 10 melanomas	Suppressed tumor growth and prolonged survival	1	Completed	NCT00004988
Clostridium novyi	NT	Hemangiosarcoma, osteosarcoma, nasal adenocarcinoma, fibrosarcoma	Reduced tumor size	1	Completed	NCT01924689
Listeria monocytogenes	ADXS11-001	Cervical cancer	Promising safety and efficacy results	1/2	Completed	NCT02164461
	ADXS31-142	Prostate cancer	Suppressed tumor growth and metastasis	1/2	Active	NCT02325557
E. coli	Nissle 191	Gastric cancer, lung adenocarcinoma	Delayed tumor growth	3	Completed	NCT02706184

cells by suppressing angiogenesis (Loeffler et al., 2008). In similar attempts, bio-engineered attenuated *Salmonella* expressing LIGHT (a TNF family cytokine) has been used to treat mice bearing syngeneic tumors.

Further, some cargo molecules like sprouty protein (SPRY1/2) are being designed to regulate tumor cell growth or inhibit angiogenesis in tumor tissues. SPRY delivered via VNP20009 strain to tumor cells has been shown to inhibit the growth of melanomas with high efficacy in vivo (Z. Liu et al., 2015). These biohybrid systems have also been employed to deliver gene-silencing RNA to inhibit the proliferation of tumors. In one such study, Zhang and co-workers (2007) employed *S. typhimurium* to deliver gene-specific shRNAs, which yields siRNAs to suppress the growth of pancreatic cancer and melanoma in mouse models. The results indicated that these siRNAs effectively triggered the degradation of the Stat3, the target RNA that is overexpressed in many kinds of cancer cells, regulating apoptosis and cellular proliferation in vivo (Zhang et al., 2007).

22.2.5 Bacteriobots

The innovative bacteriobot concept represents a new paradigm in bacteria-based hybrid drug delivery systems that combines the science of nanorobotics with bacterial therapy (Yaghoubi et al., 2020). This technology harnesses the ability of mobile bacteria to move towards a particular environmental stimulus (taxis). As discussed earlier, bacteria carry out a variety of movements like swimming, swarming, twitching, gliding, and sliding with the help of pili and flagella (Martel, 2012). These self-propulsion mechanisms make bacteria ideal for targeted drug delivery.

Typically, bacteriobots are composed of two components: the living component, i.e., the motile bacteria, and the non-living component of anti-tumor drug-containing cargo (Hosseinidoust et al., 2016). Commonly used cargo materials are liposome, PEG, alginate, cellulose, etc. Although this high-concept technology is still in its infancy, a few reports on development of these tumor-targeting bacteriobots are available. Park and co-workers developed a bacteriobot by using attenuated *S.*

typhimurium cells and combining them with Cy5.5 (near-infrared fluorescence-emitting dye)-coated polystyrene microbeads (Park et al., 2013). In the in vitro and in vivo mouse model studies, Cy5.5 signal was detected from the tumor site. The bacteriobot exhibited both chemotactic and tumor-targeting ability. In another study, non-pathogenic *E. coli* in association with naturally synthesized bio-nano-sensory systems was used to develop a bacteriobot whose movement is directed towards vascular endothelial growth factor (VEGF) (Al-Fandi et al., 2017). VEGF is overexpressed in tumor cells and it acts as chemoattractant for the *E. coli*-based bacteriobot.

One of the major advantages of bacteriobot systems is that they do not require any genetically modified bacteria and we can employ any non-pathogenic motile commensal bacteria like *E. coli* for development of the biohybrid system. However, there is a serious bottleneck in the choice and attachment of cargo material because bacterial cells often tend to lose their movement ability when they upon contact with a material (Hosseinidoust et al., 2016). Work continues to enhance efficiency of future bacteriobots.

22.2.6 Smart Bacterial Nano-Carriers

In recent times, smart bacterial nano-carriers such as bacterial-derived nanovesicles and bacterial magnetosomes have been used extensively in cancer therapy as they have good tumor penetration ability, are easy to modify and have improved drug-loading capacity. Bacterial-derived nanovesicles are made of a double lipid layer which is 20–400 nm in size. These nanovesicles are classified into different groups on the basis of their structure and sources: double-layered membrane vesicles (DMVs), cytoplasmic membrane vesicles (CMVs), outer membrane vesicles (OMVs), and outer-inner membrane vesicles (OIMVs) (Wang et al., 2018). Nanovesicles from genetically modified bacteria such as *E. coli* have been employed to deliver chemotherapeutic agents like doxorubicin (MacDiarmid et al., 2007). The researchers further modified these nanovesicles with an EGFR to target breast cancer. It was found that approximately 30% of the total EGFR-targeted nanovesicles reached the target site, approximately 20 times more than non-modified nanovesicles. In comparison to liposomal doxorubicin, a 100-fold higher dose of doxorubicin is required to show an effect equal to liposomal doxorubicin in the form of EGFR-targeted nanovesicles. The study also proved that bacteria-derived nanoplatforms have the ability to deliver siRNAs for the treatment of tumors that show drug resistance. When bacteria-derived nanovesicles were combined with the dual sequential strategy, it resulted in 100% survival up to 110 days after xenografting of MES-SA/Dx5 human uterine cells.

Bacterial magnetosomes (BMs), on the other hand, are membranous structures that are found in magnetotactic bacteria. Because of their unique characteristics such as nanoscale, high surface-to-volume ratio, uniform shape and size, low toxicity and biocompatibility, and iron crystals coated with lipoprotein membrane, they have attracted a lot of interest as alternatives to targeting drug carriers. They are used as carriers of anti-tumor substances and can also be combined with ligands that recognize molecular targets specific to tumor cells.

Bacterial magnetosomes are used for the delivery of antibiotics, anti-tumor agents, and vaccine DNA, and are also used as drug co-delivery systems. Studies showed that BM-related DOX retained its cytotoxicity against liver cancer HepG2 cells as well as breast cancer MCF-7 cells. The magnetosomes can also be modified by conjugating the drug with them for target drug delivery. For example, the conjugation of DOX molecules with BMs is one of the most promising forms of modification (Sun et al., 2009). Studies show that DOX uses bifunctional cross-linking compounds such as glutaraldehyde, disuccinimidyl suberate (DSS), and disuccinimidyl carbonate to bind to the BM membrane. DOX and BM conjugates have good dispersion in water and maintain uniform shape and magnetic properties (Liang et al., 2016). A study also showed that these conjugates are stable in a physiological pH buffer and release large amounts of drug in a pH 3.5 buffer, and this property plays an important part in terms of effective drug transport to a tumor (Guo et al., 2008). Therefore, it is expected that DOX combined with BMs will not release in the lumen of blood vessels. Furthermore, given the tumor's pH gradient, the drug is expected to be released easily from the complex specifically in the tumor's acidic microenvironment (Guo et al., 2008, Chen et al., 2021).

22.3 CHALLENGES ASSOCIATED WITH BACTERIOTHERAPY IN ONCOLOGY

The employment of bacterial vectors has widened the prospects for the biological treatment of cancer. However, like other therapies, bacterial therapy poses some challenges. The dose-limiting toxicity of bacterial delivery vehicles represents one of the major constraints for successful clinical translation of this therapy (Pálffy et al., 2005). The bacterial virulence in the immunosuppressive environment of compromised individuals may further result in potent side-effects including auto-immune reactions. Although gene manipulation approaches have attenuated toxicity to a great extent, they sometimes lower the therapeutic efficacy/invasiveness of the bacterial drug delivery system as well as disturbing the delicate microbiota of the host (Hosseinidoust et al., 2016). Another major challenge is the prophylactic effect of bacterial cancer therapy. Strains like *Mycoplasma hyorhinis, Shigella, Serratia, Klebsiella*, etc. expressing cytidine deaminase have been shown to enhance resistance to the drug gemcitabine, converting the active chemotherapeutic to its inactive form (Zhao et al., 2007).

The third major challenge is the regulation and containment of bacterial-based therapeutics (Zhou et al., 2016) to restrict post-treatment infections/possible harmful genomic mutations, considering the effects of different mutants of SARS-CoV-2, thereby limiting the chances of clinical approval of bacteriotherapy. Bacterial spores/live bacteria, unlike small molecules, cannot be sterilized effectively either by filtration or by heating. Thus, performing the production and purification of anticancer therapeutic agents according to quality protocols and adhering to the tenets of design might not be feasible. Short serum half-life, limited tropism for antigen-presenting cells (APCs), lack of predictive animal models, and good practices in the manufacturing of bacterial drug delivery vehicles are all critical to the success of bacterial cancer treatment.

22.4 FUTURE PROSPECTS AND CONCLUSION

Scientific advances in oncology over the past few decades have provided new avenues for the development of targeted precision medicine. Bacterial therapy is one such promising cancer treatment modality. With their inherent anti-tumor effects and colonization ability in the hypoxic microenvironment of tumors, oncolytic bacteria and their derivative therapies – bacteriosomes, minicells, spores, bacteriobots – have shown potential to specifically deliver therapeutic payloads, regressing both multi-drug-resistant and refractory tumors. The surface modification of these bacterial vectors via smart nano-carriers has also enhanced the pharmacokinetics and biocompatibility of anticancer agents in vivo. The vast gene-packaging ability of these systems has further allowed the expression of several anticancer agents that are amenable to genetic manipulation, enabling the development of biohybrid systems with enhanced efficacy/safety in the battle against cancer. The commonly used *E. coli*, the genetic domestication of *Salmonella sp.* and gut microbiota, i.e., *Lactobacillus, Bifidobacterium*, etc., can all be suitable vehicles for cancer treatment.

In view of the immense connectivity between host behavior and the physiology of commensal bacteria, a systemic investigation of these biochemical interactions might enable 'personalized' anticancer armamentarium to be designed. However, despite the great potential of bacterial therapy, a successful cancer treatment modality is still likely to depend on combined approaches, due to the multifactorial nature and heterogeneity of tumors at both histological and biochemical levels. Whilst bacterial therapies adapt well in hypoxic niches of tumors, cytotoxic and immunological therapies, on the other hand, are suited well for perfused areas of tumors. A synergy of these therapies with smart nano-formulations following proper research, therefore, appears optimal to target the Achilles heel of both refractory and resistant tumors.

ACKNOWLEDGMENT

AK and RD are grateful to Maharshi Dayanand University, Haryana, India for providing fellowships. This work was financially supported by the Department of Science and Technology, Science and Engineering Research Board, New Delhi, India (Grant number: SB/YS/LS-145/2014).

REFERENCES

Abdullah, M. E. Y., Alanazi, F. K., Salem-Bekhit, M. M., Shakeel, F., & Haq, N. (2019). Bacterial ghosts carrying 5-Fluorouracil: A novel biological carrier for targeting colorectal cancer. *AAPS PharmSciTech 2019 20:2*, *20*(2), 1–12. https://doi.org/10.1208/S12249-018-1249-Z

Al-Fandi, M., Alshraiedeh, N., Oweis, R., Alshdaifat, H., Al-Mahaseneh, O., Al-Tall, K., & Alawneh, R. (2017). Novel selective detection method of tumor angiogenesis factors using living nano-robots. *Sensors (Basel, Switzerland)*, *17*(7). https://doi.org/10.3390/S17071580

Ali, M. K., Liu, Q., Liang, K., Li, P., & Kong, Q. (2020). Bacteria-derived minicells for cancer therapy. In *Cancer Letters* (Vol. *491*, pp. 11–21). Elsevier Ireland Ltd. https://doi.org/10.1016/j.canlet.2020.07.024

Al-Ramadi, B. K., Fernandez-Cabezudo, M. J., El Hasasna, H., Al-Salam, S., Bashir, G., & Chouaib, S. (2009). Potent anti-tumor activity of systemically-administered IL2-expressing *Salmonella* correlates with decreased angiogenesis and enhanced tumor apoptosis. *Clinical Immunology (Orlando, Fla.)*, *130*(1), 89–97. https://doi.org/10.1016/J.CLIM.2008.08.021

Barbé, S., Van Mellaert, L., Theys, J., Geukens, N., Lammertyn, E., Lambin, P., & Anné, J. (2005). Secretory production of biologically active rat interleukin-2 by *Clostridium acetobutylicum* DSM792 as a tool for anti-tumor treatment. *FEMS Microbiology Letters*, *246*(1), 67–73. https://doi.org/10.1016/J.FEMSLE.2005.03.037

Basta, M., & Annamaraju, P. (2021). Bacterial spores. *Biological Reviews*, *7*(1), 1–23. https://www.ncbi.nlm.nih.gov/books/NBK556071/

Cao, Z. & Liu, J. (2020). Bacteria and bacterial derivatives as drug carriers for cancer therapy. *Journal of Controlled Release: Official Journal of the Controlled Release Society*, *326*, 396–407. https://doi.org/10.1016/J.JCONREL.2020.07.009

Castagliuolo, I., Beggiao, E., Brun, P., Barzon, L., Goussard, S., Manganelli, R., Grillot-Courvalin, C., & Palù, G. (2005). Engineered *E. coli* delivers therapeutic genes to the colonic mucosa. *Gene Therapy 2005 12:13*, *12*(13), 1070–1078. https://doi.org/10.1038/sj.gt.3302493

Chang, C. H., Cheng, W. J., Chen, S. Y., Kao, M. C., Chiang, C. J., & Chao, Y. P. (2011). Engineering of *Escherichia coli* for targeted delivery of transgenes to HER2/neu-positive tumor cells. *Biotechnology and Bioengineering*, *108*(7), 1662–1672. https://doi.org/10.1002/BIT.23095

Chen, Y., Liu, X., Guo, Y., Wang, J., Zhang, D., Mei, Y., Shi, J., Tan, W., & Zheng, J. H. (2021). Genetically engineered oncolytic bacteria as drug delivery systems for targeted cancer theranostics. *Acta Biomaterialia*, *124*, 72–87. https://doi.org/10.1016/J.ACTBIO.2021.02.006

Dang, L. H., Bettegowda, C., Huso, D. L., Kinzler, K. W., & Vogelstein, B. (2001). Combination bacteriolytic therapy for the treatment of experimental tumors. *Proceedings of the National Academy of Sciences of the United States of America*, *98*(26), 15155. https://doi.org/10.1073/PNAS.251543698

Dhankhar, R., Gupta, V., Kumar, S., Kapoor, R. K., & Gulati, P. (2020). Microbial enzymes for deprivation of amino acid metabolism in malignant cells: Biological strategy for cancer treatment. *Applied Microbiology and Biotechnology*, *104*(7), 2857–2869.

El-Rayes, B., & LoRusso, P. (2004). Targeting the epidermal growth factor receptor. *British Journal of Cancer*, *91*(3), 418–424. https://doi.org/10.1038/SJ.BJC.6601921

Groot, A. J., Mengesha, A., van der Wall, E., van Diest, P. J., Theys, J., & Vooijs, M. (2007). Functional antibodies produced by oncolytic clostridia. *Biochemical and Biophysical Research Communications*, *364*(4), 985–989. https://doi.org/10.1016/J.BBRC.2007.10.126

Guo, L., Huang, J., Zhang, X., Li, Y., & Zheng, L. (2008). Bacterial magnetic nanoparticles as drug carriers. *Journal of Materials Chemistry*, *18*(48), 5993–5997. https://doi.org/10.1039/b808556k

Harisa, G. I., Sherif, A. Y., Youssof, A. M. E., Alanazi, F. K., & Salem-Bekhit, M. M. (2020). Bacteriosomes as a promising tool in biomedical applications: Immunotherapy and drug delivery. *AAPS PharmSciTech*, *21*(5). https://doi.org/10.1208/S12249-020-01716-X

Hee Sam, N., Hyun Ju, K., Hyun Chul, L., Yeongjin, H., Joon Haeng, R., & Hyon E. C. (2006). Immune response induced by *Salmonella typhimurium* defective in ppGpp synthesis. *Vaccine*, *24*(12), 2027–2034. https://doi.org/10.1016/J.VACCINE.2005.11.031

Hosseinidoust, Z., Mostaghaci, B., Yasa, O., Park, B-W., Singh, A. V. & Sitti, M. (2016). Bioengineered and biohybrid bacteria-based systems for drug delivery. *Advanced Drug Delivery Reviews*, *106*(Pt A), 27–44. https://doi.org/10.1016/J.ADDR.2016.09.007

Kawatra, A., Dhankhar, R., Mohanty, A., & Gulati, P. (2020). Biomedical applications of microbial phenylalanine ammonia lyase: Current status and future prospects. *Biochimie*, *177*, 142–152. https://doi.org/10.1016/J.BIOCHI.2020.08.009

Kudela, P., Koller, V. J., & Lubitz, W. (2010). Bacterial ghosts (BGs)--advanced antigen and drug delivery system. *Vaccine*, *28*(36), 5760–5767. https://doi.org/10.1016/J.VACCINE.2010.06.087

Kuzajewska, D., Wszołek, A., Żwierełło, W., Kirczuk, L., & Maruszewska, A. (2020). Magnetotactic bacteria and magnetosomes as smart drug delivery systems: A new weapon on the battlefield with cancer? *Biology*, *9*(5). https://doi.org/10.3390/BIOLOGY9050102

Lambin, P., Theys, J., Landuyt, W., Rijken, P., Van Der Kogel, A., Van Der Schueren, E., Hodgkiss, R., Fowler, J., Nuyts, S., De Bruijn, E., Van Mellaert, L., & Anné, J. (1998). Colonisation of *Clostridium* in the body is restricted to hypoxic and necrotic areas of tumors. *Anaerobe*, *4*(4), 183–188. https://doi.org/10.1006/ANAE.1998.0161

Liang, P.-C., Chen, Y.-C., Chiang, C.-F., Mo, L.-R., Wei, S.-Y., Hsieh, W.-Y., & Lin, W.-L. (2016). Doxorubicin-modified magnetic nanoparticles as a drug delivery system for magnetic resonance imaging-monitoring magnet-enhancing tumor chemotherapy. *International Journal of Nanomedicine*, *11*, 2021–2037. https://doi.org/10.2147/IJN.S94139

Liu, S. C., Minton, N. P., Giaccia, A. J., & Brown, J. M. (2002). Anticancer efficacy of systemically delivered anaerobic bacteria as gene therapy vectors targeting tumor hypoxia/necrosis. *Gene Therapy*, *9*(4), 291–296. https://doi.org/10.1038/SJ.GT.3301659

Liu, Z., Liu, X., Cao, W., & Hua, Z.-C. (2015). Tumor-specifically hypoxia-induced therapy of SPRY1/2 displayed differential therapeutic efficacy for melanoma. *American Journal of Cancer Research*, *5*(2), 792–801.

Loeffler, M., Le'Negrate, G., Krajewska, M., & Reed, J. C. (2008). IL-18-producing *Salmonella* inhibit tumor growth. *Cancer Gene Therapy 2008 15:12*, *15*(12), 787–794. https://doi.org/10.1038/cgt.2008.48

MacDiarmid, J. A., Amaro-Mugridge, N. B., Madrid-Weiss, J., Sedliarou, I., Wetzel, S., Kochar, K., Brahmbhatt, V. N., Phillips, L., Pattison, S. T., Petti, C., Stillman, B., Graham, R. M., & Brahmbhatt, H. (2009). Sequential treatment of drug-resistant tumors with targeted minicells containing siRNA or a cytotoxic drug. *Nature Biotechnology 2009 27:7*, *27*(7), 643–651. https://doi.org/10.1038/nbt.1547

MacDiarmid, J. A., Madrid-Weiss, J., Amaro-Mugridge, N. B., Phillips, L., & Brahmbhatt, H. (2007). Bacterially-derived nanocells for tumor-targeted delivery of chemotherapeutics and cell cycle inhibitors. *Cell Cycle*, *6*(17), 2099–2105. https://doi.org/10.4161/CC.6.17.4648

MacDiarmid, J. A., Mugridge, N. B., Weiss, J. C., Phillips, L., Burn, A. L., Paulin, R. P., Haasdyk, J. E., Dickson, K.-A., Brahmbhatt, V. N., Pattison, S. T., James, A. C., Bakri, G. Al, Straw, R. C., Stillman, B., Graham, R. M., & Brahmbhatt, H. (2007). Bacterially derived 400 nm particles for encapsulation and cancer cell targeting of chemotherapeutics. *Cancer Cell*, *11*(5), 431–445. https://doi.org/10.1016/J.CCR.2007.03.012

Malmgren, R. A., & Flanigan, C. C. (1955). Localization of the vegetative form of *clostridium tetani* in mouse tumors following intravenous spore administration. *Cancer Research*, *15*(7), 473–478.

Martel, S. (2012). Bacterial microsystems and microrobots. *Biomedical Microdevices*, *14*(6), 1033–1045. https://doi.org/10.1007/S10544-012-9696-X

Mostaghaci, B., Yasa, O., Zhuang, J., & Sitti, M. (2017). Bioadhesive bacterial microswimmers for targeted drug delivery in the urinary and gastrointestinal tracts. *Advanced Science (Weinheim, Baden-Wurttemberg, Germany)*, *4*(6). https://doi.org/10.1002/ADVS.201700058

Nallar, S. C., De Qi, X., & V. D. Kalvakolanu (2017). Bacteria and genetically modified bacteria as cancer therapeutics: Current advances and challenges. *Cytokine*, *89*, 160–172. https://doi.org/10.1016/J.CYTO.2016.01.002

Pálffy, R., Gardlík, R., Hodosy, J., Behuliak, M., Reško, P., Radvánský, J., & Celec, P. (2005). Bacteria in gene therapy: Bactofection versus alternative gene therapy. *Gene Therapy 2006 13:2*, *13*(2), 101–105. https://doi.org/10.1038/sj.gt.3302635

Park, S. J., Park, S.-H., Cho, S., Kim, D.-M., Lee, Y., Ko, S. Y., Hong, Y., Choy, H. E., Min, J.-J., Park, J.-O., & Park, S. (2013). New paradigm for tumor theranostic methodology using bacteria-based microrobot. *Scientific Reports 2013 3:1*, *3*(1), 1–8. https://doi.org/10.1038/srep03394

Patyar, S., Joshi, R., Byrav, D. P., Prakash, A., Medhi, B., & Das, B. (2010). Bacteria in cancer therapy: A novel experimental strategy. *Journal of Biomedical Science 2010 17:1*, *17*(1), 1–9. https://doi.org/10.1186/1423-0127-17-21

Paukner, S., Kohl, G., & Lubitz, W. (2004). Bacterial ghosts as novel advanced drug delivery systems: antiproliferative activity of loaded doxorubicin in human Caco-2 cells. *Journal of Controlled Release: Official Journal of the Controlled Release Society*, *94*(1), 63–74. https://doi.org/10.1016/J.JCONREL.2003.09.010

Rabea, S., Alanazi, F. K., Ashour, A. E., Salem-Bekhit, M. M., Yassin, A. S., Moneib, N. A., Hashem, A. E. M., & Haq, N. (2020). *Salmonella*-innovative targeting carrier: Loading with doxorubicin for cancer treatment. *Saudi Pharmaceutical Journal*, *28*(10), 1253–1262. https://doi.org/10.1016/J.JSPS.2020.08.016

Shende, P., & Basarkar, V. (2019). Recent trends and advances in microbe-based drug delivery systems. *Daru: Journal of Faculty of Pharmacy, Tehran University of Medical Sciences*, *27*(2), 799–809. https://doi.org/10.1007/S40199-019-00291-2

Sreenivasan, Y., Sarkar, A., & Manna, S. K. (2003). Mechanism of cytosine arabinoside-mediated apoptosis: Role of Rel A (p65) dephosphorylation. *Oncogene 2003 22:28*, *22*(28), 4356–4369. https://doi.org/10.1038/sj.onc.1206486

Storz, G., & Hengge, R. (2010). Bacterial stress responses. *Bacterial Stress Responses*. https://doi.org/10.1128/9781555816841

Stritzker, J., Weibel, S., Hill, P. J., Oelschlaeger, T. A., Goebel, W., & Szalay, A. A. (2007). Tumor-specific colonization, tissue distribution, and gene induction by probiotic *Escherichia coli* Nissle 1917 in live mice. *International Journal of Medical Microbiology*, *297*(3), 151–162. https://doi.org/10.1016/J.IJMM.2007.01.008

Sun, J. B., Wang, Z. L., Duan, J. H., Ren, J., Yang, X. D., Dai, S. L. & Li, Y. (2009). Targeted distribution of bacterial magnetosomes isolated from *Magnetospirillum gryphiswaldense* MSR-1 in healthy Sprague-Dawley rats. *Journal of Nanoscience and Nanotechnology*, *9*(3), 1881–1885. https://doi.org/10.1166/JNN.2009.410

Tangney, M., van Pijkeren, J. P., & Gahan, C. G. M. (2010). The use of *Listeria monocytogenes* as a DNA delivery vector for cancer gene therapy. *Bioengineered Bugs*, *1*(4), 286–289. https://doi.org/10.4161/BBUG.1.4.11725

Wang, S., Gao, J., Li, M., Wang, L., & Wang, Z. (2018). A facile approach for development of a vaccine made of bacterial double-layered membrane vesicles (DMVs). *Biomaterials*, *187*, 28–38. https://doi.org/10.1016/J.BIOMATERIALS.2018.09.042

Warrier, V. U., Makandar, A. I., Garg, M., Sethi, G., Kant, R., Pal, J. K., Yuba, E., & Gupta, R. K. (2019). Engineering anti-cancer nanovaccine based on antigen cross-presentation. *Bioscience Reports*, *39*(10), 20193220. https://doi.org/10.1042/BSR20193220

Wiemann, B., & Starnes, C. O. (1994). Coley's toxins, tumor necrosis factor and cancer research: A historical perspective. *Pharmacology & Therapeutics*, *64*(3), 529–564. https://doi.org/10.1016/0163-7258(94)90023-X

Yaghoubi, A., Khazaei, M., Jalili, S., Hasanian, S. M., Avan, A., Soleimanpour, S., & Cho, W. C. (2020). Bacteria as a double-action sword in cancer. *Biochimica et Biophysica Acta (BBA) - Reviews on Cancer*, *1874*(1), 188388. https://doi.org/10.1016/J.BBCAN.2020.188388

Yazawa, K., Fujimori, M., Amano, J., Kano, Y., & Taniguchi, S. (2000). *Bifidobacterium longum* as a delivery system for cancer gene therapy: Selective localization and growth in hypoxic tumors. *Cancer Gene Therapy 2000 7:2*, *7*(2), 269–274. https://doi.org/10.1038/sj.cgt.7700122

Yin, L., Meng, Z., Zhang, Y., Hu, K., Chen, W., Han, K., Wu, B. Y., You, R., Li, C. H., Jin, Y., & Guan, Y. Q. (2018). *Bacillus* spore-based oral carriers loading curcumin for the therapy of colon cancer. *Journal of Controlled Release*, *271*, 31–44. https://doi.org/10.1016/J.JCONREL.2017.12.013

Zhang, L., Gao, L., Zhao, L., Guo, B., Ji, K., Tian, Y., Wang, J., Yu, H., Hu, J., Kalvakolanu, D. V., Kopecko, D. J., Zhao, X., & Xu, D. Q. (2007). Intratumoral delivery and suppression of prostate tumor growth by attenuated *Salmonella enterica* serovar typhimurium carrying plasmid-based small interfering RNAs. *Cancer Research*, *67*(12), 5859–5864. https://doi.org/10.1158/0008-5472.CAN-07-0098

Zhao, M., Geller, J., Ma, H., Yang, M., Penman, S., & Hoffman, R. (2007). Monotherapy with a tumor-targeting mutant of *Salmonella typhimurium* cures orthotopic metastatic mouse models of human prostate cancer. *Proceedings of the National Academy of Sciences of the United States of America*, *104*(24), 10170–10174. https://doi.org/10.1073/PNAS.0703867104

Zhou, H., He, Z., Wang, C., Xie, T., Liu, L., Liu, C., Song, F., & Ma, Y. (2016). Intravenous administration is an effective and safe route for cancer gene therapy using the *Bifidobacterium*-Mediated Recombinant HSV-1 Thymidine Kinase and Ganciclovir. *International Journal of Molecular Sciences 2016*, *17*(6), 891. https://doi.org/10.3390/IJMS17060891

23 Fungal Products in Medicine

Neelam Gandhi
Hansraj College, University of Delhi, New Delhi, India

CONTENTS

23.1 INTRODUCTION

For several decades, the use of antibiotics derived from bacteria and fungi as well as vaccinations against some bacterial and viral diseases have helped fight infections successfully. But our healthcare system currently faces huge challenges due to non-availability of vaccines against several pathogenic microbes and the emergence of new strains of microbes resistant to most of the existing antibiotics/drugs. Cardiovascular problems, diabetes, cancer, and autoimmune diseases are also on the rise. Hence, there is a huge demand for effective and affordable therapeutic compounds. The quest to address these health issues has led to the discovery of a large number of compounds of medicinal value from various microbes. Natural products have an advantage over synthetic molecules in being chemically very diverse, biocompatible, and more successful due to the fact that these compounds themselves have arisen in microbes by the process of natural selection operating for millions of years (Wright, 2014). Microbial metabolites have evolved to confer survival advantage in response to selection pressures faced by microbes in their habitats. People have been making use of natural products for medicinal purposes for a very long time indeed (Ji et al., 2009) and more than 50% of the drugs being used are either natural products or derived from them (Toghueo, 2019). Although extraction and identification of new potential antimicrobials or their scaffolds are beset with many difficulties, these are being overcome with new tools that have become available such as next-generation genome sequencing, bioinformatics, and analytical chemistry (Wright, 2014).

DOI: 10.1201/9781003306931-26

These approaches have enabled the pharmaceutical industry to explore a vast number of microbial species and successfully obtain a large number of compounds exhibiting antimicrobial activities.

23.2 CONTRIBUTIONS OF FUNGI IN THERAPEUTICS

Fungi have not only been the source of a large number of antimicrobials but are also a very useful expression system for production of heterologous proteins. Some fungi-derived products are proving useful in regenerative medicine, while others are being designed as vehicles for targeted delivery of drugs. These are being described below in brief.

23.3 PRODUCTS OBTAINED FROM FUNGI

The fungi kingdom includes eukaryotic organisms, both microorganisms such as yeasts and molds, and macroscopic mushrooms. They are not autotrophic, and to obtain nutrition, they secrete enzymes to digest extracellular food into simpler molecules which are then absorbed, referred to as saprophytic mode of nutrition. Since fungi naturally secrete a large variety of enzymes as well as metabolites into the extracellular medium to compete against other microbes, this has allowed pharmacologists to explore useful medical applications of these preparations. Endophytes residing within plants have also been shown to synthesize several secondary metabolites which confer protection on plants against pathogens (Ji et al., 2009), and these have been explored by researchers for novel drug discovery (de Felício et al., 2015). A large number of taxa of fungi present in the wild have been found to produce bioactive metabolites such as polysaccharides, terpenoids, and phenolic compounds like flavonoids, polyketides, and alkaloids (Jakubczyk and Dussart, 2020) which exhibit antibacterial, antiviral, anticancer, antifungal, and anti-inflammatory activities.

23.3.1 Antibacterial Activity of Fungal Products

In 1929, while working with the bacterium *Staphylococcus aureus*, Alexander Fleming observed that a fungal colony that had grown as contaminant on *S. aureus* culture plate was surrounded by a clear, transparent zone due to lysis of bacterial cells. He recognized the potential of a fungal metabolite in controlling bacterial growth and infections, and he named the substance penicillin after the name of the fungal species *Penicillium notatum* contaminating his culture plate (Gaynes, 2017). Although Fleming could not purify penicillin, he tested crude preparation of this substance from culture medium and found it to be effective against several gram-positive bacteria under laboratory conditions. Later two other scientists, Florey and Chain, succeeded in isolating the active antibacterial compound penicillin and in producing it on an industrial scale for general use (Gaynes, 2017). The medical community recognized this as a wonder drug that was to save millions of lives, and all three scientists were jointly awarded the Nobel Prize for this work in the year 1945. Bactericidal activity of penicillin occurs by inhibiting formation of new cell walls (Wright, 2014). Later, scientists chemically modified the structure of natural penicillins to produce semi-synthetic penicillins (e.g., ampicillin, carbenicillin and oxacillin) to confer additional properties on the compound such as extension of antibacterial activity to some gram-negative bacteria, resistance to penicillinase produced by some bacteria, and resistance to stomach acids to enable oral intake. Penicillins, though widely used in clinical practice to tackle bacterial infections, have been found to elicit allergic responses in some individuals and some species of bacteria have become resistant to penicillin. Hence, researchers are directing their efforts towards discovery of newer antibacterials. Various species of fungi have been reported to yield several compounds with antibacterial activity (Table 23.1). The potential of various species of *Aspergillus* to yield substances which could effectively arrest bacterial growth has been reviewed in detail by Al-Fakih and Almaqtri (2019).

Table 23.1 summarizes some fungal compounds showing their antibacterial activity, producer species, and mechanism of action.

TABLE 23.1
Bioactive Compounds of Some Species of Fungi with Antibacterial Activity

Metabolite/Compound	Fungal Species	Target Species	Effect	References
Penicillin	*Penicillium notatum*	Gram-positive bacteria (*Streptococcus, Staphylococcus, Enterococcus, Clostridium* and *Treponema* species)	Inhibits cell wall peptidoglycan synthesis	Wright (2014)
Cephalosporins	*Cephalosporium acremonium*	Gram-positive bacteria	Inhibits cell wall peptidoglycan synthesis	Wright (2014)
3,4-dimethoxyphenol and 1,3,5-trimethoxybenzene	*Aspergillus fumigatus*	*Staphylococcus aureus, Micrococcus luteus*	Inhibit growth of the bacteria tested	Furtado et al. (2002)
Extracts (high levels of 5-hydroxymethylfurfural and octadecanoic acid)	*Fusarium oxysporum* (GG008) *Fusarium oxysporum* (NFX06)	*Bacillus cereus* *S. aureus, E. coli, P. aeruginosa*	Increases permeability of cell membrane	Manganyi et al. (2019), Musavi and Balakrishnan (2014)
Xanthoradones	*Penicillium sp*	Methicillin-resistant *S. aureus*	Inhibits protein FtsZ (bacterial homolog of tubulin), preventing cell division	Schueffler and Anke (2014) as cited by Jakubczyk and Dussart (2020)
Rubellin anthraquinones	*Ramularia collo-cygni*	Gram-positive bacteria including MDR strains like *B. subtilis* (ATCC) 6633, *S. aureus* (SG) 511, S. aureus 134/94 (MRSA), *Enterococcus faecalis* 1528 (VRE) or *Mycobacterium vaccae* (IMET) 10670	Inhibit growth	Miethbauer et al. (2009)
Corollosporine 20	*Corollospora maritima*	Staphylococcus aureus	Inhibits growth	Liberra et al. (1998)

23.3.2 Antiviral Activity of Fungal Products

WHO estimates show that millions of people were infected by SARS-CoV-2 worldwide and a large number of lives had been lost. Humankind has also been very adversely affected by several other viruses, including hepatitis virus, HIV, dengue virus, influenza A virus and many more. Vaccines, though available for some viral diseases, are not available to prevent several other viral infections. Antiviral drugs currently available are either not very effective due to rapid emergence of drug-resistant strains or show serious side-effects. Therefore, a large number of studies are engaged in exploring microbial natural products to be able to fight viral infections more effectively with minimal side-effects.

Viruses can either be enveloped (lipid envelope) or non-enveloped (only the protein shell covers the genome) (Linnakoski et al., 2018), DNA or RNA viruses. The replication cycle of viruses involves several steps: attachment to the receptor on the host cell, penetration, uncoating and replication within the host cell (Carter and Saunders, 2013, as cited by Seo and Choi, 2021). Translation of viral mRNA in the host cell cytoplasm forms proteins which, together with replicated viral genomes, are packaged into new virions and released to the outside through cell lysis/budding/exocytosis.

TABLE 23.2
Bioactive Compounds of Some Species of Fungi with Antiviral Activity

Metabolite/Compound	Fungal Species	Target Species	Effect	References
Acidic polysaccharide	*Cordyceps militaris*	Influenza A virus (H1N1)	Decreases viral titer in respiratory system and increases levels of cytokines TNF-alpha and IFN-gamma in mice	Ohta et al. (2007)
Polysaccharide	*Pleurotus. abalonus*	HIV-1	Inhibits activity of reverse transcriptase and interaction of HIV-1 gp 120 with CD4	Wang et al. (2011)
Laccase	*Pleurotus ostreatus*	Hepatitis C virus (HCV)	Inhibits entry of virus into peripheral blood cells and hepatoma cells (HepG2) and inhibits HCV replication in infected HepG2 cells	EL-Fakharany et al. (2010)
Triterpenoids: Lanosta-7,9(11), 24-trien-3-one,15;26-dihydroxy (GLTA) and Ganoderic acid Y (GLTB)	*Ganoderma lucidum*	Enterovirus 71(EV71)	Interacts with the virus, preventing its adsorption to human rhabdomyosarcoma cells	Zhang et al. (2014)

Agents with antiviral activity exert their action at any of the crucial steps in the replication cycle of the virus.

Researchers report that fungi represent a very rich source of compounds/metabolites which show antiviral activity (Seo and Choi, 2021). Table 23.2 lists some fungal compounds, showing their antiviral activity along with their producer species and their mechanism of action.

23.3.3 Anticancer Activity of Fungal Products

According to a WHO report, cancer is one of the leading causes of death worldwide, and there were nearly 10 million deaths in 2020. Genetic/epigenetic changes (inactivation of tumor-suppressor genes or activation of oncogenes) may turn normal healthy cells into an altered metabolic state where they not only multiply in an uncontrolled way but also induce angiogenesis to develop new blood supply for growth and metastasis (Kumar et al., 2015).

Fungi produce a large number of structurally diverse metabolites including aromatic compounds, cytochalasans, macrolides, pyrones, naphthalenones, terpenes, etc. (Evidente et al., 2014). There are several reports of fungi-derived compounds which have the potential to serve as anticancer agents based on their in vitro inhibitory effects on human cancer cell lines or beneficial effects in mouse models of human cancer (Evidente et al., 2014). As shown in Table 23.3, one such drug is paclitaxel (generic name) (Taxol, US FDA approved), used for the treatment of different types of cancer (Kumar et al., 2019, Long, 1994). Using biotechnology and bioinformatics tools, researchers have been able to screen a large number of genera of endophytic fungi collected from *Taxus* sp. and have reported several strains capable of producing paclitaxel (Kumar et al., 2019, Xiong et al., 2013) to be able to meet the rising demand for this anticancer drug.

Table 23.3 lists some fungal compounds showing anticancer activity, with their producer species and their mechanism of action.

TABLE 23.3
Bioactive Compounds of Some Species of Fungi with Anticancer Activity

Metabolite/ Compound	Fungal sp	Mechanism/Effect	References
Paclitaxel (Taxol)	*Fusarium oxysporum, Aspergillus niger Taxomyces andreanae Aspergillus fumigatus*	Stimulates overproduction and stabilization of microtubules, arresting cell division. Exhibits antitumor activity in ovarian and breast cancers, non-small cell lung cancer, pancreatic cancer and AIDs-related Kaposi sarcoma.	Long (1994), Kumar et al. (2019), Elavarasi et al. (2012), Li et al. (2017)
Gliotoxin	*Aspergillus spp*	Induces apoptosis via the mitochondrial pathway in human cervical cancer cells and human chondrosarcoma cells	Nguyen et al. (2014)
Butenolide derivatives	*Aspergillus terreus*	Exert anticancer activity against pancreatic ductal adenocarcinoma cells	Qi et al. (2018)
L-asparaginase	*F. oxysporum, F. equiseti Aspergillus niger, A. flavus, A.fumigatus, A.oryzae, Alternaria tenuissima*	Depletes L- asparagine, an important requirement of cancer cells causing apoptosis of susceptible cells	El-Gendy et al. (2021)
Rubellins B, C, D and E and caeruleoramularins	*Ramularia collo-cygni*	Inhibit tau protein assembly and show antiproliferative, cytotoxic effects in several human tumor cell lines	Miethbauer et al. (2009)
Amanitins (amatoxin family) Amanitin derivatives	*Amanita sp*	Inhibits cell proliferation by inhibiting RNA polymerase II in pancreatic, colorectal and breast cancer cell lines in vitro	Moldenhauer et al. (2012)
Ustiloxins 26-31	*Ustilaginoidea virens*	Exert antimitotic and cytotoxic effects against several cancer cell lines of stomach, lung, breast, colon, and kidney	Koiso et al. (1994)
Beauvericin derivatives	*Beauveria bassiana Fusarium sp*	Inhibits cancer cell migration and by forming calcium channels in cell membranes changes levels of intracellular cation and results in apoptosis	Wu et al. (2018)
Antcin A, Antcin C and methylantcinate	*Antrodia camphorata*	Inhibit proliferation of human liver cancer cell lines Huh7, HepG2 and Hep3B	Hsieh et al. (2010)

23.3.4 Antifungal Activity of Fungal Products

Opportunistic fungal infections occur when the immune system is compromised due to genetic factors, to HIV infection/chemotherapy in cancer patients, or to the use of immunosuppressants post transplantation surgery (Oliveira et al., 2021). With the rise in these cases coupled with development of increased resistance to currently available antifungal drugs, serious fungal diseases are becoming a huge challenge for the healthcare system. *Candida albicans* causes invasive candidiasis and if it reaches the bloodstream, can cause high mortality rates, especially in immune-suppressed individuals. Similarly, HIV patients can develop severe meningoencephalitis caused by *Cryptococcus neoformans*. *Coccidioides* is responsible for causing life-threatening respiratory disease even in

TABLE 23.4
Bioactive Compounds of Some Species of Fungi with Antifungal Effects

Metabolite/Compound	Fungal Species	Target Species	Effect	References
Echinocandins (caspofungin, micafungin and anidulafungin) (synthetic derivatives of lipopeptides	Lipopeptide producing spp. *Aspergillus rugulovalvus, Zalerion arboricola, Papularia sphaerosperma*	Fungicidal on *Candida spp*, Fungistatic *on Aspergillus spp*	Prevent cell wall synthesis by inhibiting beta-(1,3)-D-glucan synthase needed in cell wall synthesis	Emri et al. (2013)
Amphotericin B	*Penicillium nalgiovense* Laxa	*Candida albicans*	Binds to ergosterol, creating pores in cell membrane and cell death	Svahn (2015)
Griseofulvin	*Penicillium griseofulvum*	*Microsporum sp., Epidermophyton sp., Trichophyton sp*	Interferes with spindle formation preventing mitosis	Olson and Troxell (2021)
Sordarin	*Sordaria araneosa*	*Candida albicans, Saccharomyces cerevisiae*	Inhibits protein synthesis by stabilizing ribosome/EF2 complex	Liang (2008)
Fumifungin	*Aspergillus fumigatus*	*Candida albicans Saccharomyces cerevisiae Fusarium culmorum*	NA	Mukhopadhyay et al. (1987)
Synerazol	*Aspergillus fumigatus SANK 10588*	*Candida albicans*	NA	Ando et al. (1991)
Pneumocandin B0 39	*Glarea lozoyensis Zalerion arboricola IV*	*Candida albicans and Pneumocystis carinii*	NA	Schmatz et al. (1992)

NA-Not available

immunocompetent individuals (Cole et al., 2004). In order to meet the rising demand for effective antifungal therapies, researchers have explored and reported several new antifungal agents from a number of fungal species, and griseofulvin (Olson and Troxell, 2021) and echinocandins (Emri et al., 2013) have reached the market as approved antifungal drugs. Antifungal agents bring about their action mainly by inhibiting either the synthesis of ergosterol (main fungal sterol), or other macromolecules, or by physicochemical interaction with the membrane sterols (Ghannoum and Rice, 1999). Table 23.4 lists some fungal compounds showing antifungal activity, producer species, and their mode of action.

23.3.5 Other Medically Relevant Products

Cyclosporin A, a metabolite produced by fungal species *Tolypocladium inflatum* and *Aspergillus terreus* (Anjum et al., 2012), and mycophenolic acid obtained from *Penicillium stoloniferum, P. echinulatum* and *P. brevicompactum* (Vardanyan and Hruby, 2016) have been reported to exhibit immunosuppressive activity. Both these drugs are US FDA approved and are being used in patients, either in those who have had bone marrow/organ transplants to prevent rejection, or in those suffering from inflammatory and autoimmune diseases. While Cyclosporin A prevents elicitation of an immune response by inhibiting calcineurin, activator of transcription factors (Matsuda and Koyasu, 2000), mycophenolic acid, by inhibiting inosine monophosphate dehydrogenase, blocks purine synthesis in B and T lymphocytes, suppressing the immune system (Vardanyan and Hruby, 2016).

Ergot alkaloids are produced by several species of fungi such as *Claviceps purpurea, C. africana, C. sorghi*, and being alpha-blockers, induce direct smooth muscle contraction (Ebert, 2019) and are finding applications in the management of migraines and postpartum bleeding (Shehata, 2010).

Aspergillus terreus and Monascus ruber yield lovastatin, which functions to inhibit HMG CoA reductase in the body (Hajjaj et al., 2001), helping to reduce the level of low-density lipoproteins (LDL) in the blood and thereby protecting against heart attack, stroke or diabetes, and has US FDA approval. Statins have also been shown to be anti-inflammatory and antimicrobial, and promote bone regeneration by attracting stem cells to damaged tissues so as to initiate tissue regeneration (Shah et al., 2015).

Beta -D-galactosidase obtained from *Aspergillus foetidus* is the enzyme that catalyzes breakdown of lactose into glucose and galactose (Jozala et al., 2016). Demand for it is increasing as it is important for the treatment of lactose intolerance shown by some people.

Collagenases, obtained from *Aspergillus, Cladosporium, Penicillium* and *Alternaria* (Yakovleva et al., 2006), are enzymes with several applications in the medical field, such as treatment of burns, ulcers and different types of fibrosis such as liver cirrhosis (Wanderley et al., 2017).

Leishmaniasis is responsible for many deaths worldwide, especially in immunocompromised individuals. Toghueo (2019) reported the anti-leishmanial activity of preussomerin EG1 and palmarumycin CP2 obtained from extracts of fungus *Edenia* sp. and terrenolide S, and butyrolactone VI obtained from extracts of *Aspergillus terreus*.

Tanshinones from *Trichoderma atroviride* D16, salvianolic acid from *Phoma glomerata* D14 (Toghueo, 2019), and meroterpenoids from *Talaromyces amestolkiae* YX1 (Chen et al., 2018) have been reported to show anti-inflammatory activity with the potential to be developed as anti-inflammatory drugs for the treatment of chronic inflammation-related diseases such as arthritis, cancer, and diabetes.

23.4 FUNGAL VACCINES

There has been a drastic increase in the incidence of serious fungal infections. While antifungal drugs are required to treat the infection, vaccines have the potential to prevent infection in the first place and hence are more desirable (Tesfahuneygn and Gebreegziabher, 2018). Several vaccine candidates have reached the stage of preclinical development and two vaccines have reached the clinical trial stage (Nami et al., 2019). This section briefly describes advances in the design of antifungal vaccines using a variety of strategies and new trends in vaccine development.

23.4.1 Whole Organism, Live-Attenuated Vaccines

These vaccines induce a long-lasting, strong immune response in normal healthy individuals and are quite effective in eliciting immunity against several species of pathogenic fungi such as *Histoplasma capsulatum, Blastomyces dermatitidis, Paracoccidioides brasiliensis, Pneumocystis carinii* and *Cryptococcus neoformans* (Nami et al., 2019) In one study, attenuated vaccine, produced by deletion of adhesion 1gene of *Blastomyces*, when tested on CD 4+ T cell-deficient mice, was found to elicit CD8+ T cell-mediated immune protection against *Blastomyces* (Wüthrich et al., 2002). This study suggested that vaccines may also possibly provide some measure of protection to HIV/AIDS patients as CD4+T cells were dispensable in conferring immunity against pulmonary blastomycosis.

Individuals with immunocompromised states cannot be administered these vaccines as there would be risk of infection from the vaccine itself. As these are the very individuals more prone to catching fungal infections and thus requiring protection, possible solutions sought to circumvent these issues include vaccination of these individuals before the onset of immunosuppression (i.e., before the start of chemotherapy in cancer patients, when helper T cell count is still adequate in HIV patients, or prior to planned transplant surgery) (Oliveira et al., 2021), and design of vaccines that can stimulate such immune response pathways as are relatively less compromised (Levitz and Golenbock, 2012).

23.4.2 Whole Organism, Heat-Killed or Formalin-Inactivated Vaccines

Heat-killed *Saccharomyces cerevisiae* vaccine has been reported to be effective in conferring protection against virulent strains of several fungi such as *Coccidioides posadasii, Candida albicans and Aspergillus fumigatus* (Liu et al., 2011).

23.4.3 Recombinant (Subunit) Vaccines

Two recombinant vaccines against *Candida* are reported to be under clinical trials. One is PEV7, prepared from recombinant aspartyl-proteinase 2 (Sap2), a secreted protein of *C. albicans*, and assembled into virosomes (De Bernardis et al., 2012, 2018). Rats vaccinated with this recombinant protein were protected against subsequent challenge from *C. albicans*. When the same vaccine was tested for its efficacy on healthy volunteers, specific B-cell memory responses were found to be generated in all of them (De Bernardis et al., 2018).

NDV-3 is the second vaccine, designed to contain the recombinant N-terminus of agglutinin-like sequence 3 protein (Als3p, a cell surface adhesin and invasin) of *C. albicans*, along with the adjuvant, aluminum hydroxide (Schmidt et al., 2012). It was found to generate adequate immune responses, protecting mice not only from *Candida* species but also from *Staphylococcus aureus* as surface proteins on *S. aureus* bear homology with Als3p. It was also reported that NDV-3 led to the production of high titers of antigen-specific antibodies IgG and IgA1 as well as increased levels of cytokines IFN-gamma and IL-17A in healthy volunteers (Edwards et al., 2018).

23.4.4 Conjugate Vaccines

Vaccines targeting polysaccharide epitopes, beta-glucans that are present in the cell walls of all species of fungi, may be able to confer protection against several fungi at the same time. By attaching polysaccharides (poor immunogens) covalently to a carrier protein (strong immunogens), a conjugate vaccine is produced which elicits a strong immune response targeted against both polysaccharide and the carrier protein (Xin et al., 2012). Cutler et al. (2007) observed the production of IgA and IgG antibodies in response to a vaccine produced by conjugating covalently a capsular polysaccharide, glucuronoxylomannan of *Cryptococcus neoformans*, with tetanus toxoid.

23.4.5 DNA Vaccines

These vaccines are produced by incorporating DNA, encoding the antigen from a disease-causing pathogen along with a gene which codes for co-stimulatory molecules, into a bacterial plasmid and injecting intramuscularly or subcutaneously. Gene product expressed within the host cells stimulates the generation of an immune response (Cutler et al., 2007) against the encoded antigen. Though effective antifungal DNA vaccines have been reported for murine models, none has been reported so far to be effective in humans due to their low immunogenicity.

23.4.6 Antigen-Primed Dendritic Cells

Dendritic cells (DCs), being professional antigen-presenting cells (APCs), upon recognition of fungal antigens internalize and process them to present processed antigen to naive T cells via MHC Class II molecules on their cell surface, and also secrete cytokines required for differentiation of naive CD4 T cells into various T-helper subsets. Bacci et al. (2002), working on a mouse model, have reported that DCs upon transfection ex vivo with yeast (*C. albicans*) as well as yeast RNA, exhibited higher expression of MHC class II antigens and co-stimulatory molecules upon in vivo transfer, and induced Th1-dependent protection against the fungus. Although not practical for mass immunizations, this approach may be utilized for patients who fail to respond to other modes of treatment (Ueno et al., 2019).

23.4.7 Passive Immunization

As immunocompromised individuals may not generate effective protective immunity following active immunization (Ljungman, 2012), passive immunization (by preformed antibodies) may prove to be a useful strategy. Several studies report that monoclonal antibodies (mAbs) (Ulrich and Ebel, 2020) can successfully confer protection against experimental infections caused by several pathogenic species such as *Candida, Cryptococcus, Histoplasma, Paracoccidioides, Sporothrix* or molds, *Aspergillus, Rhizopus*. The protective role of antibodies may be due either to neutralization of toxins or to inhibition of surface proteins, opsonization, or enhancement of complement-mediated phagocytosis (Cutler et al., 2007).

Although as yet there are no approved antifungal vaccines, the rapidly growing fields of genomics and proteomics may speed up the process of development of safe and effective vaccines against pathogenic fungi. Nami et al. (2019) have provided a very comprehensive account of various strategies employed so far in the development of antifungal vaccines.

23.5 PRODUCTS OBTAINED FROM FUNGI USING BIOTECHNOLOGICAL TOOLS

Since the production and approval of recombinant interferons (alpha, beta, gamma) and growth hormone in the 1980s (Jozala et al., 2016), recombinant proteins have been increasingly produced and used for the management of several medical conditions. Production of recombinant proteins requires transfection of cells with the gene of interest carried in a DNA vector. With the help of host cellular machinery, the gene translates into a protein which is obtained from either the lysate or the extracellular medium (Ahmad et al., 2014).

Yeasts, *Saccharomyces cerevisiae, Pichia pastoris, Hansenula polymorpha,Yarrowia lipolytica, Arxula adeninivorans, Komagataella sp.,Kluyveromyces lactis* and *Schizosaccharomyces pombe* (Celik and Calik, 2012) are being increasingly used for the production of heterologous proteins as they offer several advantages. Being simple and unicellular, yeasts show a fast growth rate; genetic manipulation is relatively easier than in mammalian cells; as they are eukaryotic, post-translational modification of proteins (glycosylation, acetylation, phosphorylation, etc.) occurs, unlike that of *E. coli;* and intracellular as well as secretory proteins are obtained. These systems are cost-effective and yield high levels of the desired protein by using fermenters. Researchers have also engineered strains of yeast lacking proteases (to prevent degradation of heterologous protein) (Ahmad et al., 2014) as well as showing humanized glycosylation patterns (Celik and Calik, 2012).

Saccharomyces cerevisiae has been used as a host for the production of therapeutic recombinant proteins such as insulin, insulin lispro (fast-acting insulin analog) and insulin detemir (slow-acting insulin), glucagon-like peptide1, hepatitis B surface antigen and human papillomavirus surface antigen for vaccines, hirudine, and many more products (Jozala et al., 2016).

Using a modified strain of *Aspergillus nidulans*, Yadwad et al. (1996), were able to express human interleukin-6 (IL-6). Large quantities of IL-6 are required for the production of IL-6-specific antibodies for the treatment of several inflammatory diseases including rheumatoid arthritis (Nausch et al., 2013). Rosti et al. (2014) were able to obtain human epidermal growth factor (used for promoting wound healing) from the culture medium of *Pichia pastoris*.

23.6 TISSUE ENGINEERING AND OTHER BIOMEDICAL APPLICATIONS

The polymers chitin and chitosan are present in the cell walls of fungal classes Basidiomycetes, Ascomycetes, Zygomycetes and Deuteromycetes (Elsoud and El Kady, 2019). Properties of these biopolymers, such as biocompatibility, non-toxicity, bioresorptivity, adsorption capacity, biodegradability, and antibacterial effects, confer advantages over synthetic polymers and have allowed scientists to make use of them in regenerative medicine (Elsoud and El Kady, 2019).

Although these biopolymers can be obtained from crustacean shells, fungal sources have the advantage of being non-allergenic and having consistent physico-chemical properties, as well as being easily available rather than limited to fishing sites in marine habitats.

Chitosan has been utilized as a supporting material in regenerative medicine, for instance, in wound dressing, burn treatment, artificial kidney membrane, nerve regeneration, artificial skin, artificial tendon, blood anticoagulation, and bone damage (Cheung et al., 2015). There has been a report of enhanced neural stem cell differentiation into neurons, oligodendrocytes and astrocytes by a scaffold consisting of chitosan, along with other supporting material, in rats, with the potential to be used for repair of damaged spinal cords (Jian et al., 2015). Antibacterial, antioxidant, antitumor, antidiabetic, and antiulcer properties of chitosan (Elsoud and El Kady, 2019) have enabled it to be utilized in wound dressings and several products based on chitosan are available on the market, such as Chitoflex, ChitoGauze, and Celox (Cheung et al., 2015).

23.6.1 Targeted Drug Delivery

Defects in genes or their altered expression levels can give rise to a diseased state and may be corrected more effectively by targeted introduction of a normal gene/siRNA. Serious side-effects resulting from chemotherapy against cancer can be greatly minimized by targeted delivery of anticancer drugs/immunomodulatory agents, besides improving bioavailability, specificity, and sensitivity (Sharifi-Rad et al., 2021). Chitosan-based carriers are being increasingly explored for safe, efficient and targeted delivery of the desired gene/oligonucleotides/siRNA/drugs, due to such properties of chitosan as low toxicity and low immunogenicity as well as biocompatibility and good cell membrane permeability (Cheung et al. 2015, Herdiana et al., 2021, Prabaharan, 2015). Ramana et al. (2014) reported higher efficiency of saquinavir (a drug used for the treatment of AIDS) in controlling viral proliferation in CD4+ T lymphocytes when administered as chitosan-loaded nanoparticles, due to efficient cell targeting compared to controls receiving soluble drug. Production of metallic nanoparticles using different species of fungi is contributing in nano-medicine (Moghaddam et al., 2015).

23.7 SUMMARY

Fungi represent a very rich source of a large number of biomolecules with antibacterial, antiviral, anticancer, antifungal, and anti-inflammatory properties, as well as many other medically useful substances such as immunosuppressive agents, statins, ergot alkaloids, etc. They thus contribute greatly in the management of several health issues faced by humans. Apart from approved antibiotics penicillin and cephalosporins, fungus-derived paclitaxel (Taxol) is an approved anticancer drug. Medications derived from ergot alkaloids have obtained FDA approval for the treatment of postpartum hemorrhage. The success rate of bone marrow/organ transplants has greatly improved due to the use of the immunosuppressants cyclosporine and mycophenolic acid, both fungi-derived and approved drugs. Collagenases obtained from fungi are finding an application in the treatment of burns, ulcers and different types of fibrosis. With the increase in the number of immunocompromised individuals and concomitant rise of fungal infections, fungal vaccines are being developed by scientists (two are under clinical trials) and several antifungal agents are being screened (fungi-derived griseofulvin and echinocandins are approved antifungal drugs) for protection against fungal infections.

Fungi, being eukaryotic and capable of post-translational modifications, are increasingly being used by researchers as expression systems for the production of recombinant proteins for medical use. Fungal cell wall polysaccharides with unique properties are being exploited in regenerative medicine and for various biomedical applications. Many US FDA-approved products containing chitosan for wound dressing are marketed under the names Chitoflex, Celox, Chitoseal, etc. Scientists continue to pursue the mining of more and more species of fungi to obtain natural products for biosafety purposes as well as to overcome the problem of microbial resistance to many currently used drugs.

REFERENCES

Ahmad M., Hirz M., Pichler H., and Schwab H.2014 Protein expression in Pichia pastoris: Recent achievements and perspectives for heterologous protein production (nih.gov) *Appl Microbiol Biotechnol.* 98(12), 5301–5317. doi: 10.1007/s00253-014-5732-5

Al-Fakih A.A., and Almaqtri W.Q.A. 2019 Overview on antibacterial metabolites from terrestrial Aspergillus spp, *Mycology*, 10(4), 191–209 doi: 10.1080/21501203.2019.1604576

Ando S., Satake H., Nakajima M., Sato A., Nakamura T., Kinoshita T, Furuya K., and Haneishi T. 1991. Synerazol, a new antifungal antibiotic. *The Journal of Antibiotics (Tokyo)*, 44(4), 382–389. doi: 10.7164/antibiotics.44.382

Anjum T., Azam A., and Irum W. 2012. Production of cyclosporine a by submerged fermentation from a local isolate of *Penicillium fellutanum* (nih.gov). *Indian Journal of Pharmaceutical Sciences*Jul–Aug; 74(4), 372–374. doi: 10.4103/0250-474X.107082

Bacci A., Montagnoli C., Katia Perruccio K., … and Romani L. 2002. Dendritic cells pulsed with fungal RNA induce protective immunity to Candida albicans in hematopoietic transplantation *The Journal of Immunology (jimmunol.org)* 168 (6) 2904–2913; doi: 10.4049/jimmunol.168.6.2904

Carter J., Saunders V.A. 2013. *Virology: Principles and Applications*. 2nd ed. John Wiley & Sons, Hoboken, NJ

Celik E. and Calık P. 2012. Production of recombinant proteins by yeast cells. *Biotechnology Advances* 30(5), 1108–1118. doi: 10.1016/j.biotechadv.2011.09.011. Epub 2011 Sep 22

Chen S., Ding M., … and She Z. 2018 Anti-inflammatory meroterpenoids from the mangrove endophytic fungus Talaromyces amestolkiae YX1 *Phytochemistry* 146, 8–15. doi: 10.1016/j.phytochem.2017.11.011

Cheung R.C.F., Ng T. B., Wong J.H., and Chan W.Y. 2015. Chitosan: An update on potential biomedical and pharmaceutical applications. *Marine Drugs* 14; 13(8), 5156–5186. doi: 10.3390/md13085156

Cole G.T., Xue J.M., Okeke C.N., Tarcha E.J., Hung C.Y. 2004. A vaccine against coccidioidomycosis is justified and attainable *Medical Mycology* 42(3), 189–216. doi: 10.1080/13693780410001687349

Cutler J.E., Deepe G.S., Jr, and Klein B.S. 2007. Advances in combating fungal diseases: Vaccines on the threshold (nih.gov)Nat *Revista de Microbiologia* 5(1), 13–28. Published online 2006 Dec 11. doi: 10.1038/nrmicro1537

De Bernardis F., Amacker M., Arancia S., Sandini S., … Cassone A. 2012 A virosomal vaccine against candidal vaginitis: Immunogenicity, efficacy and safety profile in animal models - PubMed (nih.gov) *Vaccine* 30(30), 4490–4498. doi: 10.1016/j.vaccine.2012.04.069. Epub 2012 May 3

De Bernardis F., Graziani S., Tirelli F., & Antonopoulou S. 2018 Candida vaginitis: Virulence, host response and vaccine prospects - PubMed (nih.gov) *Medical Mycology* 56(suppl_1), 26–31. doi: 10.1093/mmy/myx139

de Felício R., Pavão G.B., … and Debonsi H.M. 2015. Antibacterial, antifungal and cytotoxic activities exhibited by endophytic fungi from the Brazilian marine red alga Bostrychia tenella (Ceramiales) | *SpringerLinkRevista Brasileira de Farmacognosia* 25, 641–650. doi: 10.1016/j.bjp.2015.08.003

Ebert T. J. 2019. Autonomic Nervous System Pharmacology, In *Pharmacology and Physiology for Anesthesia* (pp. 282–299). doi: 10.1016/B978-0-323-48110-6.00014-4. Claviceps purpurea - an overview | ScienceDirect Topics

Edwards J.E., Jr, Schwartz M.M., Schmidt C.S., Sobel J.D. …, and Hennessey J.P. Jr. 2018. Fungal immunotherapeutic vaccine (NDV-3A) for treatment of recurrent vulvovaginal candidiasis: A Phase 2 Randomized, double-blind, placebo-controlled trial|clinical infectious diseases | Oxford Academic (oup.com), *Clinical Infectious Diseases* 66(12), 1928–1936. doi: 10.1093/cid/ciy185

Elavarasi A., Rathna G.S., and Kalaiselvam M. 2012. Taxol producing mangrove endophytic fungi *Fusarium oxysporum* from *Rhizophora annamalayana, Asian Pacific Journal of Tropical Biomedicine* 2(2), Supplement, S1081–S1085. doi:10.1016/S2221-1691(12)60365-7

El-Fakharany E.M., Haroun B.M., Ng T.B., and Redwan E.R. 2010. Oyster mushroom laccase inhibits hepatitis C virus entry into peripheral blood cells and hepatoma cells *Protein & Peptide Letters* 17(8), 1031–1039. doi: 10.2174/092986610791498948

El-Gendy M.M.A.A., Awad M.F., El-Shenawy F.S., and El-Bondkly A.M.A. 2021.Production, purification, characterization, antioxidant and antiproliferative activities of extracellular L-asparaginase produced by Fusarium equiseti AHMF4 - ScienceDirect *Saudi Journal of Biological Sciences* 28(4), 2540–2548

Elsoud M. M. A., and El Kady E. M. 2019. Current trends in fungal biosynthesis of chitin and chitosan. *Bulletin of National Research Centre*, 43, 59. doi:10.1186/s42269-019-0105-y

Emri T., Majoros L., Tóth V., and Pócsi I. 2013. Echinocandins: Production and applications. *Applied Microbiology and Biotechnology*, 97(8), 3267–3284. doi: 10.1007/s00253-013-4761-9

Evidente A., Kornienko A., Cimmino A., … and Kiss R. 2014. Fungal metabolites with anticancer activity *Natural Product Reports*, (5), 617–627. doi: 10.1039/c3np70078j. Epub 2014 Mar 20

Furtado N.A.J.C., Said S., Ito I.Y., and Bastos J. K. 2002. The antimicrobial activity of Aspergillus fumigatus is enhanced by a pool of bacteria - PubMed (nih.gov), *Microbiology Research*157(3), 207–211. doi: 10.1078/0944-5013-00150

Gaynes R. (2017). The discovery of penicillin—new insights after more than 75 years of clinical use. *Emerging Infectious Diseases*, *23*(5), 849–853. doi: 10.3201/eid2305.161556

Ghannoum M. A. and Rice L. B.,1999. Antifungal agents: Mode of action, mechanisms of resistance, and correlation of these mechanisms with bacterial resistance, *Clinical Microbiology Reviews*, 12(4), 501–517. PMID: 10515900

Hajjaj H., Neiderberger P., and Duboc P. 2001 Lovastatin biosynthesis by aspergillus terreus in a chemically defined medium. *Applied and Environmental Microbiology*, 67(6), 2596–2602. doi:10.1128/AEM.67.6.2596-2602.2001

Herdiana Y., Wathoni N., Shamsuddin S., Joni I. M., and Muchtaridi M. 2021. Chitosan-based nanoparticles of targeted drug delivery system in breast cancer treatment. *Polymers (Basel)*, 13(11), 1717. doi: 10.3390/polym13111717

Hsieh Y.-C., Rao Y.K., Chun-Chi Wu C-C., Huang C-Y.F., … and Tzeng Y.-M. 2010. Methyl antcinate A from Antrodia camphorata induces apoptosis in human liver cancer cells through oxidant-mediated cofilin-and Bax-triggered mitochondrial pathway - PubMed (nih.gov) *Chemical Research in Toxicology*, 23(7), 1256–1267. doi: 10.1021/tx100116a

Jakubczyk D. and Dussart F. 2020, Selected fungal natural products with antimicrobial properties, *Molecules*, 25(4), 911. doi: 10.3390/molecules25040911

Ji H.-F., Li X.-J., and Zhang H.-U. 2009. Natural products and drug discovery: Can thousands of years of ancient medical knowledge lead us to new and powerful drug combinations in the fight against cancer and dementia?: EMBO reports: Vol 10, No 3 (embopress.org) *EMBO Reports* 10, 194–200. doi: 10.1038/embor.2009.12

Jian R., Yixu Y., Sheyu L. Jianhong S., Yaohua Y. …, and Yilu G. 2015 Repair of spinal cord injury by chitosan scaffold with glioma ECM and SB216763 implantation in adult rats - PubMed (nih.gov) *Journal of Biomedical Materials Research Part A* 103(10), 3259–3272. doi: 10.1002/jbm.a.35466. Epub 2015 Apr 1

Jozala A.F., Geraldes D.C., …, and Pessoa A. Jr, 2016. Biopharmaceuticals from microorganisms: From production to purification. *The Brazilian Journal of Microbiology* 47(Suppl 1), 51–63 doi: 10.1016/j.bjm.2016.10.007

Koiso Y., Li Y., Iwasaki S., Hanaoka K., Kobayashi T., Sonoda R., Fujita Y., Yaegashi H., and Sato Z. (1994). Ustiloxins, antimitotic cyclic peptides from false smut balls on rice panicles caused by Ustilaginoidea virens. *The Journal of Antibiotics*, *47*(7), 765–773. doi: 10.7164/antibiotics.47.765

Kumar P., Singh B., Thakur V., Thakur A., Thakur N., Pandey D., & Chand D. (2019).Hyper-production of taxol from *Aspergillus fumigatus*, an endophytic fungus isolated from *Taxus* sp. of the Northern Himalayan region. *Biotechnology Reports (Amsterdam, Netherlands)*, *24*, e00395. doi: 10.1016/j.btre.2019.e00395

Kumar S., Ahmad M. K., Waseem M., and Pandey A. K., 2015. Drug targets for cancer treatment: An overview *Medicinal Chemistry* 5(3). doi: 10.4172/2161-0444.1000252

Levitz S.M., and Golenbock D.T., 2012 Beyond empiricism: Informing vaccine development through innate immunity research - PubMed (nih.gov). *Cell* 16; 148(6), 1284–1292. doi: 10.1016/j.cell.2012.02.012

Li D., Fu D., Zhang Y, and Zhao K. 2017. Isolation, purification, and identification of taxol and related taxanes from taxol-producing fungus *Aspergillus niger* subsp. *taxi The Journal of Microbiology and Biotechnology* 27(8), 1379–1385. doi: 10.4014/jmb.1701.01018

Liang H. 2008. Sordarin, an antifungal agent with a unique mode of action. *Beilstein Journal of Organic Chemistry*, 4 (31). doi: 10.3762/bjoc.4.31

Liberra, K., Jansen R., and Lindequist U. 1998. Corollosporine, a new phthalide derivative from the marine fungus Corollospora maritima, Werderm, 1069, *Pharmazie*, 53(8), 578–581. PMID: 9741066

Linnakoski R., Reshamwala D. and Marjomäki V. 2018. Antiviral agents from fungi: Diversity, mechanisms and potential applications *Frontiers in Microbiology*. doi: 10.3389/fmicb.2018.02325

Liu M., Clemons K. V., Bigos M., Medovarska I., Brummer E., and Stevens D. A. 2011. Immune responses induced by heat killed Saccharomyces cerevisiae: A vaccine against fungal infection (nih.gov),*Vaccine* 29(9), 1745–1753. Published online 2011 Jan 8. doi: 10.1016/j.vaccine.2010.12.119

Ljungman P. 2012. Vaccination of immunocompromised patients - Ljungman - 2012 - *Clinical Microbiology and Infection* - Wiley Online Library First published: 05 July 2012. doi: 10.1111/j.1469-0691.2012.03971.x

Long H. J., 1994 Paclitaxel (Taxol): A novel anticancer chemotherapeutic drug. *Mayo Clinic Proceedings* 69(4), 341–345. doi: 10.1016/s0025-6196(12)62219-8

Manganyi M.C., Regnier T., …and Ateba C. N., 2019, Antibacterial activity of endophytic fungi isolated from Sceletium tortuosum L. (Kougoed), *Annals of Microbiology*, 69, 659–663

Matsuda S. and Koyasu S. 2000, Mechanisms of action of cyclosporine, *Immunopharmacology*, 2000, 47, 119–125. doi: 10.1016/S0162-3109(00)00192-2

Miethbauer S., Gaube F., and Liebermann B., 2009. Antimicrobial, antiproliferative, cytotoxic and tau inhibitory activity of rubellins and caeruleoramlarin produced by the phytopathogenic fungus Ramularia collo-cygni. *Planta Medica*, 75, 1523–1525. doi: 10.1055/s-0029-1185835. Epub 2009 Jun 29

Moghaddam A.B., Namvar F., and Mohamad R. 2015. Nanoparticles biosynthesized by fungi and yeast: A review of their preparation, properties and medical applications. *Molecules*, 20, 16540–16565

Moldenhauer G., Salnikov A. V., Lüttgau S., Herr I., Anderl J., & Faulstich H. (2012). Therapeutic potential of amanitin-conjugated anti-epithelial cell adhesion molecule monoclonal antibody against pancreatic carcinoma. *Journal of the National Cancer Institute*, *104*(8), 622–634. doi: 10.1093/jnci/djs140

Mukhopadhyay T., Roy K., Coutinho L., Rupp R.H., Ganguli B.N., and Fehlhaber H.W. 1987. Fumifungin, a new antifungal antibiotic from Aspergillus fumigatus Fresenius 1863.*The Journal of Antibiotics* 40(7), 1050–1052. doi: 10.7164/antibiotics.40.1050

Musavi S.F. and Balakrishnan R.M., 2014, A study on the antimicrobial potentials of an endophytic fungus Fusarium oxysporumNFX 06. *Journal of Mathematical Biology* 3(3), 162–166

Nami S., Mohammadi R. …, and Morovati H., 2019. Fungal vaccines, mechanism of actions and immunology: A comprehensive review. *Biomedicine and Pharmacology*, 109, 2019, 3

Nausch H., Huckauf J., Koslowski R., Meyer U., Broer I., and Mikschofsky H. (2013). Recombinant production of human interleukin 6 in Escherichia coli. *PloS One*, *8*(1), e54933. doi: 10.1371/journal.pone.0054933

Nguyen V.-T., Lee J.S., Qian Z.-Ji., … and Won-Kyo J. 2014. Gliotoxin isolated from marine fungus *Aspergillus sp*. induces apoptosis of human cervical cancer and chondrosarcoma cells. *Marine Drugs*. 12(1), 69–87. Published online 2013 Dec 24. doi: 10.3390/md12010069

Ohta Y., Lee J. B., Hayashi K., Fujita A., Park D. K., and Hayashi T. (2007). In vivo anti-influenza virus activity of an immunomodulatory acidic polysaccharide isolated from Cordyceps militaris grown on germinated soybeans. *Journal of Agricultural and Food Chemistry*, *55*(25), 10194–10199. doi: 10.1021/jf0721287

Oliveira L.V.N., Wang R., … and Levitz S.M. 2021. Vaccines for human fungal diseases: Close but still a long way to go, npj, Vaccines, 6, 33, s41541-021-00294-8.pdf

Olson J.M., Troxell T. 2021. *Griseofulvin* StatPearls Publishing LLC, Treasure Island, Florida.- PubMed (nih.gov), PMID: 30726008 Bookshelf ID: NBK537323

Prabaharan M. 2015.Chitosan-based nanoparticles for tumor-targeted drug delivery. *ScienceDirect International Journal of Biological Macromolecules*, 72, 1313–1322 doi: 10.1016/j.ijbiomac.2014.10.052

Qi C., Gao W., Guan D., … and Zhang Y. 2018. Butenolides from a marine-derived fungus Aspergillus terreus with antitumor activities against pancreatic ductal adenocarcinoma cells - PubMed (nih.gov) *Bioorganic & Medicinal Chemistry*, 26(22), 5903–5910. doi: 10.1016/j.bmc.2018.10.040

Ramana L.N., Sharma S., Sethuraman S., Ranga U., and Krishnan U.M. (2014). Evaluation of chitosan nano-formulations as potent anti-HIV therapeutic systems. *Biochimica et biophysica Acta*, 1840(1), 476–484. doi: 10.1016/j.bbagen.2013.10.002

Rosti, I. A., Ramanan, R. N., Tan, J. S., Ling, T. C., & Ariff, A. B. (2014). Recovery of microquantities of human epidermal growth factor from Escherichia coli Homogenate and Pichia pastoris culture medium using expanded bed adsorption. *Separation Science and Technology*, *49*(5), 702–708.

Schmatz D.M., Abruzzo G., Powles M.A., Balkovec J.M., Black R.M., …and Bartizal K. 1992.Pneumocandin from Zalerion arboricola IV. Biological evaluation of natural and semisynthetic pneumocandin for activity against Pneumocystis carinii and Candida species. *The Journal of Antibiotics (Tokyo)*, 45, 1886–1891. doi: 10.7164/antibiotics.45.1886

Schmidt C.S., White C. Jo., Ibrahim A.S. … and Hennessey J.P., Jr. 2012. NDV-3, a *Recombinant Alum-Adjuvanted Vaccine* for *Candida* and *Staphylococcus aureus* is safe and immunogenic in healthy adults (nih.gov), *Vaccine*, 30(52), 7594–7600.Published online 2012 Oct 22. doi: 10.1016/j.vaccine.2012.10.038

Schueffler A., and Anke T. (2014). Fungal natural products in research and development. *Natural Product Reports*, 31(10), 1425–1448. doi:10.1039/c4np00060a

Seo D.J. and Choi C. 2021. Antiviral bioactive compounds of mushrooms and their antiviral mechanisms: A review. *Viruses*, 13(2), 350 doi: 10.3390/v13020350

Shah S.R., Caroline A. Werlang C.A., Kasper F.K., and Mikos A.G. 2015. Novel applications of statins for bone regeneration (nih.gov) *National Science Review* Mar 1; 2(1), 85–99. doi: 10.1093/nsr/nwu028 Published online 2014 Aug 16. doi: 10.1093/nsr/nwu028

Sharifi-Rad J., Quispe C., Butnariu M., Rotariu L. S., Sytar O., Sestito S., Rapposelli S., Akram M., Iqbal M., Krishna A., Kumar N., Braga S. S., Cardoso S. M., Jafernik K., Ekiert H., Cruz-Martins N., Szopa A., Villagran M., Mardones L., Martorell M., … Calina D. (2021). Chitosan nanoparticles as a promising tool in nanomedicine with particular emphasis on oncological treatment. *Cancer Cell International*, *21*(1), 318. doi:10.1186/s12935-021-02025-4

Shehata H., 2010. Drug and drug therapy. In Bennett, P. and Williamson, C. (Eds.) *Basic Science in Obstetrics and Gynaecology* (4th ed.), 259–277, Churchill Livingstone.

Svahn K.S., Chryssanthou E., Olsen B., Bohlin L., and Göransson U. 2015.Penicillium nalgiovense Laxa isolated from Antarctica is a new source of the antifungal metabolite amphotericin B (nih.gov). *Fungal Biology and Biotechnology* 2, 1. Published online 2015 Jan 17. doi: 10.1186/s40694-014-0011-x

Tesfahuneygn G. and Gebreegziabher G., 2018, Development of vaccination against fungal disease: A review article. *International Journal of TROPICAL DISEASE & Health* doi: 10.23937/IJTD-2017/1710005

Toghueo R. M. K. 2019. Anti-leishmanial and Anti-inflammatory agents from endophytes: A review|SpringerLink 2019. *Natural Products and Bioprospecting*, 9, 311–328.

Ueno K., Urai M., Sadamoto S., and Kinjo Y. 2019. A dendritic cell-based systemic vaccine induces long-lived lung-resident memory Th17 cells and ameliorates pulmonary mycosis - PubMed (nih.gov) *Mucosal Immunology* Jan; 12(1), 265–276. doi: 10.1038/s41385-018-0094-4. Epub 2018 Oct 2

Ulrich S. and Ebel F. 2020. Monoclonal antibodies as tools to combat fungal infections. *Journal of Fungi* 6(1), 22. doi: 10.3390/jof6010022

Vardanyan R. and Hruby V. 2016. Immunopharmacological drugs Synthesis of Best-Seller Drugs. 549–572, Academic Press, Elsevier.

Wanderley M.C., Neto J.M., Filho J.L., Lima C.A., Teixeira J.A., and Porto A.L. 2017. Collagenolytic enzymes produced by fungi: A systematic review *The Brazilian Journal of Microbiology*, 48(1):13–24. doi: 10.1016/j.bjm.2016.08.001

Wang C.R. Ng T.B., and Liu F. 2011. Isolation of a polysaccharide with antiproliferative, hypoglycemic, antioxidant and HIV-1 reverse transcriptase inhibitory activities from the fruiting bodies of the abalone mushroom Pleurotus abalonus *Journal of Pharmacy and Pharmacology* 63(6), 825–832. doi: 10.1111/j.2042-7158.2011.01274.x

Wright G. D., 2014. Something old, something new: Revisiting natural products in antibiotic drug discovery. *Canadian Journal of Microbiology*, 2014, PMID: 24588388 doi: 10.1139/cjm-2014-0063

Wu Q., Patocka J., Nepovimova E., and Kuca K., 2018. A review on the synthesis and bioactivity aspects of beauvericin, a Fusarium mycotoxin, *Frontiers in Pharmacology*, 9, 1338 doi: 10.3389/fphar.2018.01338. eCollection 2018

Wüthrich M., Filutowicz H.I., Warner T., Bruce S., and Klein B.S. 2002. Requisite elements in vaccine immunity to Blastomyces dermatitidis: Plasticity uncovers vaccine potential in immune-deficient hosts - PubMed (nih.gov). *Journal of Immunology*, 169(12), 6969–6976. doi: 10.4049/jimmunol.169.12.6969

Xin H., Cartmell J., Bailey J.J., Dziadek S., Bundle D.R., and Cutler J.E. 2012. Self-adjuvanting glycopeptide conjugate vaccine against disseminated candidiasis - PubMed (nih.gov), *PLoS One* 7(4), e35106. doi: 10.1371/journal.pone.0035106. Epub 2012 Apr 26

Xiong Z.Q., Yang Y.Y., Zhao N. et al. 2013. Diversity of endophytic fungi and screening of fungal paclitaxel producer from Anglojap yew. *Taxus x Media. BMC Microbiology*, 13, 71. doi:10.1186/1471-2180-13-71

Yadwad V.B.,Wilson S., and Ward O.P. 1996. Effect of culture conditions and induction strategies on production of human interleukin-6 by a recombinant Aspergillus nidulans strain. *Mycological Research*, 100(3), 356–360. doi: 10.1016/S0953-7562(96)80169-9

Yakovleva M.B., Khoang T.L., and Nikitina Z.K. 2006. Collagenolytic activity in several species of deuteromycetes under various storage conditions (infona.pl). *Applied Biochemistry and Microbiology*, 42(4), 431–434

Zhang W., Tao J., Yang X., … and Wu J. 2014. Antiviral effects of two Ganoderma lucidum triterpenoids against enterovirus 71 infection. *Biochemical and Biophysical Research Communications*, 449(3), 307–312. doi: 10.1016/j.bbrc.2014.05.019. Epub 2014 May 15

24 Virosomes
A Drug Delivery System

Kuntal
Gargi College, University of Delhi, New Delhi, India

Aashita Singh and Twinkle Kathuria
University of Delhi, New Delhi, India

Seema Kalra
School of Sciences, Indira Gandhi National Open University, New Delhi, India

Madhu Yashpal
Gargi College, University of Delhi, New Delhi, India

CONTENTS

DOI: 10.1201/9781003306931-27

24.1 INTRODUCTION

In the current era of drug development, delivery to the target organ is the major challenge. The three significant hurdles that need to be overcome are: (i) the drug should be able to cross the permeability barrier of the cell membrane; (ii) once inside the cell, it should be protected from degradation; and (iii) its release in the cell environment should be controlled. Virosomes can fulfill all three criteria, which is why they are a popular option for targeted delivery of drugs and a large number of other biological molecules such as proteins, peptides and genetic material inside the cell. Virosomes are reconstituted viral envelopes with viral spike glycoproteins but no viral genetic material. They retain their property to enter target cells and elicit an immune response but cannot produce disease and reproduce. Because of these properties, virosomes have diverse applications in clinical research and therapeutics. Lipid-based particles have also proved a suitable medium for mucosal vaccination (Corthésy and Bioley, 2018).

Medical Subject Headings (2001) defined virosome as a "semi-synthetic complex derived from nucleic-acid free viral particles or reconstituted viral coats. A compound of choice replaces the infectious viral nucleo-capsid. Virosomes hold on to their fusogenic activity and thus deliver antigen, drug, gene, etc., into the target cell." Almeida and co-workers synthesized the first influenza virosomes (1975) by inserting purified influenza spike proteins into preformed liposomes. Virosomes-based vaccination (virus-like particle vaccine technology) has already been approved in more than 40 countries for the elderly and infants.

The Medical Subject Headings (2001) definition points to two specific features: (i) a delivery system and (ii) an antigenic function. Since the receptor-binding and membrane-fusion properties of the parent virus are retained, virosomes are widely used in the delivery of nucleic acids, antibiotics, anticancer agents, peptides, and steroids. In addition, the viral envelope glycoproteins can stimulate a significant humoral response.

24.1.1 Structure and Composition

Virosomes are spherical, unilamellar, non-replicating, fusion-active vesicles with a mean diameter of around 150 nm (Figure 24.1). The first virosomes were formed using the influenza virus by relocating its surface glycoproteins, namely haemagglutinin and neuraminidase, within liposomal vesicles. Antigenic peptides were encapsulated to induce effective antigen-specific cytotoxic T lymphocytes (CTL) responses. Incorporating phospholipids such as di-stearoyl phosphatidylcholine and cholesterol in the membranes enhanced their lymphatic absorption and stability in vivo. This virosome could successfully elicit Class I- and Class II-specific immune responses (Almeida et al., 1975).

Influenza virus is the most commonly used virus for the formation of virosomes. It is an orthomyxovirus with a nucleocapsid with eight negative-strand RNA segments, which encode ten influenza proteins. These RNA segments form ribonucleoproteins with tightly bound nucleoprotein (NP) and viral polymerases (PA, PB1, PB2). Three subgroups of influenza have been classified to date based on the internal nucleoprotein (NP) and the matrix protein (M1): influenza A, B and C. The viral envelope has two major types of membrane proteins, hemagglutinin (HA) and neuraminidase (NA), and a small number of copies of the M2 protein (Liu et al., 2015). Fifteen different HA and nine different types of NA have been recognized to date. There are about 500 spikes composed of a HA trimer in the viral membrane, in which each monomer consists of two polypeptide chains, one each of HA1 and HA2 linked by a disulfide bridge. HA facilitates the binding of the virus to its cellular receptor, sialic acid. Attachment of the influenza virus via trimeric HA at multiple points on the host-cell surface initiates its endocytosis and internalization. In addition, 100 spikes composed of a NA tetramer are present on the viral envelope. The release of the new generation of virions is enabled by prevention of the binding of newly formed virus particles to the membrane of the virus-producing cell. NA cleaves the sialic acid residues from the cell membrane and may also permit transport of the virus through the mucin layer present in the respiratory tract, enabling the virus

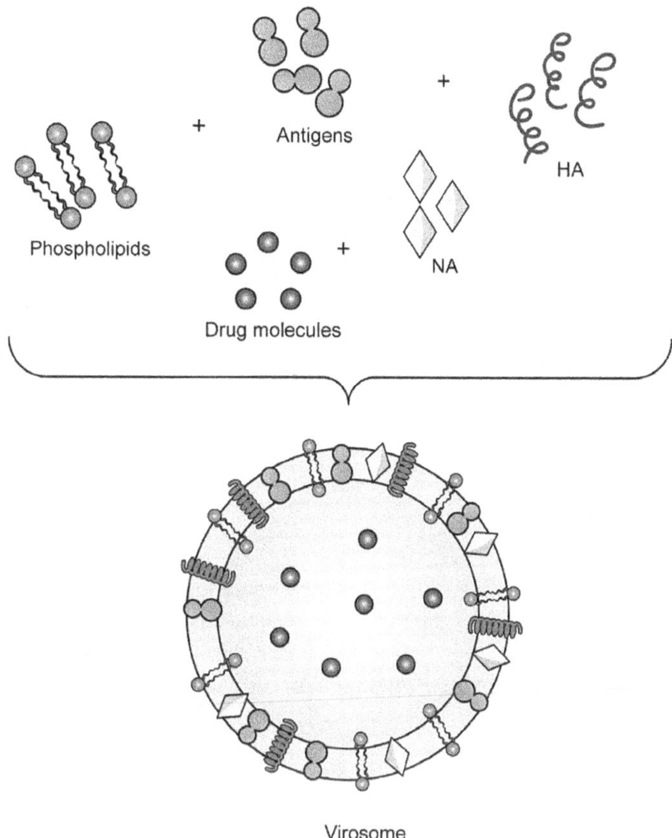

FIGURE 24.1 Structure and composition of Virosomes.

to get to the epithelial cells (Daemen et al., 2005). The membrane also has a small number of M2 proteins (Huckriede et al., 2003).

Sendai virus (HVJ, the haemagglutinating virus of Japan) is an ssRNA-containing paramyxovirus with two types of glycoproteins, fusion protein (F) and haemagglutinin-neuraminidase (HN), on the surface of its viral envelope. HN facilitates the binding of viral particles to the sialic acid residues present on the host-cell surface, and F induces fusion of the viral envelope with the host-cell membrane to enter the cell. Thus, unlike influenza viruses, these viruses do not depend on endosome-mediated uptake.

Kammer et al. (2007) developed a new generation of virosomes that can induce cytotoxic T-cell and humoral responses. These are known as IRIVs (immuno-potentiating reconstituted influenza virosomes). IRIVs contain TC- or DC-cholesterol as stabilizers which allow them to be stored by lyophilization. After reconstitution, they still retain their membrane-fusion activity and immunogenic properties.

24.1.2 Properties

Virosomes are non-toxic, biodegradable, biocompatible and non-autoimmunogenic. Their prime benefit is that they can be modified as required by replacing or including specific lipids. For example, cationic and synthetic lipids are added to DNA virosomes to enhance immunogenicity in IRIVs. Second, along with antigenicity, virosomes retain their fusion ability from their viral origin, which is essential for new-age therapeutics. In addition, hydrophobic antigens can be either adsorbed on or into the virosome membrane, whereas hydrophilic antigens can be added into the cytoplasm. Non-lipophilic compounds can also be attached to cholesterol and phospholipids such as phosphatidylserine, sphingomyelin, phosphatidylcholine and phosphatidylethanolamine.

To concentrate nucleic acids in the virosomes, cationic lipids are often added which facilitate targeted cellular delivery of genes due to a positive charge on their head group, which assists in binding to the negatively charged DNA. At present, a vast number of cationic lipids are known, including stearyl amine, DODAC (N, N-dioleyl-N, N, dimethyl ammonium chloride), DOTAP (N-[1-(2, 3-dioleoyloxy) propyl] - N, N, N-trimethylammonium chloride), etc. The most popular and promising of these is DODAC. Cationic lipids also facilitate in vivo delivery of siRNA (Sioud and Sørensen, 2003), protecting it from nuclease activity and promoting its fusion and uptake (Huckriede et al., 2003).

Several ligands such as monoclonal antibodies (MAbs), peptides and cytokines that bind cellular epitopes present on the surfaces of specific cell types can be conjugated to the virosome. Even Fab (tumor-specific monoclonal antibody fragments) can be linked to virosomes to deliver the carrier to selected tumor cells (Kalra et al., 2013).

24.1.3 Viruses Used

Various viruses have been used to develop virosomes, including human immunodeficiency virus, hepatitis B virus, influenza virus, Sendai virus, Newcastle disease virus, Sindbis virus, Rubella virus, Epstein–Barr virus, leukemia virus, Semliki Forest virus, vesicular stomatitis virus, herpes simplex virus, respiratory syncytial virus, cytomegalovirus and Chikungunya virus.

24.1.4 Advantages of Virosomes

The list of advantages of virosomes is long:

1. Virosomes are biodegradable, biocompatible, and non-lethal.
2. The FDA has approved virosomes for human use with a high safety profile.
3. There is no risk of transmission of disease.
4. There is no chance of getting anaphylaxis or auto-immunogenicity by using virosomes.
5. Virosomes can transport the drug into the cytoplasm of the target cell and lead to the amplification of the immune response.
6. Distribution, uptake and elimination of the drug in the body is extended.
7. Drugs are not degraded because they provide protection.
8. They promote fusion activity in the endolysosomal pathway.
9. A patient-specific modular vaccine regimen is allowed by virosomes.
10. Virosomes can be introduced by nasal route or injection.

These advantages support their broad applications as carriers of anticancer drugs, proteins, peptides, nucleic acids, antibiotics and fungicides.

24.1.4.1 Comparisons With Liposomes/Actual Viruses

Liposomes are a promising vehicle for delivering and targeting biologically active molecules to living cells both in vitro and in vivo, but they have limited capability to fuse with cells and thus fail to deliver encapsulated molecules appropriately to the cell cytoplasm. Virosomes have functional viral envelope glycoproteins with membrane-fusing properties and receptor binding that enables the adequate cellular delivery of encapsulated molecules.

Endosomes of host cells have low pH, which accelerates HA-mediated cell fusion where the molecule packed inside the virosome membrane is released into the cytosol of the target cell, enhancing cytosolic delivery. Virosomal technology is superior to liposomal and proteoliposomal systems

TABLE 24.1
Different Techniques Used for Biophysical and Biological Characterization of Virosomes (Singh et al., 2017)

S.No.	Property	Technique
1.	Protein Detection	SDS-PAGE
2.	Surface morphology and vesicle shape	Transmission electron microscopy Freeze break electron microscopy
3.	Size dispersion and vesicle size	Negative stain electron microscopy Transmission electron microscopy Dynamic light scattering Zeta sizer Photon connection spectroscopy Laser light diffusing Gel saturation and gel avoidance
4.	Surface charge	Free stream electrophoresis
5.	Lamellarity and phase behavior	freeze fracture electron microscopy 13-p NMR (nuclear magnetic resonance) Small edge x-beam dissipating Differential checking colorimetry.
6.	Drug release	Diffusion cell dialysis
7.	Surface chemical analysis	Static auxiliary particle mass spectrometry
8.	Pyrogenicity	Rabbit fever test Limulus ambeocyte lysate (LAL) test
9.	Fusion activity	Fluorescent resonance energy transfer assay (FRET)
10.	Percent of free medication	Mini-section centrifugation Gel avoidance and ion trade chromatography Protamine accumulation Radiolabelling
11.	Surface pH and electrical surface potential	Zeta potential estimations pH touchy tests
12.	Animal poisonous quality	Observing survival rates, histology and pathology

because these have low protection for therapeutic biomolecules from extreme microenvironments like high acid content and low pH inside organelles.

Moreover, vectors derived from enveloped viruses have an advantage as molecules delivered by such vectors escape endosomal degradation (Earp et al., 2004). Thus, the mode of action of virosomes is unlikely to result in adverse side-effects, such as cytotoxicity or non-specific inflammatory responses at the injection site.

24.1.5 Characterization

Table 24.1 summarizes the different techniques that have been used for biophysical and biological characterization of virosomes (Singh et al., 2017).

24.2 ATTACHMENT AND FUSION

The influenza virosome derived from dendritic cells and effective internalization of influenza virus is very significant for vaccines. Receptor-mediated endocytosis is induced by the binding of HA to the cell receptors, delivering virosome-encapsulated antigen-to-antigen presenting cells (APCs). Depending on their parent virus, virosomes can either fuse at a neutral pH (pH 7.4) with the plasma membrane (e.g., Sendai virus-derived virosomes) or at an acidic pH with the endosomal membrane (e.g., influenza virosomes) where fusion activity of HA is induced through a conformational change in acidic conditions (pH 4.5). If virosomes are to be used as vaccines, fusion activity is not essential for generating antibody responses against the virus they are derived from. Virosomes are liberated from their lipid envelope due to fusion of endosome and provide entry for the encapsulated drugs to the cytosol (Huckriede et al., 2005).

Microscopic analysis of macrophages incubated with virosomes and polystyrene particles reveals that virosomes can efficiently enter the cell even in the presence of cytochalasin D, a drug inhibiting actin-based phagocytosis (Hofer et al., 2009). In a study using influenza virus-encapsulated diphtheria toxin (DTA) lacking the fragment B responsible for receptor binding and entry of the molecule in the cell, successful delivery of DTA in cell cytosol was reported because of fusion-active properties of virosomes (Bron et al., 1994).

24.3 INTERACTION WITH IMMUNE SYSTEM

The antigens conjugated to the viral structures are highly stabilized and protected from degradation by preserving the B-cell epitopes. Furthermore, if integrated into the surface membrane, these antigens produce enhanced humoral response and memory (Figure 24.2). Therefore, most people have pre-existing immunity to influenza. This intensifies the action taken against influenza virosomes. Apart from this, the pathogens associated with molecular patterns (PAMP) stimulate APCs leading to TLR (toll-like receptor) activation (Rathore and Swami, 2012).

Antigen presentation is the physical presence of the membrane-bound or encapsulated antigens in the APCs (e.g., macrophages, dendritic cells, etc.), producing effective humoral and cell-mediated immunity against it by stimulation of either MHC I or MHC II presentation pathways. Surface-conjugated antigens are degraded during endosomal fusion into peptides, which are coupled to MHC Class II receptors and then these peptide/MHC Class II complexes are transported to the plasma membrane and presented to CD4+ T cells (Piccirillo & Shevach, 2001; Sakaguchi, 2005). Encapsulated antigens, on the other hand, are released into the cytosol of APCs upon endosomal fusion. These peptides are processed in the proteasomes and presented to the CD8+ T lymphocytes by MHC Class I receptors. Often encapsulated antigens, if escaped from the virosome, are degraded in the endosomes and loaded on MHC Class II receptors. The activated T helper cells mediate a

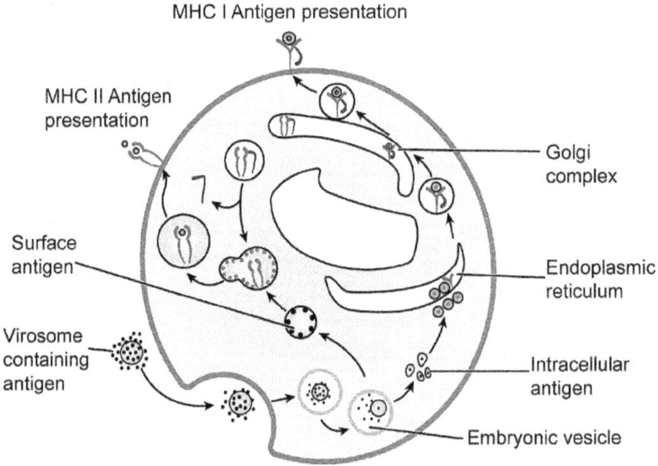

FIGURE 24.2 Interaction of virosomes with the immune system.

humoral response by stimulating B-lymphocytes leading to the production of antigen-specific antibodies. IRIVs enhance MHC Class I CTL induction through CD4+ T-cell activation, as shown by a recent study (Schumacher et al., 2004).

First successful administration of an influenza virosome in mice against the influenza virus was carried out in 1989. Several antibodies of various subclasses were elicited by intranasal (IgM, IgA, IgG) and intramuscular (IgG, IgM) immunization against HA.

24.4 PREPARATION

Influenza virosomes preparation includes three discrete steps (Figure 24.3):

i. The viral membrane is solubilized with detergent;
ii. internal viral proteins and viral RNA are removed; and
iii. membrane proteins are reconstituted in a lipid environment by stepwise removal of the detergent. (Saga and Kaneda, 2013)

The first step of IRIV formation is to inactivate the influenza virus and collapse to phospholipids by treating with detergent like beta-propiolactone. In 1987, Stegmann et al. found C12E8 to be the most suitable detergent for functional reconstitution of HA. The surface proteins (NA and HA) are then purified and incorporated into the biological membrane and virus-derived phospholipids, thus reconstituting the viral envelope. Proper reconstitution of virosomes is essential for efficient cytosolic delivery. At the same time, stepwise selective detergent removal using BioBeads is carried out (Gluck et al., 2005). In the case of DNA virosomes, after purification of the reconstituted virosomes by sucrose density gradient, they are incubated with plasmid DNA for 15 min at room temperature.

Virosomes containing immuno-stimulating complexes (ISCOMS), immuno-modulators such as muramyl dipeptide or cationic lipids are called immuno-potentiating reconstituted influenza virosomes or IRIVs.

Similarly, Sendai virus virosomes are produced by reconstitution of the Sendai fusion protein (F-protein) with or without the HA and NA proteins in viral lipids (Bagai and Sarkar, 1993, 1994). E1 and E2 envelope glycoproteins are incorporated into liposomal vesicles to prepare rubella virus virosomes. G-protein of VSV is added to preformed liposomes to generate Vesicular stomatitis virus (VSV).

A new variety of Sendai virosome called HVJ-envelope (HVJ-E) is generated, which is an inactivated HVJ particle that has been irradiated by UV light (Kaneda et al., 2002). These HVJ-E virosomes have significantly higher membrane-fusion ability but lack the potential to produce progeny virus in infected cells

24.5 APPLICATIONS OF VIROSOSMES

The property of the virus to interact with host cells and the ability to escape host immune attacks leads to improved modes of turning in bio-energetic molecules using virus-mediated generation. Characteristics such as biocompatibility and biodegradability are developed in virosomes because of membrane reconstitution. Since they are assembled in vitro, their length and structure can be modulated such that a broad range of bioactive molecules may be associated with them for focused delivery and their ability to multiply once inside the host machinery can be inhibited.

Virosomes have been used to transport molecules like proteins, nucleic acids, pills, and antibodies to diverse cell types, including gliomas cells, erythrocytes, hepatocytes, etc. (Kim et al., 2005; Kalra and Sharma, 2021; Yamada et al., 2003)

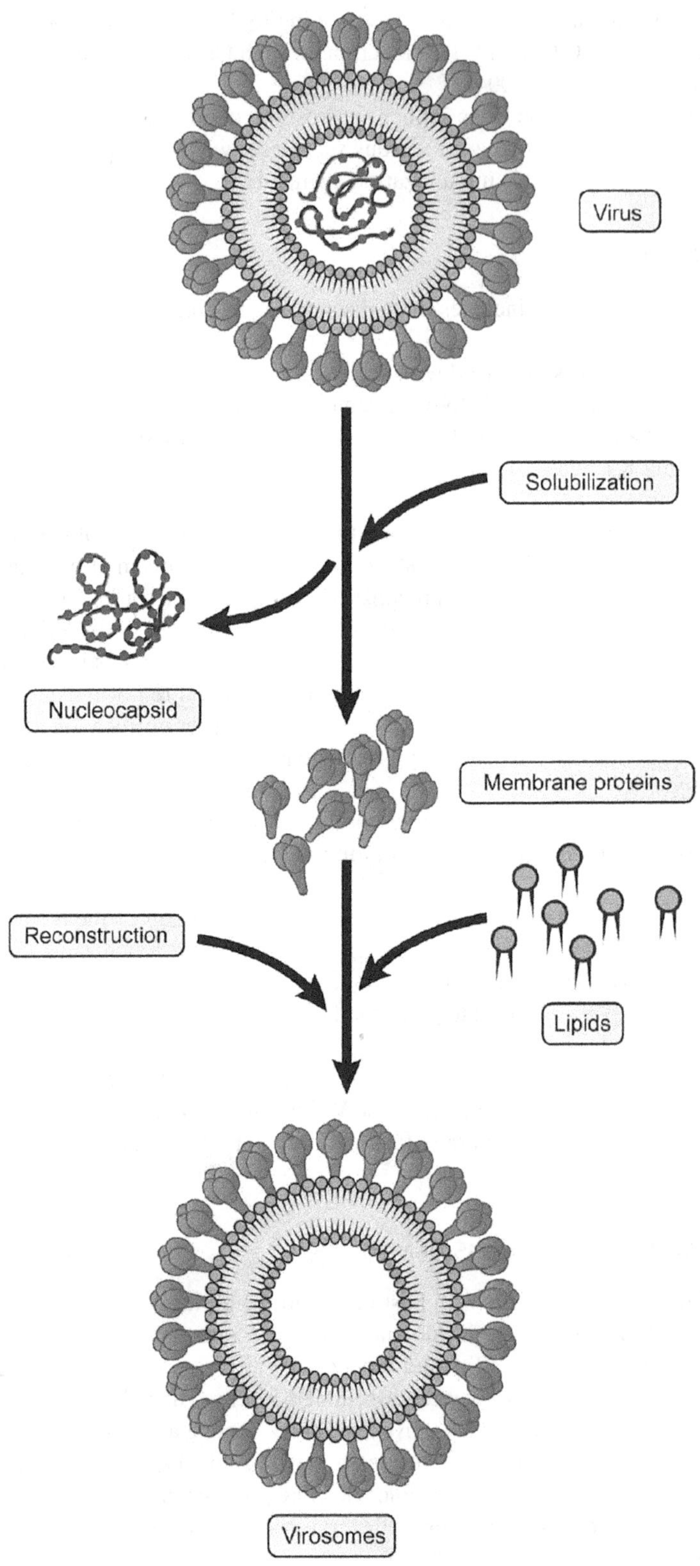

FIGURE 24.3 Preparation of virosomes from virus.

24.5.1 Virosomes for Vaccination

24.5.1.1 Virosomes for Antigen Delivery

One of the approaches is antigen delivery by virosomes, which involves antigen processing and presentation pathways. These pathways lead to the activation of the immune system involving both humoral and cellular immunity. Antibody responses are boosted by virosomes with surface-exposed antigen due to improvement of the 3D structure of the antigen, increasing antigen density, leading to greater cross-linking of B-cell receptors (Tregoning et al., 2018).

Influenza virosomes can deliver unrelated antigen of interest (AoI). This generates the immune response against AoI, which is amplified by glycoproteins present on influenza virosomes (Zhou et al., 2019). This was seen in vitro in human peripheral blood mononuclear cells (PBMC) where CD4/CD45RO+ T cells were activated and induced the cytokine profile associated with T helper 1 (Th1) stimulation, caused by virosomes. As a response to the antigen, they cause the secretion of cytokines like GM-GSF, TNF-α, IFN-γ and IL-2 in PBMC (Schumacher et al., 2004). They also result in stimulating the maturation of dendritic cells. The triple co-culture model has shown the ability of virosomes to act as an antigen carrier and delivery agent (Blom et al., 2016). A study on epithelial cells, macrophages and dendritic cells (Blom et al., 2016) concluded that virosomes do not cause any extensive inflammation and modulate the immune response in mucosal tissue of the respiratory tract.

Many virosomes have been developed and studied, leading to a supply of licensed vaccines in the market. IRIV has been used as an antigen carrier system for effective hepatitis A and influenza vaccination because of its appropriate size, structure, and composition. Various attempts have also been made to combine multiple targets like hepatitis B, diphtheria, and tetanus within a single IRIV.

24.5.1.2 Virosomes as Adjuvants and Complexed with Adjuvants

The term adjuvant is derived from a Latin word that means 'to help/enhance'. An adjuvant in a vaccine has the vital function of enhancing the host immune system. Commonly used aluminum-based salts as adjuvants have side-effects such as swelling, local inflammation and pain. In contrast, virosomes are considered safe adjuvants as they do not show any side-effects and have better adjuvant potency than aluminum salts (Figure 24.4). Virosome formulations of diphtheria and tetanus toxoids

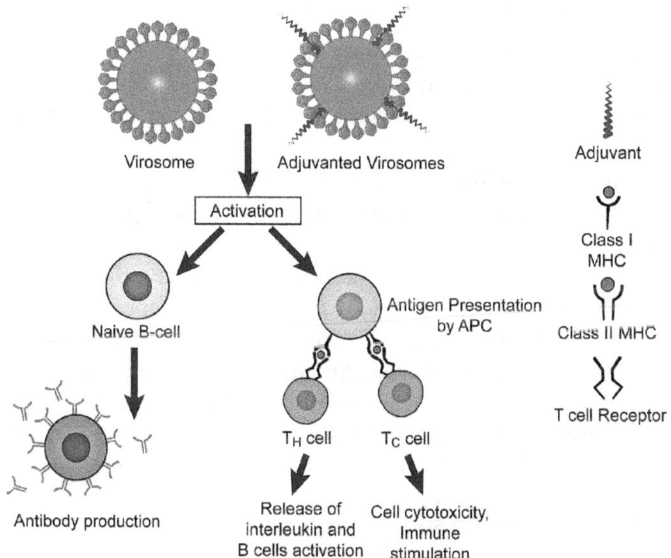

FIGURE 24.4 Mechanism of action of virosomes as adjuvants and virosomes complexed with adjuvants.

have been found to be better in terms of immunogenicity than those formed by aluminum salts (Zurbriggen, 1999).

Virosomes are also used to include adjuvants in their membrane. N-palmitoyl-S-2,3(bispalmitoyloxy)-propylcysteinyl-seryl-(lysil)3-lysine was incorporated in influenza virosomes, a lipopeptide adjuvant recognized by toll-like receptor present on B-lymphocytes and dendritic cells (Spohn et al., 2004). Adjuvant virosomes were injected in immunologically naïve Balb/C mice, causing an increased IgG response that was 150-fold more than virosomes without an adjuvant.

Cationic influenza virosomes are a preparation that involves solubilization of the viral membrane in detergent in the presence of cationic lipids that can be used for adjuvant delivery. Studies using these virosomes showed stimulation of CTLs in RAW Blue Cells, cells derived from murine RAW 264.7 macrophages/a mouse macrophage cell line (Jamali et al., 2012).

Virosomes complexed with adjuvants show immune stimulation. Vaccines with similar properties have been developed, for example, virosomal respiratory syncytial virus vaccine is adjuvanted with monophosphoryl lipid A (RSV-MPLA) (Kamphuis et al., 2012). This type of formulation shows a good response in mice and cotton rats. Studies have demonstrated that RSV-MPLA (Kamphuis et al., 2013) is a safe and immunogenic vaccine.

24.5.1.3 Some Commercial Virosome-Based Vaccines

Epaxal®

Epaxal vaccine works against hepatitis A. It is an adjuvant-based vaccine where the virosome act as an adjuvant. It was licensed in 1996. It comprises lecithin and cephalin as phospholipid components (Bovier, 2008).

Inflexal®VBerna

Inflexal® V involves the mixture of three monovalent virosome pools, which are formed from the influenza strain's specific HA and NA glycoproteins (Mischler and Metcalfe, 2002). This vaccine has been approved to treat influenza in all age groups. The influenza strains are selected annually on the recommendation of WHO and the European Medicines Agency. In Germany, Inflexal®V is also sold under the trade name InfectoVac® Flu, as Isiflu® V in Italy and as Viroflu® in the United Kingdom.

24.5.2 Virosomes for Delivery of Nucleic Acids

24.5.2.1 Virosome for DNA Delivery

With the ability to selectively bind and fuse with the cell membrane, virosomes can easily entrap DNA just like liposomes through cationic lipids, resulting in the formulation of DNA-containing (cationic) virosomes for gene delivery. Virosomes with DNA are helpful in providing a strong immune response at low doses.

Cationic lipids, mainly dioleoyldimethylammonium (DODAC) and dioleyloxypropyltrimethylammonium methyl sulfate (DOTAP), have a positively charged head group which helps in the binding of negatively charged DNA.

DODAC is responsible for the transfection activity of virosomes, which can be achieved at the highest by 30 mol % of DODAC density in the virosome membrane. The entry of DNA into the cytosol depends on the fact that HA-mediated fusion is a leaky process, allowing the DNA to enter the cytoplasm from endosomes during fusion. Another way can be the cytoplasmic entry caused by the destabilization of the endosomal membrane followed by penetration of HA fusion proteins. Some studies also show the use of the DOTAP-influenza virosomes to deliver DNA or oligonucleotides in human tumor and dendritic cells (E.R. Waelti and Glück 1998).

24.5.2.2 Delivery of siRNA by Virosomes

In eukaryotic cells, regulation of gene expression takes place by a fundamental mechanism of RNA interference. Virosomes can be used to deliver small interfering RNA (siRNA) into the cells. They are delivered as recombinant viruses having siRNA of interest (Nishimura et al., 2013). Delivery agents are essential, as free siRNA duplex molecules are difficult or poorly taken by cells in vitro. Cationic lipids and liposomes are also used for their in vivo delivery (Sioud and Sørensen, 2003) (Huckriede et al., 2007). Another method involves the use of viral capsid protein VP1 from Simian virus 40 (SV40) for packaging of siRNA into pseudovirions, which were not only able to enter target cells but also efficiently downregulated the targeted gene in vitro (Kimchi-Sarfaty et al., 2005).

Certain preparation methods are required for virosomes to deliver siRNA. These involve removal of viral genetic material and addition of solubilized membrane components and cationic lipid (DODAC or DOTAP) to the complex siRNA (De Jonge et al., 2005). The final structure of virosomes containing siRNA is confirmed by equilibrium sucrose density gradient analysis.

24.5.3 Virosomes in Cancer Treatment

An approach being used in cancer immunotherapy research, and an accepted option in advanced stages of cancer, involves increasing the innate immune response to fight against tumors. Studies show that virosomes can be regarded as a good option for the treatment of prostate cancer, breast cancer and liver cancer (Redd Bowman et al., 2020). They can be used as vectors for antigen delivery; these vectors are derived from an inactive enveloped virus. Virosomes obtained from influenza and Sendai viruses are commonly used for cancer treatment. Influenza virosomes are mostly used as adjuvants or drug delivery vectors, whereas Sendai virosomes have been used as apoptosis inducer and anticancer immune system activators.

24.5.3.1 Influenza Virosomes in Cancer Treatment

Influenza virosomes show an adjuvant effect. Studies prove that they can activate antitumour immunity. When activated systemically, immune cells destroy primary cancer cells and distant metastatic cells. Immune cells like cytotoxic T lymphocytes (CTL) target tumour-associated-antigens (TAAs), which are expressed by cancer cells. (Adamina et al., 2006) Activation of anticancer immunity requires antigen presentation of TAAs by APC. TAA should be presented with MHC-1 to activate CTLs and to gain entry to the cytoplasm (Bungener et al., 2002). Virosomes carry them in encapsulated form and release them into the cytoplasm by membrane fusion.

Studies show that influenza virosome having CD40L-expressing plasmid enhances the TAA-specific CTL stimulation. This CD40L binds to CD40 on antigen-presenting cells and increases the expression of specific molecules like B7.1 and B7.2 in the cells, which are essential for activation and amplification of naïve T cells (Cusi et al., 2005). Studies have demonstrated that when CD40L encapsulated in influenza virosomes was introduced intranasally with plasmids expressing carcinogenic antigen (CEA), is a marker of colon cancer, the co-administration of CEA and CD40L resulted in a CEA-specific CTL response which was much stronger than that shown by CEA virosomes alone, showing elevated expression of B7.1 and B7.2 on APCs.

24.5.3.2 Sendai Virosomes in Cancer Treatment

Sendai virus, also known as the haemagglutinating virus of Japan (HVJ) and its modified form HVJ-E is commonly used for virosomes production (Shimamura et al., 2003). The external surface consists of two-membrane glycoproteins, haemagglutinin-neuraminidase (HN) and fusion protein (F). Interestingly, fusion protein does not require acidic conditions as it works well in neutral condition for membrane fusion. Hence, unlike influenza virosomes, HVJ virosomes do not require the endosome to be reached for membrane fusion (Mughees et al., 2021). Studies show that HVJ-E can eradicate intradermal cancer (Kurooka and Kaneda, 2007). It was found that HVJ-E causes the release of many types of cytokines such as tumor-necrosis factor (TNF)-α, interferon (IFN)-α and (IFN)-β, and interleukin (IL)-6 by stimulating dendritic cells. Tregs are regulatory T cells that negatively control effector T cells and prevent the activation of anticancer immunity (Sasada et al., 2003). IL-6 helps inhibit the proliferation of Tregs. Therefore HVJ-E can be regarded as a powerful activator of anticancer immunity by suppressing the Treg count.

24.5.3.3 Virosomes as Drug Delivery Agents in Cancer Treatment

Virosomes can be regarded as an appropriate microbiological tool for the delivery of drugs to specific targeted cancerous cells. This requires a display of certain molecules on their surfaces such as antibodies or affibodies (Huckriede et al., 2003). Lately, virosomes have been used to deliver cancer-specific siRNA with affibody molecules of HER2 (human epidermal growth factor receptor 2), expressed on the surface of breast and ovarian cells (Nishimura et al., 2013). The latest technology also involves using magnetic forces to drive virosome-formulated anticancer medicines into the cancerous tissue. Decitabine is based on a similar technique that uses erythro-magneto-hemagglutinin virosomes (EMHVs) drug delivery system. The use of this efficient and improved delivery system resulted in a reduction of tumor mass in xenograft models of prostate cancer at low concentration (Naldi et al., 2014). A formulation of nano curcumin-based hybrid virosomes (NC-virosome) to deliver anticancerous drugs is also represented (Kumar et al., 2021).

24.5.3.4 Cervical Cancer Treatment With Virosomes

Virosomes used in cervical treatment have shown positive results, and more work is being done on vaccines based on virosome-mediated immunotherapy (Bungener et al., 2006). It was demonstrated that in the murine model system, influenza virosomes, which contain recombinant human papillomavirus, type 16 (HPV16), E7 protein antigen was helpful in treating HPV-transformed tumor cells. A combination of both humoral and cytotoxic immune responses has proven to reduce tumor growth by 70% in animals immunized by virosomes having E7 protein. This represents the efficiency of influenza-derived virosomes containing E7 protein in treating cervical cancer (Bungener et al., 2006). Another study by Walczak et al. (2010) also found that virosomes with HPV16 E7 protein caused an increase in the numbers of antigen-specific cytotoxic T lymphocytes (CTL) in mice, which helps treat cancer.

24.5.4 Virosomes in Covid Treatment

Virosomes can be used in SARS-CoV-2 vaccine formation. Recently some studies have shown that there are certain proteins on SARs-CoV that can act as targets, like E protein. Additionally, some coronaviruses also comprise hemagglutinin-esterase protein (HE). Its lectin area allows the attachment of virus and host cells. Hence only particular phospholipids and antigens of SARS-CoV-2 that can be used in virosome vaccine formation are purified. Transvac-2 is a virosome-based vaccine project currently in progress in Europe.

24.5.5 Limitations of Virosomes

1. Virosomes have viral glycoproteins on their outer surface, so their prophylactic action might induce an immunological reaction.
2. Virosomes are degraded easily and very rapidly inside the blood compartment. However, this problem can be resolved by reducing the time taken by it to reach the target site after administration.

24.6 CONCLUSION

Virosomes constitute another innovative and advanced drug-conveyance framework for biologically active molecules, including but not limited to proteins, peptides, plasmids, oligonucleotides, genes and drugs. The outer layer of virosomes can be appropriately altered to work with designated drug delivery. Virosomes can be used to transport an antigen in the host body through changed courses like intramuscular, intradermal and intranasal, taking into consideration any type of side-effect. But there is a need for comprehensive studies on bioavailability, pharmacokinetic profile, clinical effects, safety and stability to ascertain their long-term reliability as a safe, effective and affordable method of targeting and delivering the drug. The development of virosomes is restricted due to complicated assay processes for characterizing virosome products. There is a need for more efforts to speed up the development of virosome-based products.Virosomes work on receptor-mediated controlled release of biomolecules at the target site, with the extra advantages of activating both humoral and cell-mediated immunity. The ultimate successful application of RNA interference in vivo depends on the development of suitable delivery systems for siRNA.

Much work is needed, both at research and practical levels, to produce feasible models of virosomes that are safe to use, cost-effective, and have broad clinical applications.

ACKNOWLEDGMENTS

We are grateful to the many authors who provided guidance for their hard work in the completion of this chapter.

REFERENCES

Adamina, M., Guller, U., Bracci, L., Heberer, M., Spagnoli, G. C., & Schumacher, R. (2006). Clinical applications of virosomes in cancer immunotherapy. *Expert Opinion on Biological Therapy*, *6*(11), 1113–1121. https://doi.org/10.1517/14712598.6.11.1113

Almeida, J., Edwards, D., Brand, C., & Heath, T. (1975). Formation of virosomes from influenza subunits and liposomes. *The Lancet*, *306*(7941), 899–901. https://doi.org/10.1016/s0140-6736(75)92130-3

Bagai, S., & Sarkar, D. P. (1993). Reconstituted Sendai virus envelopes as biological carriers: Dual role of F protein in binding and fusion with liver cells. *Biochimica et Biophysica Acta (BBA) - Biomembranes*, *1152*(1), 15–25. https://doi.org/10.1016/0005-2736(93)90226-p

Bagai, S., & Sarkar, D. P. (1994). Effect of substitution of hemagglutinin-neuraminidase with influenza hemagglutinin on Sendai virus F protein mediated membrane fusion. *FEBS Letters*, *353*(3), 332–336. https://doi.org/10.1016/0014-5793(94)01076-5

Blom, R. A. M., Erni, S. T., Krempaská, K., Schaerer, O., van Dijk, R. M., Amacker, M., Moser, C., Hall, S. R. R., von Garnier, C., & Blank, F. (2016). A triple co-culture model of the human respiratory tract to study immune-modulatory effects of liposomes and virosomes. *PLoS ONE*, *11*(9), e0163539. https://doi.org/10.1371/journal.pone.0163539

Bovier, P. A. (2008). Epaxal®: A virosomal vaccine to prevent hepatitis A infection. *Expert Review of Vaccines*, *7*(8), 1141–1150. https://doi.org/10.1586/14760584.7.8.1141

Bron, R., Ortiz, A., & Wilschut, J. (1994). Cellular cytoplasmic delivery of a polypeptide toxin by reconstituted influenza virus envelopes (Virosomes). *Biochemistry*, *33*(31), 9110–9117. https://doi.org/10.1021/bi00197a013

Bungener, L., de Mare, A., de Vries-Idema, J., Sehr, P., van der Zee, A., Wilschut, J., & Daemen, T. (2006). A virosomal immunization strategy against cervical cancer and pre-malignant cervical disease. *Antiviral Therapy*, *11*(6), 717.

Bungener, L., Serre, K., Bijl, L., Leserman, L., Wilschut, J., Daemen, T., & Machy, P. (2002). Virosome-mediated delivery of protein antigens to dendritic cells. *Vaccine*, *20*(17–18), 2287–2295. https://doi.org/10.1016/s0264-410x(02)00103-2

Corthésy, B., & Bioley, G. (2018). Lipid-based particles: Versatile delivery systems for mucosal vaccination against infection. *Frontiers in Immunology*, *9*, 431.

Cusi, M. G., del Vecchio, M. T., Terrosi, C., Savellini, G. G., di Genova, G., La Placa, M., Fallarino, F., Moser, C., Cardone, C., Giorgi, G., Francini, G., & Correale, P. (2005). Immune-reconstituted influenza virosome containing CD40L Gene enhances the immunological and protective activity of a carcinoembryonic antigen anticancer vaccine. *The Journal of Immunology*, *174*(11), 7210–7216. https://doi.org/10.4049/jimmunol.174.11.7210

Daemen, T., de Mare, A., Bungener, L., de Jonge, J., Huckriede, A., & Wilschut, J. (2005). Virosomes for antigen and DNA delivery. *Advanced Drug Delivery Reviews*, *57*(3), 451–463.

de Jonge, J., Holtrop, M., Wilschut, J., & Huckriede, A. (2005). Reconstituted influenza virus envelopes as an efficient carrier system for cellular delivery of small-interfering RNAs. *Gene Therapy*, *13*(5), 400–411. https://doi.org/10.1038/sj.gt.3302673

Earp, L. J., Delos, S. E., Park, H. E., & White, J. M. (2004). The many mechanisms of viral membrane fusion proteins. *Current Topics in Microbiology and Immunology*, 25–66. https://doi.org/10.1007/3-540-26764-6_2

Gluck, R., Burri, K., & Metcalfe, I. (2005). Adjuvant and antigen delivery properties of virosomes. *Current Drug Delivery*, *2*(4), 395–400. https://doi.org/10.2174/156720105774370302

Hofer, U., Lehmann, A. D., Waelti, E., Amacker, M., Gehr, P., & Rothen-Rutishauser, B. (2009). Virosomes can enter cells by non-phagocytic mechanisms. *Journal of Liposome Research*, *00*(00), 090513010250098–090513010250099. https://doi.org/10.1080/08982100902911612

Huckriede, A., Bungener, L., Stegmann, T., Daemen, T., Medema, J., Palache, A. M., & Wilschut, J. (2005). The virosome concept for influenza vaccines. *Vaccine*, *23*, S26–S38. https://doi.org/10.1016/j.vaccine.2005.04.026

Huckriede, A., Bungener, L., ter Veer, W., Holtrop, M., Daemen, T., Palache, A. M., & Wilschut, J. (2003). Influenza virosomes: Combining optimal presentation of hemagglutinin with immunopotentiating activity. *Vaccine*, *21*(9–10), 925–931. https://doi.org/10.1016/s0264-410x(02)00542-x

Huckriede, A., de Jonge, J., Holtrop, M., & Wilschut, J. (2007). Cellular delivery of siRNA mediated by fusion-active virosomes. *Journal of Liposome Research*, *17*(1), 39–47. https://doi.org/10.1080/08982100601186516

Jamali, A., Holtrop, M., de Haan, A., Hashemi, H., Shenagari, M., Memarnejadian, A., Roohvand, F., Sabahi, F., Kheiri, M. T., & Huckriede, A. (2012). Cationic influenza virosomes as an adjuvanted delivery system for CTL induction by DNA vaccination. *Immunology Letters*, *148*(1), 77–82. https://doi.org/10.1016/j.imlet.2012.08.006

Kalra, A. & Sharma, S. (2021). Virosomes: A viral envelope system having a promising application in vaccination and drug delivery system. In *Nanopharmaceutical Advanced Delivery Systems*, eds Vivek Dave, Nikita Gupta, Srija Sur, ch 7 https://doi.org/10.1002/9781119711698

Kalra, N., Dhanya, V., Saini, V., & Jeyabalan, G. (2013). Virosomes: as a drug delivery carrier. *American Journal of Advanced Drug Delivery*, *1*(1), 29–35.

Kammer, A. R., Amacker, M., Rasi, S., Westerfeld, N., Gremion, C., Neuhaus, D., & Zurbriggen, R. (2007). A new and versatile virosomal antigen delivery system to induce cellular and humoral immune responses. *Vaccine*, *25*(41), 7065–7074. https://doi.org/10.1016/j.vaccine.2007.07.052

Kamphuis, T., Meijerhof, T., Stegmann, T., Lederhofer, J., Wilschut, J., & de Haan, A. (2012). Immunogenicity and protective capacity of a virosomal respiratory syncytial virus vaccine adjuvanted with monophosphoryl lipid a in mice. *PLoS ONE*, *7*(5), e36812. https://doi.org/10.1371/journal.pone.0036812

Kamphuis, T., Shafique, M., Meijerhof, T., Stegmann, T., Wilschut, J., & de Haan, A. (2013). Efficacy and safety of an intranasal virosomal respiratory syncytial virus vaccine adjuvanted with monophosphoryl lipid A in mice and cotton rats. *Vaccine*, *31*(17), 2169–2176. https://doi.org/10.1016/j.vaccine.2013.02.043

Kaneda Y. (2000). Virosomes: Evolution of the liposome as a targeted drug delivery system. *Advanced Drug Delivery Reviews*, *43*(2–3), 197–205. https://doi.org/10.1016/s0169-409x(00)00069-7

Kim, Y. D., Park, K. G., Morishita, R., Kaneda, Y., Kim, S. Y., Song, D. K., Kim, H. S., Nam, C. W., Lee, H. C., Lee, K. U., Park, J. Y., Kim, B. W., Kim, J. G., & Lee, I. K. (2005). Liver-directed gene therapy of diabetic rats using an HVJ-E vector containing EBV plasmids expressing insulin and GLUT 2 transporter. *Gene Therapy*, *13*(3), 216–224. https://doi.org/10.1038/sj.gt.3302644

Kimchi-Sarfaty, C., Brittain, S., Garfield, S., Caplen, N. J., Tang, Q., & Gottesman, M. M. (2005). Efficient delivery of RNA interference effectors via in vitro-packaged SV40 Pseudovirions. *Human Gene Therapy*, *0*(0), 050822135735001. https://doi.org/10.1089/hum.2005.16.ft-110

Kumar, V., Kumar, R., Jain, V. K., & Nagpal, S. (2021). Preparation and characterization of nanocurcumin based hybrid virosomes as a drug delivery vehicle with enhanced anticancerous activity and reduced toxicity. *Scientific Reports*, *11*(1), 1–14.

Kurooka, M., & Kaneda, Y. (2007). Inactivated sendai virus particles eradicate tumors by inducing immune responses through blocking regulatory T cells. *Cancer Research*, *67*(1), 227–236. https://doi.org/10.1158/0008-5472.can-06-1615

Liu, H., Tu, Z., Feng, F., Shi, H., Chen, K., & Xu, X. (2015). Virosome, a hybrid vehicle for efficient and safe drug delivery and its emerging application in cancer treatment. *Acta Pharmaceutica*, *65*(2), 105–116. https://doi.org/10.1515/acph-2015-0019

Mischler, R., & Metcalfe, I. C. (2002). Inflexal®V a trivalent virosome subunit influenza vaccine: production. *Vaccine*, *20*, B17–B23. https://doi.org/10.1016/s0264-410x(02)00512-1

Mughees, M. M., Ansari, M. A., Mughees, A., Farooque, F., & Wasi, M. (2021). Virosomes as drug delivery system: An updated review. *International Journal of Research in Pharmaceutical Sciences*, 12(3), 2239–2247. https://doi.org/10.26452/ijrps.v12i3.4850

Naldi, I., Taranta, M., Gherardini, L., Pelosi, G., Viglione, F., Grimaldi, S., Pani, L., & Cinti, C. (2014). Novel epigenetic target therapy for prostate cancer: A preclinical study. *PLoS ONE*, *9*(5), e98101. https://doi.org/10.1371/journal.pone.0098101

Nishimura, Y., Mieda, H., Ishii, J., Ogino, C., Fujiwara, T., & Kondo, A. (2013). Targeting cancer cell-specific RNA interference by siRNA delivery using a complex carrier of affibody-displaying bio-nanocapsules and liposomes. *Journal of Nanobiotechnology*, *11*(1). https://doi.org/10.1186/1477-3155-11-19

Piccirillo, C. A., & Shevach, E. M. (2001). Cutting edge: Control of CD8+ T cell activation by CD4+CD25+ immunoregulatory cells. *The Journal of Immunology*, *167*(3), 1137–1140. https://doi.org/10.4049/jimmunol.167.3.1137

Rathore, P. & Swami, G. (2012). Virosomes: A novel vaccination technology. *International Journal of Pharmaceutical Sciences and Research*, *2012*(10). https://doi.org/10.13040/ijpsr.0975-8232.3(10).3591-97

Redd Bowman, K. E., Lu, P., Vander Mause, E. R., & Lim, C. S. (2020). Advances in delivery vectors for gene therapy in liver cancer. *Therapeutic Delivery*, *11*(1), 833–850.

Saga, K., & Kaneda, Y. (2013). Virosome presents multimodel cancer therapy without viral replication. *BioMed Research International*, *2013*, 1–9. https://doi.org/10.1155/2013/764706

Sakaguchi, S. (2005). Naturally arising Foxp3-expressing CD25+CD4+ regulatory T cells in immunological tolerance to self and non-self. *Nature Immunology*, *6*(4), 345–352. https://doi.org/10.1038/ni1178

Sasada, T., Kimura, M., Yoshida, Y., Kanai, M., & Takabayashi, A. (2003). CD4+CD25+ regulatory T cells in patients with gastrointestinal malignancies. *Cancer*, *98*(5), 1089–1099. https://doi.org/10.1002/cncr.11618

Schumacher, R., Adamina, M., Zurbriggen, R., Bolli, M., Padovan, E., Zajac, P., Heberer, M., & Spagnoli, G. C. (2004). Influenza virosomes enhance class I restricted CTL induction through CD4+ T cell activation. *Vaccine*, *22*(5–6), 714–723. https://doi.org/10.1016/j.vaccine.2003.08.019

Shimamura, M., Morishita, R., Endoh, M., Oshima, K., Aoki, M., Waguri, S., Uchiyama, Y., & Kaneda, Y. (2003). HVJ-envelope vector for gene transfer into central nervous system. *Biochemical and Biophysical Research Communications*, *300*(2), 464–471. https://doi.org/10.1016/s0006-291x(02)02807-3

Singh, N., Gautam, S. P., Kumari, N., Kaur, R., & Kaur, M. (2017). Virosomes as novel drug delivery system: An overview. *PharmaTutor*, *5*(9), 47–55.

Sioud, M., & Sørensen, D. R. (2003). Cationic liposome-mediated delivery of siRNAs in adult mice. *Biochemical and Biophysical Research Communications*, *312*(4), 1220–1225. https://doi.org/10.1016/j.bbrc.2003.11.057

Spohn, R., Buwitt-Beckmann, U., Brock, R., Jung, G., Ulmer, A. J., & Wiesmüller, K. H. (2004). Synthetic lipopeptide adjuvants and Toll-like receptor 2—structure–activity relationships. *Vaccine*, *22*(19), 2494–2499. https://doi.org/10.1016/j.vaccine.2003.11.074

Tregoning, J. S., Russell, R. F., & Kinnear, E. (2018). Adjuvanted influenza vaccines. *Human Vaccines & Immunotherapeutics*, *14*(3), 550–564.

Waelti, E. R., & Glück, R. (1998). Delivery to cancer cells of antisense L-myc oligonucleotides incorporated in fusogenic, cationic-lipid-reconstituted influenza-virus envelopes (cationic virosomes). *International Journal of Cancer*, *77*(5), 728–733.

Walczak, M., de Mare, A., Riezebos-Brilman, A., Regts, J., Hoogeboom, B. N., Visser, J. T., Fiedler, M., Jansen-Dürr, P., van der Zee, A. G. J., Nijman, H. W., Wilschut, J., & Daemen, T. (2010). Heterologous prime-boost immunizations with a virosomal and an alphavirus replicon vaccine. *Molecular Pharmaceutics*, *8*(1), 65–77. https://doi.org/10.1021/mp1002043

Yamada, T., Iwasaki, Y., Tada, H., Iwabuki, H., Chuah, M. K., VandenDriessche, T., Fukuda, H., Kondo, A., Ueda, M., Seno, M., Tanizawa, K., & Kuroda, S. (2003). Nanoparticles for the delivery of genes and drugs to human hepatocytes. *Nature Biotechnology*, *21*(8), 885–890. https://doi.org/10.1038/nbt843

Zhou, G., Hollenberg, M. D., Vliagoftis, H., & Kane, K. P. (2019). Protease-activated receptor 2 agonist as adjuvant: Augmenting development of protective memory CD8 T cell responses induced by influenza virosomes. *The Journal of Immunology*, *203*(2), 441–452. https://doi.org/10.4049/jimmunol.1800915

Zurbriggen, R. (1999). Immunogenicity of IRIV-versus alum-adjuvanted diphtheria and tetanus toxoid vaccines in influenza primed mice. *Vaccine*, *17*(11–12), 1301–1305. https://doi.org/10.1016/s0264-410x(98)00361-2

25 Bioactive Compounds As Potential Remedy Against Coronaviruses

Kuntal
Gargi College, University of Delhi, New Delhi, India

Prabhleen Kaur
University of Delhi, New Delhi, India

Sakshi Dawer and Madhu Yashpal
Gargi College, University of Delhi, New Delhi, India

Seema Kalra
Indira Gandhi National Open University, New Delhi, India

Archna Chaudhary
SGT University, Gurugram, India

CONTENTS

DOI: 10.1201/9781003306931-28

25.1 INTRODUCTION

The term "bioactive compounds" refers to chemical compounds known to have biological entities like plants, animals or micro-organisms as their source of origin. These molecules, in turn, also affect living tissue. The two words "bio" and "active" signify any compound known to possess biological activity (Guaadaoui et al., 2014). Bioactive compounds, also abbreviated as BACs, are generally known to be present in small quantities in both edible and non-edible sources, and include polyphenols, carotenoids, glucosinolates, vitamins, omega fatty acids, organic acids, nucleosides and nucleotides, phytosterols, etc. (Kamiloglu et al., 2021). These compounds are known to have great importance in preventing a variety of illnesses and hence possess a substantial amount of health advantages. Among various diseases known to be cured by bioactive compounds, COVID, also known as coronavirus disease, is the current target in this context. Many groups of researchers worldwide are trying to determine the importance of various biologically derived compounds in treating coronaviral infections. Numerous bioactive compounds have already been proved to ameliorate multiple symptoms associated with COVID-positive individuasl.

25.2 CORONAVIRUS DISEASE (COVID-19)

Coronaviruses, in general, are known to be responsible for many zoonotic and human diseases. In Wuhan, China, a new strain of coronaviruses called Severe Acute Respiratory Syndrome Coronavirus 2 (SARS-CoV-2) has been identified. The viral disease caused by SARS-CoV-2 was declared a COVID-19 pandemic by the World Health Organization (WHO) in January 2020. Because of its spread across the globe, the disease has become of the utmost concern for the entire international community.

25.2.1 Coronaviruses

The term coronaviruses refers to a huge family of RNA viruses so named due to the crown-like spikes on their surfaces (Figure 25.1). These respiratory viruses are generally known to cause upper respiratory tract diseases and to target both animals (mostly avian) and humans as their hosts (Table 25.1). These viruses belong to the family of viruses called Coronovidae under the order Nidovirales (Figure 25.2; Suwannarach et al., 2020).

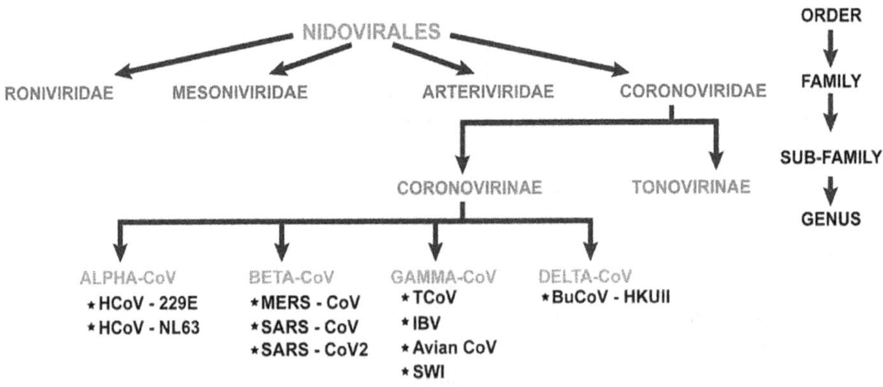

FIGURE 25.1 Classification of coronaviruses.

TABLE 25.1
Three Major Outbreaks of Zoonotic Transmissions of Coronavirus (Lockbaum et al., 2021)

Coronaviruses	SARS-CoV	MERS-CoV	SARS-CoV-2
Emergence	Emerged November 2002	Emerged September 2012	Emerged December 2019
Cause	Severe Acute Respiratory Syndrome	Middle East Respiratory Syndrome	Coronavirus Disease 2019
	Disappeared by 2004	Continues to cause local outbreaks	Declared as a pandemic by WHO, March 11, 2020

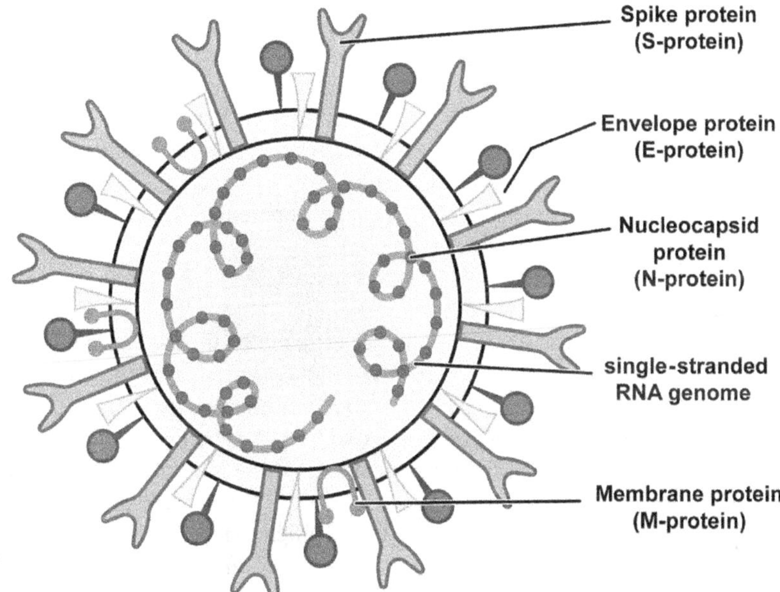

FIGURE 25.2 Generalized structure of human coronavirus.

(Haque et al., 2020; Khairan et al., 2021)

These viruses generally possess a viral envelope and a non-segmented (Suwannarach et al., 2020) positive ssRNA genome. The first human coronaviruses were identified from individuals with upper respiratory tract infections in the mid-1960s. Tyrrell and Bynoe isolated the first HCoV (B814 strain) that could cause a common cold in human volunteers following intranasal vaccination in 1965 (Tyrrell and Bynoe, 1965). Diverse coronaviruses are well known due to mutations and recombination events occurring during their genome replication. These viruses usually cause upper respiratory tract infections in humans. The outbreak of SARS-CoV-2 infection at the end of 2019 marked a global pandemic due to its ability to cause fatal pneumonia in its hosts (Xian et al., 2020). In order to validate the use of the latest information, the discussion in this chapter uses the example of SARS-CoV-2 for most of the aspects of coronaviruses.

25.2.2 Genome and Proteins

The genome of SARS-CoV-2 is organized similarly to that of other CoVs. Despite belonging to the beta group of coronaviruses, SARS-CoV-2 is different from SARS-CoV and MERS-CoV. As compared to SARS-CoV, CoV2 genes are known to have less than 80% nucleotide identity and approximately 89% of nucleotide similarity (Qamar et al., 2020). This positive-sense RNA virus is enveloped and has a complete genome of about 29.9Kb (30,000 nucleotides) (Jin et al., 2020). The genome size of SARS-CoV-2 is about 29.9Kb compared to those of MERS and SARS-CoV, which have a genome size of 30.1 and 27.9Kb, respectively. Between six and eleven open reading frames (ORFs) are found in the genome of SARS-CoV-2, which encodes more than 20 proteins (Chen et al., 2020). Of these, the first ORF (ORF1a/b) is translated to produce two polyproteins, pp1a and pp1ab, which are known to encode for 16 non-structural proteins (NSPs). Structural and other accessory proteins are, however, encoded by other ORFs. Most of the proteins encoded by SARS-CoV-2 are known to be like those of SARS-CoV. However, certain new mutations in NSPs 2 and 3 are known to be majorly responsible for the differential infectious potential of SARS-CoV-2 (Guo et al., 2020).

The virus encodes four structural and many non-structural and accessory proteins. Structural proteins encoded by the virus include nucleoprotein (N), membrane (M), envelope (E) and spike(S). All these proteins perform necessary functions, also being responsible for viral infection. Non-structural proteins, on the other hand, are generally involved in the replication and transcription of all the viral genes. Since inhibition or blocking the function of any of these NSPs will halt the replication, transcription and translation of the viral genome, it will halt viral replication and hence may combat the infection. Therefore, some of these essential NSPs like proteases and RdRp (RNA-dependent RNA polymerase, also called RNA replicase) can be employed as potential targets for creating drugs.

25.2.3 Replication and Life-Cycle

Angiotensin-converting enzyme 2 (ACE2) is the target receptor of SARS-CoV-2 for its cellular entry into the host (Figure 25.3). The spike protein, which is one of the structural proteins of the virion surface, is known to bind to ACE2, generally found on the surface of cells of the respiratory tract in humans. Spike structural protein comprises two known subunits, one of which determines the virus–host range (S1 subunit), and the other facilitates the fusion of viral and host cell membranes (S2 subunit). This is followed by the release of viral RNA into the cytoplasm of host cells, where further translation into polyproteins occurs. Viral replication inside the host then results in the pool of sub-genomic RNA, nucleocapsid and envelope proteins, all of which then assemble to form viral particles ready to bud off.

The life-cycle of a coronavirus begins as soon as the viral particle binds to the host receptor. In the case of humans, the associated host receptor for this purpose is ACE2. The spike protein of the virus particle is known to bind to ACE2 along with the host factor TMPRSS2, which stands for cell surface serine protease. Once the virus's genome, i.e., positive-sense single-stranded RNA, is released into the host cells, it is subjected to primary translation (translation of viral replication machinery) and polyprotein processing. Once the viral RNA is replicated and translated into the required structural and non-structural proteins, virion assembly takes place in the endoplasmic reticulum and Golgi apparatus. Through small vesicles, virions are transported to the cell surface and are then released from the infected host cell by the process of exocytosis.

25.2.4 Pathogenesis

SARS-CoV-2 is well known to cause the respiratory infection termed COVID-19. The virus has transmission routes including fecal, oral, direct contact, droplets, mother to child, animals to human, air-borne transmission, etc. The virus incubation period ranges from one to 14 days, with an average of about three to seven days. The most common symptoms yet known include fever, sore throat, runny nose, cough, malaise, headache, exhaustion, shortness of breath and many other flu-like symptoms. However, respiratory or even multi-organ failure is also observed in adverse situations. This very much emphasizes the urgent need to develop numerous therapeutic agents against the pathogenic virus.

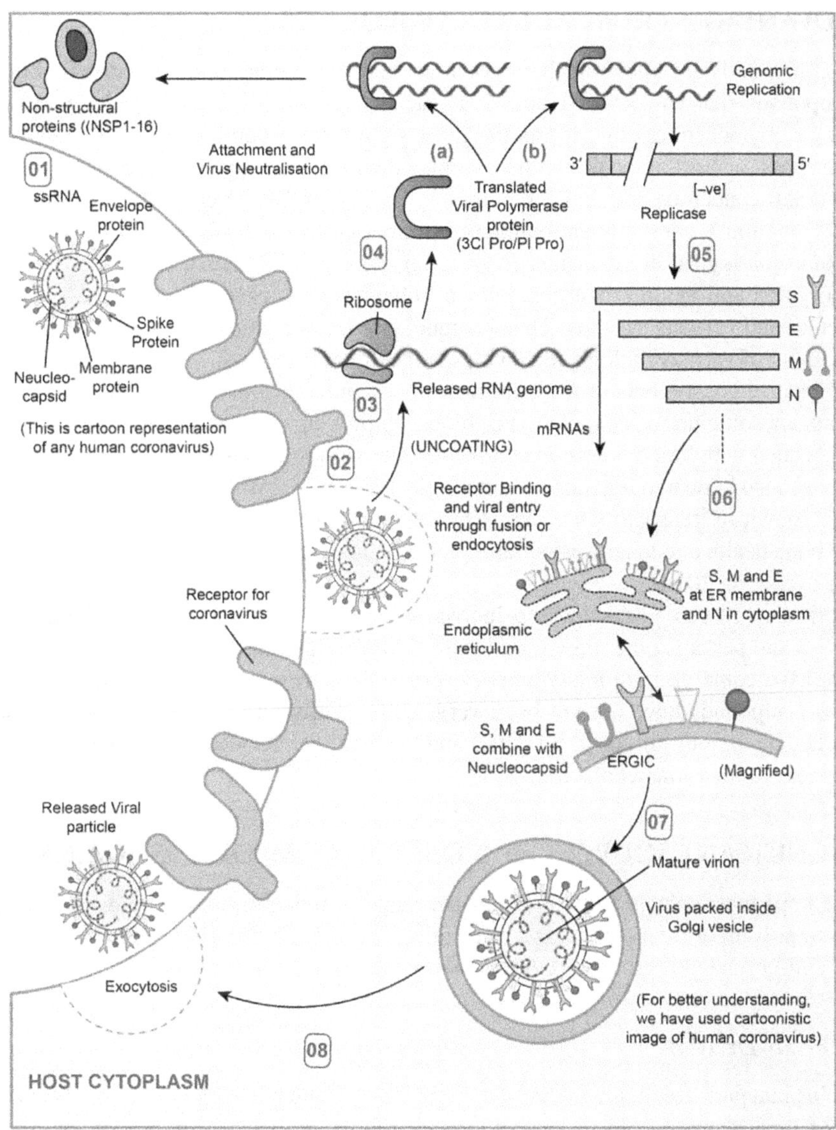

FIGURE 25.3 Life-cycle of SARS-CoV-2.

25.2.5 Epidemiology

For most known coronaviruses, bats are thought to be the general reservoir. SARS-CoV-2 has been found to have 96.2% identity in the genome sequence with Bat CoV RaTG13. The virus employs ACE2 present in the bottom region responsible for respiration of the human host as its receptor. The S protein of the virus determines the range of species of the viral host that it can affect (S1) and mediates membrane fusion of the virus and the host cells (S2). Viral RNA is then released into the cytoplasm, which is further translated into essential polypeptides, and further RNA replication is carried out in the host. The common pneumonia-like symptoms of viral infection are generally a result of an immune response called acytokine storm. The reproduction number of SARS-CoV-2 is about 2. This value, represented as R naught, gives us an estimate of how many uninfected people will catch the infection if they contact one infected individual. Such a high R naught value indicates that COVID-19 is a highly contagious disease.

25.3 ADVANTAGES OF BIOACTIVE COMPOUNDS

Bioactive compounds are known to be a potential remedy against numerous diseases. Many bioactive compounds that can prevent and treat various human illnesses have been well studied. Flavonoids, a phytochemical that forms the most significant group amongst the known bioactive compounds, possess most of the properties associated with promoting good human health, including antioxidant, anti-cancerous and anti-inflammatory properties (Walia et al., 2019). Many bioactive compounds function as powerful antioxidants, eliminating free radicals and minimizing damage to cells and other biological components. In addition, numerous bioactive compounds possess anti-inflammatory and anti-aging properties, known to have beneficial health effects, especially on the cardiovascular and nervous systems. These compounds are also known to have the ability to influence the activity of various biological enzymes and cellular receptors.

Many bioactive constituents of food promote good human health by lowering cholesterol levels in the serum, speeding up the coagulation of blood, improving circulation and reducing the complications associated with blood pressure, controlling energy homeostasis and fat storage. At the same time, some are also known to regulate immunological responses (Walia et al., 2019). Bioactive compounds are known to treat various diseases by suppressing inflammation, halting carcinogenesis, promoting bone health due to anti-osteoporotic effects, and various cardiovascular and neuroprotective effects (Teodoro, 2019).

It is critical to research and develop innovative therapy and therapeutic techniques to battle SARS-CoV-2 and perhaps other emerging future viruses (Hamoda et al., 2021). In the last decade, researchers have come up with many antiviral properties of bioactive compounds. These biologically active compounds have not yet been very well established as antiviral drugs but are being widely studied by various groups of scientists and are being put forward as novel antiviral therapeutics for the near future (Xian et al., 2020).

25.4 SIGNIFICANT SOURCES OF BIOACTIVE COMPOUNDS AGAINST COVID

Bioactive compounds isolated from an extensive range of living organisms may serve as potential candidates to prevent or treat coronaviral infections. This section discusses some of the well-known sources of anti-coronaviral bioactive compounds.

25.4.1 Plant Derived

Innumerable plant parts (including roots, stems and leaves), and many medicinal plants and herbs are a source of a large number of bioactive compounds (Table 25.2). These bioactive compounds have long been employed for curing various illnesses because of their easy availability, low cost and natural origin. Many such biologically active compounds are being used to treat viral diseases. However, the exact action mechanisms of most plant-based therapies have not yet been well described. Many ethnobotanical plants, herbs and medicinal plants have been exploited to serve as potential anti-coronaviral therapeutics (Figure 25.4). The potential of these bioactive compounds against coronaviruses is determined through in-silico, in-vitro or experimentally proven in-vivo approaches.

25.4.2 Algae Derived

Both prokaryotic cyanobacteria and many eukaryotic algal forms constitute many bioactive compounds. Cyanobacteria *Spirulina* is commonly consumed as a health food supplement due to its high protein, vitamin, mineral and lipid content and to various physiologically active molecules like chlorophyll phycocyanin and carotene. Various *Spirulina*-derived nutraceuticals and bioactive compounds, including sulfated polysaccharides, ACE inhibitor peptides, calcium spirulan, phycobiliproteins, etc., are known to possess antiviral properties and to boost both innate and adaptive

TABLE 25.2
Some Common Plant-Derived Bioactive Compounds Documented to Inhibit Coronaviral Replication by Employing Different Modes of Action

S.No.	Bioactive Compound	Plant Source	Characteristics
1	tetra-O-galloyl-β-d-glucose (TGG) (Remali and Aizat, 2021)	*Rhus chinensis*	Proposed to interfere with the viral cell fusion as it has a high binding affinity for viral S2 subunit; IC50: 4.5 micro molars
2	Luteolin (Remali and Aizat, 2021)	*Veronica linariifolia*	Proposed to interfere with viral cell fusion as it has a high binding affinity for viral S2 subunit; IC50: 10.6 micro molars
3	Aescin (Xian et al., 2020)	*Aesculus hippocastanum*	Possesses remarkable anti-SARS properties with a determined EC50 value of 3.4 micro molars per liter
4	Saikosaponin B2 (Xian et al., 2020)	*Bupleuri Radix*	Inhibition of viral entry into the host cell is thought to be its mode of antiviral action with the IC50 value of 1.7 micro molars per liter
5	Reserpine (Xian et al., 2020)	*Rauwolfia* species	Possesses significant anti-SARS properties with a determined EC50 value of 6 micro molars per liter
6	Glycyrrhizin (Remali and Aizat, 2021)	*Glycyrrhiza glabra L.* (licorice)	Documented to halt coronavirus infection by obstructing viral attachment and penetration into the host cell. It also has anti-inflammatory properties.
7	Caffeic acid (Remali and Aizat, 2021)	*Sambucus formosanaNakai*	Potent antiviral biomolecule that targets viral entry into the host cell
8	Aurantiamideacetate (Remali and Aizat, 2021)	*Artemisia annua*	Obstructs viral entry into the host by targeting inhibition of SARS Cathepsin L, which otherwise stimulates fusion of viral spike proteins and host receptors
9	Tryptanthrin (Remali and Aizat, 2021)	*S. cusia Kuntze*	Source of Plpro inhibitor tryptanthrin, and has been experimentally proved by in-vitro assays in case of HCoV-NL63

immunity. All these compounds have some or all of such properties as immuno-stimulation, immunomodulation, anti-microbial, anti-apoptotic and anti-inflammation. Spirulina supplements are widely accepted for enhancing macrophage phagocytic activity, improving the function of natural killer cells, and raising the antibody response of red blood cells. Oral consumption of *S. platensis* is proposed to have roles associated with TLR signaling in blood cells and to boost immunity (Ratha et al., 2020). Phycocyanobilins (PCBs) are a type of chromophores found in rhodophytes, a type of cyanobacteria. These blue pigments possess anti-oxidative and antiviral properties and are thought to have great potential against SARS- CoV-2 infection by inhibiting the main protease and RdRp viral proteins. Many red algae-derived lectins like Griffithsin are known to possess medicinal properties. Griffithsin is known to have both anti-HIV and anti-SARS-CoV potential (Bhatt et al., 2020).

25.4.3 Fungi Derived

Fungi serve as a rich source of biologically active compounds known to contain the therapeutic potential to curb coronaviral infections, and are involved in immunomodulation. Both filamentous fungi and mushrooms are regarded as a good reservoir of biologically active compounds.

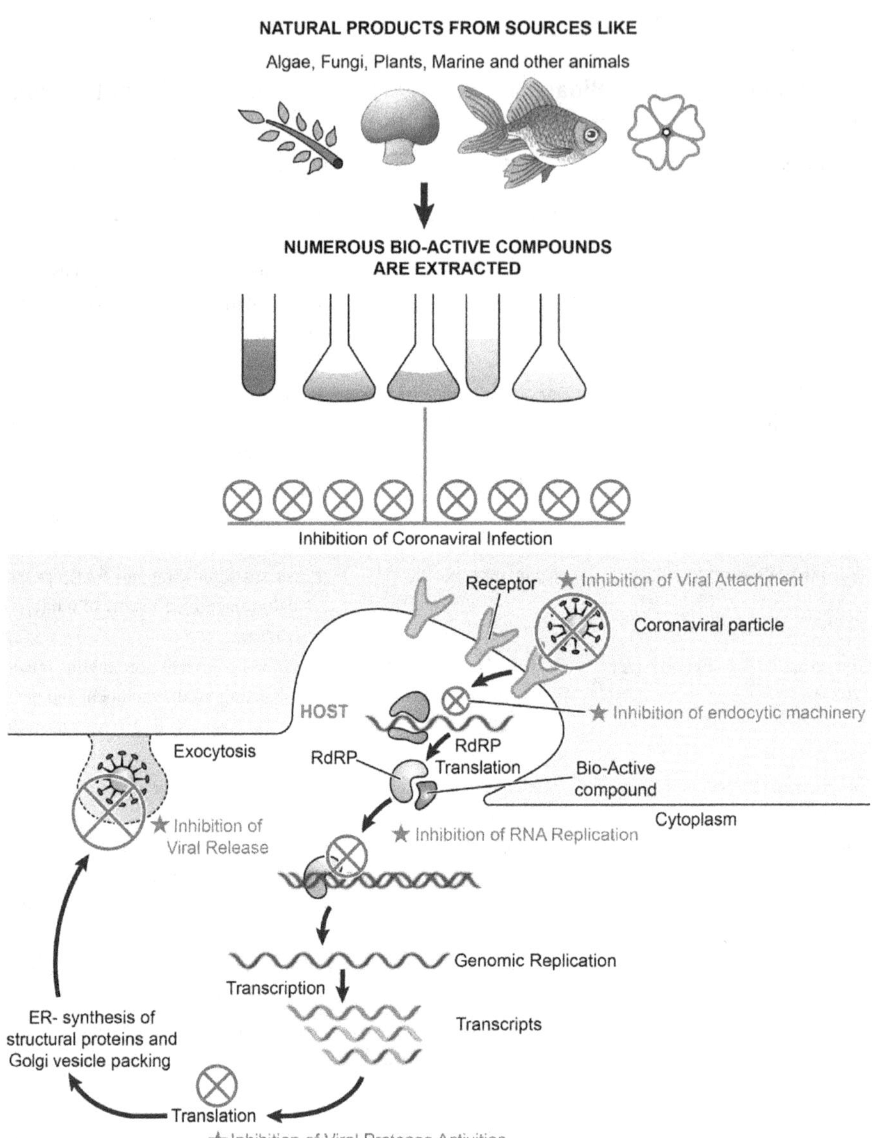

FIGURE 25.4 Bioactive compounds and their role in inhibition of coronaviral infections.

Polyketides, terpenoids, indole alkaloids, lignin derivatives and many non-ribosomal proteins possessing some sort of antiviral properties are produced by fungi. Most documented antiviral biological compounds are known to be protease inhibitors. Many fungi-derived bioactive compounds inhibit HCV NS3/4A protease. These include simeprevir, alterneroil derivatives, ω-hydroxyemodin, Griseoxanthone C and Antrodins. Different bioactive compounds are isolated from other fungal species (Suwannarach et al., 2020). Just like HCV, many HIV protease inhibitors are being repurposed to treat coronaviral infections. In-silico studies have documented Semicochliodinol B as the best lead compound for targeting the main protease of SARS- CoV-2, and it is known to have high binding affinity (Jana and Bhardawaj, 2021). Other biomolecules extracted from edible and medicinal mushrooms like colossolactone G, colossolactone VIII, colossolactone G, velutin, colossolactone E and ergosterol have been computationally shown to have potential against coronaviruses (Rangsinth et al., 2021). Many mushroom-derived bioactive compounds possess immunomodulatory activities essential for prevention and treatment of coronavirus infections. This includes T lymphocyte activation by Concanavalin A, stimulation of dendritic cells and cytokine release by Ricin-B-like lectin and many others (Suwannarach et al., 2020).

25.4.4 Animal Derived

Bioactive compounds with antiviral activities like terpenes, flavonoids, phenolic acids etc. are present in many honeybee-derived products. Many such biomolecules present in both honey and propolis are computationally revealed to target coronaviral enzymes.

Hesperetin, Kaempferol and ellagic acid have been identified as potential compounds targeting COVID-19 RdRp. In contrast, quercetin, artepillin C and the compounds listed previously have been identified as potential compounds targeting the main protease (Shaldam et al., 2021). Such molecules as have been found to have high binding scores with viral proteins can be repurposed to develop essential antiviral inhibitors. Molecular docking of bioactive chemicals extracted from honey bee products against COVID-19 targets, such as Mpro and RdRp enzymes, has aided in identification of potential biological compounds with strong binding to the COVID-19 target proteins.

25.4.5 Marine

Many marine organisms have been identified as acting as a potential source of biologically active compounds for treating coronaviral infections. Organisms like *Ecteinascidia turbinate, Petrosiacontignata, Tridideminm solidum, Pseudopterogorgiaelisabethae* and many other marine organisms are known to serve as a source of important bioactive compounds that have broadly varying mechanisms of action.

Remdesivir has been authorized by the US Food and Drug Administration (FDA) as an anti-SARS-CoV-2 drug. It is a nucleoside analog which possesses many antiviral properties. Similar to remdesivir, many nucleoside analogs have also been extracted and identified from marine sponges. These analogs, including spongouridine and spongothymidine, are employed to synthesize antiviral drugs like vidarabine and cytarabine (Hamoda et al., 2021). In addition, Pseudotheonamide C and D, Aeruginosin 98B, esculetin and ilimaquinone are marine-produced biomolecules. The former two molecules generally target host serine protease (human TMPRSS2), while the other two compounds can inhibit the activity of thre main protease of SARS-CoV-2.

Cryptotethya, a species of demosponge, has been employed to isolate nucleoside analogs, which after chemical modifications (for instance, replacement of ribose by arabinose) have helped to develop drugs like vidarabine (ara-A) and cytarabine (ara-C). The FDA has approved these drugs for their antiviral activities. Another set of nucleoside analogs employed to develop nucleoside analog inhibitors (NIs) includes Mycalisine A and B, isolated from *Mycale* species of marine sponges. Mycalisine A nucleoside analog provided the inspiration for remdesivir synthesis (Hamoda et al., 2021). It has been well established that following modest structural changes, metabolites produced from marine sponges can act as potential RdRp and other viral protein inhibitors.

25.5 DIFFERENT MODES OF ACTION OF BIOACTIVE COMPOUNDS AGAINST CORONAVIRUSES

Various studies have been conducted to confirm the anti-coronaviral properties of many biomolecules. The exact mechanism of action of most of the bioactive compounds has not yet been fully studied, but computational studies employing docking and other in-silico methods have helped determine the viral target protein (Figure 25.4). Biologically active compounds expected to act as anti-coronaviral drugs can have various mechanisms of action. Different bioactive compounds have been computationally determined to target different viral proteins or halt different steps in the process of viral replication. Some might act by inhibiting the viral entry within the host cell or viral replication or viral assembly, while others might halt viral release or interfere with the virus–host interactions. Different bioactive compounds employ different modes of action by targeting different viral proteins like spike protein (S), nucleocapsid protein (N), main protease or chymotrypsin-like protease(3Clpro), papain-like protease (Plpro), RNA-dependent RNA polymerase (RdRp), and various other viral targets. Some may even halt viral replication by targeting host protein like the host receptors that are employed by the coronaviruses for cellular entry of the virus (Figure 25.4).

25.6 DIFFERENT VIRAL TARGETS FOR BIOACTIVE COMPOUNDS AGAINST COVID

25.6.1 Spike Protein

Coronaviruses get their name from the crown of protein spikes that cover the surface of their outer membrane. The spike proteins, also known as S proteins, have been the focus of early research on the new coronavirus, since they are the key that the virus utilizes to enter host cells. It is the S protein of SARS-CoV and SARS-CoV-2 that interacts with ACE2 to enter host cells. The spike proteins on SARS-CoV-2 are nearly identical to those on SARS-CoV in terms of amino acid sequence, being 80% identical. However, the residue which interacts with ACE2 is known to be conserved in both SARS-CoV and CoV-2. It is made up of 1273 amino acid residues and is divided into three subunits, S1, S2, and S2', each of which performs a particular function during the adhesion process to the host cell (Table 25.3; Naqvi et al., 2020).

When the S protein attaches to the receptor, the host cell membrane's TM protease serine 2 (TMPRSS2) activates the S protein, allowing the virus to enter the cell (Suwannarach et al., 2020). S protein undergoes conformational modifications as a result of its entrance into the host cell's endosomes throughout this phase. Because dynamic changes in the target protein may impact immune responses, understanding these structural changes is critical for vaccine development. Despite numerous changes in SARS-S1 CoV-2's receptor-binding region, its association with ACE2 is retained in humans, pigs, civets and bats, with the exception of mice ACE2s (Naqvi et al., 2020).

Various biomolecules have been determined to have high binding affinities for the spike protein of coronaviruses (Table 25.4) and are hence proposed to interfere with viral cell fusion. Tetra-O-galloyl-β-d-glucose (TGG) isolated from *Rhus chinensis* and luteolin isolated from *Veronica linariifolia* are both documented to have high binding affinities for S2 of coronaviruses (Remali and Aizat, 2021). Bioactive compounds present in some species of plant family polygonaceae are well known to interfere in the process of interaction between spike protein of the coronavirus and ACE2 receptors of the host (Remali and Aizat, 2021). The anthraquinone glycoside compound emodin is found to considerably block the virus–host interaction by dose-dependent blockage of spike protein and host cell interaction. HCoV- 22E9 infection can be treated by saikosaponin B_2 isolated from roots of *Bupleurum chinense.* It disrupts the orientation of viral glycoproteins and hence does not allow the penetration of the virus into the host cell. *Tetrandrae radix* is a source of bis-benzylisoquinoline

TABLE 25.3
Subunits of Spike Protein (Naqvi et al., 2020)

Subunit	Function
S1	Attachment of virions with the host cell membrane by interacting with human ACE2 that subsequently initiates the infection process
S2	Works as the fusion protein that helps in the fusion of virion with the mammalian cell membrane
S2'	Remaining S2' cleaved subunit of the S protein functions as a fusion peptide

alkaloids-tetrandrine, a compound that is known to affect HCoV- OC43 viral replication by targetinss viral spike protein (Xian et al., 2020).

25.6.2 Main Protease

A major component of the viral genome is replicase gene which is translated to produce 16 NSPs. These NSPs are, however, produced after proteolytic cleavage of two overlapping proteins pp1a(450KDa) and pp1ab (750KDa). 3-chymotrypsin-like protease (3CLpro) and papain-like protease (Plpro) are known to be responsible for these proteolytic cleavages.

3Clpro is known to cleave the C-terminal end of these polyproteins while processing of N-terminal end is done by Plpro. At the N-terminal, Plpro has three cleavage sites while the next 11 sites are those for 3ClPro. Cleavage of pp1a produces 1-11 NSPs while pp1ab produces 15 NSPs, all except the 11th. A total of 16 NSPs are produced by cleavage of pp1a and pp1ab. Unless properly cleaved by Plpro and 3Clpro, several other NSPs like RNA-dependent RNA polymerases (NSP12) cannot become completely functional. In short, this enzyme is responsible for self-maturation as well as that of other viral polyproteins (Jin et al., 2020).

Main protease or 3Clpro, a non-structural viral protein of coronaviruses, is known to play a very important role in viral genome replication within the host. Mpro's functional relevance in the viral life-cycle, along with the lack of closely comparable homologues in humans, makes it an appealing choice for antiviral treatment. Computational analysis including molecular docking and simulations suggests that many marine-derived bioactive compounds can serve as SARS-CoV-2 main protease inhibitors.

Among the marine-derived compounds that have been identified as potent Mpro inhibitors are chimyl alcohol, hamigeran B, ilimaquinone, esculetin-4 carboxylic acid esters, T3, etc. The terpenoid moiety is proposed to be responsible for Mpro inhibition and is shared amongst most of the marine-derived bioactive compounds targeting 3Clpro P (Table 25.4).

Isobavachalcone, helichrysetin, pectolinarin, herbacetin, rhoifolin and quercetin 3-β-D-glucoside are plant-derived biomolecules that have been identified through in-silico screening for their potential to inhibit the activity of the main protease of coronaviruses (Remali and Aizat, 2021). A bioactive compound isolated from the lily plant *Veratrum sabadilla*, sabadinine, has been found to dock into the active site of the main protease. In addition, biomolecules such as sinigrin, beta-sitosterol and indigo isolated from plant *I. indigotica* notably curb the activity of 3Cpro (Remali and Aizat, 2021).

25.6.3 Papain-like Protease

Papain-like protease (Plpro), also called non-structural protein 3 in the case of SARS-CoV-2, plays a very important role in viral replication. Like main protease, Plpro is also involved in proteolysis. 3Clpro is known to cleave the C-terminal end of these polyproteins while processing of the

TABLE 25.4
Potential Anti-Coronaviral Bioactive Compounds Which Prevent and Treat Viral Infection by Targeting the Main Viral Protease

S.No.	Bioactive Compound	Source
1	Esculetin-4-carboxylic acid methyl ester	*Axinella cf. corrugate*
2	Hamigeran B	*Hamigeratarangaensis*
3	Chimyl alcohol (1-O-hexadecylglycerol)	*Desmapsammaanchorata*
4	Ilimaquinone	*Hippospongiametachromia*

N-terminal end is done by Plpro. At the N-terminal, Plpro has three cleavage sites, while the next 11 sites are those for 3ClPro. Cleavage of pp1a produces 1-11 NSPs, while pp1ab produces 15 NSPs, all except the 11th. Cleavage of pp1a and pp1ab produces a total of 16 NSPs.

S. cusia (Nees) extract contains a biomolecule, tryptanthrin, which has been clearly identified to inhibit the activity of coronaviral Plpro. Various herbal compounds including coumaroyltyramine, tanshinoneIIa, kaempferol, N-cis-feruloyltyramine, cryptotanshinone and quercetin have also been found to inhibit both 3Clpro and Plpro (Remali and Aizat, 2021). *S. cusia* is a source of the Plpro inhibitor tryptanthrin, experimentally proved by in-vitro assays in case of HCoV-NL63. Hirsutenone and tanshinone I isolated from *Alnus japonica* and *Salviaemiltiorrhizae* respectively are both known to suppress SARS- CoV infection by targeting Plpro (Xian et al., 2020).

25.6.4 RNA-Dependent RNA Polymerase

RNA-dependent RNA polymerase (RdRp), a non-structural protein, is required for coronavirus transcription and replication. Since RdRp plays a critical part in the virus's life-cycle, numerous polymerase inhibitors, like remdesivir, have been employed to treat a variety of viral infections. Molecules associated with inhibition of RdRp might thus lead to the development of an anti-SARS-CoV agent. *H. cordata* water extract which is safe for human consumption is known to affect RdRp activity, halting viral replication. Methanol extracts from many other plants like *Sophora subprostrata, Cimicifuga racemose, Coptis chinensis, Phellodendronchinense and Phoradendronmeliae* are also documented to inhibit viral replication by decreasing RdRp activity (Remali and Aizat, 2021). Herba and a bioactive compound, theaflavin, are known to significantly suppress SARS-CoV-2 infection by inhibiting RdRp (Xian et al., 2020)

25.6.5 Nucleocapsid (N) Proteins

The packaging of viral RNA into ribo-nucleocapsids is aided by nucleocapsid proteins (N). SARS-CoV-2's N protein is highly conserved among CoVs, sharing 90% sequence similarity with SARS-CoV. The mass of the N protein has been determined to be between 45 and 60 kDa (Thomas, 2020). The N proteins attach to viral RNA in a "bead on a string" fashion through their core's 140 amino acid-long RNA-binding domain (Naqvi et al., 2020). A predicted intrinsically disordered N-terminal domain (NTD), an RNA-binding domain (RBD), a predicted disordered central linker (LINK), a dimerization region and a predicted disordered C-terminal domain make up the SARS-CoV-2 N protein (CTD) (Cubuk et al., 2021). The NTD attaches to RNA, whereas the CTD is responsible for oligomerization. Phosphorylation is primarily controlled by the SR-rich linker. It also allows the N protein to interact with other cell components by allowing molecular mobility (Thomas, 2020). The nucleocapsid (N) protein is also abundant, making it a good candidate for vaccine and diagnostic test development.

Many natural alkaloid molecules including fangchinoline, cepharanthine, tetrandrine and other bis-benzylisoquinoline alkaloid molecules have been well documented to inhibit N and S proteins of the coronavirus (Remali and Aizat, 2021). Many plants like *Vaccinium macrocarpon, Vitis vinifera and Polygonum cuspidatum* are known to be a source of bioactive compound resveratrol which exerts its anti-coronaviral effect by removing free radicals and hence acting as an antioxidant. This biomolecule is also known to reduce inflammation by increasing nitric oxide production in tissues (Xian et al., 2020).

25.6.6 Other Viral Targets

In addition to the key targets discussed above, many other viral targets are also employed for inhibiting viral replication. Many potential leads against coronaviruses are known to target spike glycoprotein, main protein, RNA-dependent RNA polymerase, nucleocapsid proteins, etc. Some bioactive compounds which target other viral components are also known. One such example is *Glycyrrhizae radix*-extracted glycyrrhizin which targets protein Kinase C (involved in upregulation and production of nitrous synthase). Another anti-coronaviral drug target is viral helicase. Scutellarein and myricetin biomolecules are known to affect the ATPase activity of the helicase enzyme which otherwise plays an important role in viral replication (Xian et al., 2020).

25.6.7 Host ACE2 and TMPRSS2 Proteins

Coronavirus generally employs angiotensin-converting enzyme 2 (ACE2) present in the bottom region responsible for respiration of the human host as its receptor. S protein of the virus determines the range of species of the viral host that it can affect (S1) and mediates membrane fusion of the virus and host cells. Viral RNA is then released into cytoplasm which is further translated into important polypeptides, and further RNA replication is carried out in the host. IQP, VEP and various other bioactive peptides obtained from *Spirulina platensis* have been found to possess inhibitory effects on ACE (Ratha et al., 2020). Caffeic acid isolated from *Sambucus FormosanaNakai* hinders the interaction between NL63 (HCoV-NL63) human coronavirus and ACE-2 receptors and the co-receptor heparan sulfate proteoglycan present on the host cell (Xian et al., 2020).

Coronavirus generally employs a serine protease (called TMPRSS2) of the host for its entry within the cell. When the spike protein of the coronavirus attaches to the receptor, the host cell membrane's TM protease serine 2 (TMPRSS2) activates the S protein, allowing the virus to enter the cell. Antiviral drugs that target TMPRSS2 can hence inhibit viral entry within the host. Pseudotheonamide C and D isolated from marine sponge *Theonellaswinhoei*, along with aeruginosin 98B isolated from *Microcystis aeruginosa* are known to possess inhibitory activity against the host protease. The guanidino group found in both these compounds mimics the arginine substrate of the enzyme and thus inhibits TMPRSS2. It is suggested that these compounds possess bifunctional activity as their inhibitory effect on 3Clpro has also been identified (Hamoda et al., 2021).

25.7 BIOACTIVE FORMULATION AS HERBAL THERAPY

Herbal therapies have been utilized to treat numerous viral illnesses for thousands of years. Many plant extracts and ayurvedic preparations are being examined for their potential to act as antiviral therapeutics (Safet et al., 2021). These kinds of natural remedies may hence prove to be significant in curing coronaviral infections. Traditional Chinese medicine (TCM) is a very popular and effective herbal formula that was employed in 2003 to cure SARS infection. Many medicinal plants are known to act as a source of bioactive phytocompounds, nutraceuticals, traditional medicines, herbal preparations, etc. The TCM formula has been well employed in the past as a potent therapeutic against SARS-CoV. A TCM preparations called *Lianhua Qingwen* capsules (LHQWC or Clearing Pestilential Disease with Forsythiae Fructus-Lonicerae capsules) has also been documented to inhibit SARS- CoV-2 replication with 411.2 micrograms per ml as its IC50 value. It affects viral replication and morphology as well as exerting anti-inflammatory effects. Different TCM formulas are known to have different modes of action to halt viral infection. Some formulas inhibit viral replication, some regulate blood oxygen level and chemokine concentrations to damp cytokine storm formation in infected patients, while others exert anti-inflammatory and antiviral effects (Xian et al., 2020; Safet et al., 2021). *Ocimum sanctum* (tulsi), *Tinospora cordifolia* (giloy), *Zingiber officinale Roscoe* (ginger), *Elettaria cardamomum* (cardamom), *Withaniasomnifera* (Ashwagandha), *Curcuma longa* (Turmeric), *Piper nigrum* (black pepper), *Syzygiumaromaticum* (clove), *Citrus limon* (lemon) and many other Indian plant extracts and herbal drugs are used to prepare ayurvedic kadhas which are known to exert many beneficial pharmacological effects including antiviral efficacy (Safet et al., 2021).

25.8 REPURPOSING BIOACTIVE PEPTIDES FOR COVID THERAPY

Many plants and biologically active compounds isolated from different sources have been long known to have a large number of medicinal uses. Many bioactive compounds are already being employed to treat viral infections. Drug repurposing has hence become a measure for coping with acute emergencies. In this process, all the drugs that have been employed to date for treating various diseases are accessed for their potential to treat the new disease of interest. Such bioactive compounds as are already known to possess antiviral properties can be repurposed to treat coronaviral infections. Cepharanthine, a bis-benzylisoquinoline alkaloid derived from tubers of *Stephania japonica* (Qianjinteng), inhibited a 2019-nCoV-related pangolin coronavirus GX P2V infection with an EC50 value of 0.98 mol/L using a 2019-novel coronavirus-related coronavirus model in a recent study of repurposing clinically approved drugs (Xian et al., 2020).

25.9 CONCLUSION

A large number of bioactive compounds are being reviewed for their functions as anti-coronaviral drugs. These compounds have already been well known to treat various diseases due to their antioxidant, anti-inflammatory and anti-cancerous properties. Bioactive compounds not only inhibit infectious viruses but also promote physiological activities in the host that provide protection and help fight against them. Bioactive compounds are derived from a variety of sources and different compounds are known to target different viral proteins. Most of the studies that document the use of bioactive compounds as antiviral drugs are based on computational analysis. In-silico analysis including techniques like molecular docking and molecular simulations have been employed to determine the binding and mode of inhibition of bioactive compounds with various proteins of the target coronavirus. Here, we have focused on SARS-CoV-2, the most recently identified beta-coronavirus that led to the global pandemic, as a key representative of coronaviruses. Safe therapeutics against coronaviruses can be developed thanks to the effects of bioactive compounds that either directly or indirectly target viral replication and infection. Further experimental tests and biochemical assays including cell cultures can help validate the use of bioactive compounds as anti-coronaviral drugs. We look forward to the widespread use of bioactive compounds against coronavirus infections.

ABBREVIATIONS

BACs	**Bioactive compounds**
SARS-CoV-2	**Severe Acute Respiratory Syndrome Coronavirus 2**
HCoV	**Human Coronavirus**
MERS- CoV	**Middle East Respiratory Syndrome Corona Virus**
ORF	**Open reading frame**
NSP	**Non-structural proteins**
RdRp	**RNA-dependent RNA polymerase**
ACE2	**Angiotensin-converting enzyme 2**
COVID19	**Coronavirus disease 2019**
FDA	**Food and Drug Administration**
NIs	**Nucleoside analog inhibitors**
TLR	**Toll-like receptor**
PCBs	**Phycocyanobilins**
HIV	**Human immunodeficiency virus**
TMPRSS2	**Transmembrane serine 2 protease**
TGG	**Tetra-O-galloyl-β-d-glucose**

3CLpro	**Chymotrypsin-like protease**
Plpro	**Papain-like protease**
NTD	**N-terminal domain**
RBD	**RNA-binding domain**

REFERENCES

Bhatt, A., Arora, P., Prajapati, S.K. (2020) Can Algal Derived Bioactive Metabolites Serve as Potential Therapeutics for the Treatment of SARS-CoV-2 Like Viral Infection?. *Frontiers in Microbiology*, 11: 2668. doi:10.3389/fmicb.2020.596374

Chen, Y.W., Yiu, C.P.B., Wong, K.Y. (2020) Prediction of the SARS-CoV-2 (2019-nCoV) 3C-like protease (3CLpro) structure: virtual screening reveals velpatasvir, ledipasvir, and other drug repurposing candidates. *F1000Research*, 9: 129. doi:10.12688/f1000research.22457.2

Cubuk, J., Alston, J.J., Inicicco, J.J., Singh, S., Brereton, M.D.S., Ward, M.D., Zimmerman, M.I., Vithani, N., Griffith, D., Wagoner, J.A., Bowman, G.R., Hall, K.B., Soranoo, A., Holehouse, A.S. (2021). The SARS-CoV-2 nucleocapsid protein is dynamic, disordered, and phase separates with RNA. 2021. doi:10.1038/s41467-021-21953-3

Guaadaoui, A., Benaicha, S., Elmajdoub, N., Ballaoui, M., Hamal. A. (2014). What is a bioactive compound? A combined definition or a preliminary consensus. *International Journal of Nutrition and Food Sciences*, 3(3): 174–179. doi: 10.11648/j.ijnfs.20140303.16

Guo, Y.R., Cao, Q.D., Hong, Z.S., Tan, Y.Y., Chen, S.D., Jin, H.J., Tan, K.S., Wang, D.Y., Yan, Y. (2020). The origin, transmission and clinical therapies on coronavirus disease 2019 (COVID-19) outbreak – an update on the status. *Military Medical Research*, 7, 11. doi:10.1186/s40779-020-00240-0

Hamoda, A.M., Fayed, B., Ashmawy, N.S., Shorbagi, A.N.A., Hamdy, R., Soliman, S.S.M. (2021). Marine sponge is a promising natural source of anti-SARS-CoV-2 scaffold. *Frontiers in Pharmacology*, 12: 1161. doi:10.3389/fphar.2021.666664

Haque, S.K.M., Ashwa, O., Sariefl, A., Mohamed, A.K.A.J. (2020) A comprehensive review about SARS-CoV-2. *Future Virology*, 15: 9. doi:10.2217/fvl-2020-0124

Jana, P., Bhardawaj, B. (2021). Exploration of some naturally occurring fungal-derived bioactive molecules as potential SARS-CoV-2 Main Protease (MPro) inhibitors through in-silico approach. *Journal of Computational Biophysics and Chemistry*, 20, 251–266. doi:10.1142/S2737416521500113

Jin, Z., Du, X., Xu, Y. Deng, Y., Liu, M., Zhao, Y., Zhang, B., Li, X., Zhang, L., Peng, C., Duan, Y., Yu, J., Wang, L., Yang, K., Liu, F., Jiang, R., Yang, X., You, T., Liu, X., Yang, X., Bai, F., Liu, H., Liu, X., Guddat, LW., Xu, W., Xiao, G., Qin, C., Shi, Z., Jiang, H., Rao, Z., Yang, H. (2020). Structure of Mpro from SARS-CoV-2 and discovery of its inhibitors. *Nature*, 582, 289–293. doi:10.1038/s41586-020-2223-y

Kamiloglu, S., Tomas, M., Ozdal, T., Yolci-Omeroglu, P., Capanoglu, E. (2021). Bioactive component analysis. In *Innovative Food Analysis*. Galanakis, C. M. (Ed.) 41–65. Academic Press.

Khairan, K., Idroes, R., Tallei, T.E., Nasim, M.J., Jacob, C. (2021) Bioactive compounds from medicinal plants and their possible effect as therapeutic agents against COVID-19: A review. *Current Nutrition & Food Science*, 17(6). doi:10.2174/1573401317999210112201439

Lockbaum, G.J., Reyes, A.C., Lee, J.M., Tilvawala, R., Nalivaika, E.A., Ali, A., Kurt Yilmaz, N., Thompson, P.R., Schiffer, C.A. (2021). Crystal structure of SARS-CoV-2 main protease in complex with the non-covalent inhibitor ML188. *Viruses*, 13(2): 174. doi:10.3390/v13020174

Naqvi, A.T., Fatima, K., Mohammad, T., Fatima, U., Singh, I.K., Singh, A., Atif, S.M., Hariprasad, G., Hasan, G.M., Hassan, M.I. (2020). Insights into SARS-CoV-2 genome, structure, evolution, pathogenesis and therapies. *Structural Genomics Approach*, doi:10.1016/j.bbadis.2020.165878

Qamar, M.T., Alqahtani, S.M., Alamri, M.A., Chen, L.L. (2020) Structural basis of SARS-CoV-2 3CLpro and anti-COVID-19 drug discovery from medicinal plants. *Journal of Pharmaceutical Analysis* 10(4), 313–319. doi:10.1016/j.jpha.2020.03.009

Rangsinth, P., Sillapachaiyoporn, C., Nilkhet, S., Tencomnao, T., Ung, A.T., Chuchawankul, S. (2021) Mushroom-derived bioactive compounds potentially serve as the inhibitors of SARS-CoV-2 main protease: An in silico approach. *Journal of Traditional and Complementary Medicine*, 11(2): 158–172. doi: 10.1016/j.jtcme.2020.12.002.

Ratha, S.K., Renuka, N., Rawat, I., Bux, F. (2020) Prospectives of algae derived nutraceuticals as supplements for combating COVID-19 and human coronavirus diseases. *Nutrition* doi:10.1016/j.nut.2020.111089

Remali, J., Aizat, W. M. (2021). A review on plant bioactive compounds and their modes of action against coronavirus infection. *Frontiers in Pharmacology*, 2256.

Safet, A., Sarker, M.M.R., Afrin, S., Richi, F.T., Zhao, C., Zhou, J.R., Mohamed, I.N. (2021). Traditional herbal medicines, bioactive metabolites, and plant products against COVID-19: Update on clinical trials and mechanism of actions. *Frontiers in Pharmacology*, 12: 1248. doi:10.3389/fphar.2021.671498

Shaldam, M.A., Yahya, G., Mohamed, N.H., Abdel-Diam, M.M., Naggar, Y.A. (2021) In silico screening of potent bioactive compounds from honeybee products against COVID-19 target enzymes. *Environmental Science and Pollution Research*, 28, 40507–40514. doi:10.1007/s11356-021-14195-9

Suwannarach, N., Kumla, J., Sujarit, K., Pattananandecha, T., Saenjum, C., Lumyong, S. (2020). Natural bioactive compounds from fungi as potential candidates for protease inhibitors and immunomodulators to apply for coronaviruses. *Molecules (Basel, Switzerland)*, 25(8), 1800. doi:10.3390/molecules25081800

Teodoro, A.J. (2019) Bioactive compounds of food: Their role in the prevention and treatment of diseases. *Oxidative Medicine and Cellular Longevity*, Article ID 3765986, 4. doi:10.1155/2019/3765986

Thomas, L. (2020) SARS-CoV-2 Neucleo-capsid (N) protein is heavily glycosylated. https://www.news-medical.net/medical

Tyrrell, D.A., Bynoe, M.L. (1965) Cultivation of a novel type of common-cold virus in organ cultures. *British Medical Journal*, 1(5448), 1467–1470. doi: 10.1136/bmj.1.5448.1467

Walia, A., et al. (2019) Role of bioactive compounds in human health. *Acta Scientific Medical Sciences*, 3.9: 25–33.

Xian, Y., Zhang, J., Bian, Z., Zhou, H., Zhang, Z., Lin, Z., Xu, H. (2020). Bioactive natural compounds against human coronaviruses: A review and perspective. *Acta PharmaceuticaSinica B*. doi:10.1016/j.apsb.2020.06.002

Part IV

In-Silico and Mathematical Tools

26 Bioinformatics and Vaccine Development

Overview and Prospects

Swati Bajaj
Gargi College, University of Delhi, Delhi, India

Pallee Shree
Bhaskaracharya College of Applied Sciences, University of Delhi, New Delhi, India

Dileep K. Singh
University of Delhi, Delhi, India

CONTENTS

26.1 INTRODUCTION

Vaccines are "biological preparations that improve immunity to a particular disease" as defined by the World Health Organization (WHO) (Patil and Shreffler, 2019). Vaccines can halt the onset of various diseases and have significantly reduced morbidity and fatality in populations worldwide (Sunita et al., 2020). Vaccines can avert around 2 to 3 million fatalities yearly (Patil and Shreffler, 2019).

There are two arms of the immune system – the innate and the adaptive. Natural killer cells, macrophages and neutrophils provide the first line of protection and are the components of the innate immune response. This cellular network responds within 24 hours of exposure. A more targeted response is provided by the network of cells from the adaptive immune system. T- and B- lymphocytes are the participants of the adaptive immune system. B cells provide humoral immunity and produce antibodies to neutralize the antigens. However, T cells provide cell-mediated immunity, are efficient at detecting and lysing the infected cells, and also support antibody production. Adaptive immune response establishes immune memory against pathogens and performs rapidly on repeated

DOI: 10.1201/9781003306931-30

encounters. Although memory-driven secondary adaptive response is much faster and stronger, it takes 7–14 days to build an adaptive immune response after first exposure. Development of immune memory against the specific antigen (pathogen) induced by prior exposure to pathogen-driven antigens is the basis of vaccination (Terry et al., 2015).

Generation of cell-mediated and humoral immune responses is the target for effective vaccines (Munangandu et al., 2015). Antibodies from the B cells are the main immune effectors mediated by vaccines that bind specifically to a particular epitope (Oli et al., 2017; Giacomet et al., 2018). However, high-affinity antibodies and memory immune cells are induced by T cells (Rappuoli et al., 2016; Vilela Rodrigues et al., 2019). Activation of only B cells may result in the short-term effectiveness of a vaccine (Ali et al., 2014). The induction of immune memory is important for the stimulation of the immune network. The efficacy of any vaccine is directly related to the robustness of this induction. Immune memory activation, type of immune memory cells induced, and antibody persistence are the factors that influence the success of a vaccine (Sarkander et al., 2016; Oli et al., 2020).

It has taken rigorous work to provide understanding of current conventional vaccines. These vaccines empirically arrived at a time when there was a lot of ambiguity about the vaccine immune system activation. The development of vaccine-induced immunity has always been challenging in immunology (Richner et al., 2017; Goh et al., 2019; Kim et al., 2019; Oli et al., 2020).

The conventional technique for vaccine development is based on an empirical approach (Poland et al., 2018) (Figure 26.1). The outcome of this technique is poor immunogenicity and reduced specificity towards parasites that have complicated life-cycles and pathogens that show high mutation rates like RNA viruses. However, the aim in current conventional vaccines is to increase their effectiveness with specific targeting especially of pathogens that employ multiple routes of pathogenesis for their virulence in humans (Burton, 2017; Rauch et al., 2018; Oli et al., 2020). Traditional vaccines have also been driven by a '"same dose for everyone in every disease' model. To date there has been no specific vaccine to deal with the immunosenescence of aging, with inadequate understanding of how the immunogenicity of vaccines is generating inadequate vaccine response in the elderly (Poland et al., 2018). Thus, drawbacks like lack of knowledge about pathogen–host interaction, antigenic diversity, absence of animal models, and lack of permissive cell lines make the traditional approach expensive, time-consuming and less effective (Wolf et al., 2010; Kharkar et al., 2017; Sunita et al., 2020).

To overcome these challenges, a different approach is required (Wallis et al., 2019). Bioinformatics tools assist in the storage, organization, understanding, analysis and modeling of biological information from biotechnology, genetics, and molecular biology (Orozco et al., 2013;

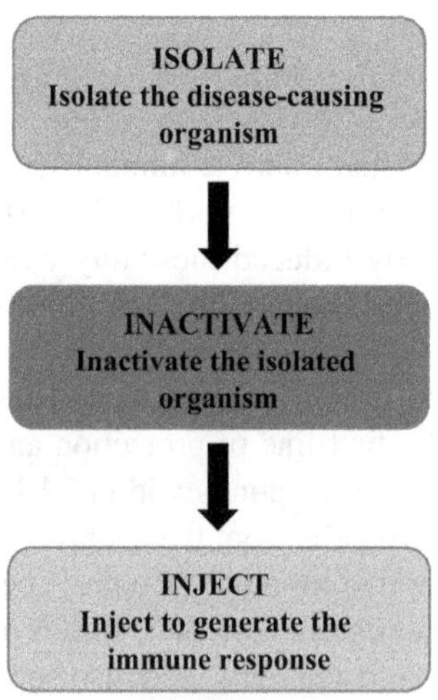

FIGURE 26.1 Empirical approach for vaccine development.

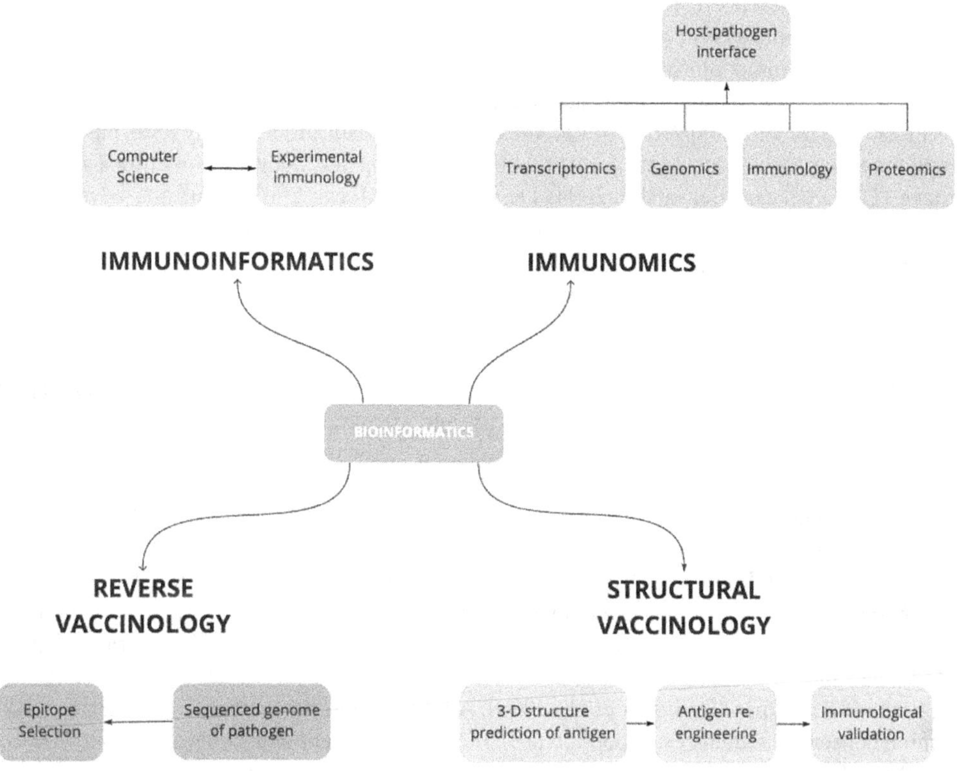

FIGURE 26.2 Bioinformatics-based approaches for vaccine design and development.

Sunita et al., 2020). Taking into account factors that influence vaccine development like genetic variability in humans, emerging and remerging infectious diseases, and antigenic diversity, personalized vaccination is the goal of bioinformatics (Oli et al., 2020). Employing bioinformatics tools (Figure 26.2) ahead of lab experiments also saves cost and time in vaccine design and development (Dubey et al., 2018; Sunita et al., 2020). Thus, it is crucial to design futuristic therapeutics that use bioinformatics tools for vaccine design.

26.2 IMMUNOINFORMATICS

The immune system works by identifying and killing pathogens, providing protection against diseases. Once recovered from certain diseases, the individual develops immunity upon re-encountering the same disease. The immune system comprises a network of thousands of molecules interacting with each other and shows high diversity both structurally and functionally (Tomar and De, 2010).

An enormous amount of clinical and epidemiological data has been generated through research experiments in immunology over a long period. A huge bank of data from immunological research is required to be managed, analyzed and stored. Immunoinformatics provide several databases in which information from all sources of immunological studies is organized and stored to provide a development resource (Oli et al., 2020).

Emerging computer and informatics technologies have led to the birth of a new outlook on vaccine design and development. The aim of immunoinformatics is to apply computational and mathematical solutions to address immunological questions (De Groot et al., 2002; He et al., 2010a). Immunoinformatics employs tools for improved understanding of host–pathogen interactions and complex immune systems. It is a technology that has added accuracy, ease, and specificity to the vaccine design pipeline (He et al., 2010a; Tomar and De, 2010).

The history of immunoinformatics dates back to the 1980s, when the first immunoinformatics tools for designing vaccines were discovered (Delisi and Berzofsky, 1985). Epitope-mapping algorithms were the most important of vaccine design informatics tools. Many tools based on T-cell epitope-mapping algorithms came into existence providing accuracy in modeling Ligang-T-cell interface (De Groot and Berzofsky, 2004; De Groot and Moise, 2007). With the advent of these tools, a complete pathogenic proteome can be examined for the quick recognition of T-cell epitopes (De Groot, 2006; He et al., 2010a).

Mapping of putative B- and T-epitopes utilizing computer algorithms has opened a window to explore potential binding sites. Immunologists can use information generated from putative antigenic proteins by analyzing pathogenic genomes to design novel vaccines. This technology greatly reduces the cost of designing vaccines compared to the conventional method which requires protein extraction and large-scale protein testing (Davies and Flower, 2007; Tomar and De, 2010). This approach, known as 'reverse vaccinology', is discussed below in detail.

The wide variability of surface antigens directly impacts the efficacy of a vaccine and the efficiency of antibodies to confer protective immunity (Rappuoli, 2007). Immunoinformatics targets such problems by enhancing the accuracy of vaccine design prognostication for a diversity of disease-causing agents (He et al., 2010a). One such algorithm for handling immune system genetic variation, called Aggregatrix, has been developed by De Groot and Martin. By identifying the best array of epitopes from a disease-causing agent, Aggregatrix allows wide HLA coverage if incorporated in a vaccine. This approach covers the broad range of variants of a strain and most of the frequent human HLA types (De Groot et al., 2008). Another algorithm, called Conservatrix, works to determine consensus sequences of peptide present within different strains of a pathogen (De Groot et al., 2004). These conserved sequences can be identified and used to design vaccines, minimizing the genetic variations between vaccine strains. This approach is important for dealing with pathogens that show high peptide sequence variability like HIV envelope proteins (Gaschen et al., 2002; De Groot et al., 2005).

T-cell activation is crucial for immune response to vaccines. In addition to their active role against infection, CD4+ helper cells are crucial for the development of memory cytotoxic T-cell and antibody-producing memory B cells. Many validated tools that depict T-cell epitope are available on the internet (Ahlers et al., 2001). Comparative analysis of these tools through performance prediction models has been possible since the early 1990s with the emergence of numerous machine-learning techniques (Wang et al., 2008; Gowthaman and Agrewala, 2008). Software that deals with T-cell epitopes includes IEDB, SVMHC, Epitopemap, PREDIVAC, KISS, BIMAS, POPI, MAPPP, and NetCTL. Epitope prediction software ABCpred, BCPPREDS, PEPITO, IgPRED specifically searches for B-cell epitopes (Maria et al., 2017). Unlike T-cell, B-cell epitope-mapping immunoinformatics tools are limited due to the complexity of in silico modeling protein targets for B cells (Enshell-Seijffers et al., 2003; Sweredoski and Baldi, 2008). However, T-cell epitope-depicting tools help accelerate the process of B-cell antigen selection. It has sometimes been found that a selected T-cell epitope also comprises a B-cell epitope (Graham et al., 1989).

The immune epitope database has provided a set of research tools for immunologists. iVax is a unified set of tools for EpiVax vaccine design comprising many algorithms like VaccineCAD, ClustiMer, EpiAssembler, EpiMatrix, BlastiMer (De Groot et al., 2009), which can be implemented to develop a multi-epitope vaccine (McMurry et al., 2005), multi-pathogen biodefence vaccine, and epitope-based vaccine (He et al., 2010a). EpiMatrix is currently in operation in the pharmaceutical industry (De Groot and Martin, 2009). iVax can also be applied to study a complete set of genes of a pathogen for vaccine candidates (He et al., 2010a).

Integrated workflows and systems for computational and informatics approaches may therefore possibly be the keys to automation of the vaccine discovery pipeline.

26.3 IMMUNOMICS

Vaccinology is a complex branch of science in which there are wide variations in disease presentation and molecular profile (Falus, 2008). Decryption of the host immunological reaction to pathogens is a challenge requiring a logical, structured, and extensive approach to vaccine development, and this can be aided by immunomics (De Sousa and Doolan, 2016).

Immunomics is the study of the immunome which refers to all the immunological reactions resulting from host–antigen interactions (Sette et al., 2005; Tomar and De, 2010). In 2001, Klysik coined the term 'immunomics,' keeping in mind that the correlation between genes and functional attributes of encoded proteins needs to be addressed by technological advances (Klysik, 2001). It is a new discipline that focuses on immune system mechanisms and uses high-throughput techniques. It provides a functional association between immunological research and its clinical importance by integrating the fields of bioinformatics, proteomics, genomics, molecular immunology, and transcriptomics (Doolan, 2011; De Sousa and Doolan, 2016).

For thousands of years, the host immune system has co-evolved with the infecting agent. The attributes of the host immune system (age, gender, immune status, etc.) and infecting agent (virulence, strain, etc.) interact with each other (Stilling et al., 2014). This interconnection further increases the complexity of the immune reactions, increases the cost of experimental verification and creates a challenge for comprehensive in-depth analysis. Immunomics aims to provide a platform for in silico high-throughput screening prior to in vitro and in vivo experimentational verification, and relies on both the host immune system and the pathogenic agents (Balls et al., 2012; Annang et al., 2015). Immunomics employs next-generation powerful computational tools and high-density microarrays for depicting pathogenic determinants and fast discovery, facilitating characterization and identification of key antigens and epitopes through a rational approach (Hecker et al., 2012; Carmona et al., 2015; De Sousa and Doolan, 2016).

Genomics, proteomics, and transcriptomics datasets available as large-scale databanks provide an information pool for immunomics-based research. TriTrypDB (http://tritrypdb.org/tritrypdb), UniProt (http://www.uniprot.org/), GenBank (http://www.ncbi.nlm.nih.gov/genbank), Innate immune database (http://www.innatedb.com), Immune epitope database (http://www.iedb.org/) are large-scale databases available for public access (De Sousa and Doolan, 2016).

Traditionally, immune response assessment methods take into account the rate of occurrence and enormity of a single individual immunological variable, e.g., antibody titre. However, a multi-dimensional approach to immunological responses with suitable biological outcomes is the focus of immunomics, for example, protective immunity. Elements like population density of T cells, evaluation of cytokine response, and epitope combinations are taken into consideration under immunomics (Quintana et al., 2004). The behavior of the immune network can be modulated through the immune response to an epitope. This has been demonstrated through immunomic microarray data for immunomic regulatory networks by Braga-Neto and Marques (2006). Moreover, immunomics can be applied to anticipate future diseases (Quintana et al., 2004).

The important determining factor for protection is the quality of the immune response generated, making its assessment crucial in immunology. Many conventional immunoassays (such as ELISA, splenic focus assay, IFN-γ ELIspot, etc.) that measure the reactivity of T- and B-cell and advance technologies (such as CyTOF technologies, Nanostring, etc.) that can assess immune responses together with phenotypic markers and cell functions at individual cell level are available. None can describe the attributes of the entire antigen-specific response if used individually; however, immunomics can do so (De Sousa and Doolan, 2016). Immunomics provide a multidisciplinary platform for integrating a distinct course of action for the development of vaccines. This bioinformatic approach provides a promising alternative for successful vaccine design using computational and mathematical algorithms to manage complex data (Raman et al., 2010), which would be difficult and expensive if done experimentally (De Sousa and Doolan, 2016).

26.4 REVERSE VACCINOLOGY

Preparation of vaccines readily accomplishes disease prevention in the most effective way as it helps in the prevention of diseases that cannot be cured (Cohen and Shoenfeld, 1996). Intervention of vaccines helped in the eradication of diseases in many developing countries, examples being smallpox and polio (Liu and Cepica, 1990; Okwo-Bele and Cherian, 2011; Dowdle et al., 2003). Until the twentieth century, vaccinations were created via the traditional approach, which used technologies such as live attenuated strains employed as vaccines, inactivated or killed strains, inactivated toxins, or subunit/conjugate vaccinations that need a microorganism culture. With the advent of the genomic age, new advances in vaccinations have been made. Advances in sequencing have provided the whole genome sequence of different microorganisms of a pathogenic nature, which has helped in the development of a new era of vaccine known as reverse vaccines. Reverse vaccinology is a technique which identify proteins expressed on the surface of a microorganism with the help of genome sequences rather than the microbe (Rappuoli, 2000; Vivona et al., 2006). In silico genomic information is utilized for vaccine development because the genome sequence is a repository of genes that code for protein which can function as potential antigens in the development of vaccine candidates (Rappuoli, 2001; Capecchi et al., 2004). Reverse vaccinology makes use of genomic information to explore and discover novel antigens for immunization. Although biochemical, microbiological, and serological methods successfully identify components beneficial for the creation of vaccines, the conventional method is often time-consuming (Rappuoli, 2000; Adu-Bobie et al., 2003; Parvizpour et al., 2020). These conventional methods often fail as some pathogenic microorganisms cannot be cultured in vitro, or the most abundant antigens may vary in sequence (Rappuoli, 2000).

Reverse vaccinology is a machine-based learning further enhanced by epitope-binding information and gene expression data addition for annotation. Earlier epitope-binding information was not utilized for reverse vaccinology due to the high likelihood of false positive results. However, as separation of noise signals using machine-based techniques has developed, epitope-binding predictions can be employed.

Reverse vaccinology has advantages over conventional approaches, as it has the potential for rapid, comprehensive assessment of the surface protein repertoire of microbes. Genomic analysis helps with antigen prediction, regardless of their immunogenicity and abundance during infection, without in vitro cultivation of the pathogenic microorganism. The basic concept of reverse vaccinology is based on in silico comparative analyses of multiple genome sequences within a heterogeneous pathogen population and commensal strains to identify antigens unique to pathogenic strains. This process helps in the development of vaccines based on non-conventional antigens and therefore exploring the non-conventional arms of the immune system. Using this approach, many of the vaccines which were previously impossible to develop have now become a reality. Reverse vaccinology procedures are now commonly used for vaccine design against several pathogens (Jaiswal et al., 2013). The first clinically registered successful vaccine based on reverse vaccinology approach was developed for the bacterial pathogen group B *Meningococcus*, the pathogen that causes 50% of meningococcal meningitis globally (Sette and Rappuoli, 2010). Finally, the discovery of the vaccine for *Meningococcus* B led to the licensed Bexsero vaccine (Folaranmi et al., 2015). It was very difficult to develop a vaccine against this disease which was a very significant cause of mortality. In a study by Masignani et al., the in-silico method of reverse vaccinology played a very significant role in development of this vaccine. The authors identified a total of 600 new genes which can code for surface protein/exported protein, 350 of which were expressed successfully and could be used further for immunization. From these, 25 unique surface proteins were demonstrated to generate bactericidal antibodies and the majority of these surface proteins were good candidates for vaccine development. None of the previously employed methodologies were used to detect them (Masignani et al., 2019).

Since the development of this vaccine against the *Neisseria meningitidis* group, based on reverse vaccinology technology, this vaccine development strategy has continuously progressed and is now recognized as an effective method that can be exploited for the development of vaccines against several kinds of pathogens (Table 26.1).

TABLE 26.1
Reverse Vaccinology Technology Aiding Design of Potential Vaccines Against Various Pathogens

S.No.	Pathogen	Species	Reference
1	*Neisseria meningitidis*	Bacteria	Rappuoli (2000), Pizza et al. (2000), Giuliani et al. (2006), Masignani et al. (2019)
2	Japanese encephalitis	Virus	Chakraborty et al. (2020)
3	H7N9 (Avian influenza A)	Virus	Hasan et al. (2019)
4	Parasite	Protozon	Goodswen et al. (2017)
5	*Streptococcus pneumoniae*	Bacteria	Talukdar et al. (2014)
6	*Staphylococcus aureus*	Bacteria	Oprea and Antohe (2013)
7	*Leptospira* spp.	Spirochetes	Dellagostin et al. (2017)
8	*Leishmania* spp.	Protozon	John et al. (2012)
9	*Listeria monocytogenes*	Bacteria	Gore and Pachkawade (2012).
10	*Acinetobacter baumannii*	Bacteria	Chiang et al. (2015); Solanki and Tiwari (2018)
11	*Campylobacter*	Bacteria	Meunier et al. (2016)
12	*Brucella melitensis*	Bacteria	Vishnu (2015)
13	SARS-CoV-2	Virus	Tahir ul Qamar et al. (2020); Ong et al. (2020); Enayatkhani et al. (2021)
14	Hepatitis	Virus	Chaudhuri et al. (2020)
15	Influenza-A (H7N9)	Virus	Hasan et al. (2019)
16	*Mycobacterium leprae*	Bacteria	Gupta et al. (2020)
17	*Bacillus anthracis*	Bacteria	Kanampalliwar et al. (2013)

With improvements in sequencing techniques, the whole genome sequence (WGS) data is easily accessible and obtainable for a wide range of pathogens. After sequence analysis, each of the unique gene set is studied to focus on prediction of epitopes for developing epitope-driven vaccines that are target specific. After that, the predicted epitope or antigen is expressed as a recombinant protein in *E. coli*. To validate surface location in vitro, the list of proteins that could be used as vaccine antigens and the animal models are further refined to evaluate immunogenicity. This process involves mice immunization and for bactericidal activity, sera of the organism are tested in vitro (Rappuoli, 2001; Mora et al., 2003). Transcriptomics and proteomics data are also used during the selection process, which helps with shortlisting of the antigens to be tested in animal models. This reduces time and cost of downstream analyses. Thus, the primary study begins by cloning each of the infectious organism's genes into an expression library. Each of the proteins expressed is isolated and in order to study the immune response of different proteins, the complex mixture is screened in a mice model. When a pool elicits a reaction, the proteins are subdivided until each protein is examined for its capacity to stimulate the immune system as well as protect the mice against the real infectious agent. Those proteins that elicit the best response of all the screened proteins can be a part of a sub unit vaccine or can be used alone (Figure 26.3).

In order to use newly discovered antigens for vaccine development, it is critical to define their biological role. Reverse vaccinology thus provides us with a very powerful tool to find and explore novel proteins that have a crucial role in pathogens. Software and tools now exist which make this reverse vaccinology a successful approach. Vaxign is one of the widely used web-based reverse vaccinology programs, and is also the first web-based vaccine design programs to predict vaccine candidates against several pathogens (Xiang and He, 2009; He et al., 2010b; Ni et al., 2017; Ong et al., 2020). In order to enhance prediction accuracy and precision, a new machine-learning approach has recently been developed using Vaxign-ML (Ong et al., 2020). Software such as Bepipred and SVMHC aids in improving the outcome of these programs (Bowman et al., 2011;

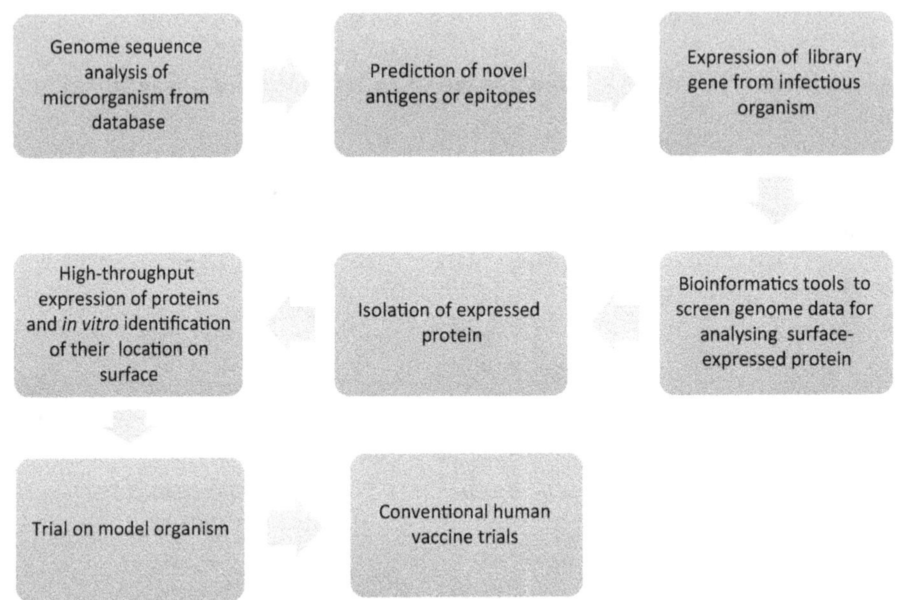

FIGURE 26.3 Workings of reverse vaccinology.

(Adapted from Kanampalliwar et al., 2013; Rappuoli, 2001).

Heinson et al., 2015; Kanampalliwar et al., 2013), DictyOGlyc[a] predicts glycosylation, BLAST is a basic local alignment search tool), the Pfam searchable database contains protein family information, and ProDOM is a database of protein domain families (Heinson et al., 2015).

There are many more tools that increase the efficacy of reverse vaccinology, and development to boost efficiency is ongoing. The reverse vaccinology technique has recently been utilized many times for designing a vaccine against COVID-19. Ong et al. (2020) applied the Vaxign and Vaxign-ML reverse vaccinology approaches along with the knowledge of existing coronavirus vaccine to predict protein candidates for the development of a COVID-9 vaccine. In another study Tahir UL Qamar et al. (2020) also applied the concept of reverse vaccinology for designing multi-epitope based subunit vaccine for COVID-19.

Reverse vaccinology helps systemically classify potential antigens that can help improve existing vaccines and create effective preparations for practically any disease whose pathogen genome sequence has been identified. It reduces the time taken to design vaccines and helps design vaccines against pathogens that are difficult to culture. Although it has several advantages, this approach also has its own limitations, as proteins can be identified by this method but it omits significant vaccine components like glycolipids and polysaccharides. Another limitation is the lack of a high-throughput system for estimating selected candidates' protective immunity. A conventional vaccinology approach will therefore always remain significant, and will always support vaccine development from antigens derived from the genome sequence. In the future, reverse vaccinology approaches will help with rapid vaccine design, but this method should be further refined by the use of different software and other bioinformatics knowledge.

26.5 STRUCTURAL VACCINOLOGY

Vaccine development of the vaccine is among the most the important events in the history of medical science, with prevention of infectious diseases achieved through vaccination drives (Ozawa et al., 2012). Conventional methods of vaccine development are not successful in the case of many diseases like human immunodeficiency virus (HIV), dengue virus, hepatitis C virus, enterovirus A71 and Chikungunya (Anasir and Poh, 2019; Rezza and Weaver, 2019). However, the advances in technology, sequencing and computational techniques of the past decade have empowered significant progress in the identification and design of potently immunogenic recombinant protein-based vaccines. With advances in structural biology and computational approaches, structure-based vaccine design enters the equation. Structure-based vaccine design is called structural vaccinology (SV) and is becoming a prominent strategy for the rational design of vaccine candidates. SV focuses on determining antigen structures and their epitopes (Dormitzer et al., 2012). It is an advanced vaccine design technique which utilizes the three-dimensional structural knowledge of antigens and epitopes to construct immunogens against the target pathogens (Nezafat et al., 2017). Structural vaccines are useful for locking antigenic proteins or envelopes in a certain structure in order to optimize epitope arrangement (Dormitzer et al., 2012; Liljeroos et al., 2015). The basic demand for SV is the molecular-level details of antigen surface that is accessible to pathogens. Recent literature has shown that using this structural knowledge could overcome some of the problems with vaccinations. Structural vaccines utilize structural protein information to design immunogens and therefore they are likely to develop novel vaccinations for tough targets.

The revolution in sequencing techniques and genome information has impacted vaccine research and led to evolution of a new technique, reverse vaccinology. in which the sequence of the protein and linear epitopes is required for vaccine design. However, there are many diseases for which vaccines cannot be developed with this protocol (Cozzi et al., 2013). In reverse vaccinology the knowledge of the pathogen genome is required, but in SV, protective determinants can be used to create antigens that are more easily redesigned and simplified for use in vaccine combinations (Kulp and Schief, 2013). SV is more effective when antigenic variation exists across closely related strains or species. SV, which is concerned with three-dimensional information about the structure, can be used to design improved vaccines (Dormitzer et al., 2008). This technology is based on the discovery of epitopes that elicit a protective immune response and are confined to certain regions within the immunogenic protein (Oscherwitz, 2016). Once these domains have been discovered and generated in recombinant form, they may be employed as efficient immunogens. Antigen crystal structures having one or more than one protection epitopes are the starting point for immunogen creation, particularly when combined with a protective antibody. SV is a rational vaccine design platform that can also aid in the development of vaccines against difficult diseases such as hepatitis C, human immunodeficiency virus, enterovirus A71, and dengue virus (Dormitzer et al., 2012; Van Regenmortel, 2016; Anasir and Poh, 2019). Hepatitis C virus is a cause for serious concern because of increasing public health problems. Despite recent advances in antiviral therapeutics, there remains a continuous increase in new hepatitis C virus infections. Recent advances in genetic and SV are now helping in the understanding of this virus envelope glycoproteins broadly neutralizing antibody responses, creating new hopes for the design of superior vaccines against the disease (McLellan et al., 2013; Yechezkel, Law, and Tzarum, 2021). Several studies on SV have been conducted proving that this technology has already been successfully used to generate vaccines against a variety of diseases and has the potential to help in the resolution of vaccine manufacturing and efficacy issues. Many previous studies have proved that SV is successful in designing novel vaccines against diseases. Attempts to design a vaccine for respiratory syncytial virus disease have been made using the concept of structural biology, which provided an immunogen that elicits enhanced protective responses compared with conventional vaccine design, and these protective responses were observed in the animal model (Liljeroos et al., 2015). Similar efforts were put into using structural biology knowledge to develop a vaccine for HIV. The structure of trimer ligand-free HIV-1–Env was selected for development of the vaccine and this conformation was fixed for determining the interactions of the receptor with the ligand. The examination of the epitope trimeric ligand-free envelope revealed that it was compatible with broadly neutralizing antibodies but not with antibodies that were poorly neutralizing (Do Kwon et al., 2015).

Meningococcus B, group B streptococcus, influenza virus, dengue virus, etc., are some of the viruses on which ongoing studies are applying the SV concept to develop vaccines. The outcomes indicate that SV has revolutionary potential for developing novel vaccines against these diseases, which are currently not amenable to conventional approaches (Nuccitelli et al., 2011; Scarselli et al., 2011; Kulp and Schief, 2013; Anasir and Poh, 2019.) Structural vaccine techniques aid in the fast discovery of monoclonal antibodies with the help of nuclear magnetic resonance, x-ray crystallography, mass spectrometry, cryo-electron microscopy, and computational methods such as computational structural modeling, scaffold designing and epitope prediction which leads to a new method for producing atomic-level precision epitope-specific immunogens (Liljeroos et al., 2015; Anasir and Poh, 2019).

Three-dimensional protein structures provide significant knowledge about the tertiary structure and position of viral epitopes, helping to solve some challenging issues impeding the development of novel vaccines against pathogens. Structural vaccinology, based on the concept of neutralization which lessens the probability of protein formation in the pathogen, provides information about the individual protein to attack as well as the regions of proteins to be targeted by the antibodies. In the neutralization process the antibodies bind at an exposed site on the surface exposed protein of the pathogen (Nuccitelli et al., 2011; Dormitzer et al., 2012). The success of neutralization is dependent upon many factors, like availability of protein binding sites, and avidity and affinity of the binding sites on the pathogens. The basis for protection against a pathogen is the prevention of viral entry, inhibition of the release of virus from infected cells, inhibition or decrease in cell-to-cell transmission, or reduced inoculum sizes.

Structural vaccinology is a methodical approach to developing an effective vaccine (Figure 26.4). The initial step involves determination of the three-dimensional atomic structure of antigens or antigen–antibody complexes (Kulp and Schief, 2013; Anasir and Poh, 2019). This can be done using structural biology tools such as cryo-electron microscopy, x-ray crystallography, nuclear magnetic resonance, etc. Remodeling of antigen or epitope is then performed based on their structural information. In order to design an effective vaccine, knowledge of bioinformatics and immunology should be combined with structural insights to design an effective antigen. The re-engineered epitope or antigen is incorporated into one of the vaccine platforms. In order to stabilize immunogenic conformations, the re-engineering mostly involves mutations of some of the amino acids sequence, which mask non-neutralizing antibodies or epitopes that are not required and optimize antigen thermostability. The re-engineered antigen may then be put into one of the vaccination platforms, such as recombinant proteins, live attenuated vaccines or virus-like particles. The final step involves checking the efficiency or immunogenicity of the candidate vaccine, which can be tested on the animal model. On the basis of the preclinical outcome, the candidate vaccine can be redesigned to enhance efficacy (Anasir and Poh, 2019; Graham et al., 2019).

Many structural vaccines are in process, with some now undergoing clinical testing (Sastry et al., 2017; Huang et al., 2018). Recent advances have demonstrated the viability of structure-based vaccinations and this concept can be utilized to develop vaccines for those diseases where conventional methods are not successful. The SV approach can also be applied to increase the efficiency of vaccines. There are certain pitfalls with this technology, such as its vulnerability to a lack of complete immunological and structural knowledge of host immune responses induced by viruses. Another drawback of this method is that the antigen antibody structures are predetermined and the

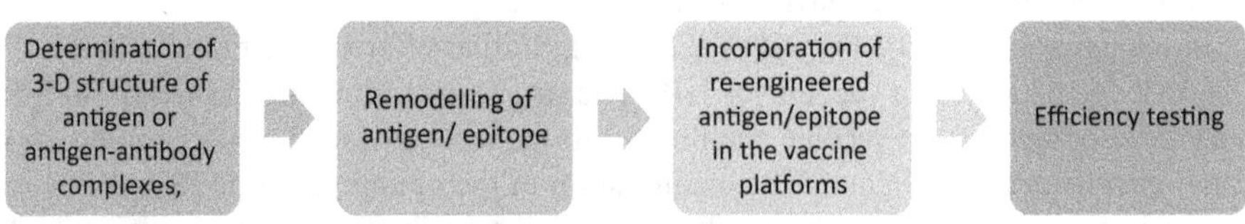

FIGURE 26.4 Structural vaccinology strategy.

(Adapted from Anasir and Poh, 2019)

structure of free molecules differ at the binding sites because of structural changes during binding. Thus, structure determined by x-ray crystallography could be different from the real immunogenic structure, which is recognized by B-cell receptors during antibody formation. Despite several drawbacks, this emerging technique is playing a significant role in development of vaccines based on 3-D knowledge of antigen antibody concepts for those diseases where conventional methods are not working.

26.6 INFECTIOUS DISEASES AND VACCINES

26.6.1 Tuberculosis

Tuberculosis is an airborne contagious illness caused by microbes pertaining to the *Mycobacterium tuberculosis* (Mtb) complex. It has been the main cause of morbidity and fatality since the bacterium was discovered in 1882 by Robert Koch (Blevins and Bronze, 2010; Sefidi-Heris et al., 2020). The disease presents diverse symptoms that range from asymptomatic or mild to severe and even fatal. Every year, around 8 million fresh cases of active Mtb are reported worldwide (Sefidi-Heris et al., 2020).

The characteristics of the bacterium, including its slow growth rate, make its study hard and cause a delay in diagnosis (Maria et al., 2017). Moreover, the inefficiency of the current vaccine, Bacillus Calmette–Guerin (BCG), is one of the major reasons for the high prevalence of the disease (Maria et al., 2017). The vaccine provides protection against extrapulmonary TB only to children and is ineffective in adults (Sefidi-Heris et al., 2020). Thus, bioinformatics tools are being used in the search for novel vaccine candidates to provide better immunity against the disease (Monterrubio-López and Ribas-Aparicio, 2015; Maria et al., 2017).

CD4+ T cells-mediated cellular immunity is extremely important to control *Mycobacterium tuberculosis* disease (Woodworth et al., 2014; De Sousa and Doolan, 2016). Thus, the study of T-cell epitopes is the basis for immunoinformatics (Vani et al., 2006; Tang et al., 2011; Sundaramurthi et al., 2012; Pandey et al., 2016; Maria et al., 2017). Bacterial genomes available from the NCBI database are screened for immunomics-based study for the identification of novel epitopes (Lindestam Arlehamn et al., 2013). Numerous epitopes have been suggested as potential candidates for vaccine design. In vivo and in vitro studies have described improved immunogenicity compared to BCG vaccine for some, including the Ag85 protein family (Mollenkopf et al., 2004) and PE, ESAT-6, and PPE protein families (Pajon et al., 2006).

26.6.2 Influenza

Infectious disease caused by any of the three RNA viruses (influenza types A, B and C) is named influenza. Influenza pandemics have become a significant socio-economic and well-being issue throughout the world. Present-day vaccines are inefficient in providing protection against fast-emerging mutations in viruses that bring antigenic shift and drift (Kumar et al., 2018; Sautto et al., 2018). Minor change in the antigenic site is referred to as antigenic shift whereas major change that results in a new virus subtype is antigenic drift (Hensley et al., 2011). This pathogenic variability accounts for the short-lasting immune response of the existing vaccines against influenza and poses a challenge in the vaccine design process (Bouvier and Palese, 2008; Oli et al., 2020).

Transmembrane proteins – neuraminidase (NA) and hemagglutinin (HA) – are the principal components in current influenza vaccines dictating the antigenic properties of the viruses (Hensley et al., 2011; Lambach et al., 2015; Hampson et al., 2017). Influenza HA has two components: the globular head and the stem region. The highly variable globular head contains the majority of antigenic sites and has remained the target for vaccine design. The stem region, previously disregarded for vaccine development, has come into prominence since the discovery of neutralizing antibodies (Bouvier and Palese, 2008, Neu et al., 2016). Successful generation of the immune response against engineered

HA stem antigens in heterosubtypic challenge models (Impagliazzo et al., 2015) and identification of numerous conserved regions in the HA stem region has opened up the possibility of developing a universal vaccine against influenza (Harris et al., 2013).

The Influenza Research Database (Zhang et al., 2017) or OpenFluDB (Liechti et al., 2010) are open databases for protein sequences of influenza virus and aids vaccine design. EpiCombFlu is another open database through which conserved motifs can be identified across the strains and targeted for vaccine design (ElHefnawi et al., 2011).

26.6.3 Zika Virus Disease

The Zika virus (ZIKV) is a mosquito-borne, single-stranded RNA flavivirus that was first isolated in Uganda's Zika forest in 1947 (Dick et al., 1952). The virus is spreading globally and symptoms range from asymptomatic to mild to severe. Severe disease is linked with congenital malformation, microcephaly and stillbirth, congenital Zika syndrome (CZS) in pregnant women (Mlakar et al., 2016; Brasil et al., 2016; de Oliveira et al., 2016; Heymann et al., 2016; Panchaud et al., 2016; Hoen et al., 2018), and an autoimmune disease of the peripheral nerves, Guillain–Barré syndrome (GBS) in adults (Méndez et al., 2017; Styczynski et al., 2017; Barbi et al., 2018). In view of the serious symptoms resulting from virus infection, the virus was declared a Public Health Emergency of International Concern by the WHO (WHO, 2016). The development of safe and effective vaccinations has become critical for infection prevention and control (Pattnaik et al., 2020).

Several vaccine contenders are in clinical trials, but as yet no vaccine is available commercially. Research for an appropriate vaccine candidate is based on many bioinformatics strategies. The best epitopes for effective vaccine design are targeted predominantly using in silico prediction models. Mirza et al. predicted antigenic CTL epitopes and B-cell epitopes with in silico studies. They applied molecular modeling and molecular docking and determined hydrophobicity, surface flexibility, surface accessibility and structure-based epitope projection for E protein, NS5 and NS3 (Mirza et al., 2016). In another study, ProPred immunoinformatics algorithms were employed to predict antigenic epitopes for HLA Class I and Class II (Dar et al., 2016). Another study also employed ProPred algorithms to predict T-cell epitopes from envelope protein (Ashfaq and Ahmed, 2016). They also applied molecular docking to investigate the interactions between HLA-B7 and B-cell epitopes. Using the *in silico* predictions, Dikhit et al. investigated nine highly conserved Class I restricted epitopes among NS2A, NS4B and NS5 envelope and capsid 1 viral proteins. The authors used PEPstr to model the tertiary structure of the epitopes (Dikhit et al., 2016). Gupta et al. (2016) created ZikaVR (http://bioinfo.imtech.res.in/manojk/zikavr/), a multi-omics platform containing information regarding the proteome, molecular biology and genome of the virus to accelerate vaccine development for ZIKV (Gupta et al., 2016).

26.6.4 COVID-19

COVID-19, an acute respiratory coronavirus 2 syndrome (SARS-CoV-2) is a highly contagious infection and was formally announced as a global pandemic on March 11, 2020 by the WHO (WHO, 2020). Every corner of the globe has been facing the challenge of keeping this outbreak in check since its first occurrence in Wuhan, China in December 2019 (Hui et al., 2020; Chatterjee et al., 2021). Symptoms may range from asymptomatic, to mild fever and common cold, to severe cases of pulmonary distress (Hui et al., 2020; Chukwudozie et al., 2021).

The arrival of the pandemic saw an upthrust in the employment of bioinformatics platforms. COVID-19 vaccine candidates demonstrate diversified technological fields including live attenuated virus vaccines, adenoviral vector-based vaccines, mRNA vaccines, immunogenic adjuvant protein vaccine, and protein subunit vaccines (Grigoryan and Pulendran, 2020; Jeyanathan et al., 2020; Kaur and Gupta, 2020; Krammer, 2020; Rawat et al., 2021).

Bioinformatics-based strategies laid the foundations for predicting antigenic epitopes and became crucial to buying time in the fight against SARS-CoV-2. Several bioinformatics software programs, web servers, and data repositories are available maintaining extensive data specific to coronavirus. Software/tools/databases used for COVID-19 vaccine development include coVdb (http://covdb.popgenetics.net), hCoronavirusesDB (http://hcoronaviruses.net/#/), coronaVIR (https://webs.iiitd.edu.in/raghava/coronavir/), SARS-CoV-2 3D (https://sars3d.com/), COVIEdb (http://biopharm.zju.edu.cn/coviedb/), GISAID (https://www.gisaid.org/), CoV3D (https://cov3d.ibbr.umd.edu/), COVID-Profiler (http://genomics.lshtm.ac.uk/), VirHostNet 2.0 (http://virhostnet.prabi.fr), and DBCOVP (http://covp.immt.res.in/) (Chatterjee et al., 2021).

26.7 CONCLUSION AND FUTURE PROSPECTS

Infectious diseases with global impact have created an urgent requirement for therapeutic vaccines. Pathogens causing such diseases are associated with complex life-cycles and antigenic variability, making the development of an effective vaccine challenging. The vaccine development pipeline has therefore called for innovative approaches to wet laboratory experimentation which will only be feasible if accompanied by bioinformatics strategies. Conventional vaccinology approaches will always remain significant and will support vaccine development through bioinformatics approaches.

Bioinformatics-based approaches such as immunoinformatics, immunomics, reverse vaccinology, and structural vaccinology are advantageous, being powerful, cost-effective, and time-saving solutions that fulfill the current need for the development of potential vaccine candidates. Continuously evolving bioinformatics platforms are being applied to solve vaccine-associated queries from experimental and clinical studies. In our opinion, this approach has great potential and is likely to play a revolutionary role in the linked challenges of 21st-century vaccine design.

REFERENCES

Adu-Bobie, J., Capecchi, B., Serruto, D., Rappuoli, R. and Pizza, M., 2003. Two years into reverse vaccinology. *Vaccine*, *21*(7-8), pp. 605–610.

Ahlers, J.D., Belyakov, I.M., Thomas, E.K. and Berzofsky, J.A., 2001. High-affinity T helper epitope induces complementary helper and APC polarization, increased CTL, and protection against viral infection. *The Journal of Clinical Investigation*, *108*(11), pp. 1677–1685.

Ali, A., Jafri, R.Z., Messonnier, N., Tevi-Benissan, C., Durrheim, D., Eskola, J., Fermon, F., Klugman, K.P., Ramsay, M., Sow, S. and Zhujun, S., 2014. Global practices of meningococcal vaccine use and impact on invasive disease. *Pathogens and Global Health*, *108*(1), pp. 11–20.

Anasir, M.I. and Poh, C.L., 2019. Structural vaccinology for viral vaccine design. *Frontiers in Microbiology*, *10*, p. 738.

Annang, F., Pérez-Moreno, G., García-Hernández, R., Cordon-Obras, C., Martín, J., Tormo, J.R., Rodríguez, L., De Pedro, N., Gómez-Pérez, V., Valente, M. and Reyes, F., 2015. High-throughput screening platform for natural product–based drug discovery against 3 neglected tropical diseases: Human African trypanosomiasis, leishmaniasis, and Chagas disease. *Journal of Biomolecular Screening*, *20*(1), pp. 82–91.

Ashfaq, U.A. and Ahmed, B., 2016. De novo structural modeling and conserved epitopes prediction of Zika virus envelop protein for vaccine development. *Viral Immunology*, *29*(7), pp. 436–443.

Balls, M., Bennett, A. and Kendall, D., 2012. Translation of new technologies in biomedicines: shaping the road from basic research to drug development and clinical application–and back again. *Pharmaceutical Biotechnology: Drug Discovery and Clinical Applications*, pp. 113–152. doi: 10.1002/9783527632909.ch6

Barbi, L., Coelho, A.V.C., Alencar, L.C.A.D. and Crovella, S., 2018. Prevalence of Guillain-Barré syndrome among Zika virus infected cases: A systematic review and meta-analysis. *Brazilian Journal of Infectious Diseases*, *22*, pp. 137–141.

Blevins, S.M. and Bronze, M.S., 2010. Robert Koch and the 'golden age' of bacteriology. *International Journal of Infectious Diseases*, *14*(9), pp. e744–e751.

Bouvier, N.M. and Palese, P., 2008. The biology of influenza viruses. *Vaccine*, *26*, pp. D49–D53.

Bowman, B.N., McAdam, P.R., Vivona, S., Zhang, J.X., Luong, T., Belew, R.K., Sahota, H., Guiney, D., Valafar, F., Fierer, J. and Woelk, C.H., 2011. Improving reverse vaccinology with a machine learning approach. *Vaccine*, *29*(45), pp. 8156–8164

Braga-Neto, U.M. and Marques Jr, E.T.A., 2006. From functional genomics to functional immunomics: new challenges, old problems, big rewards. *PLoS Computational Biology*, *2*(7), p. e81.

Brasil, P., Pereira Jr, J.P., Moreira, M.E., Ribeiro Nogueira, R.M., Damasceno, L., Wakimoto, M., Rabello, R.S., Valderramos, S.G., Halai, U.A., Salles, T.S. and Zin, A.A., 2016. Zika virus infection in pregnant women in Rio de Janeiro. *New England Journal of Medicine*, *375*(24), pp. 2321–2334.

Burton, D.R., 2017. What are the most powerful immunogen design vaccine strategies? Reverse vaccinology 2.0 shows great promise. *Cold Spring Harbor Perspectives in Biology*, *9*(11), p. a030262.

Capecchi, B., Serruto, D., Adu-Bobie, J., Rappuoli, R. and Pizza, M., 2004. The genome revolution in vaccine research. *Current Issues in Molecular Biology*, *6*(1), pp. 17–28.

Carmona, S.J., Nielsen, M., Schafer-Nielsen, C., Mucci, J., Altcheh, J., Balouz, V., Tekiel, V., Frasch, A.C., Campetella, O., Buscaglia, C.A. and Agiero, F., 2015. Towards high-throughput immunomics for infectious diseases: Use of next-generation peptide microarrays for rapid discovery and mapping of antigenic determinants. *Molecular & Cellular Proteomics*, *14*(7), pp. 1871–1884.

Chakraborty, S., Barman, A. and Deb, B., 2020. Japanese encephalitis virus: A multi-epitope loaded peptide vaccine formulation using reverse vaccinology approach. *Infection, Genetics and Evolution*, *78*, p. 104106.

Chatterjee, R., Ghosh, M., Sahoo, S., Padhi, S., Misra, N., Raina, V., Suar, M. and Son, Y.O., 2021. Next-generation bioinformatics approaches and resources for coronavirus vaccine discovery and development—A perspective review. *Vaccines*, *9*(8), p. 812.

Chaudhuri, D., Datta, J., Majumder, S. and Giri, K., 2020. In silico designing of peptide based vaccine for hepatitis viruses using reverse vaccinology approach. *Infection, Genetics and Evolution*, *84*, p. 104388.

Chiang, M.H., Sung, W.C., Lien, S.P., Chen, Y.Z., Lo, A.F.Y., Huang, J.H., Kuo, S.C. and Chong, P., 2015. Identification of novel vaccine candidates against Acinetobacter baumannii using reverse vaccinology. *Human Vaccines & Immunotherapeutics*, *11*(4), pp. 1065–1073.

Chukwudozie, O.S., Duru, V.C., Ndiribe, C.C., Aborode, A.T., Oyebanji, V.O. and Emikpe, B.O., 2021. The relevance of bioinformatics applications in the discovery of vaccine candidates and potential drugs for COVID-19 treatment. *Bioinformatics and Biology Insights*, *15*, p. 11779322211002168.

Cohen, A.D. and Shoenfeld, Y., 1996. Vaccine-induced autoimmunity. *Journal of Autoimmunity*, *9*(6), pp. 699–703.

Cozzi, R., Scarselli, M. and Ilaria Ferlenghi, I.F., 2013. Structural vaccinology: A three-dimensional view for vaccine development. *Current Topics in Medicinal Chemistry*, *13* (20), pp. 2629–2637.

Dar, H., Zaheer, T., Rehman, M.T., Ali, A., Javed, A., Khan, G.A., Babar, M.M. and Waheed, Y., 2016. Prediction of promiscuous T-cell epitopes in the Zika virus polyprotein: An in silico approach. *Asian Pacific Journal of Tropical Medicine*, *9*(9), pp. 844–850.

Davies, M.N. and Flower, D.R., 2007. Harnessing bioinformatics to discover new vaccines. *Drug Discovery Today*, *12*(9–10), pp. 389–395.

De Groot, A.S., 2006. Immunomics: Discovering new targets for vaccines and therapeutics. *Drug Discovery Today*, *11*(5–6), pp. 203–209.

De Groot, A.S. and Berzofsky, J.A., 2004. From genome to vaccine--new immunoinformatics tools for vaccine design. *Methods (San Diego, Calif.)*, *34*(4), pp. 425–428.

De Groot, A.S., Bishop, E.A., Khan, B., Lally, M., Marcon, L., Franco, J., Mayer, K.H., Carpenter, C.C. and Martin, W., 2004. Engineering immunogenic consensus T helper epitopes for a cross-clade HIV vaccine. *Methods*, *34*(4), pp. 476–487.

De Groot, A.S., Marcon, L., Bishop, E.A., Rivera, D., Kutzler, M., Weiner, D.B. and Martin, W., 2005. HIV vaccine development by computer assisted design: The GAIA vaccine. *Vaccine*, *23*(17–18), pp. 2136–2148.

De Groot, A.S. and Martin, W., 2009. Reducing risk, improving outcomes: Bioengineering less immunogenic protein therapeutics. *Clinical Immunology*, *131*(2), pp. 189–201.

De Groot, A.S. and Moise, L., 2007. New tools, new approaches and new ideas for vaccine development. *Expert Review of Vaccines*, *6*(2), pp. 125–127.

De Groot, A.S., Moise, L., McMurry, J.A. and Martin, W., 2009. Epitope-based immunome-derived vaccines: A strategy for improved design and safety. *Clinical Applications of Immunomics*, pp. 39–69.

De Groot, A.S., Rivera, D.S., McMurry, J.A., Buus, S. and Martin, W., 2008. Identification of immunogenic HLA-B7 "Achilles' heel" epitopes within highly conserved regions of HIV. *Vaccine*, *26*(24), pp. 3059–3071.

De Groot, A.S., Sbai, H., Aubin, C.S., McMurry, J. and Martin, W., 2002. Immuno-informatics: Mining genomes for vaccine components. *Immunology and Cell Biology*, *80*(3), pp. 255–269.

de Oliveira, W.K., Cortez-Escalante, J., De Oliveira, W.T.G.H., Carmo, G.M.I.D., Henriques, C.M.P., Coelho, G.E. and de França, G.V.A., 2016. Increase in reported prevalence of microcephaly in infants born to women living in areas with confirmed Zika virus transmission during the first trimester of pregnancy—Brazil, 2015. *Morbidity and Mortality Weekly Report*, *65*(9), pp. 242–247.

De Sousa, K.P. and Doolan, D.L., 2016. Immunomics: A 21st century approach to vaccine development for complex pathogens. *Parasitology*, *143*(2), pp. 236–244.

Delisi, C. and Berzofsky, J.A., 1985. T-cell antigenic sites tend to be amphipathic structures. *Proceedings of the National Academy of Sciences*, *82*(20), pp. 7048–7052.

Dellagostin, O.A., Grassmann, A.A., Rizzi, C., Schuch, R.A., Jorge, S., Oliveira, T.L., McBride, A.J. and Hartwig, D.D., 2017. Reverse vaccinology: an approach for identifying leptospiral vaccine candidates. *International Journal of Molecular Sciences*, *18*(1), p. 158

Dick, G.W., Kitchen, S.F. and Haddow, A.J., 1952. Zika virus (I). Isolations and serological specificity. *Transactions of the Royal Society of Tropical Medicine and Hygiene*, *46*(5), pp. 509–520.

Dikhit, M.R., Ansari, M.Y., Mansuri, R., Sahoo, B.R., Dehury, B., Amit, A., Topno, R.K., Sahoo, G.C., Ali, V., Bimal, S. and Das, P., 2016. Computational prediction and analysis of potential antigenic CTL epitopes in Zika virus: A first step towards vaccine development. *Infection, Genetics and Evolution*, *45*, pp. 187–197.

Do Kwon, Y., Pancera, M., Acharya, P., Georgiev, I.S., Crooks, E.T., Gorman, J., Joyce, M.G., Guttman, M., Ma, X., Narpala, S. and Soto, C., 2015. Crystal structure, conformational fixation and entry-related interactions of mature ligand-free HIV-1 Env. *Nature Structural & Molecular Biology*, *22*(7), pp. 522–531.

Doolan, D.L., 2011. Plasmodium immunomics. *International Journal for Parasitology*, *41*(1), pp. 3–20.

Dormitzer, P.R., Grandi, G. and Rappuoli, R., 2012. Structural vaccinology starts to deliver. *Nature Reviews Microbiology*, *10*(12), pp. 807–813.

Dormitzer, P.R., Ulmer, J.B. and Rappuoli, R., 2008. Structure-based antigen design: a strategy for next generation vaccines. *Trends in Biotechnology*, *26*(12), pp. 659–667.

Dowdle, W.R., De Gourville, E., Kew, O.M., Pallansch, M.A. and Wood, D.J., 2003. Polio eradication: The OPV paradox. *Reviews in Medical Virology*, *13*(5), pp. 277–291.

Dubey, K.K., Luke, G.A., Knox, C., Kumar, P., Pletschke, B.I., Singh, P.K. and Shukla, P., 2018. Vaccine and antibody production in plants: Developments and computational tools. *Briefings in Functional Genomics*, *17*(5), pp. 295–307.

ElHefnawi, M., AlAidi, O., Mohamed, N., Kamar, M., El-Azab, I., Zada, S. and Siam, R., 2011. Identification of novel conserved functional motifs across most influenza a viral strains. *Virology Journal*, *8*(1), pp. 1–10.

Enayatkhani, M., Hasaniazad, M., Faezi, S., Gouklani, H., Davoodian, P., Ahmadi, N., Einakian, M.A., Karmostaji, A. and Ahmadi, K., 2021. Reverse vaccinology approach to design a novel multi-epitope vaccine candidate against COVID-19: An in silico study. *Journal of Biomolecular Structure and Dynamics*, *39*(8), pp. 2857–2872.

Enshell-Seijffers, D., Denisov, D., Groisman, B., Smelyanski, L., Meyuhas, R., Gross, G., Denisova, G. and Gershoni, J.M., 2003. The mapping and reconstitution of a conformational discontinuous B-cell epitope of HIV-1. *Journal of Molecular Biology*, *334*(1), pp. 87–101.

Falus, A. ed., 2008. *Clinical Applications of Immunomics* (Vol. 2). Springer Science & Business Media.

Folaranmi, T., Rubin, L., Martin, S.W., Patel, M. and MacNeil, J.R., 2015. Use of serogroup B meningococcal vaccines in persons aged≥ 10 years at increased risk for serogroup B meningococcal disease: Recommendations of the Advisory Committee on Immunization Practices, 2015. *MMWR. Morbidity and Mortality Weekly Report*, *64*(22), p. 608.

Gaschen, B., Taylor, J., Yusim, K., Foley, B., Gao, F., Lang, D., Novitsky, V., Haynes, B., Hahn, B.H., Bhattacharya, T. and Korber, B., 2002. Diversity considerations in HIV-1 vaccine selection. *Science*, *296*(5577), pp. 2354–2360.

Giacomet, V., Masetti, M., Nannini, P., Forlanini, F., Clerici, M., Zuccotti, G.V. and Trabattoni, D., 2018. Humoral and cell-mediated immune responses after a booster dose of HBV vaccine in HIV-infected children, adolescents and young adults. *PloS One*, *13*(2), p. e0192638.

Giuliani, M.M., Adu-Bobie, J., Comanducci, M., Aricò, B., Savino, S., Santini, L., Brunelli, B., Bambini, S., Biolchi, A., Capecchi, B. and Cartocci, E., 2006. A universal vaccine for serogroup B meningococcus. *Proceedings of the National Academy of Sciences*, *103*(29), pp. 10834–10839.

Goh, Y.S., McGuire, D. and Rénia, L., 2019. Vaccination with sporozoites: Models and correlates of protection. *Frontiers in Immunology*, *10*, p. 1227.

Goodswen, S.J., Kennedy, P.J. and Ellis, J.T., 2017. On the application of reverse vaccinology to parasitic diseases: A perspective on feature selection and ranking of vaccine candidates. *International Journal for Parasitology*, *47*(12), pp. 779–790.

Gore, D. and Pachkawade, M., 2012. In silico reverse vaccinology approach for vaccine lead search in Listeria monocytogenes. *Biocompx*, *1*, p. 15e22.

Gowthaman, U. and Agrewala, J.N., 2008. In silico tools for predicting peptides binding to HLA-class II molecules: More confusion than conclusion. *Journal of Proteome Research*, *7*(01), pp. 154–163.

Graham, B.S., Gilman, M.S. and McLellan, J.S., 2019. Structure-based vaccine antigen design. *Annual Review of Medicine*, *70*, pp. 91–104.

Graham, C.M., Barnett, B.C., Hartlmayr, I., Burt, D.S., Faulkes, R., Skehel, J.J. and Brian Thomas, D., 1989. The structural requirements for class II (I-Ad)-restricted T cell recognition of influenza hemagglutinin: B cell epitopes define T cell epitopes. *European Journal of Immunology*, *19*(3), pp. 523–528.

Grigoryan, L. and Pulendran, B., 2020, November. The immunology of SARS-CoV-2 infections and vaccines. In Pizza, M. and Rappuoli, R. *Seminars in Immunology*. (Vol. 50, p. 101422). Elsevier.

Gupta, A.K., Kaur, K., Rajput, A., Dhanda, S.K., Sehgal, M., Khan, M.S., Monga, I., Dar, S.A., Singh, S., Nagpal, G. and Usmani, S.S., 2016. ZikaVR: an integrated Zika virus resource for genomics, proteomics, phylogenetic and therapeutic analysis. *Scientific Reports*, *6*(1), pp. 1–16.

Gupta, E., Gupta, S.R. and Niraj, R.R.K., 2020. Identification of drug and vaccine target in Mycobacterium leprae: A reverse vaccinology approach. *International Journal of Peptide Research and Therapeutics*, *26*(3), pp. 1313–1326.

Hampson, A., Barr, I., Cox, N., Donis, R.O., Siddhivinayak, H., Jernigan, D., Katz, J., McCauley, J., Motta, F., Odagiri, T. and Tam, J.S., 2017. Improving the selection and development of influenza vaccine viruses–Report of a WHO informal consultation on improving influenza vaccine virus selection, Hong Kong SAR, China, 18–20 November 2015. *Vaccine*, *35*(8), pp. 1104–1109.

Harris, A.K., Meyerson, J.R., Matsuoka, Y., Kuybeda, O., Moran, A., Bliss, D., Das, S.R., Yewdell, J.W., Sapiro, G., Subbarao, K. and Subramaniam, S., 2013. Structure and accessibility of HA trimers on intact 2009 H1N1 pandemic influenza virus to stem region-specific neutralizing antibodies. *Proceedings of the National Academy of Sciences*, *110*(12), pp. 4592–4597.

Hasan, M., Ghosh, P.P., Azim, K.F., Mukta, S., Abir, R.A., Nahar, J. and Khan, M.M.H., 2019. Reverse vaccinology approach to design a novel multi-epitope subunit vaccine against avian influenza A (H7N9) virus. *Microbial Pathogenesis*, *130*, pp. 19–37.

He, Y., Rappuoli, R., De Groot, A.S. and Chen, R.T., 2010a. Emerging vaccine informatics. *Journal of Biomedicine and Biotechnology*, *2010*, p. 218590.

He, Y., Xiang, Z. and Mobley, H.L., 2010b. Vaxign: The first web-based vaccine design program for reverse vaccinology and applications for vaccine development. *Journal of Biomedicine and Biotechnology*, *2010*, p. 297505.

Hecker, M., Lorenz, P., Steinbeck, F., Hong, L., Riemekasten, G., Li, Y., Zettl, U.K. and Thiesen, H.J., 2012. Computational analysis of high-density peptide microarray data with application from systemic sclerosis to multiple sclerosis. *Autoimmunity Reviews*, *11*(3), pp. 180–190.

Heinson, A.I., Woelk, C.H. and Newell, M.L., 2015. The promise of reverse vaccinology. *International Health*, *7*(2), pp. 85–89.

Hensley, S.E., Das, S.R., Gibbs, J.S., Bailey, A.L., Schmidt, L.M., Bennink, J.R. and Yewdell, J.W., 2011. Influenza a virus hemagglutinin antibody escape promotes neuraminidase antigenic variation and drug resistance. *PLoS One*, *6*(2), p. e15190.

Heymann, D.L., Hodgson, A., Freedman, D.O., Staples, J.E., Althabe, F., Baruah, K., Mahmud, G., Kandun, N., Vasconcelos, P.F., Bino, S. and Menon, K.U., 2016. Zika virus and microcephaly: Why is this situation a PHEIC? *The Lancet*, *387* (10020), pp. 719–721.

Hoen, B., Schaub, B., Funk, A.L., Ardillon, V., Boullard, M., Cabié, A., Callier, C., Carles, G., Cassadou, S., Césaire, R. and Douine, M., 2018. Pregnancy outcomes after ZIKV infection in French territories in the Americas. *New England Journal of Medicine*, *378*(11), pp. 985–994.

Huang, Y., Karuna, S., Carpp, L. N., Reeves, D., Pegu, A., Seaton, K., et al. (2018). Modeling cumulative overall prevention efficacy for the VRC01 phase 2b efficacy trials. *Human Vaccines & Immunotherapeutics* *14*, 2116–2127. doi: 10.1080/21645515.2018.1462640.

Hui, D.S., Azhar, E.I., Madani, T.A., Ntoumi, F., Kock, R., Dar, O., Ippolito, G., McHugh, T.D., Memish, Z.A., Drosten, C. and Zumla, A., 2020. The continuing 2019-nCoV epidemic threat of novel coronaviruses to global health—The latest 2019 novel coronavirus outbreak in Wuhan, China. *International Journal of Infectious Diseases*, 91, pp. 264–266.

Impagliazzo, A., Milder, F., Kuipers, H., Wagner, M.V., Zhu, X., Hoffman, R.M., van Meersbergen, R., Huizingh, J., Wanningen, P., Verspuij, J. and de Man, M., 2015. A stable trimeric influenza hemagglutinin stem as a broadly protective immunogen. *Science*, *349*(6254), pp. 1301–1306.

Jaiswal, V., Chanumolu, S.K., Gupta, A., Chauhan, R.S. and Rout, C., 2013. Jenner-predict server: Prediction of protein vaccine candidates (PVCs) in bacteria based on host-pathogen interactions. *BMC Bioinformatics*, *14*(1), pp. 1–11.

Jeyanathan, M., Afkhami, S., Smaill, F., Miller, M.S., Lichty, B.D. and Xing, Z., 2020. Immunological considerations for COVID-19 vaccine strategies. *Nature Reviews Immunology*, *20*(10), pp. 615–632.

John, L., John, G.J. and Kholia, T., 2012. A reverse vaccinology approach for the identification of potential vaccine candidates from Leishmania spp. *Applied Biochemistry and Biotechnology*, *167*(5), pp. 1340–1350.

Kanampalliwar, A.M., Rajkumar, S., Girdhar, A. and Archana, T., 2013. Reverse vaccinology: Basics and applications. *Journal of Vaccines & Vaccination* 4: 194.

Kaur, S.P. and Gupta, V., 2020. COVID-19 Vaccine: A comprehensive status report. *Virus Research*, 288, p. 198114.

Kharkar, P.B., Talkar, S.S., Kadwadkar, N.A. and Patravale, V.B., 2017. Nanosystems for oral delivery of immunomodulators. In *Nanostructures for Oral Medicine*. Andronescu, E. and Grumezescu, A.M. (pp. 295–334). Elsevier.

Kim, H.I., Ha, N.Y., Kim, G., Min, C.K., Kim, Y., Yen, N.T.H., Choi, M.S. and Cho, N.H., 2019. Immunization with a recombinant antigen composed of conserved blocks from TSA56 provides broad genotype protection against scrub typhus. *Emerging Microbes & Infections*, *8*(1), pp. 946–958.

Klysik, J., 2001. Concept of immunomics: A new frontier in the battle for gene function? *Acta Biotheoretica*, *49*(3), pp. 191–202.

Krammer, F., 2020. SARS-CoV-2 vaccines in development. *Nature*, *586* (7830), pp. 516–527.

Kulp, D.W. and Schief, W.R., 2013. Advances in structure-based vaccine design. *Current Opinion in Virology*, *3*(3), pp. 322–331.

Kumar, B., Asha, K., Khanna, M., Ronsard, L., Meseko, C.A. and Sanicas, M., 2018. The emerging influenza virus threat: Status and new prospects for its therapy and control. *Archives of Virology*, *163*(4), pp. 831–844.

Lambach, P., Alvarez, A.M.R., Hirve, S., Ortiz, J.R., Hombach, J., Verweij, M., Hendriks, J., Palkonyay, L. and Pfleiderer, M., 2015. Considerations of strategies to provide influenza vaccine year round. *Vaccine*, *33*(47), pp. 6493–6498.

Liechti, R., Gleizes, A., Kuznetsov, D., Bougueleret, L., Le Mercier, P., Bairoch, A. and Xenarios, I., 2010. OpenFluDB, a database for human and animal influenza virus. *Database*, *2010*, p. baq004.

Liljeroos, L., Malito, E., Ferlenghi, I. and Bottomley, M.J., 2015. Structural and computational biology in the design of immunogenic vaccine antigens. *Journal of Immunology Research*, *2015*, p. 156241.

Lindestam Arlehamn, C.S., Gerasimova, A., Mele, F., Henderson, R., Swann, J., Greenbaum, J.A., Kim, Y., Sidney, J., James, E.A., Taplitz, R. and McKinney, D.M., 2013. Memory T cells in latent Mycobacterium tuberculosis infection are directed against three antigenic islands and largely contained in a CXCR3+ CCR6+ Th1 subset. *PLoS Pathogens*, *9*(1), p. e1003130.

Liu, J.J. and Cepica, A., 1990. Current approaches to vaccine preparation. *The Canadian Veterinary Journal*, *31*(3), p. 181.

Maria, R.R., Arturo, C.J., Alicia, J.A., Paulina, M.G. and Gerardo, A.O., 2017. The impact of bioinformatics on vaccine design and development. *Vaccines*, *2*, pp. 3–6.

Masignani, V., Pizza, M. and Moxon, E.R., 2019. The development of a vaccine against meningococcus B using reverse vaccinology. *Frontiers in Immunology*, *10*, p. 751.

McLellan, J.S., Chen, M., Leung, S., Graepel, K.W., Du, X., Yang, Y., Zhou, T., Baxa, U., Yasuda, E., Beaumont, T. and Kumar, A., 2013. Structure of RSV fusion glycoprotein trimer bound to a prefusion-specific neutralizing antibody. *Science*, *340*(6136), pp. 1113 1117.

McMurry, J., Sbai, H., Gennaro, M.L., Carter, E.J., Martin, W. and De Groot, A.S., 2005. Analyzing Mycobacterium tuberculosis proteomes for candidate vaccine epitopes. *Tuberculosis*, *85*(1–2), pp. 95–105.

Méndez, N., Oviedo-Pastrana, M., Mattar, S., Caicedo-Castro, I. and Arrieta, G., 2017. Zika virus disease, microcephaly and Guillain-Barré syndrome in Colombia: Epidemiological situation during 21 months of the Zika virus outbreak, 2015–2017. *Archives of Public Health*, *75*(1), pp. 1–11.

Meunier, M., Guyard-Nicodème, M., Hirchaud, E., Parra, A., Chemaly, M. and Dory, D., 2016. Identification of novel vaccine candidates against Campylobacter through reverse vaccinology. *Journal of Immunology Research*, *2016*, p. 5715790.

Mirza, M.U., Rafique, S., Ali, A., Munir, M., Ikram, N., Manan, A., Salo-Ahen, O.M. and Idrees, M., 2016. Towards peptide vaccines against Zika virus: Immunoinformatics combined with molecular dynamics simulations to predict antigenic epitopes of Zika viral proteins. *Scientific Reports*, *6*(1), pp. 1–17.

Mlakar, J., Korva, M., Tul, N., Popović, M., Poljšak-Prijatelj, M., Mraz, J., Kolenc, M., Resman Rus, K., Vesnaver Vipotnik, T., Fabjan Vodušek, V. and Vizjak, A., 2016. Zika virus associated with microcephaly. *New England Journal of Medicine*, *374*(10), pp. 951–958.

Mollenkopf, H.J., Grode, L., Mattow, J., Stein, M., Mann, P., Knapp, B., Ulmer, J. and Kaufmann, S.H., 2004. Application of mycobacterial proteomics to vaccine design: improved protection by Mycobacterium bovis BCG prime-Rv3407 DNA boost vaccination against tuberculosis. *Infection and Immunity*, *72*(11), pp. 6471–6479.

Monterrubio-López, G.P. and Ribas-Aparicio, R.M., 2015. Identification of novel potential vaccine candidates against tuberculosis based on reverse vaccinology. *BioMed Research International*, *2015*, p. 483150.

Mora, M., Veggi, D., Santini, L., Pizza, M. and Rappuoli, R., 2003. Reverse vaccinology. *Drug Discovery Today*, *8*(10), pp. 459–464

Munangandu, H.M., Mutoloki, S. and Evensen, Ø., 2015. A review of the immunological mechanisms following mucosal vaccination of finfish. *Frontiers in Immunology*, *6*, p. 427.

Neu, K.E., Dunand, C.J.H. and Wilson, P.C., 2016. Heads, stalks and everything else: How can antibodies eradicate influenza as a human disease?. *Current Opinion in Immunology*, *42*, pp. 48–55.

Nezafat, N., Eslami, M., Negahdaripour, M., Rahbar, M.R. and Ghasemi, Y., 2017. Designing an efficient multi-epitope oral vaccine against Helicobacter pylori using immunoinformatics and structural vaccinology approaches. *Molecular BioSystems*, *13*(4), pp. 699–713.

Ni, Z., Chen, Y., Ong, E. and He, Y., 2017. Antibiotic resistance determinant-focused Acinetobacter baumannii vaccine designed using reverse vaccinology. *International Journal of Molecular Sciences*, *18*(2), p. 458.

Nuccitelli, A., Cozzi, R., Gourlay, L.J., Donnarumma, D., Necchi, F., Norais, N., Telford, J.L., Rappuoli, R., Bolognesi, M., Maione, D. and Grandi, G., 2011. Structure-based approach to rationally design a chimeric protein for an effective vaccine against Group B Streptococcus infections. *Proceedings of the National Academy of Sciences*, *108*(25), pp. 10278–10283.

Okwo-Bele, J.M. and Cherian, T., 2011. The expanded programme on immunization: A lasting legacy of smallpox eradication. *Vaccine*, *29*, pp. D74–D79

Oli, A.N., Agu, R.U., Ihekwereme, C.P. and Esimone, C.O., 2017. An evaluation of the cold chain technology in South-East Nigeria using immunogenicity study on the measles vaccines. *The Pan African Medical Journal*, *27*, (Suppl 3), p. 28.

Oli, A.N., Obialor, W.O., Ifeanyichukwu, M.O., Odimegwu, D.C., Okoyeh, J.N., Emechebe, G.O., Adejumo, S.A. and Ibeanu, G.C., 2020. Immunoinformatics and vaccine development: An overview. *ImmunoTargets and Therapy*, *9*, p. 13.

Ong, E., Wang, H., Wong, M.U., Seetharaman, M., Valdez, N. and He, Y., 2020. Vaxign-ML: supervised machine learning reverse vaccinology model for improved prediction of bacterial protective antigens. *Bioinformatics*, *36*(10), pp. 3185–3191.

Oprea, M. and Antohe, F., 2013. Reverse-vaccinology strategy for designing T-cell epitope candidates for Staphylococcus aureus endocarditis vaccine. *Biologicals*, *41*(3), pp. 148–153.

Orozco, A., Morera, J., Jiménez, S. and Boza, R., 2013. A review of bioinformatics training applied to research in molecular medicine, agriculture and biodiversity in Costa Rica and Central America. *Briefings in Bioinformatics*, *14*(5), pp. 661–670.

Oscherwitz, J., 2016. The promise and challenge of epitope-focused vaccines. *Human Vaccines & Immunotherapeutics*, *12*(8), pp. 2113–2116.

Ozawa, S., Mirelman, A., Stack, M.L., Walker, D.G. and Levine, O.S., 2012. Cost-effectiveness and economic benefits of vaccines in low-and middle-income countries: A systematic review. *Vaccine*, *31*(1), pp. 96–108.

Pajon, R., Yero, D., Lage, A., Llanes, A. and Borroto, C.J., 2006. Computational identification of beta-barrel outer-membrane proteins in Mycobacterium tuberculosis predicted proteomes as putative vaccine candidates. *Tuberculosis*, *86* (3–4), pp. 290–302.

Panchaud, A., Stojanov, M., Ammerdorffer, A., Vouga, M. and Baud, D., 2016. Emerging role of Zika virus in adverse fetal and neonatal outcomes. *Clinical Microbiology Reviews*, *29*(3), pp. 659–694.

Pandey, K., Sharma, M., Saarav, I., Singh, S., Dutta, P., Bhardwaj, A. and Sharma, S., 2016. Analysis of the DosR regulon genes to select cytotoxic T lymphocyte epitope specific vaccine candidates using a reverse vaccinology approach. *International Journal of Mycobacteriology*, *5*(1), pp. 34–43.

Parvizpour, S., Pourseif, M.M., Razmara, J., Rafi, M.A. and Omidi, Y., 2020. Epitope-based vaccine design: A comprehensive overview of bioinformatics approaches. *Drug Discovery Today*, *25*(6), pp. 1034–1042.

Patil, S.U. and Shreffler, W.G., 2019. Novel vaccines: Technology and development. *Journal of Allergy and Clinical Immunology*, *143*(3), pp. 844–851.

Pattnaik, A., Sahoo, B.R. and Pattnaik, A.K., 2020. Current status of Zika virus vaccines: Successes and challenges. *Vaccines*, *8*(2), p. 266.

Pizza, M., Scarlato, V., Masignani, V., Giuliani, M.M., Arico, B., Comanducci, M., Jennings, G.T., Baldi, L., Bartolini, E., Capecchi, B. and Galeotti, C.L., 2000. Identification of vaccine candidates against serogroup B meningococcus by whole-genome sequencing. *Science*, *287*(5459), pp. 1816–1820.

Poland, G.A., Ovsyannikova, I.G. and Kennedy, R.B., 2018. Personalized vaccinology: A review. *Vaccine*, *36*(36), pp. 5350–5357.

Quintana, F.J., Hagedorn, P.H., Elizur, G., Merbl, Y., Domany, E. and Cohen, I.R., 2004. Functional immunomics: Microarray analysis of IgG autoantibody repertoires predicts the future response of mice to induced diabetes. *Proceedings of the National Academy of Sciences*, *101*(suppl 2), pp. 14615–14621.

Raman, K., Bhat, A.G. and Chandra, N., 2010. A systems perspective of host–pathogen interactions: Predicting disease outcome in tuberculosis. *Molecular Biosystems*, *6*(3), pp. 516–530.

Rappuoli, R., 2000. Reverse vaccinology. *Current Opinion in Microbiology*, *3*(5), pp. 445–450.

Rappuoli, R., 2001. Reverse vaccinology, a genome-based approach to vaccine development. *Vaccine*, *19*(17–19), pp. 2688–2691.

Rappuoli, R., 2007. Bridging the knowledge gaps in vaccine design. *Nature Biotechnology*, *25*(12), pp. 1361–1366.

Rappuoli, R., Bottomley, M.J., D'Oro, U., Finco, O. and De Gregorio, E., 2016. Reverse vaccinology 2.0: Human immunology instructs vaccine antigen design. *Journal of Experimental Medicine*, *213*(4), pp. 469–481.

Rauch, S., Jasny, E., Schmidt, K.E. and Petsch, B., 2018. New vaccine technologies to combat outbreak situations. *Frontiers in Immunology*, *9*, p. 1963.

Rawat, K., Kumari, P. and Saha, L., 2021. COVID-19 vaccine: A recent update in pipeline vaccines, their design and development strategies. *European Journal of Pharmacology*, *892*, p. 173751.

Rezza, G. and Weaver, S.C., 2019. Chikungunya as a paradigm for emerging viral diseases: Evaluating disease impact and hurdles to vaccine development. *PLoS Neglected Tropical Diseases*, *13*(1), p. e0006919.

Richner, J.M., Jagger, B.W., Shan, C., Fontes, C.R., Dowd, K.A., Cao, B., Himansu, S., Caine, E.A., Nunes, B.T., Medeiros, D.B. and Muruato, A.E., 2017. Vaccine mediated protection against Zika virus-induced congenital disease. *Cell*, *170*(2), pp. 273–283.

Sarkander, J., Hojyo, S. and Tokoyoda, K., 2016. Vaccination to gain humoral immune memory. *Clinical & Translational Immunology*, *5*(12), p. e120.

Sastry, M., Zhang, B., Chen, M., Joyce, M. G., Kong, W. P., Chuang, G. Y., et al. (2017). Adjuvants and the vaccine response to the DS-Cav1-stabilized fusion glycoprotein of respiratory syncytial virus. *PLoS One*, 12, p. e0186854. Doi: 10.1371/journal.pone.0186854

Sautto, G.A., Kirchenbaum, G.A. and Ross, T.M., 2018. Towards a universal influenza vaccine: Different approaches for one goal. *Virology Journal*, *15*(1), pp. 1–12.

Scarselli, M., Aricò, B., Brunelli, B., Savino, S., Di Marcello, F., Palumbo, E., Veggi, D., Ciucchi, L., Cartocci, E., Bottomley, M.J. and Malito, E., 2011. Rational design of a meningococcal antigen inducing broad protective immunity. *Science Translational Medicine*, *3*(91), pp. 91ra62–91ra62.

Sefidi-Heris, Y., Jahangiri, A., Mokhtarzadeh, A., Shahbazi, M.A., Khalili, S., Baradaran, B., Mosafer, J., Baghbanzadeh, A., Hejazi, M., Hashemzaei, M. and Hamblin, M.R., 2020. Recent progress in the design of DNA vaccines against tuberculosis. *Drug Discovery Today*, *25*(11), pp. 1971–1987.

Sette, A., Fleri, W., Peters, B., Sathiamurthy, M., Bui, H.H. and Wilson, S., 2005. A roadmap for the immunomics of category A–C pathogens. *Immunity*, *22*(2), pp. 155–161.

Sette, A. and Rappuoli, R., 2010. Reverse vaccinology: Developing vaccines in the era of genomics. *Immunity*, *33*(4), pp. 530–541.

Solanki, V. and Tiwari, V., 2018. Subtractive proteomics to identify novel drug targets and reverse vaccinology for the development of chimeric vaccine against Acinetobacter baumannii. *Scientific Reports*, *8*(1), pp. 1–19

Stilling, R.M., Bordenstein, S.R., Dinan, T.G. and Cryan, J.F., 2014. Friends with social benefits: Host-microbe interactions as a driver of brain evolution and development? *Frontiers in Cellular and Infection Microbiology*, *4*, p. 147.

Styczynski, A.R., Malta, J.M., Krow-Lucal, E.R., Percio, J., Nóbrega, M.E., Vargas, A., Lanzieri, T.M., Leite, P.L., Staples, J.E., Fischer, M.X. and Powers, A.M., 2017. Increased rates of Guillain-Barré syndrome associated with Zika virus outbreak in the Salvador metropolitan area, Brazil. *PLoS Neglected Tropical Diseases*, *11*(8), p. e0005869.

Sundaramurthi, J.C., Brindha, S., Shobitha, S.R., Swathi, A., Ramanandan, P. and Hanna, L.E., 2012. In silico identification of potential antigenic proteins and promiscuous CTL epitopes in Mycobacterium tuberculosis. *Infection, Genetics and Evolution*, *12*(6), pp. 1312–1318.

Sunita Sajid, A., Singh, Y. and Shukla, P., 2020. Computational tools for modern vaccine development. *Human Vaccines & Immunotherapeutics*, *16*(3), pp. 723–735.

Sweredoski, M.J. and Baldi, P., 2008. PEPITO: Improved discontinuous B-cell epitope prediction using multiple distance thresholds and half sphere exposure. *Bioinformatics*, *24*(12), pp. 1459–1460.

Tahir UL Qamar, M., Shahid, F., Aslam, S., Ashfaq, U.A., Aslam, S., Fatima, I., Fareed, M.M., Zohaib, A. and Chen, L.L., 2020. Reverse vaccinology assisted designing of multiepitope-based subunit vaccine against SARS-CoV-2. *Infectious Diseases of Poverty*, 9(1), pp. 1–14.

Talukdar, S., Zutshi, S., Prashanth, K.S., Saikia, K.K. and Kumar, P., 2014. Identification of potential vaccine candidates against Streptococcus pneumoniae by reverse vaccinology approach. *Applied Biochemistry and Biotechnology*, *172*(6), pp. 3026–3041.

Tang, S.T., van Meijgaarden, K.E., Caccamo, N., Guggino, G., Klein, M.R., van Weeren, P., Kazi, F., Stryhn, A., Zaigler, A., Sahin, U. and Buus, S., 2011. Genome-based in silico identification of new Mycobacterium tuberculosis antigens activating polyfunctional CD8+ T cells in human tuberculosis. *The Journal of Immunology*, *186*(2), pp. 1068–1080.

Terry, F.E., Moise, L., Martin, R., Torres, M., Pilotte, N., Williams, S. and De Groot, A.S., 2015. Time for T? Immunoinformatics addresses the challenges of vaccine design for neglected tropical and emerging infectious diseases. *Expert Review of Vaccines*, *14* (1), 21–35.

Tomar, N. and De, R.K., 2010. Immunoinformatics: An integrated scenario. *Immunology*, *131*(2), pp. 153–168.

Van Regenmortel, M.H., 2016. Structure-based reverse vaccinology failed in the case of HIV because it disregarded accepted immunological theory. *International Journal of Molecular Sciences*, *17*(9), p. 1591.

Vani, J., Shaila, M.S., Chandra, N.R. and Nayak, R., 2006. A combined immuno-informatics and structure-based modeling approach for prediction of T cell epitopes of secretory proteins of Mycobacterium tuberculosis. *Microbes and Infection*, *8*(3), pp. 738–746.

Vilela Rodrigues, T.C., Jaiswal, A.K., de Sarom, A., de Castro Oliveira, L., Freire Oliveira, C.J., Ghosh, P., Tiwari, S., Miranda, F.M., de Jesus Benevides, L., de Carvalho, Ariston Azevedo, V. and de Castro Soares, S., 2019. Reverse vaccinology and subtractive genomics reveal new therapeutic targets against Mycoplasma pneumoniae: a causative agent of pneumonia. *Royal Society Open Science*, *6*(7), p. 190907.

Vishnu, U.S., Sankarasubramanian, J., Gunasekaran, P. and Rajendhran, J., 2015. Novel vaccine candidates against Brucella melitensis identified through reverse vaccinology approach. *Omics: A Journal of Integrative Biology*, *19*(11), pp. 722–729.

Vivona, S., Bernante, F. and Filippini, F., 2006. NERVE: New enhanced reverse vaccinology environment. *BMC Biotechnology*, *6*(1), pp. 1–8.

Wallis, J., Shenton, D.P. and Carlisle, R.C., 2019. Novel approaches for the design, delivery and administration of vaccine technologies. *Clinical & Experimental Immunology*, *196*(2), pp. 189–204.

Wang, P., Sidney, J., Dow, C., Mothé, B., Sette, A. and Peters, B., 2008. A systematic assessment of MHC class II peptide binding predictions and evaluation of a consensus approach. *PLoS Computational Biology*, *4*(4), p. e1000048.

Wolf, M.C., Freiberg, A.N., Zhang, T., Akyol-Ataman, Z., Grock, A., Hong, P.W., Li, J., Watson, N.F., Fang, A.Q., Aguilar, H.C. and Porotto, M., 2010. A broad-spectrum antiviral targeting entry of enveloped viruses. *Proceedings of the National Academy of Sciences*, *107*(7), pp. 3157–3162.

Woodworth, J.S., Aagaard, C.S., Hansen, P.R., Cassidy, J.P., Agger, E.M. and Andersen, P., 2014. Protective CD4 T cells targeting cryptic epitopes of Mycobacterium tuberculosis resist infection-driven terminal differentiation. *The Journal of Immunology*, *192*(7), pp. 3247–3258.

World Health Organization 2016. Emergency Committee on Zika virus and observed increase in neurological disorders and neonatal malformations 2016-3-10. http://www.who.int/mediacentre/news/statements/2016/1st-emergency-committee-zika/en

World Health Organization. 2020. WHO Director-General's opening remarks at the media briefing on COVID-19-11 March 2020. https://www.who.int/director-general/speeches/detail/who-director-general-s-opening-remarks-at-the-media-briefing-on-covid-19---11-march-2020

Xiang, Z. and He, Y., 2009. Vaxign: A web-based vaccine target design program for reverse vaccinology. *Procedia in Vaccinology*, 1(1), pp. 23–29.

Yechezkel, I., Law, M. and Tzarum, N., 2021. From structural studies to HCV vaccine design. *Viruses*, *13*(5), p. 833.

Zhang, Y., Aevermann, B.D., Anderson, T.K., Burke, D.F., Dauphin, G., Gu, Z., He, S., Kumar, S., Larsen, C.N., Lee, A.J. and Li, X., 2017. Influenza research database: An integrated bioinformatics resource for influenza virus research. *Nucleic Acids Research*, *45*(D1), pp. D466–D474.

27 Petri Net Modeling as an Aid in Bioprocess Designing

Gajendra Pratap Singh and Riddhi Jangid
School of Computational and Integrative Sciences, Jawaharlal Nehru University, New Delhi, India

Mamtesh Singh
Gargi College, University of Delhi, New Delhi, India

CONTENTS

27.1 INTRODUCTION

27.1.1 Preliminaries of Petri Net (PN) Modeling

A Petri net is a particularly directed bipartite multigraph (Murata, 1989; Peterson, 1981) that contain two disjoint set of nodes, called place (◯) and transition (▯), and these nodes are connected by directed arrows (➔) called arcs. These arcs must be used to connect places and transitions but in such a way that no two places or transitions are joined at one time, i.e., no place or transition can be connected directly. Two similar nodes can be connected via the other type of nodes. An arc which is directed from place to transition is the input place of that transition and an arc from the transition to a place is the output place of the transition. In modeling, the place node may be considered as any kind of condition of a stable thing, and the transition node may be considered as any kind of activity or event, or any action (Kansal et al., 2010, 2011; Singh, 2016; Singh and Singh, 2019).

We assign tokens (●) to the places, which we call marking, to represent any kind of resource or the availability or absence/presence of any kind of quantity. These tokens are further used for execution of Petri nets. Theoretically, we define marking [23] as $m-vector:(M_1,M_2,M_3,\ldots,M_{\hat{m}})$, where M is the number of tokens in *ith* place, $i = 1, 2, 3, \ldots, N$; m is the number of places in the Petri net structure and each $M_i \in \{0,1,2,3,\ldots,k\}$, $k \in N$, N is the natural number including zero (Peterson, 1981; Singh, 2013, 2014). Sometimes, we assign a non-negative integer to the arcs in a Petri net, which we call the weight of the arc. And if weight is not assigned, we consider it 1 by default (Petri, 1966; Singh et al., 2013a; Gupta et al., 2019b).

DOI: 10.1201/9781003306931-31

27.1.2 Definition

Formally, we define a Petri net (Murata, 1989) as a 5-tuple $C = (P, T, F, W, M_0)$ where:

$P = \{p_1, p_2, p_3, \ldots, p_m\}$: A finite set of places.
$T = \{t_1, t_2, t_3, \ldots, t_n\}$: A finite set of transitions.

$F \subseteq (P \times T) \cup (T \times P)$: A set of arcs, where F represents the flow relation representing connections from place node to transition and transition node to place node.

$W = F \rightarrow \{1, 2, 3, \ldots\}$: Weight function.
$M_0 = P \rightarrow \{0, 1, 2, 3, \ldots\}$: The initial marking.

Where $P \cap T = \emptyset$and $P \cup T \neq \emptyset$. This condition represents no place or transition can be isolate.

27.2 WHY ARE PETRI NETS NECESSARY FOR THE FERMENTATION PROCESS?

Fermentation is a process used in the manufacturing of beer, wine and similar products that occurs through physio-chemical changes in molecules including sugar or those containing starch. The process can be effectively modeled mathematically with a flowchart which helps us to understand its structure and dynamic behavior. Petri net is one of the most efficient tools for modeling dynamic processes. It is a diagrammatical representation of discrete event systems (Reddy et al., 1993; Kansal and Singh, 2008) in which the flow of tokens easily explains the movement of changes of state. It also provides information about structure and behavior of the process (Singh et al., 2020a, 2020d).

Petri net design has been used for various applications. The model described here uses the generally accepted flowchart of the fermentation process as applied in the preparation of different products.

27.3 MODELING AND NOTATION OF THE ELEMENTS OF A PETRI NET FOR FERMENTATION PROCESS

Our model of the fermentation process uses the set of places and transitions (see Table 27.1) as event and actions (Peterson, 1981).

27.4 PETRI NET MODELING OF THE FERMENTATION PROCESS

Figure 27.1 illustrates the Petri net modeling of the fermentation process. It consists of 14 places and eight transitions including one logical transition (Y/N) XOR-split as listed in Table 27.1.

27.5 MODELING DESCRIPTION AND COMPUTATIONAL STEPS

The model includes 14 places and eight transitions. The movement of tokens between the places explains the changes of state in the process. For example, the initial marking $(2, 0, 1, 0, 3, 0, 1, 0, 1, 0, 0, 1, 0, 0)$ shows that initially we have Feed 1 and Feed 2 for the input. When transitions T_1 and T_3 are enabled, they will fire two and three times respectively, and the next reachable state (Peterson, 1981) will be $(0, 2, 1, 0, 0, 3, 1, 0, 1, 0, 0, 1, 0, 0)$.

The computation part of the change of state can easily be understood as:

Initial marking (M_0): $(2, 0, 1, 0, 3, 0, 1, 0, 1, 0, 0, 1, 0, 0)$
Enabled transitions (ready to fire): T_1, T_3.

TABLE 27.1
Elements of Set of Places and Transitions

Place	Represented by	Represents	Transition	Represented by	Represents
P_1	F_1	Feed 1	**T1**	Pr_1	Preparation of 1
P_2	RW_1	Required weight for 1	**T2**	Pt_1	Pre-treatment of 1
P_3	C_1	Conditions for 1	**T3**	Pr_2	Preparation of 2
P_4	Sl_P1	Slurry Preparations 1	**T4**	Pt_2	Pre-treatment of 2
P_5	F_2	Feed 2	**T5**	BSF	Batch scale fermentation
P_6	RW_2	Required weight for 2	**T6**	A_T	Analysis test
P_7	C_2	Conditions for 2	**T7 (XOR-split)**	Y/N	Product is as desired or not.
P_8	Sl_P2	Slurry preparations 2	**T8**	Re_P	Required product
P_9	In	Inoculum			
P_{10}	Sam	Sampling			
P_{11}	Pr	Product			
P_{12}	Un_Pr	Undesired product			
P_{13}	R_Pr	Desired product			
P_{14}	Pr_Ex	Product extracted			

27.5.1 Firing Rule

Consider the firing of transition T_1 once. The reachable state will be computed using the mathematical formula (Peterson, 1981):

$$M_k(P_i) = M_{k-1}(P_i) - \#(P_i, I(T_j)) + \#(P_i, O(T_j))$$

(New marking at place P_i = Previous marking at place P_i—weight of input arc from P_i to T_j + weight of output arc from P_i to T_j)

For place P_1:

$$M_1(P_1) = M_0(P_1) - \#(P_1, I(T_1)) + \#(P_1, O(T_1))$$

$$M_1(P_1) = 2 - 1 + 0 = 1$$

Similarly, for other places:

$$M_1(P_2) = 0 - 0 + 1 = 1$$

$$M_1(P_3) = 1 - 0 + 0 = 1$$

$$M_1(P_4) = 0 - 0 + 0 = 0$$

$$M_1(P_5) = 3 - 0 + 0 = 3$$

$$M_1(P_6) = 0 - 0 + 0 = 0$$

$$M_1(P_7) = 1 - 0 + 0 = 1$$

$$M_1(P_8) = 0 - 0 + 0 = 0$$

$$M_1(P_9) = 1 - 0 + 0 = 1$$

$$M_1(P_{10}) = 0 - 0 + 0 = 0$$

$$M_1(P_{11}) = 0 - 0 + 0 = 0$$

$$M_1(P_{12}) = 1 - 0 + 0 = 1$$

$$M_1(P_{13}) = 0 - 0 + 0 = 0$$

$$M_1(P_{14}) = 0 - 0 + 0 = 0$$

Hence, after one firing of T_1, we reach $M_1 = (1,1,1,0,3,0,1,0,1,0,0,1,0,0)$. This is one step after the execution of T_1. Again, firing of T_1 will give us $(0,2,1,0,3,0,1,0,1,0,0,1,0,0)$. Further, three firings of T_3 will give us the state $(0,2,1,0,0,3,1,0,1,0,0,1,0,0)$.

27.5.2 Reachability Graph

In a similar way, the transitions will fire and we obtain the reachability graph in Figure 27.2 (Singh et al., 2013a, 2013b, 2013c; Singh and Kansal, 2016). Reachability is one of the most basic properties used in system modeling and designing. We usually are concerned with whether or not the designed system is able to reach a particular state from its initial state and if so, what firing sequence has helped us to reach that state.

27.6 WORKFLOW PETRI NET DESIGNER TOOL (WOPED 3.8.0) FOR FERMENTATION PROCESS

The model has been designed on the Workflow Petri net designer tool (WoPeD 3.8.0) and the result has been verified with the tool. In addition to the reachability graph (Figure 27.1), a qualitative analysis of the workflow net has been obtained (Figure 27.3).

The analysis shows the number of nodes (23) and edges (26) we have used in the net, arcs with more than one arc weight, whether or not our model is well-structured, the number of source and sink places in it, the number of connected components, etc. It also explains properties of the net such as boundedness and liveness. A Petri net is bounded if the number of tokens in all places of the Petri net never exceeds a finite value r, ($r \in N$). A deadlock situation occurs when no transition in the Petri net is enabled for firing. In the above example, after firing all the enabled transitions, we reach a deadlock marking where no transition is live any more. A Petri net is described as live if it is not deadlocked.

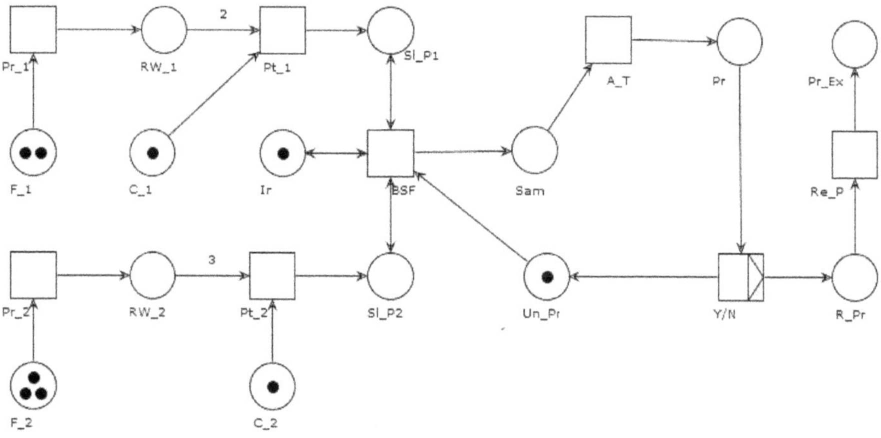

FIGURE 27.1 Petri net model of fermentation process.

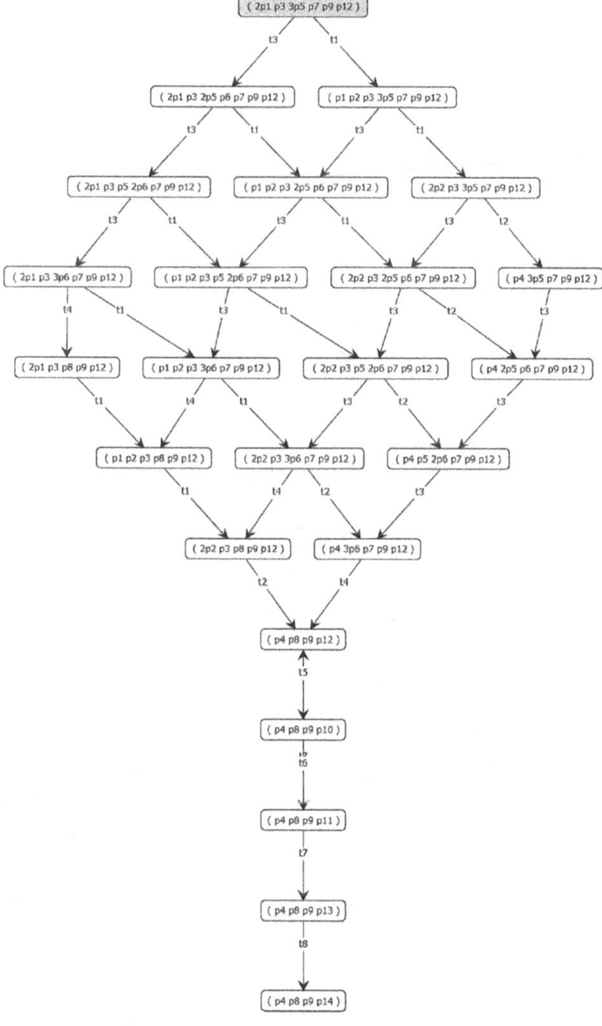

FIGURE 27.2 Reachability Graph of Figure 27.1.

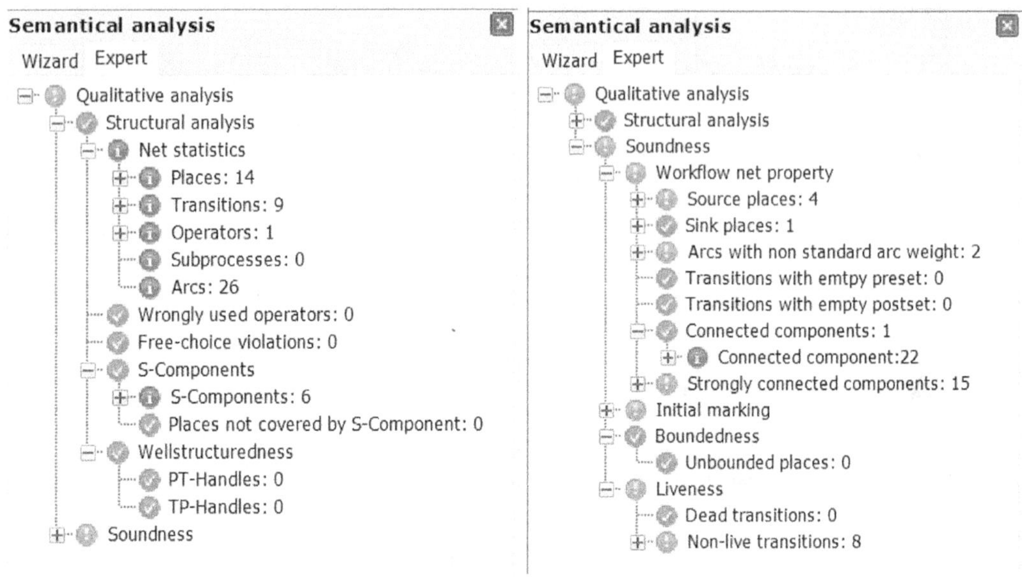

FIGURE 27.3 The qualitative analysis of the workflow net has been obtained.

The Petri net modeling thus provides all the process details with the help of this WoPeD 3.8.0 tool.

27.7 CONCLUSIONS AND SCOPE

Petri nets have been widely used in modeling of biological processes for several decades and represent one of the best and easiest ways to understand a process mathematically (Virmani et al., 2019; Singh and Gupta, 2019; Gupta and Singh, 2020; Singh et al., 2020b, 2020c). Most of the properties and behaviors can be provided by a Petri net analysis by using an reachability graph. The required pathway and details of any reachable state in the process can be found from the reachability graph. Although other analysis methods such as incidence matrix have been used, people without a mathematics background are unable to use such techniques to evaluate a process. Individuals from other disciplines, for example, biologists, can easily understand the mechanics of the net and apply it to the modeling of other processes(Chaouiya, 2007; Davidich and Bornholdt, 2008; Chen, 2010; Oyelade et al., 2016; Gupta et al., 2019a; Singh et al., 2019, 2020b, 2020c; Gupta and Singh, 2020; Afreen et al., 2021; Jha et al., 2021; Gupta et al., 2022).

ACKNOWLEDGMENTS

The authors would like to express their gratitude to the anonymous reviewers, referees and editors for their valuable guidance and suggestions. The first and second authors acknowledge the research facility provided under the Science and Engineering Research Board projects ECR/2017/003480/ PMS;MTR/2021/000378; ECR/2017/001130 and Department of Biotechnology, Ministry of Science & Technology, Govt. of India (Project ID BT/PR40251/BITS/137/11/2021).

REFERENCES

Afreen, Rukshar, Shivani Tyagi, Gajendra Pratap Singh, Mamtesh Singh, Challenges and Perspectives of Polyhydroxyalkanoate Production From Microalgae/Cyanobacteria and Bacteria as Microbial Factories:

An Assessment of Hybrid Biological System *Frontiers in Bioengineering and Biotechnology*, IF 3.644 (SCIE) Vol 9(109) 19 February 2021 https://doi.org/10.3389/fbioe.2021.624885.

Chaouiya, C., Petri net Modelling of Biological Networks. *Briefings in bioinformatics*, *8*(4), 210–219 2007

Chen, G. Q. (Ed.) Plastics completely synthesized by bacteria: polyhydroxyalkanoates. In *Plastics from Bacteria* (17–37). Springer, Berlin, Heidelberg. 2010

Davidich, M. I., S. Bornholdt, Boolean Network Model Predicts Cell Cycle Sequence of Fission Yeast. *PloS One*, 3(2), e1672 2008.

Gupta, A., Gajendra Pratap Singh, The nature of multi-blocked right artery blood flow: Computational fluid dynamics analysis. (In Hindi version) Special Volume on Mathematical-Mechanical, Vigyan Garima Sindhu, ISSN: 2320:7736 (print), Commission for Scientific and Technical Terminology, Ministry of Human Resource Development, (Department of Higher Education), Govt. of India. 112, 111–121 January–March, 2020.

Gupta, Sakshi, S. Kumawat G. P. Singh, Fuzzy Petri Net Representation of Fuzzy Production Propositions of a Rule Based System. In *International Conference on Advances in Computing and Data Sciences* (pp. 197–210). Springer, Singapore. 2019a, April

Gupta, Sakshi, Gajendra Pratap Singh, Sunita Kumawat, Petri Net Recommender System to Model Metabolic Pathway of Polyhydroxyalkanoates *International Journal of Knowledge and Systems Science* 10(2), 42–59, 2019b.

Gupta Sakshi, Gajendra Pratap Singh, Sunita Kumawat, Modeling and targeting an essential metabolic pathway of Plasmodium falciparum in apicoplast using Petri nets *Applied Mathematics-A Journal of Chinese Universities*, 37(1), 91–110, 2022 (to be published in Feb).

Jha, Madhuri, Mamtesh Singh, Gajendra Pratap Singh, Modeling of Second-Line Drug behavior in Tuberculosis using Petri net *International Journal of System Assurance Engineering and Management Accepted Int J Syst Assur Eng Manag*, 2021 https://doi.org/10.1007/s13198-021-01320-7.

Kansal, S., G.P. Singh, Petri nets and its application: A review Proceeding of National Conference on Application of mathematics in Science, Technology and Management, Shri Ram Murti Smarak College of Engineering & Technology, Bareilly, UP, September 6–7, 2008.

Kansal, S., G.P. Singh, M. Acharya, On Petri Nets Generating All the Binary n-Vectors *Scientiae Mathematicae Japonicae* 71 (2), 209–216 2010.

Kansal, S., G.P. Singh, M. Acharya, 1-Safe Petri Nets Generating Every Binary n-Vectors Exactly Once *Scientiae Mathematicae Japonicae* 74 (1), 29–36 2011.

Murata, T., Petri Nets: Properties, Analysis and Applications. *Proceedings of the IEEE*, 77(4), 541–580. 1989

Oyelade, J., I. Isewon, S. Rotimi I. Okunoren. Modeling of the Glycolysis Pathway in Plasmodium Falciparum Using Petri Nets. *Bioinformatics and Biology insights*, 10, BBI-S37296 2016.

Peterson, J. L., *Petri Net Theory and the Modeling of Systems*. Prentice Hall PTR. 1981.

Petri, C. A., Communication with automata. Dissertation, Technical University of Darmstadt 1966. http://edoc.sub.uni-hamburg.de/informatik/volltexte/2010/155/

Reddy, V. N., M. L. Mavrovouniotis M. N. Liebman, Petri net representations in metabolic pathways. *ISMB*, 93, 328–336) 1993, July

Singh, G. P., A Wheel 1-Safe Petri Net Generating all the 0; 1 n Sequences, *International Journal of Computer Applications*, 84(16), 2013.

Singh, G. P., Basic Properties of Petri Nets, *SRMS Journal of Mathematical Sciences*, 1(1), 54–71, 2014.

Singh, G. P., Applications of Petri nets in electrical, electronics and optimizations, In *2016 International Conference on Electrical, Electronics, and Optimization Techniques (ICEEOT)*, 2180–2184, IEEE, Chennai, India, 2016.

Singh, G. P., S. Kansal, Basic Results on Crisp Boolean Petri Nets, In *Modern Mathematical Methods and High Performance Computing in Science and Technology*, 83–88, Springer, Singapore, 2016.

Singh, G. P., S. Kansal, M. Acharya, Construction of a Crisp Boolean Petri net From a 1-safe Petri Net, *International Journal of Computer Applications*, 73(17), 2013a.

Singh, G. P., S. Kansal, M. Acharya, Embedding an Arbitrary 1-Safe Petri Net into a Boolean Petri Net, *International Journal of Computer Applications*, 70(6), 2013b.

Singh, G. P., S. Kansal, M. Acharya, Existence and Uniqueness of a Minimum Crisp Boolean Petri Net, *International Journal of Computer Applications*, 73(20), 2013c.

Singh, Gajendra Pratap, Aparajita Borah, Sangram Ray, A Review Paper on Corona Product of Graphs *Advances and Applications in Mathematical Sciences*, 19(10), 1047–1054, August 2020a.

Singh, Gajendra Pratap, Agraj Gupta, A Petri Net Analysis to Study the Effects of Diabetes on Cardiovascular Diseases In *6th International Conference on Computing for Sustainable Global Development (INDIACom) 481–488* IEEE Xplore, ISBN: 978-93-80544-36-6 2019.

Singh, Gajendra Pratap, Madhuri Jha, Mamtesh Singh Naina Modeling the mechanism pathways of first line drug in Tuberculosis using Petri nets *International Journal of System Assurance Engineering and Management* Springer 11(2), 313–324 2020b 10.1007/s13198-019-00940-4

Singh, Gajendra Pratap, Madhuri Jha, Mamtesh Singh, Petri-net modeling of B-cell receptor signaling pathways: A case study in CLL https://arxiv.org/abs/1906.08669 June 20, 2019

Singh, Gajendra Pratap, Madhuri Jha, Mamtesh Singh, Petri net modeling of clinical diagnosis process in Tuberculosis (In Hindi version) Special Volume on Mathematical-Mechanical, Vigyan Garima Sindhu, ISSN: 2320:7736 (print), Commission for Scientific and Technical Terminology, Ministry of Human Resource Development, (Department of Higher Education), Govt. of India. 112, 147–152 January–March, 2020c.

Singh, Gajendra Pratap, Sujit Kumar Singh, Petri Net Recommender System for Generating of Perfect Binary Tree *International Journal of Knowledge and Systems Science* 10(2), 1–12, 2019.

Singh, Gajendra Pratap, Sujit Kumar Singh, Madhuri Jha, Existence of Forbidden Digraphs for Crisp Boolean Petri Nets *International Journal of Mathematical, Engineering and Management Sciences* 5(1) 83–95 2020d.

Virmani, Ishita, Christo Sasi, Eepsita Priyadarshini, Raj Kumar, Saurabh Kumar Sharma, Gajendra Pratap Singh, Ram Babu Pachwarya, R. Paulra, Hamed Barabadi, Muthupandian Saravanan, Ramovatar Meena, Comparative Anticancer Potential of Biologically and Chemically Synthesized Gold Nanoparticles *Journal of Cluster Science*, Published on October 14, 2019 https://doi.org/10.1007/s10876-019-01695-5

28 Applications of Machine Learning in Bioprocess Development and Optimization

Dharmesh Harwani and Jyoti Lakhani
Maharaja Ganga Singh University, Bikaner, India

CONTENTS

28.1 INTRODUCTION

Bioprocess and bio-product development requires proper planning and execution throughout the product life-cycle to achieve profitable and sustainable manufacturing (Djuris and Djuric, 2017). Bioprocess products include pharmaceutical agile components (like vaccines), healthcare products (like drugs and vitamins), nutritional products (like amino acids), solvents (like alcohol), and so on. Bioprocess development is a complex process consisting of several steps: host cell screening, initializing parameters, optimization of environmental conditions, and prediction of the relationship between analytical process attributes and critical quality parameters (Djuris and Djuric, 2017; Kumar et al., 2014; Neubauer et al., 2013).

Bioprocesses are inevitable in nature. When current state and regulatory factors are known, it is possible to predict future states or products, but in fact, the biological structures are very complex and difficult to understand. For example, understanding various cell physiological mechanisms is difficult because many metabolic mechanisms are still unknown. According to Bailey (1998), bioprocess models are an approximation of actual bioprocess mechanisms. In the bioprocess model, a bioprocess control is defined, using biological components like enzymes, living organisms like yeast, and other microorganisms, to provide an optimal environment for the process to obtain the desired products. Process modeling is performed to achieve the previously stated goals of process development (Mandenius and Titchener-Hooker, 2013) for continuous assessment and evaluation of processes and their byproducts. Process modeling is important for identifying the effects of various sources of disturbance that may affect the quality of the products and delay the processes (Djuris and Djuric, 2017; Guo et al., 2017; Solle et al., 2017).

DOI: 10.1201/9781003306931-32

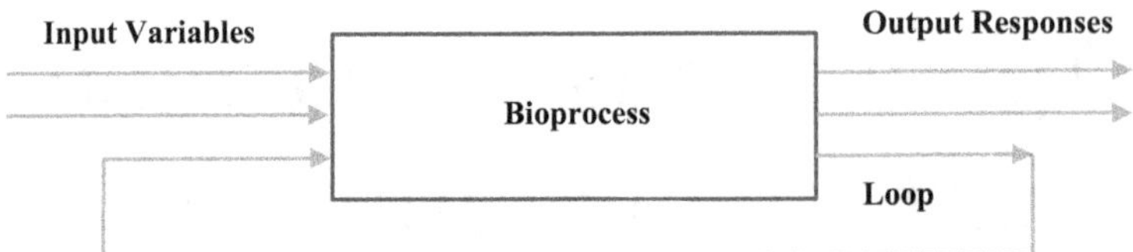

FIGURE 28.1 A block diagram for a simple bioprocess.

Most bioprocesses demonstrate dynamic non-linear behavior and some are dependent on unfamiliar chemical reactions with time-dependent attributes. This may pose challenges in bioprocess modeling, including lack of understanding of systematic biochemical reactions, online parameter measurement mechanisms, and complicated interactions between variables with their bioprocess responses (Djuris and Djuric, 2017; Rathore et al., 2017; Solle et al., 2017). A bioprocess model is a mathematical presentation of the relation between process attributes and product quality parameters. A bioprocess model can be developed as mechanical, statistical, or machine learning.

28.2 BIOPROCESS DEVELOPMENT AND MODELING

A simplified bioprocess can be illustrated as a block diagram that accepts several input variables and in response provides an output (shown in Figure 28.1 as a black box). The input variables in the initialization stages of the bioprocess provide initial parameters to the bioprocess control. These parameters trigger the bioprocess and provide output responses. Once triggered, the bioprocess control can loop the process again with output responses. In later stages, the input variables provide additional parameters and add further complexity to the bioprocess. This bioprocess control automatically stops if the desired output response is achieved. The whole process optimizes bioprocess development to achieve refinement in the output and quality of the end products. To control the quality of the outcomes, the input parameters (correct concentrations of nutrients in the medium) and other environmental factors such as pH, temperature, etc., can be optimized.

In bioprocess development, several methods are used to investigate the mathematical relationship between input parameters and output responses. Understanding the various relationships between bioprocess input parameters and output responses is important for controlling the bioprocess performance. Methods include factorial design, experimental design, and response surface methodology (Montgomery, 1984). By using these methods, researchers can scrutinize the input variables from which well-defined state-of-the-art output responses are generated. The discovery of the interaction between input variables and their output responses is difficult. For example, additional products can be developed if specific input parameters are appropriately changed. This is made possible by optimizing the bioprocess development process followed by data analysis of the experimental outputs. Various optimizing algorithms are available in the literature for providing efficient and powerful ways to optimize bioprocess development. One study suggests that artificial neural networks are excellent platforms for providing specific mathematical models for different bioprocesses, and optimal solutions can be discovered using genetic algorithms (Nagata and Chu, 2003).

28.3 BUILDING MATHEMATICAL MODELS FOR BIOPROCESSES

In bioprocess modeling, it is important to identify critical parameters and predict their values (Levin et al., 2004). Prediction can be made by applying feature selection, linear regression, and artificial neural network data analysis methods. The bioengineering and bioprocess sector is digitalizing on a vast scale and moving towards the complete automation of various bioprocesses. The new-generation bioprocess industry is heavily dependent on artificial intelligence, machine learning, computer vision, the internet of things, and other computerization network technology.

The main objectives of the bioprocess modeling methods are:

- Improvement of the action events that occurred during the process.
- Improvement of products such as secondary metabolites/enzymes/biopesticides/probiotics/hormones/aromas/organic compounds.
- Predicting estimated values of environmental factors such as pH, temperature, pressure, hydraulic retention time, substrate concentration, and bioprocess feed rates (Wang and Wan, 2009b).
- Predicting bio-parameters such as population heterogeneity, concentration of biomass, morphology, product concentration, quality, metabolic state, and media composition.
- Achieving best performance of the bioprocesses (Franco-Lara et al., 2006).

In order to improve bioprocess output, mathematical, statistical, and computational models are required for understanding, analysis, prediction, and optimization of key parameters (Escamilla-Alvarado et al., 2012). These models can be designed linearly or non-linearly depending on the bioprocess requirement. Some bioprocesses, such as microbial fermentation, are complex bioprocesses that can be dealt with by a non-linear computational model (Ahmadian-Moghadam et al., 2013).

28.4 STAGES OF BIOPROCESS DEVELOPMENT

Bioprocess development occurs in four stages:

i. Bioprocess development
ii. Bioprocess design
iii. Bioprocess optimization
iv. Industrial implementation by operational control.

The objective of the development stage is to recognize the behavior of the organism in order to optimize the parameters for the bioprocess. Process parameters targeted in the development stage include operating conditions, media composition, and feeding strategies. Mathematical modeling of the bioprocess is performed at supervision, optimization, and industrial operation control stages. After building the model, researchers select optimal conditions for operating the bioprocess. The result is called the model optimum, and the process is called bioprocess optimization (Bouaoudat et al., 2012). Various mathematical modeling methodologies are available in the literature.

28.5 BIOPROCESS OPTIMIZATION APPROACHES

There are two approaches to bioprocess optimization.

28.5.1 Sequential Optimization

This approach utilizes a model that is designed using the results of experimental outputs performed sequentially. The direct optimization approach can be applied on a single variable (parameter) or on a pair of variables at a given time. These approaches are called monothetical (or single-variable) and binomial analysis, respectively. In the monothetical approach, a single parameter of a bioprocess is used as a testing factor. Statistical and black-box approaches are commonly used in monothetical analysis. In binomial analysis, two variables (parameters) are used to design conceptual models. A subset of the factorial design is also used to exploit the sparseness and paucity of the data. The analysis is performed on this subset of the factorial design.

28.5.2 Parallel Optimization

In this approach, the model is built using the results of experimental outputs performed in parallel. Multiple critical parameters and their relationships are analyzed. For this purpose, central composite design and Box–Behnken design are used to exploit the relationship between a set of selected variables (Veljković et al., 2019). Response surface methodology (RSM) is one of the popular parallel optimization methodologies which is useful to identify, optimize, and improve critical parameters as well as bioprocess model responses. RSM uses an empirical framework that links dependent variables with other independent variables (Abdelgalil et al., 2018; Chelladurai et al., 2021). The RSM is an efficient technique for non-linear optimization. The principal limitation of RSM is that it presumes only quadratic non-linear correlation. To compensate for this limitation, machine-learning tools are used for modeling and optimization of bioprocesses, the most common of which to emerge are artificial neural networks and genetic algorithms.

28.6 MACHINE LEARNING (ML)-BASED BIOPROCESS OPTIMIZATION

ML-based bioprocess development and modeling methods have been improved, and models are based on an inference mechanism based on collected data. Using a bioprocess, a high yield of low-cost green chemicals can be obtained. Such bioprocesses can be considered green methods due to their sustainable and eco-friendly nature and the fact that they help to reduce greenhouse gas emissions and global warming (Galanie et al., 2015; Moore et al., 2017). The exploitation of microorganisms for the manufacturing of organic compounds is essential. To start the process, culture optimization is done with many repeated experimental steps to determine the optimal growth and developmental conditions for elevated biosynthesis. Culture medium optimization is a costly, laborious, ineffective, and time-consuming process. Many mathematical models have therefore been developed for a number of bioprocesses that use advanced fermentation technology. These computational models use various modeling and optimization algorithms to improve interpretation of the process input, optimum yield, and production rate. The role of ML algorithms in bioprocess development and bioprocess modeling is shown in Figures 28.2 and 28.3, respectively.

Based on the inclusion or exclusion of output data during observations, ML algorithms can be classified as supervised or unsupervised. Supervised learning algorithms are generally used for handling critical process attributes or critical quality parameters. Supervised techniques are also used to determine the interdependency of input parameters and critical process quality measures. Unsupervised learning algorithms are generally used for classification purposes.

ML algorithms are used in various applications (Table 28.1). Bioprocess development tasks in which they can be applied include genetic algorithms (Hamdia et al., 2021) used to optimize medium

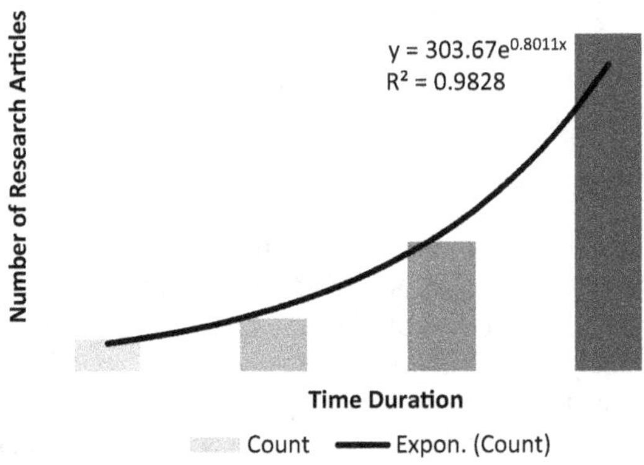

FIGURE 28.2 Research articles from the last 20 years resulting from query "bioprocess development+machine learning".

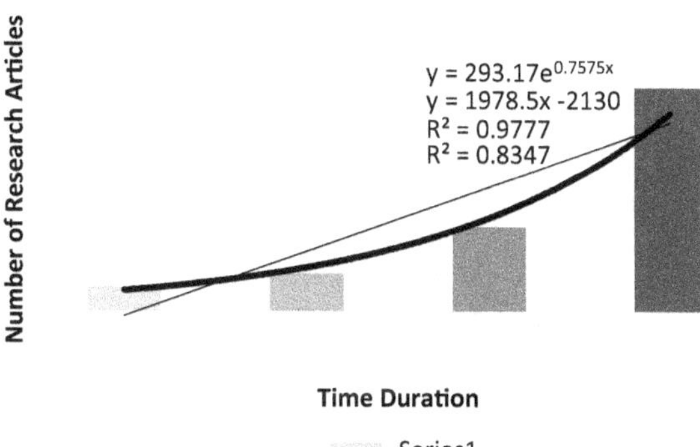

FIGURE 28.3 Research articles from the last 20 years resulting from query "bioprocess modeling+machine learning".

TABLE 28.1
Applications and Machine-Learning Approaches for Bioprocess Development

Application Area	Machine Learning Approach
Monitoring Bioprocess	• **Evolutionary Computation** McGovern (2002)
Bioprocess Modeling	• **Neural Network(NN)** Nagata et al. (2004), Kelly et al. (2005), Benini & Toffolo (2002), Jindal & Chauhan (2001), Shen et al. (2005), Ronen et al. (2002), Hu (2003) • **Genetic Algorithm(GA)** Nagata et al. (2004), Lin et al. (2004) • **Principal Component Analysis** Savery (2002) • **Fuzzy Clustering** Ronen et al. (2002) • **Parallel Genetic Algorithm** Babbar (2002), Babbar & Minsker (2002) • **Evolutionary Algorithm** Swain & Morris 2003) • **Genetic Programming** Grosman & Lewin 2004)
Prediction of Bioprocess Outcome	• **Multivariate Decision Trees** Kipling et al. (2005) • **Genetic Programming** Ellis et al. (2002) • **Artificial Neural Network** Sugimoto et al. (2005a)
Metabolomics	• **Genetic Programming** Allen et al. (2003), Kell (2002) • **Evolutionary Computation** Goodacre and Royston (2005)
Predictive Scheduling to Improve Production	• **Genetic Algorithm** Lau et al. (2003) • **Neural Networks** Pöllänen et al. (2001)
Bioanalytical Study	• **Genetic Algorithm** Jarvis (2005)
Bioprocess Optimization	• **Neural Network** Nagata et al. (2004), Nagata & Chu (2003), Naik and Bhagwat (2005) • **Genetic Algorithm** Nagata et al. (2004) (Na et al. (2002), Jarvis & Goodacre (2005), Jones & Romil (2004), Nguang et al. (2001), Nagata & Chu (2003) • **Evolutionary Algorithms** Patil et al. (2002)
Optimal Batch-Feeding	• **Artificial Neural Networks (ANNs)** Franco-Lara & Weuster-Botz (2005) • **Genetic Algorithms** Ascencio et al. (2004), Nguang et al. (2001), Sarkar & Modak (2003)
Predictive Non-Linear Modeling	• **ANN** Almeida (2002), • **Recurrent Neural Network** Malhi & Gao (2004)
Soft Sensor Modeling	• **Support Vector Machine** Lei & Sun (2005), Ascencio et al. (2004), Feng et al. (2003) • **Bayesian learning** Ascencio et al. (2004) • **Genetic Programming** Kordon et al. (2004), Kordon et al. (2003), Na et al. (2005) • **Feature Selection** Rallo et al. (2003) • **Genetic Algorithm** Na et al. (2003)

(Continued)

TABLE 28.1 *(Continued)*

Application Area	Machine Learning Approach
Reverse Engineering of Biochemical Equations	• **Genetic Algorithm** Sugimoto et al. (2005b)
Fermentation Diagnosis	• **Multivariate Analysis** Bicciato et al. (2002)
Observation of Bioprocess	• **Neural Networks** Pöllänen et al. (2001)
Parameter Estimation	• **Expectation Maximization** Xiong et al. (2005) • **Evolution Strategies** Moles et al. (2003) • **Genetic Algorithm** Roeva and Olympia (2005), Rashid et al. (2005), Choi & Park (2001)
Bioprocess Controlling	• **Dynamic Programming** Lee & Lee (2005) • **Evolutionary Algorithm** Andres-Toro et al. (2004), Cruz et al. (2003)
Fault Diagnosis	• **Support Vector Machine** Chirag et al. (2004)
Parameter Optimization	• **Genetic Algorithm** Lin et al. (2004)

components (Weuster-Botz, 2000), particle swarm optimization algorithm used to improve medium composition (Cockshott and Hartman, 2001), statistical procedures to build a model for building models where many independent variables are involved (Montgomery, 2017), and Plackett–Burman design (Kalil et al., 2000) for initial screening and optimization of the significant variables (Plackett and Burman, 1946; Abdel-Fatta et al., 2005).

There are two main reasons for the emergence of ML-based bioprocess modeling and optimization:

i. ML algorithms do not require a prior specification of the process.
ii. ML algorithms are suitable for all linear and non-linear processes.

RSM is only suitable for logarithmic variables. Machine-learning tools, however, can inherently handle any kind of nonlinearity and can easily overcome the limitations of RSM. The most popular machine-learning approach is artificial neural networks (ANNs) which deal with a search space more complex than a quadratic. ANN and RSM have been examined to design mathematical models to improve various bioprocesses (Wang and Wan, 2009a; Whiteman and Kana, 2014; Esfahanian et al., 2013). Machine-learning algorithms are well known for their generality and learning abilities. They are also efficient as bioprocess modeling tools because of their ability to deal with nonlinearity, parallelism, fault tolerance, and noise tolerance.

28.7 CONCLUSION

Since machine-learning tools and techniques work more effectively than traditional bioprocess modeling and optimization techniques, the integration of machine learning approaches is accelerating bioprocess development. All bioprocesses are dependent on indefinite time-variant reaction kinetics, and the interaction of variables with multiple degrees of correlation and time-scaling bioprocess responses. RSM is the most commonly used conventional approach for bioprocess optimization, but a major limitation is that RSM will only work for quadratic non-linear correlations. These complexities are introducing new challenges in bioprocess development, to which machine learning-based bioprocess modeling can be a reliable solution. Computational machine-learning techniques such as artificial neural networks, genetic algorithms, and evolutionary algorithms are making a considerable contribution to the estimation, monitoring, and control of non-linear and time-scaling bioprocesses. With their high predictive accuracy and optimization, these techniques are indeed suitable for a variety of bioprocess modeling applications.

REFERENCES

Abdel-Fattah, Y. R., Saeed, H. M., Gohar, Y. M., & El-Baz, M. A. (2005). Improved production of Pseudomonas aeruginosa uricase by optimization of process parameters through statistical experimental designs. *Process Biochemistry*, 40(5), 1707–1714.

Abdelgalil, S.A., Attia, A. M., Reyed, R.M., Soliman, N.A., & El Enshasy, H.A. (2018). Application of experimental designs for optimization the production of Alcaligenes faecalis Nyso Laccase. *Journal of Scientific Research*. 77(12), 713–722. http://nopr.niscair.res.in/handle/123456789/45482

Ahmadian-Moghadam, H., Elegado, F., & Nayve, R. (2013). Prediction of ethanol concentration in biofuel production using artificial neural networks. *American Journal of Modeling and Optimization*, 1(3), 31–35.

Allen, J., Davey, H.M., Broadhurst, D., Heald, J. K., Rowland, J. J., Oliver, S. G., & Kell, D. B. (2003). High-throughput classification of yeast mutants for functional genomics using metabolic footprinting. *Nature Biotechnology*, 21(6), 692–696.

Almeida, J.S. (2002). Predictive non-linear modeling of complex data by artificial neural networks. *Current Opinion in Biotechnology*, 13(1), 72–76.

Andres-Toro, B., Giron-Sierra, J. M., Fernandez-Blanco, P., Lopez-Orozco, J. A., & Besada-Portas, E. (2004). Multiobjective optimization and multivariable control of the beer fermentation process with the use of evolutionary algorithms. *Journal of Zhejiang University-SCIENCE A*, 5(4), 378–389.

Ascencio, P., Sbarbaro, D., & de Azevedo, S. F. (2004). An adaptive fuzzy hybrid state observer for bioprocesses. *IEEE Transactions on Fuzzy Systems*, 12(5), 641–651.

Babbar, M. (2002). Multiscale parallel genetic algorithms for optimal groundwater remediation design (Doctoral dissertation, University of Illinois at Urbana-Champaign).

Babbar, M., & Minsker, B. S. (2002, July). A Multiscale Master-Slave Parallel Genetic Algorithm with Application to Groundwater Remediation Design. In *GECCO Late Breaking Papers* (9–16), New York.

Bailey, J. E. (1998). Mathematical modeling and analysis in biochemical engineering: Past accomplishments and future opportunities. *Biotechnology Progress*, 14(1), 8–20.

Benini, E., & Toffolo, A. (2002, January). Axial-flow compressor model based on a cascade stacking technique and neural networks. In *Turbo Expo: Power for Land, Sea, and Air* (Vol. 3610, pp. 793–801).

Bicciato, S., Bagno, A., Soldà, M., Manfredini, R., & Di Bello, C. (2002). Fermentation diagnosis by multivariate statistical analysis. *Applied Biochemistry and Biotechnology*, 102(1), 49–62.

Bouaoudat, B. D., Yalaoui, F., Amodeo, L., & Entzmann, F. (2012). Efficient developments in modeling and optimization of solid state fermentation. *Biotechnology & Biotechnological Equipment*, 26(6), 3443–3450.

Chelladurai, S. J. S., Murugan, K., Ray, A. P., Upadhyaya, M., Narasimharaj, V., & Gnanasekaran, S. (2021). Optimization of process parameters using response surface methodology: A review. *Materials Today: Proceedings*, 37, 1301–1304.

Chiang, L. H., Kotanchek, M. E., & Kordon, A. K. (2004). Fault diagnosis based on Fisher discriminant analysis and support vector machines. Computers & Chemical Engineering, 28(8), 1389–1401.

Choi, D. J., & Park, H. (2001). Estimation of activated sludge model parameters Using respirogram and genetic algorithms. In *Bridging the Gap: Meeting the World's Water and Environmental Resources Challenges* (pp. 1–10).

Cockshott, A. R., & Hartman, B. E. (2001). Improving the fermentation medium for Echinocandin B production part II: Particle swarm optimization. *Process Biochemistry*, 36(7), 661–669.

Cruz, I. L., Van Willigenburg, L. G., & Van Straten, G. (2003). Efficient differential evolution algorithms for multimodal optimal control problems. *Applied Soft Computing*, 3(2), 97–122.

Djuris, J., & Djuric, Z. (2017). Modeling in the quality by design environment: Regulatory requirements and recommendations for design space and control strategy appointment. *International Journal of Pharmaceutics*, 533(2), 346–356.

Ellis, D. I., Broadhurst, D., Kell, D. B., Rowland, J. J., & Goodacre, R. (2002). Rapid and quantitative detection of the microbial spoilage of meat by Fourier transform infrared spectroscopy and machine learning. *Applied and Environmental Microbiology*, 68(6), 2822–2828.

Escamilla-Alvarado, C., Ríos-Leal, E., Ponce-Noyola, M. T., & Poggi-Varaldo, H. M. (2012). Gas biofuels from solid substrate hydrogenogenic–methanogenic fermentation of the organic fraction of municipal solid waste. *Process Biochemistry*, 47(11), 1572–1587.

Esfahanian, M., Nikzad, M., Najafpour, G., & Ghoreyshi, A. A. (2013). Modeling and optimization of ethanol fermentation using Saccharomyces cerevisiae: Response surface methodology and artificial neural network. *Chemical Industry and Chemical Engineering Quarterly/CICEQ*, 19(2), 241–252.

Feng, R., Shen, W., & Shao, H. (2003, June). A soft sensor modeling approach using support vector machines. In *Proceedings of the 2003 American Control Conference, 2003*. (Vol. 5, pp. 3702–3707). IEEE.

Franco-Lara, E., Link, H., & Weuster-Botz, D. (2006). Evaluation of artificial neural networks for modelling and optimization of medium composition with a genetic algorithm. *Process Biochemistry*, 41(10), 2200–2206.

Franco-Lara, E., & Weuster-Botz, D. (2005). Estimation of optimal feeding strategies for fed-batch bioprocesses. *Bioprocess and Biosystems Engineering*, 27(4), 255–262.

Galanie, S., Thodey, K., Trenchard, I. J., Interrante, M. F., & Smolke, C. D. (2015). Complete biosynthesis of opioids in yeast. *Science*, 349(6252), 1095–1100.

Goodacre, R. (2005). Making sense of the metabolome using evolutionary computation: Seeing the wood with the trees. *Journal of Experimental Botany*, 56(410), 245–254.

Grosman, B., & Lewin, D. R. (2004). Adaptive genetic programming for steady-state process modeling. *Computers & Chemical Engineering*, 28(12), 2779–2790.

Guo, W., Sheng, J., & Feng, X. (2017). Mini-review: In vitro metabolic engineering for biomanufacturing of high-value products. *Computational and Structural Biotechnology Journal*, 15, 161–167.

Hamdia, K. M., Zhuang, X., & Rabczuk, T. (2021). An efficient optimization approach for designing machine learning models based on genetic algorithm. *Neural Computing and Applications*, 33(6), 1923–1933.

Hu, C. (2003). Modeling Reaction Kinetics of Chlorine Dioxide and Volatile Organic Compounds with Artificial Neural Networks (Doctoral dissertation, University of Georgia).

Jarvis, R.M. (2005). The development of Raman spectroscopy and advanced machine learning for bioanalytical studies. The University of Manchester (United Kingdom). ProQuest Dissertations Publishing, 10903969

Jarvis, R.M., & Goodacre, R. (2005). Genetic algorithm optimization for pre-processing and variable selection of spectroscopic data. *Bioinformatics*, 21(7), 860–868.

Jindal, V. K., & Chauhan, V. (2001). Neural networks approach to modeling food processing operations. In Irudayaraj, J. M. (Ed.) *Food Science and Technology-New York-Marcel Dekker*, 305–337, CRC Press

Jones, K. O., & Romil, S. (2004). Selection of an optimum feed profile using genetic algorithms. *IFAC Proceedings*, 37(3), 577–581.

Kalil, S. J., Maugeri, F., & Rodrigues, M. I. (2000). Response surface analysis and simulation as a tool for bioprocess design and optimization. *Process Biochemistry*, 35(6), 539–550.

Kell, D. B. (2002). Metabolomics and machine learning: Explanatory analysis of complex metabolome data using genetic programming to produce simple, robust rules. *Molecular Biology Reports*, 29(1), 237–241.

Kelly, L., Lam, S. S., & Lu, S. (2005). Fermentation Process Modeling Data Preprocessing Strategies Using Neural Networks. In *IIE Annual Conference Proceedings* (p. 1). Institute of Industrial and Systems Engineers (IISE).

Kipling, K., Montague, G., Martin, E., & Morris, J. (2005). Multivariate decision trees for the interrogation of bioprocess data. In *Computer Aided Chemical Engineering* (Vol. 20, pp. 1129–1134). Elsevier.

Kordon, A., Smits, G., Kalos, A., & Jordaan, E. (2003). Robust soft sensor development using genetic programming. *Nature-Inspired Methods in Chemometrics*, 23, 69–108.

Kordon, A. et al. (2004). Biomass Inferential Sensor Based on Ensemble of Models Generated by Genetic Programming. In: Deb, K. (ed.) *Genetic and Evolutionary Computation – GECCO 2004. GECCO 2004*. Lecture Notes in Computer Science, vol 3103. Springer, Berlin, Heidelberg. https://doi.org/10.1007/978-3-540-24855-2_118

Kumar, V., Bhalla, A., & Rathore, A. S. (2014). Design of experiments applications in bioprocessing: Concepts and approach. *Biotechnology Progress*, 30(1), 86–99.

Lau, S., Willis, M. J., Montague, G. A., & Glassey, J. (2003, July). Predictive Scheduling of a Bioprocess Plant. In *Modelling, Simulation, and Optimization* (pp. 296–301).

Lee, J. M., & Lee, J. H. (2005). Approximate dynamic programming-based approaches for input–output data-driven control of nonlinear processes. *Automatica*, 41(7), 1281–1288.

Lei, L. Y., & Sun, Z. H. (2005, August). Soft sensor based on generalized support vector machines for microbiological fermentation. In *2005 International Conference on Machine Learning and Cybernetics* (Vol. 7, pp. 4305–4309). IEEE.

Levin, D. B., Pitt, L., & Love, M. (2004). Biohydrogen production: prospects and limitations to practical application. *International Journal of Hydrogen Energy*, 29(2), 173–185.

Lin, J. Q., Lee, S. M., & Koo, Y. M. (2004). Model development for lactic acid fermentation and parameter optimization using genetic algorithm. *Journal of Microbiology and Biotechnology*, 14(6), 1163–1169.

Malhi, A., & Gao, R. X. (2004, May). Recurrent neural networks for long-term prediction in machine condition monitoring. In *Proceedings of the 21st IEEE Instrumentation and Measurement Technology Conference (IEEE Cat. No. 04CH37510)* (Vol. 3, pp. 2048–2053). IEEE.

Mandenius, C. F., & Titchener-Hooker, N. J. (Eds). (2013). *Measurement, Monitoring, Modelling and Control of Bioprocesses* (132). Springer, Berlin.

McGovern, A. C., Broadhurst, D., Taylor, J., Kaderbhai, N., Winson, M. K., Small, D. A., … & Goodacre, R. (2002). Monitoring of complex industrial bioprocesses for metabolite concentrations using modern spectroscopies and machine learning: Application to gibberellic acid production. *Biotechnology and Bioengineering*, 78(5), 527–538.

Moles, C. G., Mendes, P., & Banga, J. R. (2003). Parameter estimation in biochemical pathways: A comparison of global optimization methods. *Genome Research*, 13(11), 2467–2474.

Montgomery, D. C. (1984). *Design and Analysis of Experiments: Graph*. Wiley, Darst.

Montgomery, D. C. (2017). *Design and Analysis of Experiments*. John Wiley & Sons.

Moore, R. H., Thornhill, K. L., Weinzierl, B., Sauer, D., D'Ascoli, E., Kim, J., & Anderson, B. E. (2017). Biofuel blending reduces particle emissions from aircraft engines at cruise conditions. *Nature*, 543(7645), 411–415.

Na, J. G., Chang, Y., Chung, B. H., & Lim, H. C. (2002). Adaptive optimization of fed-batch culture of yeast by using genetic algorithms. *Bioprocess and Biosystems Engineering*, 24(5), 299–308.

Na, M. G., Jung, D. W., Shin, S. H., Lee, S. M., & Sim, Y. R. (2003, July). A software sensor using a black box modeling method. In *Proceedings of the Korean Nuclear Society Conference* (pp. 100–100). Korean Nuclear Society.

Na, M.G., Lee, Y. J., & Hwang, I. J. (2005). A smart software sensor for feedwater flow measurement monitoring. *IEEE Transactions on Nuclear Science*, 52(6), 3026–3034.

Nagata, Y., & Chu, K. H. (2003). Optimization of a fermentation medium using neural networks and genetic algorithms. *Biotechnology Letters*, 25(21), 1837–1842.

Nagata, Y., Chu, K. H., & Kim, E. Y. (2004). A combined neural network/genetic algorithm technique for bioprocess modeling and optimization. In *Asian Pacific Confederation of Chemical Engineering congress program and abstracts Asian Pacific Confederation of Chemical Engineers Congress Program and Abstracts* (pp. 425–425). The Society of Chemical Engineers, Japan.

Naik, A. D., & Bhagwat, S. S. (2005). Optimization of an artificial neural network for modeling protein solubility. *Journal of Chemical & Engineering Data*, 50(2), 460–467.

Neubauer, P., Cruz, N., Glauche, F., Junne, S., Knepper, A., & Raven, M. (2013). Consistent development of bioprocesses from microliter cultures to the industrial scale. *Engineering in Life Sciences*, 13(3), 224–238.

Nguang, S. K., Chen, L., & Chen, X. D. (2001). Optimisation of fed-batch culture of hybridoma cells using genetic algorithms. *ISA Transactions*, 40(4), 381–389.

Patil, S.V., Jayaraman, V. K., & Kulkarni, B. D. (2002). Optimization of media by evolutionary algorithms for production of polyols. *Applied Biochemistry and Biotechnology*, 102(1), 119–128.

Plackett, R.L., & Burman, J. P. (1946). The design of optimum multifactorial experiments. *Biometrika*, 33(4), 305–325.

Pöllänen, J., Rousu, J., & Kronlöf, J. (2001). A Neural Network Tool for Brewery Fermentations.

Rallo, R., Ferré-Giné, J., & Giralt, F. (2003). Best feature selection and data completion for the design of soft neural sensors. In *Proceedings of AIChE 2003, 2nd Topical Conference on Sensors*. ACS San Francisco.

Rashid, R., Jamaluddin, H., & Amin, N. A. S. (2005). Parameter estimation of tapioca starch hydrolysis process: Application of least squares and genetic algorithm. *Journal-The Institution of Engineers, Malaysia*, 66(4), 51–60.

Rathore, A. S., Garcia-Aponte, O. F., Golabgir, A., Vallejo-Diaz, B. M., & Herwig, C. (2017). Role of knowledge management in development and lifecycle management of biopharmaceuticals. *Pharmaceutical Research*, 34(2), 243–256.

Roeva, O. (2005). Genetic algorithms for a parameter estimation of a fermentation process model: A comparison. *International Journal Bioautomation*, 3, 19.

Ronen, M., Shabtai, Y., & Guterman, H. (2002). Hybrid model building methodology using unsupervised fuzzy clustering and supervised neural networks. *Biotechnology and Bioengineering*, 77(4), 420–429.

Sarkar, D., & Modak, J. M. (2003). Optimisation of fed-batch bioreactors using genetic algorithms. *Chemical Engineering Science*, 58(11), 2283–2296.

Savery, J. R. (2002). *A Modular Non-Linear Approach to Empirical Principal Component Analysis Based Process Modelling*. University of London, University College London, London.

Shen, Z., Gay, R., Miao, C., Tan, T. W., Kuay, C. S., & Lee, H. M. (2005). Process modeling and automated execution for bio-manufacturing. *International Journal of Information Technology*, 12(1), 37–48.

Solle, D., Hitzmann, B., Herwig, C., Pereira Remelhe, M., Ulonska, S., Wuerth, L., … & Steckenreiter, T. (2017). Between the poles of data-driven and mechanistic modeling for process operation. *Chemie Ingenieur Technik*, 89(5), 542–561.

Sugimoto, M., Kikuchi, S., Arita, M., Soga, T., Nishioka, T., & Tomita, M. (2005a). Large-scale prediction of cationic metabolite identity and migration time in capillary electrophoresis mass spectrometry using artificial neural networks. *Analytical Chemistry*, 77(1), 78–84.

Sugimoto, M., Kikuchi, S., & Tomita, M. (2005b). Reverse engineering of biochemical equations from time-course data by means of genetic programming. *BioSystems*, 80(2), 155–164.

Swain, A. K., & Morris, A. S. (2003). An evolutionary approach to the automatic generation of mathematical models. *Applied Soft Computing*, 3(1), 1–21.

Veljković, V. B., Veličković, A. V., Avramović, J. M., & Stamenković, O. S. (2019). Modeling of biodiesel production: Performance comparison of Box–Behnken, face central composite and full factorial design. *Chinese Journal of Chemical Engineering*, 27(7), 1690–1698.

Wang, J., & Wan, W. (2009a). Optimization of fermentative hydrogen production process using genetic algorithm based on neural network and response surface methodology. *International Journal of Hydrogen Energy*, 34(1), 255–261.

Wang, J., & Wan, W. (2009b). Factors influencing fermentative hydrogen production: A review. *International Journal of Hydrogen Energy*, 34(2), 799–811.

Weuster-Botz, D. (2000). Experimental design for fermentation media development: Statistical design or global random search? *Journal of Bioscience and Bioengineering*, 90(5), 473–483.

Whiteman, J. K., & Kana, E. G. (2014). Comparative assessment of the artificial neural network and response surface modelling efficiencies for biohydrogen production on sugar cane molasses. *BioEnergy Research*, 7(1), 295–305.

Xiong, Z. H., Huang, G. H., & Shao, H. H. (2005). On-line estimation of concentration parameters in fermentation processes. *Journal of Zhejiang University. Science. B*, 6(6), 530.

Index

Page numbers in **bold** refers to tables.

For Product Safety Concerns and Information please contact our EU representative GPSR@taylorandfrancis.com
Taylor & Francis Verlag GmbH, Kaufingerstraße 24, 80331 München, Germany

www.ingramcontent.com/pod-product-compliance
Lightning Source LLC
Chambersburg PA
CBHW080122220326
41598CB00032B/4928

* 9 7 8 1 0 3 2 3 0 8 3 9 5 *